D1690202

Edited by
Ganapathy Subramanian

**Continuous
Biomanufacturing**

Continuous Biomanufacturing

Innovative Technologies and Methods

Edited by Ganapathy Subramanian

WILEY-VCH
Verlag GmbH & Co. KGaA

Editor

Prof. Dr. Ganapathy Subramanian
44 Oaken Grove
SL6 6HH Maidenhead, Berkshire
United Kingdom

Cover

Pictures being used: Coagulation factor VIII protein rendering © Molekuul.be; Vitaminpills © cst21; Measurement device containing closed vials © eGraphia.

All books published by **Wiley-VCH** are carefully produced. Nevertheless, authors, editors, and publisher do not warrant the information contained in these books, including this book, to be free of errors. Readers are advised to keep in mind that statements, data, illustrations, procedural details or other items may inadvertently be inaccurate.

Library of Congress Card No.:
applied for

British Library Cataloguing-in-Publication Data
A catalogue record for this book is available from the British Library.

Bibliographic information published by the Deutsche Nationalbibliothek
The Deutsche Nationalbibliothek lists this publication in the Deutsche Nationalbibliografie; detailed bibliographic data are available on the Internet at <http://dnb.d-nb.de/>.

© 2018 Wiley-VCH Verlag GmbH & Co. KGaA, Boschstr. 12, 69469 Weinheim, Germany

All rights reserved (including those of translation into other languages). No part of this book may be reproduced in any form – by photoprinting, microfilm, or any other means – nor transmitted or translated into a machine language without written permission from the publishers. Registered names, trademarks, etc. used in this book, even when not specifically marked as such, are not to be considered unprotected by law.

Print ISBN: 978-3-527-34063-7
ePDF ISBN: 978-3-527-69989-6
ePub ISBN: 978-3-527-69991-9
Mobi ISBN: 978-3-527-69992-6
oBook ISBN: 978-3-527-69990-2

Cover Design Bluesea Design, MacLeese Lake, Canada
Typesetting Thomson Digital, Noida, India
Printing and Binding C.O.S. Printers Pte Ltd Singapore

Printed on acid-free paper

10 9 8 7 6 5 4 3 2 1

Contents

List of Contributors *xix*

Part One: Overview of State-of-the-Art Technologies and Challenges *1*

1 Continuous Bioprocess Development: Methods for Control and Characterization of the Biological System *3*
Peter Neubauer and M. Nicolas Cruz-Bournazou
1.1   Proposed Advantages of Continuous Bioprocessing *3*
1.1.1   Introduction *3*
1.2   Special Challenges for Continuous Bioprocesses *5*
1.2.1   The Biological System in Continuous Biomanufacturing *5*
1.2.2   Inherent Changes in the Microbial System – Problem of Evolution *6*
1.2.3   Lack of Process Information *7*
1.2.3.1   Models-Based Process Development and Control for Continuous Processes *8*
1.2.3.2   Engineering Approach to Complex Systems *8*
1.2.4   Limited Control Strategies *9*
1.2.4.1   Traditional Control Strategies for Continuous Cultures *9*
1.3   Changes Required to Integrate Continuous Processes in Biotech *11*
1.3.1   A Better Physiological Understanding of the Organisms and Their Responses on the Reactor Environment *11*
1.3.1.1   Model Complexity *11*
1.3.1.2   Models *12*
1.3.2   Model-Based Process Monitoring *13*
1.3.3   Implementation of Model Predictive Control *14*
1.3.3.1   Model-Based Control *14*
1.4   Role of Iterative Process Development to Push Continuous Processes in Biotech *14*
1.4.1   Methods for Development of Continuous Processes *14*
1.4.1.1   Alternative: Fed-Batch as a System to Simulate Quasi Steady-State Conditions *16*

1.4.2	Mimicking Industrial Scale Conditions in the Lab: Continuous-Like Experiments *17*	
1.4.2.1	A Simple Model for Continuous Processes *17*	
1.4.2.2	Continuous-Like Fed-Batch Cultivations *18*	
1.4.3	Fast and Parallel Experimental Approaches with High Information Content *20*	
1.4.3.1	Computer-Aided Operation of Robotic Facilities *20*	
1.4.3.2	Model Building and Experimental Validation *21*	
1.5	Conclusions *22*	
	References *22*	
2	**Tools Enabling Continuous and Integrated Upstream and Downstream Processes in the Manufacturing of Biologicals** *31*	
	Rimenys J. Carvalho and Leda R. Castilho	
2.1	Introduction *31*	
2.2	Continuous Upstream Processes *32*	
2.2.1	Continuous Bioprocesses: With or Without Cell Recycle? *33*	
2.2.2	Early/Scale-Down Perfusion Development *34*	
2.2.3	Feeding and Operational Strategies in Perfusion Processes *35*	
2.2.4	Cell Retention Devices *36*	
2.3	Continuous Downstream Processes *41*	
2.3.1	Continuous Liquid Chromatography (CLC) *42*	
2.3.1.1	Continuous Annular Chromatography (CAC) *42*	
2.3.1.2	True and Simulated Moving Bed Chromatography (TMB/SMB) *43*	
2.3.1.3	Multicolumn Countercurrent Solvent Gradient Purification (MCSGP) *45*	
2.3.1.4	Periodic Countercurrent Chromatography (PCC) *47*	
2.3.1.5	Continuous Countercurrent Tangential Chromatography (CCTC) *50*	
2.3.2	Nonchromatographic Continuous Processes *51*	
2.3.2.1	Continuous Aqueous Two-Phase Systems *51*	
2.3.2.2	Continuous Protein Precipitation *52*	
2.3.3	Straight-Through Processes *53*	
2.3.4	Continuous Virus Clearance Processes *54*	
2.4	Integrated Continuous Processes *55*	
2.5	Concluding Remarks *59*	
	References *60*	
3	**Engineering Challenges of Continuous Biomanufacturing Processes (CBP)** *69*	
	Holger Thiess, Steffen Zobel-Roos, Petra Gronemeyer, Reinhard Ditz, and Jochen Strube	
3.1	Introduction *69*	
3.1.1	Continuous Manufacturing *69*	
3.1.2	Continuous Manufacturing of Synthetic Molecules *69*	
3.1.3	Continuous Manufacturing of Biologics *69*	
3.2	Analysis of CBP Status *71*	

3.3	Case Studies *74*	
3.4	Status and Needs for Research and Development *77*	
3.5	Engineering Challenges *79*	
3.5.1	Platform Method of QbD-Driven Process Modeling Instead of Unit Operation Oriented Platform Approaches *80*	
3.5.2	Data Driven Decisions *81*	
3.5.3	Analytics *82*	
3.5.4	QbD Methods *82*	
3.5.5	Upstream and Downstream Integration *82*	
3.5.6	Buffer Handling/Recycling *83*	
3.5.7	Process Integration of Innovative Unit Operations *84*	
3.5.8	ABC (Anything But or Beyond Chromatography) and AAC (Anything and Chromatography) *84*	
3.5.8.1	Liquid–Liquid Extraction Based on ATPE *84*	
3.5.8.2	Precipitation *86*	
3.5.8.3	Membrane Adsorbers *87*	
3.5.8.4	Innovative Materials Like Fibers or Matrices *88*	
3.5.9	Process Concepts for mAbs and Fragments *88*	
3.5.10	Single-Use Technology *91*	
3.5.11	Guided Decision for CBP *91*	
3.6	Conclusion and Outlook *96*	
	Acknowledgments *97*	
	References *97*	

Part Two: Automation and Monitoring (PAT) *107*

4 Progress Toward Automated Single-Use Continuous Monoclonal Antibody Manufacturing via the Protein Refinery Operations Lab *109*
David Pollard, Mark Brower, and Douglas Richardson

4.1	Introduction *109*	
4.2	Protein Refinery Operations Lab *111*	
4.2.1	Introduction *111*	
4.2.2	Protein Refinery Operations Lab: Design and Implementation *112*	
4.2.3	Protein Refinery Operations Lab: Process Analytical Technology (PAT) and Product Attribute Control (PAC) for the Transition to Real-Time Release (RTR) *117*	
4.2.3.1	Protein Refinery Operations Lab: Current State of PAT Technologies *118*	
4.3	Protein Refinery Operations Lab: Case Studies *122*	
4.3.1	Case Study: Perfusion *122*	
4.3.2	Case Study: Continuous Purification *124*	
4.3.3	Case Study: Proof of Concept Automated Handling of Deliberate Process Deviations *127*	
4.3.3.1	Perfusion Process Deviation Analysis (Bioreactor Temperature Shift) *127*	

4.3.3.2	Downstream Process Deviation Analysis (Viral Inactivation pH)	*128*
4.4	Summary	*129*
	Acknowledgments	*129*
	References	*129*

Part Three: Single Use Technologies and Perfusion Technologies *131*

5 Single-Use Bioreactors for Continuous Bioprocessing: Challenges and Outlook *133*
Nico M.G. Oosterhuis

5.1	Introduction	*133*
5.2	Single-Use Reactor Types	*135*
5.3	Material Aspects	*136*
5.4	Sensors	*139*
5.5	Reactor Design	*141*
5.5.1	Mass Transfer and Mixing Requirements for Continuous Processing	*141*
5.6	Scale-Up Aspects	*142*
5.7	Continuous Seed Train	*145*
5.8	New Mixer Designs	*145*
5.9	Future Outlook	*146*
	References	*147*

6 Two Mutually Enabling Trends: Continuous Bioprocessing and Single-Use Technologies *149*
Marc Bisschops, Mark Schofield, and Julie Grace

6.1	Introduction	*149*
6.2	Single-Use Technologies	*150*
6.2.1	History of Single-Use Technologies	*150*
6.2.2	Single-Use Upstream Processing	*151*
6.2.3	Single-Use Downstream Processing	*151*
6.2.3.1	Tangential Flow Filtration	*151*
6.2.3.2	Chromatography Steps	*152*
6.2.4	Early Skepticism	*152*
6.2.5	Current Trends and Future Predictions	*153*
6.3	Continuous Bioprocessing	*154*
6.3.1	Continuous Upstream Processing	*154*
6.3.2	Continuous Downstream Processing	*155*
6.3.2.1	Tangential Flow Filtration	*156*
6.3.2.2	Continuous Chromatography	*157*
6.3.3	Concerns for Continuous Bioprocessing	*158*
6.4	Integrated Single-Use Continuous Bioprocessing: Case Studies	*159*
6.4.1	Case 1: Genzyme	*159*
6.4.2	Case 2: Merck	*160*

6.4.3	Case 3: Bayer Technology Services	*161*
6.4.4	Comparison	*162*
6.4.5	Challenges and Solutions	*163*
6.4.6	Alternative Scenarios	*164*
6.5	Regulatory Aspects	*164*
6.6	Adoption Rate of Single-Use and Continuous Bioprocessing	*165*
6.7	Conclusions	*166*
	References	*167*

7 Perfusion Formats and Their Specific Medium Requirements *171*
Jochen B. Sieck, Christian Schild, and Jörg von Hagen

7.1	Introduction	*171*
7.1.1	History of Perfusion	*172*
7.1.2	Comeback of Perfusion	*172*
7.2	Characterization of Perfusion Processes	*173*
7.2.1	Productivity of Perfusion Processes	*175*
7.2.2	Cell Retention Devices	*176*
7.2.3	Steady-State Definition	*176*
7.3	Perfusion Formats	*177*
7.3.1	Innovative Perfusion Formats	*178*
7.4	Development Strategies for Perfusion Media	*179*
7.4.1	Cell Line-Specific Requirements	*181*
7.4.2	Scale-Down Models for Perfusion Processes	*181*
7.4.3	Scale-Down Cultivation Methods	*182*
7.4.4	Examples for Perfusion Scale-Down Applications	*184*
7.5	Process Development for Perfusion Processes	*185*
7.6	Case Study	*185*
7.6.1	Material & Methods	*186*
7.6.1.1	Semicontinuous Chemostat (SCC)	*187*
7.6.1.2	Repeated Batch (RB)	*187*
7.6.1.3	Semicontinuous Perfusion (SCP)	*187*
7.6.2	Results	*187*
7.6.2.1	Determination of the Starting Cell Density	*187*
7.6.3	Scale-Down Model Comparison	*188*
7.6.4	Media Screening	*189*
7.6.5	Bioreactor Confirmation	*191*
7.7	Conclusion	*192*
	Abbreviations	*193*
	References	*194*

Part Four: Continuous Upstream Bioprocessing *201*

8 Upstream Continuous Process Development *203*
Sanjeev K. Gupta

8.1	Introduction	*203*
8.2	Upstream Processes in Biomanufacturing	*205*

8.2.1	Upstream Operating Modes	206
8.2.1.1	Fed-Batch Process	206
8.2.1.2	Continuous/Perfusion Process	207
8.3	The Upstream Continuous/Perfusion Process	207
8.3.1	Upstream Process-Type Selection	209
8.3.2	Component of Continuous Upstream and Downstream Processes	209
8.3.2.1	Upstream Components: Stainless Steel and Single-Use (Su)	209
8.3.2.2	Downstream Components: Stainless Steel and Single-Use (Su)	209
8.3.3	Cell Retention Devices Used in Perfusion Process	210
8.3.3.1	Spin Filters	210
8.3.3.2	The ATF System	210
8.3.3.3	Biosep Acoustic Perfusion System	212
8.3.3.4	TFF Cell Retention Device	213
8.4	Manufacturing Scale-Up Challenges	214
8.4.1	Process Complexity and Control	214
8.4.2	Cell Line Stability	215
8.4.3	Validation	215
8.5	Single-Use Technologies: A Paradigm Change	215
8.5.1	Application of SUBs in Continuous Processing	218
8.5.2	Single-Use Continuous Bioproduction	218
8.5.3	Single-Use Perfusion Bioreactors	219
8.5.3.1	Type of Single-Use Bioreactors for Perfusion Culture	219
8.5.4	Single-Use Accessories Supporting Perfusion Culture	220
8.5.4.1	Hollow Fiber Media Exchange	220
8.5.4.2	Continuous Flow Centrifugation	220
8.5.4.3	Acoustic Wave Separation	220
8.5.4.4	Spin filters	220
8.6	FDA Supports Continuous Processing	221
8.7	Making the Switch from Batch/Fed-Batch to Continuous Processing	222
8.8	Costs and Benefits of Continuous Manufacturing	222
8.9	Costs of Adoption	223
8.10	Continuous Downstream Processing	223
8.11	Integrated Continuous Manufacturing	224
8.12	Concluding Remark	227
	Acknowledgment	228
	References	228
9	**Study of Cells in the Steady-State Growth Space**	**233**
	Sten Erm, Kristo Abner, Andrus Seiman, Kaarel Adamberg, and Raivo Vilu	
9.1	Introduction	233
9.1.1	On Physiological State of Cells: Steady-State Growth Space Analysis	234
9.1.2	Challenge of Comprehensive Quantitative Steady-State Growth Space Analysis (SSGSA)	236
9.1.3	Chemostat Culture – A Classical Tool for SSGSA	236

9.2	Advanced Continuous Cultivation Methods – Changestats	237
9.2.1	Accelerostat (A-stat)	237
9.2.2	Family of Changestats – A Set of Flexible Tools for Scanning Steady-State Growth Space	240
9.3	Review of the Results Obtained Using the Changestats	242
9.3.1	Acetate Overflow Metabolism in *E. Coli*	242
9.3.2	A-Stat in Study of Physiology of Yeast	243
9.3.3	Integration of A-Stat with High-Throughput Omics Methods and Modeling	243
9.3.4	A-Stat in Bioprocess Development	243
9.3.5	Deceleration-stat (De-stat)	244
9.3.6	Dilution Rate Stat (D-Stat)	244
9.3.7	Auxoaccelerostats	245
9.3.8	Adaptastat	246
9.4	SSGSA Using Parallel-Sequential Cultivations	247
9.5	Modeling in Steady-State Growth Space Analysis	248
	References	250

Part Five: Continuous Downstream Bioprocessing *259*

10 Continuous Downstream Processing for Production of Biotech Therapeutics *261*
Anurag S. Rathore, Nikhil Kateja, and Harshit Agarwal

10.1	Introduction	261
10.2	Continuous Manufacturing Technologies for Downstream Processing	262
10.2.1	Continuous Cell Lysis	262
10.2.2	Continuous Centrifugation	263
10.2.3	Continuous Refolding	264
10.2.4	Continuous Precipitation	267
10.2.5	Continuous Chromatography	267
10.2.6	Continuous Extraction	271
10.2.7	Continuous Filtration	272
10.3	Continuous Process Development	274
10.4	Case Studies Related to Continuous Manufacturing	276
10.5	Summary	279
	References	279

11 Evolving Needs For Viral Safety Strategies in Continuous Monoclonal Antibody Bioproduction *289*
Andrew Clutterbuck, Michael A. Cunningham, Cedric Geyer, Paul Genest, Mathilde Bourguignat, and Helge Berg

11.1	Introduction	289
11.1.1	Current Regulations and Practices	293
11.1.2	Evolving Needs: Process versus Regulatory	294
11.1.3	Current Technology Landscape	295
11.2	Batch versus Continuous: Potential Impacts on Virus Safety	297

11.2.1	Raw Material Safety/Testing	299
11.2.2	Upstream and Bioreactor Safety	301
11.2.3	Downstream Virus Removal Strategies	304
11.2.3.1	Viral Reduction by Normal Flow Filtration (NFF)	304
11.2.3.2	Chemical Inactivation (Low pH or Solvent Detergent)	308
11.2.3.3	Chromatography	311
11.2.3.4	Other Techniques	312
11.3	Validation of Viral Reduction Steps in Continuous Manufacturing Processes	313
11.3.1	Protein A Capture Chromatography	314
11.3.2	Chemical Inactivation (Low pH/Solvent Detergent)	315
11.3.3	Intermediate and Polishing Chromatography	315
11.3.4	Viral Reduction Filtration	316
11.4	Conclusion	318
	References	319

Part Six: Continuous Chromatography 321

12 Multicolumn Continuous Chromatography: Understanding this Enabling Technology 323
Kathleen Mihlbachler

12.1	Introduction	323
12.2	Modes of Chromatography	326
12.3	Interaction Mechanisms Used in Chromatographic Systems	328
12.4	Batch Chromatography	330
12.5	Semicontinuous and Continuous Batch Chromatography	331
12.5.1	Single Column	331
12.5.2	Multicolumn Parallel Operation	333
12.5.3	Multicolumn Parallel and Interconnected Operation	337
12.6	Multicolumn, Countercurrent, Continuous Chromatography	340
12.6.1	Implementing Traditional SMB Technology	341
12.6.2	SMB Technology for Biomolecules	343
12.6.3	Additional Examples of SMB Purifications	349
12.7	Risk Assessment of Continuous Chromatography	353
12.8	Process Design of Continuous Capture Step	357
12.9	Conclusion	360
	References	361

13 Continuous Chromatography as a Fully Integrated Process in Continuous Biomanufacturing 369
Steffen Zobel-Roos, Holger Thiess, Petra Gronemeyer, Reinhard Ditz, and Jochen Strube

13.1	Introduction	369
13.2	Continuous Chromatography	370
13.2.1	SMB	370
13.2.2	Serial Multicolumn Continuous Chromatography	377

13.2.3　Continuous Countercurrent Multicolumn Gradient Chromatography *378*
13.2.4　Integrated Countercurrent Chromatography *379*
13.3　Conclusion and Outlook *386*
　　　Symbols *388*
　　　References *389*

14　Continuous Chromatography in Biomanufacturing *393*
　　　Thomas Müller-Späth and Massimo Morbidelli
14.1　Introduction to Continuous Chromatography *393*
14.2　Introduction to Manufacturing Aspects of Chromatography *396*
14.3　Trade-Offs in Batch Chromatography *399*
14.4　Capture Applications *400*
14.4.1　Introduction *400*
14.4.2　Process Principle *403*
14.4.3　Application Examples *405*
14.5　Polishing Applications *406*
14.5.1　Introduction *406*
14.5.2　MCSGP (Multicolumn Countercurrent Solvent Gradient Purification) Principle *407*
14.5.3　MCSGP (Multicolumn Countercurrent Solvent Gradient Purification) Process Design *409*
14.5.4　MCSGP (Multicolumn Countercurrent Solvent Gradient Purification) Case Study *412*
14.6　Discovery and Development applications *414*
14.7　Scale-Up of Multicolumn Countercurrent Chromatography Processes *416*
14.8　Multicolumn Countercurrent Chromatography as Replacement for Batch Chromatography Unit Operations *417*
14.9　Multicolumn Countercurrent Chromatography and Continuous Upstream *419*
14.10　Regulatory Aspects and Control of Multicolumn Countercurrent Processes *419*
　　　References *421*

15　Single-Pass Tangential Flow Filtration (SPTFF) in Continuous Biomanufacturing *423*
　　　Andrew Clutterbuck, Paul Beckett, Renato Lorenzi, Frederic Sengler, Torsten Bisschop, and Josselyn Haas
15.1　Introduction *423*
15.2　Tangential Flow Filtration in Bioproduction *426*
15.2.1　Batch versus Single-Pass Tangential Flow Filtration *426*
15.2.2　Membrane Type and Format for TFF Applications *426*
15.2.3　Single-Pass Tangential Flow Filtration (SPTFF) *428*
15.2.4　Process Design *430*
15.2.5　Laboratory-Scale Process Development Example *438*

15.2.6	Consideration on Equipment Configuration and Requirements	*442*
15.3	Validation *445*	
15.3.1	Key Validation Considerations between Batch and Continuous Processing *445*	
15.3.2	Validation of Single-Pass TFF *449*	
15.4	Conclusion *453*	
	References *453*	

Part Seven: Integration of Upstream and Downstream *457*

16 Design of Integrated Continuous Processes for High-Quality Biotherapeutics *459*
Fabian Steinebach, Daniel Karst, and Massimo Morbidelli

16.1	Introduction *459*	
16.2	Perfusion Cell Culture Development *463*	
16.2.1	Objectives and Requirements *463*	
16.2.2	Bioreactor Setup *463*	
16.2.3	Physical Bioreactor Characterization *464*	
16.3	Continuous Capture Development *466*	
16.3.1	Objectives and Requirements *466*	
16.3.2	Continuous Two-Column Capture Process *467*	
16.3.3	Process Performance *468*	
16.3.4	Process Control *469*	
16.4	Operation of the Continuous Integrated Process *470*	
16.4.1	Bioreactor Operation *470*	
16.4.2	Cell Growth *470*	
16.4.3	Monoclonal Antibody Production *471*	
16.4.4	Monoclonal Antibody Capture *472*	
16.4.5	Process Performance *473*	
16.4.6	Product Quality *474*	
16.5	Conclusion *476*	
	Acknowledgment *477*	
	References *477*	

17 Integration of Upstream and Downstream in Continuous Biomanufacturing *481*
Petra Gronemeyer, Holger Thiess, Steffen Zobel-Roos, Reinhard Ditz, and Jochen Strube

17.1	Introduction *481*	
17.2	Background on Upstream Development in Continuous Manufacturing *483*	
17.3	Background on Downstream Development in Continuous mAb Manufacturing *484*	
17.4	Challenges in Process Development *485*	
17.4.1	Impact of Changing Titers and Impurities on Cost Structures *485*	

17.4.2	Impurities as Critical Parameters in Process Development	*487*
17.4.3	Host Cell Proteins as Main Problem in Process Development	*488*
17.4.4	Regulatory Aspects	*490*
17.5	Trends and Integration Approaches	*490*
17.6	Methodical Approach of Integrating USP and DSP Regarding Impurity Processing	*492*
17.6.1	Case Study: Influence of Media Components on Impurity Production	*494*
17.6.2	Case Study: Influence of Harvest Operations on Impurity Production	*495*
17.6.3	Nonchromatographic Continuous DSP Operation	*497*
17.6.3.1	ATPS	*498*
17.6.3.2	Precipitation	*499*
17.6.3.3	One Step Toward a Chromatography Free Purification Process	*500*
17.7	Conclusion and Outlook	*500*
	References	*501*

Part Eight: Quality, Validation, and Regulatory Aspects *511*

18 Quality Control and Regulatory Aspects for Continuous Biomanufacturing *513*
Guillermina Forno and Eduardo Ortí

18.1	Introduction	*513*
18.2	FDA Support for Continuous Manufacturing	*513*
18.3	PAT as a Facilitator for Continuous Manufacturing Implementation	*514*
18.4	PAT Applications in the Pharmaceutical Industry	*516*
18.5	Process Validation for Continuous Manufacturing Processes	*519*
18.6	Regulatory Documents Related to Process Validation	*520*
18.7	ICH	*520*
18.8	FDA	*520*
18.9	EMA	*521*
18.10	ASTM	*521*
18.11	Special Considerations for Continuous Manufacturing Process Validation	*521*
18.12	Scale-Down for Continuous Bioprocessing	*524*
18.13	Impact of Single-Use Systems in Process Validation	*526*
18.14	Batch and Lot Definition	*527*
18.15	Conclusion	*528*
	References	*528*

19 Continuous Validation For Continuous Processing *533*
Steven S. Kuwahara

19.1	Quality Management	*533*

19.2	Regulatory Considerations	*534*
19.3	Setting Specifications	*534*
19.4	Sequence of Events	*535*
19.5	Verification of Validated States	*536*
19.6	Choice of Test Methods	*536*
19.7	Types of Monitoring	*536*
19.8	Process Stream Analyzers	*538*
19.9	Validation/Qualification of Process Stream Analyzers	*538*
19.10	Control Charting	*540*
19.11	The Moving Range Chart	*541*
19.12	Continuous Validation	*541*
19.13	Choosing Other Control Charts	*542*
19.14	Information Awareness	*542*
19.15	Cost Issues	*543*
19.16	Revalidations	*544*
19.17	Management and Personnel	*544*
	References	*545*

20 Validation, Quality, and Regulatory Considerations in Continuous Biomanufacturing *549*
Laura Okhio-Seaman

20.1	Introduction	*549*
20.1.1	What is Continuous Biomanufacturing?	*549*
20.1.2	Improvement in Product Quality	*550*
20.1.3	Manufacturing Consistency	*550*
20.1.4	Efficient Facility and Personnel Utilization	*550*
20.1.5	Reduction in Capital Expenditure and Cost	*550*
20.2	Quality	*551*
20.2.1	Other Considerations in Quality	*552*
20.2.1.1	Contract Manufacturing Organizations (CMO's)	*552*
20.2.1.2	Good Manufacturing Practices (GMP)	*555*
20.2.1.3	Supply Chains	*555*
20.2.1.4	Change Management and Control	*556*
20.3	Validation	*557*
20.3.1	Validate to Eliminate!	*557*
20.3.2	Test Conditions for Extractable and Leachable Analysis	*560*
20.3.3	Test Solutions for Extractable and Leachable Analysis	*561*
20.3.4	Analytical Techniques for Leachables Analysis	*561*
20.3.5	Description of the Model Approach	*562*
20.3.6	Actual Formulation Approach	*563*
20.4	Regulatory	*564*
20.4.1	Current Regulatory References	*565*
20.5	Conclusion	*566*
	Further Reading	*566*

Part Nine: Industry Perspectives *569*

21 **Evaluation of Continuous Downstream Processing: Industrial Perspective** *571*
Venkatesh Natarajan, John Pieracci, and Sanchayita Ghose
21.1 Biogen mAb Downstream Platform Process *571*
21.2 Potential Platform Process Bottlenecks Pertaining to Large Scale Manufacturing *573*
21.3 Continuous Downstream Process *573*
21.3.1 Multicolumn Chromatography (MCC) for Continuous Capture *575*
21.3.1.1 Background *575*
21.3.1.2 Process Optimization *576*
21.3.1.3 Experimental Results *577*
21.3.2 Continuous Viral Inactivation *578*
21.3.3 Connected Chromatography Steps *580*
21.3.3.1 Comparison of Current and Pool-Less Process *581*
21.3.4 Continuous UF/DF Processes *582*
21.4 Productivity Comparison of Batch and Continuous Downstream Process *585*
References *585*

Index *587*

List of Contributors

Kristo Abner
Competence Center of Food and
Fermentation Technologies
Akadeemia tee 15
12618 Tallinn
Estonia

Kaarel Adamberg
Tallinn University of Technology
Department of Chemistry and
Biotechnology
Akadeemia tee 15
12618 Tallinn
Estonia

and

Competence Center of Food and
Fermentation Technologies
Akadeemia tee 15
12618 Tallinn
Estonia

Harshit Agarwal
Indian Institute of Technology
Department of Chemical
Engineering
Hauz Khas
110016 New Delhi
India

Paul Beckett
Millipore SAS
Process Solution Technologies
39 Route Industrielle de la Hardt
67124 Molsheim
France

Helge Berg
Technology Management
Millipore SAS
39 Route Industrielle de la Hardt
67124 Molsheim
France

Torsten Bisschop
Millipore SAS
Process Solution Technologies
39 Route Industrielle de la Hardt
67124 Molsheim
France

Marc Bisschops
Pall Life Sciences
Scientific Laboratory Services
Nijverheidsweg 1
1671 GC Medemblik
The Netherlands

Mathilde Bourguignat
Technology Management
Millipore SAS
39 Route Industrielle de la Hardt
67124 Molsheim
France

List of Contributors

Mark Brower
Merck & Co Inc
Biologics & Vaccines
2000 Galloping Hill Road
Kenilworth, NJ 07033
USA

Rimenys J. Carvalho
Federal University of Rio de Janeiro
COPPE
Cell Culture Engineering Laboratory
C.P. 68502
21941-972 Rio de Janeiro, RJ
Brazil

Leda R. Castilho
Federal University of Rio de Janeiro
COPPE
Cell Culture Engineering Laboratory
C.P. 68502
21941-972 Rio de Janeiro, RJ
Brazil

Cedric Geyer
Technology Management
Millipore SAS
39 Route Industrielle de la Hardt
67124 Molsheim
France

Andrew Clutterbuck
Millipore SAS
Process Solution Technologies
39 Route Industrielle de la Hardt
67124 Molsheim
France

M. Nicolas Cruz-Bournazou
Technische Universität Berlin
Department of Biotechnology
Ackerstrasse 76
ACK 24
13355 Berlin
Germany

Michael A. Cunningham
Technology Management
EMD Millipore Corporation
290 Concord Road
Billerica, MA 01821
USA

Reinhard Ditz
Clausthal University of Technology
Institute for Separation and Process Technology
Leibnizstr 15
38678 Clausthal-Zellerfeld
Germany

Sten Erm
Tallinn University of Technology
Department of Chemistry and Biotechnology
Akadeemia tee 15
12618 Tallinn
Estonia

and

Competence Center of Food and Fermentation Technologies
Akadeemia tee 15
12618 Tallinn
Estonia

Guillermina Forno
Ciudad Universitaria
Cell Culture Laboratory
UNL
FBCB
Paraje el Pozo
CC 242 Santa Fe
Argentina

and

Ciudad Universitaria
R&D Zelltek S.A.
UNL
FBCB
Paraje el Pozo
CC 242 Santa Fe
Argentina

Paul Genest
Technology Management
EMD Millipore Corporation
290 Concord Road
Billerica, MA 01821
USA

Sanchayita Ghose
Bristol-Myers Squibb
Downstream Process Development
38 Jackson Road
Danvers, MA 01923
USA

Julie Grace
Pall Life Sciences
Scientific Laboratory Services
20 Walkup Drive
Westborough, MA 01581
USA

Petra Gronemeyer
Clausthal University of Technology
Institute for Separation and Process Technology
Leibnizstr 15
38678 Clausthal-Zellerfeld
Germany

Sanjeev K. Gupta
Ipca Laboratories Ltd.
Advanced Biotech Lab
Kandivli Industrial Estate
Kandivli (west)
400067 Mumbai
India

Josselyn Haas
Millipore SAS
Process Solution Technologies
39 Route Industrielle de la Hardt
67124 Molsheim
France

Daniel Karst
ETH Zurich
Institute for Chemical and Bioengineering
Department of Chemistry and Applied Biosciences
Vladimir-Prelog-Weg 1
8093 Zurich
Switzerland

Nikhil Kateja
Indian Institute of Technology
Department of Chemical Engineering
Hauz Khas
110016 New Delhi
India

Steven S. Kuwahara
GXP BioTechnology LLC
Tucson, AZ 85741
USA

Renato Lorenzi
Millipore SAS
Process Solution Technologies
39 Route Industrielle de la Hardt
67124 Molsheim
France

Kathleen Mihlbachler
Lewa Process Technologies
Inc. Separations Development
8 Charlestown Street
Devens, MA 01434
USA

Massimo Morbidelli
ETH Zurich
Institute for Chemical and
Bioengineering
Department of Chemistry and
Applied Biosciences
Vladimir-Prelog-Weg 1
8093 Zürich
Switzerland

Thomas Müller-Späth
ChromaCon AG
Process Development
Technoparkstrasse 1
8005 Zurich
Switzerland

and

ETH Zurich
Institute for Chemical and
Bioengineering
Department of Chemistry and
Applied Biosciences
Vladimir-Prelog-Weg 1
8093 Zürich
Switzerland

Venkatesh Natarajan
Biogen
Engineering & Technology
225 Binney Street
Cambridge, MA 02142
USA

Peter Neubauer
Technische Universität Berlin
Department of Biotechnology
Ackerstrasse 76
ACK 24
13355 Berlin
Germany

Laura Okhio-Seaman
Sartorius Stedim North America
Validation Services
5 Orville Drive
Bohemia, NY 11716
USA

Nico M.G. Oosterhuis
Celltainer Biotech BV
Bothoekweg 9
7115AK Winterswijk
The Netherlands

Eduardo Ortí
Ciudad Universitaria
R&D Zelltek S.A.
UNL
FBCB
Paraje el Pozo
CC 242 Santa Fe
Argentina

John Pieracci
Biogen
Engineering & Technology
225 Binney Street
Cambridge, MA 02142
USA

David Pollard
Merck & Co Inc
Biologics & Vaccines
2000 Galloping Hill Road
Kenilworth, NJ 07033
USA

Anurag S.Rathore
Indian Institute of Technology
Department of Chemical
Engineering
Hauz Khas
110016 New Delhi
India

Douglas Richardson
Merck & Co Inc
Biologics & Vaccines
2000 Galloping Hill Road
Kenilworth, NJ 07033
USA

Christian Schild
Merck Life Science
(a business of Merck KGaA) Process Solutions
Cell Culture Media R&D
Frankfurter Strasse 250
64291 Darmstadt
Germany

Mark Schofield
Pall Life Sciences
Applications R&D
20 Walkup Drive
Westborough, MA 01581
USA

Andrus Seiman
Tallinn University of Technology
Department of Chemistry and Biotechnology
Akadeemia tee 15
12618 Tallinn
Estonia

and

Competence Center of Food and Fermentation Technologies
Akadeemia tee 15
12618 Tallinn
Estonia

Frederic Sengler
Millipore SAS
Process Solution Technologies
39 Route Industrielle de la Hardt
67124 Molsheim
France

Jochen B. Sieck
Merck Life Science
(a business of Merck KGaA)
Process Solutions
Cell Culture Media R&D
Frankfurter Strasse 250
64291 Darmstadt
Germany

Fabian Steinebach
ETH Zurich
Institute for Chemical and Bioengineering
Department of Chemistry and Applied Biosciences
Vladimir-Prelog-Weg 1
8093 Zurich
Switzerland

Jochen Strube
Clausthal University of Technology
Institute for Separation and Process Technology
Leibnizstr 15
38678 Clausthal-Zellerfeld
Germany

Holger Thiess
Clausthal University of Technology
Institute for Separation and Process Technology
Leibnizstr 15
38678 Clausthal-Zellerfeld
Germany

Raivo Vilu
Tallinn University of Technology
Department of Chemistry and Biotechnology
Akadeemia tee 15
12618 Tallinn
Estonia

and

Competence Center of Food and
Fermentation Technologies,
Akadeemia tee 15
12618 Tallinn
Estonia

Jörg von Hagen
Merck Life Science (a business of
Merck KGaA) Process Solutions
Cell Culture Media R&D
Frankfurter Strasse 250
64291 Darmstadt
Germany

Steffen Zobel-Roos
Clausthal University of Technology
Institute for Separation and Process
Technology
Leibnizstr 15
38678 Clausthal-Zellerfeld
Germany

Part One

Overview of State-of-the-Art Technologies and Challenges

1

Continuous Bioprocess Development: Methods for Control and Characterization of the Biological System

Peter Neubauer and M. Nicolas Cruz-Bournazou

Technische Universität Berlin, Department of Biotechnology, Ackerstrasse 76, ACK 24, 13355 Berlin, Germany

1.1 Proposed Advantages of Continuous Bioprocessing

1.1.1 Introduction

The change from batch to continuous processing has led to the intensification of processes in a number of industries, including steel casting, automobile and other devices, petrochemicals, food, and pharmaceuticals. Advantages include, aside from a significant increase in volumetric productivity, reduced equipment size, steady-state operation, low cycle times, streamed process flows, and reduced capital cost.

In bioengineering, continuous processing is the standard in wastewater treatment, composting, and some bioenergy processes such as biogas and bioethanol fermentations. In contrast, most production processes run as batch type operations or more specifically fed-batch processes, which is the major production technology today.

Konstantinov and Cooney provide a definition of a continuous process as "A unit operation is continuous if it is capable of processing a continuous flow input for prolonged periods of time. A continuous unit operation has minimal internal hold volume. The output can be continuous or discretized in small packets produced in a cyclic manner." [1]. They also differentiate between full continuous processes with no or minimal hold volume in the process line or hybrid processes that contain both batch and continuous process operations.

Obviously, the push in continuous manufacturing technologies was initiated by the BioPAT initiative of the Food and Drug Administration (FDA) in 2002 and the published guidance to PAT in 2004 [2], which initially aimed at a better understanding of the connections between product quality and process conditions. This lead to the need to develop quality by design (QbD), that is, the implementation of process analytical tools over the whole developmental pipeline from early product screening over the process development in the laboratory scale and during scale up. The needs for a better understanding of the impact of process parameters on the critical quality attributes (CQA) of the respective product also increased the interest in the development and implementation of novel sensors and analytical

Continuous Biomanufacturing: Innovative Technologies and Methods, First Edition.
Edited by Ganapathy Subramanian.
© 2018 Wiley-VCH Verlag GmbH & Co. KGaA. Published 2018 by Wiley-VCH Verlag GmbH & Co. KGaA.

tools. As a consequence, this better understanding of processes resulted in further process intensification and provided the instrumental basis to approach challenges in relation to continuous operation.

Aside the FDA initiative, there are several drivers for the increasing interest in continuous processing, not only in the pharmaceutical industry but also in the industrial (white) biotech industry. On one side, we see an increasing demand and thus also increasing production scale for industrial bioproducts (enzymes, small molecules, and bioenergy market) with a need for reduced costs for the products and increased competition. Considering that production scales are steadily growing and that a scale reduction close to factor 10 would be possible by continuous processing, plant sizes and the efficiency of bioprocesses could be increased significantly. On the other side, the opportunity of the selection of new biocatalysts and its implementation in the chemical synthesis for integrated chemoenzymatic processes (i.e., processes which combine chemical and enzymatic reactions) have to be competitive with the existing chemical processes and need to be integrated into the chemical production schemes. Here, continuous processes offer clear advantages.

In biopharma for recombinant proteins, antibodies, highly complex proteins, recombinant enzymes and blood factors, the efficiency of the cell factories, and production systems have dramatically increased during the last decade. Opportunities for high cell density processes with a higher volumetric product yield and quality, as well as the changing situation in view of the intellectual properties by the termination of many patents for important drugs with novel commercial opportunities for new biosimilars and biobetters are a strong driver in increasing the competition especially from emerging markets. In parallel, there is an increasing demand for establishing local production sites for defined regional markets, rather than having single production sites. Strict cost calculations as a developmental driver demand for smaller and effective, but also flexible production plants. This directs interest to evaluate continuous bioprocessing opportunities to minimize investments for production facilities, and thinking about parallelization rather than larger scales. Parallelization would also be an advantage in processes with longer plant cycle times [3] as, for example, cell culture-based products. A nice example that shows the opportunities in significantly decreasing operational and capital expenses by changing from conventional bioprocessing to continuous bioprocessing in the case of production on monoclonal antibodies (mAB) and other non-mAB processes is shown by Walther et al. [4].

However, despite the obvious opportunities of continuous processes there are many challenges to solve, mainly the demand for fast realization and risk minimization. Currently, it seems to be easier to transfer a batch process into production than to start a new, longer, and more expensive development of a continuous process even though it is expected to be more efficient.

These scenarios show that there is a big need in strategic methods concerning the development of continuous process strategies for either new products or to derive a continuous process from existing batch type processes. As early-phase product development can practically be only performed as batch processes, a key question in product development is how we can transfer a batch strategy to a continuous operation in a large process.

Specific challenges for continuous operations in the biotech industry compared to other industries are (i) the inherent nature of a natural whole cell biocatalyst, that is, a prokaryotic or eukaryotic cell, for steady evolution of its genome. (ii) Biotechnological processes generally need much higher amounts of catalyst compared to chemical industry. This biocatalyst has to be maintained by feeding. Thereby, the feed is mostly the same substance as the original substrate for the biochemical reaction and thus it competes with the yield. As a consequence, diffusion and mass transfer in the reactor have not only an effect on the efficiency of the process but also are critical for long-term operation and maintenance of the product quality.

Especially in view of the evolutionary adaptation, it is obvious that in contrast to batch procedures, continuous operation cannot be set up by trial and error but needs a fundamental basic understanding of the process in kinetic terms to allow a control of the process. This is a fundamental paradigm change in bioprocessing industries, where most operations included only a limited application of mathematical models. Also, available sensors and ways for process control are traditionally very limited.

Although current examples of so-called continuous cell culture processes in the pharma area are only semicontinuous if compared to, for example, biogas or wastewater bioprocessing, the expansion of cell cultivation processes from days to months, for example, by the perfusion technology, goes into the direction of continuous bioprocessing. However, also for these continuous bioprocess operations a big amount of labor and time is needed to establish a new process as a continuous production system. Similar as in traditional biotech, the development is currently mainly based on wet-lab experiments rather than on a systematic developmental approach. This raises the question, whether this is the only way that has proven to be successful or whether a paradigm change in the application of technologies is needed to advance the field of continuous bioprocessing. In this chapter we will discuss the methods and data which are needed to develop a continuous operation with a focus on modeling approaches.

1.2 Special Challenges for Continuous Bioprocesses

1.2.1 The Biological System in Continuous Biomanufacturing

In process engineering, there are different reactor designs typically used depending on the characteristics of the reaction system. It should be pointed, that continuous stirred tank reactors (CSTR) are known to have some disadvantages over plug flow reactors (PFR), which are typically used in the chemical industry and batch and fed-batch type of cultures – the most important being a lower concentration of product due to a constant dilution. In chemical processes, typically PFRs are preferred when high yields are required, and batches are chosen when low volumes of different product are to be produced. In addition, considering process development, continuous processes require higher investments and times at small scales. Hence, even in the chemical industry many processes with similar characteristics to protein synthesis run in PFR or batch. Still there are many process setups that allow a significant reduction in footprint, time, and costs by using continuous processes and these should be considered in bioengineering.

Now let us focus on the processes that should run continuously in bioengineering. Continuous bioprocesses are characterized by a continuous addition of substrates and simultaneous harvest, thus that the bioreactor volume stays approximately constant. If the product is released from the cell and accumulates in the culture broth, recycling by filtration units or centrifuges can be applied and this separation can be supported by immobilization of the biocatalyst. Alternatively, the bioprocess can be performed in solid-state fermentation processes where the medium runs through a static matrix where the biocatalyst is immobilized.

The biological system, which is the core component of a biotechnological production process, has many specific features that favor continuous production on one side, but also cause the specific challenges that restrict continuous processing in the biotech industries so far.

In difference to chemistry, where continuous operation became a standard, and where the change from batch to continuous operation has contributed to a significant drop in reactor sizes and investment costs and lead to modularization of production plants, bioprocesses are rather different. To understand the special challenges of continuous bioprocessing, it is important to analyze these differences between chemical and biocatalytic processes, both of which are different routes for the same outcome – the production of a chemical molecule.

A chemical reaction is characterized by the high yield of the reaction from a substrate into a product using small amounts of catalysts (order of $\mu g/l$ or mg/l). In contrast, in a bioprocess the catalyst is the (micro)-organism, which has to be produced from the substrate in a very high concentration (in industry mostly 50–100 g/l). This is mostly done in the first phase of a process (called growth phase). The product is not produced until the second phase, called production phase, which may be equal in length or even shorter than the growth phase. Interestingly, the product reaches low yields if compared to single reaction chemical processes. Highest concentrations of bioproducts are in the order between 100 and 200 g/l, or up to the order of 10–100 g/l for more complex molecules, and for many processes the yield is even lower. Generally, the yield of the product from glucose is significantly lower than 0.1 g/g since most of the glucose goes in the biomass, that is, the production of the biocatalyst. If one assumes that after biomass production the cells could be used to produce the product over unlimited time (considering that substrate required for the own maintenance is less than 10% of the amount needed for the growth phase), a significant increase of the product yield per substrate in a bioprocess would be possible. Furthermore, if turn-around times are avoided, such as harvesting of the reactor, cleaning, and preparation of the new batch including preparation of the starter cultures, time costs would also be reduced further increasing process efficiency.

1.2.2 Inherent Changes in the Microbial System – Problem of Evolution

In contrast to continuous processes in chemical engineering, bioprocesses include a perpetual evolution process, that is, a genetic change of the biocatalyst.

The rate of mutations has been extensively discussed and investigated, mainly in the context of how a continuous culture can be used for strain evolution.

However, it needs to be stated that the problem of evolution is inherent to any bioprocess. In a bioprocess the operation always starts with a very low number of cells that are multiplied during inoculum preparation and scale-up from flasks to the final size bioreactor. For a large scale bioreactor running at a scale of $10\,m^3$ (which is small in view of most common microbial bioprocesses) and a final cell number of 1.2×10^{18} for a bacterial process (i.e., 1.2×10^{11} cells/ml),[1] this makes approximately 60 generations from a selected one-cell clone and even approximately 27 generations from an inoculum stock. If the calculations concerning the accumulation of mutations as performed by Ref. [5] were assumed to be 1×10^{-3} nucleotide substitutions per genome in each generation for *Escherichia coli* [6] and 4×10^{-3} nucleotide substitutions per genome per generation in *Saccharomyces cerevisiae* [7], and would be adopted to the considered large-scale cultivation case, when assuming a homogenic frozen stock culture (which is not the case), there would be at the end of the cultivation a probability of approximately 30 mutations per each position of the nucleotide sequence of *E. coli* and 0.07 mutations per position for the yeast *S. cerevisiae*. This is an incredible diversity, which is rarely applied in molecular evolution experiments and still not exploited. These considerations may even underestimate the possible mutation frequency, as discussed in detail for chemostat environments by Ferency [8]. Therefore, it is a great challenge in continuous cultivation to direct evolution in order to guarantee continuous production of a constant product with equal quality despite the steady evolution process. While this is greatly ignored in industrial batch processes, this evolution and selection of the fittest is a critical point in continuous processing.

What can be learned from a large number of evolution experiments, especially from the extensive work of Lensky *et al.* [9], is that characteristic changes occur stepwise, that is, suddenly. Also, most importantly, fitness gain is constant with steady improvements without an upper end even in very long experiments over (tens of) thousands of generations. Experiences with natural evolution and steady fitness gain with a selection pressure for important cellular parameters, like affinity constants and maximum specific rates as described in the model below, and steady change of them by mutations [5] are a clear advantage if the continuous culture is aimed at degrading a compound which serves as a key substrate, like in degradation of wastes. However, the same process is a challenge for control in production of biomolecules where it is not possible to set a selection pressure toward the product.

1.2.3 Lack of Process Information

Historically, the low reproducibility, observability, controllability, and understanding of the biotechnological processes has driven large-scale production to an approach based on: (i) as fast as possible after inoculation, (ii) as fast as possible after induction, (iii) as fragmented as possible to avoid mixing of failed charges with high purity product. For these reasons, the fed-batch technology has become the most widely

---

1 Assuming $1g/l = 2 \times 10^{12}$ cells/l, final dry cell weight of 60 g/l.

applied standard upstream production method in large scale. Consequently, all other processes taking place before (pretreatment and medium preparation) and after (down-stream operations) were developed to meet the requirements of discrete production plants. Despite this success of the fed-batch technology, it is remarking that continuous bioprocesses have been early established [10] in parallel to the development of the fed-batch technology by Hayduck [11].

Today's advances in technology are paving the way for continuous systems with improvements in PAT, higher automation of the production, advanced control strategies, methods to direct and apply the naturally occurring evolution process to higher productivity, and most importantly regulations that promote continuous processes. This may clearly offer significant advantages by solving important challenges as are higher overall productivity, lower risk of infection, smaller reaction vessels and plants.

1.2.3.1 Models-Based Process Development and Control for Continuous Processes

In bioprocess engineering, models are used for process design, monitoring, control, and optimization [12–16]. Bioprocess complexity and restrictions have driven design and control to demand accurate and robust models [17], triggering a rapid development over the last years [18]. Advanced sensor techniques and fast computer processors enable the creation of very complex models processing enormous amounts of information [15,19]. Models are not only used to describe the behavior of living organisms but also essential to map complex systems into smaller dimension and also to obtain indirect measurements and observe non-observable events when applied as software sensors [20], for example.

The new regulations of the PAT initiative of the FDA and EMA show the importance that modeling applied to process monitoring and control is gaining in the pharmaceutical and in general in biotechnological processes [21].

Biological processes are characterized by [22–25]

- the complexity of the biological processes taking place in the bioreactor
- lack of reproducibility
- monitoring of unstable intracellular compounds at very low concentrations
- extremely fast and sensible reaction to environmental changes (offline measurements are inaccurate)
- mutations
- insufficient sensor technology
- expensive and inaccurate sensors
- highly invasive
- difficult to calibrate
- large time delays and low frequency of observations

1.2.3.2 Engineering Approach to Complex Systems

In chemical engineering, the implementation of different methods to deal with large complex systems has a long history. Engineers have developed methods like hierarchical modeling, model reusability, model inheritance, and so on. An extensive discussion of these methods and their application for the simulation of chemical plants is presented by Barton [12]. In biological systems, the modularization of separated instances of the system is not always possible. In traditional process engineering, a pump can be modeled in a modular form and then added to the flow sheet of the plant and reused as many times as needed [26]. Contrary to this, biological systems tend to show different behavior under *in vitro* conditions compared to their *in vivo* state [27]. Still, some approaches intend an analysis and modeling of biological systems with methods taken from engineering [28,29]. Kitano [30,31] emphasizes that the only possibility to understand living organisms is to consider the system as a whole. Identifying genes and proteins is only the first step, whereas real understanding can only be achieved by uncovering the structure and dynamics of the system. Kitano states the following four key properties [30]:

- *System structure*
    System structure identification refers to understanding both the topological relationship of the network components and the parameters for each relation.
- *System dynamics*
    System behavior analysis suggests the application of standardized techniques such as sensitivity, stability, stiffness, and bifurcation.
- *The control method*
    System control is concerned with establishing methods to control the state of biological systems.
- *The design method*
    System design is the effort to establish new technologies to design biological systems aimed at specific goals, for example, organ cloning techniques.

For this reason, two things are necessary in order to control a biological system and comply with the strict food and pharmaceutical regulations, namely, to observe or at least deduce the state of critical quantities and to predict to some extent the behavior of the system. The first issue is tackled using state and parameter estimation methods [32–36] in an effort to infer the conditions of the process using the information that is available. Second, the evolution of the system over time as well as its response to the control actions, characterized by nonlinear dynamics, can be predicted using mathematical models.

1.2.4 Limited Control Strategies

1.2.4.1 Traditional Control Strategies for Continuous Cultures

As the chemostat works at a preset dilution rate without any feedback control, it cannot stabilize in a process that runs close to the maximum specific growth rate. Although the chemostat is the most used continuous culture technique in research, it is rarely applied in industry. In this context other processes with a feedback control loop have developed, such as the turbidostat and the pH auxostat.

The *turbidostat* (see Ref. [37] for an excellent early review) is a continuous process where the feed rate is controlled by the online measurement of the turbidity, mostly in the outlet stream. By maximizing the flow rate at a high biomass concentration, the turbidostat can operate the process at μ_{max} and, at the same time, avoids outwashing of the biomass. Therefore, it is generally applied in processes where the flow rate should be maximized, but while maintaining the cells in the system, for example, in processes which aim the degradation of toxic compounds in wastewater treatment, or directly for the production of biomass (single cell protein), or other growth-related molecules. The turbidostat has been also a powerful tool for the selection of faster growing strains, that is, in natural selection, due to its permanent adaptation of the flow rate [38–40].

The necessity of the turbidostat to have a continuous measurement of the turbidity limits its applications to systems with no biofilm formation at the walls. Therefore, the *pH auxostat (pH stat)*, which uses the pH as a state control variable, is more robust, but is also more sophisticated in terms of the design of the medium. In the pH-auxostat the flow rate is set by the pH controller through the feeding of fresh medium to keep the pH constant. Thus, the pH auxostat can be used only for processes where biomass growth is closely correlated to changes on the pH [41]. Early pH auxostats [42,43] were typically applied for microbial processes with acidifying products, for example, in the dairy fermentation [44] or similar anaerobic processes. Such processes needed special pretreatment of the fed medium that had to be preadjusted for a certain pH and thus the biomass concentration in the reactor depended on the difference between the pH difference between the feed solution and the fermenter broth as well as on the buffer capacity of the medium, and normally not high cell densities were obtained [42].

The theoretical solution of the performance of a pH auxostat with two inlet flows, medium and a pH controlling agent, by Larsson *et al.* [45] made the pH auxostat more easy to handle and applicable as a tool also for aerobic processes. The kinetic model contains an extra function that calculates the hydrogen ions in the added base, and considers that the added new medium has the same pH as the control point for the pH in the bioreactor. The process is controlled by the definition of the inlet flow ratio of the two inlet streams, which can be easily set. This inlet flow ratio is a parameter that provides a reliable tool for process optimization. This principle was adapted to processes with an ammonia (NH_4^+) feed, for which the H^+ concentration is in good relation with the cellular uptake of NH_3 and the yield coefficient for hydrogen ions on substrate used is constant, such as for *S. cerevisiae* [41].

A third principle for control of a continuous bioprocess, namely, the nutristat, aiming at maintaining the substrate concentration at a certain level, has found less application. If online methods for the determination of the substrate are available, the nutristat is well suited to run a process at higher growth rates which are lower than μ_{max}. This is a clear advantage over the pH- or turbidostat. However, an interesting study by Rice and Hempfling [46] shows that even a pH stat can run stable at different concentrations of the growth limiting substrate, that is, specific growth rates, by variation of the substrate concentration in the feed solution or by changing the buffering capacity of the feed solution. The nutristat is clearly useful for the degradation of waste compounds, as shown, for example, by Refs [47,48]. When using glucose as substrate, the nutristat can be applied [48], however due to the low K_S value and high specific substrate uptake rate, a proper control below μ_{max} is becoming challenging, especially at high cell densities.

A special solution for culture processes with production of secreted products, such as monoclonal antibodies, at low or even zero growth rates is cell recycling. The theory of continuous cultures with cell recycling was already developed and experimentally proven in Ref. [49] based on the idea of continuous culture with cell recycling [50]. While in these processes the theory of a chemostat (see above) is valid with the cell recycling term (which can be from 0 to 1), most production processes rely on higher growth rates, as the metabolic activities at low growth rates are low and the energy supply is going to the maintenance mainly. However, the cell recycling (or cell retention, retentiostat, perfusion culture) is a practical method to increase degradation capabilities, for example, in waste treatment or maintaining organisms for which even μ_{max} is very low.

The method of perfusion culture is widely and very successfully applied in mammalian cell culture and similar processes with a low growth rate, where, for example, the production of monoclonal antibodies can be stably maintained for longer times (see Chapter 7 of this volume). Cell retention in these systems today is mainly achieved either by immobilization of the cells, by filtration, for example, use of alternating tangential flow filtration, or by centrifugation [51,52].

1.3 Changes Required to Integrate Continuous Processes in Biotech

1.3.1 A Better Physiological Understanding of the Organisms and Their Responses on the Reactor Environment

1.3.1.1 Model Complexity

The biggest challenge for modeling is to develop a general and systematic approach to find the simplest manner to describe complex systems aiming at the strictly required accuracy. The meaning of model simplification becomes more important with the increasing complexity of bioprocesses analyzed in research. The complexity of biological processes makes it very difficult to fully describe cultivations using a computer model. To name one example consider the phenotype of a microbial cell determined by >30 million macromolecules, >1000 species of small organic molecules fine-tuned in composition and number to the comprehensive set of its environmental factors [53].

In order to achieve a robust and efficient continuous process, a close monitoring of the system and a tight control are required. Due to the complexity of the microbial behavior, standard feedback control methods are not adequate and more advanced methods are required. Process control has a long tradition in development of model-based techniques, for monitoring (e.g., softsensors, observers, moving horizon estimators) and control (e.g., model predictive control, adaptive control). These methods relay on a mathematical model that fulfils some specifications as are follows:

- Ability to accurately describe the dynamic behavior of the system
- Identifiability
- Tractability

1.3.1.2 Models

A model is a poor mathematical representation of a physical system. Lack of accurate knowledge of the process to be modeled, insufficient measurement techniques, and extensive computation time hinder an exact representation of the phenomena to be described [54]. Nevertheless, models are widely used in science and their contribution to a better understanding of engineering processes and their proper design, optimization and control is out of question. Computer aided tools using model-based methods allow optimal design and operation of plants, reducing energy consumption, hazard, and environmental impact, while allowing better monitoring and control [12].

From this it can be deduced that the best model to describe a certain process is not necessarily the most accurate, but the one that describes only the relevant aspects of the system so as to get a good description with minimal effort [55].

Modeling includes a wide number of tools [56] as are principal component analysis (PCA) or partial least squares (PLS) [57], nonlinear models like neural networks [58] and also multivariate statistics [57]. Roughly said, these methods search for data correlation to reduce the dimension of the data set [59]. Also more advanced methods in knowledge discovery of data (KDD) like data mining [60] have been developed for treatment of large data sets and are applied in bioinformatics. Still, generally speaking, first principle modeling (white box) is the preferred approach to describe a complex system when mechanistic understanding (mass balances, thermodynamics, kinetics, etc.) is at hand [25,61–63]. These methods study the data characteristics to find new relations between variables and create black-box type models that describe it. By these means it is possible to look through high dimensional data and detect the most important characteristics of the system [64–66].

Contrary to black box models, mechanistic models are based on physical knowledge of the system to be described. In engineering, for example, rigorous modeling includes mass and energy balances, detailed reaction pathways, and so on. Models are the core of computer aided process engineering (CAPE) [67] and computer aided biology (CAB). The quality of every work on simulation, optimization, design, and model-based control, depends on the characteristic of the model. In engineering, models are not only used to describe the behavior of systems but also essential to map complex systems into smaller dimension more comprehensible to humans. Finally, they also serve to obtain indirect measurements of states or parameters of interest with software sensors [20], for example. Software sensors substitute measurements, which are not possible due to physical limitations, with models which predict the behavior of the nonmeasurable variable based on indirect measurements. Whenever a state of the system is to be determined, observer can be applied [35,36] or the Kalman filter [33] with its variations [34].

In system biology, various methods exist aiming at an adequate description of the dynamics of living organisms studying their gene regulatory networks [68].

Still, differential equation systems settle the standard modeling method in engineering. Systems of ordinary differential equations (ODE) have been widely applied for the description of gene regulatory networks. Usually, the system comprises rate equations of the form

$$\frac{dx_i}{dt} = f_i(x, u), \tag{1.1}$$

where x is the vector of concentration of proteins, mRNAs, or other molecules, u the vector of inputs, and f_i is a nonlinear function. Also, time delays can be added if necessary. Typical types of equations used are Monod type, switching, Heaviside, and logoid functions among others. An important advantage of nonlinear ODEs is the possibility to describe multiple steady-states and oscillations in the system [69]. Besides the requirement of testing the global convergence of the optimal solution, the bottleneck is still the state information of the parameter set creating identifiability problems. Nevertheless, many successful applications have been published showing the possibilities of ODEs to describe gene regulatory networks [70].

Although today gene regulatory network models are not applicable in industrial scales, it can be expected that systematic conversion of complex gene regulatory network models in simple tractable models will be possible in near future. Nevertheless, model complexity is closely related to instability, over parameterization, parameter correlation, and low parameter identifiability [71]. The effort required to develop and fit a model has to be justified by its application. It is useless to apply computational fluid dynamics (CFD) to the simulation of a 1 l reactor knowing that the concentration gradients can be neglected. On the other hand, simulating a reaction in a tank with 10 000 l without considering mass transfer limitations may yield in results far from reality. Summarizing, the key dynamics of a system need to be identified, isolated, and analyzed before any model is built. Currently, limitations are mainly due to the scarcity of measurement possibilities but also to the insufficiency of adequate mathematical tools.

1.3.2 Model-Based Process Monitoring

A key task in process control is to monitor the critical states and parameters of the process in order to secure proper operating conditions and desired quality even under perturbations and model mismatch [72,73]. Unfortunately, bioprocesses are characterized by their low information content caused by low concentrations, complex media, and the lack of noninvasive online sensors that can measure intracellular concentrations [74]. To overcome these problems, model-based methods to infer the conditions of the process can be applied. There is a long list of methods and applications for online state estimation [32], state observers including the classical Kalman filters [33] with their variations [34] and nonlinear observers [35,36]. Some authors use the expression software sensor or softsensor [75,76] in account of the fact that a "software" or computer-based calculation of a nonmeasurable variable which is not always a state variable, like the respiration coefficient, provides more information than the initial variable that can be directly measured.

1.3.3 Implementation of Model Predictive Control

1.3.3.1 Model-Based Control

Advances in computer capacity, sensor technology, and a better understanding of the biological system are giving place to very successful applications of advanced control strategies in continuous processes [77,78]. In the case of continuous processes, different approaches have been developed and successfully applied. Some representative examples are classical approaches to various control strategies [79–84] and state feedback control strategies [85,86]. Additionally, efforts to exploit the data generated using neural networks [64,87,88] without the need of a thorough understanding of the system have been shown. In an effort to simplify the control strategy, fuzzy logic controls [89] have also been used in bioreactors. Furthermore, advanced methods using model predictive controllers [64–66], and even based on population models [22] can be found in literature. An interesting approach is to overcome the limitations of the existing models by performing recursive estimation of its parameter estimates. By these means, models can be used also in processes that change over time. Some proves of the potential of these methods are the use of adaptive control techniques [90–92]. Finally, investigation of complex formulation for optimal control show the controllability potential of biological systems if simplified mathematical descriptions succeed to predict the dynamics of the process [93,94].

In general, the theory of monitoring and control is used under the assumption that the system to be described is time invariant and properly described by the model. Nevertheless, there exists the possibility to adapt the observer to changing system behavior by estimating the model parameters together with the states in order to adapt the model to changes in the system. This is called adaptive control [95] and proved to be very effective for many applications including bioreactors [90]. Adaptive control can be used to overcome structural deficiencies of the model as well as uncertainty in the parameters. Still an important drawback is the need of increasing information in order to find accurate estimates of both, the parameters and the states. Methods for moving horizon estimation can be used to increase the robustness of the parameter estimates if sufficient computer capacity is at hand [96]. These methods are especially relevant in biotechnology application since changes in the system (e.g., between cultivations, mutants, over time) can be observed by small variations in the parameters without the need of a change in the model structure.

1.4 Role of Iterative Process Development to Push Continuous Processes in Biotech

1.4.1 Methods for Development of Continuous Processes

In general, bioprocess development suffers from significantly longer times and costs compared to other industries [97,98]. Additionally, development of continuous processes follows different strategies than batch processes. While it is advantageous to implement the cultivation strategy at the screening or product development stage, it is difficult to implement continuous strategies in the early phases. Thus, the successful development of continuous bioprocesses depends to a bigger part than batch bioprocess development on a comprehensive understanding of the biological system. While batch processes have been traditionally developed with trial-and-error approaches, this is not possible for continuous processes so that the implementation of robust process control strategies is an important basis.

A key parameter for any bioprocess is the specific product formation rate q_p. This rate has a close connection to the specific growth rate, but this correlation is different for different products. In many cases there is no linear dependency, but q_p has an optimum below μ_{max}. Thus, to run a process with a high productivity, it is a major task in continuous bioprocess development to find this correlation between μ and q_p.

Traditionally, this is performed in chemostats, which is a long-lasting process, as the steady state must be established for each dilution rate, which takes about four to seven reactor exchanges. While generally in scientific investigations with chemostats more steady states are established in a series from an initial batch process, such consecutive long-time cultivation can lead to the selection of mutants and thus has to be performed with good controls, for example, by returning to the original dilution rate in the end of an experiment.

While for process development systems for the parallel performance of continuous bioreactors would be very interesting, the setup of such systems is a technical challenge. Parallel chemostats can minimize the problem of evolutionary selection by running experiments with different dilution rates in parallel. However, parallel experiments benefit from miniaturization, especially if big liquid volumes are handled like in the case of chemostats. Although miniaturization has made a big progress in discontinuous cultivation technologies, it is difficult to achieve in the milliliter and submilliliter scale when well defined and controlled dilution rates must be guaranteed over long time intervals.

In the past a number of approaches have been realized for parallel continuous mini and microbioreactors. Balagadde *et al.* [99] developed a microchip-based circulating loop bioreactor with a segment-wise sterilization option to avoid biofilm formation. They demonstrated this reactor for cultivations with *E. coli*. The authors observed oscillations in the cell number. No other online parameters were measured. The feed control, steady states, and so on, were not characterized. The system was developed to investigate evolution, but probably it would not be applicable in its current form for process development.

Nanchen *et al.* [100] used a 10-ml parallel continuous culture system in 17 ml Hungate tubes for 13C labeling and metabolic flux analysis. Aeration with 2 vvm, was used to also mix the system, feeding was done by a peristaltic pump, pH was measured in outlet stream and DO by a microelectrode. The system was characterized with steady states obtained over five volume changes and compared against stirred tank reactor cultivations. The system is very simple to establish and is also very valuable for parameter estimation for continuous process development. A shaken system that can be easily parallelized was developed by Akgün *et al.* [101]. The authors developed a continuous bioreactor based on shake flasks with a controlled feed, an overflow outlet channel at the side of the flask, and a top-phase aeration. While this system was extensively characterized in connection to filling volume constancy and different technical parameters and applied for a continuous culture of *S. cerevisiae*, the system has possibly so far not been applied for other studies. In view of real microbioreactors a modification to a commercial 48 bioreactor system (2 mag, Munich, Germany) was recently published by Schmieder *et al.* [102] showing first promising results for the feasibility of transferring this system to a continuous bioreactor. However, in this case so far only eight of the bioreactors were connected and the determination of key parameters, such as the K_s value had a relatively high error. While continuous cultivation until seven volume exchanges was possible, the presented data do not allow deep-going evaluation about the quality of the steady states, which even has been a problem in larger continuous cultivations.

A faster and elegant method for obtaining the necessary cellular parameters and characteristics with a lower effort compared to parallel chemostats is the A stat technology [103]. With this technology, it is possible to scan the whole growth rate space of an organism in a single experiment [104], either by continuously increasing (Acelerostat) or by decreasing (Decelerostat) the dilution rate in a way that the culture always is in a steady state. However, the technology even has wider use by applying this concept also to the continuous change of other parameters than the feed rate of the limiting substrate; therefore, the term changestat was introduced [104]. This technology of the Gradiostat was applied by various authors, but only Vilu *et al.* developed the scientific basis for it and showed the strength in view of data collection over the whole growth range (see Chapter 9 of this volume).

1.4.1.1 Alternative: Fed-Batch as a System to Simulate Quasi Steady-State Conditions

The strength of the A-stat technology, screening the whole growth space in a single experiment, starting with either a high or a low specific growth rate, is principally also possible by the application of the fed-batch technology. Here, in a similar way as in the A-stat, one can realize a feeding rate, which leads to a continuous gradual change of the specific growth rate of the system. However, in difference to a real continuous culture technology, the fed-batch is more easily to apply in high throughput approaches, as standard (micro)-reactors can be used. Simply instead of having an outgoing steam, one can live with the typical volume change of a fed-batch, which is dependent on the substrate concentration in the feed solution and the feed rate.

Although the controlled feeding to miniaturized cultures is still a challenge, first solutions are available which provide an interesting technological basis, such as 48 microplate base real feeding systems with integrated bottom channels and a hydrogel filling the capillaries [105], or even the integration of micropumps [106]. Alternatively, and more easy to use, would be systems with internal release of the substrate either from silicon as the feed-bead technology [107] or the Feed plate® (PS Biotech, Aachen, Germany). While these systems rely simply on the diffusion of the substrate into the medium, the EnBase® technology [108] (for a comprehensive review see Ref. [109]), which is based on a biocatalytic release of glucose from a polymer of glucose, allows easily varying the feed rate by the amount of the added biocatalyst. Thus, it is easy with this technology to screen for a wide range of growth rates in microplates or parallel shake flasks [108,110]. Recently, this technology was applied with the aim of finding an optimal specific growth rate for the specific production rate of a secreted heterologous enzyme in the yeast *S. cerevisiae*. In comparison to the A-stat technology the same optima was found, but in a very much shorter time [111].

As it was discussed above, the big challenge for the future is, to combine model-based and experimental approaches for continuous bioprocess development. As models in this direction must provide knowledge on the system, mechanistic models, rather than black-box approaches are needed. Therefore, it is necessary to develop design of experiment (DoE) approaches which allow the fast estimation of parameters of nonlinear models. Traditionally, in other disciplines this has been done with online optimal experimental designs (OED). In a recent study, we have applied this approach also to identify the parameters of a dynamic model for *E. coli* cultivations [112]. The strength of the application of the fed-batch approach with a model-based DoE with a sequential reoptimization of the model by a sliding windows approach succeeded in identifying the model parameters in a single parallel experiment of one day, which shows how process development can benefit from model-based and automation approaches.

1.4.2 Mimicking Industrial Scale Conditions in the Lab: Continuous-Like Experiments

1.4.2.1 A Simple Model for Continuous Processes

The challenges of continuous processes and comparison against batch or fed-batch can be better understood using a simplified description of its dynamic system. The continuous process is one of three main cultivation technologies together with batch and fed-batch. Other extensions such as sequencing batch and fed-batch also exist but will be covered later. The main difference between these three setups is the inflow and outflow streams. We can use a simple generalized dynamic model to describe the reactor in Eqs (1.2–1.5). The reader is referred to Refs [25,113] for a more detailed description of the system of equations.

The simplest system of a continuous process is the chemostat, which is characterized by a continuous flow of the incoming medium at a rate that limits one nutrient component in the bioreactor. As the rates for inflow and outflow are equal, the volume in a chemostat is constant. Through the limiting component, the specific growth rate can be controlled simply by the pump speed.

The experimental system and kinetic basis of the chemostat were developed in the groundbreaking papers by Novick and Szillard [114] and Monod [115] and further theoretically refined by Powell [116]. The chemostat applies the concept of the metabolic control, which has been earlier established by the fed-batch method, to continuous processing. By this these authors have set the basis for the wider application of the chemostat providing a detailed procedure for control, long before this was done for the fed-batch by Pirt and Kurowski [49].

A number of assumptions are necessary, the most important being: (i) species and conditions that are not described by the model (temperature, pH, trace elements, etc.) are constant, (ii) ideal mixing, (iii) monoculture. The behavior of the process can be approximately described by the following set of equations:

$$\frac{dS_j}{dt} = \frac{F_{in}}{V}(S_{j,in} - S_j) - q_{S_j} X, \tag{1.2}$$

$$\frac{dX}{dt} = \frac{F_{in}}{V}(X_{in} - X) - \frac{F_{out}}{V}(\delta X - X) + \mu X, \tag{1.3}$$

$$\frac{dO_d}{dt} = \mathrm{Kla}(O_d^* - O_d) - q_O X H, \tag{1.4}$$

$$\frac{dV}{dt} = F_{in} - F_{out}, \tag{1.5}$$

with S_j being the $j = 1\ldots N$ soluble components (substrate or product) concentrations in the medium in (g/l) ($q_{S_j} > 0$ for substrate uptake and $q_{S_j} < 0$ product secretion), X the cell dry weight of the organisms in (g/l), O_d the oxygen dissolved in the medium in (%) of saturation, V the volume of medium in the reactor in (l), and F a flow stream of the reactor in (l/h). The subindexes in and out represent the inlet and outlet streams, respectively, q_S and q_O the uptake rates of soluble components and oxygen, respectively, in (g/(g l)), Kla is the oxygen diffusion constant, O_d^* the saturation concentration of oxygen in the medium in (%), H the Henry related coefficient, and δ the cell retention (–) by membrane or perfusion systems (1 for no retention and 0 for complete recirculation).

This simple model, allows us to analyze the basic characteristics and differences of the discontinuous and continuous cultivation processes. In batch $F_{in} = F_{out} = 0$, in fed-batch $F_{in} > 0$; $F_{out} = 0$, and in continuous $F = F_{in} = F_{out} > 0$ with $D = F/V$ being the dilution rate.

1.4.2.2 Continuous-Like Fed-Batch Cultivations

It is worth stressing out that, in systems with no recirculation ($\delta = 1$), F_{out} enters only in the volume Eq. (1.5), so that equilibrium in all other states can be reached also in the fed-batch setup considering off course an infinitely large vessel.

In a continuous process, the growth rate can be easily obtained by solving Eqs (1.2) and (1.3) at steady state and considering that concentration of biomass or products in the inlet are equal to zero.

In a continuous process, the growth rate can be obtained by reformulating Eq. (1.3) to

$$0 = \frac{dX}{dt} = \left(-\frac{F}{V}\delta + \mu\right)X, \tag{1.6}$$

and further solving it to

$$\mu = \frac{F}{V}\delta = D\delta,$$

for the biomass with S_1 being the substrate and $q_{S_1} < 0$ being the specific substrate uptake rate

$$0 = \frac{dS_1}{dt} = \frac{F_{in}}{V}(S_{1,in} - S_1) + q_{S_1}X = D(S_{1,in} - S_1) + q_{S_1}X,$$

$$X = -\frac{D(S_{1,in} - S_1)}{q_{S_1}}, \quad \text{considering}(S_{1,in} - S_1) \approx S_{1,in}, \quad (1.7)$$

$$X = -\frac{DS_{1,in}}{q_{S_1}},$$

now if we consider $q_{S_1} = -\mu/Y_{X/S}$, and $\mu = D\delta$ we obtain

$$X = \frac{S_{1,in}Y_{X/S}}{\delta}, \quad (1.8)$$

and for the product S_2 or P, with $q_p > 0$ being the production rate

$$0 = \frac{dP}{dt} = -\frac{F_{in}}{V}P + q_pX = -DP + q_pX, \quad (1.9)$$

$$q_p = D\frac{P}{X}, \text{ or } P = \frac{q_pX}{D}.$$

Figure 1.1 depicts the growth rate of an *E. coli* cultivation with regard to different levels of constant feeding. Even at large feeding differences, the change in biomass concentration drives the process to a similar growth rate.

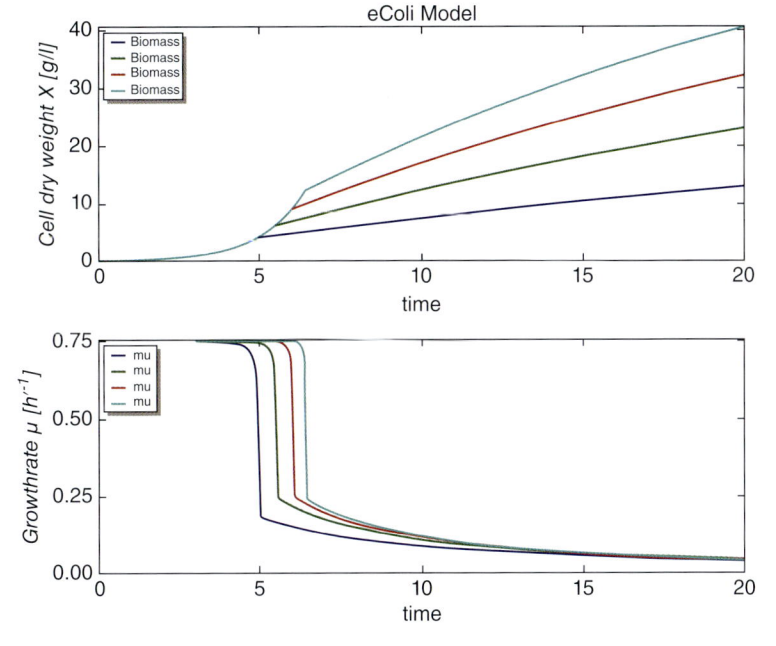

Figure 1.1 Effect of constant feeding profiles in biomass and growth rate.

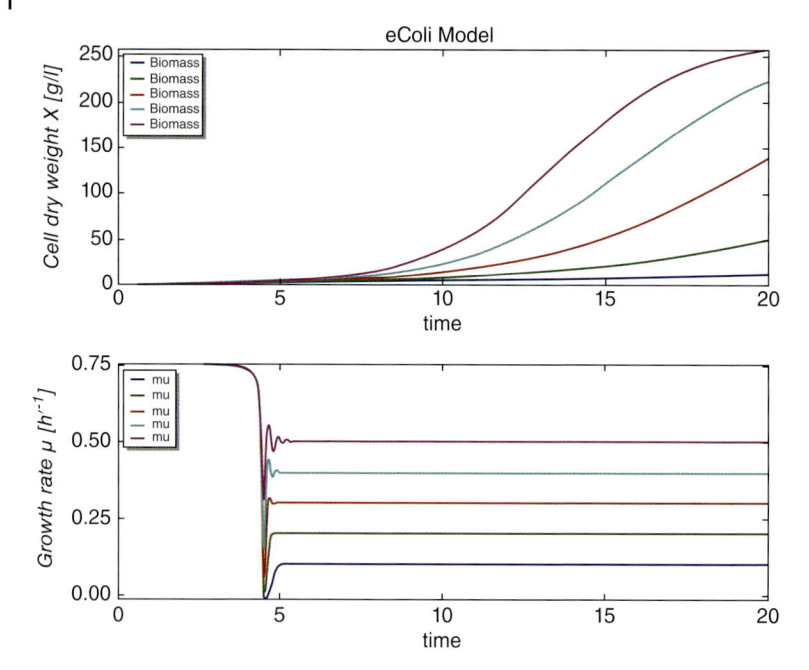

Figure 1.2 Effect of exponential feeding profiles in biomass and growth rate on the behavior of a fed-batch process.

If we apply an exponential feeding as in Figure 1.2, the dilution rate remains constant. By this, the growth rate, which is directly related to D, can be held constant so as to investigate the effect of continuous cultures in the organisms.

These exponential feeding experiments recreate conditions very similar to continuous cultures, reducing drastically material costs (experimental setup and volumes) and experimental time.

1.4.3 Fast and Parallel Experimental Approaches with High Information Content

1.4.3.1 Computer-Aided Operation of Robotic Facilities

One of the most important differences of continuous processes compared to batch or even fed-batch is its level of sophistication hence control sophistication required. The actions that can be taken to operate a batch process are limited so that control can be simple. But continuous processes require a complex control strategy to assure stability and product quality. In other words, the acceptance and profitability of continuous processes strongly depends on the progress in bioprocess monitoring, understanding, and control.

But before a model can be used for monitoring, control, and optimization purposes, it has to be build and validated with experimental data. This is not a trivial task since, as mentioned before, the reproducibility and scalability of processes is especially difficult in biological systems. The experimental efforts related are extremely high so that model building is usually left aside since scale-up and process development are carried out under high time pressure. Nowadays, advances in robotics, miniturazion, and data handling are being used to create high-throughput (HT) facilities able to perform thousands of experiments in parallel automatically. With this, new opportunities arise for a better process understanding and model building. Together with scale down techniques it is possible to create many experiments at "process like conditions" in order to rapidly fit model parameters and test the response of the system in a larger operation space.

1.4.3.2 Model Building and Experimental Validation

In biotechnology, model building is necessarily coupled to a reiterative experimental validation. Regardless of its level of complexity, models have to be constantly fitted against real observations to adjust its parameters to changes caused by variations in the environment or in the microorganism itself. On the other hand, such robotic systems require a respective control strategy, posing new challenges to experimental design and control. Process automation from product development to production requires a horizontal transfer of information. In near future, a miniaturized scale down robotic facility will be connected to the plant and run in parallel creating a miniaturized twin of the large-scale process. By this, the mathematical model traditionally used in MPC will be substituted by a more accurate description of the system with a faster response time due to the dimension difference.

Regarding the experimental planning for model validation, the efficiency in the design of multiple parallel experiments can be increased by using existing methods for data analysis and design of experiments [117–120] in order to allow an automatic evaluation of experimental results as well as the design and run of following experiments. If we managed to build a proper model to describe a process, we first need to fit the model to real data. This model will contain a set of unknown parameters that can be varied to fit the outputs of the model against observations of the real process. The aim is to find the experimental setups such that the statistical uncertainty of estimates of the unknown model parameters is minimized [121,122]. Nevertheless, there are some problems again related with the complexity and nonlinear dynamics of biological processes. First, the experiments carried out in the screening phase should emulate real process conditions [123] and generate high quality data so that systems beyond simple plates, as are mini-bioreactors [124–129] are needed. But more important is that, even for continuous processes, we have to go beyond "endpoint" or "steady-state" experiments. The dynamics of the process are essential to predict its evolution over time and the proper control strategies and for these we need dynamic experiments or at least different steady states. Because of the size and possibility of modern HT facilities, the number of factors that can be varied is very large including "actions" (pipetting, mixing, incubating, measuring, etc.) and "resources" (1-, 8-, or 96-channel pipette, shaker, photometer, flow cytometer, reaction vessels, plates, etc.). For these reasons, computer aided tools for optimal experimental design (OED) [121,130–132] are needed to maximize the efficiency of automatized laboratories. Achievements in online applications, allowing the use of the data generated to redesign the running experiment also are being developed [130,133–137] and applied in real case studies with bacterial cultivations [138], solving a number of complications as are the complexity of the biological system, the control of the experimental facilities, the low information content and long delays of the measurements, the scheduling of all actions considering resource availability, and a robust and cheap computation of the optimization.

Generally, the main factors that affect the identifiability of a model are: (i) the structure of the model, (ii) the quality of the measurements (frequency, accuracy, etc.), and (iii) the design of the experiment (inputs, conditions, etc.). OED thus, by realization of the computed experimental conditions, the information content of the measurements is maximized and the parameters are determined most accurately.

There are still important challenges that need to be solved before OED methods for optimal design and operation of robotic liquid handling stations can be reliably applied for bioprocess development. Some of the most important are the design of a robust optimization program that can assure convergence to a global solution in a limited time, the addition of nonlinear path constraints to define a more accurate search, the computation of the error propagation caused by model uncertainties, efficient methods for the solution of the scheduling problem considering all resources.

1.5 Conclusions

Continuous bioprocessing which is a standard in some bioprocesses, such as for example, waste-water treatment and biogas production, is still at its infancy in pharmaceutical production. While long-term continuous experiments are limited in view of labor and also would only provide limited knowledge in connection to the randomly occurring mutations, deeper going knowledge is needed and quality has to be implemented in the process. Therefore, modeling and advanced process control approaches form a solid basis. Especially, mechanistic and hybrid models can provide here important information on the system and the process. For their application, it has to be considered that due to variations in the process the parameters of the model will be not constant over longer time intervals but have to be regularly adjusted automatically. While such continuous optimization strategies are the state of the art in various engineering disciplines they are new in the area of bioprocessing, but can provide a significant benefit. In this context it is an advantage that various complex cellular models exist, which however by typical methods of model reduction should be adapted in a way to make the parameters identifiable. If successful, such approaches can than be a solid basis for continuous bioprocessing.

References

1 Konstantinov, K.B. and Cooney, C.L. (2015) White paper on continuous bioprocessing May 20–21, 2014 continuous manufacturing symposium. *J. Pharm. Sci.*, **104**, 813–820.
2 FDA (2004) Guidance for industry PAT – A framework for innovative pharmaceutical development, manufacturing, and quality assurance.

3 Han, C., Nelson, P., and Tsai, A. (2010) Process development's impact on Cost of Goods Manufactured (COGM). *BioProcess Int.*, **8**, 48–55.
4 Walther, J., Godawat, R., Hwang, C., Abe, Y., Sinclair, A., and Konstantinov, K. (2015) The business impact of an integrated continuous biomanufacturing platform for recombinant protein production. *J. Biotechnol.*, **213**, 3–12.
5 Gresham, D. and Hong, J. (2015) The functional basis of adaptive evolution in chemostats. *FEMS Microbiol. Rev.*, **39**, 2–16.
6 Lee, H., Popodi, E., Tang, H.X., and Foster, P.L. (2012) Rate and molecular spectrum of spontaneous mutations in the bacterium *Escherichia Coli* as determined by whole-genome sequencing. *Proc. Natl. Acad. Sci. USA*, **109**, E2774–E2783.
7 Lynch, M., Sung, W., Morris, K., Coffey, N., Landry, C.R., Dopman, E.B., Dickinson, W.J., Okamoto, K., Kulkarni, S., Hartl, D.L., and Thomas, W.K. (2008) A genome-wide view of the spectrum of spontaneous mutations in yeast. *Proc. Natl. Acad. Sci. USA*, **105**, 9272–9277.
8 Ferenci, T. (2008) Bacterial physiology, regulation and mutational adaptation in a chemostat environment. *Adv. Microb. Physiol.*, **53**, 169–229.
9 Lenski, R.E., Wiser, M.J., Ribeck, N., Blount, Z.D., Nahum, J.R., Morris, J.J., Zaman, L., Turner, C.B., Wade, B.D., Maddamsetti, R. *et al.* (2015) Sustained fitness gains and variability in fitness trajectories in the long-term evolution experiment with *Escherichia coli. Proc. Biol. Sci.*, **282**, 20152292.
10 Hayduck, F. (1923) *Low-Alcohol Yeast Process*, US Patent 1449107, Fleischmann Company.
11 Hayduck, F. (1915) Verfahren der Hefefabrikation ohne oder mit nur geringer Alkholerzeugung. German Patent 300662.
12 Barton, P.I., (1992) The modelling and simulation of combined discrete/continuous processes. Ph.D. dissertation. University of London.
13 Alvarez, M., Stocks, S., and Jørgensen, S. (2009) Bioprocess modelling for learning model predictive control (L-MPC), in *Computational Intelligence Techniques for Bioprocess Modelling, Supervision and Control* (eds M. C. Nicoletti and L.C. Jain), Springer, Berlin, pp. 237–280.
14 Lübbert, A. and Simutis, R. (1994) Using measurement data in bioprocess modelling and control. *Trends Biotechnol.*, **12**, 304–311.
15 Nicoletti, M., Jain, L., and Giordano, R. (2009) Computational intelligence techniques as tools for bioprocess modelling, optimization, supervision and control in *Computational Intelligence Techniques for Bioprocess Modelling, Supervision and Control*, (eds M.C. Nicoletti and L.C. Jain), Springer, Berlin, pp. 1–23.
16 Tabrizi, H.O., Amoabediny, G., Moshiri, B., Haji Abbas, M.P., Pouran, B., Imenipour, E., Rashedi, H., and Büchs, J. (2011) Novel dynamic model for aerated shaking bioreactors. *Biotechnol. Appl. Biochem.*, **58**, 128–137.
17 Chhatre, S., Farid, S.S., Coffman, J., Bird, P., Newcombe, A.R., and Titchener-Hooker, N.J. (2011) How implementation of quality by design and advances in biochemical engineering are enabling efficient bioprocess development and manufacture. *J. Chem. Technol. Biotechnol*, **86**, 1125–1129.
18 Preisig, H.A. (2010) Constructing and maintaining proper process models. *Comput. Chem. Eng.*, **34**, 1543–1555.

19 Li, D., Ivanchenko, O., Sindhwani, N., Lueshen, E., and Linninger, A.A. (2010) Optimal catheter placement for chemotherapy. *Comp. Aid. Chem. Eng.*, **28**, 223–228.

20 Dochain, D. (2003) State and parameter estimation in chemical and biochemical processes: a tutorial. *J. Process Control.*, **13**, 801–818.

21 fda.gov (2011) Guidance for Industry Process Validation: General Principles and Practice.

22 Henson, M.A. (2003) Dynamic modeling and control of yeast cell populations in continuous biochemical reactors. *Comput. Chem. Eng.*, **27**, 1185–1199.

23 Lei, F. and Jørgensen, S.B. (2001) Dynamics and nonlinear phenomena in continuous cultivations of *Saccharomyces cerevisiae*. Technical University of Denmark, Department of Chemical and Biochemical Engineering.

24 Lei, F., Olsson, L., and Jørgensen, S.B. (2004) Dynamic effects related to steady-state multiplicity in continuous *Saccharomyces cerevisiae* cultivations. *Biotechnol. Bioeng.*, **88**, 838–848.

25 Dochain, D. (2008) Bioprocess control. ISTE London.

26 Hady, L. and Wozny, G. (2010) Computer-aided web-based application to modular plant design. *Comp. Aid. Chem. Eng.*, **28**, 685–690.

27 Bailey, J.E. (1998) Mathematical modeling and analysis in biochemical engineering: past accomplishments and future opportunities. *Biotechnol. Progr.*, **14**, 8–20.

28 Kremling, A., Fischer, S., Gadkar, K., Doyle, F.J., Sauter, T., Bullinger, E., Allgöwer, F., and Gilles, E.D. (2004) A benchmark for methods in reverse engineering and model discrimination: problem formulation and solutions. *Genome Res.*, **14**, 1773–1785.

29 Csete, M.E. and Doyle, J.C. (2002) Reverse engineering of biological complexity. *Supramol. Sci.*, **295**, 1664–1669.

30 Kitano, H. (2002) Systems biology: a brief overview. *Supramol. Sci.*, **295**, 1662–1664.

31 Kitano, H. (2002) Computational systems biology. *Nature*, **420**, 206–210.

32 Kravaris, C., Hahn, J., and Chu, Y. (2013) Advances and selected recent developments in state and parameter estimation. *Comput. Chem. Eng.*, **51**, 111–123.

33 Welch, G. and Bishop, G. (1995) An introduction to the Kalman filter.

34 Daum, F. (2005) Nonlinear filters: beyond the Kalman filter. *IEEE Aerosp. Electron. Syst. Mag.*, **20**, 57–69.

35 Soroush, M. (1997) Nonlinear state-observer design with application to reactors. *Chem. Eng. Sci.*, **52**, 387–404.

36 Boulkroune, B., Darouach, M., Zasadzinski, M., Gillé, S., and Fiorelli, D. (2009) A nonlinear observer design for an activated sludge wastewater treatment process. *J. Process Control*, **19**, 1558–1565.

37 Watson, T.G. (1972) Present status and future prospects of turbidostat. *J. Appl. Chem. Biotechnol.*, **22**, 229–243.

38 Evdokimov, E.V. and Pecherkin, M.P. (1992) Directed autoselection of bacteria cultured in a turbidostat. *Microbiology*, **61**, 460–466.

39 Bennett, W.N. (1988) Assessment of selenium toxicity in algae using turbidostat culture. *Water Res.*, **22**, 939–942.

40 Aarnio, T.H., Suihko, M.L., and Kauppinen, V.S. (1991) Isolation of acetic acid-tolerant bakers-yeast variants in a turbidostat. *Appl. Biochem. Biotechnol.*, **27**, 55–63.

41 Pham, H.T.B., Larsson, G., and Enfors, S.O. (1999) Modelling of aerobic growth of *Saccharomyces cerevisiae* in a pH-auxostat. *Bioprocess. Eng.*, **20**, 537–544.

42 Martin, G.A. and Hempfling, W.P. (1976) A method for the regulation of microbial population density during continuous culture at high growth rates. *Arch. Microbiol.*, **107**, 41–47.

43 Büttner, R., Uebel, B., Genz, I.L., and Köhler, M. (1985) Researches with a substrate-limited Ph-auxostat .1. Bistability of growth under potassium-limited conditions. *J. Basic Microbiol.*, **25**, 227–232.

44 Pettersson, H.E. (1975) Growth of a mixed species lactic starter in a continuous "pH-Stat" fermentor. *Adv. Appl. Microbiol.*, **29**, 437–443.

45 Larsson, G., Enfors, S.O., and Pham, H. (1990) The ph-auxostat as a tool for studying microbial dynamics in continuous fermentation. *Biotechnol. Bioeng.*, **36**, 224–232.

46 Rice, C.W. and Hempfling, W.P. (1985) Nutrient-limited continuous culture in the phauxostat. *Biotechnol. Bioeng.*, **27**, 187–191.

47 Müller, R.H. and Babel, W. (2004) Delftia acidovorans MC1 resists high herbicide concentrations–a study of nutristat growth on (RS)-2-(2,4-dichlorophenoxy)propionate and 2,4-dichlorophenoxyacetate. *Biosci. Biotechnol. Biochem.*, **68**, 622–630.

48 Kleman, G.L., Chalmers, J.J., Luli, G.W., and Strohl, W.R. (1991) Glucose-stat, a glucose-controlled continuous culture. *Appl. Environ. Microbiol.*, **57**, 918–923.

49 Pirt, S.J. and Kurowski, W.M. (1970) An extension of the theory of the chemostat with feedback of organisms. Its experimental realization with a yeast culture. *J. Gen. Microbiol.*, **63**, 357–366.

50 Herbert, D. (1961) A theoretical analysis of continuous culture systems, in *Continuous Culture*, vol. 12, Society of Chemistry and Industry, London, pp. 21–53.

51 Zhang, Y., Stobbe, P., Silvander, C.O., and Chotteau, V. (2015) Very high cell density perfusion of CHO cells anchored in a non-woven matrix-based bioreactor. *J. Biotechnol.*, **213**, 28–41.

52 Kempken, R., Preissmann, A., and Berthold, W. (1995) Clarification of animal cell cultures on a large scale by continuous centrifugation. *J. Ind. Microbiol.*, **14**, 52–57.

53 Neidhardt, F.C., Ingraham, J.L., and Schaechter, M. (1990) *Physiology of the Bacterial Cell: A Molecular Approach*, Sinauer Associates.

54 Box, G.E.P. (1999) Statistics as a catalyst to learning by scientific method part II-a discussion. *J. Qual. Technol.*, **31**, 16–29.

55 Raduly, B., Capodaglio, A.G., and Vaccari, D.A. (2004) Simplification of wastewater treatment plant models using empirical modelling techniques. *Young Res.*, **2004**, 51.

56 Nomikos, P. and MacGregor, J.F. (1994) Monitoring batch processes using multiway principal component analysis. *AIChE J.*, **40**, 1361–1375.

57 AlGhazzawi, A. and Lennox, B. (2009) Model predictive control monitoring using multivariate statistics. *J. Process Control*, **19**, 314–327.

58 Kern, P., Wolf, C., Bongards, M., Oyetoyan, T.D., and McLoone, S. (2011) Self-organizing map based operating regime estimation for state based control of wastewater treatment plants. 2011 International Conference of IEEE on Soft Computing and Pattern Recognition (SoCPaR), pp. 390–395.

59 Georgescu, R., Berger, C.R., Willett, P., Azam, M., and Ghoshal, S. (2010) Comparison of data reduction techniques based on the performance of SVM-type classifiers, 2010 IEEE Aerospace Conference, pp. 1–9.

60 Witten, I.H. and Frank, E. (2005) *Data Mining: Practical Machine Learning Tools and Techniques*, 2nd edn, Morgan Kaufmann, San Francisco.

61 Barton, P.I. and Lee, C.K. (2002) Modeling, simulation, sensitivity analysis, and optimization of hybrid systems. *ACM Trans. Model. Comput. Simul. (TOMACS)*, **12**, 256–289.

62 Rippin, D.W.T. (1993) Batch process systems engineering: a retrospective and prospective review. *Comput. Chem. Eng.*, **17**, 1–13.

63 Lima, F.V. and Rawlings, J.B. (2011) Nonlinear stochastic modeling to improve state estimation in process monitoring and control. *AIChE J.*, **57**, 996–1007.

64 Ławryńczuk, M. (2008) Modelling and nonlinear predictive control of a yeast fermentation biochemical reactor using neural networks. *Chem. Eng. J.*, **145**, 290–307.

65 Zhu, G.-Y., Zamamiri, A., Henson, M.A., and Hjortsø, M.A. (2000) Model predictive control of continuous yeast bioreactors using cell population balance models. *Chem. Eng. Sci.*, **55**, 6155–6167.

66 Ramaswamy, S., Cutright, T., and Qammar, H. (2005) Control of a continuous bioreactor using model predictive control. *Process Biochem.*, **40**, 2763–2770.

67 Bayer, B. and Marquardt, W. (2004) Towards integrated information models for data and documents. *Comput. Chem. Eng.*, **28**, 1249–1266.

68 Spieth, C., Hassis, N., and Streichert, F. (2006) *Comparing Mathematical Models on the Problem of Network Inference*, ACM, pp. 279–286.

69 Heinrich, R. and Schuster, S. (1996) *The Regulation of Cellular Systems*, Springer, US.

70 Vilela, M., Chou, I.C., Vinga, S., Vasconcelos, A., Voit, E., and Almeida, J. (2008) Parameter optimization in S-system models. *BMC Syst. Biol.*, **2**, 35.

71 Stewart, W.E., Shon, Y., and Box, G.E.P. (1998) Discrimination and goodness of fit of multiresponse mechanistic models. *AIChE J.*, **44**, 1404–1412.

72 Harms, P., Kostov, Y., and Rao, G. (2002) Bioprocess monitoring. *Curr. Opin. Biotechnol.*, **13**, 124–127.

73 Schügerl, K. (2001) Progress in monitoring, modeling and control of bioprocesses during the last 20 years. *J. Biotechnol.*, **85**, 149–173.

74 Biechele, P., Busse, C., Solle, D., Scheper, T., and Reardon, K. (2015) Sensor systems for bioprocess monitoring. *Eng. Life Sci.*, **15**, 469–488.

75 de Assis, A.J. and Maciel Filho, R. (2000) Soft sensors development for on-line bioreactor state estimation. *Comput. Chem. Eng.*, **24**, 1099–1103.

76 Sundström, H. and Enfors, S.-O. (2008) Software sensors for fermentation processes. *Bioprocess Biosyst. Eng.*, **31**, 145–152.

77 Agrawal, P. and Lim, H.C. (1984) *Analyses of Various Control Schemes for Continuous Bioreactors*, Springer.

78 Henson, M.A. and Seborg, D.E. (1992) Nonlinear control strategies for continuous fermenters. *Chem. Eng. Sci.*, **47**, 821–835.
79 Andersen, M.Y., Pedersen, N.H., Brabrand, H., Hallager, L., and Jørgensen, S.B. (1997) Regulation of a continuous yeast bioreactor near the critical dilution rate using a productostat. *J. Biotechnol.*, **54**, 1–14.
80 Hu, X., Li, Z., and Xiang, X. (2013) Feedback control for a turbidostat model with ratio-dependent growth rate. *J. Appl. Math. Inform.*, **31**, 385–398.
81 Konstantinov, K.B., Tsai, Ys., Moles, D., and Matanguihan, R. (1996) Control of long-term perfusion chinese hamster ovary cell culture by glucose auxostat. *Biotechnol. Progr.*, **12**, 100–109.
82 Sowers, K.R., Nelson, M.J., and Ferry, J.G. (1984) Growth of acetotrophic, methane-producing bacteria in a pH auxostat. *Curr. Microbiol.*, **11**, 227–229.
83 Ullman, G., Wallden, M., Marklund, E.G., Mahmutovic, A., Razinkov, I., and Elf, J. (2013) High-throughput gene expression analysis at the level of single proteins using a microfluidic turbidostat and automated cell tracking. *Philos. Trans. R. Soc. B*, **368**, 20120025.
84 Yao, Y., Li, Z., and Liu, Z. (2015) Hopf bifurcation analysis of a turbidostat model with discrete delay. *Appl. Math. Comput.*, **262**, 267–281.
85 Olivieri, G., Russo, M.E., Maffettone, P.L., Mancusi, E., Marzocchella, A., and Salatino, P. (2013) Nonlinear analysis of substrate-inhibited continuous cultures operated with feedback control on dissolved oxygen. *Ind. Eng. Chem. Res.*, **52**, 13422–13431.
86 Tian, Y., Sun, K.B., Chen, L.S., and Kasperski, A. (2010) Studies on the dynamics of a continuous bioprocess with impulsive state feedback control. *Chem. Eng. J.*, **157**, 558–567.
87 Nagy, Z.K. (2007) Model based control of a yeast fermentation bioreactor using optimally designed artificial neural networks. *Chem. Eng. J.*, **127**, 95–109.
88 Sankpal, N.V., Cheema, J.J.S., Tambe, S.S., and Kulkarni, B.D. (2001) An artificial intelligence tool for bioprocess monitoring: application to continuous production of gluconic acid by immobilized *Aspergillus niger*. *Biotechnol. Lett.*, **23**, 911–916.
89 Kishimoto, M., Beluso, M., Omasa, T., Katakura, Y., Fukuda, H., and Suga, K-i. (2002) Construction of a fuzzy control system for a bioreactor using biomass support particles. *J. Mol. Catal. B Enzym.*, **17**, 207–213.
90 Bastin, G:. (2013) *On-Line Estimation and Adaptive Control of Bioreactors*, Elsevier.
91 Gresham, D. and Hong, J. (2015) The functional basis of adaptive evolution in chemostats. *FEMS Microbiol. Rev.*, **39**, 2–16.
92 Raissi, T., Ramdani, N., and Candau, Y. (2005) Bounded error moving horizon state estimator for non-linear continuous-time systems: application to a bioprocess system. *J. Process Control*, **15**, 537–545.
93 Alvarez-Vázquez, L.J. and Fernández, F.J. (2010) Optimal control of a bioreactor. *Appl. Math. Comput.*, **216**, 2559–2575.
94 Dondo, R.G. and Marqués, D. (2001) Optimal control of a batch bioreactor: a study on the use of an imperfect model. *Process Biochem.*, **37**, 379–385.
95 Åström, K.J. and Wittenmark, B. (2013) *Adaptive Control*, Courier Corporation.

96 Zavala, V.M., Laird, C.D., and Biegler, L.T. (2008) A fast moving horizon estimation algorithm based on nonlinear programming sensitivity. *J. Process Control*, **18**, 876–884.

97 Neubauer, P., Cruz, N., Glauche, F., Junne, S., Knepper, A., and Raven, M. (2013) Consistent development of bioprocesses from microliter cultures to the industrial scale. *Eng. Life Sci.*, **13**, 224–238.

98 Formenti, L.R., Nørregaard, A., Bolic, A., Hernandez, D.Q., Hagemann, T., Heins, A.L., Larsson, H., Mears, L., Mauricio-Iglesias, M., and Krühne, U. (2014) Challenges in industrial fermentation technology research. *Biotechnol. J.*, **9**, 727–738.

99 Balagadde, F.K., You, L., Hansen, C.L., Arnold, F.H., and Quake, S.R. (2005) Long-term monitoring of bacteria undergoing programmed population control in a microchemostat. *Supramol. Sci.*, **309**, 137–140.

100 Nanchen, A., Schicker, A., and Sauer, U. (2006) Nonlinear dependency of intracellular fluxes on growth rate in miniaturized continuous cultures of *Escherichia coli*. *Appl. Environ. Microbiol.*, **72**, 1164–1172.

101 Akgun, A., Maier, B., Preis, D., Roth, B., Klingelhofer, R., and Büchs, J. (2004) A novel parallel shaken bioreactor system for continuous operation. *Biotechnol. Progr.*, **20**, 1718–1724.

102 Schmideder, A., Severin, T.S., Cremer, J.H., and Weuster-Botz, D. (2015) A novel milliliter-scale chemostat system for parallel cultivation of microorganisms in stirred-tank bioreactors. *J. Biotechnol.*, **210**, 19–24.

103 Paalme, T., Elken, R., Kahru, A., Vanatalu, K., and Vilu, R. (1997) The growth rate control in *Escherichia coli* at near to maximum growth rates: the A-stat approach. *Antonie Van Leeuwenhoek*, **71**, 217–230.

104 Adamberg, K., Valgepea, K., and Vilu, R. (2015) Advanced continuous cultivation methods for systems microbiology. *Microbiology*, **161**, 1707–1719.

105 Wilming, A., Bahr, C., Kamerke, C., and Buchs, J. (2014) Fed-batch operation in special microtiter plates: a new method for screening under production conditions. *J. Ind. Microbiol. Biotechnol.*, **41**, 513–525.

106 Funke, M., Buchenauer, A., Schnakenberg, U., Mokwa, W., Diederichs, S., Mertens, A., Müller, C., Kensy, F., and Büchs, J. (2010) Microfluidic biolector-microfluidic bioprocess control in microtiter plates. *Biotechnol. Bioeng.*, **107**, 497–505.

107 Jeude, M., Dittrich, B., Niederschulte, H., Anderlei, T., Knocke, C., Klee, D., and Büchs, J. (2006) Fed-batch mode in shake flasks by slow-release technique. *Biotechnol. Bioeng.*, **95**, 433–445.

108 Panula-Perälä, J., Siurkus, J., Vasala, A., Wilmanowski, R., Casteleijn, M.G., and Neubauer, P. (2008) Enzyme controlled glucose auto-delivery for high cell density cultivations in microplates and shake flasks. *Microb. Cell Fact.*, **7**, 31.

109 Krause, M., Neubauer, A., and Neubauer, P. (2016) Scalable enzymatic delivery systems for controlled nutrient supply improve recombinant protein production. *Microb. Cell Fact.*, 15:110.

110 Siurkus, J., Panula-Perala, J., Horn, U., Kraft, M., Rimseliene, R., and Neubauer, P. (2010) Novel approach of high cell density recombinant bioprocess development: optimisation and scale-up from microliter to pilot scales while

maintaining the fed-batch cultivation mode of *E. coli* cultures. *Microb. Cell Fact.*, **9**, 35.

111 Glauche, F., Genth, R., Kiesewetter, G., Glazyrina, J., Bockisch, A., Cuda, F., Goellig, D., Raab, A., Lang, C., and Neubauer, P. (2017) New approach to determine the dependency between the specific rates for growth and product formation in miniaturized parallel robot-based fed-batch cultivations. *Eng. Life Sci.*, in press.

112 Cruz Bournazou, M.N., Barz, T., Nickel, D., Glauche, F., Knepper, A., and Neubauer, P., (2017) Online Optimal Experimental Re-Design in Robotic Parallel Fed-Batch Cultivation Facilities for Validation of Macro-Kinetic Growth Models at the Example of E. coli. *Biotechnol Bioengin*, **114**, 610–619.

113 Enfors, S.O. (2016) Fermentation Process Engineering. p. 99.

114 Novick, A. and Szilard, L. (1950) Description of the chemostat. *Supramol. Sci.*, **112**, 715–716.

115 Monod, J.L. (1950) La technique de culture continue. Théorie et applications. *Annales de l'Institut Pasteur*, **79**, 390–410.

116 Powell, E.O. (1965) Theory of the chemostat. *Lab. Pract.*, **14**, 1145–1149 passim.

117 Macarrón, R. and Hertzberg, R.P. (2011) Design and implementation of high throughput screening assays. *Mol. Biotechnol.*, **47**, 270–285.

118 Tai, M., Ly, A., Leung, I., and Nayar, G. (2015) Efficient high-throughput biological process characterization: definitive screening design with the Ambr250 bioreactor system. *Biotechnol. Prog*, **31**, 1388–1395.

119 Lutz, M.W., Menius, J.A., Choi, T.D., Laskody, R.G., Domanico, P.L., Goetz, A.S., and Saussy, D.L. (1996) Experimental design for high-throughput screening. *Drug Discov. Today*, **1**, 277–286.

120 Haaland, P.D. (1989) *Experimental Design in Biotechnology*, CRC Press.

121 Körkel, S., Kostina, E., Bock, H.G., and Schlöder, J.P. (2004) Numerical methods for optimal control problems in design of robust optimal experiments for nonlinear dynamic processes. *Optim. Method Softw.*, **19**, 327–338.

122 Korkel, D.M.S. (2002) Numerische Methoden fuer Optimale Versuchsplanungsprobleme bei nichtlinearen DAE-Modellen. Ruprecht-Karls-Universitaet Heidelberg.

123 Neubauer, P. and Junne, S. (2010) Scale-down simulators for metabolic analysis of large-scale bioprocesses. *Curr. Opin. Biotechnol.*, **21**, 114–121.

124 Hedrén, M., Ballagi, A., Mörtsell, L., Rajkai, G., Stenmark, P., Sturesson, C., and Nordlund, P. (2006) GRETA, a new multifermenter system for structural genomics and process optimization. *Acta Crystallogr. D*, **62**, 1227–1231.

125 Gill, N.K., Appleton, M., Baganz, F., and Lye, G.J. (2008) Design and characterisation of a miniature stirred bioreactor system for parallel microbial fermentations. *Biochem. Eng. J.*, **39**, 164–176.

126 Lamping, S.R., Zhang, H., Allen, B., and Shamlou, P.A. (2003) Design of a prototype miniature bioreactor for high throughput automated bioprocessing. *Chem. Eng. Sci.*, **58**, 747–758.

127 Zhang, Z., Perozziello, G., Boccazzi, P., Sinskey, A., Geschke, O., and Jensen, K. (2007) Microbioreactors for bioprocess development. *J. Assoc. Lab. Autom.*, **12**, 143–151.

128 Knorr, B., Schlieker, H., Hohmann, H.-P., and Weuster-Botz, D. (2007) Scale-down and parallel operation of the riboflavin production process with *Bacillus subtilis*. *Biochem. Eng. J.*, **33**, 263–274.

129 Huber, R., Ritter, D., Hering, T., Hillmer, A.-K., Kensy, F., Mueller, C., Wang, L., and Buechs, J. (2009) Robo-Lector: a novel platform for automated high-throughput cultivations in microtiter plates with high information content. *Microb. Cell Fact.*, **8**, doi: 10.1186/1475-2859-8-42.

130 Barz, T., López Cárdenas, D.C., Arellano-Garcia, H., and Wozny, G. (2013) Experimental evaluation of an approach to online redesign of experiments for parameter determination. *AIChE J*, **59**, 1981–1995..

131 Cruz Bournazou, M.N., Junne, S., Neubauer, P., Barz, T., Arellano-Garcia, H., and Kravaris, C. (2014) An approach to mechanistic event recognition applied on monitoring organic matter depletion in SBRs. *AIChE J.*, **60**, 3460–3472.

132 Franceschini, G. and Macchietto, S. (2007) Model-based design of experiments for parameter precision: state of the art. *Chem. Eng. Sci.*, **63**, 4846–4872.

133 Galvanin, F., Barolo, M., and Bezzo, F. (2009) Online model-based redesign of experiments for parameter estimation in dynamic systems. *Ind. Eng. Chem. Res.*, **48**, 4415–4427.

134 Galvanin, F., Barolo, M., Pannocchia, G., and Bezzo, F. (2012) Online model-based redesign of experiments with erratic models: a disturbance estimation approach. *Comput. Chem. Eng.*, **42**, 138–151.

135 Huang, B. (2010) Receding horizon experiment design with application in SOFC parameter estimation. 9th International Symposium on Dynamics and Control of Process Systems (DYCOPS 2010), July 5–7, Leuven, Belgium.

136 Stigter, J., Vries, D., and Keesman, K. (2006) On adaptive optimal input design: a bioreactor case study. *AIChE J.*, **52**, 3290–3296.

137 Zhu, Y. and Huang, B. (2011) Constrained receding-horizon experiment design and parameter estimation in the presence of poor initial conditions. *AIChE J.*, **57**, 2808–2820.

138 Cruz-Bournazou, M.N., Barz, T., Nickel, D., and Neubauer, P. (2014) Sliding-window optimal experimental re-design in parallel microbioreactors. *Chem. Ing. Tech.*, **86**, 1379–1380.

2

Tools Enabling Continuous and Integrated Upstream and Downstream Processes in the Manufacturing of Biologicals

Rimenys J. Carvalho and Leda R. Castilho

Federal University of Rio de Janeiro, COPPE, Cell Culture Engineering Laboratory, C.P. 68502, 21941-972 Rio de Janeiro, RJ, Brazil

2.1 Introduction

As stated by the deputy director of US FDA's Office of Pharmaceutical Quality in April 2016, not much has changed in pharmaceutical manufacturing since the 1960s – if a pharmaceutical scientist from that time were transported to a current manufacturing facility, he would be familiar with most production techniques [1]. Processes are still run in batch mode, with many sequential steps, each of them with starts and stops, contributing to process inefficiency, increased possibilities of human error, and long production times.

However, the batch paradigm in the synthetic drug industry started to be questioned in the beginning of the current century, and regulatory guidances encouraging innovations further pushed this discussion further [2–4]. A decade later, this stimulus for innovation took concrete shape, with two historic facts occurring from mid 2015 to mid 2016 [1]:

- In July 2015, a synthetic cystic fibrosis drug (Vertex's Orkambi®, lumacaftor/ivacaftor) was approved by US FDA, being manufactured by a fully continuous manufacturing process.
- In April 2016, US FDA approved, for the first time in history, a manufacturer's change in their production technology from batch to continuous manufacturing. The product is Janssen's synthetic HIV-1 drug Prezista® (darunavir).

In the biopharmaceutical sector, however, the change in paradigm started later. Not long ago, around 2010, considering developing a technology for a new biotech drug based on a process other than batch or fed-batch cell cultivation and batch resin chromatography purification was probably seen as insanity.

However, at that moment, in a conference of the "Cell Culture Engineering" series, one of the oral presenters asked [5]: Toward fully continuous bioprocessing – what can we learn from pharma?

Conservatism in the biopharmaceutical sector had been strong since the introduction of the first recombinant biotherapeutics into the market in the 1980s. Only incremental process improvements had been implemented over time [6]. In a market dominated by products that were unique and patent-protected, the

Continuous Biomanufacturing: Innovative Technologies and Methods, First Edition.
Edited by Ganapathy Subramanian.
© 2018 Wiley-VCH Verlag GmbH & Co. KGaA. Published 2018 by Wiley-VCH Verlag GmbH & Co. KGaA.

time-to-market concept was more important than process optimization, manufacturing efficiency, or production costs [7]. Batch and fed-batch culture processes were implemented for the vast majority of products, and continuous perfusion strategies remained confined to a niche of unstable products, such as therapeutic enzymes and blood clotting factors.

However, in the last years, the business scenario in the biopharmaceutical arena has been undergoing important changes, with growing competition from biosimilars, multiple product pipelines, and the need for greater product/scale/plant location flexibility [6,7].

These facts motivated companies and academic groups to shift research efforts toward more efficient, intensive, and flexible bioprocesses [8]. The development of single-use devices improved containment, decreased contamination risks, and thus supported the development of continuous processes. The production also of stable products by continuous perfusion processes, the evaluation of multicolumn chromatography for the continuous purification of proteins, and the integration of continuous upstream and downstream operations gained in interest and are nowadays the focus of research carried out both in industry and in academia, of many publications, and of new, specific conference series. A change of technological paradigm is currently underway in the biopharmaceutical industry, and the manufacturing facilities of the future products will certainly look different than the ones currently under commercial operation.

In the same way as it happened in 2015–2016 in the pharma industry, the current R&D efforts on continuous biomanufacturing in the biopharma sector will certainly result in historic achievements not many years away from now.

2.2 Continuous Upstream Processes

Continuous cell cultivation processes are operationally more complex than batch and fed-batch processes, since additional pumps, tanks, and controllers are needed to continuously feed medium and collect harvest. Additionally, the open-system nature of a continuous process has been traditionally claimed to offer greater potential risks of culture contamination. For these reasons, for about three decades since the approval of the first recombinant biotherapeutics, the KISS rule – "keep it simple and stupid" – prevailed when establishing a biopharmaceutical process.

Thus, the great majority of the biopharmaceutical manufacturing processes developed in the past and currently in industrial operation are based on batch and fed-batch cultivation systems. Usually 1–2 weeks after inoculation, cells are separated from the spent culture medium, and the harvested supernatant is directed to the downstream sector. Because productivities, even in fed-batch cultivation processes, are not very high, large bioreactors or sets of multiple large bioreactors are usually needed to meet the annual manufacturing demand for a given product. For example, Warikoo *et al.* [6] reported a volumetric productivity more than fivefold higher for the continuous production of a monoclonal antibody (mAb), as compared to the fed-batch process using the same cell line.

The lower productivities and larger bioreactors in turn result in large volumes of culture supernatant being directed to the downstream processing steps, requiring large equipment for clarification and for the product capture step, as well as large volumes of buffers, which need to be prepared, filtered, and stored. For all these reasons, the traditional manufacturing facilities in the biopharma industry have been characterized by large footprints of classified areas and by very high capital investment.

However, as seen in a wide range of industries, continuous processing generally increases productivity and decreases unproductive turnaround times, which means producing more in less time from smaller equipment. Therefore, implementing continuous biomanufacturing in the biopharmaceutical industry is currently seen as the way to decrease both capital and operational expenses.

Nevertheless, the drive for continuous biomanufacturing is not only economical. As also stated by regulators, continuous manufacturing presents several other advantages that also positively impact on product quality and supply [1,6]:

- Operation under steady state enables high and consistent product quality.
- Elimination of breaks between process steps and reduction of the risk of human errors make continuous manufacturing more reliable and safer.
- Continuous manufacturing allows a much quicker response to changes in demand, potentially contributing to prevention of drug shortages.

The latter point can also be important for improving readiness for countermeasures against (re)emerging pathogens, such as experienced in the recent Ebola and Zika virus outbreaks, which demonstrated the need for rapid and flexible development and manufacturing of new antibodies and vaccines, possibly in compact facilities distributed over the globe.

Based on the experience of many years of commercial industrial operation of semicontinuous and continuous cell culture facilities for the industrial production of recombinant coagulation factors, Desai [9] concluded that continuous processes do not necessarily represent increased contamination risks when compared to batch processes, and that they can be applied to many biological products, including mAbs, increasing plant productivity and potentially paving the way for lowering the overall manufacturing costs without compromising on product quality.

2.2.1 Continuous Bioprocesses: With or Without Cell Recycle?

In order to carry out a biotechnological upstream process continuously, in principle two bioreactor operation modes can be thought of: a regular continuous mode (e.g., a chemostat), or a continuous process with cell recycle, most commonly referred to as perfusion.

In the biopharmaceutical industry, considering all approved products until 2014, *Escherichia coli* is used for the production of 19% of all products, and yeast for 16.5% [10]. Animal cells (including human, other mammalian, and insect cells) are the preferred manufacturing platform, accounting for 60–65% of the approved products, and this figure is expected to increase in the future, since over the years a steadily growing prominence of animal cells over other expression systems has been observed [10–12]. This in turn is a consequence of the growing proportion, among the more recent products, of more complex proteins, especially monoclonal antibodies (mAbs), which require the ability of animal cells to correctly carry out posttranslational modifications (e.g., glycosylation) in order to meet the required product quality attributes.

The practical consequence of the dominating role of animal cells, especially mammalian cells such as CHO (Chinese hamster ovary) cells, is that, in order to establish continuous operation in the biopharmaceutical industry, perfusion is the operation mode of choice, since regular continuous processes (without cell recycle) are not adequate due to the relatively low specific growth rates of animal cells [11].

It is interesting to note, however, that perfusion has gained in popularity in the last years not only for application to the main production bioreactor in a biopharmaceutical process but also as an important tool to produce highly concentrated cell banks and to expand cells in the seed train to very high cell densities. These applications of perfusion shorten the seed train itself or accelerate the main bioreactor process due to more concentrated inocula [8,13,14].

Furthermore, perfusion has gained in popularity in the biopharmaceutical sector not only for the production of biotherapeutics, but also for developing more efficient processes for the production of viral vaccines and more efficient expansion of stem cells aiming at cell therapies [15–17].

2.2.2 Early/Scale-Down Perfusion Development

Stirred bioreactors are the workhorse in biopharmaceutical upstream processing. A wide range of such reactors are commercially available, also as single-use bioreactors, in scales ranging from minibioreactors with hundreds of microliters working volume to industrial scale bioreactors with working volumes of up to 2000 l (single-use bioreactors) or 25 000 l (stainless steel animal cell bioreactors). Most perfusion processes – either already in commercial operation for unstable products or under development for a wide range of stable and nonstable products – are based on stirred-tank reactors, but other types of perfusion bioreactor systems are possible, such as bioreactors where the cells remain entrapped, like hollow-fiber and fixed-bed bioreactors [18–20].

In order to develop a perfusion process, it is interesting to start development at small scales, in order to increase parallel operation possibilities, decrease medium consumption, and reduce handling of large volumes of liquids (feed and perfusate). A very simple small-scale system that can be used in early studies to mimic perfusion and predict performance, for example of different culture media or different medium exchange rates, are spin tubes, which consist of simple vented conical tubes placed on orbital shakers. By simply centrifuging the tubes taken from the CO_2 incubator and then carrying out a partial or total medium exchange, once or more times a day, perfusion operation under different dilution rates can be mimicked. Since medium renewal in this case is discrete, such systems operate under intermittent perfusion mode, also known as pseudoperfusion, semicontinuous, or semi-perfusion mode [21–23]. The advantage of using spin tubes under pseudoperfusion mode is that simple, regular equipment available in cell culture laboratories (CO_2 incubator, orbital shaker, centrifuge) can be employed, allowing straightforward and nonexpensive start of perfusion studies. From our own experience, for relatively high cell densities (e.g., 20E6 cells/ml), operating volumes in spin tubes can range from less than 5 ml (using regular shakers with 2.5-cm orbit) to over 20 ml (if 5-cm orbit shakers and higher speeds, e.g., 180 rpm, are used). Also, the use of spin tubes allows processing many small, low-cost disposable tubes in parallel, enabling a first quick and nonexpensive screening of multiple media and feeding strategies before moving to more sophisticated systems for further process development [23,24]. Besides spin tubes, small-scale process development can be carried out using 24-deep well plates, which allow quick medium exchange by centrifuging the whole plate and using multichannel pipettes.

However, whenever available, the best tools for perfusion process development are instrumented bioreactors that enable true continuous perfusion to be carried out and allow important culture parameters (pH, dissolved oxygen, temperature, feed/perfusate flow rates, etc.) to be controlled. Nowadays, many miniaturized bioreactor systems are commercially available, such as micro-Matrix (Applikon), ambr® 15 cell culture (Sartorius Stedim), and Micro-24 MicroReactor system (Pall), which allow very low volume operation (e.g., 5–10 ml) and/or independent operation of many parallel cultures (e.g., 24 or 48). However, for perfusion, due to the need for more ports and pumps, for a cell separation device, and ideally for a hardware/software able to control also perfusion parameters (such as feed/perfusate flow rates), studies nowadays are usually initiated in bioreactors starting at larger, yet still low working volumes (e.g., 50–250 ml), such as DASGIP® Parallel Bioreactor Systems (Eppendorf), myControl miniBioreactors (Applikon), Multifors 2 (Infors), and Wave Bioreactor™ systems (GE Healthcare). The latter three are offered by the respective manufacturers together with a cell separation device (an acoustic separator, a spin filter, and a floating filter, respectively) specifically for perfusion applications, but all four systems can be coupled to generic commercial cell retention devices, provided these are compatible with the low working volumes intended.

2.2.3 Feeding and Operational Strategies in Perfusion Processes

Perfusion processes have traditionally been fed with regular culture medium. Because in the past, perfusion processes were mainly applied to unstable products, the main concern was the mean residence time of the product inside the bioreactor, so high perfusion rates were desired. The fact that this led to rather low product concentrations in the perfusate was secondary.

However, the growing application of perfusion processes to the production of stable proteins makes it increasingly important to develop special perfusion feed media and specific feeding strategies, so that the high volumetric productivities of perfusion processes can be combined with high titers, in the same order of magnitude as those obtained in fed-batch processes.

About two decades ago, Ozturk [25] introduced the concept of cell-specific perfusion rate (CSPR). The approach of decreasing the CSPR as much as possible is known as "push-to-low" strategy [26] and can be applied in the case of stable products that do not need to be quickly removed from the bioreactor. CSPR values are usually in the range of 0.04–0.06 nl/cell/day [6,27], but optimal CSPR can vary according to what is the main goal: if to maximize the product concentration or the volumetric productivity [25].

Clincke et al. [27], for example, were able to achieve cell concentrations over 200E6 cells/ml by combining a CSPR in the range of 0.05–0.06 nl/cell/day with separate on-demand supplementation of glucose and glutamine. Warikoo et al. [6] were able to push CSPR lower, maintaining for approximately 40 days a steady-state cell density of 50–60E6 cells/ml at a CSPR of 0.04–0.05 nl/cell/day.

If very low CSPR values are desired in order to achieve higher product titers, developing special, fortified feed media should be considered. According to Konstantinov [28], a goal for the future should be to enable long-term stable processes at CSPR ranges of 0.01–0.015 nl/cell/day. A simple starting point to develop fortified perfusion feed media is to evaluate different mixtures of regular culture medium and feed media originally developed for fed-batch processes, in order to find feed mixtures that are suitable to sustain high cell densities at low perfusion rates [23,24]. However, attention should be paid to the osmolality of fortified perfusion feed media, in order to avoid osmotic stress at levels that can be deleterious to the cells.

In order to carry out "controlled-fed perfusion," some authors have combined fortification (e.g., with glucose or with both glucose and amino acids) of the medium fed to the bioreactor with feeding rates defined based on the oxygen uptake rate [29] or on the glucose consumption rate [30]. In both articles, a twofold increase in antibody volumetric productivity was obtained, when compared to a conventional perfusion. For those with further interest in this topic, more detailed discussions on perfusion feeding can be found in Refs [8,31].

A further strategy that can be adopted to improve productivity in perfusion processes is to carry out biphasic processes, where cell growth and product formation are decoupled. These processes comprise a first phase where cell growth is enhanced followed by a change in cultivation conditions in order to promote cell growth arrest, while maintaining or increasing the cell specific productivity (q_P). Examples of strategies that have been successfully used to perform biphasic perfusion are a decrease in temperature to mild hypothermic levels [32–35] or medium supplementation with short-chain fatty acids, such as butyric acid [36,37] or valeric acid [33]. In the experience of our laboratory, using statistical design of experiments to study the combination of several factors can result in optimized biphasic perfusion strategies with even greater gains in productivity [34].

2.2.4 Cell Retention Devices

In perfusion bioreactors, in order to accomplish cell recycle/retention, a cell separation device needs to be attached to the bioreactor. It can be placed either inside the bioreactor (Figure 2.1) or externally to it (Figure 2.2) [38]. Most perfusion processes use external devices, and these can either be placed within a recirculation loop (Figure 2.2b), such as in the case of cell settlers, dynamic filters, or tangential flow filters, or be connected to the bioreactor through a single line

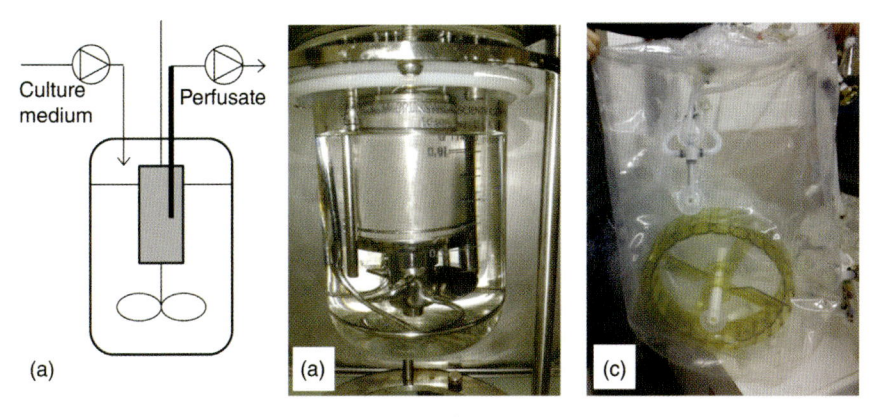

Figure 2.1 Internal cell retention devices in perfusion systems. (a) Schematic drawing (internal spin filter as example). (b) Example of small-scale bioreactor containing an internal spin filter (Bioflo 110, New Brunswick Scientific). (c) Example of an investigational development of a disposable hydrocyclone (Sartorius) placed inside a bag of a single-use Air Wheel bioreactor (PBS Biotech).

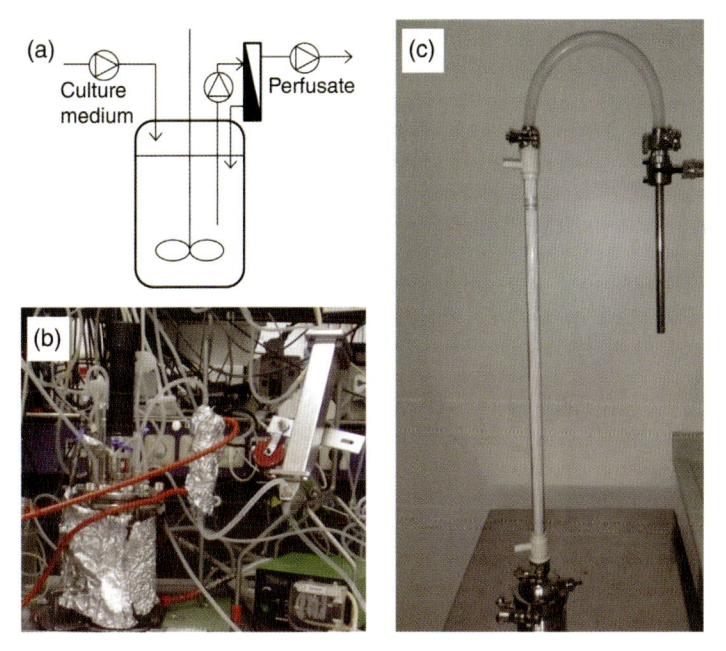

Figure 2.2 External cell retention devices in perfusion systems. (a) Schematic drawing of a TFF placed within an external recirculation loop. (b) Example of cell settler (CS20, Biotechnology Solutions) placed within a recirculation loop (bioreactor vessel and inline cooler are protected from light with aluminum foil; this photo is from a perfusion study carried out jointly by our group and the University of Bielefeld, Germany). (c) Example of an ATF™-2 system (Repligen), showing the single deep tube that connects it to the bioreactor and allows the cell suspension to be transported back and forth. Perfusate is pumped out of the system through the lateral outlets, which are connected to the extracapillary space of the hollow fiber cartridge.

(Figure 2.2c), such as in the case of the Alternating Tangential Flow (ATF™, Repligen) system. Through this single line, the cell suspension is pushed back and forth to the ATF filtration module, as a consequence of alternating cycles of pressure and vacuum that are applied to the diaphragm placed inside the bottom part of the system.

The performance of a perfusion process is highly dependent on the choice of the cell retention device, which ideally should be characterized by

- high separation efficiencies for viable cells,
- preferential removal of dead cells,
- no negative effects on cell viability,
- no fouling or clogging over time,
- stable performance and robust long-term operation,
- availability as disposable device, and
- total transmission of product to the perfusate line.

However, none among the currently available cell separation devices is able to satisfy all these features, so the choice of the cell separator will rely on a compromise, taking into account the most important performance indicators for the given biological system under study.

Several previous publications have described the different cell retention devices for animal cell perfusion cultures in detail [8,31,38–40]. Among these separators, most of them rely on the following two different separation principles:

- Filtration (e.g., tangential flow filters and alternating tangential flow systems)
- Settling in a gravitational or centrifugal field (e.g., inclined cell settlers, acoustic settlers, hydrocyclones and centrifuges)

Cell settlers and acoustic separators are based on cell sedimentation under the action of gravitational acceleration as separation principle. However, because animal cells have a relatively small size (about 15 μm in diameter) and a density close to water (about 1.06 g/cm^3), their settling velocities are quite low. In order to overcome this limitation, acoustic settlers apply an acoustic wave to the cell suspension in order to generate temporary cell aggregates at resonance nodes. The large size of aggregates allows them to quickly settle by action of gravity when the acoustic wave is periodically switched off [31,35,41].

If conventional settlers are considered, a consequence of the low settling velocity of animal cells is that for nonaggregated cells a large sedimentation area is required to ensure high separation efficiencies. Thus, vertical gravity settlers are not useful, and the settler type of choice is the inclined lamella settler, also known as cell settler (Figure 2.3a), which has a large number of parallel plates (lamellas) packed inside its body, resulting in a large packed sedimentation area, as well as in a short path between the lamellas to be overcome by the settling cells [8,42]. In process development studies performed in our laboratory, these devices, when fed with a cell suspension cooled in line down to 20 °C and subject to an intermittent vibration regime (e.g., for 15 s every 10 min), have been shown to provide very high CHO cell separation efficiencies (>99%) [33] and to enable robust, long (>30 days) perfusion runs at 50–60E6 cells/ml (unpublished results). Cell settlers are used industrially in Bayer's recombinant Factor VIII perfusion

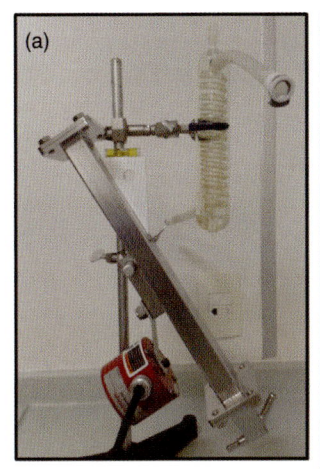

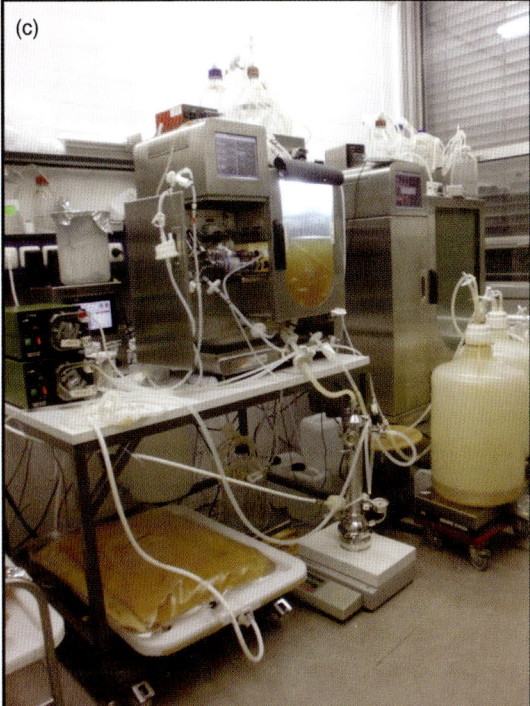

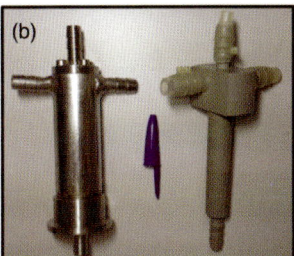

Figure 2.3 Examples of some cell retention devices. (a) Inclined cell settler (CS10, Biotechnology Solutions). (b) Hydrocyclone designed for animal cell separation, capable of processing 3 l/min. The photo compares the stainless steel prototype and the plastic version (that has been commercialized by Sartorius) to the size of a pen cover. (c) ATF™-4 system (Repligen) operating connected to an Air Wheel bioreactor (PBS Biotech). (This photo is from a perfusion study carried out jointly by our group and the University of Bielefeld (Germany).

manufacturing facility in the United States since 1993, and recent publications report their use at 3000-l perfusion scale under GMP conditions [43], as well as within an end-to-end continuous, integrated process implemented at pilot scale [44].

Settling under action of a gravitational field is the separation principle exploited by centrifuges and hydrocyclones [42]. The latter are very compact devices (Figure 2.3b), with no moving parts, that have shown separation efficiencies for CHO cells in the range of 97–99% [45,46]. In spite of being compact, hydrocyclones have a very large processing capacity, which makes their laboratory-scale evaluation under continuous perfusion mode difficult. Therefore, their use at laboratory-scale has been reported for intermittent perfusion [47], but continuous perfusion experiments performed at a 300-l production facility have indicated that this is a promising cell retention device for large-scale perfusion cultures (unpublished results). Moreover, their compact size, absence of moving parts, low cost, and the possibility of being made of plastics indicate their usefulness as single-use separators that could be placed as an internal cell retention device inside disposable bioreactor bags [8]. However, special attention must be paid to the exact internal dimensions of a hydrocyclone, since even slight variations respective to the design dimensions can significantly affect separation efficiency.

Continuous disk stack centrifuges were early on studied for use in perfusion processes, but due to the moving parts and to the potential risk of clogging of outlet ports and biomass discharge channels, their use is not very widespread in the biopharmaceutical industry [48]. On the other hand, specially designed centrifuges (e.g., Centritech® and kSep®) that contain single-use inserts to receive the cell suspension have been proposed for animal cell perfusion processes. The Centritech centrifuge is used for the industrial manufacturing of recombinant B-domain deleted Factor VIII since 2000.

Since the introduction of spin filters as the first cell retention device for animal cells in 1969 [49], filtration-based separators have always been among the most popular retention devices for animal cell perfusion processes. In the last decades, internal and external spin filters have lost in popularity, although they are still nowadays found in some industrial processes [50]. On the other hand, filtration devices based on hollow fibers, such as tangential flow filters (TFF) and alternating tangential filtration systems (ATF, Figure 2.3c) have become increasingly popular.

Clincke *et al.* [27,51] used identical hollow fiber cartridges in an ATF and in a TFF system, in order to compare both systems for perfusion of CHO cells producing a mAb. The ATF was able to give cell concentrations as high as 132E6/mL, but the alternating vacuum and pressure cycles presented difficulties to pull the highly viscous cell suspension from the bioreactor when such high cell concentrations were reached. A similar phenomenon occurred with the TFF, but at a higher cell density (>200E6/mL). Product retention in the bioreactor was observed for both filtration systems, in comparable levels.

In another comparative study, Karst *et al.* [52] evaluated the perfusion of CHO cells using ATF and TFF systems with identical hollow fiber modules. Parallel perfusion bioreactors were operated under different steady states (20E6, 40E6, and 60E6 cells/ml) and showed comparable performances in various regards, except for the fact that using the TFF a significant retention of the mAb product inside the bioreactor was observed, whereas using the ATF product transmission to the perfusate was complete.

Working with a non-mAb, large protein, Coronel et al. [53] experienced high product retention in an ATF system. The incomplete product transmission to the perfusate was confirmed for hollow fiber cartridges of different pore sizes (750 kDa, 0.2 μm, and 0.5 μm). Unfortunately, the cartridges evaluated in this work were of polysulfone and polyethersulfone, and no cartridges made of more hydrophilic materials were available to fit in the system.

Taken together, the results shown in these reports indicate that product transmission is an important issue in membrane-based cell retention devices. This phenomenon seems to be dependent on the nature of each biological system and could be related either to membrane fouling or to product adsorption of the product onto the fibers. In order to enable robust industrial application of TFF and ATF systems, a better understanding and control of this phenomenon is required.

2.3 Continuous Downstream Processes

There was a significant increase in the number of studies on continuous downstream processes in the last decade, not only for the production of labile biological products, but also for stable biomolecules due to the higher volumetric productivities that can be achieved [54–56]. The use of continuous downstream processing may present several other advantages to the process, such as faster processing, lower resin requirements, consistent product quality, reduced costs, and easy integration to continuous upstream processes. In addition, continuous processing enables more purified material to be harvested even at low capacities, which is an important advantage, considering that due to the significant increase in upstream titers achieved in the last years the DSP operations became the main bottleneck in bioprocessing [57]. However, for a purification process to be carried out under continuous mode, full automation is usually required [54].

Changing DSP to continuous mode is not straightforward, due to the higher complexity over batch mode purification. The complexity further increases when there is a need for several purification steps, as commonly required in bioprocesses. Biopharmaceutical processes usually require three or more purification steps (capture, intermediate purification, and polishing) plus virus clearance stages [58]. Thus, designing a full continuous process may be complex, but when implemented, especially when employing single-use materials, it can increase flexibility, decrease capital costs of facilities, and enable easy switch of products manufactured in campaigns [56]. The whole purification process has to guarantee that the final product has high quality, that is, is pure, active, has the correct structure, and meets the clearance criteria for critical contaminants, such as host cell DNA, host cell protein (HCP), virus, endotoxins, leached ligands, product aggregates, nonactive or degraded forms of the product, among others [59].

DSP process design is highly affected by the type of product and the upstream process. For instance, when the product is produced intracellularly, for example, in bacteria, the purification process is usually more complex due to the need for cell lysis and due to the higher amount of DNA and HCP in the starting DSP material than in the case of extracellular products produced by mammalian cells. However, in the latter case critical contaminants involve mammalian DNA, proteins, and viruses, which need to be cleared of in the purification train.

Different products will require distinct methods in order to be purified. Biopharmaceutical products as recombinant therapeutic proteins, plasmid DNA, viruses, or cells are very complex molecules or structures with different characteristics that can require totally different purification pathways. Most of the current studies on continuous purification are being carried on monoclonal antibodies, since these represent nowadays the most important class of biopharmaceutical products both approved and under development. Moreover, for mAbs, a platform process with three chromatography steps, starting with Protein A affinity chromatography as capture step, is widely used and provides high purities and high recoveries. Despite the established mAb platform purification process, due to the high costs of Protein A, alternative processes that can be carried out continuously and in large scale are currently under investigation for mAb capture from crude supernatant, such as precipitation [60] or aqueous two phase systems (ATPS) [61]. Nevertheless, most works developing continuous downstream processes focus on liquid chromatography (LC) due to its high selectivity and the extensive experience in industry using it in batch mode [62]. Therefore, several continuous LC methods have emerged in the last years, and several multicolumn systems for continuous LC have been developed and are nowadays commercially available, such as BioSC® (Novasep), ÄKTA PCC (GE Healthcare), BioSMB® (Pall Life Sciences), and Contichrom® (ChromaCon® AG), among others.

2.3.1 Continuous Liquid Chromatography (CLC)

Continuous liquid chromatography (CLC) has been applied since the mid twentieth century in the oil, sugar, and food industries, but only recently the application of such complex technologies to the purification of biologics purification has gained in popularity [63]. Semicontinuous processes, hybrid purification processes (which include both continuous and batch steps), and hybrid overall processes comprising, for example, continuous upstream and batch downstream operations are spreading in industry. Although these processes usually decrease processing time and increase overall productivity, the need to perform periodic regeneration and cleaning-in-place (CIP) operations poses difficulties to true steady-state continuous operation.

One of the major advantages of using continuous LC is to operate at full adsorption capacity of the stationary phase. In batch processes, the formation of product gradients inside the column limits loading to as low as 5–20% of the binding capacity in order to avoid loss of valuable products, whereas the CLC platforms are usually comprised by multiple columns in sequence, enabling to exploit the total binding capacity of the resins, since the flow-through product of one column is captured by the next column in the sequence [56].

2.3.1.1 Continuous Annular Chromatography (CAC)

The first annular chromatography system was proposed in 1955 and consisted of 36 tubes arranged in a slowly rotating circle, as shown in Figure 2.4a [64]. This was later developed to the current CAC design where, different from conventional chromatography, the bed has an annular-shaped bed that keeps rotating around its axis. The feed mixture is introduced continuously at a fixed point at the top of the bed, whereas the mobile phase is distributed around the rotating annulus. Due to gravitational forces, the mixture components travel through the packed stationary phase forming regular helices, with the species with stronger interactions completing a longer path within the packed bed and being collected at the bottom of the column at a position more distant to the feed point than species that weakly interact with the stationary phase [55]. Thus, important parameters of this technique are the packed bed length and the angular displacement of each component (Figure 2.4) [62].

Besides the ability to operate under true steady-state conditions [55], CAC presents other advantages, such as the possibility of being used in any chromatographic stage from capture to polishing [65] and of directly transferring batch protocols based on stepwise elution to CAC. Nevertheless, only a few reports of

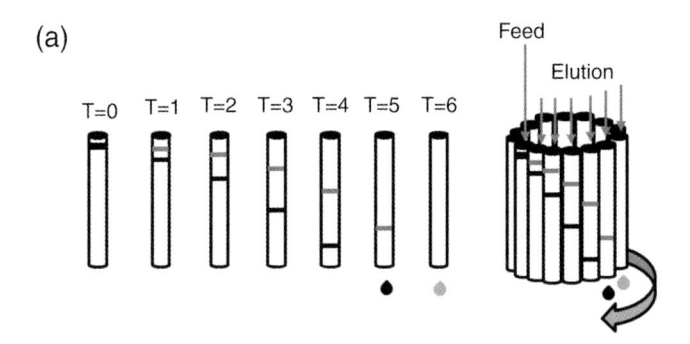

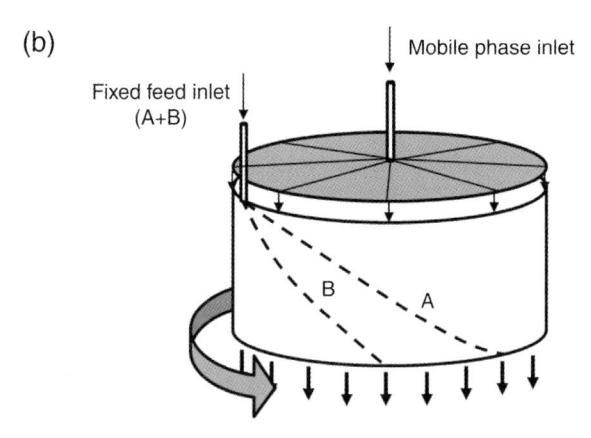

Figure 2.4 (a) Scheme of the first prototype of CAC made of packed tubes placed in a rotating circular system. (b) The annular packed-bed system established later. (Adapted from Ref. [64]. Copyright 2003, Elsevier.)

preparative CAC can be found in the literature, mainly dating back to the early 2000s [66,67]. Giovannini and Freitag [68], for instance, compared the purification of a recombinant mAb from cell culture supernatant by CAC, batch chromatography, and expanded bed chromatography (EBA), using both hydroxyapatite and Protein A resins. Yields around 90 and 80%, and purification factors of approximately 2 and 50, were obtained for hydroxyapatite and Protein A CAC processes, respectively. Hydroxyapatite CAC was further compared to EBA mode, with the latter yielding low recoveries, below 70%, as compared to 90% in CAC. In a non-mAb application, recombinant Factor VIII was continuously purified by preparative CAC using an ion-exchange resin, at 94% yield, 3.5-fold purification, and 30-fold concentration factor [69]. An autoclavable prototype was built to allow connection of the CAC system to the filtered perfusate stream. However, the authors reported peak wobbling due to inhomogeneous flow as a performance-limiting factor of the system [69].

2.3.1.2 True and Simulated Moving Bed Chromatography (TMB/SMB)

True moving bed chromatography (TMB) is a process based on countercurrent flow of both stationary and mobile phases. However, the difficulty in moving

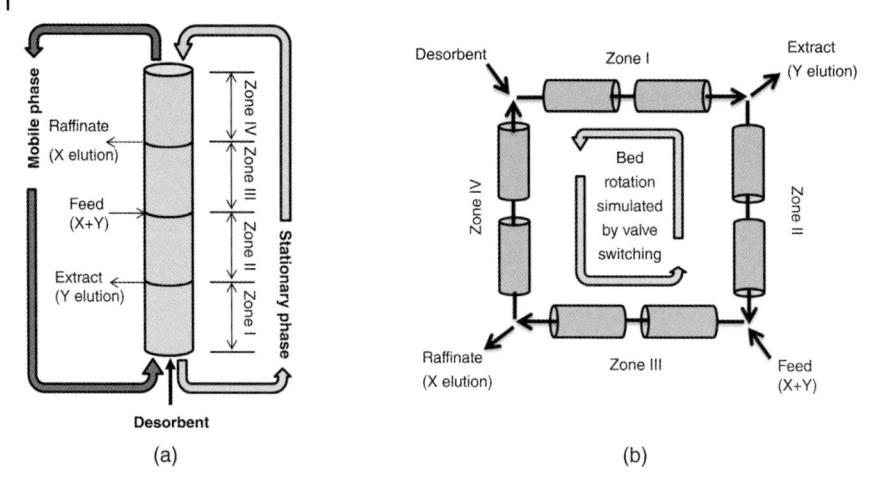

Figure 2.5 Principle of operation for binary separations. (a) True moving bed (TMB). (b) Simulated moving bed (SMB). Bed rotation is accomplished by switching mobile phase ports in the direction shown.

particles increases the complexity of this system. In order to decrease the complexity, a system with multiple columns in series receiving a countercurrent eluent flow, thus simulating the TMB concept, was designed. This process was first developed by the company Universal Oil Products under the name Sorbex and is currently known as simulated moving bed (SMB) chromatography [62]. Originally, a single rotary valve was used in the Sorbex system to switch the beds in position at defined time intervals, but the use of multiple individual valves for flow control made the system more practical.

The operation principle of TMB and SMB is based on four zones or four countercurrent columns, as shown in Figure 2.5. A scheme of TMB, with feed being injected between zones II and III, is found in Figure 2.5a. Mobile and stationary phases move in countercurrent direction, and the components with lower interactions are eluted in the raffinate stream between zones III and IV. Eluent is added at the bottom of zone I, and the components with higher interactions are eluted in the extract stream between zones I and II. In SMB (Figure 2.5b), the countercurrent movement of mobile and stationary phases is simulated by having at least one bed per zone and periodically switching the ports [62,63].

This continuous LC technique presents several advantages over batch chromatography, such as higher productivity, lower eluent consumption, and lower dilution of product when isocratic elution is performed [62]. For example, 80% reduction in eluent consumption was reported by Grill *et al.* when comparing SMB to batch chromatography to perform a binary separation to resolve a racemic pharmaceutical enantiomer [70]. Due to its binary nature, separation of enantiomers is a popular application of SMB and has also been reported at large scales [71].

In the biologics field, one interesting example of application reported the purification of adenovirus using a size-exclusion (SEC), two-column, simulated moving bed process [72]. The large size of the adenovirus enabled it to be efficiently separated from contaminants, with virus yields in the range of 86–95% [72,73]. Another interesting application consisted of using SEC-SMB to promote protein refolding and purification from solubilized inclusion bodies [74]. The refolding buffer in this system was used as eluent and tangential flow filtration was used to recycle the refolding buffer into the system, contributing to a 28-fold reduction in buffer consumption. Moreover, Kessler *et al.* [75] purified bovine IgG by a three-zone SMB process using different solvent strengths in the feed and eluent streams, establishing a gradient that allowed obtaining different internal plateaus of elution strength. Important developments that can enable a broader application of SMB in the biopharmaceutical field include going beyond binary separations. Recent examples in this regard are the development of pseudo-SMB systems for ternary and quaternary separations using solvent gradients [76,77].

The use of the term SMB can be confusing. In spite of developments toward ternary and quaternary separations, many people frequently use SMB to refer to binary fractionations. On the other hand, some other researchers use the term SMB for any type of multicolumn chromatography system in which the transport of resin is accomplished by periodically switching column inlets and outlets [78].

2.3.1.3 Multicolumn Countercurrent Solvent Gradient Purification (MCSGP)

The multicolumn countercurrent solvent gradient purification (MCSGP) is an innovative continuous chromatography technique that combines two different principles: gradient elution used in batch column chromatography and SMB chromatography [79]. This multicolumn countercurrent system relies on multiple columns that are switched opposite to the flow direction (Figure 2.6). It is possible to have linear, nonlinear, and segmented gradients, and the system is claimed to provide high yields, resolutions, and purities [55]. High resolutions are achieved, since impure fractions, which contain product (P) contaminated with stronger (S) and/or weaker (W) binders (in sections β and δ of Figure 2.6), are countercurrently reloaded to the next column by switching the ports, thus yielding pure fractions of P. As indicated in the system scheme shown in Figure 2.6, in each section in given time intervals different tasks are performed – either important tasks or minor ones (the minor ones are those indicated in brackets). For instance, the most important task in column 2 is to have all P out of the column, being directed to column 4, where it will adsorb onto the resin and get separated from W (which goes out in this section), remaining contaminated or not with S. In the next switch (opposite to flow direction), column 4 will become column 3, with the major task of not allowing S out from the column, in order to elute pure P. Section δ task is very important to keep P in the system, thus t_F and $t_δ$ are designed to recover the contaminated fractions and to eliminate W from the system, so that high purity can be then obtained in section α. Because the fractions containing overlapping contaminants and product are recycled and further processed, this process alleviates the traditional trade-off between yield and purity and is an interesting alternative for polishing steps [80].

Aumann and Morbidelli [79] applied the system for separation of peptides, having calcitonin as main product. According to these authors, 95% purity was

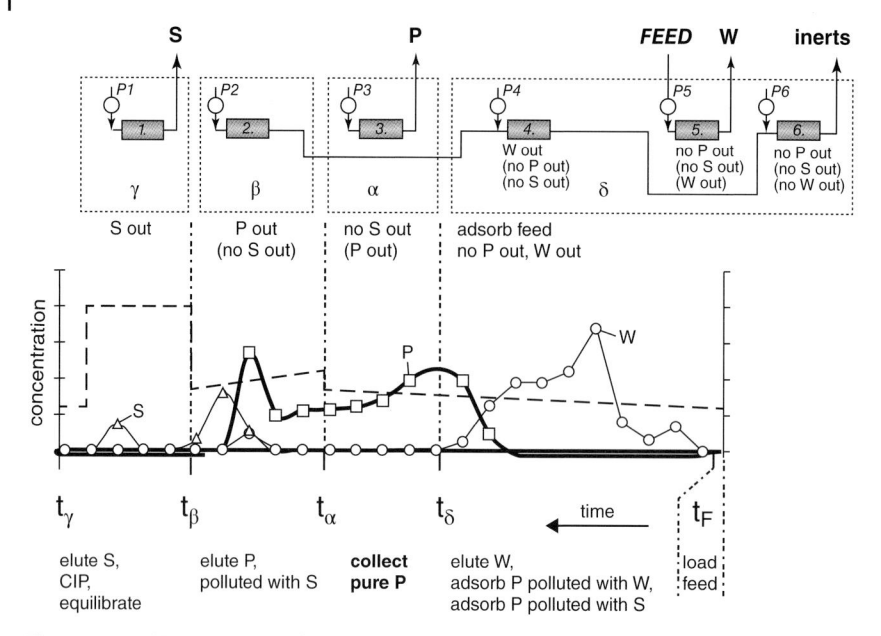

Figure 2.6 MCSGP process with six columns. The time intervals applied to the MCSGP sections are defined based on a batch chromatogram as the one shown in the lower part of the figure. (Reproduced with permission from Ref. [79]. Copyright 2007, John Wiley & Sons.)

reached at 60% product recovery in samples collected at 1200 min, operating under steady state (which was reached after 400 min of operation). Working with variants of a monoclonal antibody, Müller-Späth et al. [81] confirmed the importance of the timing to switch the columns, since it impacts process performance. In this paper, the MCSGP separation process using cation exchange chromatography was able to provide both purity and recovery above 90% for the target mAb variant. In another application focusing on purification of monoclonal antibodies, Müller-Späth et al. [80] developed a cation-exchange MCSGP process as capture step to isolate a mAb from clarified cell culture supernatant and compared it with a classical batch process using the same resin. When selected eluate fractions of the batch process were pooled (high-purity process alternative), comparable purities of 97% were obtained both for the batch process and MCSGP, but at the expense of 83% yield for the batch system as compared to 95% yield for MCSGP. On the other hand, when all eluate fractions of the batch process were pooled (high-yield process alternative), yields were more comparable (92.4% for batch process and 94.9% for MCSGP), but purity was lower for the batch alternative (88.1%, as compared to 97.4% for MCSGP). Using the high-yield process for a more complete comparison, MCSGP performed better under several aspects: two- to threefold lower buffer consumption, almost twofold higher mAb concentration in product stream, threefold larger HCP clearance, 50% higher productivity, and 83% higher loading velocity.

Different from the binary separations that characterize SMB, MCSGP enables multifraction separations, and this enlarges its applicability to bioprocesses. Krättli et al. [82], using model proteins and mAb variants, carried out processes

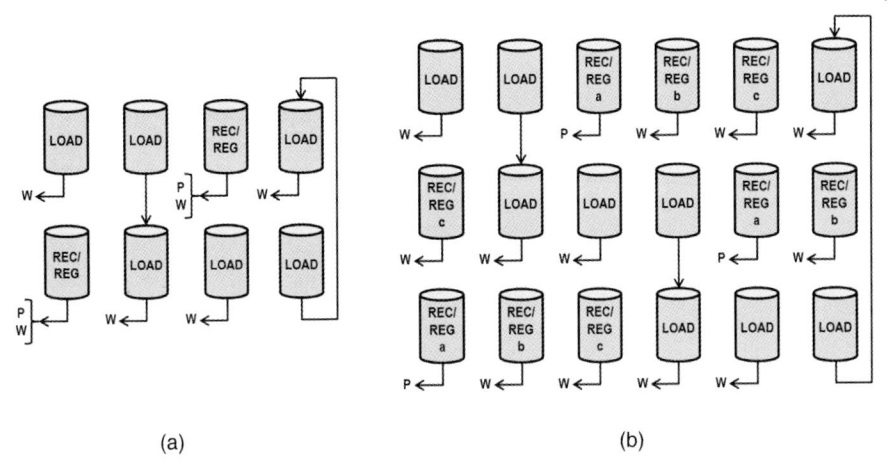

Figure 2.7 Schematic comparison of one cycle of a PCC system operating under steady state. (a) A 2-column PCC dual-flow rate strategy proposed by Ref. [83] (CaptureSMB). (Reproduced with permission from Ref. [83]. Copyright 2015, Elsevier.) (b) A 3C-PCC system. W – waste stream; P – product stream; REC/REG – comprises washing, elution, regeneration, and equilibration stages.

with three and four fractions and demonstrated that the number of columns is related to the possible number of fractions.

2.3.1.4 Periodic Countercurrent Chromatography (PCC)

In periodic countercurrent chromatography (PCC), the columns are operated in series (Figure 2.7), so that the flow-through and wash of a given column are captured by the following column, allowing for loading of the resin close to its static binding capacity [6]. Therefore, a resin utilization efficiency much higher than in conventional batch chromatography can be achieved. In PCC, each column runs the chromatographic stages (equilibration, loading, washing, elution, regeneration) in a cyclic fashion, and one cycle is considered complete when all columns have undergone all stages, so that a new cycle can start. Thus, feed can be processed continuously, whereas eluates are collected from each column periodically.

PCC systems usually consist of multiple columns with equal dimensions and the same packed material arranged in sequence, but according to the number of columns the operation of the cycles can be slightly different. Also, for a given number of columns in a PCC system, there are degrees of freedom that allow different strategies of operating the recovery/regeneration stages (i.e., washing, elution, regeneration, and equilibration). For instance, in a 3-column PCC (3C-PCC) system, washing stages interconnecting two columns at a time while the third column is being loaded can be carried out or not [6,84,85].

In order to illustrate the PCC operation principle, one possible setup for a 3C-PCC system is shown in Figure 2.7b. In each cycle, there are different steps. At the beginning (step 1), column 1 is loaded with feed and its flow-through is directed to waste until product breakthrough occurs. At this moment (step 2), the flow-through of column 1 starts to be injected in column 2, so that unbound product is adsorbed to the stationary phase in column 2. When column 1 achieves saturation, feed starts to be directed to column 2 and washing/elution starts in column 1 (step 3). Analogously to what happened in steps 1 and 2 with column 1, the flow-through of column 2 is switched from waste to column 3 when product breakthrough starts (steps 3 and 4). Column 1 can undergo regeneration in step 4 and re-equilibration in step 5, whereas column 2 undergoes washing/elution in step 5, after feed has been directed to column 3. In step 6, feed continues to be applied to column 3, but due to product breakthrough its flow-through is directed to column 1, while column 2 is being regenerated. The next step represents the start of a new cycle, with feed being applied to column 1, while column 2 is being re-equilibrated and column 3 is undergoing washing/elution. The cycles, comprising these 6 steps, are repeated, so that feed is continuously injected to the system, whereas eluates are collected from columns 1, 2, and 3 in steps 3, 5, and 1 of each cycle, respectively. The breakthrough curve observed in an individual column determines the switch time, because it is the moment when the outlet of this column is switched from waste to the next column. The way the discrete recovery/regeneration stages (washing, elution, regeneration, and re-equilibration) are carried out in a cycle is flexible and can be optimized according to process specificities. In the set-up taken as example in Figure 2.7b, washing and elution were carried out in the first step (*a*) of the recovery/regeneration operation (REC/REG), but other set-ups are possible, for example, with (*a*) consisting just of washing, and elution being carried out at step REC/REG (*b*).

Figure 2.8 illustrates the timing for switching by showing the breakthrough of an individual column: At first a baseline reflecting the breakthrough of impurities is reached, and then a second, product-containing breakthrough starts. At this moment, it is the time (t_{SW}) for the outlet stream of this given column to be switched to the next one. When the UV signal at the outlet of the first given

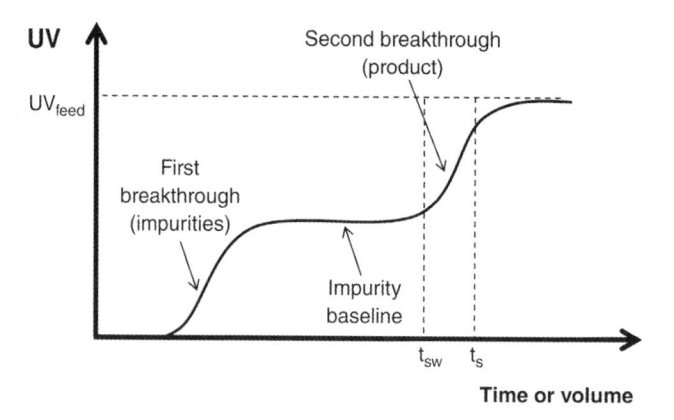

Figure 2.8 Breakthrough curve during the loading stage of a given individual column of a 3C-PCC system. The first time interval corresponds to the period that impurities are being collected in the flow-through and directed to waste. The outlet of this column must be switched to the next column immediately prior to the start of product breakthrough (t_{SW}). Feeding continues until very near to the static binding capacity of the resin. Saturation is achieved when the given column gets fully loaded at t_S, so that at this moment this column starts REC/REG and the feed starts to be injected into the next column. (Adapted from Ref. [84]. Copyright 2012, John Wiley & Sons.)

column approaches that of the feed, it is time (t_S) to start injecting the feed directly to the next column. This operation principle allows loading the resin until very close to its static binding capacity, thus enabling an increase in yield and a significant reduction in column size, as demonstrated by Godawat et al. [84].

PCC systems have been successfully applied in the purification of different products using different configurations [6,83,84,86]. Although most works use 3-column or 4-column PCC systems, Angarita et al. [83] demonstrated the feasibility of getting high performance with a 2C-PCC/CaptureSMB system (Figure 2.6a). They applied a dual-flow rate strategy, which allowed the sequential loading principle to be employed, while using the minimum required number of columns. The authors used Protein A affinity chromatography and a commercial multicolumn system (Contichrom) to compare the dual-flow rate 2C-PCC principle to batch chromatography and confirmed significant reductions in resin volume, column dimensions, and buffer consumption, with higher productivities and comparable yields and purity.

Godawat et al. [84] and Warikoo et al. [6] employed an ÄKTA™ system customized by GE Healthcare, equipped with either three or four columns and multiple pumps, valves, and UV monitors, which was later developed to the commercially available system known as ÄKTApcc. They observed high performance both for stable mAbs and labile non-mAb therapeutic products, and demonstrated the integration of the PCC systems to continuous upstream equipment, as will be discussed later in this chapter.

In order to design a PCC system, an important information that needs to be first determined is the static binding capacity of the adsorbent under operation conditions. This is accomplished by obtaining breakthrough curves by means of frontal loading experiments for specific column sizes and residence times. In order to have a continuous loading of feed (feed continuity constraint), the discrete chromatography stages within a PCC system should be scheduled such that, for a given column, the loading time (t_L) is equal or greater than the time t_{RR} needed for the recovery/regeneration stages (washing, elution, regeneration, and re-equilibration). For a N-column PCC system, t_{RR} can be calculated according to Eq. (2.1) [84]:

$$t_{RR} = \frac{t_W + t_{EL} + t_{REG} + t_{EQ}}{(N-2)}, \tag{2.1}$$

where t_W, t_{EL}, t_{REG}, and t_{EQ} are the times for washing, elution, regeneration, and re-equilibration, respectively.

The number of columns of a PCC system can be calculated according to Eq. (2.2) [84]:

$$N > 2 + \frac{t_{RR} C_P}{R_t Q_S}, \tag{2.2}$$

where Q_S is the static binding capacity determined in frontal loading experiments, C_P is the concentration of product in the feed, and R_t is the residence time during loading. R_t in turn is a function of column geometric properties (height and diameter) and operational conditions (flow rate/surface velocity). Equation (2.2) assumes that the actual binding capacity in PCC approaches the static binding capacity, otherwise Q_S should be replaced by the actual binding capacity observed in PCC operation [84].

Equation (2.2) discloses two important aspects: (i) There are multiple PPC setups that can satisfy the feed continuity constraint and fulfill a given purification task and (ii) the ratio Q_S/C_P is a key parameter to design a PCC system. Godawat et al. [84] calculated the minimum residence time needed as a function of the Q_S/C_P ratio for PCC designs with three to eight columns, and on the other hand predicted the minimum number of columns required for a range of combinations of R_t and the Q_S/C_P ratio. The authors then applied the proposed methodology to investigate the continuous Protein A capture of a mAb and the continuous capture of two enzymes using pseudoaffinity (iminodiacetate resin) or hydrophobic interaction chromatography, respectively. By applying their proposed methodology, they used a customized ÄKTA system for 3C-PCC (mAb and HIC enzyme purification) and 4C-PCC (pseudoaffinity enzyme purification) process development.

Gjoka et al. [87] also proposed a simple design strategy based on frontal breakthrough experiments to design multiple column chromatography systems, and validated it using another commercial multicolumn system (BioSMB). The predicted and measured capture efficiencies for a mAb purified by Protein A chromatography matched very well and were in the range of 93 to 99%.

2.3.1.5 Continuous Countercurrent Tangential Chromatography (CCTC)

Continuous countercurrent tangential chromatography (CCTC) was patented in 2009 [88] and relies on moving a resin slurry through a series of static mixers and hollow fiber modules in a countercurrent direction to the flow of buffers. Permeate solutions from later stages are recycled back into early stages, creating concentration gradients in the permeate solutions, saving buffer volume, and increasing process efficiency. By employing the resin in a slurry form, and not in packed beds, a true resin flow can be accomplished. Moreover, the need for time-consuming and labor-intensive packing of beds is eliminated.

The microporous hollow fiber membranes retain the resin inside the system, while nonadsorbed proteins and buffer components are removed in the permeate [56,89,90]. Each chromatography step (binding, elution, washing, regeneration, and equilibration) is carried out with the respective buffer in a stage comprising a static mixer and a hollow fiber module. The buffers for these operations are pumped into the module in a countercurrent direction to the flow of resin, and permeate streams from later stages are pumped into previous stages [88]. This countercurrent flow of buffers results in high resolutions, increases yield, and decreases buffer consumption [89]. The uniform residence times of all stages, the low operational pressure drops, and the possibility of having the whole system made of single-use components are advantages claimed for this process [56,90].

In the first experimental demonstration of purifying high-value biomolecules using CCTC, Shynkazh et al. [89] investigated setups with 1, 2, and 3 stages. Design of the CCTC system was based on binding and critical flux experiments. Using bovine serum albumin (BSA) and myoglobin as model proteins for anion exchange chromatography separation, the authors obtained BSA yields of 78, 94, and 98% for the 1, 2, and 3-stage setups, respectively. The base case system with two stages provided the BSA product with >99% purity.

Since the patent in 2009 [88], several improvements have been proposed to the system, such as the use of retentate pumps to maintain stable resin concentrations in the flowing slurry, the elimination of a slurry holding tank to improve productivity, and the introduction of an "after binder" to the binding step to increase product recovery [90]. According to the authors, these changes have resulted in an increase in productivity, a significant reduction in the equipment footprint, and a more robust process operation. Their results using the CCTC system for protein A purification of mAbs from cell culture supernatants produced both by means of low-titer and high-titer upstream processes were compared to batch column experiments. Although yields were slightly lower for the CCTC system, productivities were 2.4–3.4-fold higher than in column chromatography. Level of aggregates, distribution of charge variants, and leached Protein A concentration were compatible for both processes, while HCP in the eluates was lower for CCTC [90].

2.3.2 Nonchromatographic Continuous Processes

2.3.2.1 Continuous Aqueous Two-Phase Systems

Aqueous two-phase systems (ATPS) represent a type of liquid–liquid extraction (LLE) process that is not based on the use of organic solvents, as most LLE processes are. Being aqueous in nature, the immiscible phases used to promote separation are biocompatible and do not affect the biomolecule to be purified. The separation is based on the differential partitioning of the biomolecules present in a mixture between two immiscible aqueous phases, which can be two different polymer solutions (e.g., dextran and polyethylene glycol, PEG) or a polymer and a salt solution (e.g., PEG and phosphate) [91]. It is a simple purification technique that is easy to apply and to scale up – in several industrial sectors there is a great experience in implementing continuous multistage countercurrent LLE processes at large scale [57,58].

In spite of the aforementioned advantages of this process, the lower resolution that ATPS usually presents made this approach less attractive for high-value-added biotherapeutics. In particular, in the case of monoclonal antibody purification, the comparison to the high performance of Protein A affinity chromatography hampered in the past this technique from being more widely employed.

However, due to the higher titers currently obtained in the upstream processing and due to the rapidly changing biopharmaceutical business scenario, marked by the trend for continuous processing, by more competition, and by the need to decrease manufacturing costs [8], the interest in the use of this technology to purify biopharmaceutical products increased significantly in the last few years.

Rosa *et al.* [61] developed a continuous multistage countercurrent process to purify mAbs from cell culture supernatants (Figure 2.9). They employed a PEG/phosphate system comprising three different steps: (i) extraction, for removal of most high-molecular mass (MM) contaminants; (ii) back-extraction of product; and (iii) washing, for removal of lower MM contaminants and polymer-rich phase, and for polymer recycling. The IgG was recovered with

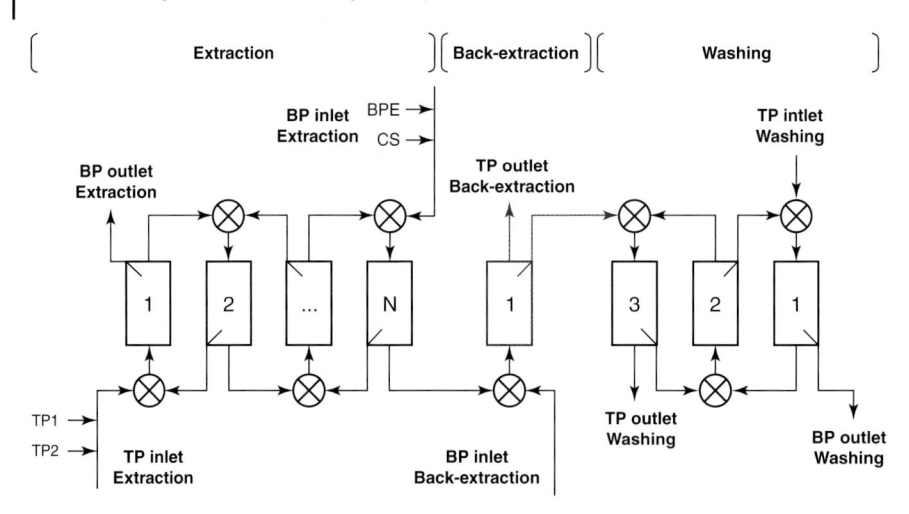

Figure 2.9 Schematic representation of countercurrent mixer-settler battery used for ATPS separation of IgG from cell culture supernatant. (Reproduced with permission from Ref. [61]. Copyright 2013, John Wiley & Sons.)

yields of 80 and 100% from CHO and PER.C6® supernatants, respectively, and purities were higher than 99% in both cases. In another application focusing on mAb purification, multistage extraction and washing provided purity of 75%, yield of 95%, and HCP clearance of 73% [92].

Recently, van Winssen et al. [93,94] proposed a novel approach called Tunable Aqueous Polymer Phase Impregnated Resins (TAPPIR®). In this process, in order to improve separation efficiency and speed, one aqueous phase is immobilized inside the pores of solid particles, which are then suspended in the second aqueous phase containing the biomolecules. This allows phase emulsification and partitioning of biomolecules between the two aqueous phases, with no need for phase separation after the extraction step. However, so far only reports showing the proof of principle of this technology for dye separation [93] and a model mixture of lysozyme and myoglobin [94] can be found in literature.

2.3.2.2 Continuous Protein Precipitation

Fractional precipitation is one of the simplest processes applicable to the separation of proteins. It relies on causing part of a mixture of proteins to precipitate through alteration of some property of the solvent, changing protein solubility under those new conditions. The distribution of hydrophobic and hydrophilic amino acid residues at the surface of a protein determines its solubility in a solvent and, although hydrophobic residues tend to be located in inner parts of the protein, there are always hydrophobic groups on the surface, exposed to the solvent, which are as important as charged and other polar groups in determining the protein behavior [95].

The proteins to be isolated are usually found in aqueous solvents, and protein solubility can be manipulated by changing properties of water, for example, through changes in pH, ionic strength, addition of water-soluble polymers or miscible organic solvents, temperature variations, or combinations of these.

Although the process is simple and inexpensive, the lower resolution traditionally achieved and the risk of applying conditions that might affect protein structure are some of the reasons why this process has not been so often applied for the isolation of high-value-added biologics. However, new and innovative approaches have been developed in the last few years, demonstrating the feasibility of applying the long-existing technology of protein precipitation to efficient biotherapeutics purification processes, also under continuous mode [95–98]. These recent advances enable combining the low cost, scalability, and ease of operation of precipitation processes to the higher productivities of continuous processes.

Besides manipulating conditions to decrease protein solubility, in continuous precipitation processes shear and mixing conditions are also critical conditions for process performance [55]. Different mixers have been employed in continuous precipitation, such as continuous stirred tank reactors (CSTR), mixed suspension and mixed product removal reactors (MSMPR), and tubular reactors [55,60,98,99].

As with other purification techniques, the recent publications on protein precipitation focus on monoclonal antibodies. One of the reports proposed a 4-step process combining ethanol, $CaCl_2$, pH, and temperature to induce precipitation [100]. In another paper, the same group investigated the combination of $CaCl_2$ with polyethylene glycol (PEG) precipitation for the purification of recombinant mAbs from cell culture supernatants. Yields were in the range of 80–95%, and purities were comparable to Protein A affinity purification [101]. Another publication of the group studied the further combination of the three aforementioned precipitation techniques ($CaCl_2$, PEG, and ethanol) with caprylic acid (CA) precipitation, proposing either a 2-step process ($CaCl_2$, CA) that performs well enough to replace a Protein A step, or a 4-step process ($CaCl_2$, CA, PEG, ethanol) that gives final yield and purity approaching drug substance values [96].

These works were then followed by reports on the adaptation of the precipitation methods to continuous mode. Hammerschmidt *et al.* [98] used tubular reactors to carry out a two-stage $CaCl_2$/ethanol continuous precipitation process, which resulted in yields higher than 90%, 79% reduction in HCP, 98% reduction in DNA, aggregate levels below 1%, with no effects on antibody binding or isoform distribution. Hammerschmidt *et al.* [97] studied sequential precipitation adding PEG and manipulating the pH. The continuous process was carried out in tubular reactors, and gave results comparable to the batch precipitation process in terms of yield, HCP, DNA, and aggregates. Precipitate recovery was successfully carried out using a tangential flow filtration system (instead of centrifugation), thus allowing for the use of single-use devices.

The four types of mAb precipitation techniques ($CaCl_2$, CA, PEG, ethanol) were combined in the economic analysis carried out by Hammerschmidt *et al.* [60], which showed that the precipitation-based process enabled cost reductions over Protein A-based conventional technology at all stages (clinical phases I–III and commercial production) of the life cycle of a therapeutic antibody.

2.3.3 Straight-Through Processes

Some processes are unable to operate continuously, but can provide fast processing and thus can be a useful tool for integrated bioprocessing strategies.

Straight-through processes (STP) are based on matching flow rates and conditions (pH, conductivity) of outbound and inbound streams of sequential purification steps, allowing direct feeding of one eluate to the next column and thus reducing the need for holding tanks [102–104]. These processes usually present several advantages over traditional batch processes, such as higher throughput, lower buffer consumption, and/or higher productivity.

The development of straight-through processes requires the use of chromatography systems with multiple pumps and valves, so that multiple buffers can be dealt with and streams can be pumped from one step (column, monolith, membrane adsorber) to another. Hughson *et al.* [102] have adapted an ÄKTA Explorer 10 system to develop a 3-step STP process for the purification of an unstable recombinant biotherapeutics. Hold steps were eliminated, and a direct comparison of six different combinations of purification sequences was carried out, involving two ion exchange options as first step, a hydrophobic membrane adsorber as second step, and three affinity chromatography options as third step. The final STP process enabled the three steps to be completed in less than 5 h with starting material volumes of up to 400 ml. Overall yields for replicate STP runs were 65–70% (calculated based on biological activity), and HCP levels in the final eluate were 31.8 ng per projected dose.

Shamashkin *et al.* [103] developed an STP platform for mAb purification based on an ÄKTA FPLC system that included three processing steps: a Protein A capture step, an anion exchange chromatography step in flow-through mode, and an in-line virus filter. Product quality (HCP, leached Protein A, high molecular mass aggregates, and antibody recovery) in the STP final eluates was comparable to that of the standard purification process, and gains were claimed in terms of processing speed, reduced number of holding tanks, decreased amount of in-process samples, and lower operating costs.

Other tools that can increase process throughput and are useful for integrated, continuous bioprocessing are resins operated in negative or flow-through mode [103] and convective-flow adsorbents, such as monoliths, nanofibers, and membrane adsorbers [59,102,105–107].

2.3.4 Continuous Virus Clearance Processes

Viral clearance is a critical issue in biopharmaceutical production, since any risk of virus transmission, as occurred in the 1980s through contaminated blood products [62], must be avoided. Viral clearance can be accomplished by removal or by inactivation of virus [108]. The current regulatory requirements establish that at least two orthogonal steps for viral clearance should be included in the downstream process, and that each of these steps should provide a log reduction value (VLR) of at least 4 [56]. In most processes, the viral clearance steps comprise one virus inactivation and one virus removal technique.

Several operations are routinely used for viral clearance, such as membrane filtration [109] and chromatographic separations [110] for viral removal, and UV irradiation [56,111,112], low pH inactivation [113,114], and solvent/detergent incubation [115] for viral inactivation. However, the need to apply these techniques under continuous mode is a recent requirement that has arisen from the trend for continuous biomanufacturing in the biopharmaceutical industry.

Low pH inactivation can be easily integrated into mAb purification processes that comprise a Protein A chromatography step, since the virus inactivation step can be accomplished by holding the Protein A eluate for a given incubation time at the low elution pH (usually around 3.0). This strategy was applied within an end-to-end continuous production and purification process, where the eluates of the Protein A columns of the 3C-PCC capture system were collected and held at low pH for a residence time of 1 h [113]. In order to warrant the efficiency of a low pH virus inactivation process, Klutz *et al.* [114] proposed the use of a coiled flow inverter (CFI) as a suitable incubation vessel, and developed mathematical models to design the system, based either on a logarithmic reduction value (LRV) or on a minimum residence time (MRT) approach.

Orozco *et al.* [116] have designed a chamber that incubates a continuous product stream for a desired incubation time, typically for 1 h. Since typical incubation times and flow rates do not result in plug flow conditions, one of the biggest challenges to ensure efficient continuous viral inactivation is to address dispersion of the product stream. Therefore, in order to meet the multilog viral clearance requirements for a given step, the authors based their design work on the constraint that only 1–10 ppm of the original product should exit the chamber before the specified time.

Salm *et al.* [117] developed a fully automated, 100-l scale continuous downstream process including an in-line viral inactivation/conditioning step, and also demonstrated feasibility of multiday virus reduction filtration.

In general, in-line viral conditioning [117,118] and continuous incubation chambers [116] can be used to convert batch viral inactivation processes already well established in industry, such as low pH and solvent/detergent methods, but can also be applied to alternative inactivation techniques, such as incubation in neutral, concentrated arginine solutions, as proposed by McCue *et al.* [108].

Continuous flow UV irradiation is another effective method for virus inactivation [56,111,112]. Caillet-Fauquet *et al.* [111] developed a continuous flow system that was shown to inactivate some viruses and bacteria without affecting the biological activity of bloodborne material. This system could be easily adapted to continuous purification of biopharmaceutical products.

Regarding virus removal steps, membrane-based filtration could be operated in pseudocontinuous mode by using two or more membrane modules arranged in parallel, with automatic switching of devices whenever filter regeneration or replacement be needed [44,56]. Furthermore, the various options of continuous chromatography systems already discussed are expected to continue significantly contributing to the viral clearance robustness of biopharmaceutical purification processes, such as batch chromatography processes do.

2.4 Integrated Continuous Processes

Although perfusion cultivation has been an industrial reality for several unstable products for over two decades, the processes as a whole were hybrid ones, since the downstream steps for these molecules remained batch processes. So, even if the upstream process enabled a low residence time of the molecule inside the bioreactor, thus reducing the risk of product degradation, the long-lasting sequence of batch purification steps and intermediate holding tanks remained a potential source of damage to the product. Moreover, the benefits of continuous chromatography, such as decrease in resin volume needed and higher productivities, were not taken advantage of.

However, the advances experienced in the last years in the development of tools for improving continuous upstream processing and enabling continuous downstream processing (e.g., commercial continuous countercurrent chromatography systems) are making the development of fully integrated continuous bioprocesses possible, operating end-to-end under continuous mode.

The earlier works (2001–2012) that investigated integrated bioprocessing setups focused on integrating perfusion cultivation with the first purification step, that is, product capture [6,59,84], whereas more recent works (2015) show interesting results from end-to-end integrated, continuous operation [44,113]. In an early study from 2001–2002 to carry out product capture simultaneously to perfusion cultivation, Castilho et al. developed Protein A membrane adsorbers and used computational fluid dynamics to design a rotating disk filter (Figure 2.10a), which jointly resulted in a filtration/adsorption system that was coupled to a bioreactor (Figure 2.10b) in order to simultaneously promote cell retention and product capture [59,119–121]. Protein A ligand from a sterile filtered solution was immobilized *in situ* onto cellulose microporous membranes placed inside the previously autoclaved system prior to process start, warranting aseptic operation. The perfusion/capture system was fully automated, so that the captured mAb was recovered through automated, short, periodic elution cycles, during which bioreactor feeding and perfusate withdrawal were shortly interrupted. The results demonstrated that IgG concentration in the eluate was 14-fold higher than inside the bioreactor (Figure 2.10c), and silver-stained, reducing SDS-PAGE analysis showed that a high-purity IgG could be directly recovered, during cell cultivation, from a very impure bioreactor material (Figure 2.10c). Interesting aspects that could be studied to further develop this concept could be the use of twin rotating disk filters operating alternately, in order to avoid short interruptions of perfusion operation to carry out elution, as well as the replacement of the flat-sheet membrane adsorbers and the rotating disk filter by hollow fiber membrane adsorbers that could be fitted either in ATF or TFF systems.

In 2012, Warikoo et al. [6] and Godawat et al. [84] reported an integrated perfusion/capture process using the ATF cell retention device and a PCC system. Proof of concept runs for a mAb and a non-mAb product (an enzyme) were carried out, with viable cell densities in the bioreactors for both products maintained in the range of 50–60E6/mL.

The mAb capture system consisted of a Protein A 3C-PCC setup with 2.5 min residence time in the columns. Over a period of 30 days, 114 column operations, with a total of 228 automated column switches, were carried out. The Protein A resin capacity utilized in PCC was confirmed to be comparable to the static binding capacity (~49 mg/ml resin), and mAb critical quality attributes, including

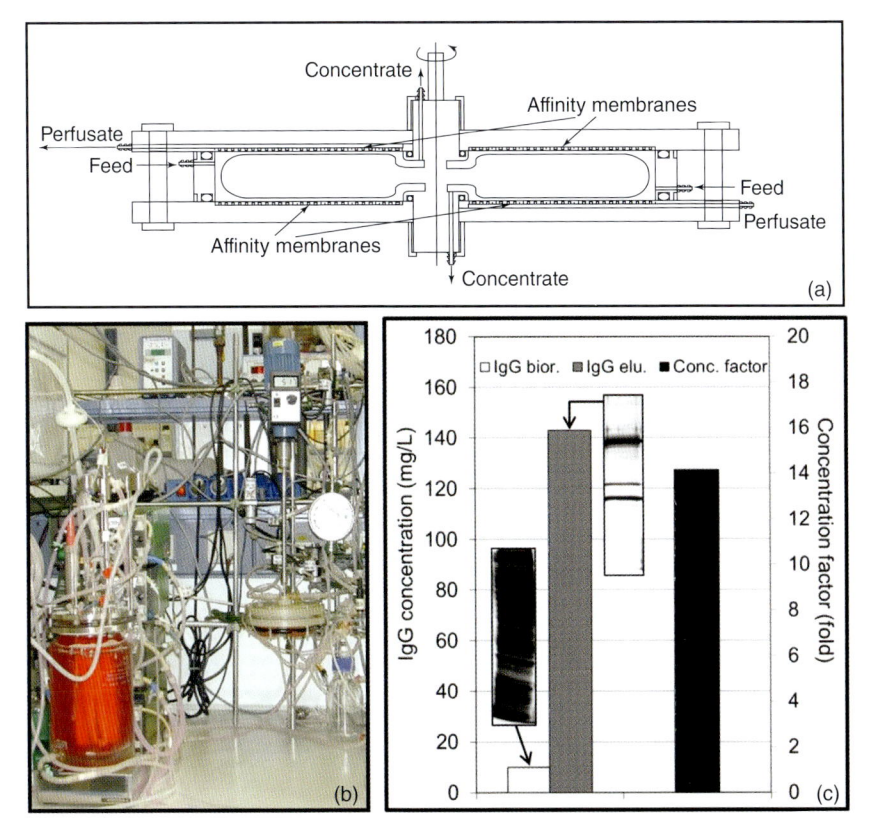

Figure 2.10 Integrated perfusion/capture of a mAb produced by CHO cells. (a) Rotating disk filter (RDF). (Reproduced with permission from Ref. [121]. Copyright 2003, John Wiley & Sons.) (b) System for integrated perfusion/capture, comprising the RDF prototype and Protein A membrane adsorbers. (Adapted from Refs [59,119].) (c) IgG concentration measured by nephelometry in bioreactor and eluate samples, showing a concentration factor of 14-fold. High purity of the eluate is evidenced by reducing silver-stained SDS-PAGE analysis, showing the heavy and light chains of the eluted mAb and a band of a contaminating protein at approximately 22 kDa.

potency, purity, HCP, residual Protein A, and aggregation levels, were comparable over the entire duration of PCC operation [84].

For the enzyme capture using a HIC resin, the feed continuity constraint implied that either a residence time in the column beyond 4.8 min had to be applied or a 4th column had to be added to the PCC [84]. This latter option was chosen by the authors, and 166 column operations were carried out during 9 days of integrated continuous cultivation/capture operation [84]. The resin capacity utilized in the 4C-PCC was comparable to the static binding capacity (4 mg/ml resin), and the recovery and all critical quality attributes were comparable across the four columns and throughout the entire period [6]. Resin capacity utilization was increased by 20 and 50% for the mAb and the enzyme, respectively. Further improvements were a decrease of 25 (mAb) and 46% (enzyme) in buffer usage, and a 25-fold (mAb) or 13-fold (enzyme) reduction in the total resin volume used [84].

In a report on a fully integrated, end-to-end continuous bioprocess for a mAb produced by a CHO/DHFR cell line, Godawat et al. [113] considered that their design goals were achieved: (a) process train simplification; (b) uninterrupted, automated purification to the drug substance over an extended period of time; (c) steady-state operation with respect to process flows and product quality; and (d) no change with time in the quantities held up in the unit operations. Two sequential PCC systems were used. The first one consisted of a Protein A 3C-PCC step fed with cell culture supernatant from a 12-l bioreactor. A 2-l disposable surge bag was placed in the stream leading the perfusate to the 3C-PCC. After a low pH viral inactivation step carried out in a stirred tank with 1 h residence time, there was a second 3C-PCC with strong cation exchange resin. The second 3C-PCC was designed so that the mAb was eluted directly in the formulation buffer at the required drug substance concentration. This eluate was directly fed into an anion-exchange membrane adsorber polishing step operated in flow-through mode that kept formulation buffer and protein concentration unaltered. Overall yield was approximately 80%, and the purified drug substance was generated at a rate of 8 g/day. The steady-state operation resulted in consistent product quality over ~30 days, and this additionally indicated the feasibility of the CHO/DHFR system for long continuous cell culture processes. The process performance was comparable to the batch process, with HCP clearance of 4 logs, removal of residual protein A to 1.0 ppm, and <5% aggregation. No time-dependent performance change was observed and critical quality attributes of the mAb product were comparable to release specifications of the traditional batch process.

A pilot plant fully based on single-use (SU) equipment and performing an automated continuous process up to the formulation step was designed and operated by Klutz et al. [44]. The SU units interconnected by flexible tubing resulted in a completely closed system, which allowed the implementation of the "ballroom concept": lower segregation and lower room classification requirements. This in turn resulted in lower HVAC needs and decreased capital and operational expenditures (CAPEX and OPEX). All process units from cell culture to viral filtration were placed in a class D production "ballroom," and only the seed laboratory and the postviral filtration operations were placed in segregated class C rooms. The pilot plant consisted of two perfusion bioreactors operated at 8-l working volume in parallel, each one connected to an inclined cell settler. At steady-state, viable cell density was 60–70E6/mL, product titer was 0.12 g/l, and specific productivity was 5.3 pcd. A perfusion rate of 3 vvd was applied, corresponding to a cell-specific perfusion rate of approximately 0.05 nl/cell/day. Downstream of the perfusion bioreactor, the process consisted of membrane filtration steps for clarification and concentration, respectively, followed by a Protein A BioSMB system for mAb capture, a coiled flow inverter reactor for low pH viral inactivation, two flow-through mode chromatography steps (with mixed-mode and anion exchange resins, respectively) in an additional BioSMB system, sterile and viral filtration steps, and two membrane filtration steps (ultrafiltration and dialysis, respectively) to accomplish product formulation. All filtration steps consisted of two modules in parallel to achieve continuous processing by alternate operation. Low bioburden was warranted by gamma sterilization or autoclaving for 95% of all parts that came in contact with product, or by sanitization with 40%

IPA + 0.5 M NaOH of the other materials. A short, proof of concept run lasting 2.5 days processed 120 l of concentrated harvest material (corresponding to approximately 840 l of raw perfusate), with an overall yield of 75%. The bulk drug substance obtained met the specifications for an industrial mAb product, with <35 ppm HCP and >98.5% monomer content. Although robust operation of the pilot plant for several weeks is still to be done, Klutz *et al.* [44] show concepts that could translate into a commercial facility [122].

2.5 Concluding Remarks

Following the historic approval, by the US FDA, of a new synthetic pharmaceutical product fully manufactured under continuous mode in 2015 and of a change from batch to continuous in the manufacturing process of another synthetic drug in 2016, similar developments can be anticipated for the next years in the manufacturing of biologics.

In the last 15 years, important advances have been achieved in enabling technologies. Single-use technologies have been developed that allow upstream and downstream processes to be easily interconnected and to be carried out in a closed system, reducing potential contamination risks, increasing plant design flexibility, and decreasing room segregation and classification requirements. Much progress has been achieved in cell engineering regarding stability and specific productivity of recombinant cell lines, encouraging long-term upstream operation and increasing titers obtained in perfusion in spite of the continuous dilution of bioreactors. Optimized chemically defined, animal-derived component free media have been developed, supporting higher cell densities. New or improved cell retention devices have been designed and launched commercially, reducing the historical bottlenecks related to cell separation. Improved online, in-line, and at line monitoring techniques have been developed, contributing to the establishment of fully automated, PAT-compliant continuous processes. Continuous multicolumn countercurrent chromatography techniques for biologics have significantly advanced and complete systems are nowadays commercially available. Multimodal resins and highly convective adsorbents, such as membrane adsorbers, monoliths, and nanofibers, have been developed and can be used to perform flow-through purification steps, fitting well into continuous downstream processes. Efficient continuous, nonchromatography-based processes that are easily scalable and lower in cost have given promising results for the purification of mAbs. Continuous viral clearance systems have been proposed and proven to be feasible.

All these technologies have contributed to the development of several continuous integrated processes that have been reported in laboratory or pilot scale in the last few years [44,113]. As observed in the past for other industries, continuous biomanufacturing has meanwhile been proven to yield high-quality products, lot-to-lot consistency, smaller footprint requirements, shorter processing times, and higher productivities. When implemented side-by-side with single-use technology and advanced process automation, lower room segregation and classification are needed, labor requirements are decreased, and lower risks of human errors are expected. Altogether, process intensification is under way and this will have as consequence that the biofacilities of the future will be more flexible regarding scale, location, and multiproduct manufacturing, and that capital investment and operational expenditures will decrease, which in turn will hopefully translate into lower final selling costs and larger patient coverage, also in less favored regions of the world.

In order to be fully mature for industrial implementation, some issues still need to be addressed in order to have robust, fully continuous integrated processes operating at commercial scale. For example, compatibility of interfaces between individual unit operations still needs to be improved and, for some of the unit operations, systems are not yet available for industrial implementation [54]. However, considering the large ongoing research effort that academy, biopharmaceutical industry, and equipment suppliers are putting into integrated continuous biomanufacturing, along with the encouraging actions that have been taken by regulators [1,123,124], it can be expected that these bottlenecks will not take long to be overcome. Thus, as a result of the confluence of the emerging technologies and of the business need to innovate, the regulatory authorization of biologics manufactured by fully continuous integrated bioprocesses might be just a few years away from now.

References

1 Yu, L. (2016) Continuous manufacturing has a strong impact on drug quality. In FDA Voice, April 12, 2016. Available at http://blogs.fda.gov/fdavoice/index.php/2016/04/continuous-manufacturing-has-a-strong-impact-on-drug-quality/ (accessed April 19, 2016).

2 Kossik, J. (2002) Think small: pharmaceutical facility could boost capacity and slash costs by trading in certain batch operations for continuous versions. Pharmama IR .com, article ID/DDAS-SEX 52B. Available at http://www.pharmamanufacturing.com/articles/2002/6/ (accessed April 19, 2016).

3 US Food and Drug Administration (2003) Guidance for industry PAT – A framework for innovative pharmaceutical manufacturing and quality assurance, Center for Drug Evaluation and Research. Available at http://www.fda.gov/cder/guidance/5815dft.htm (accessed April 19, 2016).

4 Abboud, L. and Hensley, S. (2003) Factory shift: new prescription for drug makers: update the plants. Wall Street Journal, Available at http://www.wsj.com/articles/SB10625358403931000 (accessed April 19, 2016).

5 Konstantinov, K. (2010) Towards fully continuous bioprocessing: what can we learn from pharma? Presentation at Cell Culture Engineering XII Conference, Engineering Conferences International, Canada, April 25–30, 2010.

6 Warikoo, V., Godawat, R., Brower, K., Jain, S., Cummings, D., Simons, E., Johnson, T., Walther, J., Yu, M., Wright, B., McLarty, J., Karey, K.P., Hwang, C., Zhou, W., Riske, F., and Konstantinov, K. (2012) Integrated continuous production of recombinant therapeutic proteins. *Biotechnol. Bioeng.*, **109**, 3018–3029.

7 Gottschalk, U. (2013) Biomanufacturing: time for change? *Pharm. Bioprocess.*, **1**, 7–9.

8 Castilho, L.R. (2014) Continuous animal cell perfusion processes: the first step toward integrated continuous biomanufacturing, in *Continuous Processing in Pharmaceutical Manufacturing* (ed. G. Subramanian), Wiley-VCH Verlag GmbH, New York, pp. 306–326.
9 Desai, S.G. (2015) Continuous and semi-continuous cell culture for production of blood clotting factors. *J. Biotechnol.*, **213**, 20–27.
10 Walsh, G. (2014) Biopharmaceutical benchmarks 2014. *Nat. Biotechnol.*, **32**, 992–1000.
11 Castilho, L.R. (2016) Biopharmaceutical products: an introduction, in *Biotechnology in Human and Animal Health* (eds V. Thomaz-Soccol, R.R. Resende, and A. Pandey), Elsevier, Philadelphia.
12 Jesus, M. and Wurm, F.M. (2011) Manufacturing recombinant proteins in kg-ton quantities using animal cells in bioreactors. *Eur. J. Pharm. Biopharm.*, **78**, 184–188.
13 Tao, Y., Shih, J., Sinacore, M., Ryll, T., and Yusuf-Makagiansar, H. (2011) Development and implementation of a perfusion-based high cell density cell banking process. *Biotechnol. Prog.*, **27**, 824–829.
14 Pohlscheidt, M., Jacobs, M., Wolf, S., Thiele, J., Jockwer, A., Gabelsberger, J., Jenzsch, M., Tebbe, H., and Burg, J. (2013) Optimizing capacity utilization by large scale 3000L perfusion in seed train bioreactors. *Biotechnol. Prog.*, **29**, 222–229.
15 Genzel, Y., Vogel, T., Buck, J., Behrendt, I., Ramirez, D.V., Schiedner, G., Jordan, I., and Reichl, U. (2014) High cell density cultivations by alternating tangential flow (ATF) perfusion for influenza A virus production using suspension cells. *Vaccine*, **32**, 2770–2781.
16 Baptista, R.P., Fluri, D.A., and Zandstra, P.W. (2012) High density continuous production of murine pluripotent cells in an acoustic perfused bioreactor at different oxygen concentrations. *Biotechnol. Bioeng.*, **110**, 648–655.
17 Cunha, B., Aguiar, T., Silva, M.M., Silva, R.J., Sousa, M.F., Pineda, E., Peixoto, C., Carrondo, M.J., Serra, M., and Alves, P.M. (2015) Exploring continuous and integrated strategies for the up- and downstream processing of human mesenchymal stem cells. *J. Biotechnol.*, **213**, 97–108.
18 Tapia, F., Vogel, T., Genzel, Y., Behrendt, I., Hirschel, M., Gangemi, J.D., and Reichl, U. (2014) Production of high-titer human influenza A virus with adherent and suspension MDCK cells cultured in a single-use hollow fiber bioreactor. *Vaccine*, **32**, 1003–1011.
19 Lesch, H.P., Heikkilä, K.M., Lipponen, E.M., Valonen, P., Müller, A., Räsänen, E., Tuunanen, T., Hassinen, M.M., Parker, N., Karhinen, M., Shaw, R., and Ylä-Herttuala, S. (2015) Process development of adenoviral vector production in fixed bed bioreactor: from bench to commercial scale. *Hum. Gene. Ther.*, **26**, 560–571.
20 Zhang, Y., Stobbe, P., Silvander, C.O., and Chotteau, V. (2015) Very high cell density perfusion of CHO cells anchored in a non-woven matrix-based bioreactor. *J. Biotechnol.*, **213**, 28–41.
21 Henry, O., Kwok, E., and Piret, J.M. (2008) Simpler noninstrumented batch and semicontinuous cultures provide mammalian cell kinetic data comparable to continuous and perfusion cultures. *Biotechnol. Prog.*, **24**, 921–931.

22 Vergara, M., Becerra, S., Díaz-Barrera, A., Berrios, J., and Altamirano, C. (2012) Simultaneous environmental manipulations in semi-perfusion cultures of CHO cells producing rh-tPA. *Electron. J. Biotechnol.*, **15** (6). doi: 10.2225/vol15-issue6-fulltext-2

23 Bettinardi, I.W. (2016) Desenvolvimento de Estratégias de Alimentação com Meios Concentrados para Cultivo de Células Animais em Perfusão. M.Sc. dissertation, COPPE - Federal University of Rio de Janeiro, Brazil.

24 Bettinardi, I.W. and Castilho, L.R. (2016) Small-scale comparison of pseudoperfusion feeding strategies using basal and concentrated feed media. Presentation at the Cell Culture Engineering XV Conference, Engineering Conferences International, USA, May 8–13, 2016.

25 Ozturk, S.S. (1996) Engineering challenges in high density cell culture systems. *Cytotechnology*, **22**, 3–16.

26 Konstantinov, K.B., Tsai, Y., Moles, D., and Matanguihan, R. (1996) Control of long-term perfusion Chinese hamster ovary cell culture by glucose auxostat. *Biotechnol. Prog.*, **12**, 100–109.

27 Clincke, M.F., Mölleryd, C., Zhang, Y., Lindskog, E., Walsh, K., and Chotteau, V. (2013) Very high density of CHO cells in perfusion by ATF or TFF in WAVE bioreactor. Part I. Effect of the cell density on the process. *Biotechnol. Prog.*, **29**, 754–767.

28 Konstantinov, K.B. (2013) The promise of continuous bioprocessing. Presentation at Integrated Continuous Biomanufacturing I Conference, Engineering Conferences International, Spain, October 20–24, 2013.

29 Feng, Q., Mi, L., Li, L., Liu, R., Xie, L., Tang, H., and Chen, Z. (2006) Application of "oxygen uptake rate-amino acids" associated mode in controlled-fed perfusion culture. *J. Biotechnol.*, **122**, 422–430.

30 Yang, J.D., Angelillo, Y., Chaudhry, M., Goldenberg, C., and Goldenberg, D.M. (2000) Achievement of high cell density and high antibody productivity by a controlled-fed perfusion bioreactor process. *Biotechnol. Bioeng.*, **69**, 74–82.

31 Chotteau, V. (2015) Perfusion processes, in *Animal Cell Culture*, Cell Engineering vol. 9 (ed. M. Al-Rubeai), Springer International Publishing Switzerland, Basel, pp. 407–443.

32 Chuppa, S., Tsai, Y.S., Yoon, S., Shackleford, S., Rozales, C., Bhat, R., Tsay, G., Matanguihan, C., Konstantinov, K., and Naveh, D. (1997) Fermentor temperature as a tool for control of high density perfusion cultures of mammalian cells. *Biotechnol. Bioeng.*, **55**, 328–338.

33 Coronel, J. (2015) Desenvolvimento de um Processo de Cultivo Contínuo com Reciclo Celular para Produção de um Biofármaco Lábil, Ph.D. thesis. COPPE–Federal University of Rio de Janeiro, Brazil.

34 Coronel, J., Klausing, S., Heinrich, C., Noll, T., Figueredo-Cardero, A., and Castilho, L.R. (2016) Valeric acid under mild hypothermia presents advantages over butyric acid as productivity enhancer in CHO cell cultivations. *Biochem. Eng. J.* **114**, 101–109.

35 Rodriguez, J., Spearman, M., Tharmalingam, T., Sunley, K., Lodewyks, C., Huzel, N., and Butler, M. (2010) High productivity of human recombinant beta-interferon from a low-temperature perfusion culture. *J. Biotechnol.*, **150**, 509–518.

36 Kim, J.S., Ahn, B.C., Lim, B.P., Choi, Y.D., and Jo, E.C. (2004) High-level scu-PA production by butyrate-treated serum-free culture of recombinant CHO cell line. *Biotechnol. Prog.*, **20**, 1788–1796.
37 Hong, J.K., Lee, G.M., and Yoon, S.K. (2011) Growth factor withdrawal in combination with sodium butyrate addition extends culture longevity and enhances antibody production in CHO cells. *J. Biotechnol.*, **155**, 225–231.
38 Castilho, L.R. and Medronho, R.A. (2008) Animal cell separation, in *Animal Cell Technology: From Biopharmaceuticals to Gene Therapy* (eds L.R. Castilho, A.M. Moraes, E.F.P. Augusto, and M. Butler), Taylor & Francis, New York, pp. 273–294.
39 Voisard, D., Meuwly, F., Ruffieux, P.A., Baer, G., and Kadouri, A. (2003) Potential of cell retention techniques for large-scale high-density perfusion culture of suspended mammalian cells. *Biotechnol. Bioeng.*, **82**, 751–765.
40 Castilho, L.R. and Medronho, R.A. (2002) Cell retention devices for suspended-cell perfusion cultures. *Adv. Biochem. Eng. Biotechnol.*, **74**, 129–169.
41 Dalm, M.C., Jansen, M., Keijzer, T.M., van Grunsven, W.M., Oudshoorn, A., Tramper, J., and Martens, D.E. (2005) Stable hybridoma cultivation in a pilot scale acoustic perfusion system: long term process performance and effect of recirculation rate. *Biotechnol. Bioeng.*, **91**, 894–900.
42 Medronho, R.A. (2003) Solid–liquid separation, in *Isolation and Purification of Proteins* (eds B. Mattiasson and R. Hatti-Kaul), Marcel Dekker, New York, pp. 131–190.
43 Yang, W.C., Lu, J., Kwiatkowski, C., Yuan, H., Kshirsagar, R., Ryll, T., and Huang, Y.M. (2014) Perfusion seed cultures improve biopharmaceutical fed-batch production capacity and product quality. *Biotechnol. Prog.*, **30**, 616–625.
44 Klutz, S., Magnus, J., Lobedann, M., Schwan, P., Maiser, B., Niklas, J., Temming, M., and Schembecker, G. (2015) Developing the biofacility of the future based on continuous processing and single-use technology. *J. Biotechnol.*, **213**, 120–130.
45 Deckwer, W.D., Medronho, R.A., Anspach, F.B., and Lübberstedt, M. (2001) Method for separating viable cells from cell suspensions. US Patent 6,878,545 B2.
46 Pinto, R.C.V., Medronho, R.A., and Castilho, L.R. (2008) Separation of CHO cells using hydrocyclones. *Cytotechnology*, **56**, 57–67.
47 Pinto, R.C.V. (2007) Separação de células CHO utilizando hidrociclones, Ph.D. thesis. COPPE–Federal University of Rio de Janeiro, Brazil.
48 Björling, T., Dudel, U., and Fenge, C. (1995) Evaluation of a cell separator in large scale perfusion culture, in *Animal Cell Technology: Developments Towards the 21st Century* (eds E.C. Beuvery, J.B. Griffiths, and W.P. Zeijlemaker), Kluwer Academic Publishers, Dordrecht, pp. 671–675.
49 Himmelfarb, P., Thayer, P.S., and Martin, H.E. (1969) Spin filter culture: the propagation of mammalian cells in suspension. *Science*, **164**, 555–557.
50 Figueredo-Cardero, A., Martínez, E., Chico, E., Castilho, L.R., and Medronho, R.A. (2014) Rotating cylindrical filters used in perfusion cultures: CFD simulations and experiments. *Biotechnol. Progr.*, **30**, 1093–1102.
51 Clincke, M.F., Möllery, C., Samani, P.K., Lindskog, E., Fäldt, E., Walsh, K., and Chotteau, V. (2013) Very high density of Chinese hamster ovary cells in perfusion by alternating tangential flow or tangential flow filtration in WAVE

BioreactorTM. Part II: Applications for antibody production and cryopreservation. *Biotechnol. Prog.*, **29**, 768–777.

52 Karst, D.J., Serra, E., Villiger, T.K., Soos, M., and Morbidelli, M. (2016) Characterization and comparison of ATF and TFF in stirred bioreactors for continuous mammalian cell culture processes. *Biochem. Eng. J.*, **110**, 17–26.

53 Coronel, J., Heinrich, C., Figueredo-Cardero, A., Northoff, S., and Castilho, L.R. (2013) Process evaluation for the production of a labile recombinant protein. Presentation at the 23rd ESACT Meeting, European Society for Animal Cell Technology (ESACT), France, June 23–26, 2013.

54 Konstantinov, K.B. and Cooney, C.L. (2015) White paper on continuous bioprocessing. *J. Pharm. Sci.*, **104**, 813–820.

55 Rathore, A.S., Agarwal, H., Sharma, A.K., Pathak, M., and Muthukumar, S. (2015) Continuous processing for production of biopharmaceuticals. *Prep. Biochem. Biotechnol.*, **45**, 836–849.

56 Zydney, A.L. (2016) Continuous downstream processing for high value biological products: a review. *Biotechnol. Bioeng.*, **113**, 465–475.

57 D'Souza, R. and Azevedo, A. (2013) Emerging technologies for the integration and intensification of downstream bioprocesses. *Pharm. Bioprocess.*, **1**, 423–440.

58 Moraes, A.M., Castilho, L.R., and Bueno, S.M.A. (2008) Product purification processes, in *Animal Cell Technology: From Biopharmaceuticals to Gene Therapy* (eds L.R. Castilho, A.M. Moraes, E.F.P. Augusto, and M. Butler), Taylor & Francis, New York, pp. 295–328.

59 Castilho, L.R., Anspach, F.B., and Deckwer, W.D. (2002) An integrated process for mammalian cell perfusion cultivation and product purification using a dynamic filter. *Biotechnol. Prog.*, **18**, 776–781.

60 Hammerschmidt, N., Tscheliessnig, A., Sommer, R., Helk, B., and Jungbauer, A. (2014) Economics of recombinant antibody production processes at various scales: industry-standard compared to continuous precipitation. *Biotechnol. J.*, **9**, 766–775.

61 Rosa, P.A.J., Azevedo, A.M., Sommerfeld, S., Mutter, M., Bäcker, W., and Aires-Barros, M.R. (2013) Continuous purification of antibodies from cell culture supernatant with aqueous two-phase systems: from concept to process. *Biotechnol. J.*, **8**, 352–362.

62 Carta, G. and Jungbauer, A. (2010) Introduction to protein chromatography. *Protein Chromatography: Process Development and Scale-Up*, Wiley-VCH Verlag GmbH, New York.

63 Müller-Späth, T. and Morbidelli, M. (2015) Multicolumn countercurrent gradient chromatography for the purification of biopharmaceuticals, in *Continuous Processing in Pharmaceutical Manufacturing* (ed. G. Subramanian), Wiley-VCH Verlag GmbH, New York, pp. 227–254.

64 Hilbrig, F. and Freitag, R. (2003) Continuous annular chromatography. *J. Chromatogr. B.*, **790**, 1–15.

65 Wolfgang, J. and Prior, A. (2002) Continuous annular chromatography. *Adv. Biochem. Eng. Biotechnol.*, **76**, 233–255.

66 Giovannini, R. and Freitag, R. (2002) Continuous separation of multicomponent protein mixtures by annular displacement chromatography. *Biotechnol. Prog.*, **18**, 1324–1331.

67 Iberer, G., Schwinn, H., Josić, D., Jungbauer, A., and Buchacher, A. (2001) Improved performance of protein separation by continuous annular chromatography in the size-exclusion mode. *J. Chromatogr. A*, **921**, 15–24.

68 Giovannini, R. and Freitag, R. (2001) Isolation of a recombinant antibody from cell culture supernatant: continuous annular versus batch and expanded-bed chromatography. *Biotechnol. Bioeng.*, **73**, 522–529.

69 Vogel, J.H., Nguyen, H., Pritschet, M., Van Wegen, R., and Konstantinov, K. (2002) Continuous annular chromatography: general characterization and application for the isolation of recombinant protein drugs. *Biotechnol. Bioeng.*, **80**, 559–568.

70 Grill, C.M., Miller, L., and Yan, T.Q. (2004) Resolution of a racemic pharmaceutical intermediate. A comparison of preparative HPLC, steady state recycling and simulated moving bed. *J. Chromatogr. A*, **1026**, 101–108.

71 Hsu, L.C., Kim, H., Yang, X., and Ross, D. (2011) Large scale chiral chromatography for the separation of an enantiomer to accelerate drug development. *Chirality*, **23**, 361–366.

72 Nestola, P., Silva, R.J.S., Peixoto, C., Alves, P.M., Carrondo, M.J.T., and Mota, J.P.B. (2014) Adenovirus purification by two-column, size-exclusion, simulated countercurrent chromatography. *J. Chromatogr. A*, **1347**, 111–121.

73 Mota, J.P., Silva, R., Nestola, P., Peixoto, C., and Carrondo, M. (2015) Robust design and operation of quasi-continuous adenovirus purification by two-column, simulated moving-bed, size-exclusion chromatography. Presentation at Integrated Continuous Biomanufacturing II Conference, Engineering Conferences International, USA, November 1–5, 2015.

74 Wellhoefer, M., Sprinzl, W., Hahn, R., and Jungbauer, A. (2014) Continuous processing of recombinant proteins: integration of refolding and purification using simulated moving bed size-exclusion chromatography with buffer recycling. *J. Chromatogr. A*, **1337**, 48–56.

75 Kessler, L.C., Gueorguieva, L., Rinas, U., and Seidel-Morgenstern, A. (2007) Step gradients in 3-zone simulated moving bed chromatography. Application to the purification of antibodies and bone morphogenetic protein-2. *J. Chromatogr. A*, **1176**, 69–78.

76 Jiang, C., Huang, F., and Wei, F. (2014) A pseudo three-zone simulated moving bed with solvent gradient for quaternary separations. *J. Chromatogr. A*, **1334**, 87–91.

77 Wei, F., Shen, B., Chen, M., and Zhao, Y. (2012) Study on a pseudo-simulated moving bed with solvent gradient for ternary separations. *J. Chromatogr. A*, **1225**, 99–106.

78 Bisschops, M., Frick, L., Fulton, S., and Ransohoff, T. (2009) Single-use, continuous countercurrent, multicolumn chromatography. *Bioprocess Int.*, 7, 18–23.

79 Aumann, L. and Morbidelli, M. (2007) A continuous multicolumn countercurrent solvent gradient purification (MCSGP) process. *Biotechnol. Bioeng.*, **98**, 1043–1055.

80 Müller-Späth, T., Aumann, L., Strohlein, G., Kornmann, H., Valax, P., Delegrange, L. et al. (2010) Two step capture and purification of IgG2 using multicolumn countercurrent solvent gradient purification (MCSGP). *Biotechnol. Bioeng.*, **107**, 974–984.

81 Müller-Späth, T., Aumann, L., Melter, L., Ströhlein, G., and Morbidelli, M. (2008) Chromatographic separation of three monoclonal antibody variants using multicolumn countercurrent solvent gradient purification (MCSGP). *Biotechnol. Bioeng.*, **100**, 1166–1177.

82 Krättli, M., Müller-Späth, T., and Morbidelli, M. (2013) Multifraction separation in countercurrent chromatography (MCSGP). *Biotechnol. Bioeng.*, **110**, 2436–2444.

83 Angarita, M., Müller-Späth, T., Baur, D., Lievrouw, R., Lissens, G., and Morbidelli, M. (2015) Twin-column CaptureSMB: a novel cyclic process for protein A affinity chromatography. *J. Chromatogr. A*, **1389**, 85–95.

84 Godawat, R., Brower, K., Jain, S., Konstantinov, K., Riske, F., and Warikoo, V. (2012) Periodic counter-current chromatography – design and operational considerations for integrated and continuous purification of proteins. *Biotechnol. J.*, **7**, 1496–1508.

85 Baur, D., Angarita, M., Müller-Späth, T., Steinebach, F., and Morbidelli, M. (2016) Comparison of batch and continuous multi-column protein A capture processes by optimal design. *Biotechnol. J.* doi: 10.1002/biot.201500481.

86 Carta, G. and Perez-Almodovar, E.X. (2010) Productivity considerations and design charts for biomolecule capture with periodic countercurrent adsorption systems. *Sep. Sci. Technol.*, **45**, 149–154.

87 Gjoka, X., Rogler, K., Martino, R.A., Gantier, R., and Schofield, M. (2015) A straightforward methodology for designing continuous monoclonal antibody capture multi-column chromatography processes. *J. Chromatogr. A*, **1416**, 38–46.

88 Shinkazh, O. (2009) Countercurrent tangential chromatography methods, systems, and apparatus. US Patent No. US7,988,859B2.

89 Shinkazh, O., Kanani, D., Barth, M., Long, M., Hussain, D., and Zydney, A.L. (2011) Countercurrent tangential chromatography for large-scale protein purification. *Biotechnol. Bioeng.*, **108**, 582–591.

90 Dutta, A.K., Tran, T., Napadensky, B., Teella, A., Brookhart, G., Ropp, P.A. et al. (2015) Purification of monoclonal antibodies from clarified cell culture fluid using Protein A capture continuous countercurrent tangential chromatography. *J. Biotechnol.*, **213**, 54–64.

91 Molino, J.V.D., Viana Marques, DdeA., Junior, A.P., Mazzola, P.G., and Gatti, M.S.V. (2013) Different types of aqueous two-phase systems for biomolecule and bioparticle extraction and purification. *Biotechnol. Prog.*, **29**, 1343–1353.

92 Eggersgluess, J.K., Richter, M., Dieterle, M., and Strube, J. (2014) Multi-stage aqueous two-phase extraction for the purification of monoclonal antibodies. *Chem. Eng. Technol.*, **37**, 675–682.

93 van Winssen, F.A., Merz, J., and Schembecker, G. (2014) Tunable aqueous polymer-phase impregnated resins – a novel approach to aqueous two-phase extraction. *J. Chromatogr. A*, **1329**, 38–44.

94 van Winssen, F.A., Merz, J., Czerwonka, L.M., and Schembecker, G. (2014) Application of the tunable aqueous polymer-phase impregnated resins-technology for protein purification. *Sep. Purif. Technol.*, **136**, 123–129.

95 Scopes, R.K. (1994) *Protein Purification: Principles and Practice*, 3rd edn, Springer, New York.

96 Sommer, R., Tscheliessnig, A., Satzer, P., Schulz, H., Helk, B., and Jungbauer, A. (2015) Capture and intermediate purification of recombinant antibodies with combined precipitation methods. *Biochem. Eng. J.*, **93**, 200–211.

97 Hammerschmidt, N., Hobiger, S., and Jungbauer, A. (2016) Continuous polyethylene glycol precipitation of recombinant antibodies: sequential precipitation and resolubilization. *Process Biochem.*, **51**, 325–332.

98 Hammerschmidt, N., Hintersteiner, B., Nico Lingg, N., and Jungbauer, A. (2015) Continuous precipitation of IgG from CHO cell culture supernatant in a tubular reactor. *Biotechnol. J.*, **10**, 1196–1205.

99 Raphael, M., Rohani, S., and Sosulski, F. (1995) Isoelectric precipitation of sunflower protein in a tubular precipitator. *Can. J. Chem. Eng.*, **73**, 470–483.

100 Tscheliessnig, A., Satzer, P., Hammerschmidt, N., Schulz, H., Helk, B., and Jungbauer, A. (2014) Ethanol precipitation for purification of recombinant antibodies. *J. Biotechnol.*, **188**, 17–28.

101 Sommer, R., Satzer, P., Tscheliessnig, A., Schulz, H., Helk, B., and Jungbauer, A. (2014) Combined polyethylene glycol and $CaCl_2$ precipitation for the capture and purification of recombinant antibodies. *Process Biochem.*, **49**, 2001–2009.

102 Hughson, M., Carvalho, R.J., Cruz, T.A., and Castilho, L.R. (2015) Laboratory scale continuous linear purification as a development tool for recombinant blood protein processing, using chromatographyic resins and membranes. Presentation at Integrated Continuous Biomanufacturing II, Engineering Conferences International, USA, November 1–5, 2015.

103 Shamashkin, M., Godavarti, R., Iskra, T., and Coffman, J. (2013) A tandem laboratory scale protein purification process using Protein A affinity and anion exchange chromatography operated in a weak partitioning mode. *Biotechnol. Bioeng.*, **110**, 2655–2663.

104 Blom, H., Eriksson, C., Forss, A., and Skoglar, H. (2015) Continuous downstream processing of a monoclonal antibody using periodic counter current chromatography (PCC) and straight through processing (STP). Presentation at Integrated Continuous Biomanufacturing II, Engineering Conferences International, USA, November 1–5, 2015.

105 Orr, V., Zhong, L., Moo-Young, M., and Chou, C.P. (2013) Recent advances in bioprocessing application of membrane chromatography. *Biotechnol. Adv.*, **31**, 450–465.

106 Nascimento, A., Rosa, S.A.S.L., Mateus, M., and Azevedo, A.M. (2014) Polishing of monoclonal antibodies streams through convective flow devices. *Sep. Purif. Technol.*, **132**, 593–600.

107 Hardick, O., Dods, S., Stevens, B., and Bracewell, D.G. (2015 Nanofiber adsorbents for high productivity continuous downstream processing. *J. Biotechnol.*, **213**, 74–82.

108 McCue, J.T., Selvitelli, K., Cecchini, D., and Brown, R. (2014) Enveloped virus inactivation using neutral arginine solutions and applications in therapeutic protein purification processes. *Biotechnol. Progr.*, **30**, 108–112.

109 Jackson, N.B., Bakhshayeshi, M., Zydney, A.L., Mehta, A., van Reis, R., and Kuriyel, R. (2014) Internal virus polarization model for virus retention by the Ultipor((R)) VF Grade DV20 membrane. *Biotechnol. Prog.*, **30**, 856–863.

110 Roberts, P.L. (2014) Virus elimination during the purification of monoclonal antibodies by column chromatography and additional steps. *Biotechnol. Progr.*, **30**, 1341–1347.

111 Caillet-Fauquet, P., Di Giambattista, M., Draps, M.-L., Sandras, F., Branckaert, T., de Launoit, Y. *et al.* (2004) Continuous-flow UVC irradiation: a new, effective, protein activity-preserving system for inactivating bacteria and viruses, including erythrovirus B19. *J. Virol. Methods*, **118**, 131–139.

112 Li, Q., Macdonald, S., Bienek, C., Foster, P.R., and Macleod, A.J. (2005) Design of a UV-C irradiation process for the inactivation of viruses in protein solutions. *Biologicals*, **33**, 101–110.

113 Godawat, R., Konstantinov, K., Rohani, M., and Warikoo, V. (2015) End-to-end integrated fully continuous production of recombinant monoclonal antibodies. *J. Biotechnol.*, **213**, 13–19.

114 Klutz, S., Lobedann, M., Bramsiepe, C., and Schembecker, G. (2016) Continuous viral inactivation at low pH value in antibody manufacturing. *Chem. Eng. Process*, **102**, 88–101.

115 Roberts, P.L. (2008) Virus inactivation by solvent/detergent treatment using Triton X-100 in a high purity factor VIII. *Biologicals.*, **36**, 330–335.

116 Orozco, R., Guillen, N., Chung, L., Godfrey, S., and Coffman, J. (2015) Considerations for an incubation chamber for continuous viral inactivation. Presentation at Integrated Continuous Biomanufacturing II Conference, Engineering Conferences International, USA, November 1–5, 2015.

117 Salm, J., Fiadeiro, M., Orozco, R., Kublbeck, J., Noyes, A., Horne, J., LaCasse, D., Sacramo, A., Gupta, S., Coffman, J., and Fahrner, R. (2015) Enabling technologies for integrated/continuous downstream processing of biologics. Presentation at Integrated Continuous Biomanufacturing II Conference, Engineering Conferences International, USA, November 1–5, 2015.

118 Xenopoulos, A. (2015) A new, integrated, continuous purification process template for monoclonal antibodies: process modeling and cost of goods studies. *J. Biotechnol.*, **213**, 42–53.

119 Castilho, L.R. (2001) *Development of a Dynamic Filter for Integrated Perfusion Cultivation and Purification of Recombinant Proteins from Mammalian Cells*, VDI Verlag, Düsseldorf, Germany.

120 Castilho, L.R., Anspach, F.B., and Deckwer, W.D. (2002) Comparison of affinity membranes for the purification of immunoglobulins. *J. Membr. Sci.*, **207**, 253–264.

121 Castilho, L.R. and Anspach, F.B. (2003) CFD-aided design of a dynamic filter for mammalian cell separation. *Biotechnol. Bioeng.*, **83**, 514–524.

122 Goudar, C., Titchener-Hooker, N., and Konstantinov, K. (2015) Integrated continuous biomanufacturing: a new paradigm for biopharmaceutical production. *J. Biotechnol.*, **213**, 1–2.

123 Woodcock, J. (2015) Introducing new technology in pharmaceutical manufacturing. Presentation at Integrated Continuous Biomanufacturing II Conference, Engineering Conferences International, USA, November 1–5, 2015.

124 Lee, S.L., O'Connor, T.F., Yang, X., Cruz, C.N., Chatterjee, S., Madurawe, R.D., Moore, C.M.V., Yu, L.X., and Woodcock, J. (2015) Modernizing pharmaceutical manufacturing: from batch to continuous production. *J. Pharm. Innov.*, **10**, 191–199.

3

Engineering Challenges of Continuous Biomanufacturing Processes (CBP)

Holger Thiess, Steffen Zobel-Roos, Petra Gronemeyer, Reinhard Ditz, and Jochen Strube

Clausthal University of Technology, Institute for Separation and Process Technology, Leibnizstraße 15, 38678 Clausthal-Zellerfeld, Germany

3.1 Introduction

3.1.1 Continuous Manufacturing

Continuous processing is in general associated with utilizing "economy of scale" potentials in the industrial manufacturing of products. The transfer of any manufacturing technology from batch to continuous mode is widely considered as an indicator of maturity. This was observed in car manufacturing by Ford in the 1920–1930s, petrol refining and bulk chemical manufacturing in the 1920–1950s, steel casting in the 1950–1960s, food processing in the 1950–1980s, and so on, and now pharmaceuticals. In addition, innovative management and quality control methods like lean management, simultaneous engineering, rapid prototyping, six-sigma, and so on, have been transferred from branches already under cost pressure like automotive or IT-business to other branches, to prevent cost/competition pressure [1].

3.1.2 Continuous Manufacturing of Synthetic Molecules

Following the announcement of the Novartis/MIT development cooperation for continuous manufacturing strategies for synthetic APIs [2], many studies have recently been published, trying to transfer such activities into biologics manufacturing [3–8].

3.1.3 Continuous Manufacturing of Biologics

In biopharmaceuticals production this has not created major attention until recently, although the use of multi-thousand liter reactors for animal cell culturing have reached large-scale dimension. At first theoretical studies on cost structures were conducted [4–8] pointing out benefits in lower CAPEX and lower OPEX at higher titers. Some of the studies focused on widening the general

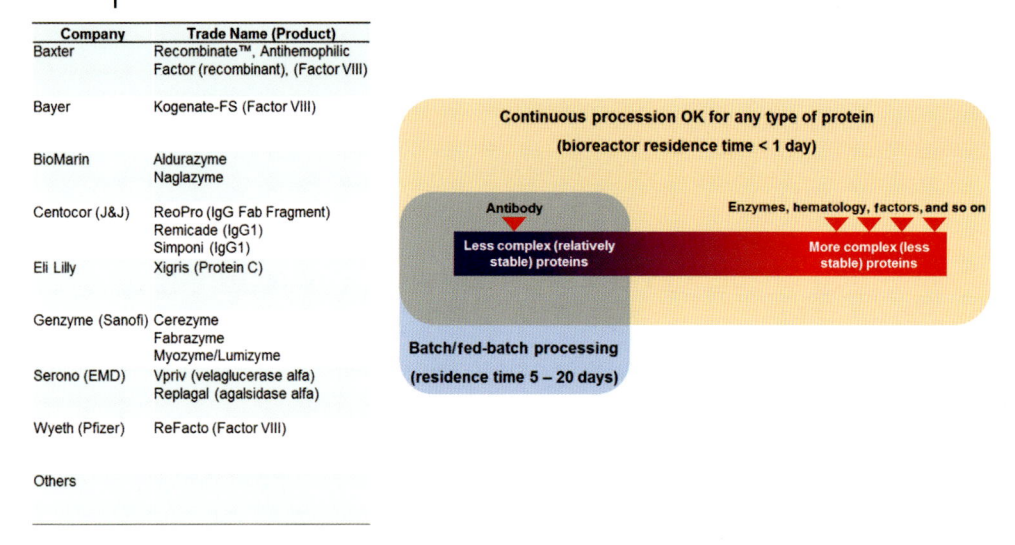

Company	Trade Name (Product)
Baxter	Recombinate™, Antihemophilic Factor (recombinant), (Factor VIII)
Bayer	Kogenate-FS (Factor VIII)
BioMarin	Aldurazyme
	Naglazyme
Centocor (J&J)	ReoPro (IgG Fab Fragment)
	Remicade (IgG1)
	Simponi (IgG1)
Eli Lilly	Xigris (Protein C)
Genzyme (Sanofi)	Cerezyme
	Fabrazyme
	Myozyme/Lumizyme
Serono (EMD)	Vpriv (velaglucerase alfa)
	Replagal (agalsidase alfa)
Wyeth (Pfizer)	ReFacto (Factor VIII)
Others	

Figure 3.1 Perfusion fermentation for biologics. (Reproduced with permission form Ref. [1].)

bottleneck in downstream processing for high-volume products like mAbs in 1000–2000 kg/annum scale by transfer to a continuous operation [4–6,9]; others looked at the special benefits from low-volume products like fragments and stratified medicines at 10–100 kg/annum amount scale [7,8,10].

Riske [1] points out that biologics are fermented about 90% by batch or fed-batch and at high titers are mainly cost-driven by downstream processing. The business is traditionally conservative in order to guarantee product quality by process robustness in very complex manufacturing steps for quite complex bioactive large molecules. Therefore, innovation is more product than process/technology driven. However, new business scenarios create strong needs for improvement, preferentially by disruptive innovation.

Literature search shows that more products on the market and even more candidates in clinical trials are already fermented in continuous perfusion mode than earlier estimated [1,11–16], see Figure 3.1.

Therefore, the general idea to simply transfer downstream into continuous operation mode as well seems a quite logical conclusion. Nevertheless, sound downstream solutions have not yet made their way into industrial use.

The combination of continuous flow, small reactors and skids, and steady-state cell cultures in upstream have to be combined with innovative solutions in downstream in order to generate minimized hold-up times, less process steps, closed systems, and maybe disposable technology. In continuous operation, with pilot scale already being manufacturing scale, there are no scale-up issues; therefore, robust processing/product quality can be reached. In addition, high productivity in manufacturing can be achieved by choosing defined chemical media [1,2,12,13].

However, above and beyond the economic gains to be drawn from continuous processing of biologics, another aspect might/will prove its value in future, which is the process stability and reproducibility improvements gained from a continuous operation.

3.2 Analysis of CBP Status

Besides all the obvious advantages, GE summarized the following 10 reasons for why sponsors still hesitate [17]:

1) Economic justifications and investment risks
2) Performance reliability (incidence of failure)
3) Implementation: process control and QS procedure
4) Regulatory body dispositions/filing concerns
5) Equipment fitness for extended usage
6) Platform fitness for extended times
7) Operational concerns, like
   - lot designations
   - establishing the means of achieving the required robust process throughput balance
   - fears that equipment cleaning may be more difficult and complicated than in batch
   - the fact that when any unit operation in CB is down for any reason, the whole process can be down
   - supply chain concerns
8) How robust/flexible the process is for new platforms
9) How robust/flexible the process is in accommodating new entity types
10) How robust/flexible the process is in accommodating new business models

The paper concludes that most of the reasons are emotionally conservative and not data driven. Some concerns are obviously similar for batch and continuous operation mode.

Nevertheless, most of the reasons can be explained as typical by general innovation cycles and their obstacles as Nygaard points out in Figure 3.2 [18,19].

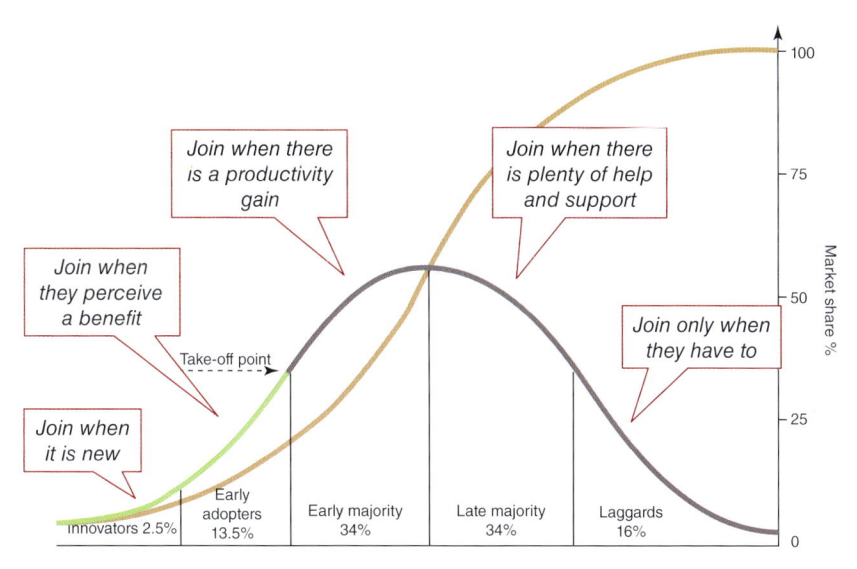

Figure 3.2 Law of diffusion of innovations [18].

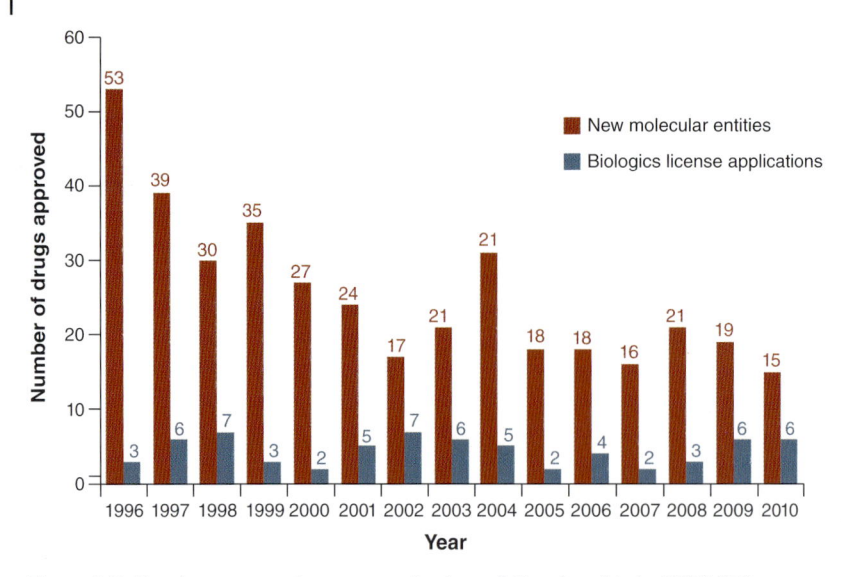

72 | *3 Engineering Challenges of Continuous Biomanufacturing Processes (CBP)*

Figure 3.3 Big pharma new drug approvals at an all-time low (six in 2010) [22].

However, hesitation is not denial of the needs. And industrial needs change due to increasing regulatory demands, business competition, and healthcare system budget limitations, which become obvious from Figures 3.3–3.5.

These aspects combined have caused dozens of research sites to be closed and about 100,000 jobs in the United States as well [20–22].

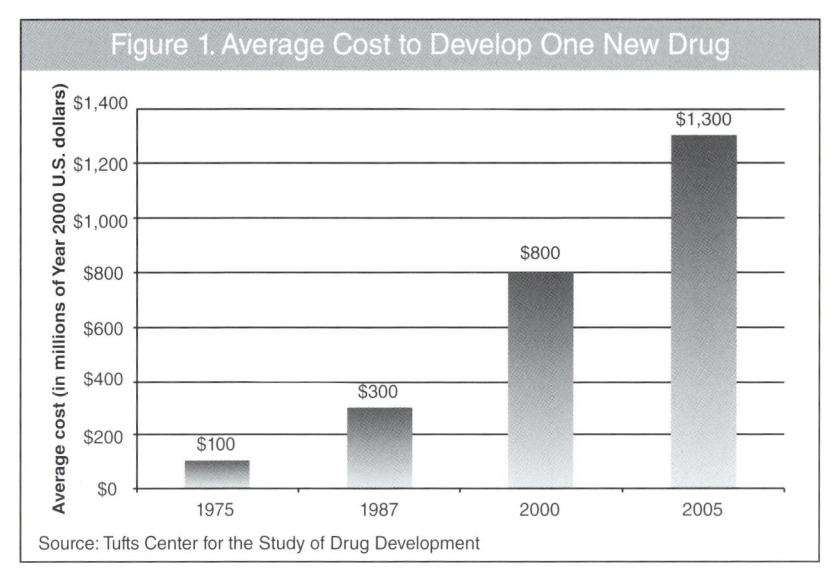

Figure 3.4 R&D spending flat since 2005 due to exponential cost increase. (Reproduced with permission form Ref. [21]. Copyright 2012, Nature Publishing Group.)

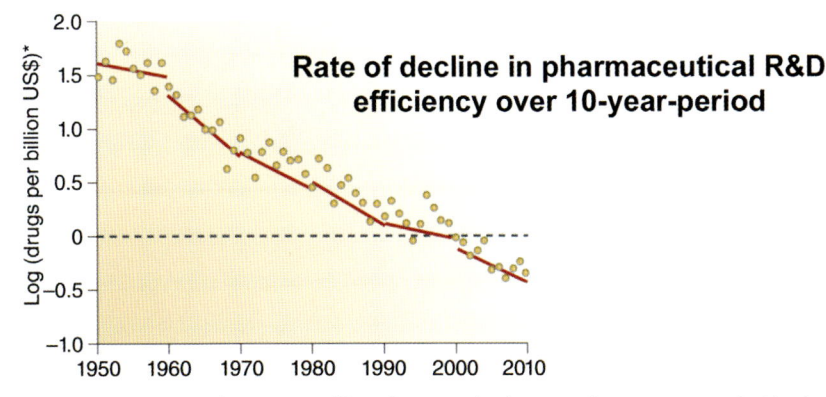

Figure 3.5 Ninety-eight percent of big pharma sales from products 5 years and older (average patent life: 11 years). (Reproduced with permission form Ref. [20]. Copyright 2012, Nature Publishing Group.)

Still, the main decision criteria are whether a *product* or product group exists that can *only* be successfully brought to the market and whether the innovative therapy is needed and therefore process innovation is unavoidable.

An example of taking a strategic decision was UCB's decision in favor of the existing innovative SMB technology for enantiomer discrimination and separation in large scale [23–25], industrialized with the aid of NOVASEP, a small vendor company, backed up by Merck/Darmstadt for reliability.

Additionally, customer surveys made by NNE Pharmaplan [18,19] describe the actual situation as driven by

- Regulatory concerns like GMP issues, quality systems, and control aspects dominated with
  - consistency and reproducibility issues,
  - comparative analysis of product quality for CBP versus batch,
  - the regulator's view somewhat unclear as there are differences between FDA, EMA, and RoW, and
  - unclear analytical requirements.
- Production management and staffing issues at 24/7 operation are minor questions with details on
  - how successful CBP could be implemented and
  - end user studies; how CBP would look like in detail need further clarification.

As a first conclusion, all these statements refer to lack of experience and concern from authority's point of view.

The concerns about regulatory authorities and lot definitions can be dealt in Refs [26,27]. Continuous manufacturing is seen officially as "a key enabler in modernizing pharmaceutical manufacturing." In contrast, FDA "has been working to stimulate development of novel manufacturing technologies" that enables "continuous manufacturing" as there are a "multitude of advantages." QbD is a key method for success: "while QbD is catching up on development, manufacturers have been reluctant to modernize manufacturing" in order to gain "increased robustness and lower costs" [28]. Potential for reduced cost is gained by integrated processes with less steps, smaller equipment and facilities as well as online-monitoring and control for increased product quality assurance in real time.

Quality advantages are gained due to shorter contact time at 37 °C in fermentation, shorter processing time, real-time process control, and generation of data that can be data-mined for higher sophisticated process control, which opens up the option for real-time process release. Enhanced reproducibility and control is a consequence.

Likewise, the definition of batch and lot is consistent at 21 CFR 210.3 and applicable for continuous manufacturing as defined "as a unit produced by continuous processes in a unit of time or quantity." SMB technology in enantio-separation of pharmaceuticals integrated in batch operation synthesis is a good example [23].

A useful approach of Frank Nygaard from NNE Pharmaplan compares single-use and CBP by risk analysis. Nevertheless, CBP is much more in an early stage of new technologies, whereas single-use is at the stage of productivity gain proven. Lessons learned are discussed in detail later.

General technical questions in process design and engineering are as follows:

- To decide on innovate unit operations and equipment
- Integration of USP and DSP
- Integration of formulation as well as fill and finish
- Integration of innovate units in total processes
- Integration of buffer handling and supply or recycling
- Single use or stainless steel
- PAT and QbD methods
- Process modeling to develop total/complete/full processes with innovate units integrated and to generate platform methods instead of platform processes
  - that is, the actual platform process approach predefines a sequence of unit operations with set operation parameter windows
  - whereas a platform method like total process modeling with predictive rigorous physicochemical models based on model parameters determined in miniaturized laboratory-scale for each new component system applying QbD-sound statistical means for experimental model parameter measurement error determination and model validation as well as risk-based analysis approach that allows fast and efficient prediction of the best process sequence of unit operations and optimal robust operation parameter
- To decide on disposables or steel or both mixed
- To specify appropriate bioanalytics for process control in contrast to QA

3.3 Case Studies

A first and quite comprehensive case study of complete continuous operation in pilot scale is described in Ref. [28].

Conclusions from case studies from different working groups are as follows:

- Continuous cell separation is some challenge to be overcome by settlers, centrifuges, UF systems [29], or filtration aids [30]

- Disposables are mainly chosen to reduce potential microbial contamination risk during long-term operation [31]
- As operation takes place in a closed system, S1 environment is sufficient and no clean rooms are needed [31]
- Efforts and risks in scale-up engineering are avoided by numbering up, ball-room park plant type [2,31] as shown in a MIT-Novartis study on NCEs
- Temperature only 37 °C for three instead of 12 h fermentation time [28] causes higher product stability
- Still small hold-ups are needed for appropriate reliable continuous operation by definition and validation of hold-times [28,30], for example, use for VI at pH hold [31]
- VF blocking, tests for sudden breakthrough are needed to be more predictive than any reaction to pressure increase

The most active group at the moment from Genzyme has published laboratory-scale studies and even piloting trials – Figure 3.6 flowchart with parameters and results.

Another study from the German "invite group" [31] summarizes their activities within some funded projects, operating USP in 28 day and DSP in 2.5 day continuous mode in laboratory scale. In contrast to Genzyme's activities, some questions on long-term continuous operation are still open. Based on Pall/Tarpon, two flow-through chromatography plants, that is, Prot A capture with eight as well as combined IEX and Mixed Mode with six columns, are applied. For two column-MCSGP processes, no full process scale studies are available at the moment, only combination of perfusion and protein A [32].

Chromatography is still seen as "best practice" choice to be integrated at least in one to two steps by an efficient total DSP process with regard to selectivity, capacity, robustness, and costs; continuous chromatography is a key challenge to

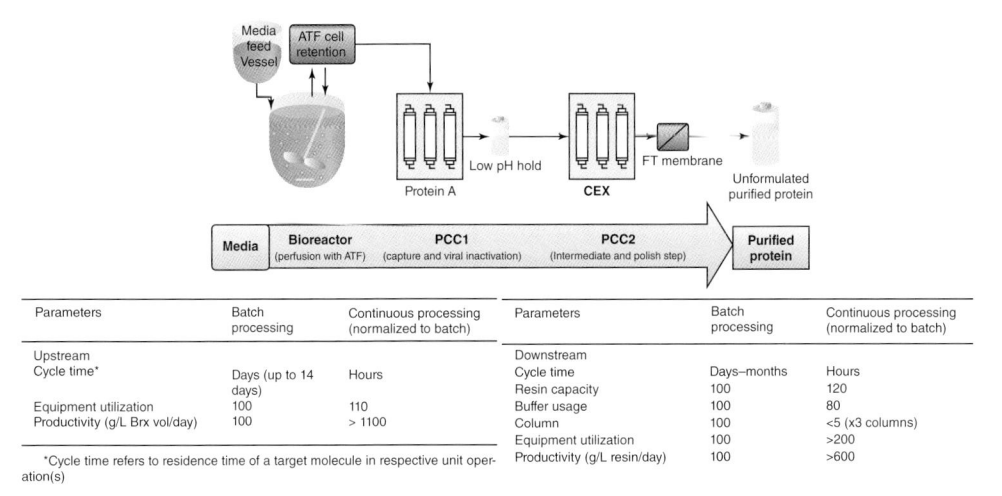

Parameters	Batch processing	Continuous processing (normalized to batch)	Parameters	Batch processing	Continuous processing (normalized to batch)
Upstream Cycle time*	Days (up to 14 days)	Hours	Downstream Cycle time	Days–months	Hours
			Resin capacity	100	120
Equipment utilization	100	110	Buffer usage	100	80
Productivity (g/L Brx vol/day)	100	> 1100	Column	100	<5 (x3 columns)
			Equipment utilization	100	>200
			Productivity (g/L resin/day)	100	>600

*Cycle time refers to residence time of a target molecule in respective unit operation(s)

Figure 3.6 Integrated fully continuous production of mAbs. (Reproduced with permission form Ref. [28]. Copyright 2015, Elsevier.)

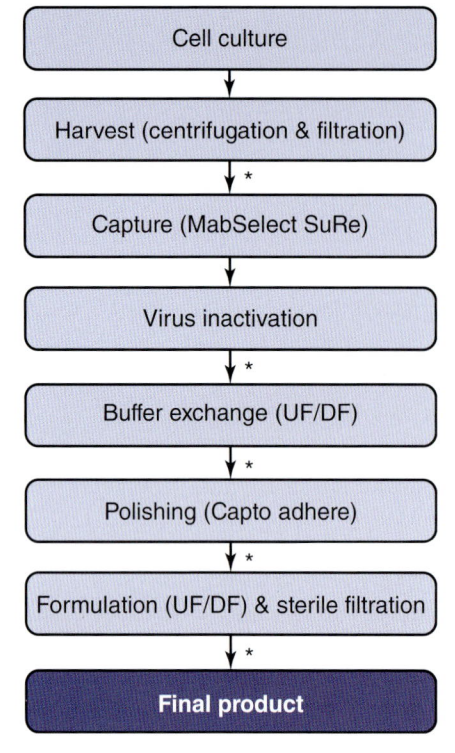

```
┌─────────────────────────────┐
│        Cell culture         │
└──────────────┬──────────────┘
               ▼ *
┌─────────────────────────────┐
│ Harvest (centrifugation & filtration) │
└──────────────┬──────────────┘
               ▼ *
┌─────────────────────────────┐
│   Capture (MabSelect SuRe)  │
└──────────────┬──────────────┘
               ▼
┌─────────────────────────────┐
│      Virus inactivation     │
└──────────────┬──────────────┘
               ▼ *
┌─────────────────────────────┐
│   Buffer exchange (UF/DF)   │
└──────────────┬──────────────┘
               ▼ *
┌─────────────────────────────┐
│    Polishing (Capto adhere) │
└──────────────┬──────────────┘
               ▼ *
┌─────────────────────────────┐
│ Formulation (UF/DF) & sterile filtration │
└──────────────┬──────────────┘
               ▼ *
┌─────────────────────────────┐
│       Final product         │
└─────────────────────────────┘
```

Figure 3.7 Single-use process concept [34].

be addressed [33]. Therefore, Chapters 13 and 17 in this book deal with possible solutions in detail and describes status and design principles.

GE Healthcare proposes a flexible antibody purification process based on ReadyToProcess™ products in flow-through mode (Figure 3.7) [34].

Disposable devices are chosen, and preferentially flow-through mode is established instead of bind-and-elute, to enable direct loading within continuous process scheduling [34,35].

Merck Millipore has published a concept together with Sandoz [36] in flow-through operation as being continuous as well, Figure 3.8. Platform-based protein A capture step is substituted by precipitation followed by CIX and AIX in flow through mode. A now started EU-funded project will enable experimental results in pilot scale.

Taking a step back and considering unbiased nonplatform unit operations as well as a generalized DSP scheme may help in the choice of efficient units and their suitable sequence.

Jungbauer published a review on continuous downstream processing [9]. His group has published quite some time ago the generalized DSP schema, Figure 3.9 [37]. Within this chapter all the following ABC (anything but or beyond chromatography) units are arranged along such a general scheme.

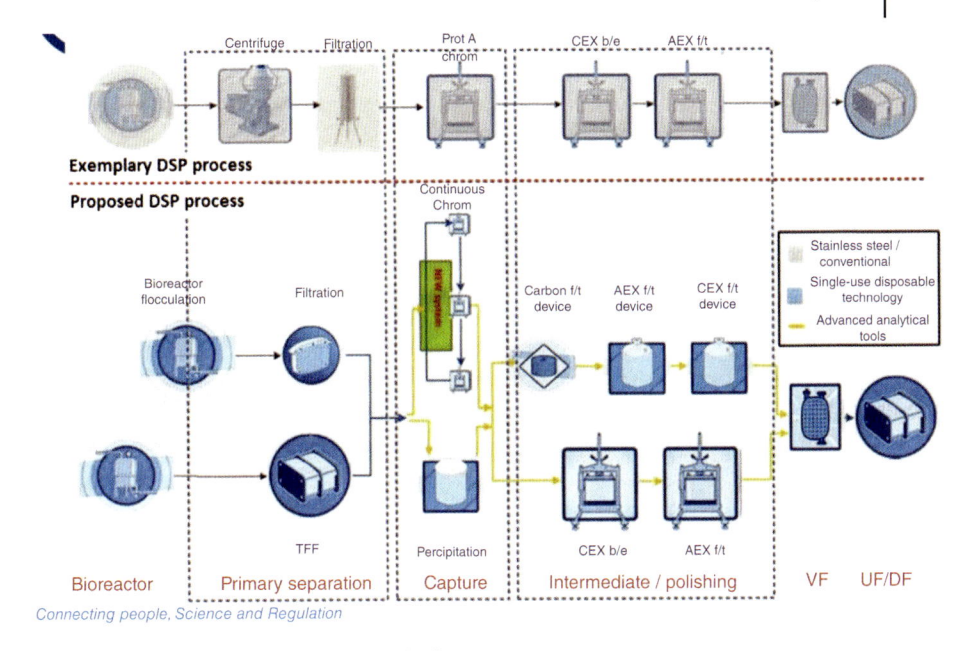

Figure 3.8 Flow-through process concept [36].

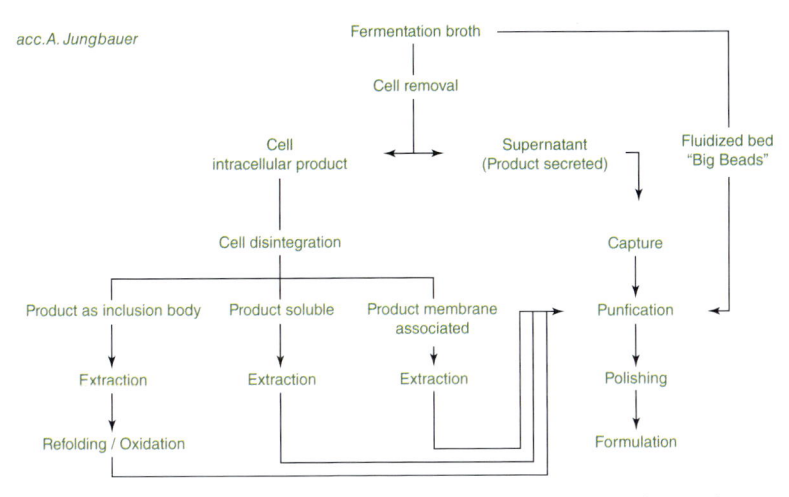

Figure 3.9 Generalized DSP scheme. (Reproduced with permission form Ref. [37]. Copyright 2010, John Wiley & Sons, Inc.)

3.4 Status and Needs for Research and Development

Figure 3.10 visualizes a key message of this chapter. Currently, standard product lifecycle (scenario 1) results in price decay due to generics/biosimilars competition for products running out of patent protection. Any potential reaction by

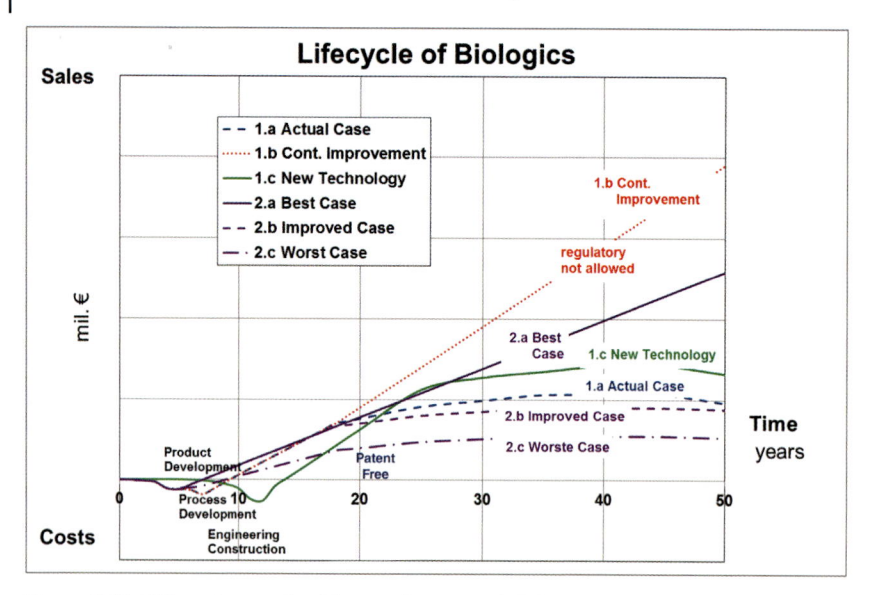

Figure 3.10 Different scenario of innovation toward CBP.

ongoing process improvement (scenario 1.b) is forbidden by regulation, because it involves so called "major change," which typically requires additional clinical studies to prove product equivalence. Due to the still limited knowledge on the outcome of changing to new technology in order to maintain profitability is actually not foreseeable. Therefore, known company studies, for example, by Sanofi [27] and GE [32], Invite [31] and Pall/Tarpon [31] or Sandoz/Millipore [36] aim to transfer existing platform unit operations into continuous operation.

Of course, these platform-based unit approaches could be easier and faster filed than any major unit operation changes, because the platform unit operations are already accepted by the authorities. Major process changes generating different side product profiles in larger than 0.1% concentration would require new clinical trials.

The benefits (scenario 2.a) are seen in lower capex and flexibility. But, the drawback of sticking to platform units could be seen in 2.a or 2.c, because generic/biosimilar competition will cause price decay as well.

Due to these logics, the proposal is to aim at any scenario of best-case improvement, which will modify the unit operations in continuous operation as well and generate minimal COGs. This could not be beaten by any generics/biosimilar competition. Therefore, any reduction in early margins would be compensated later by maintaining competitiveness.

Only obstacles are the efforts and methods needed for data-driven decisions in process development of fully integrated innovative unit operations without compromising product quality and patient safety.

Appropriate piloting and long-term operation of a prototype is needed to prove this to regulatory authorities.

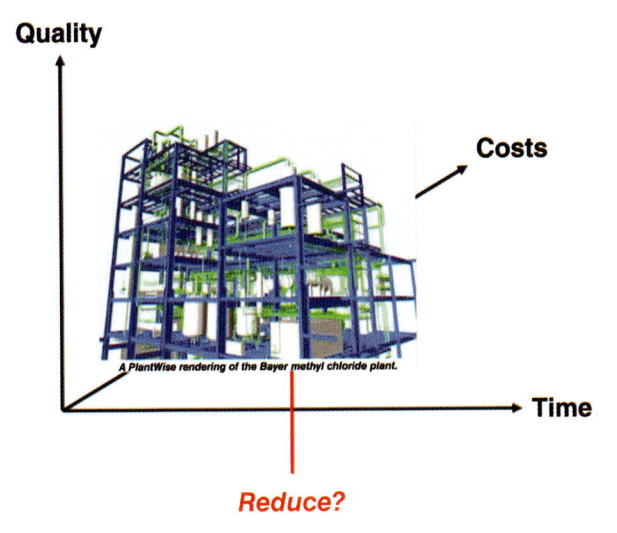

Figure 3.11 Fundamental objective function in engineering projects. (Reproduced with permission form Ref. [10]. Copyright 2014, John Wiley & Sons, Inc.)

3.5 Engineering Challenges

A recently published analysis on the impact of different new engineering approaches in various chemical–pharmaceutical industry branches [10,38–40] outline, that the central objective function in engineering is the best combination of all three parameters, quality, time, and budget, Figure 3.11, – and not just one or two of those alone. While customers like to define all three parameters, in order to arrive at a win–win situation between supplier and user, with two parameters defined by the customer, the third one needs to remain under control of the supplier.

In the strictly regulated world of pharmaceutical drug manufacturing quality of the equipment is directly related to quality attributes of the product, and therefore controlled by regulatory authorities. The second parameter budget, here CAPEX and OPEX as TOTEX, needs – as described before – to be minimized. Consequently, only project time is open for adaptation to the other two set parameters. An approach, from which vendor and customer may benefit, is modular engineering [10,38–43].

Discussing objective function parameters versus time, a glance at typical project execution schedule curves point out potential savings, Figure 3.12.

Either 1. increase the s-shape incline or 2. parallelize different tasks. Parallelization is limited due to technical reasons of the information available at each project stage. Due to actual simultaneous engineering approaches based on 3D-construction methods, only minor savings are possible. On the other side, the steepness of the different project task curves open up only minor optimization potential, either keeping in mind that, for example, simple pumps have delivery times of approximately 6 months and whole package units are delivered within about 9 months after receiving orders [10].

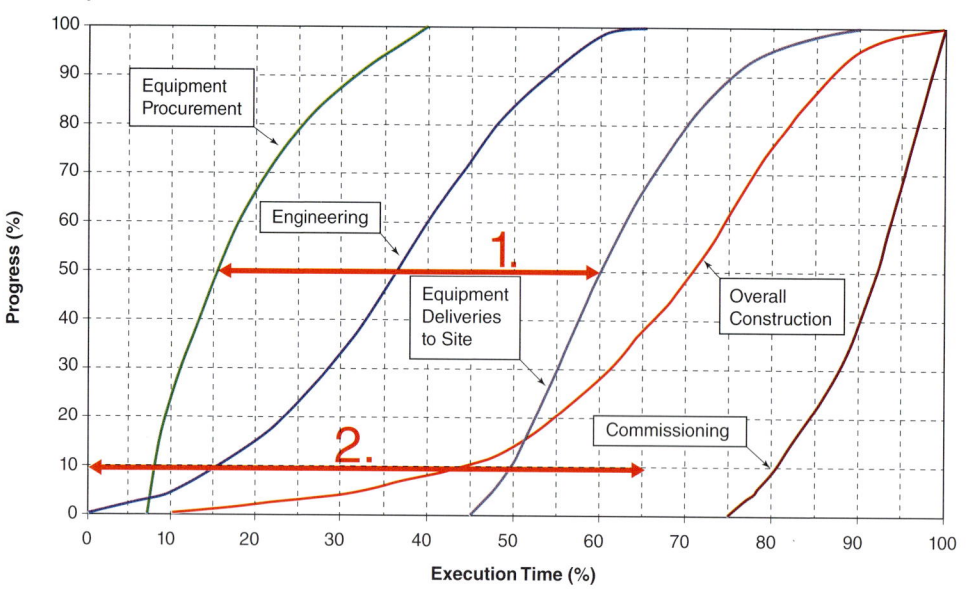

Figure 3.12 Conventional project execution schedule. (Reproduced with permission form Ref. [10]. Copyright 2014, John Wiley & Sons, Inc.)

3.5.1 Platform Method of QbD-Driven Process Modeling Instead of Unit Operation Oriented Platform Approaches

These scenarios indicate that instead of sticking to platform process approaches, more innovation is needed. Optimal unit operation sequences and the valid operation window for new drug candidates are fixed by process modeling with aid of predictive rigorous models, which are based on separated effects of fluid-dynamics, phase equilibrium, and mass transfer. Model parameters are efficiently determined in miniaturized laboratory-scale experiments with sound measurement error calculation, and model validation is performed, for example, by Monte Carlo simulation studies as described in Ref. [8].

In contrast, the full optimization potential of a specific continuous operation is not realized if the unavoidable safety margins of a platform process approach – with the main objective to be fast and efficient in limited development resources – stay in place.

Nevertheless, when working only on typical products such as mAbs and willing to accept fast success, but compromising on the performance side, the idea to transfer an accepted unit operation scheme like mAb platform, typically follows Figure 3.13.

However, assessing the broad variety of new candidates in clinical trials (see Figure 3.14: How many platforms would you need?) results in at least 5–10, which is way outside a standardized platform process approach.

Therefore, the "conservative" biomanufacturing platform approach needs to be revisited [46]. The authors have proposed for years a platform *approach* to be able

OFF THE SHELF PLATFORM FOR RAPID PROCESS TRAIN ASSEMBLY

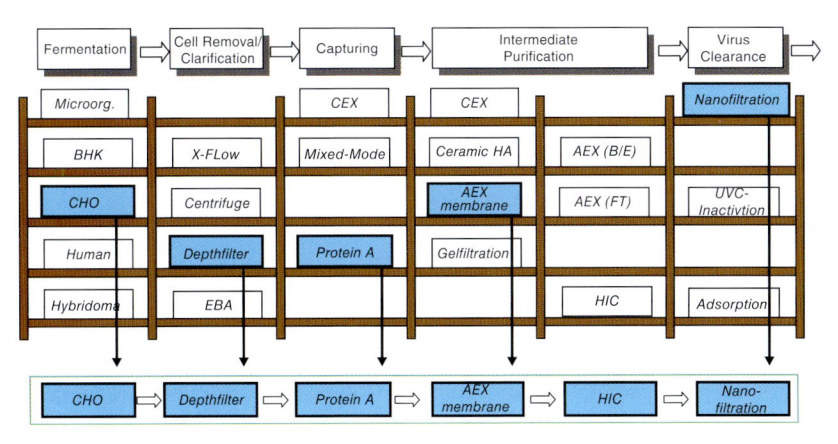

Figure 3.13 Platform for rapid process development [44].

to adapt to molecular variety of candidates needed, and to address the time pressure in process development as well.

The platform approach is process modeling in combination with experimental determination of model parameters for each molecular new candidate system in laboratory scale. The experimental setup and methods need to be standardized [47–49].

3.5.2 Data Driven Decisions

The general aim is to file for approval on the basis of data driven decisions. This coincides with FDA/EMA demands on PAT and QbD approaches.

This approach is also quite in-line with approaches from other working groups [50,51].

Company	Total # of Molecules	Drug Format (Class)					% of Drug Formats
		NCE	mAb	Recomb Protein.	Oligo-nucleotide	Vaccine	Others e.g. ICK	
A	10	4	2		1	2	1	83
B	7	4		1			2	50
C	16	12	3	1				50
D	19	15	3			1		50
E	8	5	1		2			50
F	15	10	2			3		50
G	14	10	3	1				50

Figure 3.14 Drug formats – diversity in existing oncology pipelines [45].

The future will show whether a database of physical properties for biomolecules can be developed. It must be kept in mind, that, for example, overlapping peaks are present in all large-scale chromatographic processes, which cannot be resolved by classical process scale detection methods. However, the diode array detector (DAD), well known in analytical applications, can be applied to perform an online peak deconvolution. In general, there are two different approaches to the peak deconvolution idea for biomanufacturing. Both ideas can be enhanced by mathematical peak identification methods, as well as model assumptions. Spectroscopic parameters of each molecule are not as exact as necessary to be determined by theory with aid of molecular modeling or by HTS measurements, because the detection quantification is finally sensitively dependent on the lifetime and status as well as type of the detector applied in manufacturing, and moreover, the buffer system used as well. Online peak deconvolution has the potential to shift pooling decisions from empirical to data-based decisions and might, therefore, be considered for all chromatographic processes (Chapters 13 and 17).

3.5.3 Analytics

Data-driven decisions need, of course, reliable data. Data from process parameters like flow, temperature, pressure, pH, and conductivity are easily accessible. Product quality-related data in different complex molecular matrices due to impurities, contaminants, side component amounts, and different salt types and concentrations cause significantly more problems.

A major obstacle at the moment is the analytical effort required [51–53] to generate data in appropriate quality, as summarized by Jungbauer [52]. In addition, Ottens review [50] clarifies some aspects as well. These reviews point out the analytical needs and efforts from an engineering point of view. The regulatory authorities have already described their demands [54–59].

At the moment, robotics [60,61] or quantitative MS [62,63] may provide options. Currently under research is an approach for peak deconvolution [64] by DAD UV/VIS detectors, in which the authors see the most promising approach for application in manufacturing, combined with classical existing QA methods, for example, Prot A, SEC, Bradford, 2D-GE [65] (Chapters 13 and 17).

3.5.4 QbD Methods

Although PAT and QbD approaches are already quite well developed, they are far from being applied as a broad standard [48,66,67]. In addition, production operation robustness analysis is quite simple to be applied [68].

3.5.5 Upstream and Downstream Integration

The more recent recognition that the total process optimum is only by chance the sum of the single unit operation optima, holds especially true for biologics. If the titer is pushed up further, [65] the Golgi apparatus generates under stress exponentially more and different side components/impurities, which due to their

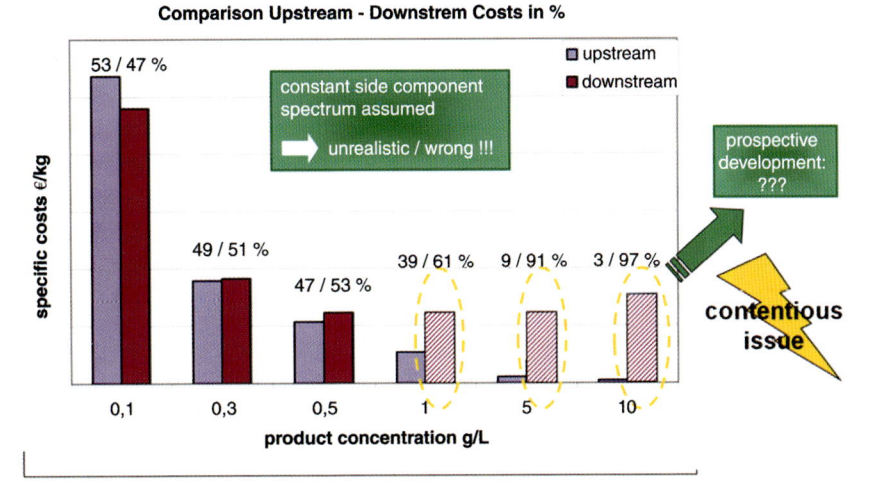

Figure 3.15 Downstream COG at higher titers with higher side components. (Adapted from References [8,65].) COG minimum is a function of cost-dominant DSP release at medium/high titers in USP [3–7].

growing similarity with the target compound are increasingly difficult to separate [69,70]. This is one of the reasons, why the productivity bottleneck is currently shifting more and more toward DSP (Figure 3.15).

Ref. [65] proposes an approach to integrate USP and DSP, and in doing so, finds the total process optimum at medium high titers. Konstantinov's group stated that as an advantage of CBP [1], medium high titers are achieved by perfusion fermentation at high cell densities in contrast to classical fed-batch.

3.5.6 Buffer Handling/Recycling

As one of the issues identified in the discussion about common industrial concerns in Chapter 2, the supply chain in any continuous and flexible manufacturing needs to be dealt with.

Jungbauer discussed in a review the buffer recycling options [71]. Unmentioned in this context, already some years ago this approach was quite consequently proposed in Refs [8,72] for mAb platforms. Figure 3.16 depicts by aid of a Sankey diagram the amounts of buffers at the different unit operation steps, with their potential for recycling.

It was shown that buffer recycling is economically feasible, gaining COG reductions of about 30% at ROIs within 1 year, as proven and validated in laboratory scale. Scale-up to manufacturing should not cause technical problems, as all the units employed are state of the art.

Under most company sustainability guidelines, this is considered appropriate [73]. In addition, there is a growing water shortage around the world even in fast growing industrial regions; especially, access to quality drinking water used as feed for making purified water for injection (WFI) manufacturing presents often a major issue.

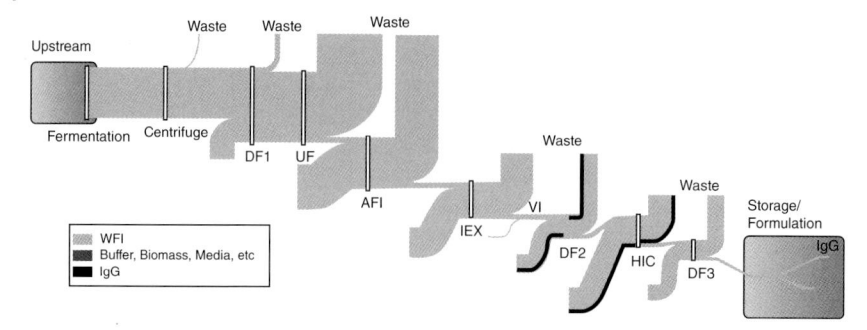

Figure 3.16 Sankey diagram of water deployment for a typical state-of-the art monoclonal antibody production process. (Reproduced with permission form Ref. [8]. Copyright 2012, John Wiley & Sons, Inc.)

3.5.7 Process Integration of Innovative Unit Operations

Addition or substitution of unit operations in an established and maybe already approved process sequence requires special attention, care, and includes a cost-benefit assessment for the whole process. Simple replacement of one function by another will almost certainly end in corrupting the whole process. Process parameters have to be carefully adjusted before and after the alternative unit operation in order to generate the full optimization potential.

Executing such an activity on an experimental trial and error basis is practically impossible due to the efforts in time and cost, which is one of the major reasons, why bioprocess community has been extremely hesitant to follow this route. Here again, process modeling under QbD approaches is the key enabling method of choice.

3.5.8 ABC (Anything But or Beyond Chromatography) and AAC (Anything and Chromatography)

3.5.8.1 Liquid–Liquid Extraction Based on ATPE

An alternative to the classical nonmiscible organic up-taking phase allows aqueous two-phase systems (ATPS). One of the first research groups, reporting such a solution in Germany, was found in 1980s [74], but industrialization was not accepted due to the manifold of centrifuges required, which was not economic. ATPS is generated, when a water-based solution from a polymer (e.g., polyethyleneglycole, PEG) and a solution of a second polymer (e.g., dextrane) or inorganic salts (e.g., phosphate [75]) respective organic salts [76]) in the corresponding two-phase region are contacted/mixed. A major challenge for continuous operation is the recycling of those phases after recovering the target component (e.g., by precipitation [77] or back-extraction/-washing [78]). The implementation of the process in suitable equipment is not simple, because ATPS phases separate quite awkwardly, and due to their low surface tension and small density differences tend to easily form emulsions. Also, the high viscosities (up to 40 mPas) of the PEG phase hinder efficient mass transfer and settling behavior that makes it difficult to process. Therefore, in search for alternatives, ATPS on the basis of alcohols [79,80], ionic liquids [76], starch [81], are investigated, with the objective to make hydrodynamics and cost efficiency acceptable, because not all components of the ATPS are easy to run in cycle.

Liquid–liquid extraction by aid of ATPS has meanwhile gained some acceptance in downstream processing of mAbs. This is based on the special component system, where both phases contain about 80% water. Thereby, the product damage, which is often caused by organic systems, is prevented. Besides, in scientific literature, a number of publications deals with these applications and have proven good mAb yield/recovery and purity. The ATP extraction is not yet applied in large scale.

Practically, mAb manufacturing is divided in upstream and downstream processing. In upstream, the mAb and numerous other components are generated as metabolic products by the cells during fermentation. Normally, cell separation is followed by purification and respective isolation of the mAb from solution, which is named downstream processing. Some activities deal with the integration of ATP extraction in downstream or for cell separation [82–84]. A recently developed membrane extractor [85,86] combines classical extraction with membrane-based system, which allows cell separation and purification successfully.

Normally, in chemistry – where the usual organic counter-phase of any water-based feed systems is quite different in physical properties – phase separation is straightforward without additives. But, in any biotechnological ATP system, appropriate phase separation of two water-based phases with quite similar physical properties is challenging. One potential solution is to reach efficient phase separation with aid of membranes. Because, both phases contain a high amount of water, a separation by hydrophilic or hydrophobic membrane types is possible for both phases. In order to gain an efficient separation nevertheless, a method proposed by Riedl [87] could be chosen. To the aqueous system, a biocompatible surfactant is added and distributes between the phases according to its distribution coefficient. Due to the tenside, Riedl *et al.* achieve a higher hydrophobicity of the lighter phase, which leads to a reduction of contact angle of lighter phase on a hydrophobic membrane. This generates a selectivity of the membrane regarding the phases [86].

Based on miniaturized shaking experiments to gain phase equilibrium parameters and kinetic measurements on single droplets via a standardized single-droplet measurement cell as well as a phase-equilibrium stage model, the chosen ATPS was proven experimentally feasible – and was successfully purified both in standard extraction equipment like mixer–settler batteries (MSB) as well as with packing- and Kühni-columns [78]. Potential saturation of the large surface areas of the internals by product and side components as well as product stability was quantified.

Additional research actions were focused on single-use/disposable equipment dealing with the development and optimization of a membrane type extractor. The general principle of function is explained in Figure 3.17. At first the structure of the mixing region for the two phases involved was optimized. Due to strong mixing and high surface to volume ratio, extremely fast mass transfer is achieved. Directly afterward, both phases pass through a separation zone, which draws off either the more organic or the more aqueous phase due to its tendency for solvation. Driving force in this type of phase separation device is the

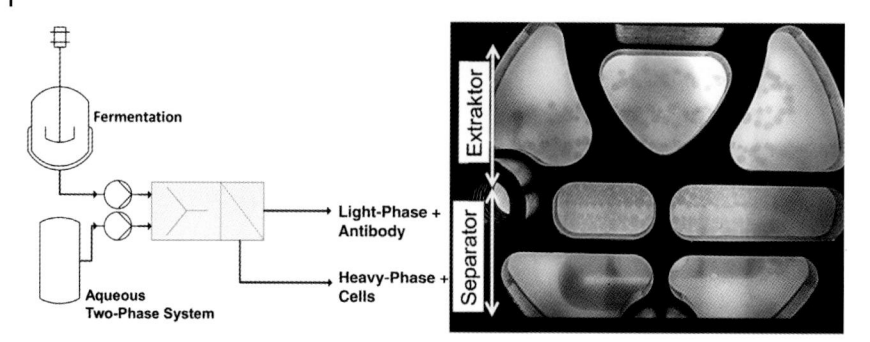

Figure 3.17 Miniaturized LL-extractor device. (Reproduced with permission form Ref. [88].)

transmembrane pressure, which is the average pressure difference between feed and permeate chamber (Figure 3.17).

Characterization of this innovative membrane extractor device is at first based on the standard ECFE test system, water/acetone/butyl-acetate. At the moment, the not yet totally optimized device could be operated with very flexible phase ratios at throughputs of about 40 ml/min [i.e., 2.4 l/h]. A few 10 l/h are considered sufficient for a scale of 2.000 l fermenter. Above this throughput range, longer separation distances are necessary to achieve a complete phase separation. A feasibility study has applied this membrane extractor successfully within an ATPE process. Fermentation broth with CHO cells is mixed following the described principle of aqueous two-phase systems and afterward the heavier phase is separated from the lighter phase by aid of the membrane. To enhance phase separation, a biocompatible tenside is added at the beginning, which increases hydrophobicity of the heavier phase. At the moment optimal setups of tenside and membrane type are under investigation by screening, in order to enlarge the currently quite low break through pressures of less than 300 mbar [88].

3.5.8.2 Precipitation

Today, antibody concentrations of 3–5 g/l are regularly reached [89–93]. By applying newest cell retention devices (ATF perfusion modules), even higher titers of up to 25 g/l can be achieved [94,95]. These developments are answers of upstream processing to ever increasing demands for new and innovative low-cost monoclonal antibodies [89–94].

The changes of USP manufacturing lead to 15–100 kg mAb/batch at titers of 5 g/l in 20–25 kl bioreactors [92,93]. The equipment of downstream processes was originally designed for much lower concentrations of antibodies, resulting in capacity limitations of the DSP equipment. These limitations not only lead to an increase of processing time and material consumption but also of costs in downstream processing [65].

In addition to these limitations, new problems, accompanying process development trends in USP toward high product titers, arise. Type and concentration of impurities vary strongly between different fermentations due to process changes [52,96–99]. The resulting impurities, apart from their increasing amount, become more similar to the product in their isoelectric point, molecular weight, and hydrophobicity [65].

These significant problems in advancing mAb manufacturing require new optimization approaches. Examples could consist in new optimization strategies, technical solutions, and methods in process development. Technical solutions consists in the application of innovative technologies like aqueous two-phase extraction (ATPE) [78,100–104], precipitation [105–107], and membrane-based adsorption [108–110], among others. New methods in process development could include extended use of mini-plant facilities and a stronger integration of modeling [89,111,112].

ATPE has been discussed before, but precipitation may also present possible technology for protein purification. It is considered as low-cost technology resulting in high yields [65,107]. Precipitation can be used in different ways. It already is applied in industrial scale for protein purification [113–118]. Another application consists in the purification and clarification step of process volumes prior to chromatographic steps. HCP and media components could be separated before a chromatographic capture would take place resulting into a heightened protection of chromatographic resins [65,119,120]. Latest research results demonstrate the possible application of precipitation as selective mAb manufacturing tool [109]. It can be used for high titer broth and is much less expensive than protein A chromatography [105]. The scale up only depends on the processed volumes [105]. The precipitation of the antibodies themselves can be carried out using ammonium sulfate [21,120,121]. Positively charged polymers precipitate impurities like acidic HCP, DNA, and media components [65,116,120,122]. Antibodies can also be separated from impurities by negatively charged polymers [120], polyelectrolytes, organic solvents (ethanol), and polymers such as polyethylene glycol of varying lengths [109].

Recent work at the Institute of Separation and Process Technology at TU Clausthal presents a combination of ATPE and precipitation as possible alternative to centrifugation and protein A chromatography [123].

ATPS as cell harvest operation was optimized and data are being published [124]. For subsequent precipitation development, a small number of reagents was identified which lead to a selective precipitation of IgG or HCP. The most promising results were achieved by applying citric acid or cold ethanol. A DoE-based process optimization delivered high yields from high titer broth. Appling this ATPE-precipitation process to different fermentation broths of different titers and impurity compositions showed varying results. The exact composition of impurities strongly influences the process performance.

At the moment, an ATPE-precipitation process alone cannot substitute all chromatographic steps in the benchmark process. The resulting yields and purities, however are promising to present a more cost-effective alternative to centrifugation, protein A and a second chromatography step. In order to design a chromatography-free mAb-process, an additional integration of upstream and downstream processing (Chapters 13 and 17) could provide a useful approach in process development to achieve high product quality at lower cost [123].

3.5.8.3 Membrane Adsorbers

In the meantime, IEX, HIC, and even salt tolerant HIC materials for membrane chromatography are available from different vendors in pilot or large scale. The first case studies have applied those materials in a final polishing step in flow through mode at pilot scales. Binding capacity is still too low for an efficient bind-and-elute mode and will remain to be assessed in future for large volume products due to geometry aspects. In a fast flow operation scheme at the nearly pure product level, those devices may show considerable advances in process time, cleaning efforts, or disposability [125–128].

3.5.8.4 Innovative Materials Like Fibers or Matrices

The search for innovative materials as auxiliaries in chromatography, which is still the work-horse and also a major cost driver, and membranes is still of high importance, but also risk. From idea to industrial transfer of a new synthesis takes easily up 10 years with costs in the million Euro range, especially, if classical full experimental trial and error approaches are applied. More systematic knowledge-based approaches are therefore under fundamental research [129–131]. Still first new products need to prove their efficiency and performance in a process environment.

Another (old) idea got new drive by these simulation studies, the short and quick loading and elution phases of core shell materials [132], with beads being operated in membrane style [133].

Disposable, single-use, that is, cost-effective adsorbents were proposed by Millipore [134] based on fibers, not beads. New approaches [135] provide easy packing properties and relatively high loadings. Also, old patents on woven fabrics are still around [136].

Additionally, rod type columns, possible up to ID about 50 mm and larger, enable fast flow at low pressure and no mass transfer kinetic limitations [137], some of which can be applied on a process scale.

Also (old) ideas of radial flow column equipment are under consideration for high volume, that is, fast flow at low pressures combined with product concentration.[1] In this portfolio, expanded bed and magnetized fluidized bed matrices and equipment rank as well for direct loading with unclarified broth [138], in order to reduce one unit operation in the process scheme.

3.5.9 Process Concepts for mAbs and Fragments

Searching for the holy grail in bioprocessing, that is, implementing the general idea of being best in class with an optimized process from the beginning and integrating any unit operation fit for manufacturing, then a continuous processing concept like shown in Figure 3.18, might result.

After ATPE extraction, a one- or two-step precipitation may be followed by multistep UF-filter modules. Continuous UF is still a challenge to adopt. Final polishing is done either with iCCC, described in Chapters 13 and 17 in this book in detail, or membrane adsorbers, depending on the volume scale and purification selectivity needed.

Statistical DOE analyses, described in detail in Ref. [123] result in the conclusion that HCl, a shorter feeding time (0.01 h; HCl precipitation), medium temperature (15 °C; HCl precipitation), a low energy input (300 rpm; HCl

---

1 http://www.sepragen.com/page/Products-Chromatography-Columns.aspx (access January 22, 2016).

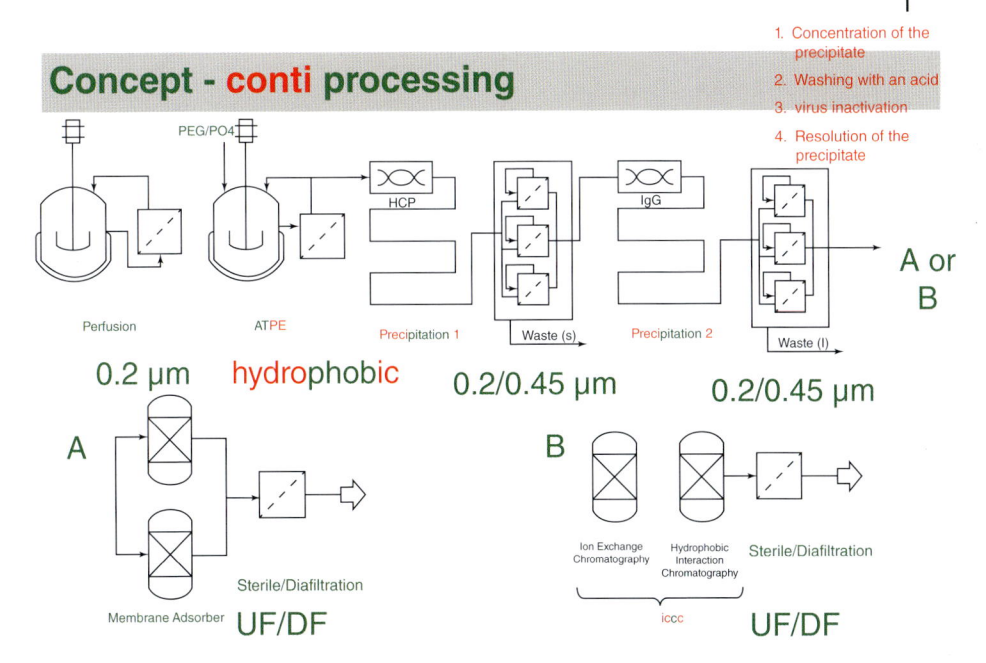

Figure 3.18 Conti biomanufacturing process concept with alternative DSP units. (Reproduced with permission form Ref. [123]. Copyright 2016, Tekno Scienze Publisher.)

precipitation), a high pH (4.75; HCl precipitation), a high EtOH concentration (30 wt%), a low feeding rate (0.5 ml/min; EtOH precipitation), a low temperature (−10 °C), a low energy input (300 rpm; EtOH precipitation), and a high pH value (5.75–6.75; EtOH precipitation). Parts of these final parameters were chosen in order to make a compromise of antagonistic effects. Parameter ranges of the most important parameters can be viewed in Figure 3.19.

Additional experiments were carried out using different fermentation broths. The IgG concentrations ranged from 0.2 g/l of up to 5.4 g/l including different compositions of HCP. The ATPE/precipitation purification process of the high titer fermentation broth resulted into a yield of 94%, a SEC-based purity of 79%, and an Elisa-based purity of 99.8% (only regarding proteins) at a titer of 8.3 g/l and an HCP concentration of 10.96 mg/l. The actual HCP concentration is probably higher because only a generic ELISA assay was used. The HCP content was reduced by 81.4% and no DNA was detectable in the resolved IgG pellet [114]. In Figure 3.20, a SEC diagram of the top phase, the supernatant of the HCl precipitation and of the resolved IgG is presented.

In Figure 3.20, the monomer peak of the HCl supernatant is slightly reduced compared to the other two phases. Probably, some of the aggregates were dissolved into monomers during EtOH precipitation and pellet dissolution resulting into a high product yield. But, there is a significant reduction compared to the IgG peak of the fermentation broth. Keeping the high yield in mind, this reduction can be explained by a reduction of other components that elute at the same time due to their size and structure. A significant reduction of the aggregate

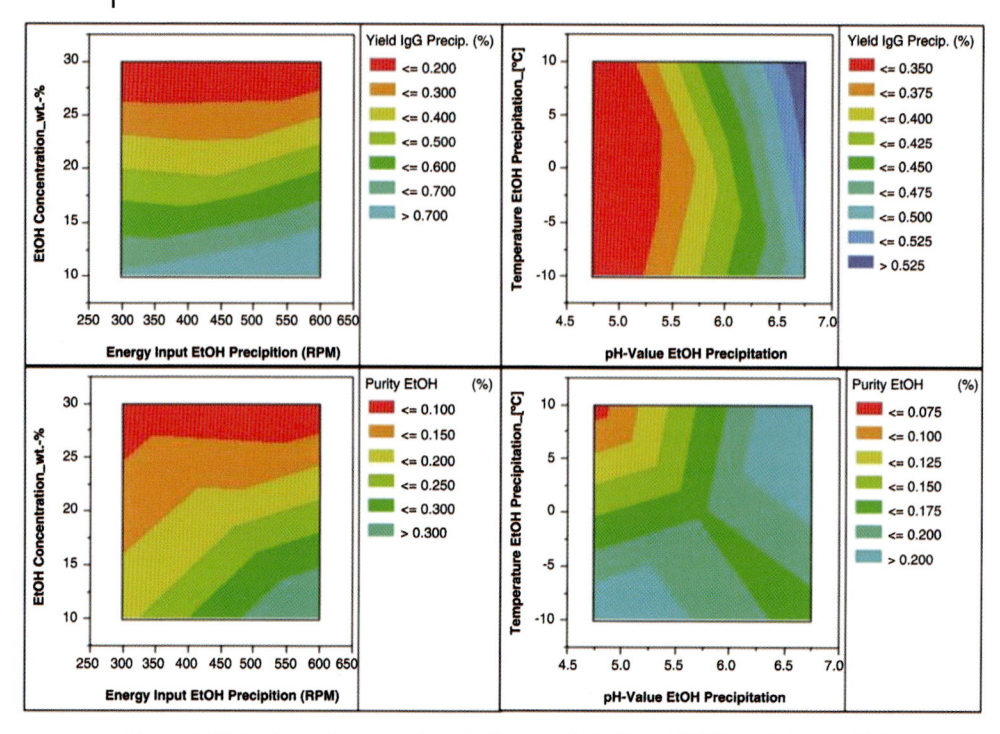

Figure 3.19 Contour diagrams of results from optimization of EtOH precipitation. (Adapted with oermission from Refs [123,124].)

peak is also visible. The peak areas of HCP and of smaller molecules (starting around 10.5 min) are reduced compared to the peaks of the top phase. The overlap of the IgG and HCP peak generate an error in the exact concentration determination. Therefore, a sufficient separation of both parameters by SEC is not possible but the qualitative difference remains clear [123].

Figure 3.21 depicts the advantages of the alternative concept described with regard to the following platform-benchmark:

- Adaptation of capacity and therefore adaptation to high titers is easily possible by PEG400 concentration in ATPE as well as concentration of precipitation-media in precipitation without significant additional COGs
- CAPEX of a centrifuge set is spared
- If needed for process yield and purity, multistage operation of ATPE is easily adaptable by combining the single devices
- FTIR analysis shows no modifications of the IgG structure
- With the aid of precipitation, virus inactivation could be integrated in a washing step
- During redissolution of the precipitate, the process volume could be adjusted with regard to the subsequent process steps
- Solid–liquid separation could be fulfilled by a set-up of UF-filters in parallel operation

The COG analysis depicts at the moment a factor of 3–5 reduction at optimized parameter setups [139].

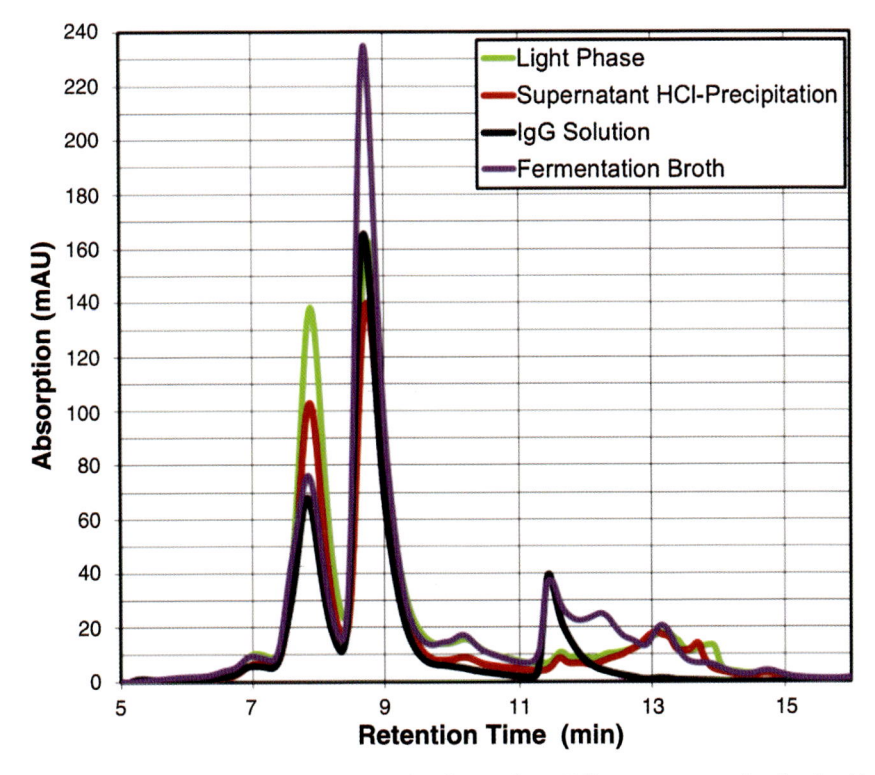

Figure 3.20 SEC diagram of fermentation broth, top phase, HCl supernatant and redissolved IgG pellet resulting from a fermentation broth with a product concentration of 5.4 g/l. (Reproduced with permission form Ref. [123]. Copyright 2016, Tekno Scienze Publisher.)

3.5.10 Single-Use Technology

Single-use technology is broadly applied in process development to reduce man power and time efforts and for CMOs to be most flexible in pilot-scale, and fast between applications [140–142].[2]

Working groups have established a helpful scheme to guide a decision for SUS applications, see Figure 3.22.

3.5.11 Guided Decision for CBP

It seems a quite clever idea [18] to transfer this scheme for any guided decision for CBP, as shown in Figure 3.23.

Design space definition via product CQAS, prior knowledge, and risk assessment leads to preliminary parameter identification. Based on that, an experimental DOE including risk assessment results in an experimental parameter screening. Combined with modeling, a control strategy can be derived which leads to batch record definition. Manufacturing data is evaluated in comparison to the experimental results as a design space, which is being verified and updated.

---

2 http://www.biopharma-reporter.com/Upstream-Processing/Roche-expanding-capacity-40-across-network-but-will-still-use-CMOs (access January 22, 2016).

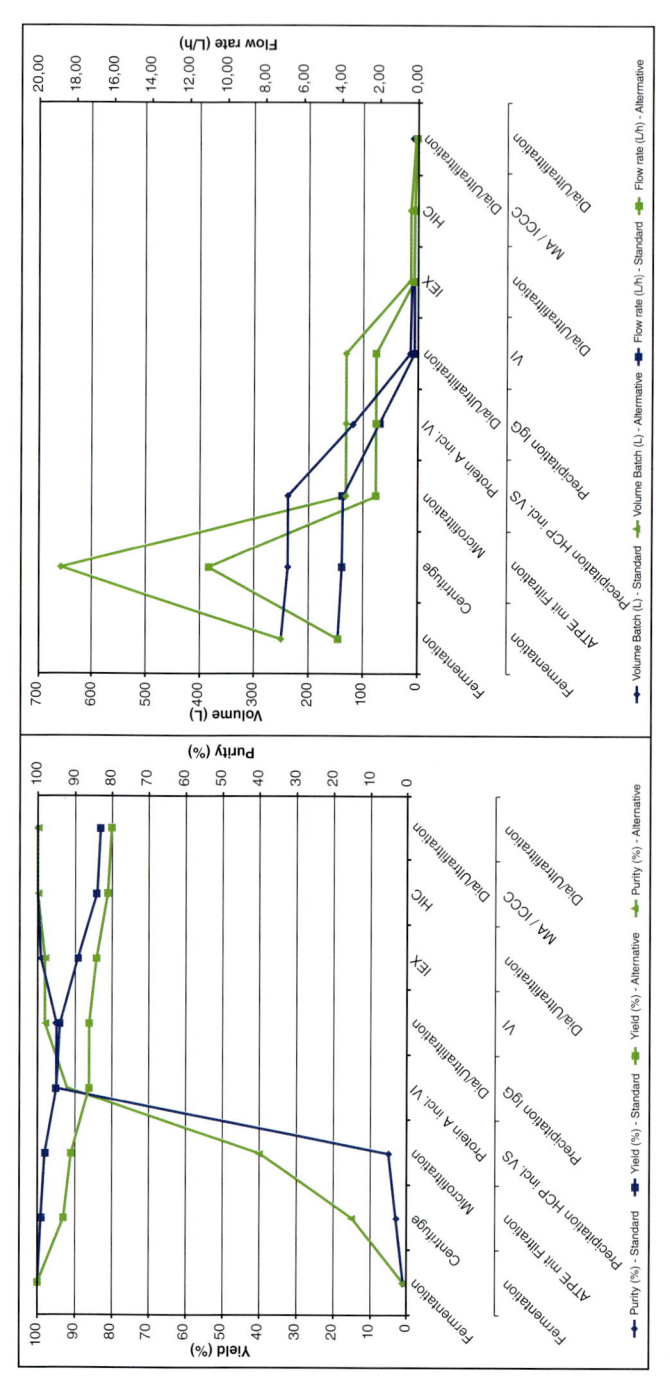

92 | *3 Engineering Challenges of Continuous Biomanufacturing Processes (CBP)*

Figure 3.21 Comparison between platform and alternative process concept.

- Example from PDA TR 66: Application of Single-Use Systems (SUS) in Pharmaceutical Manufacturing

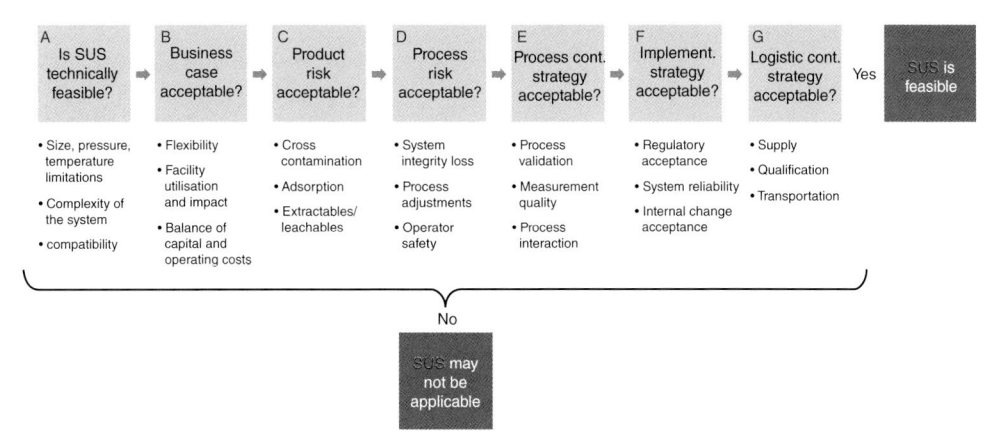

A Is SUS technically feasible?	B Business case acceptable?	C Product risk acceptable?	D Process risk acceptable?	E Process cont. strategy acceptable?	F Implement. strategy acceptable?	G Logistic cont. strategy acceptable?	Yes → SUS is feasible
• Size, pressure, temperature limitations • Complexity of the system • compatibility	• Flexibility • Facility utilisation and impact • Balance of capital and operating costs	• Cross contamination • Adsorption • Extractables/ leachables	• System integrity loss • Process adjustments • Operator safety	• Process validation • Measurement quality • Process interaction	• Regulatory acceptance • System reliability • Internal change acceptance	• Supply • Qualification • Transportation	

No → SUS may not be applicable

Figure 3.22 Guided decision process for SUS [18].

- Guide based on the decision strategy for SUS presented in PDA TR 66

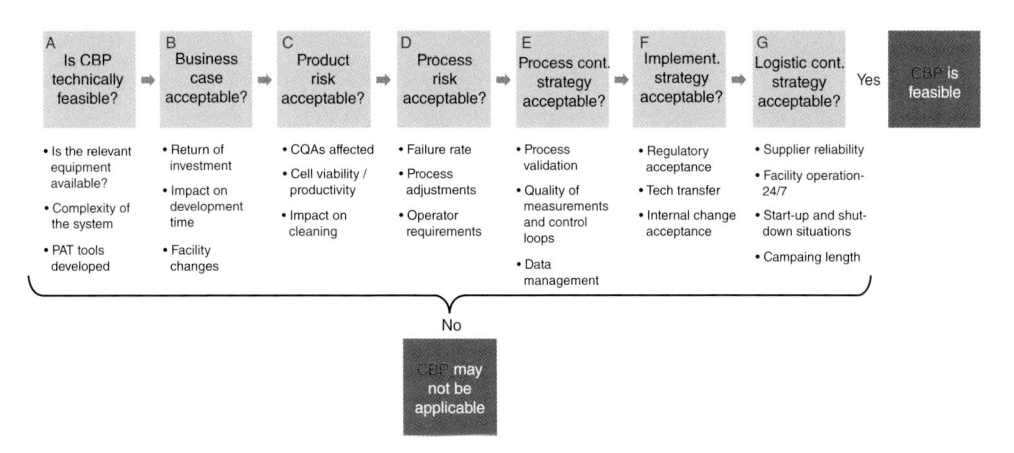

A Is CBP technically feasible?	B Business case acceptable?	C Product risk acceptable?	D Process risk acceptable?	E Process cont. strategy acceptable?	F Implement. strategy acceptable?	G Logistic cont. strategy acceptable?	Yes → CBP is feasible
• Is the relevant equipment available? • Complexity of the system • PAT tools developed	• Return of investment • Impact on development time • Facility changes	• CQAs affected • Cell viability / productivity • Impact on cleaning	• Failure rate • Process adjustments • Operator requirements	• Process validation • Quality of measurements and control loops • Data management	• Regulatory acceptance • Tech transfer • Internal change acceptance	• Supplier reliability • Facility operation-24/7 • Start-up and shut-down situations • Campaing length	

No → CBP may not be applicable

Figure 3.23 Guided decision process for CBP – analogy to SUS [18].

Figure 3.24 depicts the contribution of statistical tools in use: screening DOE approaches with sensitivity analyses lead to statistical DOE plans and models to determine parameter ranges. Scale-up correlations, uncertainty analysis, and performance monitoring with regard to PAT technologies result in verification and updating of the design space. In addition, multivariate statistical process control (MSPC) is used for a sound control strategy enabling real-time release [143].

At the start of process development, taking advantage of CBP in development does not require to run manufacturing in CBP mode, column life time studies and testing parameter ranges in one set up is possible. General knowledge of relationships between CQAs and CPPs is developed. Basics for feed-forward and feed-backward controls like the impact of perfusion rate on viable cell concentration

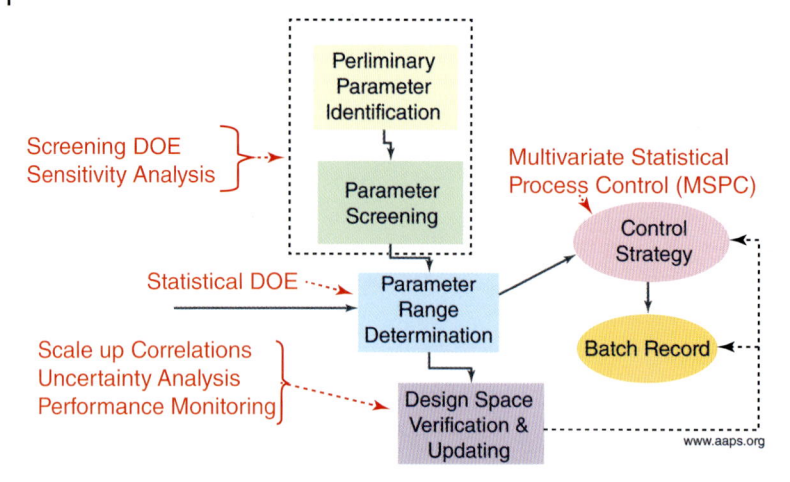

Figure 3.24 Statistical tools in design space considerations. (Reproduced with permission form Ref. [144].)

Table 3.1 Impact on column loading [18].

Process step	Yield range	Amount of product	Column loading range (g/l resin)	Number of runs on a 2 l column	Column loading range (g/l resin)
500 lcell culture yield	2.7–3.3 g/l	1350–1650			
Harvest	80–90%	1080–1485			
Prot A	75–85%	810–1262	16.2–25.2 (50 l)	20–32	20
IEX	85–95%	689–1119	6.89–11.2 (100 l)	35–60	10
HIC	80–90%	551–1079	5.51–10.8 (100 l)	34–58	8

and elution conductivity on pool volume. Finally, critical process indicators CPI are identified.

A typical impact on column loading is demonstrated by a short comparison [18], Table 3.1.

During the discussion of "quasicontinuous," one sophisticated definition of "continuous" is provided by Konstantinov: "everything is continuous, even batches, if they are split and repeated in cycles"; therefore, the often already applied splitting of lots/batches into sub-batches is already an intermediate step toward CBP! It has an impact on residence time, pool volume, process time, resolution, and total yield as well as an impact on CQAs – which has to be experimentally proven by DOE under QbD. For example, a typical solution in industry is to run a number of cycles on each column, in order to decrease the column size and schedule the operation into the whole manufacturing flow of buffer handling and staff availability: Does the longer hold time before loading and after pooling have an impact on CQAs? And what are the criteria for pooling the sub-batches? These experimentally defined results could directly be transferred to CBP – nothing new and nothing mysterious.

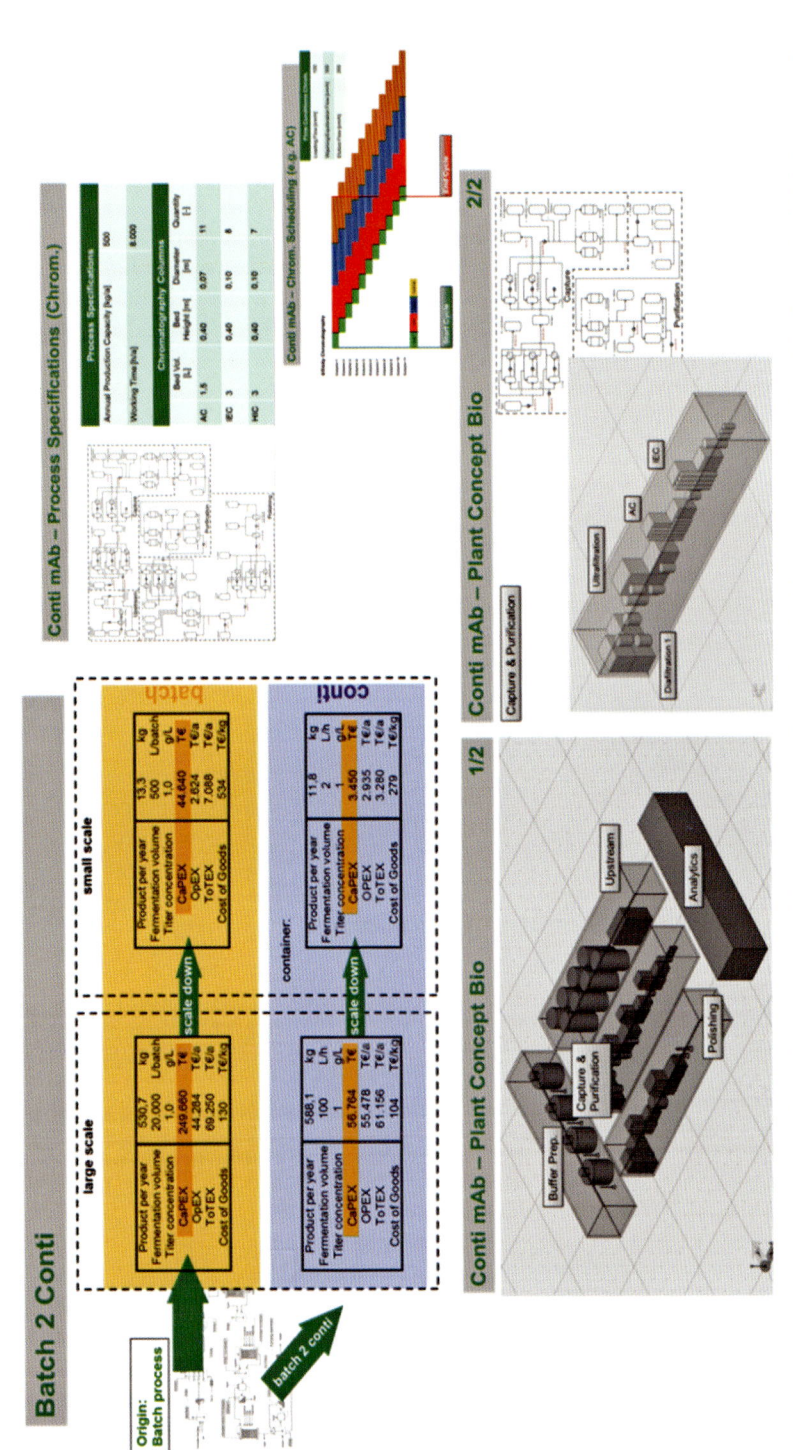

Figure 3.25 Batch to continuous industrialization by modular engineering. (Reproduced with permission form Ref. [33]. Copyright 2014, American Chemical Society.)

Studies that demonstrated these impacts have been published [7,8,10,33,40] quite a while ago, Figure 3.25. Pointing out that at best, with low CAPEX by dedicated modular equipment engineering and low OPEX at appropriate life time of auxiliary media, reduced cleaning buffer consumption and process control to reduce man-power of operation with the existing platform approaches operated continuously the economy of scale is not needed for small volume products in the tens to hundreds kg per year range. In large scale, even a COG reduction by a factor of five is possible.

Additionally, recent studies based on total process integration of alternative unit operations– which are not the actual platforms – hint at a potential factor of 10 in COG reduction becoming possible [65,123].

3.6 Conclusion and Outlook

A detailed analysis of perceived obstacles for a broader implementation of CBP shows, that executing such a concept is not rocket science, it is just sound technical transfer, appropriately combining existing methods and equipment parts. However, in doing so, analytics as a provider of critical information in the decision-making will have to play a much larger than earlier thought, see case study data in Section 17.6.2 in this book.

Some references are already there, that is, others have already implemented, successfully continuous bioprocessing facilities: "Amgen CEO Robert Bradway hinted several years ago that the company was on the "cusp" of a new manufacturing process for making cell-based drugs that would upend the industry, being faster and cheaper. Today, Amgen says that this time has arrived with the completion of a $200 million plant in Singapore, which incorporates continuous processing. The Thousand Oaks, CA-based biotech manager said, the plant in Tuas uses single-use bioreactors, disposable containers, continuous purification processing, and real-time quality analysis for monoclonal antibody manufacturing. Started in early 2013, it was built in less than two years." [145].

"With cost pressures weighing heavily on drug-makers, most are looking for ways to do more with less. In manufacturing, that often means cutting jobs and closing plants. However, Genzyme, the biotech arm of Sanofi, is experimenting with continuous processing, "a new manufacturing approach that potentially can save time, space, and equipment costs by eliminating the batch approach production." [146].

These approaches aimed at fast success based on platform unit operations being transferred from batch to continuous. This objective was reached. First case studies have been successfully established. Others could learn from this, with the risk of having to be the first taken away from them.

A recent review on continuous biomanufacturing describes the different unit operations in DSP as not total process integrated, and ends with stating the need for further research [147]. Of course this is (always) correct, but industrialization is quite straightforward, if wanted/needed.

Nevertheless, to become long-term sustainable "best in class at low COG," "new" (i.e., actual nonplatform) unit operation technologies with platform method process simulation need to become more widely applicable, which also means "accessible" for biopharmaceutical SMEs, which have the scientific capabilities for innovative drug concepts, but lack the resources to establish and integrate the different modules mentioned in highly efficient, yet cost effective integrated and continuous operation. Therefore, a kind of prototype approach is needed for an unbiased assessment and integration of all tools, that is, hardware, software, and mindset, to gain faith with decision makers, and convince operational staff by comprehensive training as well. This will include, for example, demonstration of stable operation of continuous chromatography columns over the time span equivalent to handling feed from 200 batches or 1 year of operation and appropriate scheduling. Likewise, membrane units for filtration, to be operated as workhorses to determine and document reliable function, including necessary replacement before breakthrough.

Production scheduling and integration of online analytical data for process control into operational decisions will have to be trained under real life conditions to make operating personnel familiar with the new tools, assisting in more robust and reliable operations.

In particular, regulatory authorities need to become involved as early as possible the QbD approach practiced, and the quality of the data generated for filing, as well as the criteria set for real-time release decisions in the manufacturing environment. The products still have to be generated by the companies themselves, of course. But, getting rid of any short- or long-term cost pressure, having the best in class process and technology, without the threat potential biosimilars/generics competition waiting over the horizon, will be the overall sustainable benefit of such action.

The proposed approach toward a modular continuous plant that will have about a factor of 10 lower CAPEX and about factor 5–10 lower OPEX will enable even the large number of SMEs on the market to become fully integrated manufacturers.

Innovative drug candidates – with or without the whole SME or startup company – may not need to be sold to big pharma for manufacturing and marketing, but could directly be industrialized by the developers themselves. Fast track filing and operational excellence is possible due to platform methods provided to the SMEs and implemented with suitable efforts. As an additional effect for society, such a procedure would increase the innovation cycle for new drugs on the market at moderate costs.

Acknowledgments

The authors would like to thank their colleagues at the Institute for Separation and Process Technology at Clausthal University of Technology for inspiring discussions and mutual work.

References

1 Riske, F. (2012) BioProcess International European Conference & Exhibition, Düsseldorf.

2 Novartis-MIT Center for Continuous Manufacturing (2015) novartis-mit.mit.edu (access December 12, 2015).
3 Sinclaire, A. and Brown, A. (2013) Continuous processes economic evaluation. Available at www.continuous-bioprocessing.com (access December 1–2, 2015).
4 Johnson, T. (2015) Vision: integrating upstream and downstream in a fully continuous facility. Available at www.continuous-bioprocessing.com (access December 1–2, 2015).
5 Xenopoulous, A. (2015) A new, integrated, continuous purification process template for monoclonal antibodies: process modelling and cost of goods studies. *J. Biotechnol.*, **213**, 42–43.
6 Petrides, D. *et al.* (2014) Biopharmaceutical process optimization with simulation and scheduling tools. *Bioengineering*, **1**, 154–187.
7 Strube, J., Sommerfeld, M., and Lohrmann, M. (2007) Processes development and optimization for biotechnology production – monoclonal antibodies, in *Bioseparation and Bioprocessing*, 1st edn (ed. G. Subramanian), John Wiley & Sons, Inc., pp. 65–99.
8 Strube, J., Grote, F., Helling, C., and Ditz, R. (2012) Bio-process design and production technology for the future; and modeling and experimental model parameter determination with quality by design (QbD) for bioprocesses, in *Biopharmaceutical Production Technology*, vol. 1+2 (ed. G. Subramanian), John Wiley & Sons, Ltd, Weinheim, Chapters 11 and 20.
9 Jungbauer, A. (2013) Continuous downstream processing of biopharmaceuticals. *Trends Biotechnol.*, **31** (8), 479–492.
10 Strube, J., Ditz, R., Fröhlich, H., Köster, D., Grützner, T., Koch, J., and Schütte, R. (2014) Efficient engineering and production concepts for products in regulated environments – dream or nightmare? *Chem. Ing. Tech.*, **86** (5), 1–9.
11 Chu, L. and Robinson, D.K. (2001) Industrial choices for protein production by large-scale cell cultures. *Curr. Opin. Biotechnol.*, **12**, 180–187.
12 Li, F., Amanullah, A. *et al.* (2010) Cell culture processes for monoclonal antibody production. *mAbs*, **2** (5), 466–477.
13 Butler, M. and Menses-Acosta, A. (2012) Recent advances in technology supporting biopharmaceutical production for mammalian cells. *Appl. Microbiol. Biotechnol.*, **96**, 885–894.
14 Tan, Z. and Shirwaiker, R.A. (2012) A review of emerging industrial and systems engineering trends and future directions in biomanufacturing In *Proceedings of the 2012 Industrial and Engineering Conferences*, (eds G. Lin and J.W. Hermann), Institute of Industrial Engineers.
15 Symphogen (2016) Dechema: Single Cell Technology, Frankfurt am Main, Germany, June 2–3, Available at www.dechema.de (access December 07, 2015).
16 Health Network Communications (2016) Biosimilar Drug Development World Meeting, Barcelona, February 10–11, Available at http://www.healthnetworkcommunications.com/biosimilar (access December 07, 2015).

17 Whitford, W. and G.E. Healthcare (2015) Continuous Biomanufacturing: 10 Reasons Sponsors Hesitate, Bioprocess online white paper, Available at www.bioprocess.com (access December 07, 2015).

18 Nygaard, F. and NNE Pharmaplan (2015) What holds industry back from broad implementation of continuous processing? PDA Meeting Continuous Manufacturing, Berlin Germany, September 21–23, Available at www.pda.org (access December 07, 2015).

19 Munk, M. and Langer, E. (2015) What is holding industry back from implementing continuous processing: can Asia adopt more quickly? *BioPharma Asia*, **4** (6), 16–22.

20 Scannell, J.W. (2012) Diagnosing the decline in pharmaceutical R&D efficiency. *Nat. Rev. Drug Discov.*, **11**, 191–200.

21 Allison, M. (2012) Reinventing clinical trials. *Nat. Biotechnol.*, **30**, 41–49.

22 Avik, R. (2012) Stifling new cures: the true cost of lengthy clinical drug trials, Manhattan Institute Project FDA Report No. 5, March, 1–13.

23 Schulte, M. and Strube, J. (2001) Preparative enantioseparation by simulated moving bed chromatography (review). *J. Chromatogr. A*, **906** (1), 399–416.

24 Schulte, M., Britsch, L., and Strube, J. (2000) Continuous preparative liquid chromatography in the downstream processing of biotechnological products. *Acta Biotechnol.*, **20** (1), 3–15.

25 Schulte, M., Wekenborg, K., and Strube, J. (2007) Continuous chromatography in the downstream processing of products of biotechnology and natural origin, in *Bioseparation and Bioprocessing*, 1st edn (ed. G. Subramanian), Wiley-VCH Verlag GmbH & Co.KGaA, Weinheim, pp. 225–255.

26 Woodstock, J. (2014) FDA, MIT-CMAC International Symposium on Continuous Manufacturing of Pharmaceuticals. Cambridge MA, May 20.

27 Woodstock, J. (2013) FDA, US House of Representatives, December 12.

28 Godwat, R., Konstatinow, K., Rohani, M., and Warikoo, V. (2015) End-to-end integrated fully continuous production of recombinant monoclonal antibodies. *J. Biotechnol.*, **213**, 13–19.

29 Duvar, S., Heine, M., Hecht, V., and Ziehr, H. (2015) Evaluation of different cell culture retention systems for perfusion. Braunschweig International Symposium on Pharmaceutical Engineering Research SPhERe, October 19 and 20.

30 Singh, N., Arunkumar, A., Chollangi, S., Tan, Z.G., Borys, M., and Li, Z.J. (2016) Clarification technologies for monoclonal antibody manufacturing processes: current state and future perspectives. *Biotechnol. Bioeng.*, **113** (4), 698–716.

31 Klutz, S. *et al.* (2015) Developing the biofacility of the future based on continuous processing and single-use technology. *J. Biotechnol.*, **213**, 120–130.

32 Müller-Späth, T., Ströhlein, G., and Bavand, M. (2013) Productivity boost for biopurification: twin-column ultra-high resolution chromatography. *Genet. Eng. Biotechnol. News*, **33** (10), 34–35.

33 Zobel, S., Helling, C., Ditz, R., and Strube, J. (2014) Design and operation of continuous countercurrent chromatography in biotechnological production. *Ind. Eng. Chem. Res.*, **53** (22), 9169–9185.

34 GE Healthcare (2015) A flexible antibody purification process based on ReadyToProcess™ products, Application note 28-9403-48 AB, Available at www.gehealthcare.com (access December 07, 2015).
35 Pollard, D. (2015) Merck talk ppt, Advances towards automated continuous mAb processing, Available at www.merck.com (access December 07, 2015).
36 Hribar, G. and Gillespie, C. (2015) Next generation biopharmaceutical downstream processing – continuous bioprocessing. PDA meeting on Continuous Manufacturing, Berlin, Germany, September 21–23, Available at www.pda.org (access December 07, 2015).
37 Carta, G. and Jungbauer, A. (2010) *Protein Chromatography: Process Development and Scale-Up*, Wiley-VCH Verlag GmbH, Weinheim.
38 Lang, J., Stenger, F., and Schütte, R. (2012) Chemieanlagen der zukunft – unikate und/oder module. *Chem. Ing. Tech*, **84** (6), 883–884.
39 Helling, C., Fröhlich, H., Eggersglüß, J., and Strube, J. (2012) Fundamentals towards a modular microstructured production plant. *Chem. Ing. Tech.*, **84** (6), 892–904.
40 Rottke, J., Grote, F., Fröhlich, H., Köster, D., and Strube, J. (2012) Efficient engineering by modularization into package units. *Chem. Ing. Tech.*, **84** (6), 885–891.
41 Kielburger, G. (2015) Kosten senken und Geschäftsmodelle überdenken, Industrie 4.0 im Großanlagenbau, November 17, Available at www.process.vogel.de (access December 07, 2015).
42 Stump, B. and Badurdeen, F. (2012) Integrating lean and other strategies for mass customization manufacturing: a case study. *J. Intell. Manuf.*, **23**, 109–124.
43 Stephan, D. (2015) Vorläufiger Stopp für Crackerneubau: Personalengpass in Folge des Shalegas-booms, September 26, Available at www.process.vogel.de (access December 07, 2015).
44 Kaiser, K. (2010) Dechema Trainings Course on Downstream Processing, Clausthal, October 2010, BHC Wuppertal.
45 Merck KGaA (2012) Drug formats – diversity in existing oncology pipelines,, Darmstadt.
46 Noaiseh, G. and Moreland, L. (2013) Current and future biosimilars, potential practical application in rheumatology. *Biosimilars*, **3**, 27–33.
47 Helling, C., Borrmann, C., and Strube, J. (2012) Optimal integration of directly combined hydrophobic interaction and ion exchange chromatography purification processes. *Chem. Eng. Technol.*, **35** (10), 1786–1796.
48 Helling, C. and Strube, J. (2012) Modeling and experimental model parameter determination with Quality by Design (QbD) for bioprocesses, in *Biopharmaceutical Production Technology*, 1st edn (ed. G. Subramanian), John Wiley & Sons, Ltd, Weinheim, pp.
49 Strube, J., Grote, F., Josch, J.P., and Ditz, R. (2011) Process development and design of downstream processes. *Chem. Ing. Tech.*, **83** (7), 1044–1065.
50 Hanke, A.T. and Ottens, M. (2014) Review – purifying biopharmaceuticals: knowledge-based chromatographic process development. *Trends Biotechnol.*, **32** (4), 210–220.
51 Nfor, B.K. *et al.* (2012) Multi-dimensional fractionation and characterization of crude protein mixtures: towards establishment of a database of protein

purification process development parameters. *Biotechnol. Bioeng.*, **109** (12), 3070–3063.

52 Tscheliessnig, A.L. *et al.* (2013) Review – host cell protein analysis in therapeutic protein bioprocessing – methods and applications. *Biotechnol. J.*, **8**, 655–670.

53 Westoby, M. *et al.* (2011) Effects of solution environment on mammalian cell fermentation broth properties – enhanced impurity removal and clarification performance. *Biotechnol. Bioeng.*, **108** (1), 50–58.

54 Simmermann, H. and Donelly, R.P. (2005) Defining your product profile and maintaining control over it part 1. *BioProcess Int.*, **3** (6), 32–40.

55 Champion, K. *et al.* (2005) Defining your product profile and maintaining control over it, part 2. *BioProcess Int.*, **3** (8), 52–57.

56 Boerner, R. and Clouse, K. (2005) Defining your product profile and maintaining control over it, part 3. *BioProcess Int.*, **3** (9), 50–56.

57 Brorson, K. and Philipps, J. (2005) Defining your product profile and maintaining control over it, part 4. *BioProcess Int.*, **3**, 50–54.

58 EMA (1999) ICH Topic Q 6 B, Note for guidance on specifications: test procedures and acceptance criteria for biotechnology/biological products (cmp/ich/365/96), Available at www.ema.eu.int (access December 07, 2015).

59 EMA (2012) Report on the expert workshop on setting specifications for biotech products, EMA/CHMP/BWP/30584/2012, March 01, Available at www.ema.eu.int (access December 07, 2015).

60 Cambridge Healthtech Institute (2016) Biotherapeutics Analytical Summit Conference, Bethesda, MD, March 14–18, Available at www.biotherapeuticsAnalyticsSummit.com (access December 07, 2015).

61 Siemens Healthcare (2015) Automated ELISA processing BEP 2000 Advance System, Available at http://www.healthcare.siemens.com/infectious-disease-testing/systems/bep-2000-advance-system/features-benefits (access December 07, 2015).

62 Hughes, C.J., Vissers, J.P.C., and Langridge, J.I. (2012) Performance of ACQUITY UPLC M-Class in Proteomics Nanoscale Applications, Available at http://www.waters.com/webassets/cms/library/docs/720005244en.pdf (December 07, 2012).

63 Schulze-Wierling, P., Hubbuch, J. *et al.* (2007) High-throughput screening of packed-bed chromatography coupled with SELDI-TOF MS analysis: monoclonal antibodies versus host cell protein. *Biotechnol. Bioeng.*, **98** (2), 440–450.

64 Brestrich, N., Hubbuch, J. *et al.* (2014) A tool for selective inline quantification of co-eluting proteins in chromatography using spectral analysis and partial least squares regression. *Biotechnol. Bioeng.*, **111** (7), 1365–1373.

65 Gronemeyer, P., Ditz, R., and Strube, J. (2014) Trends in upstream and downstream process development for antibody manufacturing. *Bioengineering*, **1**, 188–212.

66 Ljunglöf, A. *et al.* (2011) Rapid development for purification of a Mab. *BioProcess Int.*, **June**, 62–68.

67 Meitz, A. *et al.* (2014) An integrated downstream process development strategy along QbD principles. *Bioengineering*, **1**, 213–230.

68 Helling, C., Dams, T., Gerwat, B., Belousov, A., and Strube, J. (2013) Physical characterization of column chromatography: stringent control over equipment performance in biopharmaceutical production. *Trends Chromatogr.*, **8**, 55–71.

69 Wang, X., Hunter, A.K., and Mozier, N.M. (2009) Host cell proteins in biologics development: identification, quantitation and risk assessment. *Biotechnol. Bioeng.*, **103**, 446–458.

70 Grzeskowiak, J.K., Tscheliessnig, A., Wu, M.W., Toh, P.C., Chusainow, J., Lee, Y.Y., Wong, N., and Jungbauer, A. (2010) Two-dimensional difference fluorescence gel electrophoresis to verify the scale-up of a non-affinity-based downstream process for isolation of a therapeutic recombinant antibody. *Electrophoresis*, **31**, 1862–1872.

71 Jungbauer, A. and Walch, N. (2015) Buffer recycling in downstream processing of biologics. *Curr. Opin. Chem. Eng.*, **10**, 1–7.

72 Grote, F., Ditz, R., and Strube, J. (2012) Downstream of downstream processing: development of recycling strategies for biopharmaceutical processes. *J. Chem. Technol. Biotechnol.*, **87** (4), 481–497.

73 Sun, H. (2015) The impact of upstream supply and downstream demand integration on quality management and quality performance, Available at http://www.emeraldinsight.com/0265-671X.htm (access December 12, 2015).

74 Hustedt, H. *et al.* (1985) Protein recovery using two-phase systems. *Trends Biotechnol.*, **3** (6), 139–144.

75 Rogers, R.D. *et al.* (1995) *Aqueous Biphasic Separations: Biomolecules to Metal Ions*, Springer US, Boston, pp. 191.

76 Zheng, Y. *et al.* (2015) Mechanism of gold (III) extraction using a novel ionic liquid-based aqueous two phase system without additional extractants. *Sep. Purif. Technol.*, **154**, 123–127.

77 Magamichi, K., Rajni, H.K., and Bo, M. (2000) Aqueous two-phase systems: methods and protocols. *Methods Technol.*, **11**, 371–379.

78 Eggersglüß, J.K., Both, S., and Strube, J. (2012) Process development for the extraction of biomolecules application for downstream processing of proteins in aqueous two-phase systems. *Chim. Oggi*, **30** (4), 32–36.

79 Chen, X. *et al.* (2013) Extraction of Tryptophan enantiomers by aqueous two-phase systems of ethanol and $(NH_4)_2SO_4$. *J. Chem. Technol. Biotechnol.*, **88** (8), 1545–1550.

80 Montalvo-Hernández, B. *et al.* (2012) A recovery of crocins from saffron stigmas (*Crocus sativus*) in aqueous two-phase systems. *J. Chromatogr.*, **1236**, 7–15.

81 Mutalib, A. *et al.* (2014) Characterisation of new aqueous two-phase systems comprising of Dehypon®LS54 and K4484®Dextrin for potential cutinase recovery. *Sep. Purif. Technol.*, **123**, 183–189.

82 Merchuk, J.C. *et al.* (1998) Biomedical sciences and applications, aqueous two-phase systems for protein separation. *J. Chromatogr. B*, **711** (1–2), 285–293.

83 Tsukamoto, M. *et al.* (2009) Cell separation by an aqueous two-phase system in a microfluidic device. *Analyst*, **134** (10), 1994–1998.

84 Veide, A. *et al.* (1983) A process for large-scale isolation of beta-galactosidase from *E. coli* in an aqueous two-phase system. *Biotechnol. Bioeng.*, **25** (7), 1789–1800.

85 Wellsandt, T., Stanisch, B., Helling, C., Fröhlich, H., and Strube, J. (2015) Characterization method for separation devices based on micro technology. *Chem. Ing. Tech.*, **87** (1–2), 1–10.

86 Wellsandt, T. *et al.* (2015) Micro separation technology – part 1: LL-extraction. *Chem. Ing. Tech.*, **87** (9), 1198–1206.

87 Riedl, W. *et al.* (2008) Membrane-supported extraction of biomolecules with aqueous two-phase systems. *Desalination*, **224** (1–3), 160–167.

88 Wellsandt, T. (2016) Miniaturization of LL-extraction devices, Dissertation, Clausthal University of Technology.

89 Shukla, A.A. and Thömmes, J. (2010) Recent advances in production of monoclonal antibodies and related proteins. *Trends Biotechnol.*, **28** (5), 253–261.

90 Jain, E. and Kumar, A. (2008) Upstream processes in antibody production: evaluation of critical parameters. *Biotechnol. Adv.*, **26**, 46–72.

91 Li, F., Vijayasankaran, N., Shen, A., Kiss, R., and Amanullah, A. (2010) Cell culture processes for monoclonal antibody production. *mAbs*, **2**, 466–479.

92 Kelley, B. (2007) Very large scale monoclonal antibody purification: the case for conventional unit operations. *Biotechnol. Prog.*, **23**, 995–1008.

93 Kelley, B. (2009) Industrialization of mAb production technology: the bioprocessing industry at a crossroads. *mAbs*, **1**, 443–452.

94 Chon, J.H. and Zarbis-Papastoitsis, G. (2011) Advances in the production and downstream processing of antibodies. *New Biotechnol.*, **28**, 458–463.

95 Butler, M. and Meneses-Acosta, A. (2012) Recent advances in technology supporting biopharmaceutical production from mammalian cells. *Appl. Microbiol. Biotechnol.*, **96**, 885–894.

96 Tait, A.S., Hogwood, C.E.M., Smales, C.M., and Bracewell, D.G. (2012) Host cell protein dynamics in the supernatant of a mAb producing CHO cell line. *Biotechnol. Bioeng.*, **109** (4), 971–982.

97 Hogwood, C.E.M., Tait, A.S., Koloteva-Levine, N., Bracewell, D.G., and Smales, C.M. (2013) The dynamic of the CHO cell protein profile during clarification and protein A capture in a platform antibody purification process. *Biotechnol. Bioeng.*, **110** (1), 240–251.

98 Jin, M., Szapiel, N., Zhang, J., Hickey, J., and Ghose, S. (2010) Profiling of host cell proteins by two-dimensional difference gel electrophoresis (2D-DIGE): Implications for downstream process development. *Biotechnol. Bioeng.*, **105**, 206–316.

99 Liu, H.F., Ma, J., Winter, C., and Bayer, R. (2010) Recovery and purification process development for monoclonal antibody production. *mAbs*, **2**, 480–499.

100 Eggersglüß, J.K., Richter, M., Dieterle, M., and Strube, J. (2014) Multi-stage aqueous two-phase extraction for the purification of monoclonal antibodies. *Chem. Eng. Technol.*, **37** (4), 1–9.

101 Eggersglüß, J.K., Wellsandt, T., and Strube, J. (2014) Integration of aqueous two-phase extraction into downstream processing. *Chem. Eng. Technol.*, **37** (10), 1–12.

102 Silva, MartaF.F., Fernandes-Platzgummer, A., Aires-Barros, M., and Raquel; Azevedo, A.M. (2014) Integrated purification of monoclonal antibodies directly from cell culture medium with aqueous two-phase systems. *Sep. Purif. Technol.*, **132**, 330–335.

103 Rosa, P.A.J., Ferreira, I.F., Azevedo, A.M., and Aires-Barros, M.R. (2010) Aqueous two-phase systems: a viable platform in the manufacturing of biopharmaceuticals. *Extract. Tech.*, **1217**, 2296–2305.

104 Rosa, P.A.J., Azevedo, A.M., Sommerfeld, S., Mutter, M., Bäcker, W., and Aires-Barros, M.R. (2013) Continuous purification of antibodies from cell culture supernatant with aqueous two-phase systems: from concept to process. *Biotechnol. J.*, **8**, 352–362.

105 Hammerschmidt, N., Tscheliessnig, A., Sommer, R., Helk, B., and Jungbauer, A. (2014) Economics of recombinant antibody production processes at various scales: industry-standard compared to continuous precipitation. *Biotechnol. J.*, **9**, 766–775.

106 Sommer, R., Tscheliessnig, A., Satzer, P., Schulz, H., Helk, B., and Jungbauer, A. (2015) Capture and intermediate purification of recombinant antibodies with combined precipitation methods. *Biochem. Eng. J.*, **93**, 200–211.

107 Hammerschmidt, N., Hintersteiner, B., Lingg, N., and Jungbauer, A. (2015) Continuous precipitation of IgG from CHO cell culture supernatant in a tubular reactor. *Biotechnol. J.*, **10** (8), 1196–1205.

108 Fröhlich, H., Villian, L., Melzner, D., and Strube, J. (2012) Membrane technology in bioprocess science. *Chem. Ing. Tech.*, **84**, 905–917.

109 Chenette, H.C.S., Robinson, J.R., Hobley, E., and Husson, S.M. (2012) Development of high-productivity, strong cation-exchange adsorbers for protein capture by graft polymerization from membranes with different pore sizes. *J. Membr. Sci.*, **423–424**, 43–52.

110 Weaver, J., Husson, S.M., Murphy, L., Wickramasinghe, S., and Ranil, S. (2013) Anion exchange membrane adsorbers for flow-through polishing steps: part II. Virus, host cell protein, DNA clearance, and antibody recovery. *Biotechnol. Bioeng.*, **110**, 500–510.

111 Bhambure, R., Kumar, K., and Rathore, A.S. (2011) High-throughput process development for biopharmaceutical drug substances. *Trends Biotechnol.*, **29**, 127–135.

112 del Val, I.J., Kontoravdi, C., and Nagy, J.M. (2010) Towards the implementation of quality by design to the production of therapeutic monoclonal antibodies with desired glycosylation patterns. *Biotechnol. Prog.*, **26**, 1505–1527.

113 Azevedo, A.M., Rosa, P.A.J., Ferreira, I.F., and Aires-Barros, M.R. (2009) Chromatography-free recovery of biopharmaceuticals through aqueous two-phase processing. *Trends Biotechnol.*, **27**, 240–247.

114 Rosa, P.A.J., Azevedo, A.M., Sommerfeld, S., Bäcker, W., and Aires-Barros, M.R. (2012) Continuous aqueous two-phase extraction of human antibodies using a packed column. *J. Chromatogr. B*, **880**, 148–156.

115 Azevedo, A.M., Rosa, P.A.J., Ferreira, I.F., and Aires-Barros, M.R. (2009) Chromatography-free recovery of biopharmaceuticals through aqueous two-phase processing. *Trends Biotechnol.*, **27**, 240–247.

116 Giese, G., Myrold, A., Gorrell, J., and Persson, J. (2013) Purification of antibodies by precipitating impurities using polyethylene glycol to enable a two chromatography step process. *J. Chromatogr. B*, **938**, 14–21.

117 Oelmeier, S.A., Ladd-Effio, C., and Hubbuch, J. (2013) Alternative separation steps for monoclonal antibody purification: combination of centrifugal partitioning chromatography and precipitation. *J. Chromatogr. A*, **1319**, 118–126.

118 Kuczewski, M., Schirmer, E., Lain, B., and Zarbis-Papastoitsis, G. (2011) A single-use purification process for the production of a monoclonal antibody produced in a PER.C6 human cell line. *Biotechnol. J.*, **6**, 56–65.

119 Kuczewski, M., Schirmer, E., and Lain, B. (2014) PEG precipitation: a powerful tool for monoclonal antibody purification, Available at www.biopharminternational.com (accessed August 1, 2014).

120 Gagnon, P. (2012) Technology trends in antibody purification. *J. Chromatogr. A*, **1221**, 57–70.

121 Grodzki, A. and Berenstein, E. (2010) Antibody purification: ammonium sulfate fractionation or gel filtration, in *Immunocytochemical Methods and Protocols*, 1st edn (eds C. Oliver and M.C. Jamur), Humana Press, New York, pp. 15–26.

122 Ma, J., Hoang, H., Myint, T., Peram, T., Fahrner, R., and Chou, J. (2010) Using precipitation by polyamines as an alternative to chromatographic separation in antibody purification processes. *J. Chromatogr. B*, **878**, 798–806.

123 Gronemeyer, P. *et al.* (2016) Implementation of aqueous two-phase extraction combined with precipitation in a mAb manufacturing process. *Chim. Oggi*, **34**, 66–70.

124 Gronemeyer, P. *et al.* (2016) DoE based integration approach of upstream and downstream processing regarding HCP and ATPE as harvest operation. *Biochem. Eng*, **113**, 158–166.

125 Josch, J.P. and Strube, J. (2012) Characterization of feed properties for conceptual process design involving complex mixtures. *Chem. Ing. Tech.*, **84** (6), 918–931.

126 Jungbauer, A. and Kaltenbrunner, O. (1996) Fundamental questions in optimizing ion-exchange chromatography of proteins using computer-aided process design. *Biotechnol. Bioeng.*, **52**, 223–236.

127 Schwellenbach, J., Kosiol, P., Sölter, B., Taft, F., Villain, L., and Strube, J. (2016) Preparation and characterization of high capacity, strong cation-exchange fiber based adsorbents. *J. Chromatogr.*, **1447**, 92–106.

128 Helling, C., Dams, T., Gerwat, B., Belousov, A., and Strube, J. (2013) Physical characterization of column chromatography: stringent control over equipment performance in biopharmaceutical production. *Trends Chromatogr.*, **8**, 55–71.

129 Ndocko Ndocko, E., Ditz, R., Josch, J.P., and Strube, J. (2011) New material design strategy for chromatographic separation steps in bio-recovery and downstream processing. *Chem. Ing. Tech.*, **83** (1–2), 113–129.

130 Osberghaus, A., Drechsela, K., Hansen, S., Hepbildikler, S.K., Nath, S., Haindl, M., von Lieres, E., and Hubbuch, J. (2012) Model-integrated process development demonstrated on the optimization of a robotic cation exchange step. *Chem. Eng. Sci.*, **76** (9), 129–139.

131 Salvalaglio, M., Paloni, M., Guelat, B., Morbidelli, M., and Cavallotti, C. (2015) A two level hierarchical model of protein retention in ion exchange chromatography. *J. Chromatogr. A.*, **1411**, 50–62.

132 Qin, W., Silvestre, M., and Franzreb, M. (2014) Magnetic microparticles@UiO-67 core–shell composites as a novel stationary phase for high performance liquid chromatography. *Appl. Mech. Mater.*, **703**, 73–76.

133 McGlaughin, M.S. (2012) An emerging answer to the downstream bottleneck. *BioProcess Int.*, **10** (5), 58–61.

134 Yavorsky, D. (2012) MERCK-MILLIPORE, Bedford, United States, New Disposable Technology for Chromatographic Purification of Biopharmaceuticals, SPICA, September 30–October 3, Bruessels.

135 Schwellenbach, J., Taft, F., Villain, L., and Strube, J. (2016) Preparation and characterization of high capacity, strong cation-exchange fiber based adsorbents. *J. Chromatogr.*, **1447**, 92–106.

136 Li, C., Ladisch, C.M., Yang, Y., Hendrickson, R., Keim, C., Mosier, N., and Ladisch, M.R. (2002) Optimal packing characteristics of rolled, continuous stationary-phase columns. *Biotechnol. Prog.*, **18**, 309–316.

137 Oksanen, H.M., Domanska, M., and Bamford, D.H. (2012) Monolithic ion exchange chromatographic methods for virus purification. *Virology*, **434**, 271–277.

138 Mullick, A. and Flickinger, M.C. (1999) Expanded bed adsorption of human serum albumin from very dense *Saccharomyces cerevesiae* suspensions on fluoride-modified zirconia. *Biotechnol. Bioeng.*, **65** (3), 282–290.

139 Huter, M. (2016) Biologics Precipitation. Master thesis, ITVP at Clausthal University of Technology.

140 Wagner, R. (2012) 2. Laupheimer Zelltage, Rentschler Biotechnologie GmbH, June 11–12, Laupheim.

141 Geipel-Kern, A. (2009) Einweglösungen als Zukunftstrend in der Arzneimittelherstellung: Flexible Abfüllsysteme, November 28, Available at www.process.vogel.de (access December 07, 2015).

142 Stephan, L. *et al.* (2015) Developing the biofacility of the future based on continuous processing and singel-use technology. *J. Biotechnol.*, **213**, 120–130.

143 Sharmista, C. (2012) Advantages of CBP, IFPAC Annual Meeting, Baltimore, January.

144 Sharmista, C. (2012) Design space considerations, AAPS Annual Meeting, October 11, Chicago.

145 Palmer, E. (2014) Amgen opens $200M continuous purification plant in Singapore November 20, Available at http://www.fiercepharmamanufacturing.com/story/amgen-opens-200m-continuous-purification-plant-singapore/2014-11-20 (access January 22, 2016).

146 Palmer, E. (2013) Sanofi's Genzyme looking hard at continuous manufacturing, January 31, Available at http://www.fiercepharmamanufacturing.com/story/sanofis-genzyme-looking-hard-continuous-manufacturing/2013-01-31 (access January 22, 2016).

147 Zydney, A.L. (2016) Continuous downstream processing for high value biological products: A Review. *Biotechnol. Bioeng.*, **113, 3**, 465–475.

Part Two

Automation and Monitoring (PAT)

4

Progress Toward Automated Single-Use Continuous Monoclonal Antibody Manufacturing via the Protein Refinery Operations Lab

David Pollard, Mark Brower, and Douglas Richardson

Merck & Co Inc, Biologics & Vaccines, 2000 Galloping Hill Road, Kenilworth, NJ 07033, USA

4.1 Introduction

Therapeutic proteins and vaccines continue to gain momentum in the market place where biologically derived molecules own a predominant position among the top 10 selling drugs. Biotherapeutics remain the largest and fastest growing class of novel pharmaceuticals. In this category, the FDA approved 45 new drugs in 2015, up from 28 in 2012 [1] and over 400 biomolecules currently in clinical development [2]. This success has been dominated by full-length humanized monoclonal antibodies (mAbs), yet expansion into new modalities is accelerating, including antibody drug conjugates, multispecific antibodies, glycan engineered proteins, novel fragments and scaffolds.

Despite this growing success, the biopharmaceutical industry continues to face strict competitive challenges from multiple pressures, including economic, regulatory, and political, as outlined in Table 4.1. As a result, there is a significant drive to boost the overall productivity of biotherapeutic programs by shortening development timelines and lowering both development and production costs, while maintaining product quality [3]. Because of the large number of full-length antibodies in development, and their relatively high dosing requirements, low-cost mAb production solutions must be aggressively pursued. Supply chains need to be streamlined in order to respond to the needs of patients in a more personalized and agile approach. This synchronization of manufacturing should shorten the time between the start of manufacturing and providing drug to the patient. The drug on demand concept, supported by continuous processing, should provide a responsive approach to changing drug demand and minimize inventory costs. These cost reduction initiatives, in combination with regional manufacturing, should help to expand patient accessibility to biologics and vaccines. In addition, the expansion from traditional mAbs to a portfolio of heterogeneous multimodality products will require flexible low-cost manufacturing capacity.

These issues are pressuring the industry away from the traditional fixed 10–20 kl stainless steel manufacturing facilities, as built in the last two decades, toward more flexible modular approaches (Figure 4.1). These new facilities of the future can easily accommodate both large and small drug demands from the

Continuous Biomanufacturing: Innovative Technologies and Methods, First Edition.
Edited by Ganapathy Subramanian.
© 2018 Wiley-VCH Verlag GmbH & Co. KGaA. Published 2018 by Wiley-VCH Verlag GmbH & Co. KGaA.

Table 4.1 Challenges impacting the biotech and biopharma industry.

Pressures	Potential requirements
• Increasing pipeline of heterogeneous Products • Greater range of product demands with increased uncertainty • Cost pressures 　– Evolving reimbursement environment 　– Global competition 　– Lost drug exclusivity • Need for global presence	First in class pipeline Agility and flexibility to adapt to change Proactively prepare for upcoming needs Develop technologies for new business growth and drive down costs Coordinate technology to drive R&D continuum to manufacturing World class clinical and synchronized manufacturing supply to the patient

diverse modality platforms, each of which may require specialized technologies in the production process. These lower cost, single-use-enabled modular facilities are based upon closed processing principles that enable reduced air handling requirements and multiple production facility processing (Figure 4.1). The elimination of using steam and clean in place reduces the facility footprint and provides a lower capital cost facility.

Net present cost estimates, that include capital, depreciation, and overhead, show multi $100 million savings to be made from implementing single-use modular facility processing [4]. It is envisaged that the facility of the future can accommodate both single-use fed-batch processing, such as for a highly potent low demand mAb (50 kg/yr), and an automated continuous process for a high-demand product (>300 kg/yr). The continuous processing of mAbs by the integration of perfusion to multicolumn chromatography and remaining purification steps has been previously described [4]. The combination of efficient multicolumn countercurrent chromatography [4,5], elimination of hold steps,

Current state

mAb focus, rigid batch size
Stainless steel fed batch (4-6 g/L)
High cost to reconfigure facilities
Separate DS & DP
Long lead time supply chain

Commercial media
Fed batch (4-6g/L)
Conventional QC release

Fixed costs focus

Future state

Range of drug modalities & batch sizes
Toolbox of robust high titer integrated single-use platforms
Flexible, low cost, agile capacity
Responsive short lead time supply chain

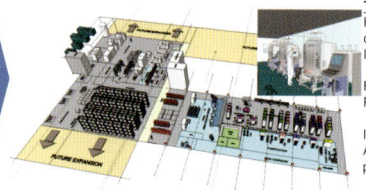

Toolbox
High producer cell line
In house media

Fed batch > 10g/L
Perfusion > 3g/Lday

Integrated processing
Automated continuous processing

Real time release
Integrated DS & DP

Variable costs focus

Figure 4.1 CHO mAb manufacturing improvement. Transitioning from stainless steel to modular single-use facility of the future with flexible synchronized manufacturing supply. This marks a transition from fixed cost to a variable cost focus.

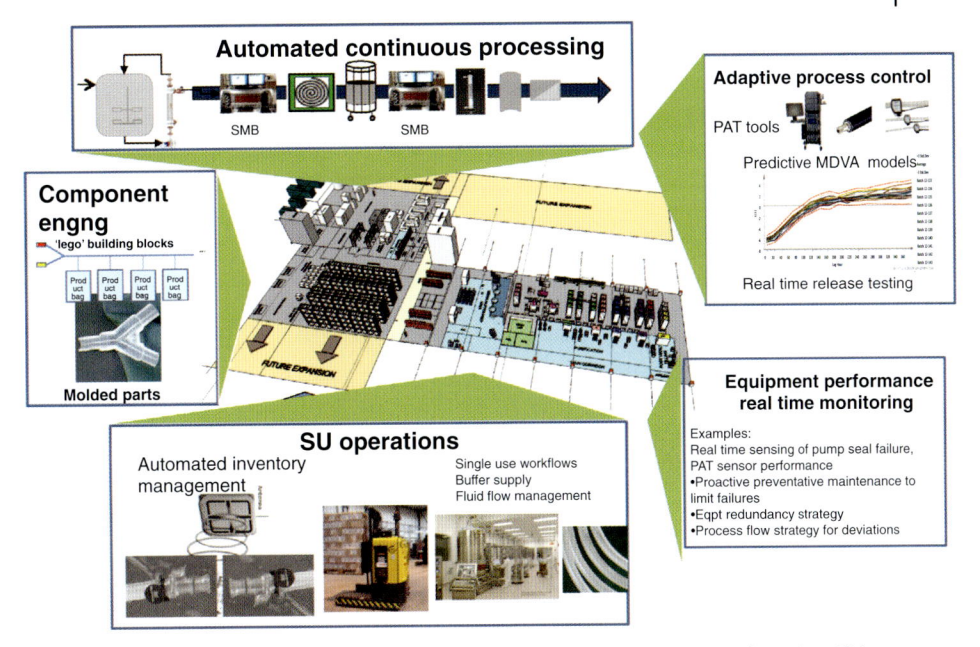

Figure 4.2 Vision of the modular facility of the future for bioprocessing manufacturing. This incorporates the automated continuous processing in an open ballroom facility combined with adaptive process control, single-use operations including disposable component engineering.

and high-throughput purification membranes [6] can lead to significant efficiency and cost improvements (3×) [4]. It is envisaged that automated continuous processing can be combined with process analytical monitoring and product attribute control to enable real-time release from the bioprocessing manufacturing floor (Figure 4.2). This should reduce the QC and QA burden and shorten the total manufacturing duration. Ultimately this "drug on demand" approach should reduce inventory costs and provide a more synchronized approach that will shorten the time from the start of manufacturing to the patient. This will enable the manufacturing supply to more easily adapt to changing demand from a multiple modality product portfolio. This chapter will describe the progress made to develop this fully automated protein refinery facility.

4.2 Protein Refinery Operations Lab

4.2.1 Introduction

Typical batch purification for monoclonal antibodies is comprised of processing the entire bioreactor volume of material through discrete unit operations in succession with large pool tanks between them to hold intermediate volumes. Initially the crude cell culture broth is processed through clarification (centrifugation) and/or depth filtration to remove cells and reduce particulates followed by low bioburden filters to clarify the stream from any remaining particulates (Figure 4.3). The clarified stream is further processed through an affinity capture chromatography

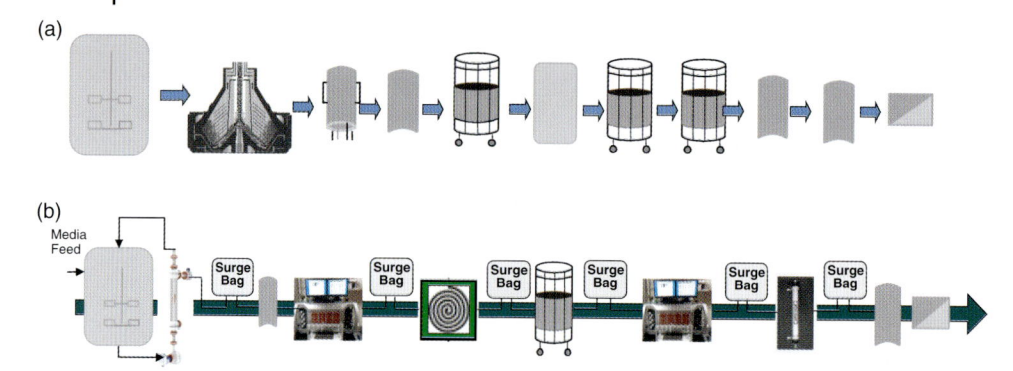

Figure 4.3 mAb production: Comparison of traditional batch processing versus integrated continuous processing. (a) Batch processing incorporating discrete batch operations moving from one unit operation to the next with product hold steps in between. (b) Continuous processing is fully integrated single-use operations with small surge bags between unit operations for priming and gradient control between unit operations.

step (protein A) for bulk purification and the effluent treated to a low pH viral inactivation step for a specified duration prior to loading on an anion exchange membrane or resin. The subsequent polished product is passed through nanofiltration for virus removal followed by concentration and buffer exchange by ultrafiltration. In contrast, continuous processing is the complete integration of upstream such as perfusion, through multicolumn chromatography (such as simulated moving bed chromatography (BioSMB)), viral inactivation via pH retention loop, and subsequent membrane purification (Figure 4.3) without stopping for significant product pooling at any location. The continuous processing boundary can be further expanded to the remaining unit operations with single pass tangential flow filtration and low bioburden filtration. The ultrafiltration product is then diafiltered into formulation buffer into discrete lots that can ultimately define batch definition criteria. Previous proof of concept work using manual control incorporated single-use operations and disposable fluid path for closed processing [4]. This demonstrated the feasibility to reduce the processing time in plant for clinical production from 3 to 5 days for batch processing, down to less than 1 day for the continuous option. This continuous process is assumed to operate at pseudo steady state for the flows into and out of each unit operation. The properties of each stream, such as pH, conductivity, concentration, are not constant and gradients in these properties are expected. Small volume surge bags between each continuous unit operation act to dampen the gradients and are used as sampling and adjustment points for the process streams (Figure 4.3).

4.2.2 Protein Refinery Operations Lab: Design and Implementation

The completion of the experimental proof of concept, alongside the significant cost reductions realized in the facility of the future process economic modeling, provided the business case to create a dedicated lab facility for continuous bioprocessing: Protein Refinery Operations Lab (PRO lab). The elements of this automated testing lab are defined in Figure 4.4 that include process

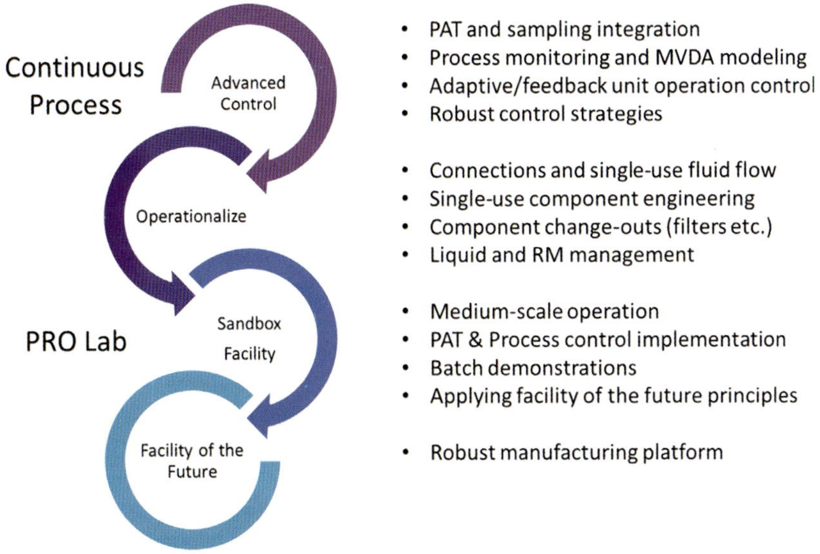

Figure 4.4 Protein Refinery Operations Lab (PRO lab). Sandbox evaluation lab to develop automated continuous processing for clinical and commercial facilities of the future.

monitoring and multivariate data analysis (MVDA) integrated with process analytical tools (PAT) and adaptive control strategies. The PRO lab provides an environment for the operationalizing of the procedures and workflows of continuous operations that would ultimately be transferred to clinical and commercial facilities of the future, as well as enabling evaluation of robustness for the single-use skids and manifold components. Control procedures will be implemented that allow for automated handling of process deviations. This sandbox capability will allow execution of planned process deviations and shows that process automation and control can rectify the majority of potential processing issues in a "lights out" approach. For example, a pH deviation after protein A capture would lead to increased aggregates. In this instance, the flow of high-aggregate material is either diverted out of the process to a waste vessel or the aggregates are cleared by automatic method adjustment at the subsequent polishing membrane step. The facility must accommodate line of sight to commercial manufacturing that will be scaled up to around 500–2000 l scale.

The design of the lab incorporates a perfusion bioreactor (10–50 l scale) integrated to stream clarification, BioSMB protein A chromatography, to viral pH inactivation, to polishing purification, via membrane or resin, followed by nano- and ultrafiltration. This equipment is arranged in a "horseshoe"-type approach (Figure 4.5) with the at line PAT tools centered in the middle of the lab. Buffer and media prep space and waste accumulation is provided outside of the process area. The total lab area is 460 ft^2 (43 m^2) and ultimately a total of 10–100 g mAb/day could be processed via the 10–50 l upstream capacity. The continued productivity improvement of high producing cell lines enables commercial perfusion to be at the 200–500 l scale with the purification using similar sized hardware already installed in the PRO lab. Therefore, it is anticipated that

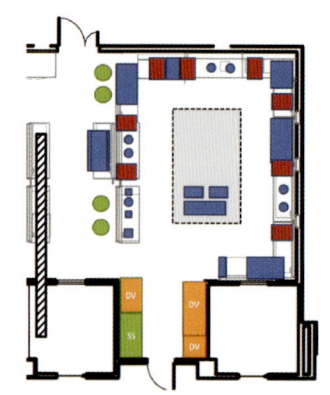

Figure 4.5 The Protein Refinery Lab (PRO lab) designed in a "horseshoe" layout from perfusion to ultrafiltration. Aseptic sampling from different unit operations brings samples to the at line analytics centered in the middle of the room. Key: blue: process equipment, red: surge bag, orange: Delta V control, green: buffers and single use supplies, gray: PAT.

scale-up from this lab will not be a significant hurdle. For example, the BioSMB hardware will be a similar scale but the prepacked columns will increase diameter from 1.1 cm in the PRO lab to 8 cm in the Facility of the Future. Currently the surge bags between the unit operations are 250 ml gamma-irradiated vessels equipped with single-use pH and conductivity probes. For commercial scale, it is anticipated to utilize currently commercially available 50-l SU mixer vessels. The increase of the surge vessels from 250 ml benchtop units to 50 l standalone mixer vessels represents the largest expansion needed in the process area to accommodate the larger daily processing volumes.

The automation strategy for the continuous process from the lab to clinical and commercial manufacturing scales is for the distributed control system (DeltaV) to provide supervisory control to each individual unit operation/skid. Individual skids will call upon preloaded methods and utilize local control through existing supplier software. This equipment communicates to the DeltaV controller via an OPC, Profibus, RS-232, or similar interface. The automation of the lab took around 12 months to design, build, and implement. This included writing the base control strategies and also the communication drivers to connect the overarching supervisory control to the bioprocessing skids such as perfusion auto TFF (Spectrum Labs) and multicolumn chromatography (BioSMB, Pall). Automated actuator systems were also defined and implemented to control fluid flow management between skids and for unit operations that are operated without a vendor supplied skid such as depth filtration. A subsequent 3 months was needed for debugging of automation procedures. The design of the automation

4.2 Protein Refinery Operations Lab

Table 4.2 Design and operating principles of the PRO Lab.

Each unit operation of the process is represented	Surge vessels to collect and feed next unit operation
Each unit operation operates/controlled by its original supplier software	All unit operations integrated via an overarching supervisory control automation (Delta V) connected to the unit operations via OPC
Redundant filters implemented with automated switching	The automation procedures code must be directly transferable to a clinical or commercial FoF
SU component engineering based upon Merck catalog of common Lego building blocks with closed processing methodologies	Automation incorporates speed compensated methods with break points
Each unit operation and surge bag amenable to accept PAT monitoring and control enabled and aseptic sampling enabled	Automated start-up and shutdown sequences are required.

protocols and hardware were coordinated as an interdivisional initiative between Merck Research Labs and Merck Manufacturing Division in order to have line of sight to commercial manufacturing. This allows the direct transfer of the supervisory control protocols from the lab to the clinical or commercial manufacturing Facility of the Future. This approach was part of the guiding design and operating principles used for developing the automation protocols (Table 4.2).

A set of process dashboard layouts was created to support user interaction with the automation system. These included an overarching supervisory layout of the complete integrated process from perfusion through chromatography to ultrafiltration highlighting key process parameters specific to each unit operation. An example is shown in Figure 4.6a that shows the end-to-end "horseshoe" setup of the processing suite providing an instant summary of the progress status of the process and the key outputs from each unit operation as well as alarm status. From this layout, an automated startup sequence can be selected wherein the unit operations are cascaded on starting from the inoculation of the bioreactor. For example, when the upstream cell concentration reaches its desired target, 50 million cells/mL as measured by online capacitance, then the permeate feed is initiated and directed to the protein A capture system that then triggers the cascade of startup to the next unit operation. Similarly, an automated shutdown sequence was also created to shutdown each unit operation in a cascaded approach when filtrate stops flowing out of the perfusion membrane.

An expanded layout of each unit operation has also been created to see all of the relevant process parameters and equipment modules associated with that step. An example of the detailed unit operation layout is shown in Figure 4.6b for depth filtration. The strategy employs dual paths for redundant filtration trains. For each filter set, the single-use pressure sensors monitor the backpressure indicating fouling. The flow can be switched over to the new depth filter train automatically based on Boolean logic on either a high delta pressure reading from the sensors or at a specific volumetric filter loading. Automatic alerts are then sent to end users to

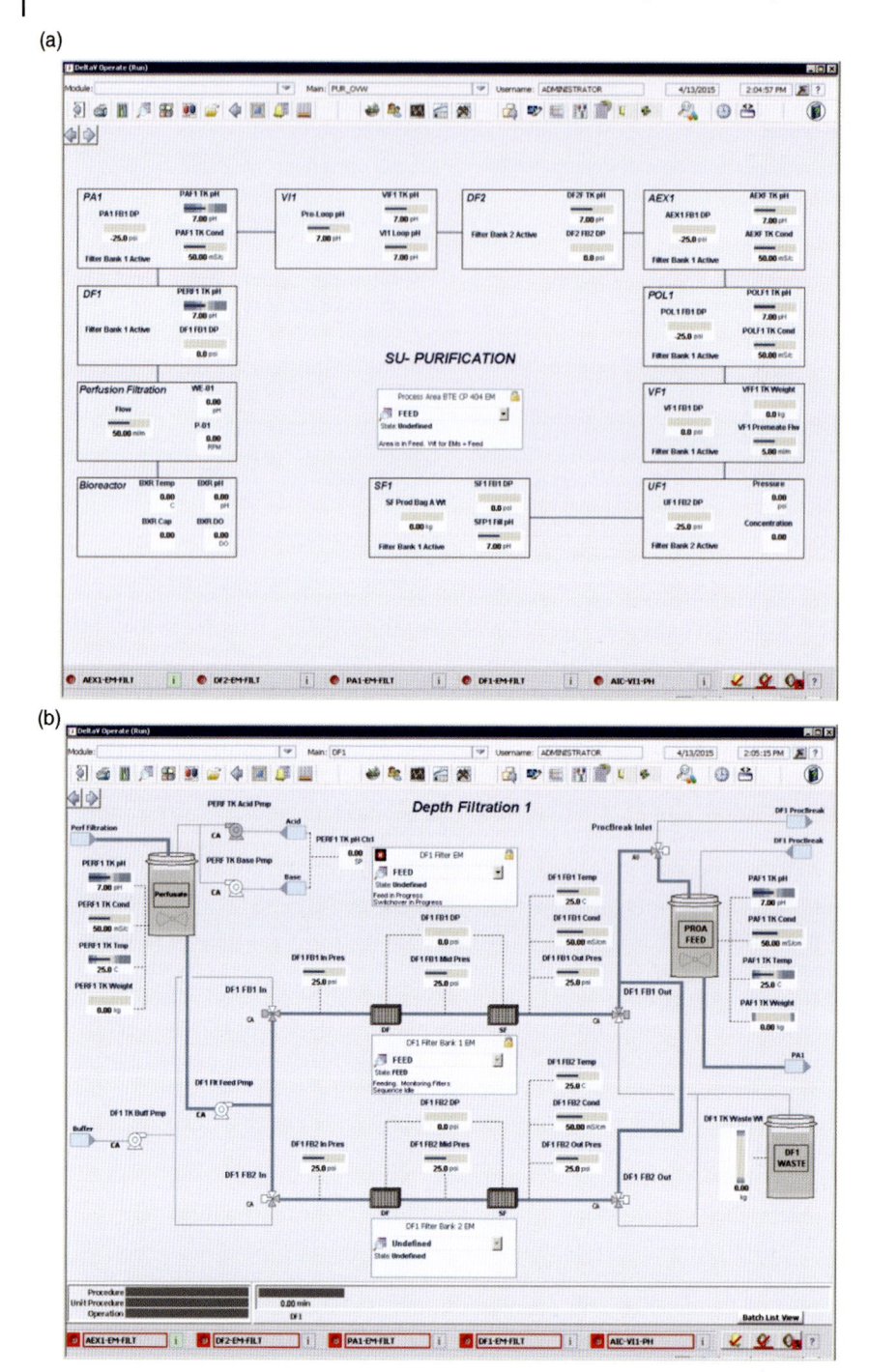

Figure 4.6 Examples of the automation (Delta V) user control layouts: overarching process control (a) and the depth filtration (b).

replace the fouled filter set. This approach has been implemented for all dead-end filtration needs of the process such as depth filtration, guard filters, and bioburden reduction filtration.

The overarching control system has speed compensation methodology that gives flexibility to the process stream for accumulation in the surge bags. For example, if accumulation occurs at a particular surge bag stage then the flow downstream of that surge vessel can be increased to reduce the volume. Also, if the system has a short interruption in a unit operation, the flow is automatically held at the previous surge bag, upstream of the interrupted unit operation, until the issue is rectified. If the disruption is lengthier, material can be sent out of the process to a process break (waste) bag. The ability to adjust the variable flow control allows the overall total productivity of the process to be varied. It is anticipated that this capability will be used in a commercial process to provide a quick response to changing product demands needs. The supervisory automation has the ability to swap or bypass unit operations to allow flexibility in a lab setting to try different approaches and new experimental process flows. These features can be locked out in a GMP environment when flexibility is less desirable.

4.2.3 Protein Refinery Operations Lab: Process Analytical Technology (PAT) and Product Attribute Control (PAC) for the Transition to Real-Time Release (RTR)

Process analytical technology is being developed to enable product attribute control during the continuous manufacturing of biologics. Combining these tools with parametric modeling using both uni- and multivariant analysis will provide more robust and reproducible bioprocessing that incorporates the key issues of variability such as lot-to-lot quality differences in raw materials (Figure 4.7). In the PRO lab, we have begun to integrate the process analytical tools to use for feed-forward and feedback control. Examples for perfusion include in-line capacitance for controlling of biomass levels and in-line Raman for controlling glucose concentrations. New

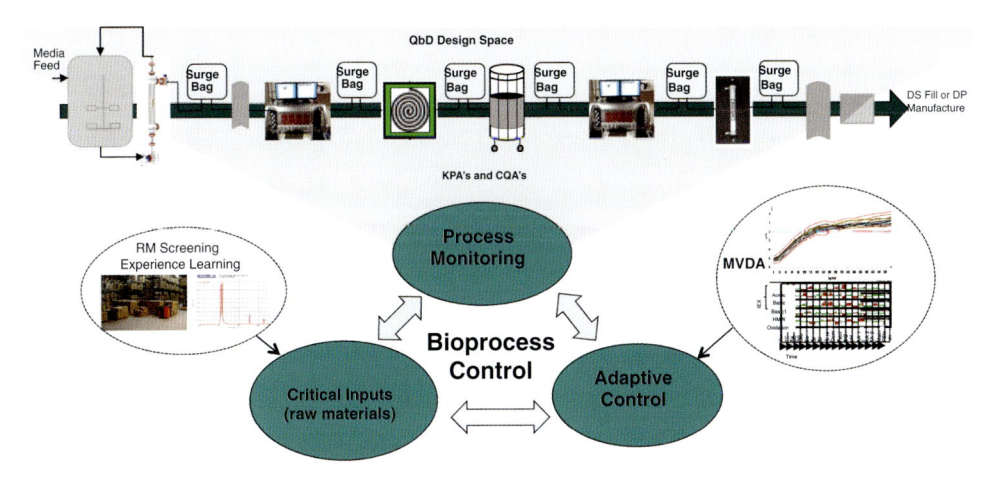

- End product testing transition to real-time release testing
- Real-time automated control: process responds to variability and disturbances
  - End-to-end prediction models of the complete process
  - RM control ➡ Process input ➡ Product quality and yield

Figure 4.7 Real-time release via PAT and parametric models with multivariant analysis.

approaches from suppliers are now making it feasible to use the multivariate analysis tools for feed-forward and feedback control. These approaches using SIMCA batch online (Umetrics), SIPAT (Siemens), and Pi (OSisoft) for data historian integration into DeltaV have been implemented into the PRO lab. This transition of analytical testing and process control into the processing area should reduce the burden for both QC and QA resources in the manufacturing environment. Recent implementation of handheld RAMAN spectroscopy in large molecule clinical manufacturing has allowed for release of raw materials in 2 days compared to up to 30 days with conventional testing. In addition, combining the raw material release with drug substance, and drug product rapid online and at line testing should provide a path for real time release. Januvia, Merck's blockbuster small molecule diabetes medication, has been released in real time via NIR spectroscopy for more than 10 years with no tablet ever tested in a QC lab. Building from this success, the plan is to leverage the regulatory experience to build the capabilities for the more complex large molecule real-time release.

4.2.3.1 Protein Refinery Operations Lab: Current State of PAT Technologies

To develop the real time release capabilities for biologics, the analytical toolkit needs to enable real-time data via online or at line methods to cover the main characterization areas of primary structure (such as glycosylation, charge heterogeneity, sequence), biological potency, impurities (product and process related), and quaternary structure (aggregation, sub visible particulates). These tools are being developed and implemented into the PRO lab with a goal of aligning process and product data (Figure 4.8). A status of the current technology is outlined in Figure 4.9. For the upstream process, the spectroscopy approaches for key analytes such as glucose and lactate are feasible via spectroscopy methods such as Raman. The demonstration at Merck has shown feasibility to measuring glucose down to 0.5 g/l sensitivity for perfusion runs of >30 days. Capacitance has been implemented for an online measure of viability and cell density. This has been implemented during perfusion to enable controlled cell bleeding to maintain the cell concentration; the case study in Section 4.3.1 provides more detail. Direct critical attribute monitoring in both upstream and downstream has advanced with the development of new online UPLC tools with UV and mass spectrometric detection (Waters Patrol). For purification, online UPLC has shown to provide useful multiattribute approaches for monitoring of titer, aggregates, and excipients (Figure 4.10). The profiling of the upstream perfusion process allows both online titer and a profiling of the media and host cell protein profiles with UPSEC analysis on permeate over a 20-day period. Similar demonstrations include titer and aggregate control during UF/DF for high-concentration purification and real-time peak cutting during AEX purification. Analysis of process impurities such as residual DNA, HCP are currently feasible via at line approach to rapid ELISA methods [7]. Significant online technology gaps remain to be addressed that include robust aseptic sampling, online osmolality, potency, particulates, color, and bioburden.

Aseptic sampling is a key enabler for PAT by supporting the sample connection between the process and analyzers without introducing new contaminants. The PRO lab is evaluating the BEND Mast aseptic sampling system with cell removal system and automated sample management. Autoclavable sample pilots with sanitized

4.2 Protein Refinery Operations Lab | **119**

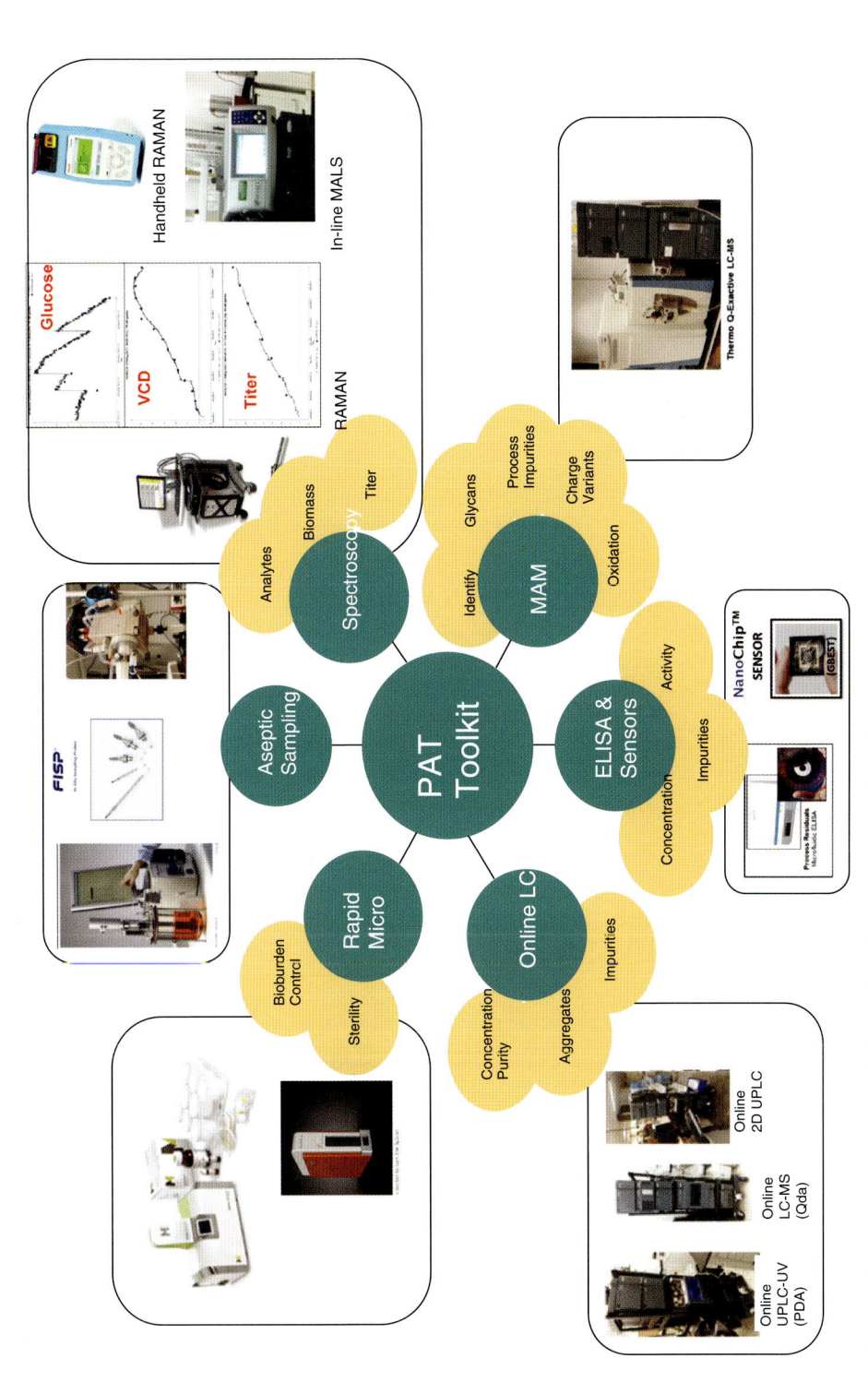

Figure 4.8 Expanding biologics PAT toolkit under evaluation in the PRO lab.

CPP's & CQA's

Parameter	PAT
Glucose/Lactate	Raman
Biomass/Viability	Capacitance
Titer	Waters UPLC
Osmolarity	At line
Quality(glycosylation)	Peptide Mapping/MS
Sterility	Off line

Parameter	PAT	
Concentration	Inline UV280 or Waters UPLC	
Purity	IEX: Waters Patrol, SEC: Waters Patrol	Peptide Mapping / MS
Quality Endotoxin	At line	
Quality: n glycans	At line	
Impurities	At line (Gyro lab ELISA's)	
Quality Bioburden	Offline Microbial Monitoring	
Conductivity	Exising Commerical Sensors	
pH	Solid State pH (Senova)	
Flow	Single Use Ultrasonic Flow (Transonics)	

Figure 4.9 Summary of Process Analytical Tools to support key bioprocess parameters for upstream and downstream continuous processing. Key: green refers to technology already commercially available and implementation demonstrated, yellow: potential technology in development with proof of concept feasibility or at line alternative, red: technology gap that lacks technology solution.

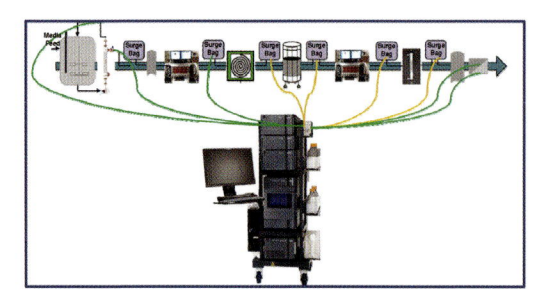

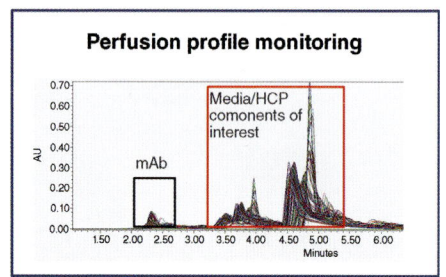

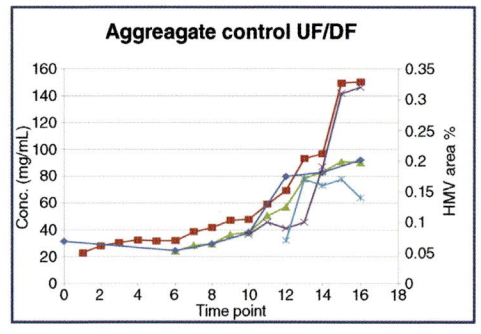

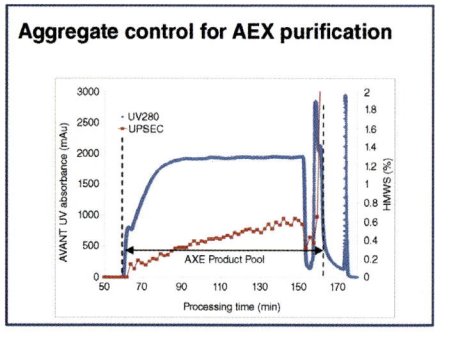

Figure 4.10 Examples of UPSEC multiattribute process monitoring across a continuous processing from perfusion and the unit operation surge bags. Examples include online perfusion monitoring, aggregate control during ultrafiltration, and chromatography profiling during AEX purification using a 5 min UPSEC method.

sampling points are located at the bioreactor and at the multiple surge bag stations of the purification. Aseptic cell-containing or cell-free samples are pneumatically transported to the sample tubes on the liquid handling sample hub. Initial evaluation has shown feasibility with bioreactor sterility remaining for >40 days allowing automatic sampling during steady-state production. Further improvements remain including sampling volume, analyzer integration, and robustness of cell removal. A disposable solution with reduced fouling would be a welcome technology advancement. A number of technology approaches are being developed for rapid microbiology testing that include BacT/Alert, Milliflex Quantum, and MuScan (Figure 4.8). In the PRO lab, the control of bioburden needs to be better understood for continuous bioprocessing. Typical continuous processing batches could be up to 100 days in duration, so assurance of bioburden is an important issue with the closed single-use systems approach. Samples are routinely aseptically removed and evaluated for viable microbial content with MuScan (Innosieve Diagonostics). This provides rapid detection of microorganisms in less than 60 min via concentration, staining, and patented fluorescence scanning methodology. Currently this seems to be the only commercially available technology that provides a 1 h turnaround time to support bioburden testing for continuous manufacturing. Figure 4.11 shows that different concentrations of microorganisms can be determined by the MuScan approach. This technology is being used to characterize the bioburden during the development work for continuous processing.

The improvement of sensitivity and data analysis is driving an expansion of the applications for LC-MS into the characterization of biological process streams and QC environments. Recently, the combination of peptide mapping with LC-

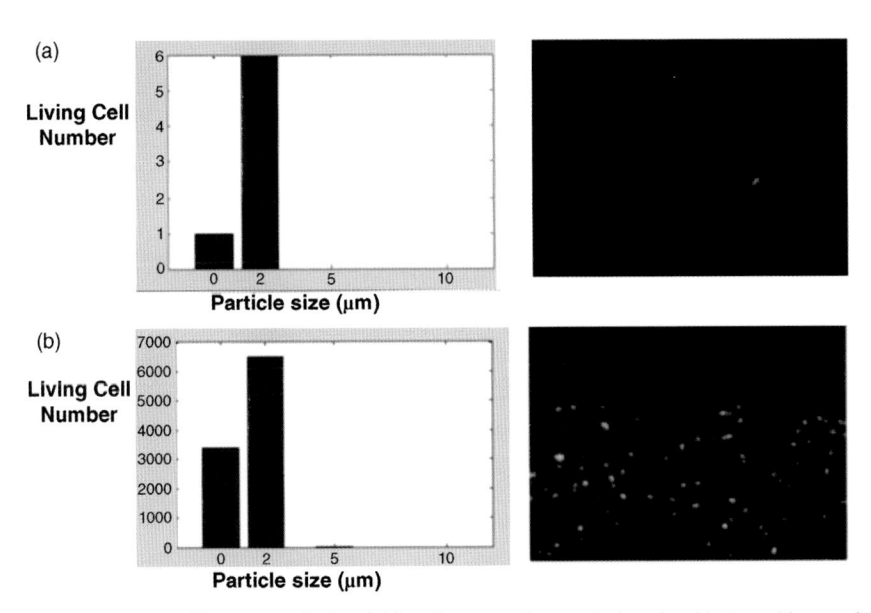

Figure 4.11 An offline example for viable microorganism analysis using MuScan. The result summary provides a total number of objects, living cells, size, and fluorescence image. Process samples were used to generate two examples. (a) Low bioburden sample with 7 living cells/sample. (b) A high bioburden sample with a total of 9913 living cells.

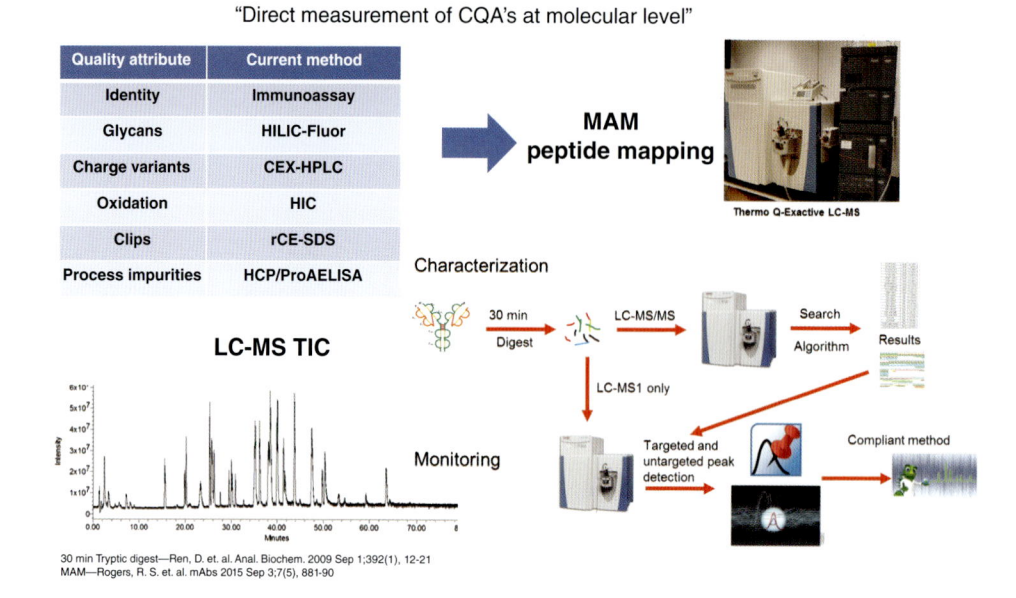

Figure 4.12 Multiattribute method via peptide mapping/LC-MS.

MS has enabled the development of multiattribute methodology [8]. A single peptide mapping LC-MS method has shown to evaluate multiple attributes (identity, glycans, charge variants, oxidation) that potentially eliminate six separate assays from the QC lab (Figure 4.12). These methodologies are now being added to Merck's analytical toolbox and compared against current methods. During 2016, the PRO lab worked to implement this methodology into online sampling of the continuous processing, such as perfusion, and purification stages. The automated aseptic sampling will deliver a sample to an automated liquid handler for protein A purification and automated digestion and then fed to the LC/MS. Interfacing these tools with SIPAT, DeltaV, and SIMCA batch, online will enable data feedback and product attribute control during the bioprocessing in real time. This product attribute control can be the first step in continuous product quality monitoring throughout production and providing key attribute data ahead of final fill and potential real-time product release [9].

4.3 Protein Refinery Operations Lab: Case Studies

4.3.1 Case Study: Perfusion

The PRO lab incorporates perfusion from 10 to 50 l scale. Performance at and above 2 g/(L day) has been achieved using a 1 vvd media throughput. Reaching this target is an important milestone as process cost analysis shows favorable economics above this target. As the facility of the future is envisaged to be predominately single use, the automated disposable tangential flow filtration (auto TFF, Spectrum labs) was

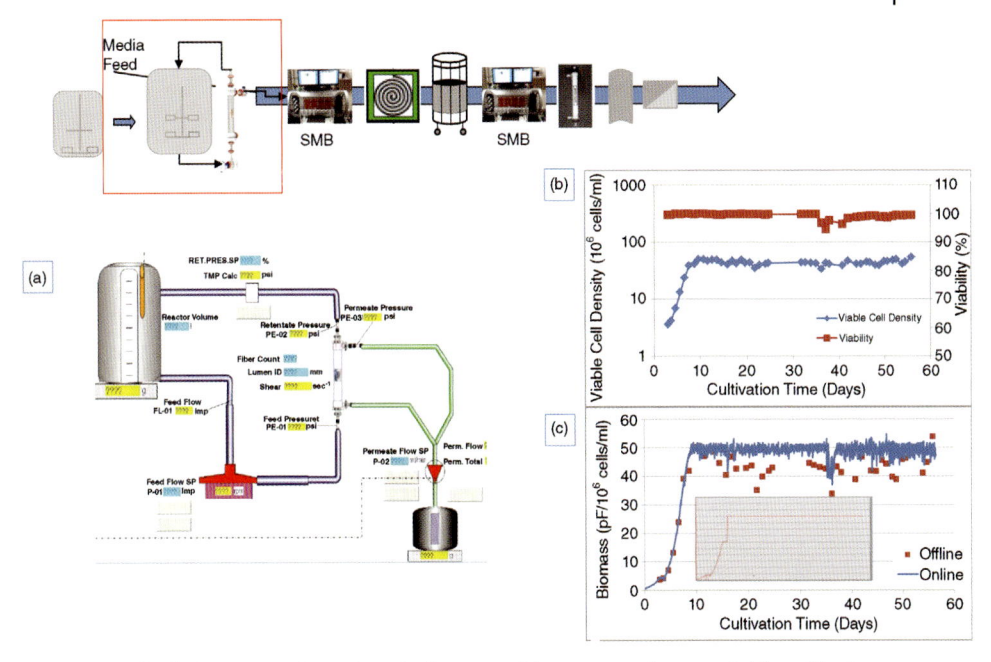

Figure 4.13 Perfusion processing using single use enable automated tangential flow filtration (Spectrum Labs). Steady-state cell concentration was achieved for >40 days using real-time online capacitance control. (a) Single-use configuration of the auto TFF (Spectrum labs). (b) Control of cell concentration from automated cell blending using capacitance for cell concentration. (c) Correlation between offline and on line cell concentration.

developed and implemented (Figure 4.13). The auto TFF is a fully automated standalone skid for perfusion bioreactors that has been incorporated into the distributed control system in the PRO lab. The flow diagram of the auto TFF system is shown in Figure 4.13a with the key components represented (bioreactor, membrane, pressure sensors, scales). Control features offered through the skid include the media feed gravimetric control, automated flux control, permeate flow using single-use flow and pressure sensors. These are all accessible through the recipes in the Supervisory Pro lab control system. Optimization experiments were conducted using common membrane materials (PS, mPES, mixed ester) as well as common geometries (0.5, 1 mmdia × 20, 40, 60 cm tall). The mPES membrane with 1 mm diameter and 20 cm tall were found to be optimal [10]. To date, 60-day perfusion proof of concept studies have been completed with membranes of this configuration and has yielded stable high-cell viability (>98%) and densities (70 MM cells/mL) during this extended period of cultivation. Continued development of new membranes is expected to further understand and minimize fouling. The biomass concentration is a critical process parameter to maintain constant during perfusion cell culture that may influence critical quality attributes such as glycosylation and charged variant profiles. An automated cell bleed maintains the cells in the exponential growth phase at all times and minimizes shifts of metabolism. In scale-down experiments where automation is minimal, perfusion bioreactors are typically bled in bolus aliquots from the bioreactors at a constant interval (such as every 12 or 24 h). This results in a cyclic behavior where the cell mass fluctuates between pre and post bleed states. To design a cell bleeding strategy, an online signal for biomass is made available to the DeltaV control system. Biomass levels in the bioreactor are measured online via a capacitance probe in the bioreactor (Hamilton/Fogale). A good correlation between online capacitance cell density and offline CEDEX cell counter measurements were obtained (Figure 4.13c). With this correlation, the cell bleed in the PRO lab has been fully automated with feedback control using the online capacitance signal. Here, a cell bleed pump is activated when the capacitance is above the desired set point, and the cells are removed from the bioreactor. The bioreactor volume is maintained at a constant mass through the auto TFF skid control resulting in a dilution of cell mass. By using tuning parameters with a long time constant, the cell bleed can be considered a continuous process. The result of this automation strategy can be seen in Figure 4.13b where the online signal (inset) shows an exponential growth period over a growth period followed by an extended time at a constant cell density (70 MMcells/mL +/−2 MMcells/mL).

4.3.2 Case Study: Continuous Purification

The key enabler for the transition to continuous biologics production is continuous multicolumn chromatography (CMCC). Here, the inherently batch process is run over multiple columns simultaneously allowing for continuous antibody mass flow into the step and a periodic antibody elution out of the step. The BioSMB CMCC system is comprised of a single-use flow path and valve block enabling gamma sterilization and ultimately closed processing from the bioreactor to the drug substance bag. The closed process designation has many advantages for Facility of the Future concepts, including decreased turnaround time between campaigns in a multiproduct facility, and potentially parallel processing of multiple products within a single suite. The closed process designation also indicates that the process streams in that closed process are in a low bioburden environment that would eliminate concern for bioburden contamination of chromatographic steps. In the future, the BioSMB will incorporate gamma irradiated flow paths but in the meantime the wetted path for continuous BioSMB system was sanitized by conventional means using caustic solution. Therefore, bioburden ingress into the process through the columns themselves or inefficient sanitization of the valve cassette are primary concerns to the continuous process. An example of the bioburden impact is shown in Figure 4.14a and b. Two perfusion bioreactor runs were carried out to validate sanitization procedures as well as the performance of the continuous chromatography. One of the primary indicators of bioburden growth in the protein A step is blinding of the 0.22 μm guard filters ahead of the columns. An example is pressure spikes shown in Figure 4.14a needing frequent automated filter switching in batch 1. This bioburden either entered the process through the protein A columns or through the postcaustic wash. An expanded sanitization procedure was utilized in batch 2 adding a hydrogen peroxide wash solution in addition to the caustic wash. This more rigorous wash procedure significantly reduced the pressure fluctuations from filtration fouling as seen in Figure 4.14b. This shows the promise that a fully gamma-irradiated single-use flow path as well as gamma-sterilized protein A

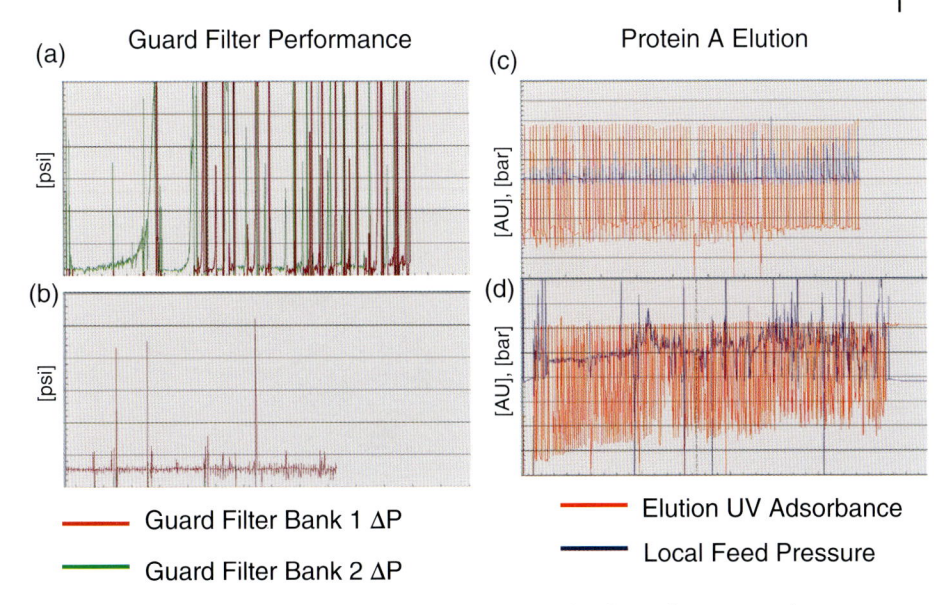

Figure 4.14 Continuous mAb capture using continuous multicolumn chromatography over 50 days. (a) Guard filter differential pressure from automated perfusion run 1 and (b) automated perfusion run 2. (c and d) Elution UV and feed pressure from automated perfusion 1.

columns can enable a low bioburden control throughout a continuous chromatography and subsequent linked continuous steps.

The UV elution profiles shown in Figure 4.14c and d show consistent performance over a period of 40 days and a total of 160 cycles per column.

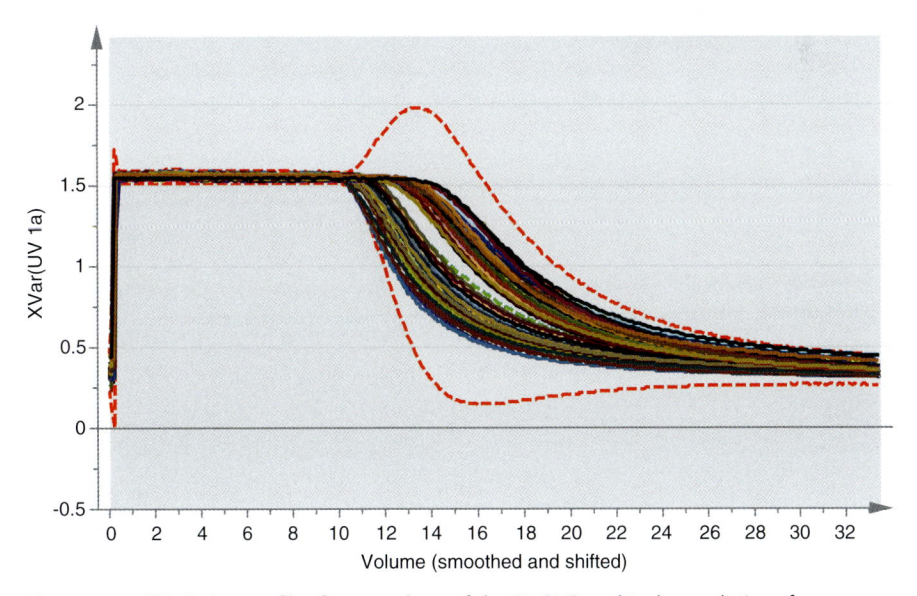

Figure 4.15 UV elution profiles from a subset of the BioSMB multicolumn elutions from automated perfusion run 2 over 50 days.

126 | *4 Progress Toward Automated Single-Use Continuous Monoclonal Antibody Manufacturing*

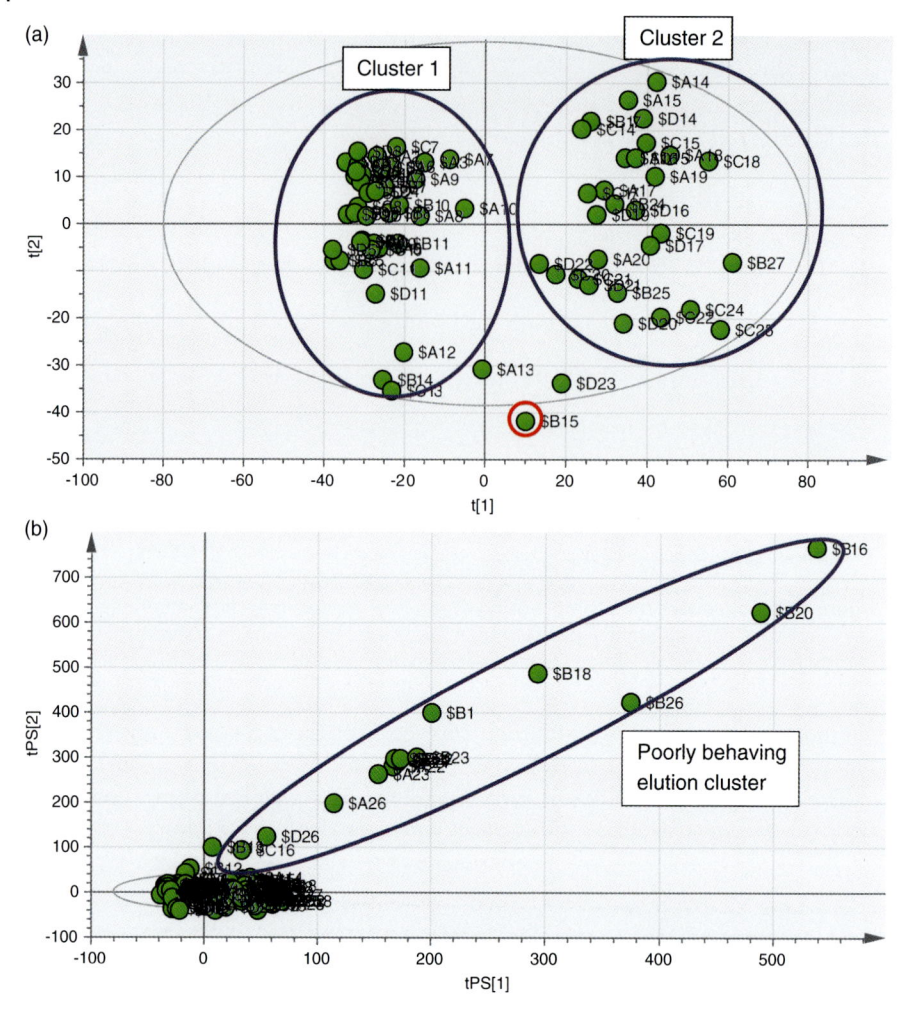

Figure 4.16 (a) Principal Component Analysis of selected elution peak UV data from automated perfusion run 1. (b) Expanded data set encompassing bad elution profiles.

An easier way to interpret the UV data from column elution is through principal component analysis (PCA), a form of multivariate data analysis. PCA is a technique where the large data sets, described by several inter-correlated dependent variables are reduced to a set of new orthogonal variables called principal components [11]. These principal components are then displayed visually in an area plot to easily identify outliers. In Figure 4.15, a subset of the UV elution profiles of Figure 4.14d are plotted with elution volume. Here, you can observe some scatter in the peak width according to small variations in the titer in the perfusate stream being loaded onto the columns, but still it is difficult to identify outliers in the data.

Upon analyzing the same data set after PCA transformation, the data are reduced down to the area plot, as shown in Figure 4.16a. Here, all elutions lying inside of the lightly colored oval area are statistically similar in their first two principal components. There is one elution profile that is statistical outlier (B15). Further, the data falls into two clusters inside of the oval (left and right), representing elutions from columns that were loaded at a slightly higher titer than the other cluster, thus shifting the length of the elution peak. By expanding the data set to include known abnormal elution peaks (as identified from conductivity plots), a new area plot shown in Figure 4.16b was constructed. Here, the poorly behaving elution peaks formed a new cluster along a common trajectory outside of the initial oval created in by the original PCA model. It can be seen that most of these irregular profiles originated from column "B." Although no action was taken during this particular experiment to segregate these peaks, one could imagine control scenarios that once column elutions start trending outside of the oval, one of two actions could be taken automatically: (1) Column "B" would automatically be taken out of service and replaced with a preinstalled column in a different position on the valve block or (2) the product stream is sent to waste until a time when corrective action is taken by the operators. This example highlights the power of MVDA modeling in continuous processes. By interfacing MVDA software (SBOL) with the control system (DeltaV), one can easily automate responses to abnormal situations as long as there is confidence that the planned corrective action will remedy the observed deviation.

4.3.3 Case Study: Proof of Concept Automated Handling of Deliberate Process Deviations

Other continuous processes such as oil refining are highly automated facilities. The chemical reactions and separation processes are characterized by perturbing steady-state operation and observing the subsequent impact on the process. This high degree of automation concept is being developed for the biologics Facility of the Future and are being undertaken in the PRO lab for continuous protein production. Ultimately, the goal of the PRO lab is to perturb the upstream process and observe how that deviation propagates through the remaining continuous process. Initial studies to study the impact of process perturbations to drug molecule critical quality attributes using the online and at line process analytical tools in both the upstream and downstream processes have begun. In this work, a continuous perfusion bioreactor was operated for 30 days. Permeate was fed directly to the continuous multicolumn chromatography skid (protein A affinity chromatography) after day 10 of cultivations. The effluent of the protein A columns was fed through an automated viral inactivation step using an in-line pH adjustment and a tubular reactor for residence time. Samples were taken from the bioreactor and from the viral inactivation product stream after pH adjustment for product stability, with the process flow shown in Figure 4.17a.

4.3.3.1 Perfusion Process Deviation Analysis (Bioreactor Temperature Shift)
Upon reaching steady cell density and product titer, permeate was directed into the continuous downstream apparatus consisting of continuous protein A chromatography and continuous viral inactivation. Figure 4.17b shows that after day 10 of continuous operation, the charge variant profile remained constant until day 23 (13 days). On day 22, 2 °C temperature was made in the perfusion

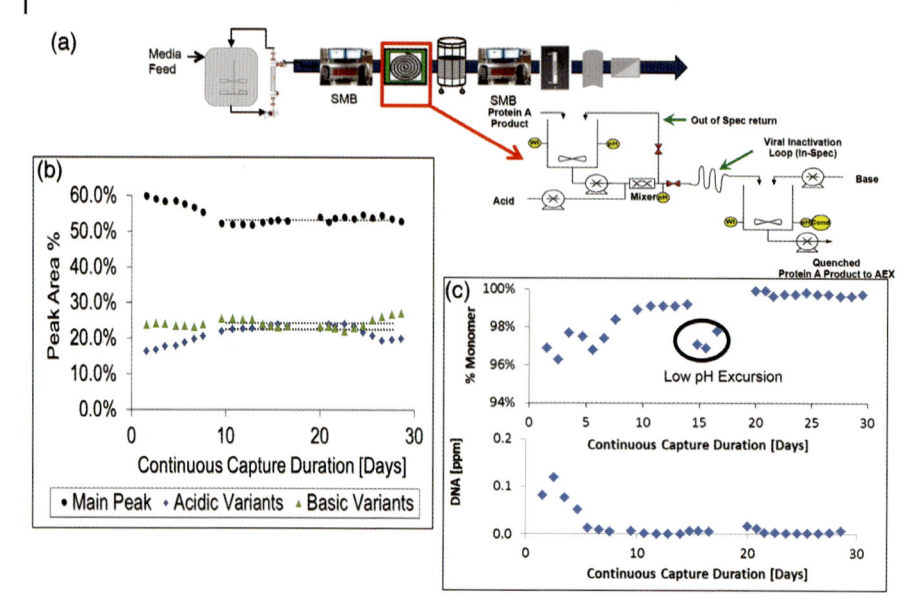

Figure 4.17 CHO mAb continuous processing in the PRO Lab. (a) Continuous protein production process flow diagram from perfusion bioreactor through continuous protein A chromatography and continuous viral pH inactivation. (b) Charge variant profile showing response to bioreactor temperature shift on day 22. (c) Aggregate response to low virus inactivation pH on day 14.

bioreactor. The change in bioreactor temperature could them be directly correlated with the shift in charge variant profile. Looking forward to the fully automated facility of the future, robust process knowledge could be used to feedback process set points (such as temperature) in response to measured changes in critical quality attributes of the product (such as charge variant profile).

4.3.3.2 Downstream Process Deviation Analysis (Viral Inactivation pH)

The process stream was sampled after the virus inactivation step, post pH adjustment to a stable hold point. As with the previous example with charged variants, the aggregation profile required 10 days postcontinuous operation to reach a steady-state value >99% monomer, as seen in Figure 4.17c. After day 10, there were 4 days of consistent monomer levels postviral inactivation. On day 14, the pH set point was lowered 0.8 pH units. The response to this process perturbation was the percent monomer dropped from >99 to 96% showing a significant change. After three days of operation, with increased operation the pH set point was returned to its initial value, and the percent monomer rebounded back to >99%. It is envisioned that the parametric multivariant models will allow automated decision-making as to the disposition of the fluid flow with the high aggregates. For example, can the aggregate level be removed at subsequent polishing step or should it be segregated from the batch at the surge bag. The PRO lab will be used to build these foundational concepts to provide a toolbox of algorithms to deal with the majority of common process deviations.

4.4 Summary

This work has shown that significant strides toward implementation of continuous bioprocessing integrated with sophisticated automation strategies are within reach. The protein refinery lab has shown to provide the necessary capability to develop and operationalize continuous bioprocessing. This was built upon a clear set of guiding principles for design and operation for the evaluation and implementation of automated continuous processing. This included the key foundational decision of the early engagement between R&D and Manufacturing to establish directly transferable methodologies and automation code for the line of site to clinical and commercial manufacturing. Work in the PRO lab has so far shown feasibility of automated continuous processing up to 60 days. The PRO lab shows that true commercial lights out manufacturing with long duration (60–100 days) is within reach during the next 5 years. Efforts continue to expand the integration of PAT and drive toward product attribute control. Key technology gaps were outlined in this work that include analytical needs such as online rapid microbiology and potency that could be solved by industry consortia approaches. Conquering these challenges will enable a complete real-time release solution to be available. Work in the PRO lab will continue to show that automated approaches to process deviations are feasible and will be incorporated into risk assessments for continuous processing. This will be benefited by improved robustness of single-use technologies and operation practices that are built upon close collaboration with end users and technology suppliers. Implementing continuous processing will require new quality and regulatory approaches. This will require expanding the collaboration between end users, equipment suppliers, and regulators. This should ultimately deliver a lower cost agile solution that provides a flexible supply chain to the varying drug demands with a global presence.

Acknowledgments

The authors wish to acknowledge multiple scientists at Merck & Co.Inc that assisted with providing knowledge and data sharing. A special thanks includes Chris Kistler, Bill Napoli, Hao Chen, Kristen O'Neill, Huijuan Li, Jun Heo, Maria Khouzam, John Higgins, Dave Moyle, Jeff Johnson, and Merck Single-Use Network, including Mark Petrich single-use component engineering team. We also wish to thank our many external collaborators, including Dave Serway (Spectrum Labs), Rich Rogers (JUST Biotherapeutics), Marc Biscchops (Pall), Jim Stout *et al.* (Natrix), Maury Bayer (Proconex).

References

1 Thayer, A.M. (2016) *Chem. Eng. News*, **94** (10), 22–23.
2 Reichert, J.M. (2012) *MAbs*, **4** (1), 1–3.

3 Kaitin, K.I. (2010) *Clin. Pharmacol. Ther.*, **87** (3), 356–361.
4 Brower, M., Hou, Y., and Pollard, D. (2015) in *Continuous Processing in Pharmaceutical Manufacturing* (ed. G. Subramanian), Wiley VCH Verlad GmbH, Weinheim, pp. 255–296.
5 Bisschops, M. (2013) *Pharm. Bioprocess.*, **1** (4), 361–372.
6 Ying, H. *et al.* (2015) *Biotechnol. Prog.*, **31** (4), 974–982.
7 Heo, J.H. *et al.* (2014) *Pharm. Bioprocess.*, **2** (2), 129–139.
8 Rogers, R.S. *et al.* (2015) *MAbs*, **7** (5), 881–890.
9 Richardson, D.D. (2016) Real time release for biologics: current state future targets. Oral presentation at IFPAC 2016.
10 Pinto, N. (2015) ACS San Diego Presentation.
11 Abdi, H. and Williams, L.J. (2010) *Wiley Interdiscip. Rev. Comput. Stat.*, **2** (4), 433–459.

Part Three

Single Use Technologies and Perfusion Technologies

5

Single-Use Bioreactors for Continuous Bioprocessing: Challenges and Outlook

Nico M.G. Oosterhuis

Celltainer Biotech BV, Bothoekweg 9, 7115AK Winterswijk, The Netherlands

5.1 Introduction

After introduction of the wave-type single-use bioreactors at the end of the nineties of the last century [1], single-use bioreactors are fully adapted by industry now. The biopharmaceutical industry has widely accepted single-use technology in general and single-use bioreactors particularly. Single-use bioreactors nowadays are applied in (GMP) production, process development and at research level [2]. The advantages of applying single-use technologies are such that nowadays most new facilities apply single-use bioreactors even up to volumes of 1000 l and bigger [3]. In an industry survey by BioPlan Associates [4] it has been found that almost 50% of the industry expects to see a 100% single-use facility in operation within the next 5 years. The bioreactor is only one unit-operation in the full (single-use) train of process steps for biopharmaceutical processes. However, it is a key step, which not only determines the product quantity produced (and therefore the scale of the process) but also determines the product quality.

By introducing continuous processing, due to the potentially higher volumetric productivity, the operational scale can decrease a factor 5–10. Product costs per gram of for an average antibody are in the same range when a 14–21-day fed-batch process, carried out at $10\,m^3$-scale is compared with a 40-day continuous perfusion process at $1\,m^3$-scale [5]. Such, the application of single-use equipment even becomes more and more attractive. When the scale of operation is decreased with a factor 10 and the upstream part of the process is fully disposable, the direct, equipment related investment costs will be decreased by a factor 5–10. Significant savings (factor 2–3) are made as well in investment levels of the infrastructure (both scale related as due to simplification).

A single-use bioreactor typically consists of two–three separate parts given as follows:

1) The single-use (plastic) bag
2) The support to the bag, including motor or shaking platform and means for heat transfer
3) The control unit for process control

In many cases the control unit is comparable to that of traditional steel or glass bioreactors and a separate part of the system. Some suppliers have integrated the control part with the support structure of the bag. The parts 1 and 2 serve as the bioreactor, which replaces a steel or glass reactor. After processing, the bag is emptied (harvest) and the empty bag is thrown away. Bags come fully sterile; γ-sterilization at 25–40 (or even 50) KGy is the standard way of sterilization of these bags that ensures full sterility before use.

Usually, bags are filled using filter sterilization of media, or in some cases, also preautoclaved medium is used. Connections to the bags can be made either under laminar flow, using sterile connectors or using tube-welders.

Key aspect in introducing single-use bioreactors in GMP manufacturing (but also in process development and R&D) is the absence of cleaning in between runs. Cleaning requires extensive validation and documentation [6]. The absence of these cleaning procedures and the absence of steam-sterilization result in a much more simple infrastructure leading to much lower investment costs of a new facility. Other arguments of applying single-use bioreactors are: lower risk of infection; easy handling and simple or hardly any maintenance (no O-rings, bearings, valve seals, etc.); reduced energy demand (no steam); fast turn-around (also leading to less capital expenditures); higher flexibility (exchange of equipment is easier); more simple installation and operational qualification (IQ/OQ) procedures.

The high flexibility of single-use equipment is clearly illustrated in Table 5.1.

As mentioned above, it's obvious that single-use bioreactors today take an important role in bioprocessing. For biopharmaceutical production, especially the easy validation leads to significant cost savings. Also in combination with cell separation devices (filters, settlers, centrifuges), single-use bioreactors are applied for continuous (perfusion) processes nowadays.

Another recent breakthrough of single-use bioreactors is into the application of microbial (fermentation) processes. Where in first instance only mammalian processes were carried out in single-use bioreactors, nowadays also microbial fermentations, like *Escherichia coli* and yeast-based processes for biopharmaceuticals are running in these types of bioreactors [8].

Table 5.1 Flexibility single-use equipment compared to traditional stainless steel equipment.

Aspect	Single-use equipment	Traditional stainless steel equipment
Time from investment decision to mechanical complete	6 mo	18 mo
Commissioning and qualification	<1 mo	3–6 mo
Capital cost	Lower	>3× higher
Operational cost	1.5–3× higher	Lower
Turnaround time	<1 d	1 wk

Source: Adapted with permission from Ref. [7]. Copyright 2015, BioProcess International.

Hence, single-use bioreactors have a clear benefit not only in biopharmaceutical processes but also in more traditional fermentation processes, like the (bulk) fermentation of lysine, where they are applied in the seed-train. This avoids the usage of shakers and simplifies the seed-train. Where in first instance steel – fixed-piped – reactors were used, nowadays more flexible single-use bioreactors are applied as well. The advantages are less contamination risks, higher flexibility, easy implementation, less infrastructure (no autoclaves or additional piping), and so on.

5.2 Single-Use Reactor Types

Despite the success of the "wave-bioreactor" in this respect, during the last decade also the stirred single-use systems became more and more popular. Main argument to introduce more complex single-use stirred systems is the 1:1 comparability with existing stainless steel or glass bioreactors.

A single-use bioreactor can be classified into the type of mixing technology that is applied. This can be a "shaking platform" at which the bag is placed at a platform where the movement of the platform results in mixing of the liquid in the bag, as well as dispersion of the gas into the liquid to ensure gas–liquid mass transfer (Figure 5.1).

The alternative, as mentioned, is to apply a fully disposable stirring device, rotating by means of a disposable bearing mechanism or based on a magnetic drive (Figure 5.2).

Furthermore, there are some specific designs, applied in some more specific processes.

As a result a very wide range of bioreactor types have been introduced into the market, all having their specific benefits and drawbacks (Table 5.2).

Nowadays, rocking type bioreactors are mostly used in seed processes or to produce small batches of product for investigation purposes. Especially in those processes where single-use bioreactors replace existing equipment, stirred systems are applied to avoid lengthy validation procedures. Product comparability of the product as was produced into a (steel) stirred bioreactor has to be proven.

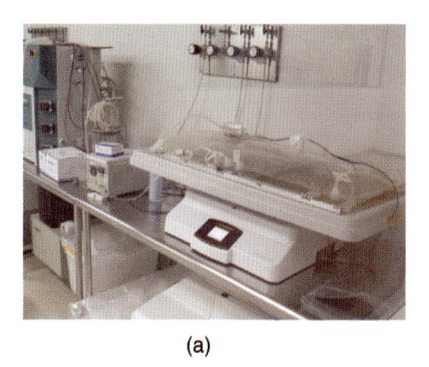

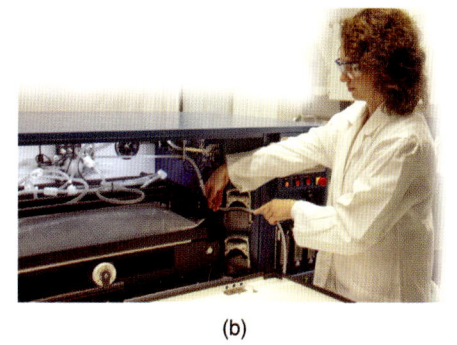

(a)   (b)

Figure 5.1 Different types of rocking platform-based single-use bioreactor. (a) BioStat Cultibag RM, 20 l (Sartorius Stedim Biotechnology); (b) CELL-tainer 20 l (Celltainer Biotech).

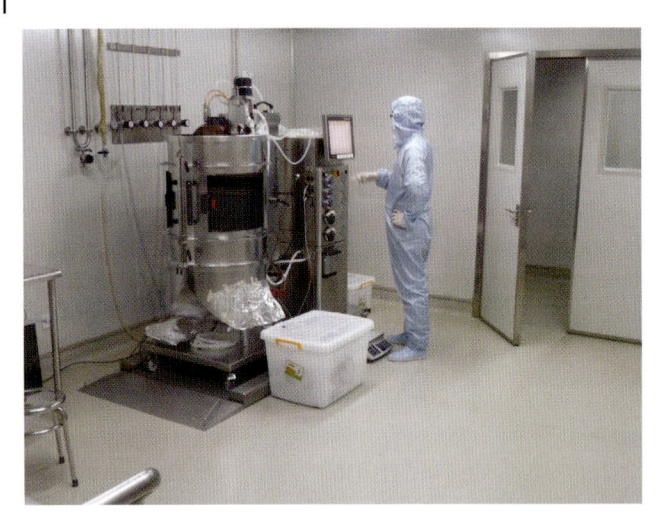

Figure 5.2 Single-use bioreactor (stirred) (200 l Biostat Cultibag STR (Sartorius Stedim Biotechnology)).

In newly designed products or processes, the reactor type chosen might be of less importance. At the end, the reactor needs to create optimal process conditions. Especially, for continuous processing this might be a challenge (see below). Also aspects of easy handling, simple design, and cost of bags have to be taken into account when designing a process. As stirred bags are more costly, and cost of production becomes more and more an important issue, a simple bag design can be an important factor in final choice of the bioreactor.

5.3 Material Aspects

A continuous biopharmaceutical process, running in average 30 days, requires a robust and reliable reactor design. Not only the single-use bag should last the full process without getting mechanical or wearing issues but also sensors, filters, tubing, connectors, and so on need to withstand a lengthy process.

The bag material aspects and sensor criteria are discussed below. Other materials like tubing and filters are comparable to materials used with glass and steel bioreactors. However, precautions are required to avoid wetting of the off-gas filter (both peltier cooling as filter heaters are applied). To avoid wear of the tubing (especially thermoplastics are sensitive for mechanical wear), proper guiding of the tubing is recommended.

Many investigations have been done to prove the safety of the materials used. Up to date it clearly has been demonstrated based on extensive leachables and extractable studies that materials used are fully safe [9]. Testing of materials and full validation is the responsibility of the vendors. Regulatory institutions like EMEA (Europe) and FDA (USA) have published many guidelines, which have to be fulfilled [10]. Finally, it is the user responsibility to check and document the validation guidelines as been followed by the vendors.

Table 5.2 Overview of available single-use bioreactors (using single-use bags).

Reactor name	Volume (total l)	Bag type	Mixing	$k_L a$ (h^{-1})	Supplier
Wave	1–500	Pillow	Rocking	<10	GE Healthcare
Biostat Cultibag RM	1–100	Pillow	Rocking	<10	Sartorius Stedim Biotech
XRS-20	2–20	Pillow	Biaxial rocking	73	Pall
Appliflex	1–25	Pillow	Rocking	<24	Applikon Biotechnology
CELL-tainer®	0.2–35	Pillow or square	Two-dimensional rocking	550	Celltainer Biotech
CELL-tainer®	5–275	Pillow or square	Two-dimensional rocking	400	Celltainer Biotech
HyClone™	30–2000	Tankliner (round)	Stirrer	<15	Thermo Fisher
Hyperforma SUF	30 and 300	Tankliner	Stirrer	>200	Thermo Fisher
Biostat Cultibag STR	50–1000	Tankliner (round)	Stirrer	<25	Sartorius Stedim Biotech
XDR™	10–2000	Tankliner (round)	Stirrer	<25	GE Healthcare
XDR-50 MO	50	Tankliner (round)	Stirrer	na	GE Healthcare
Mobius Cellready	50–2000	Tankliner	Stirred	<60	Merck - Millipore
Allegro™	20–200	Tankliner (round)	Stirred	<30	Pall
Nucleo™	50–100	Tankliner (square)	Paddle	<20	Pall
SBX	10–200	Tankliner (round)	Orbital shaker	<30	Kühner

Modern materials for manufacturing single-use bags are based on multilayer films. These films, produced by coextruding (blown-film, cast film, lamination, etc.) are applying for the product-side of the film polyethylene type materials (LDPE, low-density; LLDPE, linear low-density; HDPE, high-density; etc.). The outer side of the film can be made of PA (Nylon6) or PET. Such materials show to be a proper barrier for gas, solvents, vapor, and odor. Also the strength of these materials is such that large-volume bags can be constructed.

Both physical and vapor resistance tests are performed to ensure the suitability and functionality of the bag film materials [11]. Also the absence of toxic materials is crucial. Therefore, tests are performed to detect extractables after exposure of the bags to acid (pH<2) and alkaline (pH>9) conditions, as well as to water for injection and pure ethanol. After exposure, volatile, semivolatile, and nonvolatile components are analyzed. Also, detection of TOC, pH value, conductivity, and weight loss analysis are performed. Guidelines are available from regulatory instances like FDA (USA) and EMEA (Europe). Several testing procedures are available to prove the absence of leachables and extractables. Testing of the materials is the vendors' responsibility and has to be documented in a validation dossier.

For ports and fittings, usually materials like polyethylene (PE), Engage® (a polyolefin elastomer), and Evatane® (copolymerized ethylene-vinyl acetate) are used. The last two materials show good γ-resistance. All these materials can be heat welded to the bag films. Connectors, to ensure sterile connection of the bag with other process equipment like medium and storage bags but also with filters, are usually made of polycarbonate (like Makrolon® from Bayer). Connections to the bag can be done under a laminar flow regime or, more practically, by using a so-called sterile tube-welder which heat-seals the tubing together. To close tubing's, heat sealers are used. These heat sealing requires special tubing material made of thermoplastic elastomers (e.g., C-flex, PharmaPure, Solmed, etc.). Also quick fittings are available, making it possible to make sterile connections without any further tools (e.g., Kleenpak® sterile connectors). An extensive overview of couplings is given by Rothe and Eibl [12].

All materials used have to be stable against γ-irradiation at levels up to 40 kGy (or even up to 50 kGy). This also means that all tests of materials, including leachables and extractables have to be performed after γ-irradiation and (rapid-) aging of the materials for at least 4 months. It's not obvious that all materials resist γ-irradiation at such levels. Also the alternative method, electron beam irradiation (β-irradiation) may lead to changes in material properties [13]. Materials like PE can oxidize due to γ-irradiation in the presence of air or oxygen. Changed material properties after irradiation may lead to unwanted cracking of the polymers that can finally lead to a leak in the bag or breaking of the single-use component under stress conditions.

Before release of the single-use bags many validations have to be done by the vendor, not only to prove the sterility of these bags but also especially to prove the suitability and safety of the materials used. Manufacturing of the bags is done under cleanroom conditions to reduce dust, particle levels, and to reduce the starting bioburden level before irradiation (Figure 5.3).

Figure 5.3 Production and assembly of single-use bags. (Photo: courtesy of Haemotronic S.p. A., Italy.)

Most vendors have their propriety film material and also use propriety components. Such approach makes the end users dependent on the vendor and potentially it raises challenges in terms of continuity of supply and reduces flexibility. A debate has started, also including vendors to get to a standardization of disposable design, including materials applied [14].

By standardization of designs, materials, and components vendors of disposables can run leaner manufacturing, which finally results in lower cost to the end users. Material consistency is key in reducing validation efforts. Nowadays, when the end user wants to change to another supplier of single-use bags, for whatever reason, a lengthy revalidation of materials and even of the process (including new consistency runs) is needed to fulfill the requirements of the regulatory authorities like FDA or EMEA. Such efforts potentially hold back the introduction of single-use technologies. Therefore, standardization of materials and application of materials that are freely available (like certain film materials) by vendors are strongly recommended.

5.4 Sensors

For continuous processes, it's essential that the online sensors for measuring parameters like temperature, pH, dissolved oxygen (DO), and so on, need to have a long life-time but moreover the sensor needs to be gamma resistant as well as the sensor should be preassembled into the bag before (gamma-) sterilization. Another requirement is that the sensors are (relatively) cheap, so they can be discarded with the bag after running the process.

Conventional sensors are mostly made of steel or glass. Although technologies exist to introduce a conventional steel electrode into the bag (e.g., Kleenpak™ connector from Pall), the big disadvantage is that reuse of the electrode requires additional handling and especially also additional validation procedures. This makes the bioreactor not fully single-use.

New promising sensor types for application into single-use bioreactors, mixing devices, and down-stream operations are those developed by Gymetrics SA[1], a Swiss startup founded in 2011. First, commercial products will be available by 2016 and are pH sensors based on electrochemical technology for both upstream application (based on a conductive polymer) and downstream application (based on a metal-oxide).

The sensor for upstream application is composed of a three solid-state electrode configuration. The working electrode is modified with a conducting polymer layer. Auxiliary and reference electrodes are glassy carbon and Ag/AgCl, respectively.

The sensor requires an aqueous media with a minimum conductivity of 10 mS/cm, which is well below the level typically found in physiological solutions. The electrodes are sensitive to high temperature and the probes cannot be

---

1 www.gymetrics.com.

autoclaved. However, they are insensitive to gamma irradiation and show no alteration in the sensor performance before and after sterilization at irradiation dose of 40 kGy.

The conditions of the sensor storage are relatively easy as long as the sensors are not light or heat sensitive, providing that they are not exposed to temperatures above 70 °C. The limiting factor for shelf life will be the sterile packaging rather than the sensor itself, which theoretically has an unlimited shelf life when stored at ambient temperature. A shelf life guarantee of 24 months is foreseen and real-time testing at 6 months so far has shown no impact on sensor performance before or after sterilization.

In order to validate and to test the sensor performance, extensive laboratory pH testing was carried out in a variety of buffers and with CHO cells. The sensors are calibrated at one or three points 4, 7, and 9. Tests were carried out at pH = 7 with CHO cells at the Zürcher Hochschule für Angewandte Wissenschaften, Life Science und Facility Management (ZHAW). The sensor showed a maximum drift of 0.06 pH per 24 h with a sensitivity of 80 mV/pH (Figure 5.4). The pH of the culture medium was also followed and controlled in real time using commercial pH meter. The tests were done using Gymetrics pH prototype material on CHO culture (XM 111-10) used to produce SEAP as a model protein. The potential drift was 49 mV or 0.84 pH after 14 days (the sensor sensitivity is ~ 80 mV/pH). The point "a" shows a drop of temperature of 6% after a media change.

It may be concluded that these "dry" sensors show a potential better performance in terms of gamma stability, life time during dry storage and drift compared with the presently used fluorophoric sensors, which are light sensitive and also less stable under gamma sterilization conditions.

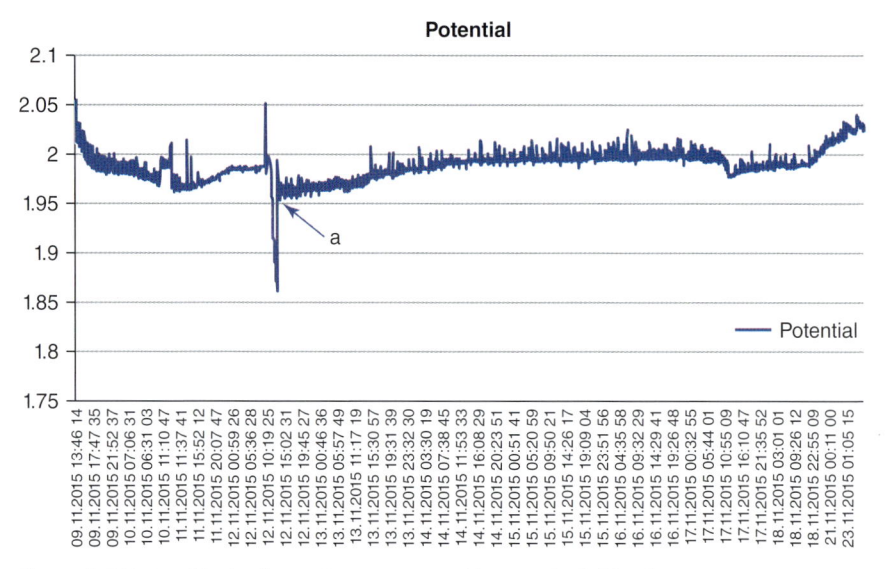

Figure 5.4 Tests with the Gymetrics up-stream pH sensor in CHO culture.

5.5 Reactor Design

5.5.1 Mass Transfer and Mixing Requirements for Continuous Processing

Continuous (perfusion) processes commonly result in higher cell densities. Where, in average in a fed-batch CHO-based process cell densities of 10–15×10^6 cells per ml are reached, in a continuous process, cell densities of 100×10^6 cells per ml are not unusual [15]. The specific oxygen uptake rate $\left(q_{O_2}^{max}\right)$ of CHO-cells is within the range of 5–10 pmoll per cell per day [16]. Assuming the cell density $C_{cell} = 10 \times 10^6$ cells/ml, the oxygen uptake rate, OUR, can be calculated as follows:

$$\mathrm{OUR} = q_{O_2}^{max} \cdot C_{cell},$$

being approximately 2–4 mmol/(l h).

When an optimal dissolved oxygen concentration of 30% of air saturation is assumed, the mass transfer coefficient, $k_l a$, can be calculated from the equilibrium equation for mass transfer

$$\mathrm{OTR} = k_l a \cdot (c_{O_2,L^{sat}} - c_{O_2,L}).$$

From the above equation, it follows the required oxygen mass transfer coefficient to support a fed-batch type of cell culture process, $k_l a \approx 15$–$30\,\mathrm{h}^{-1}$.

Most single-use bioreactors suited for mammalian cell culture have a oxygen mass transfer coefficient, $k_l a$, which is in the range of 20–$50\,\mathrm{h}^{-1}$ and therefore such reactors are well suited to support the cell densities as been obtained in fed-batch culture.

However, when the cell density becomes a factor 10 higher, as been obtained in perfusion cultures, aeration with normal air is not sufficient to support the oxygen demand of those types of cultures. At a liquid oxygen concentration of 0% of air saturation (which in practice should be avoided to be that low), the driving force for oxygen transfer is maximal. In case of a cell density of 10^8 cells/ml, under such conditions, the required oxygen mass transfer coefficient should be: $k_l a \approx 200\,\mathrm{h}^{-1}$.

Marine-type impellers are not suited to achieve such high $k_l a$-values. Application of other impeller types, like gas dispersing impellers (e.g., Rushton-type impellers), is very limited due to the higher energy dissipation of such impellers and therefore higher shear rates into the liquid which may damage the cells [17]. Another approach, usually followed, is the application of a microsparger, through which pure oxygen is sparged into the reactor. An axial pumping stirrer (like a marine impeller) ensures the distribution of the oxygen through the reactor. However, the disadvantage is that such a sparger makes the bag design more complex. Also additional process control (gas flow controller) is required to avoid too high oxygen concentrations into the liquid and an additional inlet gas filter is required as well. The use of pure oxygen requires more strict safety precautions as well.

Besides the transfer of oxygen, also the transfer of carbon dioxide been produced by the cells has to be realized. With a respiration quotient, $RQ = 1$, the same molar amount of carbon dioxide as the consumed amount of oxygen has to be removed from the culture liquid. However, the mass transfer rate for CO_2 is 89% of that of O_2 [18]. As the CO_2 concentration in the gas phase can reach values of 20–40% (due to the CO_2 production of the cells), the carbon dioxide mass transfer coefficient needs to be in general a factor 1.5 higher than the oxygen mass transfer coefficient. This means that for CO_2 in perfusion culture, at a cell density of 10^8 cells/ml, the $k_l a^{CO2} \approx 300\,h^{-1}$. To remove CO_2 from the culture, a microsparger is not suited. Therefore, the strategy is that oxygen is sparged through the microsparger and air (sometimes even nitrogen) is sparged to a bubble or tube sparger. The larger bubbles will take-up a part of the CO_2. However, a high gasflow with larger bubbles should be avoided as coalescence of the bubbles, including bursting of the bubbles at the liquid surface, can lead to unwanted high shear stresses on the cells [19].

Moreover, for larger scale single-use bioreactors (>100 l), the mixing time can be rather high. Mixing times of 30–60 s are not unusual [3]. A simple regime analysis [20] shows that the time constant for oxygen transfer:

$$t_{OT} = 1/k_l a,$$

will be in the order of magnitude of 100 s (at a $k_l a \approx 40\,h^{-1}$). As the specific time for oxygen mass transfer and the mixing time are close together, this potentially may lead to the occurrence of oxygen concentration gradients in the liquid.

Although stirred single-use reactors are a more or less logical follow-up of what's already installed and in use in biomanufacturing, the introduction of continuous bioprocessing can lead to serious issues in applying these types of reactors. Instead of applying classical stirred systems an alternative can be the application of the CELL-tainer® bioreactor, which can achieve high $k_l a$-values under mild shear forces [21]. As been demonstrated with a highly shear sensitive microalgae in comparison to a stirred bioreactor, the CELL-tainer® shows excellent oxygen mass transfer with decreased cell damage and less foam formation. For working volumes up to 200 l, the CELL-tainer can support high cell densities at limited rocking rates.

In Table 5.3, a comparison is made between a fed-batch process and a continuous perfusion process at 200 l scale. From this table it can be concluded that assuming average product titres for an antibody, a 200 l fed-batch reactor results in a yearly production of 24 kg nonpurified protein and a 200 l continuous perfusion reactor a yearly production has of 139 kg of nonpurified protein.

5.6 Scale-Up Aspects

Although, as explained above, the maximum production scale can be considerably reduced when applying continuous processing, a thorough scale-up procedure is needed to scale a process from the laboratory scale (<5 l) to the final production scale ($\approx$ 200 l). As media and disposables can be rather costly even at 10 l scale, and high-throughput screening and process development nowadays can be carried out at milliliter scale, a proper translation of these conditions to the final production scale is required.

Table 5.3 Comparison of the volumetric productivity of a fed-batch and perfusion process at 200 l scale.

	Fed-batch	Perfusion
Viable cell density ($\times 10^6$ c.ml^{-1})	17	60
Total cell density ($\times 10^6$ c.mL^{-1})	34	90
Viability (at end of process)	50	70
Product titer (mAb) (g l^{-1})	6	1.1
Medium exchange rate (l l^{-1} d^{-1})	0.05	2
Volumetric productivity (g l^{-1} d^{-1})	0.43	2.2
Runs per year (d)	20 (14)	7 (45)
Total yearly amount (at 200 l) (kg)	24	139

A popular method for scale-up is geometric similarity. However, as can be explained from Table 5.4, such method may create unwanted conditions. It's obvious that when scaling-up parameters like the stirrer speed (assuming geometric similarity) has to change. A parameter, which influences the mixing as well as the shear forces in the liquid, is the tip speed of the impeller. However, when keeping the tip speed constant under geometric similar conditions, the power-input/volume (P/V) is more than a factor three reduced at 2000 l scale. As the oxygen mass transfer is directly related to the power input

$$k_l a = f(P \cdot V^{-1})^a \cdot (v_s)^b,$$

where $a = 0.4$–0.7 and $b = 0.2$–0.5 (depending on the medium properties) [17], the effect of a factor three reduction in volumetric power input may result in a drop of the oxygen mass transfer coefficient with a factor 1.5–2.2. As mass transfer can be critical for both oxygen and carbon dioxide, keeping the tip speed constant may result in gas–liquid mass transfer issues.

Table 5.4 Variation of critical process parameters when scaling-up from 50 to 2000 l using geometric similarity.

Process parameter	N (min^{-1})	v_{tip} (ms^{-1})	Re (−)	$P V^{-1}$ (W m^{-3})
50 l reactor	150	1.1	49 420	22.3
Equal[2] N	150	3.9	605 406	293.1
Equal v_{tip}	42	1.1	171 936	6.7
Equal Re	12	0.31	49 420	0.15
Equal $P V^{-1}$	63	1.6	254 270	22.3

Source: Adapted with permission from Ref. [3]. Copyright 2014, BioProcess International.

---

2 Equal parameter at 2000 l scale.

Another frequently used parameter to keep constant when applying geometric similarity is the volumetric power input. As explained above, this will ensure comparable gas liquid mass transfer conditions. However, in such case, the Reynolds number increases with approximately a factor five. This may result in a change of the flow regime from transient (at 50 l scale) to highly turbulent at 2000 l scale. A change in flow regime may effect the liquid distribution in the reactor (mixing profiles) as well as the shear stress to the cells due to the liquid flow.

Scale-up of a mixed regime can lead to process differences between the different scales as expressed in Table 5.4.

When scaling-up, it's most important to keep conditions of the cells – the microenvironment – constant. However, as is expressed in a simplified scheme, Figure 5.5, many parameters have either a direct impact on the cell, or can be of influence to other parameters.

For scale-up, also for single-use bioreactors and even for scale-up from 1 l up to 1000 l, it's recommended to do a regime analysis (estimation of the critical time constants) for the large-scale reactor. Knowing and understanding these essential time constants (in most cases this are the time constants for gas/liquid mass transfer; oxygen consumption and CO_2-production; mixing), make it possible to carry out small-scale simulation experiments (by high-throughput process development) and to investigate the impact on biological parameters like growth of the cells, viability, productivity, and product quality. Also, shear effects can be simulated on small scale: Flow cytometry can be used to elucidate the impact of shear stress on the cells [21], but also cell viability is a proper parameter to estimate the impact of shear stress on cells.

Scale-up of a mixed regime can lead to process differences between the different scales. A geometric similarity of reactors at different scales, as been promoted by some suppliers of single-use bioreactors [3] is therefore not useful.

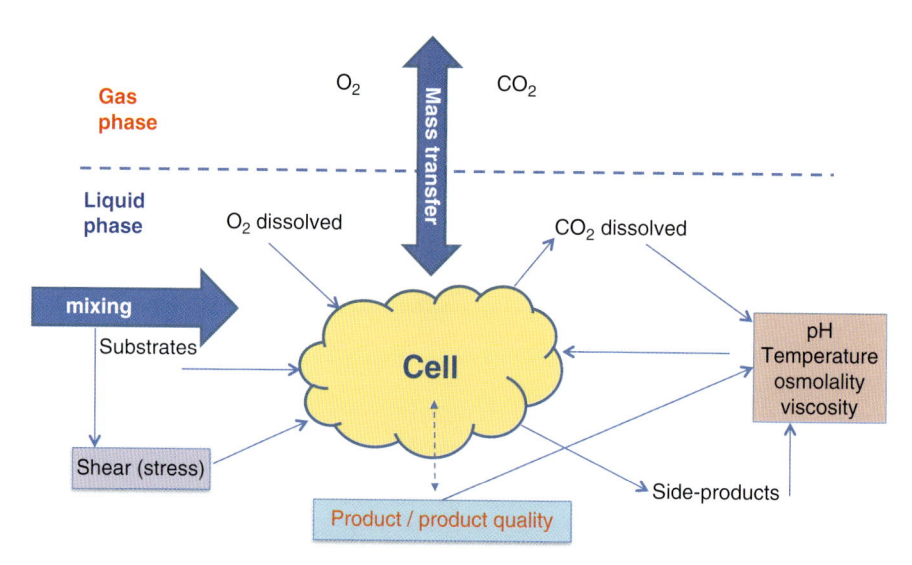

Figure 5.5 Simplified scheme expressing the interaction of various process parameters on the microenvironment of the cell. (Adapted from Ref. [22].)

A more fundamental approach of the impact of shear stress to CHO-cells is described by Sieck et al. [23]. Although under different stress regimes, characterized by the maximum hydrodynamic stress, τ^{max}, cell viability was affected by both extreme sparging as well as agitation, the cellular metabolism (productivity, product quality) did not change significantly. This means that determination of cell growth and viability under stressed conditions might be a suitable parameter to determine the impact of shear. Especially, stress caused by sparging created the highest damaging effect on the cells.

5.7 Continuous Seed Train

In order to seed a larger scale bioreactor (100 l and above), many steps have to be performed to generate sufficient viable cell material for inoculation. Usually, seeding is done in mammalian cell culture at starting densities of $0.2–0.5 \times 10^6$ cells/ml. Where, in a batch growth maximum cell densities are reached of $2–5 \times 10^6$ cells/ml, an expansion factor of 10 is the maximum achievable. In many cases, a factor seven is applied. Such, in order to seed a 200 l bioreactor, starting with a frozen vial of 1–5 ml, at least six steps are required. Despite the flexibility of single-use equipment and the reduced risk of contamination when transferring a culture to a next stage, it's still very laborious and much equipment is needed (shakers, various size wave reactors).

When using continuous (perfusion) processing in the step before the final production size reactor, higher cell densities can be reached. Assuming a cell density of 30×10^6 cells/ml in a perfusion culture, in order to seed a 200 l reactor, a preculture of only 2–3 l is required. This already saves two steps.

Another saving is possible by applying a CELL-tainer single-use bioreactor, including the expansion set [24]. By using this expansion set, which reduces the bag volume without interfering sterility, a starting volume of $\approx$150 ml is feasible. Such a volume could be inoculated with a cryovial as has been demonstrated[3] for an *E. coli* culture. Expansion within the same bag, by removing the expansion blocks (placed underneath the bag), provides the possibility to increase the volume within the same bag up to 25 l. When running in perfusion mode, at least 150×10^9 cells can be produced even in a 5 l volume. Such amount of cells is sufficient to inoculate even a 1000 l bioreactor in one step. This results in a one-step seeding process, thus saving labor, equipment, and consumables.

5.8 New Mixer Designs

A new, promising, mixer design for shear sensitive applications is the "Bachellier"-impeller by Enevor[4]. This mixer design provides a laminar and defined flow, Figure 5.6, into the liquid instead of the turbulent flow as been created by standard

---
3 www.celltainer.com.
4 www.enevor.com.

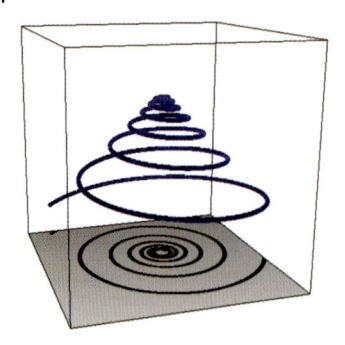

Figure 5.6 Mixing profile of the "Bachellier"-impeller. (Courtesy of Cellmotions Inc. and Carl Bachellier.)

Figure 5.7 Prototype design of the "Bachellier"-impeller. (Photo's courtesy of Enever, Inc., USA).

turbine or axial pumping stirrers. Due to these laminar flow conditions, shear sensitive cells (including microcarriers and stem cells) are evenly distributed over the reactor without having an impact on the cell or particle itself. Scale-up of the impeller is showing comparable conditions in terms of flow profile and shear. Some different scales of this impeller are shown by Figure 5.7.

5.9 Future Outlook

The application of single-use bioreactors has reopened a new era of bioreactor design. Where in the past bioreactors where more or less using a standardized design (except for some very large-scale operations), since the introduction of the wave-type bioreactor it has clearly been demonstrated that a complete different design of bioreactors can be very successful in supporting the cells for optimal growth and production.

Introduction of flexible materials being the "culture vessel" gives the opportunity to redesign the bioreactor such that the reactor is able to create optimal conditions for the cell. Where in the past, most processes had to be adapted to the reactor or series of reactors due to need to apply the installed (stainless-steel) base, now the reactor can be much easily adapted to the process. This will have an impact on the bioreactor landscape in future. Although many suppliers still stick to the traditional stirred tank, new and very effective bioreactor designs come-up. Newly designed types of impellers might exchange traditional impellers. Reactor shapes (vessel versus pillow-shaped bags) will depend on application and cost of materials. The costs of disposables become a more important factor.

Single-use – applying gamma irradiation for sterilization – results in special requirements for the construction materials used in terms of leachables and extractables, but also for strength and life-time of materials. New sensor designs may be needed as well. It is expected that in the coming 5–10 years the bioreactor landscape will be fully dominated by single-use in biomanufacturing as well as in R&D and process development environment [4].

References

1 Singh (1999) Disposable bioreactor for cell culture using wave-induced agitation. *Cytotechnology*, **30** (1–3), 149–158.
2 Langer, E.S. (2012) Single-use bioreactors get nod. *Genet. Eng. Biotechnol. News*, **32** (14), 16.
3 De Wilde, D. et al. (2014) Superior scalability of single-use bioreactors. *BioProcess Int.*, **12** Suppl (8), 14–24.
4 BioPlan Associates (2015) 12th Annual Report and Survey of Biopharmaceutical Manufacturing Capacity and Production, Rockville, MD, April, Available at http://www.bioplanassociates.com/12th.
5 Wagner, R. (2015) Value added perfusion-based manufacturing. *Eur. Biotechnol.*, **14**, 86–87.
6 FDA (2006) Guide to Inspections Validation and Cleaning Processing.
7 Thomas, A. and Munk, M. (2015) Meeting the demand for a new generation of flexible and agile manufacturing facilities. *BioProcess Int.*, **13** (Suppl 11), 16–23.
8 Oosterhuis, N.M.G., Neubauer, P., and Junne, S. (2013) Single-use bioreactors for microbial cultivation. *Pharm. Bioproc.*, **1** (2), 167–177.
9 Bestwick, D. and Colton, P. (2009) Extractables and leachables from single-use disposables. *Bioprocess Int.*, **7** Suppl (1), 88–94.
10 Uetwiller, I. (2006) Testing and validation of disposable systems. *Gen. Eng. News*, **26** (3), 54–57.
11 Vanhamel, S. and Masy, C. (2011) Production of disposable bags: a manufacturer's report, in *Single-Use Technology in Biopharmaceutical Manufacture* (eds R. Eibl and D. Eibl), John Wiley & Sons, Inc., pp. 113–134.
12 Rothe, S. and Eibl, D. (2011) Systems for coupling and sampling, in *Single-use Technology in Biopharmaceutical Manufacture* (eds R. Eibl and D. Eibl), John Wiley & Sons, Inc., pp. 54–65.
13 Woo, L. and Sandford, C.L. (2002) Comparison of electron beam irradiation with gamma processing for medical packaging materials. *Radiat. Phys. Chem.*, **63**, 845–850.
14 Wolton, D. et al. (2015) Standardization of disposables design: the path forward for a potential game changer. *BioProcess Int.*, **13** (11), 21–23.
15 Chotteau, V., Zhang, Y., and Clincke, M.-F. (2015) Very high cell density in perfusion of CHO cells by ATF, in *Continuous Processing in Pharmaceutical*

Manufacturing, 1st edn (ed. G. Subramanian), John Wiley & Sons, Inc., pp. 339–356.

16 Goudar, C.T., Piret, J.M., and Konstantinov, K.B. (2011) Estimating cell specific oxygen uptake and carbon dioxide production rates for mammalian cells in perfusion culture. *Biotechnol. Progr.*, **27** (5), 1347–1357.

17 van 't Riet, K. and Tramper, J. (1991) *Basic Bioreactor Design*, Marcel Dekker Inc..

18 Royce, P.N.C. and Thornhill, N.F. (1991) Estimation of dissolved carbon-dioxide concentrations in aerobic fermentations. *AIChE J.*, **37** (11), 1680–1686.

19 Tramper, J. et al. (1986) Shear sensitivity of insect cells in suspension. *Enzyme Microb. Technol.*, **8**, 33–36.

20 Kossen, N.W.F. and Oosterhuis, N.M.G. (1985) Modelling and scale-up of bioreactors, in *Biotechnology*, vol. 2 (eds H.J. Rehm and G. Reed), Wiley-VCH Verlag GmbH, pp. 572–605.

21 Hillig, F., Porscha, N., Junne, S., and Neubauer, P. (2014) Growth and DHA production performance of the heterotrophic marine microalgae *Crypthecodinium cohnii* in the wave-mixed single use reactor CELL-tainer. *Eng. Life Sci.*, **14**, 254–263.

22 Moser, A. (1981) *Bioproceßtechnik*, Springer Verlag.

23 Sieck, J.B. et al. (2014) Adaptation for survival: phenotype and transcriptome response of CHO cells to elevated stress induced by agitation and sparging. *J. Biotechnol.*, **189**, 94–103.

24 Oosterhuis, N.M.G. (2015) Perfusion process design in a 2D rocking single-use bioreactor, in *Continuous Processing in Pharmaceutical Manufacturing*, 1st edn (ed. G. Subramanian), John Wiley & Sons, Inc., pp. 155–163.

6

Two Mutually Enabling Trends: Continuous Bioprocessing and Single-Use Technologies

Marc Bisschops,[1] Mark Schofield,[2] and Julie Grace[3]

[1]*Pall Life Sciences, Scientific Laboratory Services, Nijverheidsweg 1, 1671 GC Medemblik, The Netherlands*
[2]*Pall Life Sciences, Applications R&D, 20 Walkup Drive, Westborough, MA 01581, USA*
[3]*Pall Life Sciences, Scientific Laboratory Services, 20 Walkup Drive, Westborough, MA 01581, USA*

6.1 Introduction

Biopharmaceutical manufacturing has relied on batch processing in stainless steel systems for decades. This mode of operation has served the industry well; it has been key to the success of many biopharmaceuticals and over the past several years manufacturing efficiency has significantly improved. These successes have suppressed the need for a paradigm shift in biopharmaceutical manufacturing. However, the biopharmaceutical landscape is rapidly changing. The introduction of biosimilars, multiple drugs targeting the same indication, a desire for regional rather than centralized manufacturing, and the possibility of blockbuster biopharmaceuticals that require manufacturing on the multiple ton scale supply additional challenges to bioprocess. Biopharmaceutical companies have recognized these trends and are preparing themselves for a future in which manufacturing efficiency becomes a key to success.

In other industries, process intensification has been employed for more than a century to improve efficiency of manufacture and reduce cost. Possibly, the best known example is the story of the Ford Model T. Henry Ford was a pioneer in introducing manufacturing efficiency. Ford combined two strategies together to realize his vision of a car for the masses. The first strategy, which was not a new concept even in automobile manufacture, was to develop parts that were interchangeable. They could be assembled directly without additional modification. The second strategy was to develop a continuously moving assembly line so cars could be bolted together as they moved down the line.

Through these advances Ford was able to reduce the price for the Model T by 71%. With that, Ford made cars available to a much wider audience than before. In less than a decade, the number of cars registered in the US increased from 9 million (in 1920) to 26 million (1929).

At this point, perhaps, we can see some parallels between car manufacturing and bioprocess. Single-use systems are "plug and play," they do not need to be sterilized before and after a process. As such they mirror the interchangeable parts in car manufacture. There are additional parallels in continuous processing: In the automobile production line, the car is gradually pieced together as it moves down the line; in bioprocess, the target molecule also moves from one step to the next, but for the most part contaminants are being removed as it progresses through the process.

Ford's success lay in combining the two strategies and realizing that interchangeable parts facilitated the move to continuous production. In bioprocess, we may also see that the synergies between single use and continuous processing also yield dramatic improvements in efficiency. However, the relationship between the two might be somewhat different. The move to continuous bioprocesses means that instead of each step being performed once or twice on the very large scale, unit operations can be greatly reduced in size and each step performed hundreds or perhaps even thousands of times. This reduction in scale further facilitates the move to single use technologies.

This chapter reviews the synergies and analogies between the rise of single-use technologies in the biopharmaceutical industry and the advent of continuous bioprocessing.

6.2 Single-Use Technologies

6.2.1 History of Single-Use Technologies

Single-use technologies have been used in the pharmaceutical industry since the early 1980s. At this time, filter manufacturers started producing small capsule filters to replace filters that previously needed stainless steel housing assemblies for use. The capsules could be sterilized through autoclaving, thus eliminating the need to sterilize in place. Additionally, smaller lab-scale syringe filters became available presterilized through gamma irradiation. The success of these initial formats led to the manufacturing of larger scale capsules by the end of the 1980s. By early 1990s, higher area 10 in. modular filters were common and these are still being used to this day. Presterilized production-scale filter capsules were beginning to see adoption around this time as well [1].

In the 1980s, polymer films were developed into 2D format biocontainers commonly referred to as "pillow" bags. These bags were commonly used for the storage of up to 1600 l of media, serum, and buffer. Through the 1990s, 3D bags were being manufactured in increasing volumes up to 3000 l and corresponding totes for containment and support soon followed.

As users became more comfortable with these early single-use options, requests surfaced for "systems" that integrated filter capsules and biocontainers via tubing and corresponding single-use connections (hosebarbs, "Y"s, "T"s, etc.). These pre-constructed systems were provided with sterile claims through the continued use of gamma irradiation. Totes and containment systems had to be improved during this time to address initial concerns over leaking and potential damage to the flexible and delicate systems.

The adoption of single use was furthered in 2007 with the introduction of the first sterile connection device (Pall Corporation, Port Washington, NY) to facilitate secure fluid transfer preserving product integrity and reducing product waste. Today, we see single-use equipment being used in almost all stages of biopharmaceutical production, the technology sees applications from bioreactors all the way to fill/finish needles.

6.2.2 Single-Use Upstream Processing

Single-use technologies such as T-flasks and roller bottles have been used for preculture operations since the 1970s. The first single-use bioreactor was launched by Vijay Singh (who established the company Wave Biotechnology) in 1996. The Wave bioreactor consists of a pillow-shaped single-use bioreactor bag that is placed on a heated rocker. The headspace in the bag is inflated to allow it to maintain its shape and aeration relies on the liquid–gas interface. The wave bioreactor remains particularly popular for cell expansion/preculture operations. However, the limited scalability has kept it from becoming the dominant design for large scale manufacturing.

The success of the Wave, despite its limited scalability, led to the development of single-use stirred tank bioreactors and the first 250 l stirred tank bioreactor was launched in 2004. These bioreactors rely on a stainless steel housing that provides the rigid structure for a 3D bag mounted with an integrated (single-use) agitation device. This can either be a paddle or a more traditional stirrer. In 2009, the first single-use bioreactor with 2000 l capacity was launched, enabling single-use commercial manufacturing. Currently, various (including Sartorius, Millipore, GE, Thermo, and Pall) single-use bioreactor designs are available with working volumes up to 2000 l.

The current expression levels for monoclonal antibodies in CHO platforms have increased by one order of magnitude over the past 15 years. As a result, one single-use bioreactor nowadays has the same manufacturing capacity as a traditional 20 000 l stainless steel bioreactor 15 years ago. This has driven the adoption of the single-use stirred tank bioreactor, which has become the favored design for single-use bioprocess, finding a wide range of applications in clinical manufacturing for various biopharmaceuticals since its initial launch.

6.2.3 Single-Use Downstream Processing

6.2.3.1 Tangential Flow Filtration
The first single-use capsule for tangential flow filtration (TFF) was launched in 2007 by Pall Corporation. Since then, single-use solutions for TFF have become available at various scales. A TFF system essentially consists of pumps, valves, pressure sensors (+ optional flow sensors), a product container, a recirculation container, and a membrane module. All components that are directly in touch with the process solutions are currently available in single-use format.

6.2.3.2 Chromatography Steps

The first, and still the most successful, single-use applications in the field of chromatography involve membrane adsorbers. The best known example is probably the use of the membrane adsorber with Q-functionality (strong anion exchange, based on quaternary ammonium group) for removal of impurities such as endotoxins, host cell proteins, and viruses.

More recently, the introduction of the first single-use chromatography system, the Äkta ready (GE Healthcare), in 2008 was a milestone in the use of single-use technologies for downstream processing, offering ready to use disposable flow paths eliminating the need for cleaning. The launch of this system coincided with the introduction of prepacked chromatography columns by GE Healthcare (Uppsala, Sweden) and Repligen Corporation (Waltham, MA). Although the technologies became available, single-use chromatography remains cost prohibitive for large-scale applications in manufacturing. For routine manufacturing, the chromatography column needs to be run for many cycles in order to depreciate the costs of the expensive chromatography media. As a result, one should expect that the use of single-use chromatography in its traditional operating mode will be limited to (early stage) clinical manufacturing, where reuse of the chromatography media is highly uncommon.

6.2.4 Early Skepticism

Despite the many declared and possibly "perceived" benefits to single use, there have been skeptics. Early on, single-use adoption was limited to small-scale and development activities. The debut of larger bioprocess containers (up to 2,000 l and beyond), accompanied by the increase in protein titers from cell cultures expanded the scope of disposables. The main objections against the use of single-use components for biopharmaceutical manufacturing were costs and environmental concerns. In addition to this there were concerns, which still linger, related to leachables and extractables.

Most initial skeptics of single-use manufacturing acknowledged the fact that single-use manufacturing strategies would reduce the capital expenses associated with building or expanding manufacturing facilities. It was, however, repeatedly pointed out that the recurring costs of the single-use components would have a significant impact on the operational expenses of manufacturing [2]. Various studies were performed to support or reject the hypothesis that single-use bioprocessing has a positive impact on the overall costs of goods. It is beyond the scope of this chapter to present a thorough review of the economic impact [3,4]. Notably, Langer and Ranck outlined some aspects to consider in evaluating the return on investment for single-use bioprocessing [5]. It seems fair to state that there is no generic answer and that a case-to-case evaluation would be required to determine the economic impact of single-use bioprocessing.

Environmental concerns about the disposal of single-use were expressed by various experts [2]. These effects, however, are balanced by the sanitization, cleaning, and/or sterilization necessitated by stainless steel equipment. This involves chemicals and high-purity water that also carry a significant carbon footprint. Many companies, both vendors and end users, have engaged in complex modeling activities to determine the environmental impact of using single-used devices in bioprocessing compared to traditional stainless steel systems. The current consensus can be summarized by the findings of Whitford and Scott [6] "most advanced studies have concluded that for most installations, disposables reduced the environmental footprint (ecological stress) and impact of a biomanufacturing facility. Very rigorous comparative analyses indicate that single-use bioprocess technologies exhibit lower environmental impact than reusable bioprocessing technologies in all impact categories examined. From terrestrial ecotoxicity to marine eutrophication to ozone depletion – in the long run, SU-based manufacturing has been determined to be more environmentally friendly."

The most highly cited quality concern over single-use technologies is extractables and leachables. Langer and Nader reported extractables and leachables to be the highest among the top four current concerns regarding single-use [7]:

1) Leachables and extractables (75.9%)
2) Breakage of bags and loss of production material (67.5%)
3) Vendor dependence and/or single-source issues (63.8%)
4) Material incompatibility with process fluids (63.8%)

Extractables are compounds that can migrate from a material into a solvent under exaggerated test conditions, and leachables are materials that are shown to migrate into the drug product under normal test conditions. Safety risks associated with extractables and leachables are real: Replacing human serum albumin with polysorbate 80 in the drug epoetinum alfa led to an increased incidence of antibody positive pure cell aplasia [8]. Additionally, leachables may be toxic in their own right, but it is possible that they may also affect product efficacy and stability.

Evaluating leachables and extractables is new territory for many users and that has supplied a barrier to adoption. To address this, suppliers have invested heavily to allay these concerns and nowadays, more and more single-use materials (and components) have been tested to provide assurance that product quality and patient safety is not at risk. As a result, many skeptics have turned around over the last few years to acknowledge that single-use biomanufacturing is a viable alternative to stainless steel facilities.

6.2.5 Current Trends and Future Predictions

The pinnacle of single use bioprocessing is the Amgen facility in Singapore that opened in 2014. This "next generation biomanufacturing facility" deploys single-use technologies throughout the whole process, including single-use bioreactors and disposable plastic containers. Productivity is enhanced through the use of these technologies combined with continuous purification processing and real-time quality analysis. Through the use of single use technologies, Amgen expects a 60% reduction in the cost of the bioprocessing [9]. The perceived success of this facility is currently generating a lot of interest in the bioprocessing community and is likely to be imitated.

6.3 Continuous Bioprocessing

6.3.1 Continuous Upstream Processing

Large-scale production of recombinant proteins is normally performed by either a fed-batch or a perfusion process. Fed-batch relies in the gradual addition of nutrients to improve productivity and growth. Product is only harvested at the end of the process. In a perfusion process, fresh media is constantly added to the bioreactor while product and growth inhibiting substances are constantly removed. As such, perfusion can be seen as a continuous process and maybe considered an important driver for operating a continuous downstream process as well.

Most commercialized biopharmaceutical products are manufactured through fed-batch processes, but perfusion cell culture technology has always been of interest for the industry as it offers high specific productivity, allowing a larger manufacturing capacity in a smaller bioreactor. Some biopharmaceutical products are manufactured through perfusion cell culture at substantial scale. Of the documented examples, the most notable are infliximab (Remicade, Johnson Biologics) and recombinant factor VIII (Kogenate FS, Bayer Healthcare).

That said, most companies have tended to avoid perfusion cell culture when possible. Only in situations where fed-batch processes were less effective because of product stability, for example, have perfusion systems been implemented in manufacturing. The main challenge for perfusion has been the complexity of cell retention. Initially, spin filters were proven to be the most reliable and effective cell retention system. However, the main limitation of spin filters is their scalability: bioreactor volumes scale with the cube of the radius, whereas the surface area of the spin filter scales with the square of the radius. As a result, an internal spin filter can occupy a significant part of the bioreactor at manufacturing scale [10].

Gravity settlers have been used as an alternative to spin filters. The main drawback of gravity settlers is the relatively long residence time of the cells in the gravity settler under conditions that could be suboptimal for the cells. In addition to this, gravity settlers are less robust in terms of handling significant variations in perfusion rates.

Many of the concerns associated with cell retention in perfusion bioreactors have been addressed by the Alternating Tangential Flow filtration (ATF) technology developed by Refine (now part of Repligen Corporation, Waltham, MA). The ATF technology employs hollow fibers filters to retain the cells in the perfusion bioreactor. This has contributed to revive interest in perfusion bioreactors. However, despite the alternating flow, product is often seen to be increasingly retained by the hollow fiber filter over time due to membrane fouling.

A promising future alternative for cell retention is the acoustic wave separation technology developed by FloDesign Sonic (commercialized through Pall Life Sciences). Here, a standing 3D sound wave is employed to retain cells, but this approach has not yet been commercialized for larger scales.

Even though the cell retention technology has significantly improved over the past decades, one major drawback of perfusion bioreactors remains: high volume

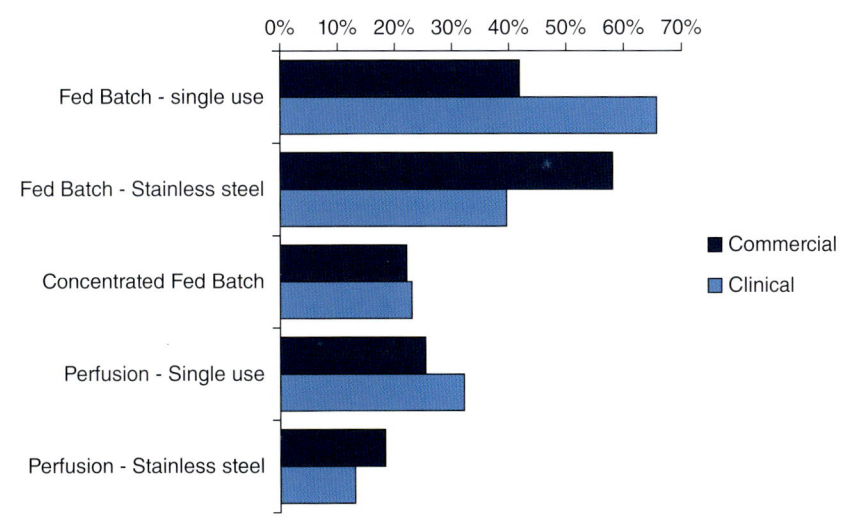

Figure 6.1 Preference for specifying certain bioreactor technologies for clinical or commercial scale manufacturing. The bars indicate the number of respondents who indicated the likelihood of specifying that bioreactor technology as "likely" or "very likely." (Reproduced with permission from Ref. [11]. Copyright 2013, Bioplan Associates.)

of media. Perfusion rates in the range of 1–3 bioreactor volumes per day are fairly common. With specific productivities on the order of 1 g/l/day, the product concentration coming from the bioreactor is limited to 0.3–1 g/l. In the near future, improvements in cell culture technologies could potentially increase this to 3 g/l/day, but at this time fed-batch expression has already advanced to produce 3–5 times higher titers.

In a recent report, Bioplan Associates reported the results of a survey among over 300 industry professionals on their opinion on continuous bioprocessing [11]. One of the topics addressed was preference for bioreactor technologies for clinical and commercial scale manufacturing (Figure 6.1).

The study suggests that single-use bioreactors are a mature technology, being the first choice for clinical scale manufacturing and approaching the importance to stainless steel bioreactors for commercial scale manufacturing. The study also shows that perfusion bioreactor implementation is considered "likely" or "very likely" with around half the frequency of fed-batch bioreactors.

6.3.2 Continuous Downstream Processing

Here, unit operations can be separated into two classes. The first class involves flow through of the product containing solution. Examples are centrifugation, normal flow filtration, and flow-through chromatography. These types of processes lend themselves naturally to be implemented in a continuous bioprocessing platform. Filters will be clogged after a certain time, but in such case they can be configured in an alternating configuration so that the clogged filter can be regenerated or exchanged while a backup filter is in service. A similar scheme can be implemented for flow-through chromatography columns that will eventually be saturated. As long as the time for exchanging or regeneration is (significantly) shorter than the service time, a continuous flow operation can be maintained.

The second class of unit operations is that process in which the product is retained within the unit operation. This can – for instance – involve a traditional tangential flow filtration system, in which the product is recirculated and retained while buffer and smaller molecular weight components permeate through the membrane. Other examples are chromatography steps in which the product is captured on the media and virus inactivation processes that require a certain contact time. This "second class" of unit operations is not so easily adopted to continuous processing. They can still be employed, but their semicontinuous mode of operation normally requires the periodic output of these processes to be collected in a surge vessel to smooth the flow to the next step.

6.3.2.1 Tangential Flow Filtration

Tangential flow filtration traditionally involves recirculation of the process fluid across a recirculation tank. As a result, this process step does not lend itself readily to continuous bioprocessing.

As a response to this, the single-pass TFF system was developed and commercialized by Pall Corporation (Port Washington, NY). Here, the process fluid enters the TFF module and is distributed over multiple cassettes in parallel (Figure 6.2). As liquid permeates the membranes, the volume of the process fluid decreases. This is compensated by distributing the process fluid over fewer cassettes in parallel as the process proceeds. This allows a high transmembrane flow velocity, and hence optimal transmembrane flux throughout the entire flow path [12].

More recently, the in line concentrator (ILC) was launched (Pall Corporation, Port Washington, NY) as a single-use cassette for single-pass TFF. Contrary to the traditional SPTFF system, this module does not rely on external instrumentation for controlling volumetric concentration factor (the ratio of initial volume to retentate volume).

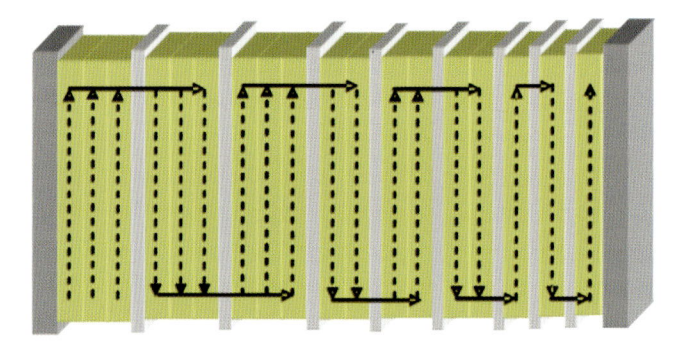

Figure 6.2 Typical flow path in the Single-Pass TFF system.

6.3.2.2 Continuous Chromatography

Capture chromatography by nature tends to be operated as a batch process since the product solution does not continuously flow through the system. In large-scale chemical manufacturing industries, including food processing and fine chemicals, ion exchange processes were converted to continuous unit operations by implementing a multicolumn approach. This has resulted in carrousel-type continuous ion exchange processes and simulated moving bed technology.

The concept of a multicolumn capture chromatography process is schematically shown in Figure 6.3. On the left-hand side, this diagram shows a batch chromatography column operated in down-flow. As the process solution is applied on the column, the chromatography material gradually attains equilibrium with the feed solution. Inside the column, a mass transfer zone is developed above which the chromatography media is fully saturated and below which all protein of interest has captured from the solution. As a result, the product is only actively being captured in the mass transfer zone. The diagram in the middle and on the right-hand side of Figure 6.3 shows how a continuous multicolumn chromatography process only focuses on that mass transfer zone, eliminating the idle zones above and below. In addition to this, the column that is taken from the top of the load zone is saturated well beyond its dynamic binding capacity. Any material that breaks through the first column in the load step is captured on one or more subsequent columns. This allows a continuous chromatography process to be operated at a much higher capacity utilization than batch chromatography processes.

The potential of multicolumn chromatography for biopharmaceuticals was recognized relatively early for transgenic biopharmaceutical products [13]. Idec Pharmaceuticals (now Biogen) was one of the pioneers providing proof of concept using a carrousel-type SMB for monoclonal antibodies [14]. This approach confirmed the potentials of multicolumn chromatography for biopharmaceutical applications in terms of specific productivity gains and reduction of buffer consumption.

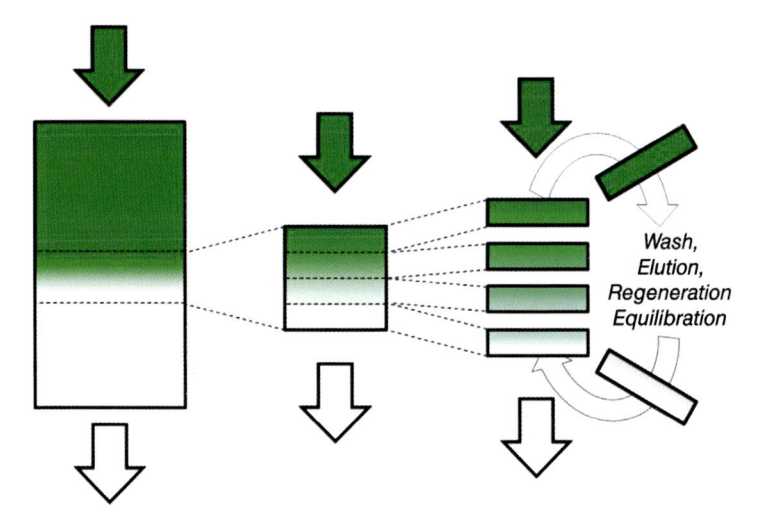

Figure 6.3 Conceptual diagram of a continuous multicolumn chromatography process.

It was not until 2006 when suppliers launched suitable hardware solutions that the bioprocessing industry seriously started exploring the potentials for a variety of chromatography applications. Most early studies involved the transformation of the Protein A chromatography process into a continuous alternative [15]. Subsequent case studies demonstrated the capability of continuous multicolumn chromatography for other modes as well, including aggregate removal through cation exchange chromatography and hydrophobic interaction chromatography [16–18]. In addition to this, the purification of vaccines (virus-like particles) was explored using ion exchange and size exclusion [19,20].

6.3.3 Concerns for Continuous Bioprocessing

Over the past few years, the biopharmaceutical industry has witnessed a growing interest in (integrated) continuous bioprocessing solutions. Nonetheless, quite a few concerns within the industry remain. Most of them relate to cGMP aspects and quality systems.

Munk recently presented results from a survey among industry professionals in which opportunities, challenges, and concerns on continuous bioprocessing were addressed [21]. The top three operational concerns are listed in Figure 6.4. The concerns were rated on a scale ranging from 0 to 5 (critically important). The maximum ranking of around three suggests that concerns exist, but the expectation is they should be manageable.

Additional concerns include the complexity of continuous bioprocess equipment, sampling strategies, and process control strategies [22]. On the other hand, various opportunities for continuous bioprocessing were identified, particularly with respect to potential improvement of product quality control [23]. The following are the most prominent promises of continuous bioprocessing:

- Improved product quality through improved process consistency
- Improved process understanding, facilitated by mathematical modeling
- Advanced process control and PAT
- Real-Time Release Testing (RTRT)

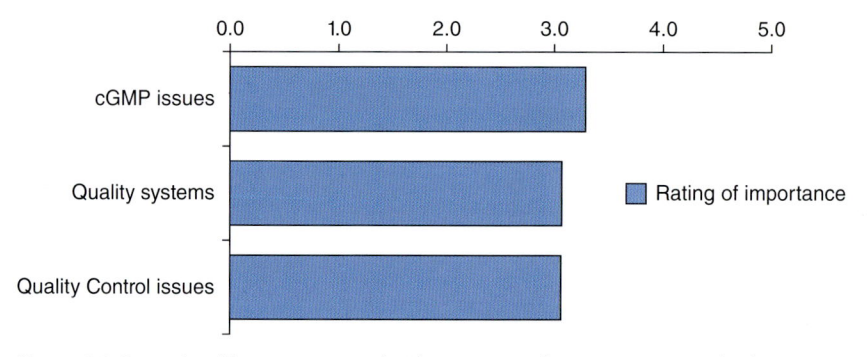

Figure 6.4 Operational issues associated with continuous bioprocessing, results from a survey among bioprocess professionals [21]. The importance was rated on a scale from 0–5 (5 being critically important).

It may be clear that many of these opportunities will require a lot of development before they can be realized. A few studies in this area, however, seem to support the hypothesis that the amount of process data that is generated by continuous bioprocessing will help in providing evidence for consistent manufacturing and will help in identifying potential process deviations before they would become problematic [24].

6.4 Integrated Single-Use Continuous Bioprocessing: Case Studies

With continuous bioprocessing technologies becoming available for various steps in the overall downstream processing cascade, it makes sense to explore the impact of combining multiple steps into a cascade. Preliminary work that combines multiple continuous processing steps into a small island of continuous manufacturing showed significant gain-specific productivity due to the synergy between the two steps [25].

However, several companies are exploring completely integrated continuous bioprocessing from cell culture to drug substance. The most advanced examples to date have been demonstrated by Sanofi-Genzyme (Framingham, MA), Merck (Kenilworth, NJ), and Bayer Technology Services (Leverkusen, Germany). The most recent work by these companies is reviewed below.

6.4.1 Case 1: Genzyme

Genzyme has commercialized various recombinant human enzymes. These proteins are intrinsically unstable and lose activity during fed-batch bioreactor processes. Because of challenges related to product stability, upstream manufacturing for these recombinant human enzymes has been implemented on a perfusion bioreactor. This is because the perfusion rate is normally around 1 bioreactor volume per day, so the product is present in the perfusion bioreactor for a much shorter time than in the fed-batch process that may be operated for more than 14 days until harvested.

The perfusion process may be operated for 60 days or more. This makes a compelling argument to couple this upstream continuous production with continuous downstream purification. As such, this served as a starting point for developing a fully continuous manufacturing platform. In one of the earlier papers, the combination of a perfusion bioreactor and a continuous capture chromatography step was described for the production of recombinant human enzymes and monoclonal antibodies [26]. More recently, proof of principle for a fully integrated continuous manufacturing process for monoclonal antibodies was presented [27].

The upstream part of the proof of concept study involved a 12 l bioreactor with an ATF technology for cell retention. The bioreactor was operated at 50–60 million cells/ml and perfusion rates were 2–3 bioreactor volumes per day (0.04–0.05 nl/cell/day). The specific productivity stabilized around 0.8 g/l/day, yielding 10 g/day product (feed concentration in the effluent was in the range of 0.28–0.42 gm/l).

The feed solution was passed through a 0.2 μm filter into a small 2 l surge bag before entering the capture chromatography step. This step was implemented on a 3C-PCC system (GE Healthcare) operating three Protein A columns. The 3C-PCC system was operated such that a column was eluted approximately every 4 h. As a result, the periodicity of the eluted antibody provided sufficient time for repetitive batch inactivation time in a mixed container. The virus inactivation involved a 1 h incubation at lower pH. The elution peaks were thus inactivated one by one before the intermediate product solution was further purified.

Before polishing, the inactivated product was diluted to reduce conductivity by in-line addition of dilution buffer. The intermediate product solution was then applied on another 3C-PCC system with three cation exchange columns (Capto S). The eluted product was directly applied on a membrane adsorber that was directly connected to the 3C-PCC system. The cation exchange process was operated such that product was eluted every 2 h and the membrane adsorber was washed and regenerated every 24 h. The elution conditions of the cation exchange process were designed such that no further ultrafiltration/diafiltration was needed before formulation.

In the proof of principle study, bioburden control was addressed by adding azide to the solution prior to entering the first capture chromatography step. No integrated process control system was used to ensure that volumetric flow rates between the subsequent unit operations were automatically balanced.

6.4.2 Case 2: Merck

Merck & Co. Inc. (Kenilworth, NJ) has been one of the pioneers of integrated continuous bioprocessing. Where Genzyme took a step-by-step approach [26,27] toward proof of principle of the integrated biomanufacturing platform, Merck started integrating unit operations from the very beginning, using single-use elements where possible.

The laboratory is designed in a horseshoe alignment, starting with a 10 l perfusion bioreactor on one end and going all the way to a drug substance on the other outer end [28]. In the center of the horseshoe, there is a centralized analytical island performing various assays on samples that can be taken from nearly any point in the process. Figure 6.5 shows a panoramic view of the integrated bioprocessing laboratory in Kenilworth, NJ.

Figure 6.5 Panoramic view of the horseshoe lay-out of the continuous monoclonal antibody laboratory at Merck & Co. Inc. (Kenilworth, NJ). The process runs from the bioreactors on the left-hand side to the polishing and final filtration steps at the right-hand side. On the very far right-hand side, the analytical island is shown.(Reproduced with permission from Ref. [28]. Copyright 2015, Mark Brower.)

The perfusion bioreactor is operated around 1 g/l/day specific productivity and the product titers coming from the perfusion bioreactor are approximately 1 g/l. Cell retention is performed using a single-use format tangential flow filtration system by Spectrum Labs. Protein A chromatography is implemented on a BioSMB system (Pall Life Sciences). The sterility of the BioSMB system is protected using a guard filter that is replaced daily. The eluted product is collected in a small surge vessel to manage the intermittent flow coming from the Protein A multicolumn chromatography system. The virus inactivation is then performed in a plug flow reactor that is sized to provide the required incubation time. After addition of base in another small surge container, the inactivated and quenched product is applied to anion exchange column. The flow through of which is directly applied on to a second BioSMB system operating the cation exchange chromatography step (Poros HS, Thermo) in capture mode to remove aggregated product [17].

The approach Merck took is unique in the way the entire bioprocess system is operated continuously from perfusion to bioreactor and all unit operations are fully controlled so that the flow rates coming from one process system into the next process system are automatically balanced. In addition to this, all samples for product characterization and in process control are automatically taken and analyzed using an LC-MS system and a UPLC. All unit operations, as well as the analytical systems, are controlled through a centralized process control system based on DeltaV. The platform is designed to support the implementation of multivariate data analysis tools, allowing adaptive control of the unit operations.

6.4.3 Case 3: Bayer Technology Services

Bayer Technology Services (Leverkusen, Germany) has taken an approach that targets a platform that is both continuous and disposable [29]. Once fully assembled, the platform aims at a flow path that is a (technically) closed system, and hence it would be conceivable to operate the entire manufacturing in a ballroom facility. With this approach, Bayer has estimated that the entire manufacturing from perfusion bioreactors to drug substance can be done in a ballroom with 19.2×9.6 m area only. This facility of the future would include two perfusion bioreactors (200 l), all downstream processing systems and tanks for buffer preparation and media preparation.

A scale-down pilot plant for such biomanufacturing facility has been established and operated in a (non-cGMP) laboratory space at Bayer Technology Services with 7×14 m floor area. The proof of concept facility was operated in two sections that were disconnected. The first section covered the upstream process that was performed using two 10 l perfusion bioreactors with inclined gravity settlers for cell retention. The bioreactors were operated at a perfusion rate of 3 bioreactor volumes per day. The stationary viable cell density was in the range of 60–70 million cells/ml and the cell specific productivity was 5.3 pg/cell/day. The clarified harvest had a product concentration of approximately 0.115 mg/ml.

The clarified harvest was filtered via two Sartoguard NF filters. After a few days, the maximum pressure on one filter was reached after which the product flow would automatically be directed to the second filter. Meanwhile, the first filter was replaced by a new filter, using a tube welder to maintain aseptic conditions. The clarified harvest was collected in a reservoir in which it was preconcentrated, using a hollow fiber ultrafiltration module. This also enabled concentration of the harvest from the bioreactors to 0.70 g/l.

Since the total manufacturing capacity of the upstream section and the downstream section was not fully aligned, these parts were operated as two disconnected cascades. The downstream processing operations were however, linked together to establish a fully continuous purification platform, consisting of the following main elements.

The Protein A chromatography step was performed using a BioSMB. This process was designed such that both the feed solution and eluted product run continuously (uninterrupted).

The virus inactivation is performed using a coiled flow inverter that ensures close-to-plug-flow behavior, even at lower linear velocities. As a result, the residence time distribution during incubation is reasonably well controlled. Prior to entering the plug flow reactor, acid is dosed into the product solution to reach the required incubation pH. At the end of the virus inactivation loop, the solution is neutralized by adding TRIS buffer to the product solution. The dose pump for the acidification and the neutralization are both operated with a feedback control loop with a pH probe.

The polishing consists of two chromatography steps. In the pilot plant operated in Leverkusen, this was performed via a mixed mode (Capto Adhere, GE Healthcare) and an anion exchange chromatography step (Hypercel STAR AX, Pall Life Sciences). Both steps were operated in flow-through mode. The mixed mode chromatography step was configured as a multicolumn chromatography process with a countercurrent load step and the anion exchange process was operated in flow-through mode with two units alternating. Both steps combined in one BioSMB® system. Final downstream processing involved additional filtration, virus filtration, formulation, and bioburden reduction steps.

The entire process was automated using a Siemens PCS7 process control system. Except for the upstream section (bioreactors and cell retention), the entire continuous downstream processing cascade was established in a single-use format.

6.4.4 Comparison

All three case studies provide proof of principle for integrated continuous bioprocessing and are compared in Table 6.1. It is interesting to recognize some of the similarities as that may give us an understanding of the future direction of continuous processing. For instance, all the processes rely on perfusion upstream and Protein A-based multicolumn chromatography for the capture step. If we look at differences, we can see where there might be some remaining challenges, these include cell retention and virus inactivation.

Table 6.1 Comparison of the integrated continuous processes implemented by Merck, Genzyme, and Bayer.

Bioreactor	Genzyme	Merck	Bayer
	Perfusion (12 l)	Perfusion (10 l)	Perfusion (2 × 10 l)
Cell retention	ATF	TFF	Inclined gravity settler
Capture	3C-PCC MabSelect SuRe	BioSMB MabSelect SuRe	BioSMB MabSelect SuRe
Virus inactivation	Stirred vessel, one elution peak at a time	Repetitive batch or plug flow reactor	Coiled flow inverter (plug flow reactor)
Polishing (1)	3C-PCC Capto-S	Anion exchange	BioSMB (mixed mode)
Polishing (2)	Sartobind Q	BioSMB Cation IX	Anion exchange
Automation	None	DeltaV	Siemens DCS
Reported capacity PoC	8 g/day	10 g/day	22.7 g/day
Analytics	Offline	Integrated	Offline

6.4.5 Challenges and Solutions

All three groups report bioburden challenges. Many of the bioburden issues relate to the fact that aseptic fluid handling in a development laboratory is inherently more difficult than in a controlled manufacturing environment (clean room area). Pumps and connectors at laboratory scale are generally not designed for closed and/or aseptic operations.

Bioburden is being addressed by Genzyme by introducing azide in the clarified cell supernatant. This approach helps in demonstrating proof of concept at laboratory scale. For larger-scale operation, Genzyme appears to depend upon a combination of sanitization and sterilization in place.

Merck uses filters directly after the chromatography pumps to reduce bioburden in the product solution and the use of single-use technologies where applicable. Bayer Technology Services relies heavily upon the use of (gamma irradiated) single-use components even at laboratory scale.

Process control and process automation becomes more important as multiple unit operations are interconnected. Even with (small) interstage surge bags as described by Merck and Bayer Technology Services, the time-averaged flow rate going into a next unit operation needs to be equal to the flow rate of the intermediate product solution coming from the previous unit operation. As a result, local control of the individual process skids is no longer sufficient and some sort of supervisory process control needs to be implemented. The integrated continuous bioprocess platforms established by Merck and Bayer Technology Services seem to be most advanced in this aspect, relying on DeltaV and Siemens PCS7 process control systems, respectively. As a result, these companies also seem to be most advanced in terms of the integration of analytical technologies, moving toward adaptive process control and eventually real-time release testing.

6.4.6 Alternative Scenarios

All three case studies already presented focus on a combination of a perfusion bioreactor in combination with a continuous downstream process. As Konstantinov and Cooney suggested, one could also consider hybrid models in which batch and continuous unit operations are combined to establish the most efficient manufacturing platform [30]. In an earlier occasion, Merck already presented proof of concept for the combination of a fed-batch bioreactor and a continuous downstream process [17]. This case study showed how the timelines for purifying a batch of monoclonal antibodies could be reduced from 5 or 6 days to approximately 24 h by using continuous downstream processing. This would have a significant impact on facility throughput.

In traditional manufacturing industries, this combination of batch upstream and continuous downstream is actually one of the dominant designs. Many antibiotics, amino acids, and carboxylic acids that are produced through bacterial fermentation are manufactured in a fed-batch bioreactor process combined with continuous downstream processing. This is a result of the fact that these large-scale manufacturing processes are mainly driven by operating expenses rather than capital expenses. As a result, specific productivity is not necessarily the target function for optimization. Instead, minimizing the use of auxiliary materials such as water and chemicals has become the main driver.

In this respect, (intensified) fed-batch processing is (still) superior to perfusion processes, simply because the amount of product produced per unit volume of cell culture media is higher in a fed-batch process. In spite of the recent revival of perfusion cell culture processes, this may well become the more attractive platform for more stable recombinant proteins and monoclonal antibodies.

It is also this scenario in which the combination with a single-use format would have significant added value. The impact of the changeover time between subsequent manufacturing batches or campaigns becomes more significant as the net manufacturing time decreases. As a result, the impact of shorter turnover times between campaigns will have a larger impact.

6.5 Regulatory Aspects

In the early discussions of continuous bioprocessing technologies, regulatory aspects were often presented as the biggest hurdle on the road toward implementation. Over the past few years, however, regulatory agencies have repeatedly expressed their support for exploring continuous manufacturing in the pharmaceutical industry [31–33]. This support was driven by the ambition to modernize pharmaceutical manufacturing into a more flexible and agile industry, with an enhanced capability for process control and product quality control.

Some of the questions related to regulatory aspects still need to be addressed, particularly for continuous downstream processing. The most often addressed question relates to the batch definition. For hybrid processes, it seems natural to define a batch based on the noncontinuous operations. For a hybrid process consisting of a fed-batch cell culture process in combination with a continuous downstream process, it seems to be an obvious choice to have the bioreactor schedule dictating the batch definition.

For fully integrated continuous processes (perfusion campaigns), various options have been mentioned. The batch definition could either be based on a certain processing time, a certain amount of product produced or it could even be related to the use of raw materials (including the time of a single-use assembly). Eventually, one could argue that the batch definition should be based on a risk-based decision. Larger batches will reduce the costs associated with quality control testing, whereas smaller batches will reduce the impact of a batch rejection.

Some other challenges on the road toward validating continuous bioprocessing technologies for downstream processing could relate to reliable scale-down models for process validation (including viral clearance studies) and quality by design studies. Some of these challenges can be (partially) mitigated by *in silico* process studies, but these will never replace experimental process characterization studies.

6.6 Adoption Rate of Single-Use and Continuous Bioprocessing

Munk [21] gave an estimate where the various manufacturing technologies are in the Rogers' technology adoption lifecycle curve. This curve represents the adoption of new technologies by various groups of customers during the course of time. According to Munk, the early majority of customers are currently implementing single-use bioprocessing technologies at a fast rate, whereas continuous bioprocessing is still in its early adoption stage, somewhere in between the innovators and the early adopters (see also Figure 6.6).

The adoption of single-use bioreactor technologies for actual manufacturing was enabled by the significant increase in expression levels observed over the past decade. Titers for monoclonal antibodies increased from below 1 gm/l to an average of 3.5 gm/l nowadays, and some companies are consistently in the 5–10 gm/l range. This allows manufacturing the same amount of antibodies in a much smaller bioreactor. This increase in titer coincided with the launch of single-use bioreactors with working volumes up to 2000 l, allowing production batches with 5–20 kg of product.

The increase in titer also caused the manufacturing bottleneck to shift from the bioreactor toward the downstream processing and the primary capture step in particular. Even though the binding capacity of affinity chromatography media has nearly doubled over the past 10–15 years, the chromatographic capture steps still lag behind with the throughput that has been realized in upstream processing. The limitations were not only caused by limitations

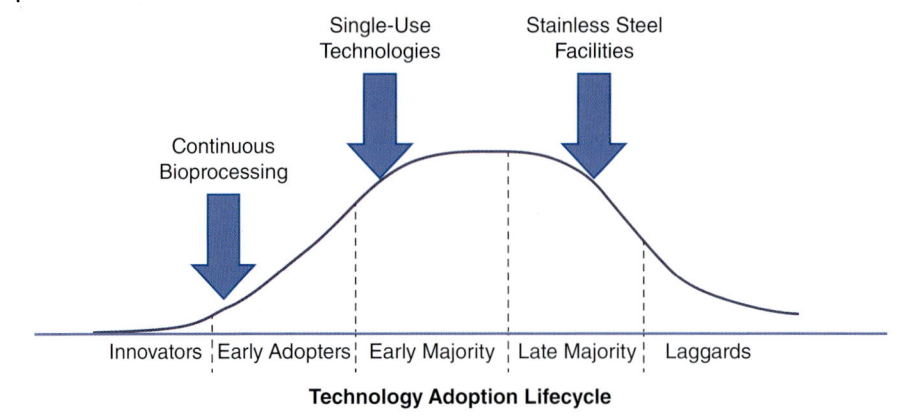

Figure 6.6 Rogers' bell curve describing the technology adoption lifecycle for technologies, adapted from Munk [21].

column sizes and media volume. The challenges related to the increasing buffer consumptions were equally important and in some facilities even the most problematic limitation (e.g., [34]).

6.7 Conclusions

Over the past decade, the downstream processing throughput has become one of the key challenges in the biopharmaceutical industry. These challenges were mainly caused by the success of biopharmaceutical medicines on the market in combination with the significant increase in expression levels in the upstream process.

The increasing titers have also enabled the use of single-use bioprocessing technologies for late-stage clinical and commercial manufacturing in upstream processing. The higher concentrations allowed manufacturing relevant amount in bioreactors that then became small enough to be designed in a viable single-use format.

Most of the downstream processing steps have traditionally been sized around the mass of protein that was to be purified. As a result, these could not benefit from the process intensification successes that were achieved in the upstream process. On the contrary: with increasing titers, the downstream processing unit operations became the limiting steps.

The required process intensification that is needed to resolve the downstream processing bottleneck that was created, however, can be realized by continuous bioprocessing. In addition to mitigating the throughput bottleneck, continuous bioprocessing will also allow the downstream processing unit operations to be designed in a more compact format, and hence a single-use design for the entire downstream processing platform becomes a viable option. The technologies that are required for this transition are currently available or will be launched in the next 12 months.

The development and scale-up of single-use bioreactor technologies followed the titer increase reasonably well, but for continuous downstream processing, the need for a more efficient manufacturing strategy is immediate. As a result, one should expect that the adoption of continuous downstream processing would not necessarily have to take the same time that it took for single-use bioreactors. The conceptual designs for continuous downstream processing technologies were presented in 2007. Over the past 5 years, the interest and (initial) adoption of continuous bioprocessing technologies has really taken off and viable commercial scale equipment will be implemented within 10 years after its initial launch.

References

1. Martin, J. (2011) A brief history of single-use manufacturing. Biopharm International Supplement, November.
2. DePalma, A. (2009) Single-use systems make headway with skeptics. *Genet. Eng. News*, **29**, 19.
3. Monge, M. and Sinclair, A. (2009) Disposables cost contributions: a sensitivity analysis. *Biopharm. Int.*, **22**, 4.
4. Sinclair, A. (2008) How to evaluate the cost impact of using disposables in biomanufacturing. *Biopharm. Int.*, **21**, 6.
5. Langer, E. and Ranck, J. (2005) The ROI case: economic justification for disposables in biopharmaceutical manufacturing. Bioprocess International, October Supplement, pp. 46–50.
6. Whitford, W.G. and Scott, C. (2014) Single-use and sustainability, Bioprocess International, April Supplement.
7. Langer, E. and Nader, R.A. (2014) Single-use technologies in biopharmaceutical manufacturing: a 10-year review of trends and the future. *Eng. Life Sci.*, **14**, 238–243.
8. Pang, J., Blanc, T., Brown, J., Labrenz, S., Villalobos, A., Depaolis, A., Gunturi, S., Grossman, S., Lisi, P., and Heavner, G.A. (2007) Recognition and identification of UV-absorbing leachables in EPREX pre-filled syringes: an unexpected occurrence at a formulation-component interface. *PDA J. Pharm. Sci. Technol.*, **61** (6), 423–432.
9. Amgen (2014) Amgen Outlines Strategy, Growth Objectives And Capital Allocation Plans. Press release, October 28.
10. Bonham-Carter, J. and Shevitz, J. (2011) A brief history of perfusion biomanufacturing. *Bioprocess Int.*, **9** (9), 24–30.
11. Langer, E.S. (2013) 10th Annual Report and Survey of Biopharmaceutical Manufacturing. Bioplan Associates, April.
12. Casey, C., Gallos, T., Alekseev, Y., Ayturk, E., and Pearl, S. (2011) Protein concentration with single-pass tangential flow filtration. *J. Membr. Sci.*, **384** (1–2), 82–88.
13. Fulton, S. (2001) Ton-scale production of recombinant protein pharmaceuticals. Presentation held at Recovery of Biological Products X, Cancun, June 7.

14 Thömmes, J., Sonnenfeld, A., Conley, L., and Pieracci, J. (2003) Protein A affinity simulated moving bed chromatography. Presented at Recovery of Biological Products XI, Banff, September 15.

15 Bisschops, M. (2012) BioSMB™ technology: continuous countercurrent chromatography enabling a fully disposable process, in *Biopharmaceutical Production Technology* (ed. G. Subramanian), Wiley-VCH Verlag GmbH, Weinheim, Germany, pp. 769–791.

16 Allen, L. (2011) Developing purification unit operations for high titre monoclonal antibody processes. Presented at IBC Antibody Development and Production Conference, Bellevue WA, March.

17 Brower, M. (2011) Working towards an integrated antibody purification process. Presented at IBC Biopharmaceutical Manufacturing and Development conference, San Diego CA, September.

18 Pieracci, J., Mao, N., Thömmes, J., Pennings, M., Bisschops, M., and Frick, L. (2010) Using simulated moving bed chromatography to enhance hydrophobic interaction chromatography performance. Presented at Recovery of Biological Products XIV, Lake Tahoe, August.

19 Jiang, H. (2010) Purification of H5N1 and H1N1 VLP based vaccines. Presented at IBC Single-Use conference, La Jolla CA, June 15.

20 Bisschops, M. (2011) Disposable SEC in vaccine purification using BioSMB technology. Presented at PREP Conference, Boston, July 12.

21 Munk, M. (2015) Continuous bioprocessing: what is holding the industry back from implementing continuous bioprocessing more widely. Presented at PDA Manufacturing Science Workshop, Washington DC, October 01.

22 Kaltenbrunner, O. (2015) Design and control of continuous processing for biopharmaceutical manufacturing. Presented at PDA Manufacturing Science Workshop, Washington DC, October 01.

23 Lee, S.L. (2015) Regulatory initiatives for supporting innovation in pharmaceutical manufacturing. Presented at PDA Manufacturing Science Workshop, Washington DC, September 30.

24 Bisschops, M., Strawn, M., To, B., and Coffman, J. (2014) Multivariate data analysis: managing large amounts of data coming from continuous biomanufacturing processes. Presented at Recovery of Biological Products XVI, Rostock, Germany, July 28.

25 Rogler, K., Gjoka, X., Martino, A., Ayturk, E., Gantier, R., and Schofield, M. (2015) Productivity and economic advantages of coupling single-pass tangential flow filtration to multi-column chromatography for continuous processing. Presented at PREP Conference, Boston, MA, July 29.

26 Warikoo, V., Godawat, R., Brower, K., Jain, S., Cummings, D., Simons, E., Johnson, T., Walther, J., Yu, M., Wright, B., McLarty, J., Parey, K.P., Hwang, C., Zhou, W., Riske, F., and Konstantinov, K. (2012) Integrated continuous production of recombinant therapeutic proteins. *Biotechnol. Bioeng.*, **109** (12), 3018–3029.

27 Godawat, R., Konstantinov, K., Rohani, M., and Warikoo, V. (2015) End-to-end fully integrated continuous production of recombinant monoclonal antibodies. *J. Biotechnol.*, **213**, 13–19.

28 Brower, M. (2015) Protein Refinery Operations Lab (PRO Lab): a sandbox for continuous protein production & advanced process control. Presented at Integrated Continuous Bioprocessing (ICB2) Conference, Berkeley (CA), November.
29 Klutz, S., Magnus, J., Lobedann, M., Schwan, P., Maiser, B., Niklas, J., Temming, M., and Schembecker, G. (2015) Developing the biofacility of the future based on continuous processing and single-use technology. *J. Biotechnol.*, **213**, 120–130.
30 Konstantinov, K.B. and Cooney, C.I. (2014) White Paper on Continuous Bioprocessing. Presented at International Symposium on Continuous Manufacturing of Pharmaceuticals, Cambridge, MA, May 20.
31 Moore, C.M.V. (2011) Continuous manufacturing – FDA perspective on submissions and implementation. Presented at PQRI Workshop, Bethesda, MD, September 13.
32 Chatterjee, S. (2012) FDA perspective on continuous manufacturing. Presented at IFPAC Annual Meeting, Baltimore, January.
33 Woodcock, J. (2014) Modernizing pharmaceutical manufacturing – continuous manufacturing as a key enabler. Presented at International Symposium on Continuous Manufacturing of Pharmaceuticals, Cambridge, MA, May 20.
34 Conley, L. (2015) Improving manufacturing network productivity: overcoming the downstream bottlenecks. Presented at the 4th International Conference on Accelerating Biopharmaceutical Development, Scottsdale, AZ, March 3.

7

Perfusion Formats and Their Specific Medium Requirements

Jochen B. Sieck, Christian Schild, and Jörg von Hagen

Merck Life Science (a business of Merck KGaA), Process Solutions, Cell Culture Media R&D, Frankfurter Strasse 250, 64291 Darmstadt, Germany

7.1 Introduction

Perfusion refers to a bioreactor process featuring continuous harvesting of supernatant and concurrent supply of fresh medium, while cells are retained in the bioreactor by means of a cell retention device. Perfusion processes are highly productive due to the ability to cultivate cells at high densities by continuous addition of nutrients and parallel removal of metabolites [1–4]. Cell densities up to 2×10^8 cells/ml depending on the perfusion rate, the cell-line, and the chosen medium have been reported [5,6]. Highly controlled perfusion processes operate for months and produce fragile proteins with a reproducibly high quality [7]. There is also a lower level of impurities due to the high cell viability, the possibility to integrate a continuous downstream process, and the compatibility with Quality by Design (QbD) and monitoring by process analytical technology (PAT) [8]. Continuous protein production in perfusion mode with integrated affinity chromatography has been applied with an uninterrupted run time of 30 days [9]. Typical process durations are in the range of 2 weeks up to several months. In order to achieve steady-state at high cell densities, partial removal of cells is necessary to improve viability by preventing the accumulation of dead cells [10].

Advantages of perfusion are a high volumetric productivity, steady-state product quality, and low residence time of product in the bioreactor. On the other hand, perfusion is more complex in process control and equipment requirements, with potentially higher failure rates [11]. Perfusion also has high medium consumption as one of the main cost factors and the resulting volumes of harvest containing relatively low concentration of product needs to be handled in downstream [11].

7.1.1 History of Perfusion

First industrial perfusion manufacturing appeared in the 1980s. Even using the delicate cell retention devices and low yielding cell lines available then, relatively high cell concentrations and productivity compared to batch processes were reported [12]. The most common retention devices used at this time were gravimetric settlers, which showed limited cell retention capabilities, and spin filters, which were challenging in terms of overall reliability, fluctuating cell retention performance, and scale-up. The change of spin filter units was impractical and perfusion was repeatedly interrupted [13]. The need of a continuous supply of fresh media required more reliable retention devices to achieve higher cell densities and volumetric productivities [14].

In the 1990s, improvement of cell lines resulted in fed-batch processes that were now able to achieve product concentration around 1 g/l with lower technical complexity and higher ease of operation in up to $20\,m^3$ bioreactors [15]. The success of fed-batch cultivation decelerated the development of spin-filter technology and other cell retention devices. Fed-batch cultivation became the dominant cultivation method in upstreamin the last decades [16]. Within the last 25 years, the volumetric yields have increased 20-fold in industrial processes [17]. The foundations for these developments were improved cell lines, stronger expression systems, improved cell culture media, and better process controls [7].

Nevertheless, perfusion continued to be used for instable molecules like factor VIII, interferons, and therapeutic enzymes, and gradually cell retention technology improved as more reliable technologies became available [11].

7.1.2 Comeback of Perfusion

Perfusion has made a comeback in the last decade because of the high volumetric productivity and improved quality [16]. One of the main advantages compared to fed-batch cultivation is to reach high viable cell densities with a high productivity in relatively small bioreactors with long term continuous cultivation [14]. The trend for using single-use technology has advantages in low initial investments, operating costs, and allows for high flexibility at lower risk of contamination. However, the maximum working volume is currently limited to 2000 l [18,19]. In perfusion mode, these relatively small single-use systems are able to reach yields that are comparable to fed-batch processes in $10-20\,m^3$ scale [11]. With the capacity to produce both stable and nonstable biopharmaceuticals, and the combination with single-use technology, perfusion is developing into an enabling technology to implement manufacturing in smaller and more flexible production facilities of the future [11].

Thus there is also an increasing demand for improved perfusion media that allow for higher volumetric productivity and lower medium demand. Medium development targets will, therefore, typically be related to process performance (e.g., increased viable cell density or cell-specific productivity), steady-state maintenance (e.g., control growth rate, improve viability, avoid limitations), or improving product quality [20–22]. As the focus for medium development has been on fed-batch media and feeding strategies for the last decades, this chapter seeks to review the state of the art in development of animal origin free, chemically defined perfusion media.

7.2 Characterization of Perfusion Processes

In this section, the most important parameters characterizing perfusion processes will be reviewed, starting with the fundamental variables to describe perfusion processes.

The perfusion rate (P) is defined as the rate at which fresh medium is added to the bioreactor. To achieve a constant volume, the harvest rate (H) should be equal to P. Typically, these rates are expressed as specific rates (volume of medium per volume of bioreactor per day (vvd)). If a cell discard strategy (bleeding) of cell culture broth is applied, the bled volume (B) needs to be resupplied as well, increasing the perfusion rate P (day^{-1}) compared to the harvest rate H (day^{-1}).

$$P = B + H, \tag{7.1}$$

where P = perfusion rate (day^{-1}), B = bleed rate (day^{-1}), H = harvest rate (day^{-1}) [21].

For "open" cell retention devices (CRDs) with a cell retention <100%, the harvest rate can be mathematically split into a cell-free harvest and a bleed stream equal to the bioreactor cell density. This allows simulating the performance of "open" CRDs using "closed" CRDs using an additional cell discard line to realize the representative cell loss of the "open" CRD system, facilitating a process transfer from one CRD technology to another.

The cell-specific perfusion rate (CSPR) is an important parameter to describe the nutrient supply per cell in a continuous bioprocess. Elegant perfusion operations can be realized using CSPR control [2,4]. Perfusion rate and the viable cell density are required for CSPR calculation, shown in Eq. (7.2).

$$\text{CSPR} = \frac{P}{X}, \tag{7.2}$$

where CSPR = cell-specific perfusion rate (nl/cell/day), P = perfusion rate (day^{-1}), X = cell density (10^6 cells/ml) [21]. (*Remark:* In this chapter, CSPR will be expressed in pl/cell/day.)

The CSPR describes the available volume of media per cell and day and is an indicator for the nutrient supply per cell. The optimum CSPR and minimum CSPR vary for each cell culture medium and cell line, typically ranging from 50 to 500 pl/cell/day [21]. CSPR$_{min}$ is defined herein as the minimum flow rate of fresh medium for cultivation of a given cell line. When exceeding CSPR$_{min}$ even temporarily through uncontrolled growth of the culture or interrupted medium supply, growth rate, viability, and specific productivity may decrease. Controlling CSPR at a constant level (CSPR-stat) allows for consistent nutrient concentrations at changing cell densities, for example, during the exponential growth phase [14]. A reduction of CSPR to perfusion rates near CSPR$_{min}$ allows achieving product concentrations similar to fed-batch processes through lowering the volumetric flow rate. This reduces the volumetric medium demand and increases medium utilization.

Medium utilization refers to the amount of nutrients added to the perfusion process that is consumed during the process. A high medium utilization is indicated by low residual concentrations of nutrients (e.g., glucose, amino acids, etc.) in bioreactor supernatant and harvest. One objective for perfusion media development is increasing medium utilization; however, some nutrients may need to be maintained at minimum concentrations to avoid cellular responses of limitation.

Medium depth (Eq. (7.3)) describes the maximum cell density that can be cultivated with a given medium flow rate and is the reciprocal of the minimum CSPR. The minimum CSPR can be determined by allowing cells to grow in perfusion at a fixed absolute flow rate. As cell density increases, the CSPR will decrease until growth stops when $\text{CSPR}_{\min}$ is reached [21]. Perfusion medium development typically targets to increase medium depth and decrease CSPR so that either a constant cell density can be maintained using less medium, or to increase the cell density of the process to improve productivity. Thus these values are important characteristics for the comparison of perfusion media [23].

$$\text{Medium depth} = \text{CSPR}_{\min}^{-1} = X_{\max} \times P^{-1}, \tag{7.3}$$

where CSPR = cell-specific perfusion rate (nl/cell/day), P = perfusion rate (day^{-1}), X = cell density (10^6 cells/ml) [21].

If steady-state processes are targeted, perfusion processes should be operated with a certain offset (safety margin, S) to $\text{CSPR}_{\min}$ to facilitate optimum growth rate, high viability, and high specific productivity. This value is defined here as $\text{CSPR}_{\text{crit}}$.

$$\text{CSPR}_{\text{crit}} = (1 + S) \times \text{CSPR}_{\min}, \tag{7.4}$$

where CSPR = cell-specific perfusion rate (pl/cell/day), P = perfusion rate (day^{-1}), S = offset constant or "safety margin" (−).

The safety margin required depends on the process, controls, cell line, medium composition, and operational handling. In general, decreasing the safety margin will increase productivity as well as medium utilization, but also results in higher risk of process failure, for example, due to variations in the process parameters or interruptions of perfusion. Increasing the safety margin will result in a process more robust against small excursions resulting in higher medium flow rate and reduced medium utilization.

For the determination of metabolic rates, mass balancing is applied considering constant flow rates and volume [24]. The specific consumption rate can be determined from medium exchange rate, VCD, and the difference of substrate concentration in fresh medium and supernatant [24,25]. The derivative for a substrate in perfusion processes is shown exemplary in Eq. (7.5):

$$\begin{aligned}
\frac{dc_S \times V}{dt} &= F_{\text{in}} \times c_{S,\text{in}} - F_{\text{out}} \times c_{S,\text{out}} - r \times V, \\
F_{\text{in}} &= F_{\text{out}} = F, \\
P &= \frac{F}{V}, \\
\frac{dc_S}{dt} &= P \times (c_{S,\text{in}} - c_{S,\text{out}}) - r, \\
r &= q_S \times X, \\
\frac{dc_S}{dt} &= P \times (c_{S,\text{in}} - c_{S,\text{out}}) - q_s \times X,
\end{aligned} \tag{7.5}$$

where V = bioreactor volume (l), F = flow (l/day), P = perfusion rate (day^{-1}), c_S = substrate concentration (g/l), r = reaction rate (g/l/day), q_S = specific substrate consumption rate (g/cell/day), X = cell density (10^6 cells/ml) [24].

Equation (7.5) is used to determine the consumption and production rates of substrates, metabolites, and products in a perfusion bioreactor without substrates or metabolite retention.

7.2.1 Productivity of Perfusion Processes

Higher productivity over fed-batch processes is often claimed as a major advantage of perfusion processes. However, due to the constant dilution of the bioreactor content, the titers in perfusion processes are typically lower compared to fed-batch processes, unless the product is retained in the bioreactor. This leads to different definitions of productivity, which are reviewed in the following section.

The volumetric productivity (VP (g/l/day)) is defined as the yield of product per bioreactor volume per time. It is calculated from the current product concentration multiplied with the harvest rate and describes the current productivity at a given time.

$$VP_i = c_{P,i} \times H_i = X_i \times q_{P,i}, \tag{7.6}$$

where VP = volumetric productivity (g/l/day), $c_{P,i}$ = concentration of product on day i (g/l), H = harvest rate (day^{-1}), X = cell density (10^6 cells/ml), q_P = cell specific productivity (pg/cell/day) [21].

The yield (Y (g)) of upstream production is defined as the mass of accumulated product. For fed-batch processes, yield is equal to product concentration of the harvest multiplied by harvest volume. In the case of "concentrated fed-batch" processes, the product is retained in the bioreactor and is harvested similar to a fed-batch process. For high cell densities, the solid biomass (packed cell volume (%)) is a significant fraction of the total broth volume and needs to be considered [26].

In perfusion, the daily yield equals the accumulated mass of product since process start.

$$Y_i = \int_0^i c_{P,i} \times H_i \tag{7.7}$$

where Y = yield (g), c_P = product concentration (g/l), H = harvest rate (day^{-1}).

Space time yield (STY (g/l/day)) is equal to the yield Y (Eq. (7.7)) divided by the process duration and bioreactor volume and sometimes confused with volumetric productivity, but is a more accurate definition of the overall productivity of an upstream production process. The difference in productivity between fed-batch and perfusion processes is most suitably explained in terms of STY [16].

$$STY_i = \frac{Y_i}{V_{BR} \times (t_i - t_0)}, \tag{7.8}$$

where STY (g/l/day), Y = yield (g), t = process time (days) [27,28].

For the calculation of cellular productivity, Eq. (7.9) can be used.

$$q_{P,i} = ((c_{P,i} - c_{P,i-1}) \times (t_i - t_{i-1})^{-1} + P \times c_{P,i}) \times X^{-1}, \qquad (7.9)$$

where q_P = specific cellular productivity (pg/cell/day), P = perfusion rate (day^{-1}), c_P = product concentration (g/l), X = cell density (10^6 cells/ml) [24].

It should be noted that in contrast to the yield, q_P has to be calculated based on the perfusion rate P, not the harvest rate H if a bleed strategy is applied. In this case, according to Eq. (7.1), P is higher than H due to the bleed rate B. The product produced by the cells but lost through bleeding needs to be taken into account to avoid an underestimation of q_P.

7.2.2 Cell Retention Devices

To achieve perfusion, a cell retention device (CRD) needs to be attached to or included in the bioreactor. The selection of a CRD system has a strong impact on the perfusion process, its viable cell density, and productivity. CRDs are classified as "closed" systems if a physical barrier (e.g., a filtration membrane) provides 100% cell retention, while "open" CRDs will allow a fraction of cells to pass through the system and leave the bioreactor [29]. While closed systems will allow for very high cell densities, dead cells and debris accumulate in the bioreactor, resulting in lower culture viability and membrane fouling, as indicated by partial product retention ("sieving") or blocking of the filter [30]. The aspects of the most commonly used systems, ATF and TFF, were studied by Clincke et al. and Karst et al. [5,31].

In contrast, "open" retention devices may be easier to operate over long process durations as dead cells and debris are constantly removed from the bioreactor [29,32–35]. However, parameters of open CRDs require additional optimization and their effective cell densities, and thus the possible volumetric productivity of the resulting perfusion process are ultimately limited. In addition, cell retention efficiency has to be considered in scale-up and tech transfer to achieve comparable performance. Finally, the harvest will not be cell-free and seamless coupling to downstream operations [9,36] would require an additional cell removal step.

7.2.3 Steady-State Definition

Continuous processes offer the advantage that they can be operated in steady state, indicating that no changes in critical process parameters occur as long as the steady state persists. In mammalian cell perfusion processes, steady-state duration is typically limited due to the genetic instability of organisms and limited operational stability of cell retention devices.

For a full steady state, it is required that all inputs and outputs as well as all bioreactor parameters remain constant over time, for example, viable cell density, viability, product concentration and quality, metabolite and nutrient concentrations, pH, and DO. However, some practitioners use the term steady state to describe that some controlled parameters are in steady state, for example, a (online) controlled biomass steady state.

It may be required to observe the process for several days to determine if a steady state was reached because some nutrients might be consumed at only slightly higher rate than the rate at which fresh nutrients are added, such that a limitation will only become apparent after some time. In summary, a full steady state in cell culture perfusion is difficult to achieve, to maintain, and to demonstrate [8].

7.3 Perfusion Formats

In this section, differentiation between perfusion processes and their resulting impact on the perfusion medium requirements will be discussed.

Perfusion processes can be segmented in different sequential phases: after inoculation, the cells will typically grow exponentially for several days (growth or ramp-up phase). Perfusion will be initiated and perfusion rate gradually increased to avoid nutrient limitations and to achieve high cell densities. When the target VCD is achieved, the process will be directed toward maintaining this VCD for a certain period (transition phase), ideally resulting in a full steady state after several days (steady-state phase). It is worth noting that the cellular nutrient requirements differ in each phase (e.g., growth versus steady-state phase) and the media need to be optimized accordingly [20].

A critical aspect when designing the perfusion process is the stability of the product. Some products (e.g., blood factors, therapeutic enzymes, and some fusion proteins) are prone to biochemical modification in the bioreactor, for example, enzymatic cleavage, oxidation, deamination, and so on, which negatively impact the products therapeutic efficacy. For such products, the residence time in the bioreactor becomes a critical process parameter. These products were the traditional field where perfusion was consistently applied for manufacturing despite the rise of fed-batch processing, as the fed-batch concept does not allow for controlling the residence time. The maximum residence time RT_{max} can be assessed experimentally [21] and the perfusion process designed accordingly. As long as perfusion rate is selected to maintain $RT<RT_{max}$, medium and process optimization can target achieving higher cell densities and/or productivities to increase the overall process productivity while sustaining product quality and efficacy [21]. Residence time distribution in CSTRs should be considered [37].

Another important differentiator between perfusion processing formats currently discussed is if long-term steady state is actually desired. Steady-state operation typically requires active control of the cell density, for example, by discarding cell culture broth ("bleeding") [21]. As this strategy also includes discarding product from the bioreactor, the overall yield of the process is reduced. Strategies to reduce growth rate can be employed, for example, by controlled nutrient limitations, addition of growth inhibitors, or by temperature reduction [26,38–40], to reduce the product loss through bleeding. To avoid product loss, some process formats are performed without bleeding to maximize productivity by sacrificing the steady state [26,41]. For these processes, typically no steady state can be achieved, as the process is highly dynamic. For example, if the harvest rate is set constant and the cells are growing unrestricted, the CSPR will constantly change.

It should be noted that introducing nutrient limitations as well as temperature reduction may alter cellular metabolism [20,42,43]. For example, for the extension of the antibody production phase in batch processes, temperatures between 28 and 37 °C were investigated with hybridoma cells [44]. At 31 °C, higher viabilities and lower glucose uptake rate were found. Other results showed a positive influence of temperature reduction to 33 °C on the cell density and glucose uptake rate [45]. The specific production rate was constant in a range between 34 and 39 °C [46]. With CHO cells it was shown that a decrease of temperature from 37 to 34 °C had no influence on cell density and viability but on the growth rate. The temperature in this range had no influence on the production rate of mAb [38,47]. A transfer of these outcomes from batch processes to perfusion processes could result in lower perfusion rates and savings in medium consumption and volumes that have to be handled in downstream operations. Another advantage of lower fermentation temperatures is the decreased oxygen uptake rate, which can be limiting in large-scale production processes [14]. Through lowering the perfusion bioreactor temperature from 37 to 34 °C the OUR decreased up to 25% resulting in higher volumetric productivity in OTR-limited reactors [38]. To summarize, process temperature may have a strong impact on cellular metabolism and, as a consequence, medium requirements, which needs to be taken into account during process and medium development. If metabolic uptake rates are determined with the goal to optimize nutrient composition, the target process temperature should be applied to ensure that suitable uptake rates are considered.

Nutrient uptake and metabolite production rates also change with CSPR [21]. As already stated, achieving steady state requires a safety margin from $CSPR_{min}$ ($CSPR_{min} < CSPR_{crit}$). In contrast, for processes without cell discard, cell growth will automatically stop when $CSPR_{min}$ and X_{max} are reached, indicating a full exploitation of the available medium depth. Such "dynamic" process schemes allow for simplified process control and shorter process durations at higher cell densities and medium utilization.

7.3.1 Innovative Perfusion Formats

Perfusion is the production process of choice for unstable products. However, there are examples where stable molecules such as monoclonal antibodies are produced in perfusion [11] and practitioners are evaluating various other innovative applications of perfusion processes [48]. As both stable and nonstable molecules can be produced using perfusion, it could be applied as single upstream platform for different biopharmaceutical products in flexible multiproduct manufacturing facilities [11,26,49].

One direction of development is the intensification of bioprocesses using perfusion. As cost pressure increases for the biopharma industry, perfusion is revisited because of the possible gains in productivity over standard fed-batch processes [11]. The required bioreactor sizes can be reduced, allowing for smaller, less capital intensive facilities. Smaller required bioreactor scales also allow using disposable bioreactors for production, which are limited in terms of volume today [11,18]. If increasing productivity for robust molecules is targeted, a format called "concentrated fed-batch" (CFB) has been proposed, featuring bioreactors attached to an ATF system equipped for nanofiltration to not only retain the cells in the bioreactor, but also the product [26,41]. In this scheme, all the spent medium can be discarded as it does not contain any product, which greatly simplifies the perfusion harvest handling. As no bleed should be performed due to the high product concentration in the reactor, no steady state is achieved and product quality needs to be characterized as function of process time or at the end of the process [26]. Harvest and purification are performed batchwise. However, it was reported that the overall yield of these processes is reduced by a high fraction of solids in the bioreactor as well as a lower specific productivity compared to "traditional" fed-batch [26], which may be due to the decreased equivalent CSPR [21].

In contrast, a steady-state strategy is required to implement an integrated continuous biologics manufacturing process, directly coupling upstream production and downstream purification [9]. Using an ATF in microfiltration mode, Warikoo et al. realized a direct coupling of steady-state perfusion production of antibodies as well as therapeutic enzymes to a semicontinuous affinity chromatography step, operated with several columns to allow for continuous loading of the bioreactor harvest on the columns [9,36]. An advantage of long perfusion process durations is that these will improve a manufacturing plant's utilization rate, because fewer downtimes will occur for cleaning and preparation of the bioreactors. In this example, the integrated process could be operated autonomously for several weeks in steady state [9,50].

In addition to the production step of a manufacturing process, perfusion can be applied already in the inoculation train, which is normally performed in batch mode in a series of bioreactors of increasing scale. In these batch processes, cells need to be maintained in exponential growth. To avoid nutrient limitations, cell densities are typically very low. Using perfusion, much higher cell densities can be achieved without limitation, which allows bypassing several expansion steps and improve manufacturing flexibility [51,52]. Cells can also be withdrawn from such a perfused expansion bioreactor and frozen as high cell density, large volume process intermediate [53]. Decoupling the thaw and expansion steps from the production step may greatly increase the scheduling flexibility of a biomanufacturing plant. Furthermore, as all these process intermediates should have the same cultivation history, variability due to small differences in the expansion can be minimized.

7.4 Development Strategies for Perfusion Media

The heterogeneity of perfusion processes already described indicates that defining a perfusion medium formulation to suit all the different types of processes is challenging. Thus, optimal results are typically achieved by medium formulations customized for the given process and cell line, as offered by specialized media developers. Alternatively, medium formulations from literature, available in-house medium platforms, or commercially available media can be screened for promising candidates [20]. Examples are available showing that media designed for batch production processes, which typically provide high overall nutrient concentrations, can be applied for perfusion [54,55]. Alternatively, fed-batch media can be used after fortifying them with their specifically designed companion feeds. Component solubility has to be taken into consideration as well, in particular those of amino acids as they may be interdependent [56]. Once a starting formulation for the given cell line and the intended process has been identified, several routes for development and optimization can be followed.

Many proven methods to improve cell culture media are based on studying cellular metabolism [20]. Generally most straightforward approach is spent medium analysis. Depending on the analytical capabilities consumption of several nutrients as well as production of metabolites can be studied. This data is used to eliminate limitations for the cells on single-component level. One step further are metabolomics studies, taking the whole of the nutrient and metabolite concentrations into account to give a more detailed picture of the cellular metabolism under given conditions [57–61]. Transcriptome analysis can be performed on CHO cells as well [62–66]. The transcriptome gives a full picture of environmental conditions that a cell population responds on transcriptome level at the sampling time point. Schaub et al. have successfully applied next-generation sequencing (NGS) to characterize the transcriptome of high and low producing CHO cells [67]. This allowed to improve titers through supplementation of potentially limiting components. Transcriptomics using CHO microarrays has been applied successfully to validate the scale-up of a perfusion process [63] and study the stress response of CHO cells confronted with hydrodynamic stress from agitation, sparging, and their combination [68]. These "omics" methods have a great potential to study the cellular behavior on a holistic level. However, experimental design needs to take many aspects into account that can cause variation, and the obtained results need to be evaluated carefully to identify the specific responses to the experimental conditions [68,69]. Besides improving media, these methods could be applied to generate advanced cell lines as well [70].

Empirical experimentation can be effectively enhanced using multivariate analysis and experiment design, allowing for dramatically reduced numbers of experiments required to screen large numbers of parameters, for example, components and their concentrations. In many case studies, practitioners are combining multivariate methods and high throughput experimentation to study a wide range of parameters in relatively short timeframes [71–73]. For such approaches, high throughput systems as well as software tools for experimental design and evaluation of results are required. On the other hand, great advantages can be achieved without an extended demand for metabolic understanding or analytical technologies, as the formulations are initially selected and optimized based on easily measurable parameters like viable cell density and productivity.

A general limitation in cell culture medium development is that substrate consumption rates of a given cell line in a new medium composition are unknown. Even if a limiting substrate for one medium and one cell line has been determined, due to the differences in cell lines and clone-to-clone variation, the limiting substrate may differ from cell to cell [20]. One strategy to overcome this issue is the deliberate introduction of a limiting compound to the medium. Ideally, this would be a nutrient that can be measured quickly and easily, for example, glutamine or glucose. If the limiting substrate reaches low concentrations, perfusion rate would be increased to avoid any other limitation. Such a scheme may simplify process development, scale-up, and tech transfer. In addition, the establishment of scale-down models is somewhat simplified, because a controllable limitation can be implemented independent of bioreactor scale.

7.4.1 Cell Line-Specific Requirements

The first Chinese hamster ovary (CHO) cell line was isolated from *Cricetulus griseus* by Theodore T. Puck in the late 1950s [74]. In the 1960s, the CHO cell line was the target for further virus infections and mutagenesis and in this process various cell line adaptions were generated. In 1963, the proline-dependent CHO-K1 cell line was derived from the original one [75–77]. The CHO cell line can be grown in suspension, is easy to handle in scale-up and continuous reactors, and can grow in chemically defined media without supplements like sera or proteins. In 1987, the tissue plasminogen activator was approved by the Food and Drug Administration as the first biopharmaceutical produced in CHO [65]. Today, the cultivation of CHO cells in suspension is widely used in the biopharmaceutical industry to produce monoclonal antibodies and other drug proteins such as growth factors, hormones, and cytokines because of their ability to perform posttranslational modifications and human-like glycosylation [78]. CHO cells reach high viable cell densities over 10^7 cells/ml in batch and fed-batch processes in large volumes up to 20 m^3 [79], and over 10^8 cells/ml in perfusion processes [5,80]. Moreover, they regularly exceed yields of 2 g/l of recombinant proteins by means of selecting host cells, expression vectors, L-glutamine synthetase (GS) engineering, and selection strategies like dihydrofolate reductase deletion (dhfr$^{-/-}$) by mutagenesis [5,6,81–83]. The original CHO cells were the target of modifications and mutagenesis whereby many different lineages exist today. Figure 7.1 illustrates the lineage of CHO-K1, CHO-S, and CHO-DG44 [84].

Each cell line has own requirements regarding selection of medium and included components that are critical for the production of biopharmaceuticals. For some CHO cell lines, specific medium platforms have been developed that take typical metabolic needs of the cells' lineage into account, for example, Cellvento™ or Ex-Cell® media and feeds [84].

Despite the cell line itself, the transfection method plays an important role. CHO cells are frequently transfected using the DHFR or the GS system, using different selection markers typically added supplementary to basal media [64]. The GS system uses the selection of cells that can grow without the addition of glutamine, as the GS is part of the selection system, allowing for selection by eliminating glutamine from the basal medium.

7.4.2 Scale-Down Models for Perfusion Processes

Scale-down model development is a semiempirical method based on reproducing the prevailing conditions of large scale in the scale-down model [85]. Similar to classical scale-up methodology, a scale-down model needs to simulate the

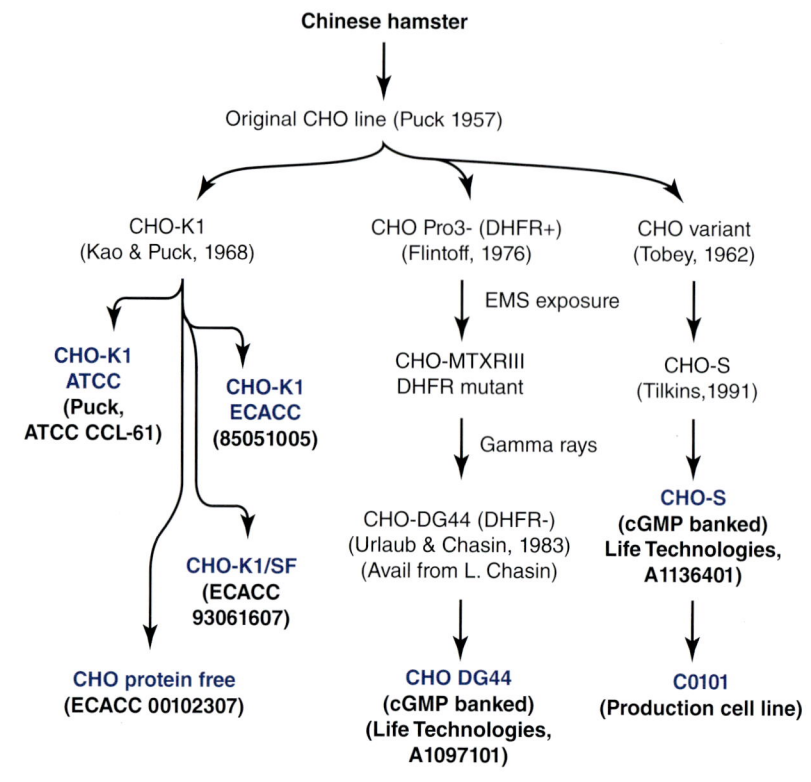

```
                        Chinese hamster
                              ↓
              Original CHO line (Puck 1957)
              ↙               ↓               ↘
         CHO-K1         CHO Pro3- (DHFR+)      CHO variant
     (Kao & Puck, 1968)  (Flintoff, 1976)      (Tobey, 1962)
           ↓              ↓ EMS exposure          ↓
      CHO-K1           CHO-MTXRIII             CHO-S
       ATCC      CHO-K1  DHFR mutant        (Tilkins, 1991)
      (Puck,    ECACC        ↓                   ↓
   ATCC CCL-61)(85051005) Gamma rays
                              ↓                CHO-S
                     CHO-DG44 (DHFR-)      (cGMP banked)
                     (Urlaub & Chasin, 1983) Life Technologies,
        CHO-K1/SF    (Avail from L. Chasin)  A1136401)
         (ECACC              ↓                   ↓
        93061607)
           ↓
    CHO protein free    CHO DG44              C0101
   (ECACC 00102307)   (cGMP banked)      (Production cell line)
                     (Life Technologies,
                         A1097101)
```

Figure 7.1 Lineage and development history of prominent CHO cell lines. Modifications and selection after mutagenesis are the root cause of different cell line characteristics. (Reproduced with permission from Ref. [84]. Copyright 2013, Nature Publishing Group.)

limiting parameter of a process as accurately as possible, for example, the oxygen transfer for many bacterial processes [86–88]. For mammalian cell culture processes, a constant "microenvironment as experienced by the cells" has been proposed as scaling strategy, including pH, temperature, and nutrient supply, but also bioreactor design, specific energy dissipation rates, mixing time, superficial gas velocity, oxygen transfer, gas hold-up volume, and pCO_2 stripping [89]. In perfusion, the most characteristic phenomenon is the continuous medium exchange. Several authors defined the CSPR as the supply of a single cell with fresh media [14,21]. The CSPR is independent of the chosen scale and the cultivation system. Similar CSPRs can be realized in chemostat and perfusion. If all systems have the same CSPR, the cells supposedly have a similar metabolic behavior [21]. This makes the CSPR a suitable scale-down criterion for perfusion processes.

7.4.3 Scale-Down Cultivation Methods

There are many possible cultivation modes to perform perfusion scale-down based on the CSPR criterion. The smallest bioreactor volume that can be operated with cell retention is probably in the range of 0,25 to 1 l. These are normally

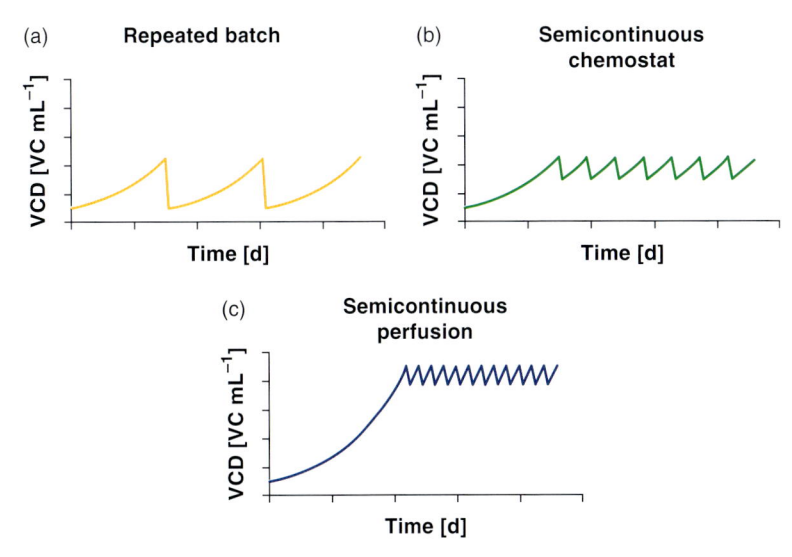

Figure 7.2 Illustration of VCD time course of (a) typical repeated batch, (b) semicontinuous chemostat, and (c) semicontinuous perfusion processes.

individual benchtop systems, not suitable for high throughput experimentation like basal medium screening. For such applications, shaken systems or automated microbioreactors can be applied. These systems need the possibility to be operated, for example, in semicontinuous perfusion with cell retention through centrifugation, semicontinuous chemostat without cell retention, or repeated batch as a simplification of a chemostat process. In Figure 7.2, a comparison of three possible cultivation methods for a scale-down model of perfusion are depicted.

Semicontinuous cultivation is a simplification of a fully continuous process where the medium exchange will be performed in discrete steps like once per day, resulting in dynamic changes of nutrient concentrations over 24 h.

It should be noted that the applicability of scale-down models depends on the medium development stage. For early development, for example, screening for a basal medium for the specified cell line, high throughput is of primary importance and should optimally be performed in automated high-throughput systems. At this stage, the culture systems can be simplified and only need to mimic perfusion to a degree. Once one or few basal perfusion medium formulations have been identified, the demand in scale-down model representativeness increases. For example, if component optimization should be performed on metabolic data (e.g., spent medium analysis), the cells in the scale-down model need to have a representative environment resulting in a comparable metabolism in scale-down model and bioreactors.

Scale-down models are simplifications and should represent the most critical parameters of the full scale process [85]. As it is not possible to reproduce all parameters in a scale-down model at the same time [90,91], typically a confirmation through well controlled bioreactors is necessary.

7.4.4 Examples for Perfusion Scale-Down Applications

In this section, different case studies are described where scale-down models of perfusion processes have been applied. SpinTubes (TubeSpin® Bioreactors, TPP Techno Plastic Products AG, Trasadingen, CH) are 50 ml conical tubes that can be applied for cultivation instead of classical Erlenmeyer shake flasks at lower footprint and higher throughput. These tubes can be used as high-throughput tools for medium and process development if a suitable operating mode is selected [92–94]. Particularly, oxygen transfer needs to be considered [95,96]. In our own case study (see Section 6.4), we present results for applying SpinTube bioreactors in semi-continuous perfusion (SCP), semicontinuous chemostat (SCC), and repeated batch (RB) as simplification of semicontinuous chemostat. It was found that the SCP and SCC modes were comparable in terms of growth and metabolism in perfusion bioreactors. However, the SCP model provided the best comparability with the full-scale perfusion bioreactor in terms of specific productivity [23]. The SCP mode was overestimating the $CSPR_{min}$ values for the medium and cell line combination by about 20% compared to bioreactor results, which is probably due to the daily dynamics if only one exchange operation per day is performed.

To reduce such dynamics, chemostat cultures can be operated using a continuous exponential feed regime followed by sequential medium withdrawal. This was first realized in shake flasks and later transferred to the ambr15™ [97]. Due to the high degree of automation, the ambr system allows for 24–48 parallel cultivations in microbioreactors. Ho et al. have successfully applied the ambr in semicontinuous chemostat in combination with multivariate experimental design for medium screenings, but did not present bioreactor confirmation data [98]. Combining an automated bioreactor platform in chemostat with multivariate experiment design is a powerful combination for medium development [71]. However, due to the regular dilution of the bioreactor chemostat operation will not allow for high cell densities. No cell retention mechanism is available despite sedimentation, which allows for high cell densities but resulted in one case in differences in metabolism and productivity [99]. Ram reviews the available systems for perfusion scale-down models and comes to the conclusion that a mature perfusion microbioreactor platform is currently still a technology gap for the biopharmaceutical industry [100].

Despite the fact that no single ideal platform for perfusion medium and process development exists today, a combination of the available tools at different development stages can be applied. In the early development stages, high throughput is required, which justifies compromises in comparability. Later, more representative scale-down models are required to facilitate optimizing metabolism, productivity, or product quality. A good example for the application of different scale-down models of gradually increasing complexity has been recently presented by Yang, who applied deep well plates for high-throughput screening of media in which the cells grow and produce [50]. Later, they applied SpinTube bioreactors operated in SCP to optimize the medium, because this model provided a good representation of the cellular metabolism and product quality, compared to real perfusion bioreactors [96]. Targeting media providing high q_P and low CSPRs, the group could achieve a fourfold increase in titers and volumetric productivity compared to their starting condition based on commercially available batch or fed-batch media [50,96].

7.5 Process Development for Perfusion Processes

A comprehensive methodology for development and optimization of high cell density perfusion bioreactor processes for a given medium and cell line combination was published by Konstantinov *et al.*, including the "push-to-low" approach [21]. The authors include comments on online and offline parameters requiring implementation and optimization, for example, for controlling viable cell density.

First, the optimization space needs to be determined [21]. The limiting factors of a perfusion process may be the following:

1) The stability of the product in the bioreactor environment may require to remove the product from the bioreactor after a certain residence time (RT_{max}). Residence time distribution in CSTRs must be considered [37].
2) The perfusion rate may be limited by the cell retention device to a given maximum (P_{max}), especially for larger bioreactor scales.
3) A maximum cell density that a bioreactor can support (X_{max}), for example, in terms of OTR. Increasing OTR can be achieved be increasing oxygen flow rate or agitation speed. Both parameters may cause shear stress responses at some point [68]. However, it has been demonstrated that "shear sensitivity" is often overstated and that quite harsh conditions are required to damage modern industrial CHO cell lines in general [101–106]. It was also discussed that achievable cell density may be limited in terms of possible cell volume fraction in cell culture broths [5]. Disposable bioreactors have been applied successfully for high-density perfusion as well [54].
4) Most important in terms of medium development is that $CSPR_{min}$ or $CSPR_{crit}$ will limit the possible X_{max} because of the P_{max} given by the cell retention device. Optimizing the medium for a given perfusion process and cell line would allow for an expansion of the optimization space.

Konstantinov *et al.* outline a complete experimental framework to determine the boundaries of the optimization space as well as subsequent optimization experiments. Following this framework should allow developing the most productive process possible for a given product, medium, and cell line combination.

7.6 Case Study

In the following section, a case study for perfusion scale-down model development will be discussed. The aim of the study was the selection of a scale down model representative of high-density perfusion bioreactors but with increased experimental throughput, intended for medium screening and optimization. CSPR was chosen as the scale-down criterion. In particular, the goal was to increase medium depth (Eq. (7.3)) and to reduce $CSPR_{min}$ (Eq. (7.2)) to allow for reduced medium flow rates at constant or improved q_P (Eq. (7.9)). To determine medium depth, cells were grown in perfusion at a given perfusion rate until the maximum VCD (X_{max}) was reached. For fast and simultaneous screening, 50 ml SpinTube bioreactors were chosen [96]. To increase throughput and reproducibility by reduction of manual handling steps, the growth phase was performed batchwise in 600 ml SpinTubes, centrifuged, and resuspended to achieve a high starting VCD for the semicontinuous perfusion experiment, allowing to reach X_{max} within 5 days.

7.6.1 Material & Methods

Several CHO-S and CHO-K1-derived proprietary CHO cell lines were used in this study, engineered using either the GS or the DHFR transfection system [84]. All cell lines produced monoclonal antibodies.

All experiments used several proprietary dry powder cell culture media that were dissolved in Milli-Q® water. The formulations were chemically defined, animal origin free, and contained no hydrolysates. They were mixed with 1.2–2 g/l sodium bicarbonate and hypoxanthine-thymidine supplement. pH was adjusted to 7.1 ± 0.1 using 2 M sodium hydroxide solution. Osmolality was at 330 ± 20 mOsmo/kg. Final cell culture media were sterile filtered using Stericups®. MSX was used for L-glutamine synthetase-engineered cell lines to select cells [107]. MSX inhibits the endogenous L-glutamine synthetase so that GS-engineered cells are selected in L-glutamine deficient medium [108].

Cryopreserved cells with a volume of 1 ml and a cell density of 10×10^6 cells/ml were thawed and diluted in 29 ml medium. After 3 days of cultivation in an orbital shaker at 37 °C, 5% CO_2, 80% relative air moisture, and 320 rounds per minute with a shaking diameter of 50 mm, the cells were diluted to 0.3×10^6 viable cells/ml for 2 days culture duration or 0.2×10^6 cells/ml for 3 days of cultivation. The cultivation vessels were 50 ml SpinTubes with a nominal filling volume of 30 ml. After four passages, cells were transferred into 600 ml SpinTubes with a nominal filling volume of 400 ml. Figure 7.3 illustrates the preparation of the experiments. A cell density of 0.4×10^6 cells/ml at 37 °C, 5% CO_2, and 200 rounds per minute (A). After 3 days, the vessel (B) was centrifuged with $800 \times g$ at 4 °C for 10 min. After centrifugation (C), the supernatant was discarded and the pellet suspended in fresh medium to achieve target inoculation VCD (D). Then the cell suspension

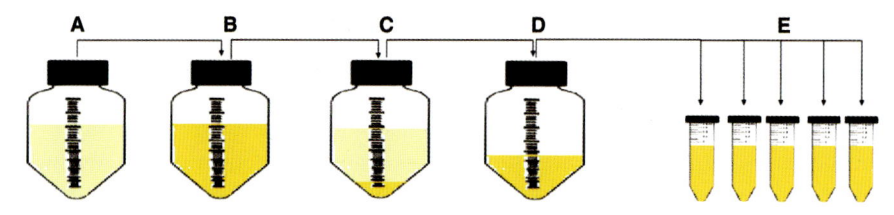

Figure 7.3 Illustration of the cultivation process for media screening. Precultivation using 600 ml SpinTubes, centrifugation, discarding the supernatant, resuspension in less volume, and aliquotation in 50 ml SpinTubes (see text for description of A–E).

was aliquoted into several 50 ml orbital shaken spin tubes with a filling volume of 20 ml (E).

The scale-down criterion was constant CSPR in comparison to the bioreactor process. In the different scale-down modes, the starting cell densities were chosen due to the estimated growth rate and the resulting CSPRs. In semicontinuous chemostat, the media exchange rate was adapted to the growth rate and should be around 50% medium exchange per day (i.e. for a doubling time of 24 h). In repeated batch mode, one media exchange every two days was performed. In semicontinuous perfusion, 100% medium was exchanged per day.

7.6.1.1 Semicontinuous Chemostat (SCC)

In semicontinuous chemostat, every day an aliquot of the cell suspension was discarded and the tubes refilled with fresh media. The daily media exchange thus depended on the growth rate. In this process mode, there was always a minimum concentration of by-products and less nutrients compared to repeated batch.

For semicontinuous chemostat mode, the starting cell density was 9×10^6 cells/ml. After every 24 h of cultivation, viable cell density, viability, and metabolite and substrate concentration were analyzed. The suspension was diluted by removing an aliquot one time per day resulting in a dilution rate between 25 and 40% depending on the current cell density and growth rate.

7.6.1.2 Repeated Batch (RB)

The repeated batch was the most simplified cultivation method of the mentioned processes. The idea was to simulate semicontinuous chemostat with less operator interaction to increase throughput. The cells grew 2 days in fresh media until they were limited. For repeated batch mode, the starting cell density was 5.5×10^6 cells/ml. After 24 h of cultivation, viable cell density, viability, and metabolite and substrate concentration were measured. After 48 h of cultivation, viable cell density, viability, and metabolite and substrate concentrations were determined. An aliquot of about 75% was discarded and the culture diluted with fresh medium to reach the starting cell density of 5.5×10^6 cells/ml.

7.6.1.3 Semicontinuous Perfusion (SCP)

In SCP, once per day the cell suspension was centrifuged, the supernatant discarded, and resuspended with fresh media. The daily media exchange removed by-products and refreshed nutrients. In semicontinuous perfusion mode, the cell suspension with 15×10^6 cells/ml was cultivated for 24 h. After every 24 h of cultivation, viable cell density, viability, and metabolite and product concentrations were measured. The vessel was centrifuged with $800 \times g$ at 4 °C for 10 min. After centrifugation the supernatant was discarded and the pellet resuspended in fresh medium.

7.6.2 Results

7.6.2.1 Determination of the Starting Cell Density

For semicontinuous perfusion, the determination of the required starting cell density to reach a steady state or the maximum turning point within 5 days is

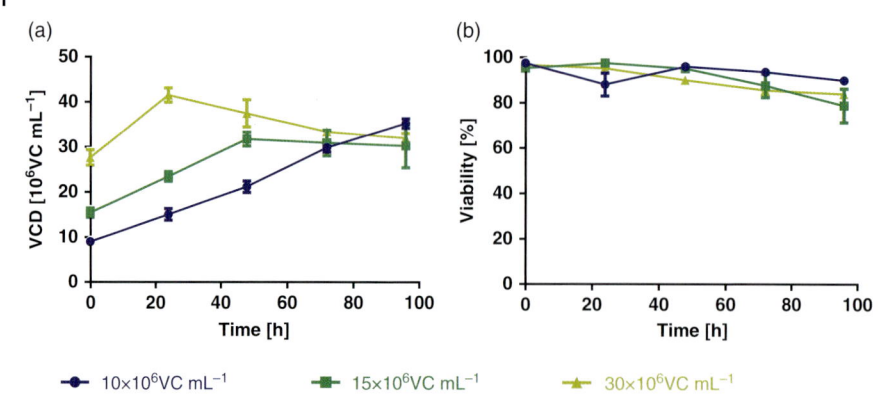

Figure 7.4 VCD (a) and viability (b) in semicontinuous perfusion in SpinTubes. Dependency of starting cell densities (10, 15, 30 × 10⁶ cells/ml) in a five-day cultivation process with one media exchange per day through centrifugation. Cultivation conditions: CHO-S-derived cell line, Cellvento™ CHO-100 Medium, 50 ml SpinTubes, 30 ml working volume, 37 °C, 320 RPM, 5% CO_2, 80% relative air moisture. Error bars represent 1 standard deviation ($n = 3$).

needed. Three different starting conditions were tested with the CHO-S-derived cell line in Cellvento CHO-100 Medium. The starting cell densities were 10, 15, 30 × 10⁶ cells/ml at day 0. Once per day, the SpinTubes were centrifuged and cells resuspended in fresh medium. For results see Figure 7.4.

It was observed that inoculation at 10 × 10⁶ cells/ml did not reach the maximum VCD in 96 h, but both other experimental conditions did. For the SCP protocol of this particular cell line, a starting cell density of 15 × 10⁶ cells/ml was chosen because of the stable viable cell density between 48 and 96 h, which facilitates an approximate metabolic characterization in steady state. Furthermore, it needs less cells than starting at 30 × 10⁶ cells/ml and this increases the potential throughput.

7.6.3 Scale-Down Model Comparison

Perfusion bioreactor experiments were analyzed to select a representative scale-down model. The bioreactor runs were performed with the same CHO-S-derived cell line used in the following experiments using Cellvento CHO-100 Medium [54,55].

In Figure 7.5a, the cell densities in the three operating modes studied are presented. The red line indicates the X_{max} of the reference bioreactor experiment, operated at 2 vvd with the same cell line and Cellvento CHO-100 Medium. At 1 vvd in semicontinuous perfusion, VCD was about 50% of respective bioreactor VCD. In Figure 7.5b, the calculated CSPRs are depicted. It can be observed that after about 48 h, all the CSPRs were comparable and on the target value of 33 pl/cell/day.

Figure 7.5c indicates the impact of the cultivation mode on specific cellular productivity. The SCP mode best represented the bioreactor conditions of 9 pg/cell/day. SCP was selected as scale-down model operating mode because it best represented specific productivity and $CSPR_{min}$ of the perfusion bioreactor process.

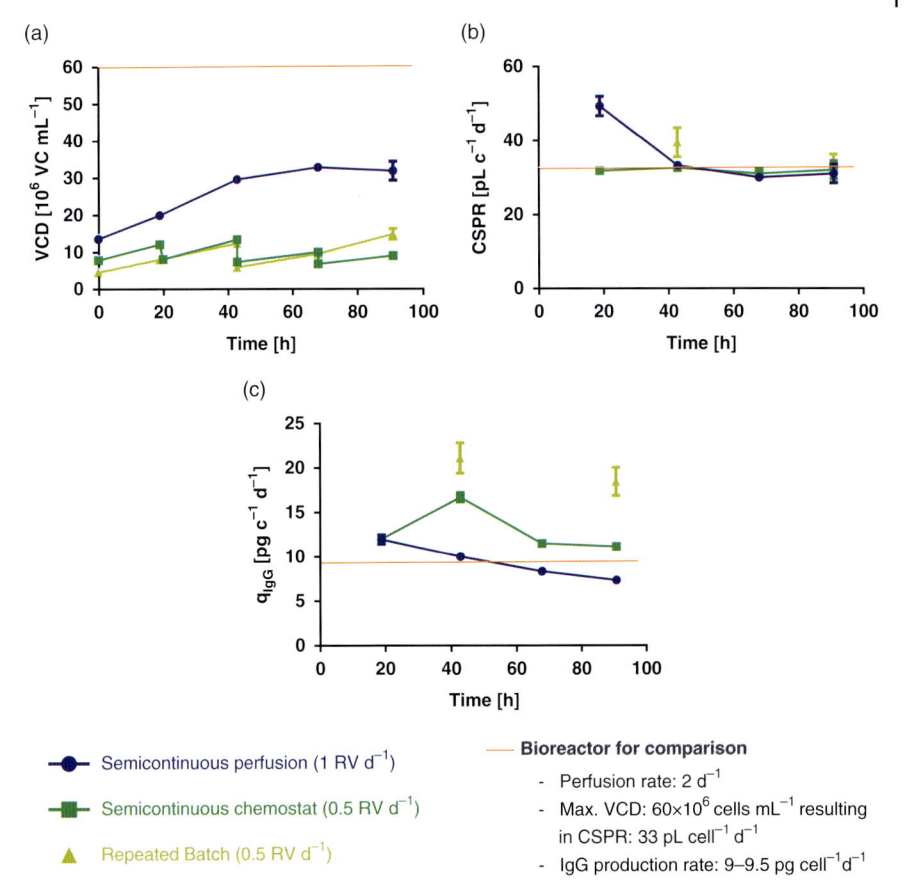

Figure 7.5 VCD (a), CSPR (b), and q_{IgG} (c) in semicontinuous perfusion, semicontinuous chemostat, and repeated batch in SpinTubes. Comparison of different cultivation methods to optimize a scale-down model for perfusion media screening in a five-day cultivation process. Error bars indicate one standard deviation ($n = 5$).

7.6.4 Media Screening

Cellvento CHO-100 Medium was chosen as the standard perfusion medium and as reference because of the prework in the bioreactors and scale-down model. All following media compositions were compared with this reference medium. After scale-down model optimization, the following experiments were performed with a standard laboratory CHO-K1 cell line without GS activity. In Figure 7.6, a selection of results is shown. During the screening, 24 different cell culture media and mixtures were screened. The excluded results showed one of the following characteristics: low VCD, low viability, low titer, or low cell-specific IgG production rate.

Duration of experiments was extended up to 6 days in order to reach the maximum VCD. The experiments started with a VCD between 29 and 32×10^6 cells/ml (see Figure 7.6a). The highest VCD was reached with media

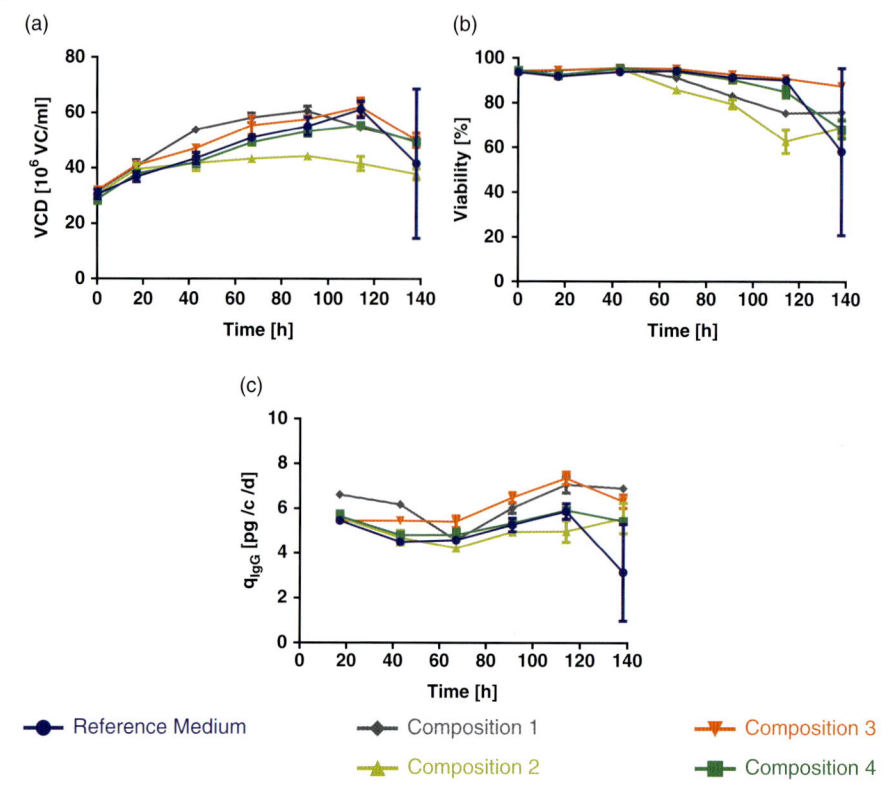

Figure 7.6 VCD (a), Viability (b) and q_{IgG} (c) in semicontinuous perfusion. Performance comparison screening different media compositions in six days semicontinuous perfusion. Cultivation conditions: CHO-K1 derived cell line, 20 ml working volume. Error bars indicate one standard deviation ($n = 4$).

composition 3 and 62×10^6 cells/ml, while reference medium and composition 1 reached 61×10^6 cells/ml. Composition 4 and composition 2 had a lower VCD with 55×10^6 cells/ml and 44×10^6 cells/ml, respectively. The viability of reference medium, composition 4, and composition 3 were stable until 91 h, while the viability of composition 2 dropped to 63%. The highest viability at 138 h of cultivation time had composition 3 with 88%. The highest q_{IgG} was reached with composition 3 and composition 1 with 7.4 pg/cell/day and 7.1 pg/cell/day, respectively. The highest q_{IgG} with the reference medium was 5.9 pg/cell/day with this cell line. The semicontinuous perfusion system allows for an insight to the metabolic behavior of the cells in the different cell culture media close to $CSPR_{min}$, which can directly be used to identify components that may be limiting for further cell growth. Figure 7.7 is an example for the different specific uptake and production rates of the cells in the described conditions.

Due to depletion of glucose and L-glutamine (data not shown), the consumption rates were similar with all media compositions. Lactate production rate also depended on the glucose concentration in the medium. For the reference medium as well as compositions 2 and 4, the lactate production rate was nearly zero,

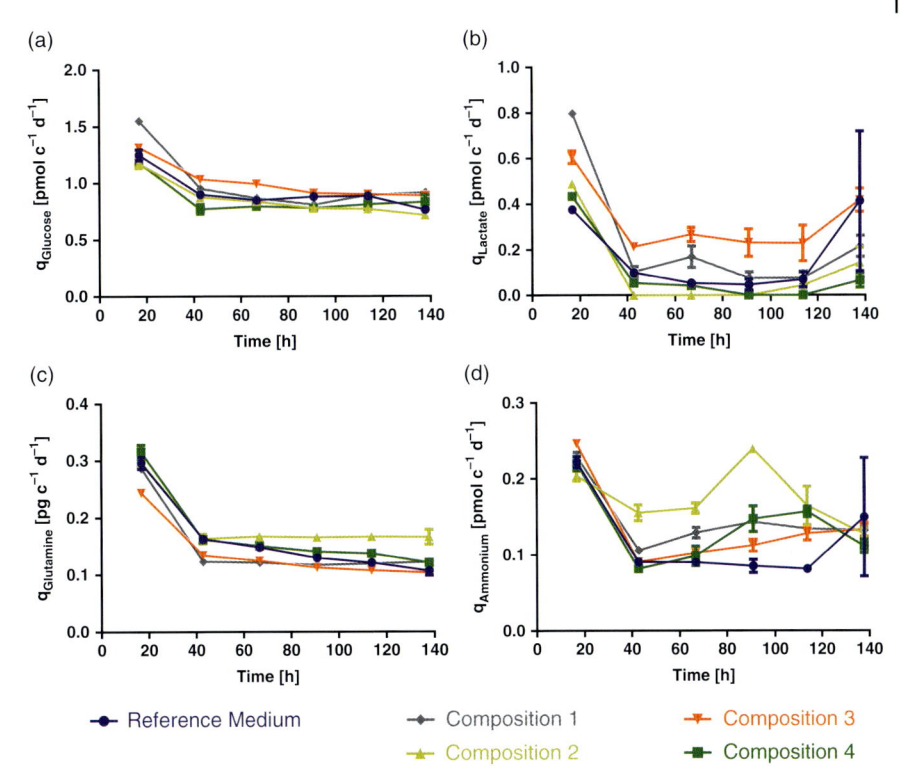

Figure 7.7 Metabolic behavior in the scale-down model. (a) $q_{Glucose}$, (b) $q_{Lactate}$, (c) $q_{Glutamine}$, and (d) $q_{Ammonium}$ in semicontinuous perfusion in SpinTubes. Performance comparison screening different media compositions in six-day semicontinuous perfusion. Cultivation conditions: CHO-K1 derived cell line, 20 ml working volume. Error bars indicate one standard deviation ($n = 4$).

indicating a favorable lactate metabolism. The ammonium production rate was at highest with composition 2. The metabolic results from the scale-down model for the reference medium were similar to the metabolic results in the bioreactor process, indicating that the SCP scale-down model is partly predictive for cellular metabolism in fully controlled, continuous perfusion bioreactors.

7.6.5 Bioreactor Confirmation

As mentioned, scale-down models can only represent the larger systems in some aspects, not all, and thus confirmation runs have to be performed. In our confirmation of scale-down model results, two bioreactor runs were performed. The media were the reference Cellvento CHO-100 Medium and composition 3. The aim was the investigation of $CSPR_{min}$, maximum VCD, q_{IgG}, and metabolic rates.

With Cellvento CHO-100 Medium a maximum VCD of 99×10^6 cells/ml with a viability of 76% was achieved. Viability is shown in Figure 7.8b. The media composition 3 was able to reach 158×10^6 cells/ml with a final viability of 89%. Cell-specific perfusion rate, calculated from the point of increasing the perfusion

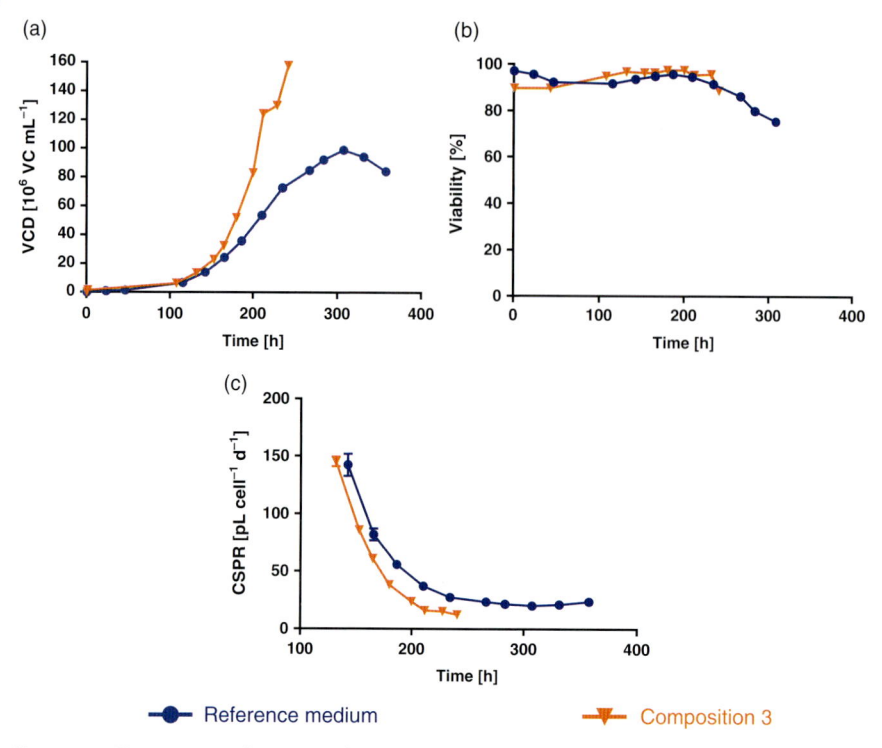

Figure 7.8 Bioreactor confirmation of SCP results. (a) VCD, (b) Viability. Comparison of standard medium and new perfusion medium in perfusion bioreactor processes. Cultivation conditions: CHO-K1-derived cell line, perfusion rate 2 vvd, cell retention through alternating tangential flow filtration, 4200 ml working volume, 37 °C, 40% air saturation, DO cascade controlled. Error bars indicate standard deviation of double measurements ($n = 1$).

rate to two bioreactor volumes per day, is given in Figure 7.8c. The $CSPR_{min}$ for the reference was 20 pl/cell/day and 13 pl/cell/day for the composition 3. This indicates a better suitability of the investigated new medium for perfusion with this cell line, which confirms the findings from the scale-down model.

7.7 Conclusion

Only a few experiments regarding high-throughput screening methods for perfusion have been reported yet and also validations of experiments in perfusion bioreactors were rare [34,96]. Fernandez *et al.* showed that simulation of perfusion is a more accurate cultivation method to analyze metabolic behavior in perfusion than batch or continuous processes [34]. Most high-throughput systems in literature had been implemented for fed-batch media development and not for perfusion systems.

The key parameter for the scale-down development was to reach a comparable CSPR to get similar metabolic rates and IgG production. If all systems have the same CSPR, the cells should have a similar metabolic behavior [21]. In these experiments, the cultivation system has an influence on the metabolic behavior despite providing similar CSPR values. In semicontinuous perfusion, the cells were limited in nutrient supply, indicated by reaching X_{max} and $CSPR_{min}$. A similar limitation occurred in the well-controlled bioreactors [54,55]. It is possible that the cells were not limited in repeated batch and semicontinuous chemostat, indicated by different substrate concentrations before media exchange. The metabolic behavior of this cell line was not comparable at lower cell densities despite the comparable CSPR.

After determining cultivation method and conditions, a CHO-K1 cell line was used for medium screening and optimization. In the following screening, composition 3 showed highest VCD, highest titer, and highest q_{IgG}. After this medium composition was identified, bioreactor runs were performed to confirm the results of scale-down experiments. The product concentration showed significant differences between reference and new medium resulting from both increased q_{IgG} and VCD.

Due to the bioreactor validation, an estimation of accuracy of the scale down could be performed. A qualitative prediction regarding VCD, titer, and metabolic behavior was achieved. The comparison of the scale-down model and the bioreactor run shows that Konstantinov's prediction of similar metabolic behaviors at the same CSPR was correct except for the scale-down model in repeated batch mode. In addition, the statement of Fernandez et al. was confirmed that simulation of perfusion is a better model than batch cultivation to predict metabolic behavior [34]. The CSPR of 13 pl/cell/day in the bioreactor experiment was lower than other reported CSPRs in high cell density systems, using perfusion bioreactors [5,9,21].

Abbreviations

ATF	alternating tangential flow filtration
B	bleed rate (day^{-1})
c	concentration (g/L)
CFB	concentrated fed-batch
CHO	Chinese hamster ovary
CRD	cell retention device
CSPR	cell-specific perfusion rate (pl/cell/day]
CSTR	continuous stirred-tank reactor
D	day
DHFR	dihydrofolate reductase
F	flow rate (l/day)
G	g-force (gravity)
GS	glutamine synthetase
H	harvest rate (day^{-1})
IgG	immunoglobulin G
k_L	oxygen transfer coefficient
$k_L a$	volumetric oxygen transfer coefficient
mAb	monoclonal antibody
OTR	oxygen transfer rate

OUR   oxygen uptake rate
P   perfusion rate (day^{-1})
PAT   process analytical technology
q   cell specific production/consumption rate (pg/cell/day]
QbD   Quality by Design
RPM   revolutions per minute (min^{-1})
RT   room temperature
S   safety margin (−)
STY   space time yield (g/l/day)
t   time (h)/(day)
V   volume (l)
VCD   viable cell density (10^6 cells/ml)
VP   volumetric productivity (g/l/day)
vvd   volume per volume per day
X   viable cell density (10^6 cells/ml)
Y   yield (g)

References

1 Ozturk, S.S. (2006) Cell culture technology: an overview, in *Cell Culture Technology for Pharmaceutical and Cell-Based Therapies* (eds S. Ozturk and W.S. Hu), Taylor & Francis, New York, pp. 1–14.

2 Ozturk, S.S. (2004) Development of highly effective cell culture perfusion bioreactors for commercial production of biologicals: past, present, and future, in 3rd Recombinant Protein Production Meeting, Tavira, Algarve, Portugal.

3 Xie, L. and Zhou, W. (2006) Fed-batch cultivation of mammalian cells for production of recombinant proteins, in *Cell Culture Technology for Pharmaceutical and Cell-Based Therapies* (eds S. Ozturk and W.S. Hu), Taylor & Francis, New York, pp. 349–386.

4 Ozturk, S.S. (1996) Engineering challenges in high density cell culture. *Cytotechnology*, **22**, 3–16.

5 Clincke, M.F. *et al.* (2013) Very high density of CHO cells in perfusion by ATF or TFF in WAVE bioreactor. Part I. Effect of the cell density on the process. *Biotechnol. Prog.*, **29** (3), 754–767.

6 Clincke, M.F. *et al.* (2013) Very high density of Chinese hamster ovary cells in perfusion by alternating tangential flow or tangential flow filtration in WAVE Bioreactor-part II: applications for antibody production and cryopreservation. *Biotechnol. Prog.*, **29** (3), 768–777.

7 Wurm, F.M. (2004) Production of recombinant protein therapeutics in cultivated mammalian cells. *Nat. Biotechnol.*, **22** (11), 1393–1398.

8 Subramanian, G. (2014) *Continuous Processing in Pharmaceutical Manufacturing*, Wiley-VCH Verlag GmbH, Weinheim.

9 Warikoo, V. *et al.* (2012) Integrated continuous production of recombinant therapeutic proteins. *Biotechnol. Bioeng.*, **109** (12), 3018–3029.

10 Deschênes, J.S. *et al.* (2006) Use of cell bleed in a high cell density perfusion culture and multivariable control of biomass and metabolite concentrations. *Asia Pac. J. Chem. Eng.*, **1** (1–2), 82–91.

11 Pollock, J., Ho, S.V., and Farid, S.S. (2013) Fed-batch and perfusion culture processes: economic, environmental, and operational feasibility under uncertainty. *Biotechnol. Bioeng.*, **110** (1), 206–219.

12 Himmelfarb, P., Thayer, P.S., and Martin, H.E. (1969) Spin filter culture: the propagation of mammalian cells in suspension. *Supramol. Sci.*, **164** (3879), 555–557.

13 Woodside, S.M., Bowen, B.D., and Piret, J.M. (1998) Mammalian cell retention devices for stirred perfusion bioreactors. *Cytotechnology*, **28** (1–3), 163–175.

14 Ozturk, S.S. (1996) Engineering challenges in high density cell culture systems. *Cytotechnology*, **22** (1–3), 3–16.

15 Popović, M.K. and Pörtner, R. (2012) Processing, consumption and effects of probiotic microorganisms. *Encyclopedia of Life Support Systems*, Eolss Publishers, Oxford.

16 Langer, E.S. and Rader, R.A. (2014) Continuous bioprocessing and perfusion: wider adoption coming as bioprocessing matures. *Bioprocess. J.*, **13** (1), 43–49.

17 Hacker, D.L., De Jesus, M., and Wurm, F.M. (2009) 25 years of recombinant proteins from reactor-grown cells: where do we go from here? *Biotechnol. Adv.*, **27** (6), 1023–1027.

18 Eibl, R. *et al.* (2010) Disposable bioreactors: the current state-of-the-art and recommended applications in biotechnology. *Appl. Microbiol. Biotechnol.*, **86** (1), 41–49.

19 Eibl, R. *et al.* (2009) *Bioreactors for Mammalian Cells: General Overview*, in *Cell and Tissue Reaction Engineering*, Springer, pp. 55–82.

20 Vijayasankaran, N. *et al.* (2009) *Animal Cell Culture Media*, in *Encyclopedia of Industrial Biotechnology*, John Wiley & Sons, Inc., New York.

21 Konstantinov, K. *et al.* (2006) *The "Push-to-Low" Approach for Optimization of High-Density Perfusion Cultures of Animal Cells, Cell Culture Engineering* (ed. W.-S. Hu), Springer, Berlin, pp. 75–98.

22 Rouiller, Y. *et al.* (2014), Modulation of mAb quality attributes using microliter scale fed-batch cultures. *Biotechnol. Progr.*, **30** (3), 571–583.

23 Schild, C. *et al.* (2015) Design criteria and requirements for the development of perfusion media. Integrated Continuous Bioprocessing II. Claremont Hotel, Berkeley, CA.

24 Adams, D., Korke, R., and Hu, W.S. (2007) Application of stoichiometric and kinetic analyses to characterize cell growth and product formation, in *Animal Cell Biotechnology* (ed. R. Pörtner), Humana Press, pp. 269–284.

25 Chmiel, H. (2005) *Bioprozesstechnik*, 3rd edn, Spektrum Akademischer Verlag, Heidelberg.

26 Yang, W.C. *et al.* (2016) Concentrated fed-batch cell culture increases manufacturing capacity without additional volumetric capacity. *J. Biotechnol.*, **217**, 1–11.

27 Trummer, E. *et al.* (2006) Process parameter shifting: part I. Effect of DOT, pH, and temperature on the performance of Epo-Fc expressing CHO cells cultivated in controlled batch bioreactors. *Biotechnol. Bioeng.*, **94** (6), 1033–1044.

28 Trummer, E. *et al.* (2006) Process parameter shifting: part II. Biphasic cultivation – a tool for enhancing the volumetric productivity of batch

processes using Epo-Fc expressing CHO cells. *Biotechnol. Bioeng.*, **94** (6), 1045–1052.

29 Woodside, S.M., Bowen, B.D., and Piret, J.M. (1998) Mammalian cell retention devices for stirred perfusion bioreactors. *Cytotechnology*, **28** (1), 163–175.

30 Kelly, W. et al. (2014) Understanding and modeling alternating tangential flow filtration for perfusion cell culture. *Biotechnol. Progress*, **30** (6), 1291–1300.

31 Karst, D.J. et al. (2016) Characterization and comparison of ATF and TFF in stirred bioreactors for continuous mammalian cell culture processes. *Biochem. Eng. J.*, **110**, 17–26.

32 Gorenflo, V.M. et al. (2002) Scale-up and optimization of an acoustic filter for 200L/day perfusion of a CHO cell culture. *Biotechnol. Bioeng.*, **80** (4), 438–444.

33 Baptista, R.P., Fluri, D.A., and Zandstra, P.W. (2013) High density continuous production of murine pluripotent cells in an acoustic perfused bioreactor at different oxygen concentrations. *Biotechnol. Bioeng.*, **110** (2), 648–655.

34 Fernandez, D. et al. (2009) Scale-down perfusion process for recombinant protein expression, in *Animal Cell Technology: Basic & Applied Aspects: Proceedings of the 19th Annual Meeting of the Japanese Association for Animal Cell Technology (JAACT), Kyoto, Japan, September 25–28, 2006* (eds. S. Shirahata et al.), Springer, Dordrecht, pp. 59–65.

35 Castilho, L.R. and Medronho, R.A. (2002) Cell retention devices for suspended-cell perfusion cultures, in *Tools and Applications of Biochemical Engineering Science*, Springer, pp. 129–169.

36 Godawat, R. et al. (2012) Periodic counter-current chromatography: design and operational considerations for integrated and continuous purification of proteins. *Biotechnol. J.*, **7** (12), 1496–1508.

37 Bailey, J.E. and Ollis, D.F. (1986) *Biochemical Engineering Fundamentals*, McGraw-Hill.

38 Chuppa, S. et al. (1997) Fermentor temperature as a tool for control of high-density perfusion cultures of mammalian cells. *Biotechnol. Bioeng.*, **55** (2), 328–338.

39 Fox, S. et al. (2003) Maximizing interferon-gamma production by Chinese hamster ovary cells through temperature shift optimization: experimental and modeling. *Biotechnol. Bioeng.*, **85** (2), 177–184.

40 Du, Z. et al. (2015) Use of a small molecule cell cycle inhibitor to control cell growth and improve specific productivity and product quality of recombinant proteins in CHO cell cultures. *Biotechnol. Bioeng.*, **112** (1), 141–155.

41 Zijlstra, G. et al. (2013) High cell density XD cultivation of CHO cells in the BIOSTAT cultiBag STR 50L single-use bioreactor with novel microsparger and single-use exhaust cooler.

42 Godia, F. and Cairo, J. (2006) Cell Metabolism, in *Cell Culture Technology for Pharmaceutical and Cell-Based Therapies* (eds S. Ozturk and W.S. Hu), Taylor & Francis, New York, pp. 81–112.

43 Miller, W.M., Blanch, H.W., and Wilke, C.R. (2000) A kinetic analysis of hybridoma growth and metabolism in batch and continuous suspension culture: effect of nutrient concentration, dilution rate, and pH. *Biotechnol. Bioeng.*, **67** (6), 853–871. (Reprinted from *Biotechnol. Bioeng.*, 32, 947–965, 1988.)

44 Reuveny, S. et al. (1986) Factors affecting cell growth and monoclonal antibody production in stirred reactors. *J. Immunol. Methods*, **86** (1), 53–59.
45 Sureshkumar, G.K. and Mutharasan, R. (1991) The influence of temperature on a mouse–mouse hybridoma growth and monoclonal antibody production. *Biotechnol. Bioeng.*, **37** (3), 292–295.
46 Bloemkolk, J.W. et al. (1992) Effect of temperature on hybridoma cell cycle and mAb production. *Biotechnol. Bioeng.*, **40** (3), 427–431.
47 Weidemann, R., Ludwig, A., and Kretzmer, G. (1994) Low temperature cultivation: a step towards process optimisation. *Cytotechnology.*, **15** (1), 111–116.
48 Gottschalk, U., Brorson, K., and Shukla, A.A. (2012) The need for innovation in biomanufacturing. *Nat. Biotechnol.*, **30** (6), 489–492.
49 Gottschalk, U., Brorson, K., and Shukla, A.A. (2013) Innovation in biomanufacturing: the only way forward. *Pharm. Bioprocess.*, **1** (2), 141–157.
50 Yang, Y. (2016) Media development toward high cell density, high productivity and low perfusion rates for continuous biomanufacturing platforms. Cell Culture World Congress, Munich.
51 Yang, W.C. et al. (2014) Perfusion seed cultures improve biopharmaceutical fed-batch production capacity and product quality. *Biotechnol. Progress*, **30** (3), 616–625.
52 Pohlscheidt, M. et al. (2013) Optimizing capacity utilization by large scale 3000L perfusion in seed train bioreactors. *Biotechnol. Progress.*, **29** (1), 222–229.
53 Seth, G. et al. (2013) Development of a new bioprocess scheme using frozen seed train intermediates to initiate CHO cell culture manufacturing campaigns. *Biotechnol. Bioeng.*, 110, 1376–1385.
54 Clutterbuck, A. (2015) Evaluation of Single-Use Bioreactors for Continuous Processing, in *InnovativeTtechnologies and Methods in Supporting Continuous Biomanufacturing*, Robinson College, Cambridge, UK.
55 Cunningham, M.A. (2014) Webinar: Integrated solution for high cell density cell culture from bench to pilot scale.
56 Salazar, A., Keusgen, M., and von Hagen, J. (2016) Amino acids in the cultivation of mammalian cells. *Amino Acids*, **48**, 1–11.
57 Ahn, W.S. and Antnoniewicz, M.R. (2013) Parallel labeling experiments with [1, 2–13C]glucose and [U-13C]glutamine provide new insights into CHO cell metabolism. *Metab. Eng.*, **15** (0), 34–47.
58 Goudar, C. et al. (2010) Metabolic flux analysis of CHO cells in perfusion culture by metabolite balancing and 2D [13C, 1H] COSY NMR spectroscopy. *Metab. Eng.*, **12** (2), 138–149.
59 Mulukutla, B.C., Gramer, M., and Hu, W.S. (2012) On metabolic shift to lactate consumption in fed-batch culture of mammalian cells. *Metab. Eng.*, **14** (2), 138–149.
60 Quek, L.E. et al. (2010) Metabolic flux analysis in mammalian cell culture. Metabolic Engineering: Metabolic Flux Analysis for Pharmaceutical Production, Special Issue. 161–171.
61 Tsao, Y.S. et al. (2005) Monitoring Chinese hamster ovary cell culture by the analysis of glucose and lactate metabolism. *J. Biotechnol.*, **118** (3), 316–327.

62 Hammond, S. et al. (2011) Chinese hamster genome database: an online resource for the CHO community at www.CHOgenome.org. *Biotechnol. Bioeng.*, 109, 1353–1356.

63 Jayapal, K. and Goudar, C. (2014) Transcriptomics as a tool for assessing the scalability of mammalian cell perfusion systems, in *Mammalian Cell Cultures for Biologics Manufacturing* (eds W. Zhou and A. Kantardjieff), Springer, Berlin, pp. 227–243.

64 Jayapal, K.P. et al. (2007) Recombinant protein therapeutics from CHO cells: 20 years and counting. *Chem. Eng. Prog.*, **103** (10), 40.

65 Kim, Y.G. (2012) Omics-based CHO cell engineering: entrance into post-genomic era. *Adv. Genet. Eng. Biotechnol.*, **1**, 1.

66 Xu, X. et al. (2011) The genomic sequence of the Chinese hamster ovary (CHO)-K1 cell line. *Nat. Biotechnol.*, **29** (8), 735–741.

67 Schaub, J. et al. (2010) CHO gene expression profiling in biopharmaceutical process analysis and design. *Biotechnol. Bioeng.*, **105** (2), 431–438.

68 Sieck, J.B. et al. (2014) Adaptation for survival: phenotype and transcriptome response of CHO cells to elevated stress induced by agitation and sparging. *J. Biotechnol.*, **189**, 94–103.

69 Polizzi, K.M. and Kontoravdi, C. (2015) Genetically-encoded biosensors for monitoring cellular stress in bioprocessing. *Curr. Opin. Biotechnol.*, **31**, 50–56.

70 Le, H. et al. (2015) Cell line development for biomanufacturing processes: recent advances and an outlook. *Biotechnol. Lett.*, **37** (8), 1553–1564.

71 Lyons, D. et al. (2015) Modeling perfusion at small scale using ambr15™. Integrated Continuous Bioprocessing II, Claremont Hotel, Berkeley, CA.

72 Rouiller, Y. et al. (2013) A high-throughput media design approach for high performance mammalian fed-batch cultures. *mAbs*, **5** (3), 501–511.

73 Jordan, M. et al. (2013) Cell culture medium improvement by rigorous shuffling of components using media blending. *Cytotechnology*, **65** (1), 31–40.

74 Tjio, J.H. and Puck, T.T. (1958) Genetics of somatic mammalian cells. II. Chromosomal constitution of cells in tissue culture. *J. Exp. Med.*, **108** (2), 259–268.

75 Ham, R.G. (1963) An improved nutrient solution for diploid Chinese hamster and human cell lines. *Exp. Cell. Res.*, **29**, 515–526.

76 Rapp, F. and Hsu, T.C. (1965) Viruses and mammalian chromosomes IV. Replication of herpes simplex virus in diploid Chinese hamster cells. *Virology*, **25** (3), 401–411.

77 Waubke, R., Hausen, H.Zur., and Henle, W. (1968) Chromosomal and autoradiographic studies of cells infected with herpes simplex virus. *J. Virol.*, **2** (10), 1047–1054.

78 Walsh, G. and Jefferis, R. (2006) Post-translational modifications in the context of therapeutic proteins. *Nat. Biotechnol.*, **24** (10), 1241–1252.

79 Wurm, F. (2004) Production of recombinant protein therapeutics in cultivated mammalian cells. *Nat. Biotechnol.*, **22** (11), 1393–1398.

80 Kompala, D.S. and Ozturk, S.S. (2006) Optimization of high cell density perfusion bioreactors, in *Cell Culture Technology for Pharmaceutical and Cell-Based Therapies* (eds S. Ozturk and W.S. Hu), Taylor & Francis, New York, pp. 155–224.

81 Wood, C.R. *et al.* (1990) High level synthesis of immunoglobulins in Chinese hamster ovary cells. *J. Immunol.*, **145** (9), 3011–3016.
82 Barnett, R.S. *et al.* (1995) Antibody production in Chinese hamster ovary cells using an impaired selectable marker. *Antibody Expression and Engineering*, ACS Symposium Series, American Chemical Society, Washington, DC.
83 Daramola, O. *et al.* (2014) A high-yielding CHO transient system: coexpression of genes encoding EBNA-1 and GS enhances transient protein expression. *Biotechnol. Prog.*, **30** (1), 132–141.
84 Lewis, N.E. *et al.* (2013) Genomic landscapes of Chinese hamster ovary cell lines as revealed by the *Cricetulus griseus* draft genome. *Nat. Biotechnol.*, **31** (8), 759–765.
85 Palomares, L., Lara, A., and Ramirez, O.T. (2010) Bioreactor Scale-Down, in *Encyclopaedia of Industrial Biotechnology: Bioprocess, Bioseparation, and Cell Technology* (ed. M.C. Flickinger), John Wiley & Sons, Inc., New York, pp. 1–13.
86 Oosterhuis, N.M.G. and Kossen, N.W.F. (1984) Dissolved oxygen concentration profiles in a production-scale bioreactor. *Biotechnol. Bioeng.*, **26** (5), 546–550.
87 Bylund, F. *et al.* (1999) Scale down of recombinant protein production: a comparative study of scaling performance. *Bioprocess Biosyst. Eng.*, **20** (5), 377–389.
88 Enfors, S.O. *et al.* (2001) Physiological responses to mixing in large scale bioreactors. *J. Biotechnol.*, **85** (2), 175–185.
89 Berridge, J., Seamon, K., and Venugopal, S. (2009) A-Mab: a case study in bioprocess development.
90 Zlokarnik, M. (2005) *Scale-Up*, 2nd edn, Wiley-VCH Verlag GmbH, Weinheim.
91 Hempel, D.C. (1986) Grundlagen des Scale-Up für biotechnologische Prozesse in Rührfermentern *Jahrbuch Biotechnologie*, Carl Hanser Verlag, München.
92 Strnad, J. *et al.* (2010) Optimization of cultivation conditions in spin tubes for Chinese hamster ovary cells producing erythropoietin and the comparison of glycosylation patterns in different cultivation vessels. *Biotechnol. Progress*, **26** (3), 653–663.
93 Jordan, M. and Jenkins, N. (2007) Tools for high-throughput medium and process optimization. *Animal Cell Biotechnology: Methods and Protocols*, Humana Press, 193–202.
94 Strnad, J. *et al.* (2011) Optimal process mode selection for clone screening. *Acta Chim. Slov.*, **58** (2), 333–341.
95 Zhang, X. *et al.* (2009) Efficient oxygen transfer by surface aeration in shaken cylindrical containers for mammalian cell cultivation at volumetric scales up to 1000L. *Biochem. Eng. J.*, **45** (1), 41–47.
96 Villiger-Oberbek, A. *et al.* (2015) Development and application of a high-throughput platform for perfusion-based cell culture processes. *J. Biotechnol.*, **212**, 21–29.
97 Poulsen, B.R. (2012) Scale-down tools for the evaluation of perfusion processes. Cell Culture World 2013, Munich.
98 Ho, D. *et al.* (2013) A simplified microbioreactor culture model to mimic perfusion. Integrated Continuous Biomanufacturing, Castelldefels, Spain.

99 Kreye, S. (2016) Live Webinar: ambr15 as a sedimentation-perfusion model for cultivation characteristics and product quality prediction.
100 Ram, R.J. (2016,) Tools for continuous bioprocess development. *BioPharm. Int.*, **29** (1), 18–25.
101 Nienow, A.W. (2006) Reactor engineering in large scale animal cell culture. *Cytotechnology*, **50** (1), 9–33.
102 Nienow, A.W. (2009) Scale-up considerations based on studies at the bench scale in stirred bioreactors. *J. Chem. Eng. Jpn.*, **42** (11), 789–796.
103 Hu, W., Berdugo, C., and Chalmers, J.J. (2011) The potential of hydrodynamic damage to animal cells of industrial relevance: current understanding. *Cytotechnology*, **63**, 445–460.
104 Sieck, J.B. *et al.* (2012) Development of a scale-down model of hydrodynamic stress to study the performance of an industrial CHO cell line under simulated production scale bioreactor conditions. *J. Biotechnol.*, **164** (1), 41–49.
105 Nienow, A.W. (2014) Re "Development of a scale-down model of hydrodynamic stress to study the performance of an industrial CHO cell line under simulated production scale bioreactor conditions"[Sieck, J.B., Cordes, T., Budach, W.E., Rhiel, M.H., Suemeghy, Z., Leist, C., Villiger, T.K., Morbidelli, M., and Soos, M. (2013) J. Biotechnol., 164, 41–49]. *J. Biotechnol.* 171 (1), 82–84.
106 Chalmers, J.J. (2015) Mixing, aeration and cell damage, 30+ years later: what we learned, how it affected the cell culture industry and what we would like to know more about. *Curr. Opin. Chem. Eng.*, **10**, 94–102.
107 Cockett, M.I., Bebbington, C.R., and Yarranton, G.T. (1990) High level expression of tissue inhibitor of metalloproteinases in Chinese hamster ovary cells using glutamine synthetase gene amplification. *Biotechnology (NY)*, **8** (7), 662–667.
108 Brown, M.E. *et al.* (1992) Process development for the production of recombinant antibodies using the glutamine synthetase (GS) system. *Cytotechnology*, **9** (1–3), 231–236.

Part Four

Continuous Upstream Bioprocessing

8

Upstream Continuous Process Development

Sanjeev K. Gupta

Ipca Laboratories Ltd, Advanced Biotech Lab, Kandivli Industrial Estate, Kandivli (west), 400067 Mumbai, India

8.1 Introduction

Upstream process is a key step of any biopharmacuetical manufacturing process, the batch and fed-batch processes are predominantly used by biopharma industries for the manufacturing of most of the recombinant therapeutic proteins used for critical human diseases. However, the continuous/perfusion process has also been employed for the manufacturing of some of the biopharmaceutical products available in the market. It is now proven by many industries with gradual evolution that transition from batch to continuous process can yield significant benefits. A critical assessment of these benefits is required for the implementation of continuous processing in manufacturing.

Over the past two decades, the biopharmaceutical industries grew quickly and focused on bringing innovative products to the market. This era of product innovation led to generation of high turnover and profit margins leading to setting up a modern manufacturing technology without much concern about the cost and manufacturing assets. With the growth of biopharma industry, it is even more realized that there is major issues with the design and cost of their manufacturing approaches [1]. Evolution of various novel technologies as well as extensive research has improved understanding the costs of goods (COGs) for the production of recombinant therapeutic protein several folds, which ultimately resulted reductions of operating cost via process improvements and enhanced operational efficiencies [2–6]. Increase in the product titer in upstream cell culture [7] and improved purification yield are the key examples of manufacturing process improvement [8]. Furthermore, examples of operational efficiencies include developing platform processes [9,10] and operational improvement strategy [11] that allows better utilization of existing resources. In the face of this changing landscape, two common needs for future biomanufacturing are emerging: increased flexibility and reduced cost of goods (COGs). Manufacturing flexibility allows companies to manage a complex and evolving portfolio where product numbers, volumes, and types are always in flux due to scientific and market uncertainties, and mergers and acquisitions.

Continuous Biomanufacturing: Innovative Technologies and Methods, First Edition.
Edited by Ganapathy Subramanian.
© 2018 Wiley-VCH Verlag GmbH & Co. KGaA. Published 2018 by Wiley-VCH Verlag GmbH & Co. KGaA.

Table 8.1 Characteristics of fed-batch and continuous/perfusion upstream process.

Characteristics	Fed-batch	Continuous (perfusion)
Bioreactor type	Single-use or stainless steel	Single-use or stainless steel
Scale	Single use up to 2000 l Stainless steel up to 20 000 l	Single use or stainless steel 50–2000 l
Cell density	$5–30 \times 10^6$ cells/ml	$30–120 \times 10^6$ cells/ml
Culture duration (run cycle)	12–20 d	20–60 d
Product concentration (titer)	0.5–8.0 g/l	~20% of fed-batch
Harvest	At the end of the batch	Daily/continuously
Waste product accumulation	Yes/high	No/very less
Product residence time	Up to entire culture duration	2–4 d
Product stability	High	Low
Clone stability issues/requirement	Moderate/up to 30–40 generation	High/70–100 generation
Process complexity	Moderate	High
Process control need	Moderate	High
Operation skill	Moderate	High/sophisticated
Product quality issue	High	Low
Contamination risk	Moderate	Moderate
Media consumption	Low	Very high
Operation cost	Moderate to high	Moderate
Purification capacity	Small to medium	Very large
Capex investment cost	High	Moderate

Source: Adapted from Refs [29,32].

Several biopharmaceutical industries successfully switched from batch or fed-batch to continuous process to maximize manufacturing flexibility without compromising with the product quality and operational excellence, resulted reducing the cost of goods [12–14]. In recent past, several such benefits have been qualitatively described and explored for the cost-effective biotherapeutic development by various companies [15–18].

Upstream process intensification through adaptation from batch/fed-batch to continuous manufacturing has been useful effectively in industries such as steel casting [12], petrochemical, chemical, food, and pharmaceuticals with huge efficiency [19–23]. Continuous manufacturing offers various advantages over batch/fed-batch manufacturing. Table 8.1 includes stable operation, reduced equipment size, high-volumetric productivity, streamlined process flow, low-cycle times, and reduced capital cost [24]. An example of biopharma industry is the ongoing project at the Novartis-MIT Center for Continuous Manufacturing.

Until now, a significant number of industrial and academic researchers are aggressively occupied in the development of continuous processing platforms [25,26]. The development efforts are further encouraged by US-FDA in a recent conference presentation [27,28].

Continuous manufacturing results in shorter process time mainly due to oversight of hold steps, higher productivity, and lower manpower obligation. Over the past decade, there has been a rising interest in the continuous processing for those industries involved in production of biological products. Continuous operations in upstream process/perfusion have been used for production of biologics for more than two decades now. However, recent advances of continuous downstream process have led the industry to visualize developing an integrated bioprocessing platform for an efficient continuous biomanufacturing [29,30].

The advantages of continuous manufacturing over batch and fed batch processes includes continued process with consistent product quality, reduced equipment size, high-volumetric productivity, smooth process flow, low-process cycle times, and reduced capital and operating cost (Table 8.1). This technology, however, offers various challenges, which need to be addressed much ahead of the implementation [29]. Recent research has focused on implementation of continuous upstream, continuous downstream [31], and continuous processing for production of several biologics, primarily monoclonal antibodies [30].

In this chapter, we describe various approaches used for upstream process development with more emphasis on continuous manufacturing platform being developed and practiced in biomanufacturing industries for the production of an affordable and cost-effective therapeutic protein. Features of several cell retention devices used so far by industries for continuous upstream process is described in detail. How would single use devices support an efficient, simpler, less footprint, and cost-effective continuous upstream process is also discussed with few suitable examples. How single-use technology including bioreactors, holders, filters, and bags can be used in continuous manufacturing is also described. Last but not the least, also touched upon implementation of an integrated continuous processing (upstream and downstream processes) for the production of biotherapeutics with robust product quality. Overall, this chapter constitutes current updates on various platform technologies potentially used for continuous upstream process development for the manufacturing of potential therapeutic products.

8.2 Upstream Processes in Biomanufacturing

There are mainly three types of upstream processes practiced largely by biopharmaceutical industries: batch, fed-batch, and continuous processing. As mentioned in earlier paragraph, among these processes, the fed-batch process has been the preferred choice by biopharma industries, mainly due to its ease of development resulting optimal yield and desirable product quality. However, recent advancement of continuous processing has attracted many biopharma companies for the exploration of the existing process with the continuous manufacturing process, as this process offers various advantages over batch and fed-batch processes (described above) [32].

8.2.1 Upstream Operating Modes

I) Batch process
II) Fed-Batch Process
III) Continuous process

For the production of any biotherapeutic protein, industries are practicing mainly two modes of operations, Fed-batch and continuous/perfusion culture. Both the modes start with vial/clone thaw; revival and expansion of the cells for the seed culture are used for scale-up and finally inoculation of the production bioreactor. The main features of this process are described below (Figure 8.1) [32]:

8.2.1.1 Fed-Batch Process

The fed-batch process of an upstream process majorly includes the following steps:

- Inoculation of bioreactor at ~75–90% of the final working volume.
- Intermittent feeding of the culture with feed supplements up to the final working volume during the fed-batch process.
- The in-process sample analysis is done for the cell count, viability, product titer, and biochemical parameter (glucose, lactate, etc.) analysis.
- Bioreactor is terminated and the culture is harvested based upon a predetermined criteria (e.g., percentage of viable cells)
- The batch duration may range from 10–20 days.
- The culture is usually harvested using centrifuge or depth filtration devices.
- The product is purified using chromatography-based purification process.

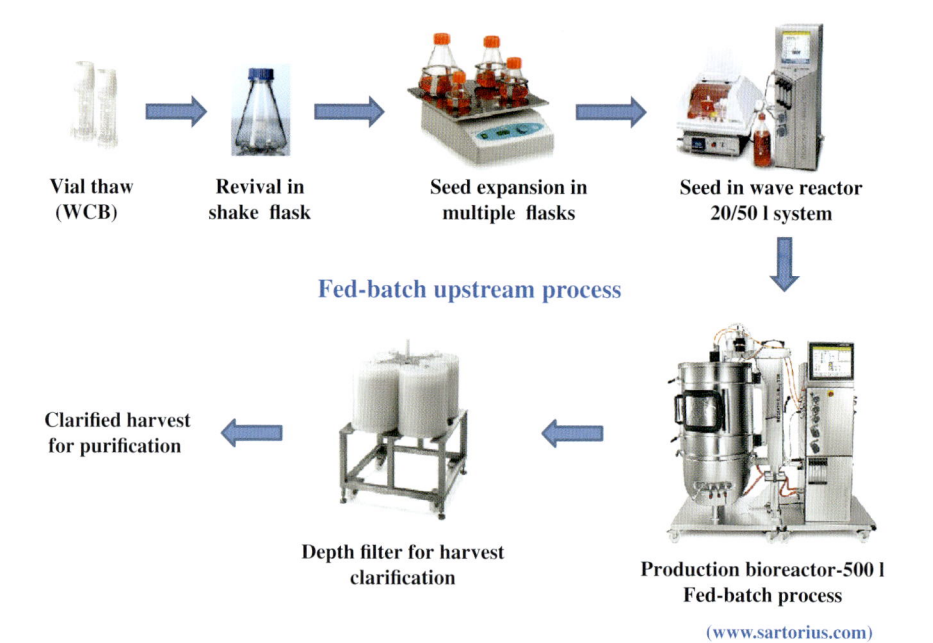

Figure 8.1 Process flow for a typical fed-batch/batch upstream production process.

- The product is characterized using various Bio-analytical tools.
- The purified product is called drug substance (DS) and used for the formulation and fill and finish called drug product (DP).
- A series of fed-batch runs are performed in succession and product is accumulated.

8.2.1.2 Continuous/Perfusion Process

The continuous/perfusion process of an upstream process majorly includes the following steps [32]:

- Cell vial thaw, expansion, and scale-up
- Seed inoculation of a seed bioreactor for large-scale production
- Achieving high-cell density for the perfusion start
- Fresh media is fed into the bioreactor in continuous mode
- Culture supernatant containing the desired protein continuously harvested through a cell retention device
- Cell retention device retains the cells and circulate to the culture vessel/bioreactor
- The culture volume is kept constant by replenishing and harvesting media at the same rate
- The process run cycle typically ranges from 20 to 60+ days
- Perfusion process leads to high volumetric productivity as compared to fed batch

Advantage of Continuous Upstream/Perfusion [32]

Continuous upstream offers following advantages over fed-batch cell culture process:

- *Product quality issue-process type*
- Bioreactor is harvested in continuous mode
- Bioreactor operating conditions are constant and can be accurately controlled
- Instable products are removed continuously
- Some products may require a constant production environment
- *Higher volumetric productivity*
- Volumetric productivity: 5–10 times higher than the fed-batch culture
- Production requires smaller culture vessels/bioreactors
- Much higher peak cell density compared to fed-batch
- The bioreactor can be operated continuously for extended periods of time at high cell density
- Reduced capital cost
- *Longer production campaigns compared to fed-batch*
- Reduced labor cost

8.3 The Upstream Continuous/Perfusion Process

The continuous upstream process is now gaining importance in biopharmaceutical development including vaccines. The major benefit of the continuous/perfusion upstream process is high cell density and high productivity in a relatively lesser size bioreactor as compared to the batch/fed-batch processes. In order to sustain high cell number and productivity, there is need to feed fresh medium during the cell propagation and production phases. The secondary metabolites accumulated in the batch and fed-batch processes are not removed during the process, in contrast, the metabolites and toxic substances are replenished at an optimized perfusion rate with feeding of fresh medium. Upstream continuous or perfusion is possible with a good separation device for retention of the cells in the bioreactor [33]. Several such cell retention devices are being used for the perfusion culture that performs well to a greater or lesser degree. These devices include, spin filters, gravity-based cell settlers, centrifuges, cross-flow filters, alternating tangential-flow filters (ATF), vortex-flow filters, acoustic settlers (sonoperfusion), and hydrocyclones. However, the spin filters, cell settlers, ATF, and acoustic settlers are potentially used with a small to large-scale bioreactors (Figure 8.2).

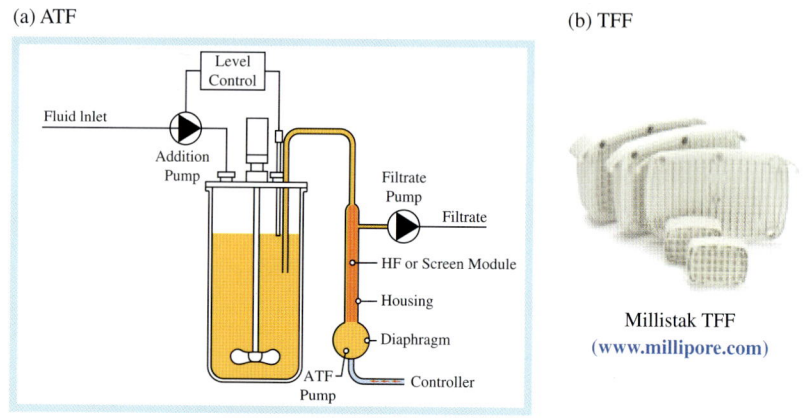

(a) ATF

(b) TFF

Complete perfusion process assembly with ATF
(www.refinetech.com)

Millistak TFF
(www.millipore.com)

(c) BioSep

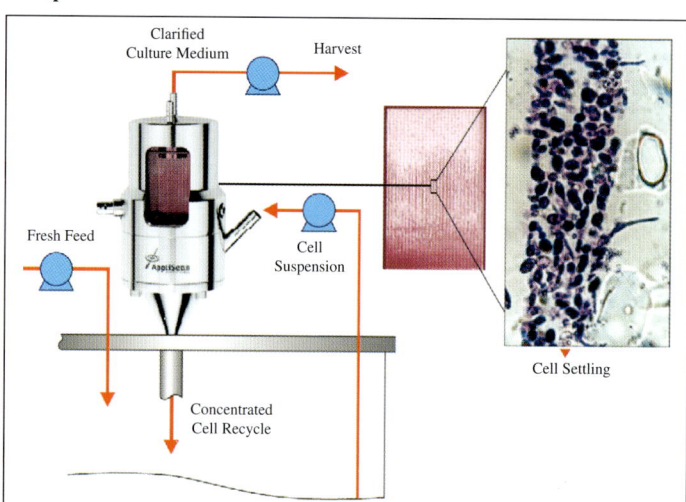

A typical configuration of acoustic cell retention device (BioSep) (www.applikon-bio.com)

Figure 8.2 Schematic diagram of "cell retention devices" used for continuous upstream process: (a) ATF, (b) TFF, and (c) BioSep.

The continuous upstream is becoming as the manufacturing method of choice due to availability of more sophisticated equipment that offers (1) improved sterile filtration, (2) refined pumps with better control, (3) improved understanding with ample data of perfusion technology, and (4) availability of cost effective and rich medium favorable for perfusion culture. However, the key drivers for adapting perfusion process are reducing "cost of goods" (COGs) and capital investment (Capex cost) that ultimately leads production of affordable biotherapeutic drugs to a large population [34].

8.3.1 Upstream Process-Type Selection

For a decade or so, major transition in biopharmaceutical manufacturing occurred due to significant improvement in the productivity. Further, advancements in development of novel expression vector, modified host, clone selection and screening devices, process development approaches, and media and feed formulations resulted in significantly enhanced cell growth to a very high cell concentration and impressive expression yield as compared to early days. In early 1990s with fed batch process, attainable cell concentration were about $5-10 \times 10^6$ cells/ml with considerably high product titer of 1.0–2.0 g/l; however, today those are greater than 20×10^6 cells/ml with product concentrations of up to 10 g/l. Those results are further amplified by the use of perfusion process, through which substantially higher cell concentrations and product titer can be achieved [35,36].

8.3.2 Component of Continuous Upstream and Downstream Processes

8.3.2.1 Upstream Components: Stainless Steel and Single-Use (Su)

- Medium, feed, and buffer preparation vessels/tank
- Medium and feed holders
- Different size bioreactors with control tower
- Tubing and connector (different sizes)
- Welder and sealers for aseptic connection (optional)
- Cell retention devices for perfusion process
- Harvest collection vessels/tanks (temperature controlled)
- Harvest hold vessels

8.3.2.2 Downstream Components: Stainless Steel and Single-Use (Su)

- Buffer preparation vessels/tank
- Purification resins
- Different size columns (single to multiple)
- Chromatography skids
- UF/DF and virus purification skids
- Harvest hold tanks/vessels
- Buffer preparation and hold tanks/vessels (multiple size)
- Different pore size filters
- Tubings and connectors (different sizes)
- Welder and sealers for aseptic connection (optional)
- Product/eluate hold tank

8.3.3 Cell Retention Devices Used in Perfusion Process

Cell retention devices that can be used for the perfusion processes are as follows [32]:

- Cell settlers-conical and inclined settlers
- Centrifuges
- Cell immobilization
- Spin filters, internal, or external
- Hollow fiber filtration
- Acoustic resonance
- Microfiltration
- Alternating tangential filtration (ATF)

Some of the key devices are described as below:

8.3.3.1 Spin Filters

Perfusion offers more volumetric productivity from a low producer clone as perfusion process can increase the cell concentration many folds by means of cell retention devices such as spin filters, which allows retention of cells inside the bioreactor and harvesting the culture supernatant for the purification of the expressed proteins [37]. The spin filter was the most frequent cell retention device used those days as it was the best device available for perfusion process. However, the spin filters have been dropped out largely due to limitation in the scalability and unreliability. The spin filters are used internally and externally in a bioreactor; however, external filters have drawbacks with respect to cost, maintenance, and sterilization. In addition, success of batch and fed-batch process not only repressed the wider use of spin filters but also of other potential cell retention technologies. The process scalability and operation difficulties associated with the spin filters stigmatized the process and the fed-batch process dominated well into the next decades.

Furthermore, despite the dominance of fed-batch as an industry standard, perfusion continued to be choice of cell culture process. Perfusion offered an exceptional solution for the production of labile products (unstable) that could not sustain in the toxic environment of a fed-batch culture. With use of perfusion devices, such toxic metabolites or products could be easily removed and replaced with fresh medium quickly from the bioreactor and stored after harvest properly to maintain their stability. Although the use of spin filters for perfusion process declined but other cell separation devices based on gravity settler, centrifugation, and alternate tangential flow (ATF) slowly emerged.

8.3.3.2 The ATF System

ATF is the most widely used system now a day as a cell retention device. It offers nearly linear scale-up for simplicity of operation and validation. Generally, conventional filtration systems will fail rapidly when used to separate media from a complex suspension of a cell culture with a high bioburden. By contrast, the ATF system does not foul due to its unique flow dynamics and self cleaning capability available with wide range of pore size that performs longer compared to other devices. This device contains a standard hollow-fiber unit used for separation of the

Table 8.2 ATF details for different bioreactor sizes and scale-up.

Process type	ATF 2	ATF 4	ATF 6	ATF 8	ATF 10
Continuous or concentrated Fed batch	0–4 l	4–25 l	25–150 l	150–400 l	400–1000 l
Microcarrier processes	01–10 l	10–50 l	50–250 l	250–1000 l	Up to 5000 l

Source: www.refinetech.com.

cells and product. However, unlike other systems that recirculate a culture through a filter in one direction, the alternating tangential-flow action persistently cleans the fibers every 5–10 s with a back flush action. This system requires only a single connection to the culture vessel; cells and media circulate to the ATF system, flowing reversibly through the hollow fibers. The diaphragm controls the flow by moving up and down in the ATF system's pump. This generates a rapid low-shear flow between vessel and pump, ensures fast exchange and rapid return of the cells to the bioreactor, and minimizes their residence outside the bioreactor. The choice of pore size for the hollow fiber determines what elements are retained and which ones pass through to the permeate [34] (www.refinetech.com).

ATF is designed to increase the productivity of cell-derived recombinant biopharmaceuticals, provides a more reliable and efficient process of cell separation with the inherent ability to support cell growth to extreme concentrations. The ATF system allows increased volumetric productivity and reduced bioreactor size. The ATF system has ability to scale the process on a linear basis from 1 l to >1000 l and can be used with both traditional stainless steel and single-use bioreactors and with all cell types including adherent lines. The ATF system is used in both clinical and commercial manufacturing. A wide range of ATF systems are available for different bioreactor size and volume (Table 8.2).

Advantages of the ATF System

- Cell concentrations in the range 40–150 million/ml with suspension cells such as CHO and PER.C6®
- Continuous protein production of 1 g/l/day and higher
- Cells are maintained in a healthy state providing a more consistent product quality
- Rapid media removal or exchange at over 1 vv/h
- A filtered product stream ready for purification
- No osmolarity increase, constant removal of waste/toxic molecules
- Faster, simpler virus production with a reduced cost of goods

  High volumetric productivity in continuous culture

Application of ATF

- Concentrated fed-batch
- High productivity continuous upstream
- Virus production and filtration

- Rapid media removal and exchange
- Harvest clarification
- Cell and protein concentration
- Cell banking and quick manufacturing
- Seed transfers and reduction of seed train steps
- Microcarrier wash, cell retention, and perfusion culture

Example/Case Study-Perfusion Using ATF Cell Retention Device
A continuous upstream/perfusion process is developed in a 2 l small bioreactor using ATF filtration device for the production of CHO-based monoclonal antibody. The medium used was animal component free for the suspension CHO cell line cultivation. The peak cell density achieved over 40×10^6 cells/ml with a perfusion rate of two reactor volume/day and reached up to 48×10^6 cells/ml during the perfusion process. The major challenge was to provide required aeration at such high concentration of the cells, pure oxygen was supplied by an open-tube sparger resulting in satisfying oxygenation until $25-30 \times 10^6$ cells/ml but became limiting at higher cell densities due to the low k_la of these bubbles and the small liquid height. A porous sparger was used either alone or in combination with the open tube for pure oxygen supply. In terms of operation, the perfusion process operated using the ATF device was satisfactory, without filter fouling, easy to operate, and to adjust in comparison with other separation devices by filtration or acceleration [38].

8.3.3.3 Biosep Acoustic Perfusion System
The Applikon BioSep system is a unique, cell retention device developed for high cell density perfusion processes. High frequency resonant ultrasonic waves are used for separation of cells from product instead of a physical mesh or filters or hollow fibers, it offers all the benefits of conventional cell retention devices but without their inherent problems and limitations.

Applications of the Biosep System

- Cell concentration for perfusion process
- Cell washing and cell culture harvest.

The BioSep device is based on the acoustic resonance technology, a nonfouling/nonclogging retention system. BioSep uses SonoSep Technology. This system can be used in a continuous mode for several months, which makes the most favorable equipment for the continuous/perfusion upstream cultures. The BioSep can be useful in both R&D (minimum 1 l/day), process optimization/development, and on production scale (1000 l/day) with linear scale-up. Furthermore, the BioSep can be seen as harvest clarification the first step in the Downstream Processing. The Biosep has a capacity of perfusing 1–1000 l/day culture, divided over five different models (www.applikon-bio.com).

Example/Case Study
A novel temperature-controlled and larger-scale acoustic separator was evaluated for perfusion culture in a 100 l bioreactor where cells were perfused up to 400 l/day

for a 10^7 CHO cell/ml that produced up to 34 g/day/400 l recombinant protein. The maximum performance of 96% separation efficiency at 200 l/day was obtained by setting the separator temperature to 35.1 °C, the recirculation rate to three times the harvest rate, and the power to 90 W. While there was no detectable effect on culture viability, viable cells were selectively retained, especially at 50 l/day, where there was fivefold higher nonviable washout efficiency. Overall, the new temperature-controlled and scaled-up separator design performed reliably in a way similar to smaller scale acoustic separators [39]. However, this system is not favored by most of the industries due to scalability issues after certain scale.

8.3.3.4 TFF Cell Retention Device

The continuous process has been practiced traditionally by industries for the production of unstable proteins in cell culture platform. The use of continuous process for the production of stable protein is limited due to low protein concentration, process complexity, cell culture medium, and feed costs. The new single use technology is gaining attention from various industries for the implementation of continuous upstream process as the single use technology reduces the process complexity compared to the conventional stainless steel technology. Moreover, the improved yield of continuous process compared to fed-batch process offers use of smaller single-use upstream systems, both for clinical as well as commercial scale production. The new cell retention devices allow developing relatively less complex perfusion process and enables intensification of fed-batch processes. The cell retention device used for the perfusion process can be used to radically boost the cell density of the seed ($N-1$) bioreactor to speed up the production process and achieve gains in efficiency [40].

The critical components of a continuous/perfusion system are the bioreactor vessel, controller, and cell retention device. The cell retention device is frequently used to describe the type of perfusion system with gravity-based and filtration devices being the most common. Recently, a new cell retention device called tangential flow microfiltration called Prostak (EMD Millipore) is being tested for small-scale perfusion cell culture process. This prototype device is derived from EMD Millipore Prostak™-microfiltration series of products typically used in the primary clarification of cell culture (www.millipore.com).

In a recent study, Chris *et al.* have used a new prototype Prostak TFF system for perfusion cell culture study [40]. The TFF Prostak is a 3 channel with 0.2 um 640 cm^2 membrane prototype small-scale devices used for continuous system. This device can be presterilized by autoclave. The recombinant protein producing CHOS cell line used for the evaluation of Prostak TFF system as perfusion device. A high cell density up to 70×10^6 cells/ml is achieved in 12 days culture when used Prostak TFF cell retention device. These high cell densities would be enabling in seed ($N-1$) bioreactor applications to shorten seed train and overall production times. Additionally, the FogaleTM biomass sensor is confirmed as an accurate method to track cell viability online. Subsequently, the same cell retention device was used for continuous upstream process in the production bioreactor up to 30 days and achieved up to 40×10^6 cells/ml with the final cell viability of >85% during the perfusion culture [40].

Above described 30 day perfusion process is an example of how TFF Prostak system could be used in a continuous production process where the protein of interest was continuously harvested from the permeate. The data suggest that the new TFF device Prostak can also be used for the continuous upstream cell culture process both for high-density seed generation and production of recombinant protein in a production bioreactor.

8.4 Manufacturing Scale-Up Challenges

The perfusion process could be used for a quick production of preclinical material generation for the companies requiring increased protein production as numerous perfusion technologies can quickly deliver the required quantity of recombinant protein. One general approach is to select a small-scale cell-retention device that offers a high degree of assurance for scaling to a commercial manufacturing process. Since scaling up processes in large bioreactors introduces several challenges, therefore engineers do not prefer perfusion devices to add additional complications. Several technologies have evolved for the large-scale production, and each method brings its own limitations. The biggest challenges include performance of the cell line at various scales, optimizing the fresh medium feed rate and constant filtrate flow rate keeping the product titer and quality uniform. Other parameters that would usually need consideration are selection of appropriate perfusion device and its performance for optimal production. Additionally, unlike the older systems, a failure in the perfusion devices does not mean failure of the run (e.g., failure of ATF, spin filters). In such instances, the perfusion device can be easily exchanged with another in a sterile way to continue the process. Another big issue is the continuous process is selection of appropriate size of bioreactor and its design and configuration as high cell density in perfusion process demands high mass transfer as well as oxygen flow. Furthermore, when greater perfusion rates are required, increasing recirculation flow rates can result into an inefficient cell separation and significant cell and viability loss, which lowers output and increases costs, significantly. Centrifuges have been used successfully for several continuous processes at larger scale, often to very high flow rates. However, the high level of fine-tuning needed to retain the reproducibility of such systems mainly during process scale-up. Despite those issues, skilled and experienced individuals are able to handle the cell retention device and learn with the experience that how to handle the scale-up challenges. They thoroughly evaluate and get better scale-up and scale-down performance [34]. Some of the common challenges with the continuous processes are summarized as follows.

8.4.1 Process Complexity and Control

Continuous upstream processing is a complex process, however the challenges are managed through engineering, skills, and experience.

- Process development and consistency in scale-up processes
- Selection of appropriate cell retention device
- Meeting requirement of cells at high cell density culture
- Maintaining cell viability on higher side
- Optimizing perfusion rate
- Minimizing contamination risk

- Reliable scale-down perfusion models
- Need to model hold times and fluid handling challenges
- Experienced manpower with diverse technical capability
- Scheduling and raw material logistics
- Robust manufacturing systems with appropriate automation
- Developing cost-effective processes (e.g., media and feed cost)

8.4.2 Cell Line Stability

As the continuous upstream process lasts longer compared to the fed-batch process, the cell line stability plays major role in scale-up studies, titer, and maintaining uniformity of the product quality. Unlike in fed-batch process, the continuous processes runs up to 60–90 days, therefore, it is very important to develop a robust cell line which can sustain till the end of the perfusion batch. Generally, the cell line stability is checked up to 120 generations and those clones are selected for perfusion process, which shows stability up to 100–120 generations. The major criteria for cell line selection are described as follows [32]:

- The cell line must stably express the product of interest for the duration of the continuous process.
- Selection of an expression system and cell line development platform that allows developing a productive and stable cell line as well as compatible to the continuous process.
- Cell line and process should be able to produce uniform and acceptable product quality.
- Stability of cell line should be up to 100–120 generation or even more depending upon the duration and continuous culture process. For example, CHEF1 cell line shown to have a stable expression out to 120 generations ($\sim$ 120 days).

8.4.3 Validation

Another manufacturing challenge in the continuous manufacturing is validation of the process; this demands the following [32]:

- The continuous manufacturing process must be validated to show consistency and reproducibility of the entire process run.
- Sophisticated analytical techniques should be applied to monitor the process and product quality.

The product quality is assessed using major techniques discussed in the following sections.

8.5 Single-Use Technologies: A Paradigm Change

Continuous manufacturing process for cell–culture-derived biotherapeutics offers the potential of superior daily productivity and hence smaller facility footprints than batch and fed-batch manufacturing processes. However, their use has been held up in the past by perceived greater logistical and validation complexity as well as higher likelihoods of technical failures. In recent past, continuous processing upstream development aspires to overcome some of these obstacles with the promise of superior productivities and lower failure rates. With the advent of single-use technologies for cell culture as well as purification operations has revolutionized importance in the potential of bioprocess development based on continuous cell culture systems. It is also important to consider less tangible operational factors such as ease of development and flexibility as well as the environmental burden of these strategies.

Furthermore, continuous innovations in single-use bioreactor designs have stirred the biopharmaceutical industry toward their increased use. A large number of companies, for example, Sartorius, Thermo, GE, Pall, and Millipore are currently supplying single use bioreactors (SUBs) with different configuration and sizes. The innovative single use bioreactors are available today are as small as 1–2000 l. Recently, ABEC, USA-based company has launched 4300 l world's largest single use bioreactor with a working volume of 3500 l (BioPharma reporter.com, November 10, 2015). However, the upstream process scale-up cost and handling challenges for fed batch and continuous processes are yet to be proven. In addition to single use bioreactor, remarkable improvements have been made in other components such as performance of sensors, bag components, and so on keeping disposability in mind [31].

Table 8.3 highlights top therapeutic products commercialized using perfusion upstream process for their manufacturing. These include mainly recombinant monoclonal antibodies, blood factor, and enzymes. The perfusion culture has been selected as the preferred mode of upstream process in cases of labile products (e.g., Xigris, Kogenate and Cerezyme), where as in case of mAb perfusion process it is selected due to company experience, manufacturing set up, and low titres. For example, Centocor's (now Janssen Biotech) has developed two blockbuster products such as "Reopro" and "Remicade" using perfusion platform process (Table 8.3) [31].

Table 8.3 also reveals that the most common perfusion systems adopted in commercial processes are spin filters and gravity settlers. However, Janssen Biotech has switched to a newer perfusion technology, alternating tangential flow (ATF) perfusion (Refine Technology, Edison, NJ) for the production of more recent mAbs, such as Simponi (Centocor, 2006).

The single-use (SU) technology provides multiple important features, including easy handling, lower initial investment, manufacturing footprint, and operating costs. Many biopharmaceutical industries are now adopting the single-use technology in their manufacturing for the process development applications (cell culture, purification, etc.). A good number of upstream operations are now supported by single-use devices, including media, feed, and buffer preparation, cell culture seed generation, and finally production (Figure 8.3). A number of diverse sparging (mass-transfer), impeller design, with and without shaft and cell suspension designs are available. These include various packed bed reactors and rocker-style, top- or bottom-mounted impellers, and orbitially shaken mechanical agitation devices [41,42].

Table 8.3 The biotherapeutics produced by perfusion process.

Product brand (generic)	Clinical indication	Company	Approval date (US)	Perfusion device	Bioreactor size (l)
Reopro (Abciximab)	PCTA	Janssen Biotech	1994	Spin filter	500
Remicade (Infliximab)	RA + other AI diseases	Janssen Biotech	1998	Spin filter	500 and 1000
Simulect (Basiliximab)	Transplant rejection	Novartis	1998	Rotational sieve filtration	250
Simponi (Golimumab)	RA + other AI diseases	Janssen Biotech	2009	ATF	500 and 1000
Stelara (Ustekinumab)	Psoriasis	Janssen Biotech	2009	ATF	500
Xigris (Rec. activated Protein C)	Sepsis	Eli lilly	2001	Gravity settler	1500
Rebif (Interferon beta-1a)	Multiple sclerosis	Merck-Sereno	1998	Fixed bed	75
Kogenate-FS (rec. Factor-VIII)	Hemophilia-A	Bayer	2000	Gravity settler	200
Fabrazyme (Agalsidase Enz)	Fabry disease	Genzyme	2003	Gravity settler	2000
Myozyme (Aglucosidase alfa)	Pompe disease	Genzyme	2006	Gravity settler	4000

Source: Reproduce with permission from Ref. [31]. Copyright 2013, John Wiley & Sons, Inc.

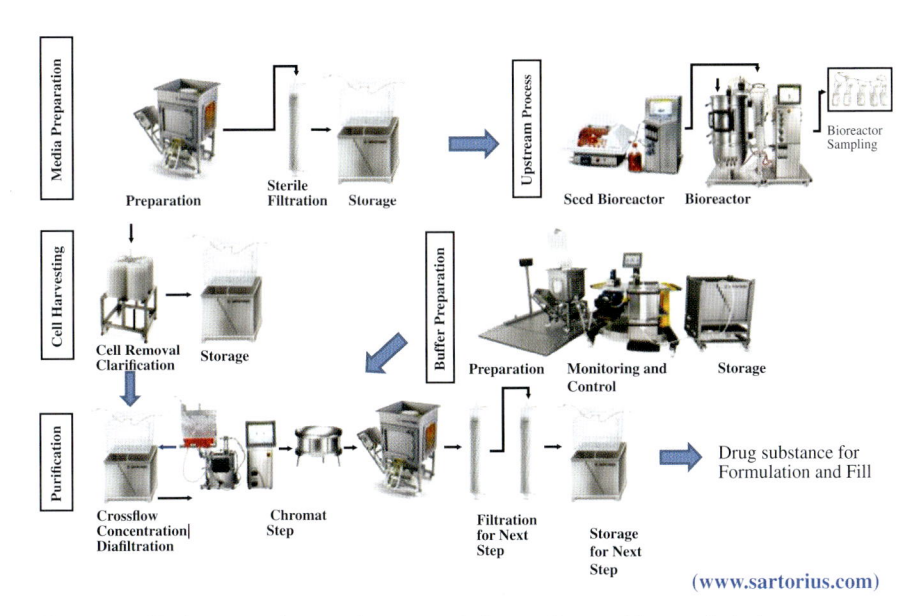

Figure 8.3 Single-use equipments/components for continuous biomanufacturing.

8.5.1 Application of SUBs in Continuous Processing

Continuous processing (CP) can provide valuable advantages such as better product quality and reductions in manufacturing construction and operating costs [43]. In continuous processing, materials continuously flow between series of equipment linked with the process module as they are processed uninterrupted into intermediates or final products. The operation goes this way for various length of time from days to months. Batch or fed-batch upstream production is generally segmented into several individual steps and even performed at separate suites, buildings, or sites and often requires large intermediate hold and transport processes. In contrast, CP occurs without interruption in continuous mode at a single location.

8.5.2 Single-Use Continuous Bioproduction

Continuous processing offers various benefits when applied to biomanufacturing or continuous bioproduction (CB). They include reductions in classified area, manpower requirements, and operational steps. Many modern upstream approaches already use fundamental technologies that can support simple adaptation to continuous bioproduction. It is also encouraged by regulatory agencies (US-FDA) and provides numerous tangible benefits. Recently, continuous processes are being used by a growing number of biopharmaceutical manufacturers in several unit operations [44]. Upstream, a variety of manufacturing scale perfusion bioreactors are currently becoming available. Successful continuous/perfusion process even exists in GMP manufacturing including various approved products (Table 8.3). Genzyme's new Geel, Belgium manufacturing plant houses 4000 l perfusion bioreactors.

Implementation of single-use continuous process offers numerous advantages in biomanufacturing including heightens process flexibility, efficiency, capability, and product consistency [45]. Single-use continuous bioproduction supplies comprehensive process flexibility due to its equipment being so easy to clean, inspect, and maintain. It supports ease of product changeover because it is very modular, reconfigurable, and transportable. Single-use continuous bioproduction supports a reduced process development time and operation cost as it eliminate clean-in-place (CIP) and steaming service development time. SU flexibility is very complementary here and includes open architecture and "hybrid" implementations. The chance of contamination can be minimized by the use of single-use equipment promoting the closed and highly integrated operations.. This enables such approaches as the growing "Factory of the Future" initiative of manufacturing even different product types in single gray space (or ballroom) of reduced classification.

Use of single-use integrated system offers several benefits such as a reduction in (1) facility construction utilities and costs, (2) manufacturing suite area and classification, (3) manufacturing operation steps and costs, (4) validation requirements, (5) quality systems maintenance, and (6) operations personnel. In fact, there really are a few real financial, engineering, or regulatory concerns in the consideration of single-use continuous production biopharmaceutical manufacturing. Many envision closed, disposable, integrated, and continuous bioproduction systems for biopharma in the near future.

8.5.3 Single-Use Perfusion Bioreactors

The single use hollow fiber perfusion bioreactor is available for over 40 years now. However, single-use production scale perfusion equipment become available very recently. Single-use and hybrid production-scale perfusion equipment appeared to support majority of the process formats and the platforms. Single-use perfusion equipment currently available support mechanically agitated suspension, hollow fiber, floating filter, and packed bed reactors. These single-use equipment are integrated with proper ancillary equipments and nearly all large-scale single-use bioreactors are competent of providing perfusion process. The bioreactor accessories commercially available to support single-use perfusion culture include hollow fiber exchange, continuous flow centrifugation, and acoustic wave separation.

8.5.3.1 Type of Single-Use Bioreactors for Perfusion Culture

Stirred Tank Suspension Reactors
Mechanically agitated (stirred-tank) bioreactors, containing either suspension or anchorage-dependent cultures on microcarrier, have conquered the biomanufacturing industry. The stirred tank bioreactor (STR) is the most accepted suspension systems as it is well understood and allows simple operation and easy scale-up. A suitable support matrix such as microcarriers are dextran-based, glass and plastic microcarriers are used for the adherent cell culture process.

Fixed/Floating Filter Bioreactors
Presterilized single-use wave-action (or rocking) bioreactors available from different sources provide another perfusion-capable solution. In this case, the rocking motion of disposable bags provides oxygen transfer to the cultured cells. Also available with integral perfusion culture facility and utilize an innovative floating filter that is kept cleaned and unclogged by the wave motion. This is a simple, disposable perfusion bioreactor that can be used for various bioproduction processes.

Packed Bed Bioreactors
Packed bed (PB) bioreactors are a type of entrapment culture that can maintain a variety of cell types for long periods of time. PB reactors providing such features as extremely low shear due to their macroporous matrix immobilization of cells and are currently employed in a number of culture applications.

Hollow Fiber Perfusion Bioreactors
Hollow Fiber Perfusion Bioreactors (HFPB) allow perfusion of high-density culture. Cells are usually seeded within the cartridge body, but outside of the hollow fibers in the "extracapillary space" (ECS). In this configuration, fresh medium is pumped through the hollow fibers allowing both recombinant proteins

and culture nutrients to diffuse through the fiber walls in each direction. Culture medium from the cartridge can be oxygenated within this loop. The basic features of a HFPB system include a very high culture binding surface-to-volume ratio, immobilization of cells at very high cell density, and a selectable porosity in the fibers.

8.5.4 Single-Use Accessories Supporting Perfusion Culture

8.5.4.1 Hollow Fiber Media Exchange
Several external hollow fiber-based perfusion-supporting devices are now available providing manufacturing-scale operation in single use. The advantages of hollow fiber media exchanger include almost linear scale-up, simplicity of operation and validation, and choice of filter materials and pore size. However, the major challenge is the optimization of operation parameter in a hollow fiber perfusion as cell aggregates due to high cell density causes filter clogging. In addition, maintaining high cell viability, optimizing pump speed and measuring filter membrane pressure are three key parameters to consider for a successful hollow fiber perfusion culture.

8.5.4.2 Continuous Flow Centrifugation
An array of rotor designs available for centrifuges presenting a high capacity continuous flow that enables a perfusion mode of cell culture. The major advantage of continuous centrifuges includes providing consistent performance, easy to transfer, and ease of scale-up. Moreover, the single use components can also resist the centrifuge forces throughout the perfusion culture. The centrifuge device in a perfusion process becomes closed system when aseptic single use tubing is used for the complete circuit. Although this device provides low shear isolation of the cells with minimum reduction in cell viability, absence of filters leads cell loss to the permeate stream.

8.5.4.3 Acoustic Wave Separation
As described earlier in this chapter, separation of cells by means of ultrasound provides for the isolation of cells or small particles from fluids without using invasive materials demanded in other devices such as spin filters or centrifugal approaches. These cell retention devices are often referred to as ultrasonic separators, filters, resonators or acoustic wave separators (AWS). Ultrasonic separators have a chamber where the original signal is converted into a standing acoustical wave field that severely limits dispersion of the cells flowing through the chamber. The acoustic energy establishes a "virtual" screen or mesh providing a low-shear, noncontact, nonfouling, nonmoving means of cell separation. This device is used nowadays for the separation of cells in animal perfusion culture for medium to large scale continuous biomanufacturing.

8.5.4.4 Spin filters
The spin filters have been used for the production of a couple of biological products. Both internal and external filters of many designs have been successfully employed for decades. This prevents filter clogging by developing a tangential flow of medium

across a screen during perfusion process. Some suppliers now provide single-use spin filter units of varying sizes for use with many cell culture platforms. The main body of the disposable unit is polycarbonate and features a filtering open mesh of highly specialized monofilament fabric.

8.6 FDA Supports Continuous Processing

Industry leaders such as Genzyme, Bayer, Janssen, Merck-Serono, Novartis, and Lonza for Eli Lilly have manufactured approximately 19 marketed recombinant proteins including monoclonal antibody (mAb) so far using continuous upstream or elements of continuous processing, and these products are predominantly blockbusters with annual revenues totaling approximately $20 billion [46]. The industry giant Amgen has taken initiative for continuous biomanufacturing in Singapore plant is an example of changing the processing paradigm [47]. Many other companies such as GlaxoSmithKline, Johnson & Johnson, Genzyme, Bristol-Myers Squibb, AstraZeneca, Samsung BioLogics, and Novartis AG are among the pharmaceutical companies and contract manufacturing organizations growing biological facilities to manufacture drugs in innovative ways. Continuous upstream processing offers productivity improvement as well as biologics standardization [48]. Implementation of continuous processing in downstream process of a mAb can improve the throughput significantly with a much smaller footprint. Other advantages of continuous processing include consistent product quality, smaller equipment, smooth processes, low process cycle times, reduced operating costs, increased flexibility, high equipment utilization rates, high volumetric productivity, more automation coupled with less human interaction, effective use of single-use equipment, and reduced inventory and storage needs [46,48,49].

FDA has been a strong supporter of continuous processing as early as 2004 when it released *Pharmaceutical cGMPs for the 21st century–A Risk-Based Approach*. FDA strongly believes that the manual handling of a product can be potentially reduced by use of continuous manufacturing and also it allows a better process control. In addition, online monitoring of continuous processing can make possible real-time testing approaches and can support FDAs quality-by-design (QbD) initiatives [50]. The risk of contamination encountered with continuous processing does not require discard of entire batch as continuous processing allows regular sampling and analysis of recombinant products though online monitoring system. Conventional sampling and offline testing normally done for a batch process in a continuous system is challenging "as it only represents a point in time, not the complete "batch," given the continuous nature." For this reason, online analytical testing methods are recommended. "Critical process parameters (CPP)" are also needed to control the process and maintain the steady-state condition.

Although, there are no regulatory hurdles for implementing continuous manufacturing [50] however, there are some issues that continue with the biological process, for example, performing quality assurance/quality control with products manufactured using continuous system as well as *defining lots and batches* [46]. The main concern lies in the fact that a batch can be defined based on quantity manufactured or duration of the process [51]. However, the actual regulatory issue is traceability of the batch when there are consequences of a recall for a manufacturer due to some quality issues. FDA is in the position to offer incentives to instrument supplier for the developing required continuous unit operations (particularly downstream) and PAT (process analytical technology) [48]." These technologies could provide improved information about product quality attributes, such as charge heterogeneity, bioactivity, degree of aggregation, glycosylation profile, and impurity levels.

8.7 Making the Switch from Batch/Fed-Batch to Continuous Processing

Batch and fed-batch processes have been the preferred mode of upstream process but, however, it involves multiple steps and does not have enough operation flexibility due to limitation in process control, therefore, industry seeks more efficiency [47]. Batch processing is becoming a process bottleneck, and the industry is looking for more efficient way to manufacture biopharmaceuticals. Implementation of continuous manufacturing is advantageous than batch process as a continuous operation can be processed in continuous mode, and a continuous unit operation can process a continuous flow input for an expanded period of time [48].

Resistance to acceptance of continuous upstream processing in the past has been based on the unreliability and complexity of available resources such as production bioreactors, cell retention devices, and associated control equipment, but however, due to advancement in the technology many of these concerns are no longer pertinent. "Adoption of continuous downstream is limited by technological immaturity," for example, a sterile chromatography column is required for continuous capture in downstream. In addition, lack of multicolumn chromatography systems in polishing step is a limitation.

A change from batch or fed-batch process to continuous manufacturing may necessitate new equipment, process control parameters, and control strategies to establish product equivalency. Another challenge is regulatory filing and approvals; if a manufacturing change occurs postapproval, changes should be filed to regulatory authority with a proper justification and bioequivalence studies (Biosimilarty with reference to Innovator drug and own product in case of Biosimilars) – changes should be summarized and justified with bioequivalence (Biosimilarity) studies, for postchange approval and relaunch.

8.8 Costs and Benefits of Continuous Manufacturing

The complexity of continuous manufacturing could make the switch unpleasant to potential users, and there are also concerns about process development control, contamination risks, and scale-up potential of traditionally batch or fed-batch based process [46]. Moreover, there is some nervousness and challenges surrounding the integration of upstream and downstream processes and the interaction between unit operations, as it is well documented that changes upstream affect other steps down the line [52].

There are also some worries about the overall novelty of the technology in the continuous space, as not all equipment needed to perform continuous processes may yet be available for GMP manufacturing. Furthermore, both traditional stainless steel and single-use equipment were not necessarily meant for constant use and this type of wear and tear may accelerate product breakdown [46].

8.9 Costs of Adoption

Continuous biomanufacturing is thought to offer cost savings within the range of 30–50% [46,53]. As informed by analyst Marcus Ehrhardt from PwC to *The Wall Street Journal*, a traditional manufacturing facility is estimated to cost $150 million to construct, while a continuous plant would cost significantly less,; however, the price of construction depends on capacity plan, technology used, manufacturing location, and so on. Biologics manufacturing facilities are much more costlier than small molecule pharma drugs, it is estimated that the initial cost for a continuous manufacturing construction could surpass $30 million; same cost incurred for the setting up recent Vertex continuous manufacturing plant. GlaxoSmithKline's has established a new hybrid continuous-batch manufacturing facility in Jurong, Singapore expected to be functional by 2016 cost approximately $25 million. For existing facilities looking to retrofit, the cost of new continuous/perfusion equipment is not that high, but the real cost is adapting the facility and the validation workload. The operation costs of a continuous manufacturing depend on whether single-use or stainless steel operations are used by plant head. According to data published by Refine technology, the stainless steel continuous upstream operation would cost approximately $44.1 per 500 kg/year, whereas the single-use technologies would cost approximately $11 million [46].

8.10 Continuous Downstream Processing

The continuous downstream process of a continuous manufacturing has been recognized as the most difficult step by many biopharmaceutical industry representatives, and experience in this area is limited [48]. Although continuous upstream bioreactors have been running for decades, and the technology understanding, supply chain management, and operating experience already exists, in contrast for downstream continuous manufacturing, the supplier side of the technology is partially developed and immature. Despite these limitations, continuous downstream processing is much more expedient, fast as processing takes hours instead of days, potentially leading to improved product quality. Recently, Pall life science and GE healthcare have developed continuous chromatography system that can be potentially used for continuous downstream process optimization and further process scale-up to the manufacturing scale.

In continuous upstream process, as the volume of harvest to be purified increases, required buffer and elution volumes in batch-capture chromatography increases [53]. A continuous protein capture option could be a good alternative to reduce column size and lower resin utilization, thus potential cost saving. To reduce consumption of chromatography buffers, resins, and solvents, several companies now a days are attempting multicolumn chromatography process, specifically, multicolumn countercurrent solvent gradient purification (MCSGP). In batch chromatography, a resin is often not loaded at maximum capacity, and capture columns generally have to be cycled multiple times. A column in batch is typically loaded approximately 80% of its capacity, in contrast multicolumn chromatography provides complete solution and can enhance working chromatography capacity [46]. Furthermore, multicolumn chromatography is ultimate for those wanting to maximize productivity and improved resin capacity utilization. "The general way for continuous chromatography is to install multiple columns in series so that the columns can be loaded to 100% of its capacity and any product flowthrough will be captured by the next column in series. Implementation of continuous columns results in a 20–30% increase in resin utilization as compared to a batch process. In continuous mode of operation, column size can be reduced up to 20 times smaller than normal as low-flow rates allow for small column size. A multicolumn chromatography system starts with a complex mixture, which is loaded into the system and goes through a series of washing, elution, regeneration, and equilibration steps. These procedures repeat in subsequent columns, helping to balance downstream productivity with upstream titers. Many downstream equipment manufacturers, such as Pall, Novasep, Tarpon Biosystems, GE Healthcare, Semba, and ChromaCon, offer continuous chromatography systems that can be used for continuous manufacturing. In addition to continuous chromatography skids, the in-line dilution system could also be potentially used for buffer preparation and dilution. The dilution system is helpful in handling the large volume buffer requirement. Further, this reduces the equipment size and footprint. The in-line dilution system is commercially sold by several manufactures, such as Pall, GE healthcare, Asahi Kasei, and so on.

In addition to multicolumn and in-line dilution technology, downstream operations could benefit from novel continuous viral inactivation and ultrafiltration/diafiltration unit operations. To standardize downstream operations, the rational design of highly specific ligands for the capture step analogous to protein-A in mAb purification could be a beneficial approach [48].

8.11 Integrated Continuous Manufacturing

In the current scenario of a strong product pipelines, quick fluctuation in market demands and rising competition from biotechnology and biosimilar companies are gradually more driven to develop innovative solutions for highly flexible and cost-effective manufacturing. Implementation of integrated continuous processing a universal manufacturing platform can address these challenging demands that comprise high-density upstream continuous/perfusion cell culture and a directly coupled continuous purification protein capture step. Recently, Alex

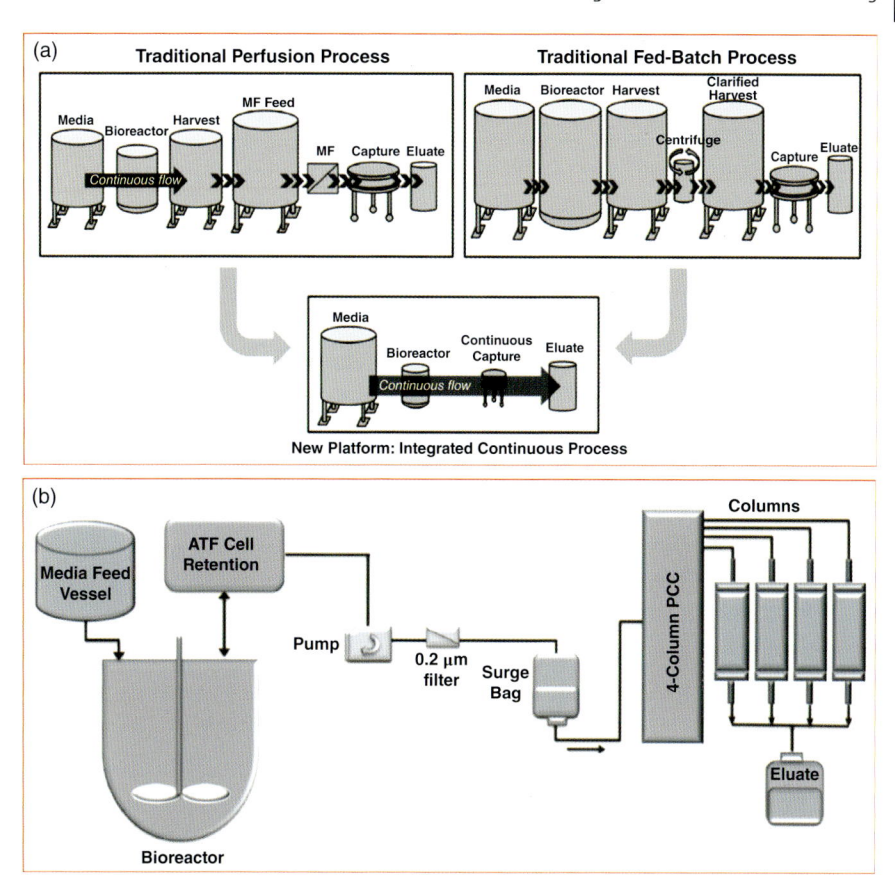

Figure 8.4 (a) Traditional fed-batch, traditional perfusion culture, and new platform integrated continuous processing. (b) Integration of four column PCC system with a perfusion cell culture bioreactor. (Reproduced with permission from Ref. [44]. Copyright 2012, John Wiley & Sons, Inc.)

Xenopoulos has implemented a new purification format for monoclonal antibody purification comprised of flocculation-based clarification, capture by continuous multicolumn protein-A chromatography followed by flowthrough polishing step. The new platform process allows a robust, single-use manufacturing solution as well as reduced over-all cost of goods, significantly. Modeling studies prove that the individual clarification, capture, and polishing chromatography offer significant advantages as stand-alone unit operations [54].

The recent study on integrated continuous processing reported by Warikoo *et al.*, demonstrate successful integration of a continuous/perfusion bioreactor and a four-column periodic counter-current chromatography (PCC) system for the continuous capture of desired therapeutic protein (Figure 8.4a and b). Two experiments were performed: production of (1) a monoclonal and (2) a recombinant human enzyme. Both the proteins were expressed in CHO expression platform, with a high density perfusion cultures at a quasi-steady state of $50-60 \times 10^6$ cells/ml for more than 60 days, led to much higher volumetric

Table 8.4 Design objective for integrated continuous processing.

Category	Design objectives
General	Closed operation, less environmental control Universal platform for any therapeutic proteins Less manual intervention, high level of automation High volumetric productivity through process intensification
Quality	High and consistent product quality Lower impurities due to high cell viability in upstream Less intermediate testing and stability study due to elimination of hold step Minimized bioburden load due to closed operation
Cost	Low capital cost due to reduced facility footprint (e.g., elimination of harvest clarification and hold step) Lower buffer volume and chromatography media Reduced QC environmental testing cost due to close operation
Speed	Fast process development, scale-up study, and tech transfer Fast clinical production and commercialization
Flexibility	Flexibility, mobility due to reduced equipment size Low cycle time Rapid capacity increase/decrease Simplified transfer to new facility

Source: Reproduced with permission from Ref. [44]. Copyright 2012, John Wiley & Sons, Inc.

productivities than current perfusion or fed-batch processes. The directly integrated and automated PCC system ran nonstop for 30 days without indications of time-based performance decline. The product quality observed for the continuous capture process was comparable to that for a batch-column operation. Furthermore, the integration continuous approach including cell culture and PCC led to a remarkable decrease in the equipment footprint and elimination of several nonvalue-added unit operations, such as clarification and intermediate hold steps [44].

This work also explores a novel continuous process technology that addresses the need for flexibility, consistent product quality, high-process output, and low cost (Table 8.4). The primary intent of developing this process is to introduce a universal continuous biomanufacturing platform, capable of handling various types of recombinant products, while significantly reducing the size of the manufacturing footprint and capital cost.

For development of integrated continuous process platform, multiple bioreactors with 12 l working volume (Broadley-James Corp., Irvine, CA) were run in continuous/perfusion mode using the ATF (Refine Technologies) cell retention device with polyethersulfone 0.2 mm filters. Cell density was measured offline using Vi-CELL (Beckman Coulter) cell counter. The recombinant CHO cells capable of growing in chemically defined medium used for the continuous production of monoclonal antibody or rhEnzyme proteins. The bioreactors were inoculated with viable cell of 0.5×10^6 cells/ml and allowed to grow up to $50–60 \times 10^6$ cells/ml. Perfusion began 24 h post inoculation at one reactor volume/day with the rate increasing proportional to cell concentration. A steady-state cell-specific perfusion rate of 0.04–0.05 nl/cell-day was maintained. DO was kept above 30% of air saturation. pH was maintained above 6.8 through sodium carbonate addition, but not exceeding 6.95. The harvest obtained from the bioreactors was directly loaded onto the periodic countercurrent (PCC) system without additional clarification.

The PCC system was integrated to the bioreactor for continuous protein capture. In PCC, the retention time of a protein on the column can be reduced without increasing the column size because the breakthrough from the first column can be captured on the second column installed in series. To achieve continuous capture of the recombinant protein, the PCC was directly connected to the bioreactor as shown in Figure 8.4a. The harvest from the bioreactor/ATF was pumped into a 2 l disposable bag using a peristaltic pump (Masterflex). A 0.2 mm filter (Millipack 40, Millipore) was added between the bioreactor and the surge bag as an additional sterile barrier. MabSelect SuRe (GE Healthcare) and a Hydrophobic Interaction Chromatographic (HIC) media in a XK16TM, 1.6 cm and 6 cm (GE Healthcare) column were used to capture MAb and rhEnzyme, respectively. Each column operation consisted of equilibration, load, wash, elution, and regeneration steps. The directly integrated and automated PCC system ran nonstop for 30 days without indications of time-based performance decline. The product quality observed for the continuous capture process was comparable to that for a batch-column operation. Furthermore, continuous processing of the harvest confers significant advantages with respect to the protein quality. Specifically, elimination of the harvest and other hold steps decreases target protein exposure to enzymatic, chemical, and physical degradation and thereby mitigates product stability risks [54].

8.12 Concluding Remark

Continuous processing is now gaining more importance for a successful and cost effective manufacturing of a recombinant product. The continuous upstream/perfusion process allows improved product quality consistent process, increased volumetric productivity, significant reduced cost, and handling of intermediates. The operational efficiencies and flexibility of single-use solutions further promotes adoption of continuous manufacturing. Recent bioproduction trends, such as modular architectures, synergize with both single-use and benefits, we anticipate growing interest in continuous upstream process supporting continuous bioproduction initiatives. The continuous upstream leads to a higher volumetric productivity as compared to the batch and fed-batch processes. However, this is more complex with respect to development, standardization, scale-up and manufacturing as well as regulatory approvals. The availability of improved cell retention devices such as ATF and TFF can address the upstream continuous process challenges. Implementation of Single-use bioreactor and other components also allows easy and cost effective manufacturing of a therapeutic product as compared to stainless steel and batch/fed-batch culture process, respectively. Continuous operation also has the potential to positively influence the downstream processing, as there is greater flexibility in meeting various market demands. Furthermore, a novel solution-integrated continuous bioprocessing offers exclusive advantages over traditional approaches for manufacturing recombinant products including monoclonal antibodies and human enzymes. The flexibility of the system can be further improved by incorporating single-use disposable solutions in both upstream and downstream processes. In spite of multiple advantages offered by continuous processing, due to lack of experience, technical challenges and confidence, adaptability and switching from batch/fed-batch process remains challenge for biopharma companies. However, the adaptation of continuous processes in manufacturing can be advantageous in developing cost effective and affordable recombinant therapeutic drugs.

Acknowledgment

Author would like to thank Mr. Vivek Yadav, Manager (Upstream), Ipca Laboratories Ltd., Mumbai, India for providing articles and assistance. Author is also thankful to Dr. Ashok Kumar, President-CRD and Ipca Laboratories Ltd., Mumbai, India for providing the necessary moral support.

References

1 Farid, S.S. (2007) Process economics of industrial monoclonal antibody manufacture. *J. Chromatogr. B Analyt. Technol. Biomed. Life Sci*, **848** (1), 8–18.
2 Sinclair, A. and Monge, M. (2002) Quantitative economic evaluation of single use disposables in bioprocessing. *Pharm. Eng.*, **22** (3), 20–34.
3 Rathore, A.S., Levine, H., Curling, J., Kaltenbrunner, O., and Latham, P. (2004) Costing issues in the production of biopharmaceuticals. *BioPharm. Int.*, **17** (2), 46–55.
4 Werner, R.G. (2004) Economic aspects of commercial manufacture of biopharmaceuticals. *J. Biotechnol.*, **113**, 171–182.
5 Rajapakse, A., Titchener-Hooker, N.J., and Farid, S.S. (2005) Modelling of the biopharmaceutical drug development pathway and portfolio management. *Comput. Chem. Eng.*, **29**, 1357–1368.
6 Farid, S.S. (2013) Cost-effectiveness and robustness evaluation for biomanufacturing. *BioProcess Int.*, **11** (11), 20–27.
7 Croughan, M. (2008) The silver anniversary of clinical protein production from recombinant CHO cells. ITQB Seminar, Oeiras, Portugal.
8 Gronemeyer, P., Ditz, R., and Strube, J. (2014) Trends in upstream and downstream process development for antibody manufacturing. *Bioengineering*, **1** (4), 188–212.
9 Kelley, B. (2009) Industrialization of mAb production technology: the bioprocessing industry at a crossroads. *mAbs*, **1** (5), 443–452.
10 Shukla, A.A. and Thömmes, J. (2010) Recent advances in large-scale production of monoclonal antibodies and related proteins. *Trend Biotechnol*, **28** (5), 253–261.

11 Han, C., Nelson, P., and Tsai, A. (2010) Process development's impact on cost of goods manufactured (COGM). *BioProcess Int.*, **8** (3), 48–55.
12 Tanner, A.H. (1998) *Continuous Casting: A Revolution in Steel*, Write Stuff Syndicate, Fort Lauderdale, FL.
13 Thomas, H. (2008) Batch-to-continuous–coming out of age. *Chem. Eng.*, **805**, 38–40.
14 Reay, D., Ramshaw, C., and Harvey, A. (2013) *Process Intensification*, Elsevier, 27–55.
15 Baker, J. (2013) Matching flows: the development of continuous bioprocessing, new initiatives in the approval of bioproducts, and assurance of product quality throughout the product lifecycle. Integrated Continuous Biomanufacturing, Barcelona, Spain.
16 Weintraub, K. (2013) Biotech firms in race for manufacturing breakthrough. MIT Technology Review.
17 Whitford, W.G. and Sargent, B. (2013) Continuous processing: from cookie preparation to cell-based production. The Cell Culture Dish.
18 Konstantinov, K.B. and Cooney, C.L. (2014) Continuous bioprocessing. International Symposium on Continuous Manufacturing of Pharmaceuticals (p. White Paper4), Available at http://iscmp.mit.edu/white-papers/white-paper-4.
19 Reay, D., Ramshaw, C., and Harvey, A. (2008) *Process Intensification: Engineering for Efficiency, Sustainability and Flexibility*, Butterworth-Heinemann, Amsterdam.
20 Anderson, N.G. (2001) Practical use of continuous processing in developing and scaling up. *Org. Process Res. Dev.*, **5** (6), 613–621.
21 Thomas, H. (2008) Batch-to-continuous coming out of age. *Chem. Eng.*, **805**, 38–40.
22 Fletcher, N. (2010) Turn batch to continuous processing. Manufacturing Chemist, pp. 24–26.
23 Laird, T. (2007) Continuous processes in small-scale manufacture. *Org. Process Res. Dev.*, **11** (6), 927.
24 Utterback, J.M. (1994) *Mastering the dynamics of innovation*, Harvard Business School Press, Boston, MA.
25 Integrated Continuous Biomanufacturing Conference. October 20–24 (2013) Barcelona, Spain.
26 International symposium on continuous manufacturing of pharmaceuticals, May 20–21 (2014) MIT-Cambridge, Massachusetts.
27 Moore, C. (2011) Continuous manufacturing-FDA perspective on submissions and implementation. 3rd Symposium on continuous flow reactor Technology for industrial applications, Lake Como.
28 Baker, J. (2013) Matching flows: The development of continuous bioprocessing, new initiatives in the approval of bioproducts, and assurance of product quality throughout the product lifecycle. Integrated Continuous Biomanufacturing Conference, Oct 20–24, 2013, Barcelona, Spain.
29 Konstantinov, K.B. and Cooney, C.L. (2015) White paper on continuous bioprocessing. May 20–21, 2014 continuous manufacturing symposium. *J. Pharm. Sci.*, **104** (3), 813–820.

30 Rathore, A.S., Agarwal, H., Sharma, A.K., Pathak, M., and Muthukumar, S. (2015) Continuous processing for production of biopharmaceuticals. *Prep. Biochem. Biotechnol.*, **45** (8), 836–849.

31 Pollock, J., Ho, S.V., and Farid, S.S. (2013) Fed-batch and perfusion culture processes: economic, environmental, and operational feasibility under uncertainty. *Biotechnol. Bioeng.*, **110** (1), 206–219.

32 Bioprocess International. (2009) Online education series.

33 Voisard, D. et al. (2003) Potential of cell retention techniques for large-scale high-density perfusion culture of suspended mammalian cells. *Biotechnol. Bioeng.*, **82** (7), 751–765.

34 Bonham-Carter, J. and Shevitz., J. (2011) A brief history of perfusion biomanufacturing how high-concentration cultures will characterize the factory of the future. BioProcess International, 9 (9)

35 Carstens, J. (2009) Perfusion! Jeopardy or the ultimate advantage? BioProcess International., webinar

36 Bonham-Carter, J. et al. (2010) Which Process Option Is Right for Me? Bioresearch Online.

37 Shevitz, J. et al. (1989) Stirred tank perfusion reactors for cell propogation and monoclonal antibody production, in *Advances in Biotechnological Processes*, vol. 11 (ed. A. Mizrahi), Alan R. Liss., Inc., New York, p. 81.

38 Chotteau and Veronique (2010) Study of Alternating Tangential Flow filtration for perfusion and harvest in Chinese Hamster Ovary cells cultivation. KTH.

39 Gorenflo, W.M., Smith, L., Dednisky, B., Persson, B., and Piret, J.M. (2002) Scale-up and optimization of an acoustic filter for 200L/day perfusion of a CHO cell culture. *Biotechnol. Bioeng.*, **80**, 438–444.

40 Martin, C.S., Padilla-Zamudio, J., Rank, D., McInnis, P., Kozlov, M., Reynolds, S., Parella, J., and Madrid, L. (2015) Novel small scale TFF cell retention device for perfusion cell culture systems. *BMC Proc.*, **9** (Suppl 9), P25.

41 Bsargent, B. and Whitford, W.G. (2014) Single-use perfusion culture enables continuous bioproduction. The Cell Culture Dish, Available at (http://cellculturedish.com/2014/07/single-use-perfusion-culture-enables-continuous-bioproduction/).

42 Galliher, P. and Pralong, A. (2013) When the process becomes the product: single-use technology and the next biomanufacturing paradigm. *BioPharm Int. Suppl.*, **26** (4), s27–s30.

43 Schaber, S.D., Gerogiorgis, D.I., Ramachandran, R., Evans, J.M., Barton, P.I., and Trout, B.L. (2011) Economic analysis of integrated continuous and batch pharmaceutical manufacturing: a case study. *J. Ind. Eng. Chem.*, **50** (17), 10083–10092.

44 Warikoo, V., Godawat, R., Brower, K. et al. (2012) Integrated continuous production of recombinant therapeutic proteins. *Biotechnol. Bioeng.*, **109** (12), 3018–3029.

45 Whitford, W.G. (2013) Supporting continuous processing with advanced single-use technologies. *BioProcess Int.*, **11** (4 Suppl.), 46–52.

46 Langer, E. (ed.) (2014) Continuous bioprocessing and perfusion: increased adoption expected. *11th Annual Report and Survey of Biopharmaceutical Manufacturing*, BioPlan Associates, pp. 103–116.

47 Ghose, S., Nordberg, R., and Forss, A. (2015) "The Future of Continuous Downstream Processing," on-demand webcast on *BioPharm International*, Feb, 20 https://event.on24.com/eventRegistration/EventLobbyServlet?target=regist...
48 Konstantinov, K.B. and Cooney, C.L. (2014) White paper on continuous bioprocessing. International Symposium on Continuous Manufacturing of Pharmaceuticals: Implementation, Technology & Regulatory, Cambridge, MA.
49 Hernandez, R. (2015) Continuous manufacturing: a changing processing paradigm. *BioPharm. Int.*, **28** (4), 20–41.
50 Chatterjee, S. (2012) FDA perspective on continuous manufacturing. IFPAC Annual Meeting, Baltimore, MD.
51 Allison, G. *et al.* (2014) Regulatory and quality considerations for continuous manufacturing. The International Symposium on Continuous Manufacturing of Pharmaceuticals: Implementation, Technology & Regulatory, Cambridge, MA.
52 Challener, C. (2014) Bioprocessing & Sterile Manufacturing. Pharmaceutical Technology, 38.
53 Bisschops, M. *et al.* (2009) Single-use, continuous-countercurrent, multicolumn chromatography. Supplement to BioProcess International, Available at http://www.bioprocessintl.com/downstream-processing/chromatography/singl (accessed February 15, 2015).
54 Xenopoulos, A. (2015) A new, integrated, continuous purification process template for monoclonal antibodies: process modeling and cost of goods studies. *J. Biotechnol.*, **213**, 42–53.

9

Study of Cells in the Steady-State Growth Space

Sten Erm,[1,2] Kristo Abner,[2] Andrus Seiman,[1,2] Kaarel Adamberg,[1,2] and Raivo Vilu[1,2]

[1]Tallinn University of Technology, Department of Chemistry and Biotechnology, Akadeemia tee 15, 12618 Tallinn, Estonia
[2]Competence Center of Food and Fermentation Technologies, Akadeemia tee 15, 12618 Tallinn, Estonia

9.1 Introduction

Experimental investigations and modeling of steady-state metabolism is the foundation for the quantitative characterization of cell physiology (see below; [1,2]). Steady-state of cell culture can be defined as the state of unchanging concentrations of different molecules inside and outside of the cells, which is the consequence of one-to-one relationship between steady-state flux patterns of biochemical processes and environmental conditions maintained constant. Classical cultivation method for the study of steady-state physiology is chemostat [3,4]. Steady-states of growing cells are investigated in chemostat one by one – after achieving a steady-state, the data are collected for the analysis, and the growth conditions, usually growth rate, is changed, and a new steady-state is achieved. As the time needed for the stabilization of steady-states is usually hours, a systematic study of different growth conditions and corresponding physiological states takes long time and requires remarkable resources. This is the reason why growth characteristics of cells are generally studied in batch cultures. Steady-state metabolism could, in theory, be studied also in batch cultures, but transient frequently uncontrolled changes of substrate and product concentrations in batches do not allow the data obtained be considered fully applicable for accurate quantitative characterization of the steady-states [1].

The basic drawback of chemostat cultures – high time and resource need together with difficulties in parallelization of experiments (low throughput), is even more aggravated taking into account the number of different parameters determining the steady-state growth conditions – concentrations of growth substrates, pH, temperature, and so on. Multidimensionality of steady-state growth space analysis (SSGSA), however, is unavoidable if the aim is a more comprehensive understanding of cells physiology.

In this chapter, we show that the noted disadvantages of chemostat cultivation in studies of steady-state metabolism can be at least partially compensated using

Continuous Biomanufacturing: Innovative Technologies and Methods, First Edition.
Edited by Ganapathy Subramanian.
© 2018 Wiley-VCH Verlag GmbH & Co. KGaA. Published 2018 by Wiley-VCH Verlag GmbH & Co. KGaA.

advanced continuous cultivation methods changestats, such as accelerostat (A-stat; [5]), dilution rate stat (D-stat; [6]), and sequential-parallel (mother–daughter) cultivation schemes [7], and so on.

Metabolic flux analysis (MFA) and flux balance analysis (FBA), essentially modeling methods of steady-state intracellular metabolism [8], are widely used together with high-throughput experimental methods currently with great success for the optimization of biotechnological processes [9–18]. However, these models do not include cell cycle mechanisms, cell geometry, and so on, and they are not suitable for the high-throughput *ab initio* cell design. We shall indicate some avenues in the development of the single cell models, where noted drawbacks of the MFA/FBA models would be overcome, and we argue that combination of these novel models together with advanced steady-state cultivation methods could open new perspectives in high-throughput cell design complementing highly efficient array of omics methods and recombineering methods (MAGE, CRISPR Cas9, etc.).

9.1.1 On Physiological State of Cells: Steady-State Growth Space Analysis

Definition of the physiological state of cells is essential for reliable quantitative description, analysis, and successful reproduction of their phenotypes *in silico*. The physiological states of cells are determined not only by molecular characteristics of cell components and structure of metabolic network ultimately coded by DNA but also by intracellular concentrations of metabolites and values of fluxes determined by the external environmental parameters. Environmental parameters determining the physiological states of cells are not as numerous as intracellular parameters, but among them nutrient availability and concentrations, temperature, pH, and oxygen concentration form the backbone of the environmental part of the multidimensional growth space of cells. It should be emphasized separately that as cells are complex dynamic systems, history of the growth processes is also important in determining physiological states in nonsteady-state conditions. Needless to say that need to take into account history of the cultures makes the task of characterization of nonsteady physiological states much more complicated and difficult in comparison with steady-states. This makes steady-state growth space a unique subsystem of the growth space. The steady-state growth analysis is analogous to the classical thermodynamics, which is theory of equilibrium processes. There are quite different theories for the analysis of nonequilibrium thermodynamic processes and new branches of thermodynamics are being developed for them also now. However, the equilibrium thermodynamics has been and is in the foundation of this fundamental branch of physics. The same seems to hold also for theories of cell physiologies – theories and tools of steady-state cell physiology is the foundation.

"Physiological state" and "growth space" are concepts, which have not been addressed exhaustively in the earlier research, despite the fact that classics of steady-state research Herbert, Málek, Tempest, and others published their first fundamental papers on the topic already in the 1950s [19]. The concept of physiological state has been revisited recently by Konstantinov [20], Stephanopoulos *et al.*, [21], and growth and nongrowth spaces have been defined and studied in batch cultures by Refs. [22,23]. However, as we specifically find great value in detailed analysis of the combinations of environmental parameters supporting growth and patterns of intracellular parameters associated with these combinations we define, in this chapter, the sets of aforementioned combinations where cells show growth (specific growth rate $\mu > 0$/h), as the growth space. Importantly, as explained above, only in steady-state growth space the physiological state of cells can accurately be determined and studied due to one-to-one relationship between patterns of biochemical processes and combinations of environmental parameters [1]. We propose that a more complete and quantitative understanding of cell metabolism can be achieved by SSGSA – systematic high-resolution characterization of steady-state growth space of cells through collecting systems-level data in different physiological states and *in silico* analysis of quantitative relationships between the combinations of environmental parameters and characteristics of cell metabolism. The data, results of modeling and understanding of steady-state physiology of growing cells form a fundamental core of know-how for the cell design.

Ideally, all the necessary cell characteristics (e.g., μ, carbon flow, cellular composition) necessary to understand their functioning and physiological states form surfaces in the growth space. Figure 9.1 illustrates an example of a three-dimensional growth space presenting μ as a function of pH and residual substrate

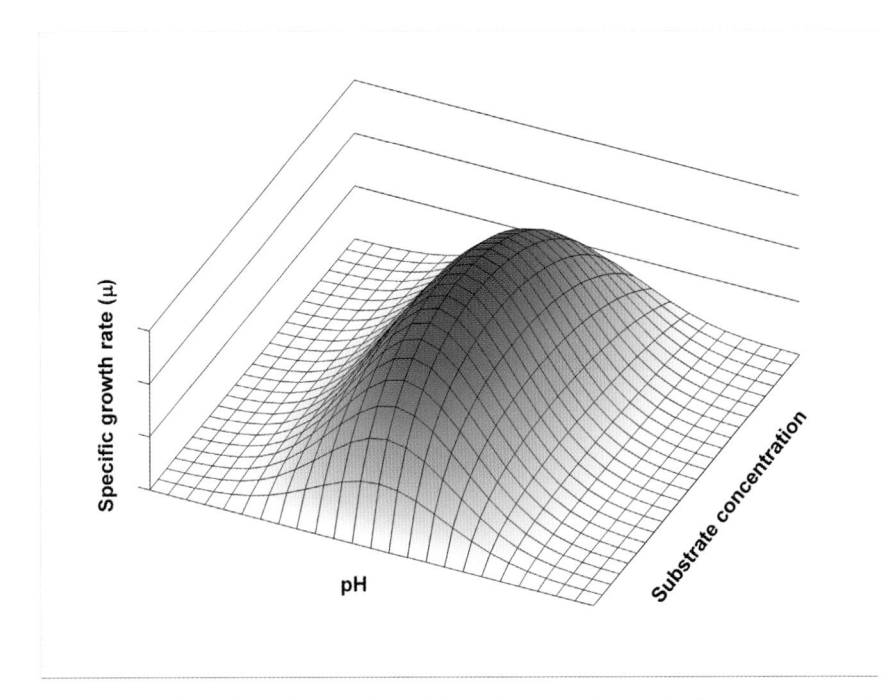

Figure 9.1 Surface of specific growth rate (μ) as a function of residual substrate concentration and pH in three-dimensional projection of steady-state growth space. The surface was generated based on the Monod equation [4] and near-to-real general relationships of μ and pH. Residual substrate concentration starts inhibiting μ_{max} above a certain concentration due to osmotic stress [2].

concentration generated using the Monod equation (see below; [4]) and near-to-real relationships of pH and μ. The illustration is just one possible case out of many and certainly more growth space surfaces can be envisaged. Although the multidimensional SSGSA is out of scope of the metabolic models used currently, ultimately, systematic SSGSA should contribute decisively to the acceleration of systems biology-based metabolic engineering of organisms into cell factories.

9.1.2 Challenge of Comprehensive Quantitative Steady-State Growth Space Analysis (SSGSA)

Continuous cultivation methods, for example, chemostat and also others, which enable to grow cells at strictly defined physiological steady-states with unchanging concentrations of intra- and extracellular molecules, metabolic flux patterns, and so on, are the basis for the SSGSA. While continuous cultivation methods have been around for a while now, analytical methods for the measurement of the molecules relevant for quantitatively defining steady-states of metabolism have only recently emerged. These high-throughput omics methods make possible measurements of mRNAs (transcriptomics), proteins (proteomics), intracellular metabolites and extracellular compounds (metabolomics), and so on. They have been generally used for relative comparison of different growth conditions, but absolute quantification of abundances/concentrations of intracellular molecules is also now possible (24–26), and these methods together with appropriate modeling tools allow in principle comprehensive quantitative SSGSA if coupled with steady-state continuous cultures. However, as noted and analyzed below, different models currently used are not the most suitable for this. Development of models (describing cell cycle mechanisms, cell geometry besides metabolic networks and allowing SSGSA) for analysis of steady-state metabolism of cells is currently among the greatest challenges of systems and synthetic biology.

9.1.3 Chemostat Culture – A Classical Tool for SSGSA

Chemostat is very well suited for the investigations of steady-state physiological states – it allows conducting experiments in strictly defined physiological steady-states through controlling and maintaining constant environmental conditions, for example, temperature, pH, limiting substrate concentration, and so on. Chemostat cultivation method was developed simultaneously by Novick and Szilard [3] and Monod [4]. Since then chemostats have been widely used and the results obtained have been excellently reviewed several times [1,27,28]. Steady-state of metabolism in chemostat is achieved and maintained by adding fresh media and removing cultivation broth from the cultivation vessel at a constant rate defined as the dilution rate (D) ($D = F/V$, where F is the feed rate (l/h) and V the volume of cultivation broth (L) in the reactor). Growth rate of cells μ can be controlled by D and in the steady-state conditions $\mu = D$, which is the most important relationship theory of chemostat. According to the Monod's equation $\mu = \mu_{max} \times \tilde{S}/(\tilde{S} + K_S)$, where $\tilde{S}$ is the residual limiting substrate concentration in the cultivation broth and K_S constant characterizing affinity of cells toward the growth-limiting substrate. This is an exemplary, simple relationship of concentration of limiting substrate (an environmental parameter) and growth rate characterizing the physiological state of the cells [4]. Numerous, mainly, empirical relationships of environmental parameters and intracellular parameters characterizing physiological state of cells have been discovered and proposed, and basically all these can be studied using chemostat in combination with omics methods, data analysis, and modeling. However, there is a clear need for models that could explain these relationships transparently and comprehensively.

Chemostats can be run stepwise but also in parallel. As mentioned above, stepwise chemostats are very time-consuming due to the need to stabilize the culture after each step change, making them also prone to the emergence of unwanted mutations (3,27,29–31). Both of these issues can be circumvented by using remarkably shorter changestat cultivation methods (e.g., A-stat, D-stat – see below) that enable high-resolution SSGSA through the continuous change of one or several environmental parameters in a single experiment, thus not needing the long stabilization phases after each stepwise change of the growth rate. Importantly, if the experiment is carried out properly, cells grown in changestats are in quasi steady-states, which are representatives of steady-states in chemostats.

Reduction of time of experiments can be achieved carrying out chemostat experiments in parallel. A special case of sequential-parallel cultivations of cells in multiple reactors set-up is described below [7]. In this case, cell cultures are cultivated and transferred from a "mother" reactor to one or several "daughter" reactors in the controlled steady-states increasing effectively the range of environmental conditions scanned in acceptable time (Figure 9.1). The throughput of the experiments can be increased even further using different schemes of multiplexed microscale continuous culture platforms [32–34].

9.2 Advanced Continuous Cultivation Methods – Changestats [2]

9.2.1 Accelerostat (A-stat)

The first changestat – accelerostat (A-stat) – was developed in the early 1990s at the National Institute of Chemical Physics and Biophysics in Tallinn, Estonia by the research group of Prof. Raivo Vilu while studying the growth of E. coli [5]. The idea of A-stat is based on the stepwise chemostat technique, but instead of increasing dilution rate D stepwise after achieving initial steady-state in chemostat, continuous increase of D at constant acceleration is started in A-stat according to algorithm: $D = D_0 + a_D \times t$ where D_0 is the initial D, a_D is acceleration of D, and t is the time since acceleration of D is started. Identical to the chemostat, the experimenter can control μ of the cells in A-stat through D under steady-state representative conditions. A-stat produces higher resolution data and is much more time- and resource-efficient compared to chemostat. A-stat has been the most used method among changestats (Table 9.1) mainly because it enables to study the effects and dynamics of μ, one of the most important physiological parameters of growing cells (Figure 9.2).

Table 9.1 List of changestat experiments with main results [2].

Main results	Reference
Accelerostat (A-stat)	
Escherichia coli	
Saturation of respiration capacity proposed to trigger acetate overflow at $\mu = 0.38/h$	[5]
Disruption of acetate recycling in the PTA-ACS node proposed to trigger acetate overflow at $\mu = 0.27/h$; A-stat and chemostat comparable at transcriptome level; μ-dependent metabolome, transcriptome, and proteome	[35]
Acetate overflow postponed and reduced fourfold in a mutant strain with coordinated activation of PTA-ACS and TCA cycles	[36]
Faster growth is achieved by increasing catalytic and translation rates of proteins	[37]
Saccharomyces cerevisiae	
Interlaboratory study determined μ_{crit} values for various different laboratory yeasts to select a reference *S. cerevisiae* strain	[38]
Mutant strain with derepressed glucose control displays 5% higher μ_{crit}	[39]
Supplementing mineral medium with oleic acid increases μ_{crit} by 8%	[40]
Hanseniaspora guilliermondii	
Two-phase switch from fully respiratory to respirofermentative growth with acetate excretion preceding ethanol; optimal range of μ for high biomass yield	[41]
Zygosaccharomyces rouxii	
Crabtree effect detected under aerobic conditions; effect of acceleration on growth characteristics also a function of the rate of change in environmental substrate concentrations	[42]
Lactococcus lactis	
Continuous shift from mixed acid fermentation to homolactic fermentation with faster growth; μ-dependent metabolome, transcriptome, and proteome	[43]
Faster growth is achieved by increasing catalytic and translation rates of proteins	[44]
Corynebacterium glutamicum	
Direct coupling between μ and demeton-S-methyl biodegradation in cometabolism with fructose; A-stat comparable to De-stat	[45]
Dunaliella tertiolecta	
Optimized kinetic parameters for photobioreactor design and bioprocess conditions for the production of vitamin and carotenoids	[46,47]
Deceleration-stat (De-stat)	
Escherichia coli	
Determination of maintenance energy requirements near zero-growth conditions	[48]
Stress response was less apparent during a smooth change in nutrient availability compared to nutrient shifts in fed-batch conditions	[49]
Rhodobacter capsulatus	
Optimization of photosynthetic efficiency for the production of hydrogen from acetate and light energy by the determination of important components of the light energy balance	[50]

Table 9.1 (Continued)

Main results	Reference
Saccharomyces cerevisiae	
Stronger stress response toward rapid changes in substrate concentration and D in chemostat compared to gradual changes in De-stat	[51]
Thalassiosira pseudonana and *Phaeodactylum tricornutum*	
Up to 94% of time can be saved compared to chemostats using 10 times faster decelerations than required for maintaining quasi steady-state while losing only 5% in accuracy to estimate maximal biomass productivity rate	[52]
Dilution rate stat (D-stat)	
Zygosaccharomyces rouxii	
Decreasing the threonine/methionine ratio in the ingoing medium proved that methionol synthesis occurs only in the Ehrlich pathway	[53]
Saccharomyces uvarum and *Saccharomyces cerevisiae*	
Dynamic effects of temperature on μ, biomass yield, and by-product profiles; metabolic events accompanying the transition of metabolism from carbon to nitrogen limitation	[6]
Lactic acid bacteria	
By-product patterns of fermentative metabolism for various LAB depend on the galactose to arginine ratio in the cultivation medium	[54]
Escherichia coli	
Maximal acetate coutilization with glucose maintained up to $D = 0.2/h$ but totally lost between $D = 0.45/h$ and $0.5/h$	[35]
Yarrowia lipolytica	
Best ratio of nitrogen to carbon in the cultivation medium for lipid production without carbon loss into citric acid is between 0.021–0.085 N-mol/C-mol	[55]
Auxoaccelerostats	
Lactic acid bacteria	
Water activity is the most important environmental parameter affecting growth in cheese-like conditions	[56]
Determination of pH and temperature optima for achieving fastest growth of different lactic acid bacteria	[57]
Saccharomyces cerevisiae	
Dose effect curves (IC_{50}) five-times higher when the inhibitory aliphatic monocarboxylic acid is added smoothly to the growth environment compared to rapidly	[58]
Similar threshold levels for stress responses toward increased temperature, ethanol, salt, or organic acid concentration causing the decrease of μ_{max} for a recombinant and laboratory strain	[51,59]
Adaptastat	
Escherichia coli	
Enables to study cell metabolism near μ_{max} under substrate-limitation and the first to produce nutrient-limited growth on two substrates	[60]

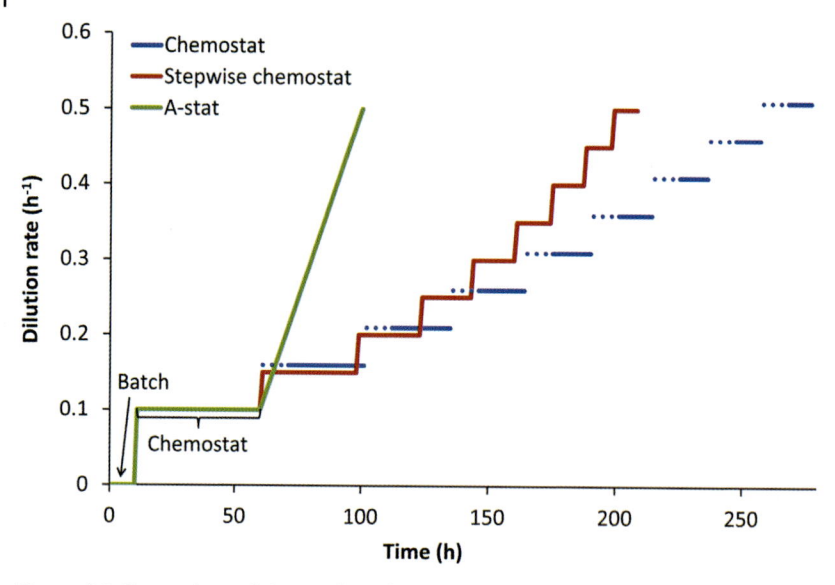

Figure 9.2 Comparison of time and resolution of accelerostat, stepwise chemostat, and chemostats carried out separately as independent cultures to scan through the specific growth rate region 0.1–0.5/h. Acceleration in A-stat $a_D = 0.01/h^2$. Five working volumes were assumed sufficient to reach steady-state in chemostats. Separate chemostat experiment (blue lines) take more time than stepwise chemostats as they all are started as batch cultures (dotted line), while only one batch is needed for the whole stepwise chemostat experiment

9.2.2 Family of Changestats – A Set of Flexible Tools for Scanning Steady-State Growth Space

The "family" of changestat methods can be divided into two groups: chemostat-based and turbidostat-based methods. The former methods enable to study cells at various μ in nutrient-limiting conditions, and the latter in substrate excess conditions at μ_{max}.

The most important benefit of using changestats for SSGSA is that they enable to scan through a region of the growth space within one experiment, and in reasonable time in most cases. This feature is illustrated in Figure 9.3. In addition to the significant save in time and resources when using changestats instead of separately performed chemostats, changestats enable to monitor dynamic changes of steady-state metabolism with high resolution, and accurately detect metabolic switch-points and optimal (maximum, minimum) growth conditions. For instance, metabolic shift from respiratory to respirofermentative growth (e.g., start of acetate overflow metabolism in *E. coli*), optimal environmental conditions for the highest productivity of target product formation and inhibition effects on substrate consumption can be reliably determined (see below for examples). These features of changestats make them unique for both fundamental studies of metabolism and bioprocess optimization.

It was stated above that changestats allow describing steady-state cell physiology equally well to chemostats if the experiments are carried out

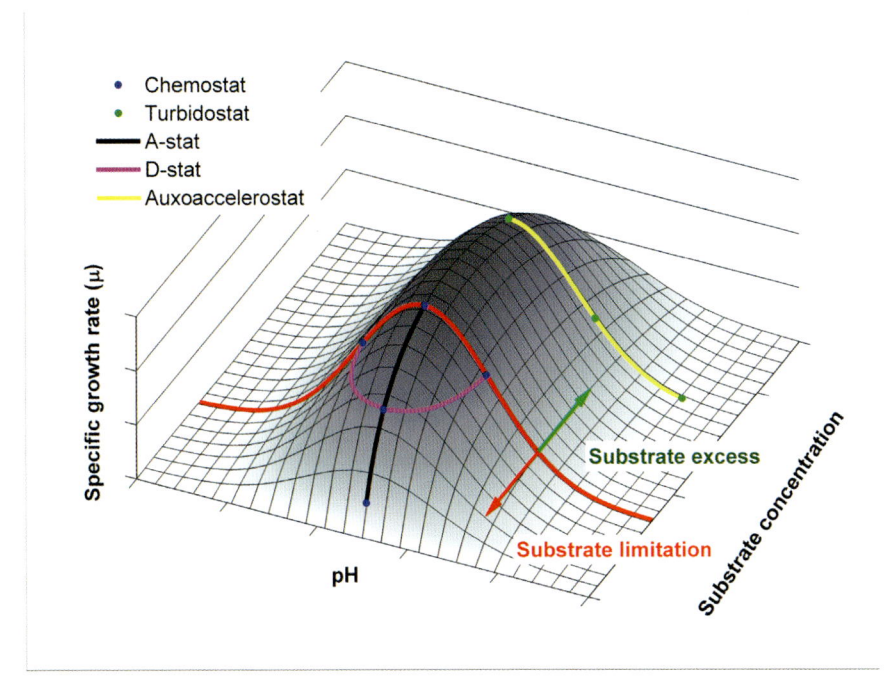

Figure 9.3 High-resolution steady-state growth space analysis (SSGSA) using different changestats (A-stat – accelerostat; D-stat – dilution rate stat). Blue and green points represent chemostat and turbidostat experiments, respectively, which can be carried out during the A-stat or auxoaccelerostat experiments. Refer to text for details of the methods [2].

properly. This means that the rate of change (acceleration/deceleration) of the environmental parameter of interest studied allows the cells to be maintained in a steady-state representative state (quasi steady-state), that is, cells are able to adapt to changing environmental conditions and remain despite the continuous change of the environmental parameters in the steady growth state corresponding to their current values. This can be validated during a changestat experiment most easily by stopping the acceleration/deceleration and setting the culture into chemostat mode: if growth characteristics do not change, then changestat data is equal to that of chemostat. If the rate of change of controlling environmental factor is too fast, metabolism will be disrupted and quasi steady-state lost.

Generally, the rate of change of the environmental parameter used for the scanning of the growth space depends on μ_{max} of the organism under study: the higher it is, the faster the environmental conditions can be changed. For an A-stat experiment, appropriate acceleration rate can be estimated based on μ_{max} of the organism to be in the range of 0.01–$0.04 \times \mu_{max}$ [6]. This is supported by several studies where μ_{max} was reached for different microorganisms in A-stats [5,35,46,48,61–64]. A careful study of the effect of acceleration rates on growth of *Lactococcus lactis* showed that $\mu_{max} = 0.59/h$ coinciding with batch data was reached in A-stat using accelerations <0.005/h^2 [63]. It was concluded in Ref. [42] that acceleration of 0.001/h^2 is the fastest, which can be used for maintaining steady-state growth during A-stats with yeasts. Generally, A-stat data collected from experiments using accelerations in the range of 0.007–0.05 × μ_{max} are comparable to chemostat data. However, the exact rate of change for each case should be determined and tested carefully, where detailed characterization of a novel organism is required (e.g., last step in optimization of growth conditions) [65].

The best evidence for proving that changestat and chemostat data on steady-state metabolism coincide quantitatively is found by the results of a number of experiments with different microorganisms where practically identical numerical values were obtained for major growth characteristics [35,38,41,42,45–47,66,67]; for transcriptome [35] and proteome [68] expression in E. coli. It has been shown that comparable results of A-stat and D-stat experiments have been well reproducible [35,38,42,43,47,67].

9.3 Review of the Results Obtained Using the Changestats [2]

The results obtained applying changestats for fundamental research of metabolism, bioprocess optimization and other purposes are summarized in Table 9.1 [2]. The main results presented in Table 9.1 are also commented in brief sections below.

9.3.1 Acetate Overflow Metabolism in E. coli

Since A-stat allows determining precisely μ-dependent metabolic switch points and follows the metabolic events around the switch point with high resolution, A-stat has been applied in a number of papers for studying acetate overflow metabolism in E. coli: excretion of acetate observed above a certain growth rate μ_{crit} on glucose.

Saturation of respiratory chain capacity was proposed as a reason for causing acetate accumulation in the cultivation broth at high μ by Paalme et al. in the 1990s (5,48). The results of the first transcriptome study using A-stat suggested that genes of the main acetate producing (Pta, AckA, PoxB) and consuming enzymes (Acs) might be involved [69]. The following comprehensive systems biology study, also using A-stat, showed that acetate overflow was triggered by down-regulation of the enzyme acetyl-CoA synthetase (Acs), causing the disruption of acetate recycling in the PTA-ACS node [35]. This hypothesis was proven in Ref. [36] as coordinated activation of the PTA-ACS and TCA cycles postponed the start of acetate overflow up to $q_S = 6.0$ mmol/(gDCW h) and reduced carbon flow to acetate fourfold at μ_{max} [36]. Application of A-stat was instrumental in these studies for proposing a novel hypothesis and engineering more efficient strains since it enabled to precisely detect the switch point of acetate overflow and monitor the metabolic events at high resolution.

9.3.2 A-Stat in Study of Physiology of Yeast

A-stat has been used due to its noted above advantages in study of μ-dependent metabolic switch-points for detailed characterization of the Crabtree effect in yeast [70] where cells switch from fully respiratory to respirofermentative growth above a certain μ (in Crabtree-positive yeasts), termed the critical μ (μ_{crit}) [71].

A large interlaboratory study of μ_{crit} values for laboratory yeasts to select a reference *Saccharomyces cerevisiae* strain amenable to experimental techniques used in genetic, physiological and biochemical engineering research was carried out using A-stat [38]. As seen in Table 9.1, A-stat has also been used for several studies of different physiological effects and media optimization in case of different yeasts. A detailed μ-dependent characterization of a non-*Saccharomyces* yeast *Hanseniaspora guilliermondii* [41] and *S. cerevisiae* [6] revealed a two-phase switch from fully respiratory to respirofermentative growth with the excretion of acetic acid preceding that of ethanol. Notably, the same two-phase switch was observed in a series of 20 chemostats of *S. cerevisiae* [72].

9.3.3 Integration of A-Stat with High-Throughput Omics Methods and Modeling

Integration of high-resolution steady-state cultures with omics measurements and metabolic modeling has strong potential for leading to more accurate and quantitative understanding of metabolism at whole-cell level. A comprehensive systems biology approach was applied for *E. coli* [35] and *L. lactis* [43] by coupling A-stat cultivation with transcriptome, proteome and metabolome analyses, and metabolic flux analysis. Both studies represent a unique data set of global regulation of metabolism with rising μ. Further computational analysis of these data combined with absolute quantitative proteome analysis [24] revealed that both microorganisms with very different metabolism – *E. coli* and *L. lactis* – use the same principles in the regulation of gene expression levels (mRNA, protein, metabolic fluxes) in order to achieve faster growth: (i) post-transcriptional control of protein abundances (changes in protein levels are not strictly determined by changes in mRNA levels) and (ii) post-translational control of metabolic flux rates (higher flux throughput is achieved through the increase of apparent *in vivo* catalytic rates of enzymes) [37,44]. The latter studies demonstrate the power of combining high-resolution A-stats with high-throughput omics analyses.

9.3.4 A-Stat in Bioprocess Development

The Dutch group led by Prof. R. H. Wijffels has used A-stat for the optimization of kinetic parameters of photobioreactors for microalgae cultivations [46] and bioprocess conditions for vitamin and carotenoid synthesis by *Dunaliella tertiolecta* [47]. It was shown that A-stat yields comparable results with chemostats even in such systems where the limiting nutrient is not evenly distributed within the bioreactor [46,47]. Notably, the benefits of A-stat were evident in optimization of production yields of vitamins C and E, and carotenoids lutein and β-carotene in microalgae *D. tertiolecta*, as optimal yields were achieved at different light intensities for each product [47]. Prof. R. H. Wijffels' group has also used another changestat method for bioprocess optimization (Section 9.3.5).

9.3.5 Deceleration-stat (De-stat)

The changestat method deceleration-stat (termed De-stat in this review) is in principle an A-stat with the only difference that D is decreased at a constant rate (opposite to A-stat). Although the first study to implement De-stat was [48] for determination of maintenance energy requirements in *E. coli* [73], De-stat has been most extensively used by Prof. R. H. Wijffels' group [74]. An important advantage of De-stat compared to A-stat is the save of time through shorter stabilization times needed to attain the initial steady-state at higher μ and use of faster rate of change of D [52]. Notably, De-stat has been shown to yield comparable results with A-stat in the study of demeton-*S*-methyl biodegradation capability by *C. glutamicum* [45].

Wijffels and coworkers first used De-stat for optimizing the photosynthetic efficiency of the purple nonsulfur bacterium *Rhodobacter capsulatus* for the production of hydrogen from acetate and light energy by determining important components of the light energy balance: biomass growth and maintenance, generation of hydrogen, and photosynthetic heat dissipation [50]. Additionally, Prof. R. H. Wijffels' group has developed a simulation for estimating maximal productivity of algal biomass based on De-stat data including even cases where deceleration exceeded the rate at which quasi steady-state could be maintained [52]. They show that up to 94% of time can be saved compared to chemostats using 10 times faster decelerations than required for maintaining quasi steady-state while losing only 5% in accuracy to estimate maximal biomass productivity rate.

Two studies have utilized De-stat also to specifically study stress responses. First, Neubauer and coworkers studied the stringent and general stress response during entry of *E. coli* into glucose starvation [49]. Second, response of *S. cerevisiae* to sudden or gradual decrease in glucose concentration and D has been followed based on the expression of the general stress response protein Hsp12p [51]. Similarly in Ref. [49], they detected a stronger stress response, through protein Hsp12p expression, toward rapid changes in substrate concentration and D in chemostat compared to gradual changes in De-stat. Application of De-stat in both studies led to the indication that stress response mechanisms in cells have adapted to rapid changes of environmental conditions.

9.3.6 Dilution Rate Stat (D-Stat)

The third widely used changestat method is dilution rate stat (D-stat). Although it was first used to study the Ehrlich pathway in the yeast *Zygosaccharomyces rouxii* [53], D-stat characteristics were first thoroughly formulated in Ref. [6] and later most extensively practiced by the groups of Prof. R. Vilu and Prof. T. Paalme in Tallinn, Estonia [75]. The concept of D-stat is simple: one environmental parameter (e.g., temperature, pH, cultivation medium composition) is smoothly changed while D and other parameters are kept constant according to algorithm:

$N = N_0 + a \times t$, where N is the parameter being changed, N_0 is the initial value of the parameter being changed, a is the rate of change of parameter N, and t is time [6]. The main advantage of D-stat compared to chemostat is that D-stat allows one to study the impact of an environmental parameter (including cultivation medium composition) on cell physiology at controlled μ over a range of values within one experiment at high-resolution, thus more potentially detecting optimal growth conditions (Figure 9.1).

D-stat is suitable for determining critical values and limits of both environmental (e.g., temperature, pH) and cultivation medium (e.g., relative nitrogen or vitamin concentration) factors that affect μ. For instance, D-stats with smooth increase in temperature detected the dynamic effects of temperature on μ, biomass yield, and by-product profiles of *Saccharomyces uvarum* [6].

D-stat is possibly most useful regarding bioprocess development for studying the consumption patterns of different substrates. In such cases, after steady-state in chemostat has been achieved on one medium, feeding of another medium with a different composition (added, removed or modified concentration of component(s)) is started and increased in time while simultaneously decreasing the feeding rate of the initial medium to keep D constant. This D-stat approach has yielded the following main results: (1) decreasing the ratio of threonine to methionine in the ingoing medium of *Z. rouxii* cultivations revealed that methionol synthesis occurs in the Ehrlich pathway [53]; (2) by-product patterns of fermentative metabolism for various LAB depend on the galactose to arginine ratio in the cultivation medium [54]; (3) maximal acetate consumption of ~35 mmol/gDCW by *E. coli* under glucose-limitation is maintained up to $D = 0.2/h$ while coutilization is totally lost between $D = 0.45$ and $0.5/h$ [35]; (4) the best ratio of nitrogen to carbon in the cultivation medium for producing lipids without carbon loss into citric acid by *Yarrowia lipolytica* is between 0.021–0.085 N-mol/C-mol [55].

9.3.7 Auxoaccelerostats

A-stat and D-stat are suitable for high-resolution study of steady-state cell physiology in nutrient limiting conditions. The classical method for studying cells in steady-state under substrate excess conditions is turbidostat [76] where cells are forced to grow with μ_{max} by adjusting D based on biomass concentration (OD as the indicator) using the following simple logic: if OD is below or above the set point value, D is decreased or increased, respectively. In addition to culture turbidity, cells can be forced to grow with μ_{max} through feedback from pH (pH-auxostat; [77]), CO_2 (CO_2-auxostat; [78]), electrical capacitance of the culture (permittistat; [79]), or other parameters. However, turbidostat and auxostats are not suitable for high-resolution study of steady-state growth space under nutrient excess since, similar to chemostat, the culture has to be stabilized after every shift in environmental conditions. Therefore, auxoaccelerostats were developed [6] for achieving higher resolution through smoothly changing one environmental parameter under nutrient excess conditions. An auxoaccelerostat culture is controlled online by the experimenter through a parameter that is either directly (e.g., OD) or indirectly (e.g., pH, percentage of dissolved oxygen concentration in bioreactor (pO_2%), percentage of CO_2 in the off-gas) related to cell growth. The algorithm for operating an auxoaccelerostat is the same as for D-stat.

Similar to A-stat and D-stat, auxoaccelerostats are suitable for the determination of metabolic switch-points and optimal growth conditions, but under nutrient excess. Applying various auxoaccelerostats for studying the effects of pH, temperature, salt concentration, and water activity on the growth of the cheese bacterium *Lactobacillus paracasei* revealed that water activity is the most important environmental parameter affecting the growth of cells in cheese-like conditions [56]. Another study utilizing pH-, pO_2-, and CO_2-auxoaccelerostats to determine the effects of pH, temperature, pO_2%, water activity, inhibitory substances, or medium components (tryptone or yeast extract) on the growth of *S. cerevisiae* and *L. lactis* determined, for instance, the temperatures and tryptone concentrations inhibiting the growth yield [6]. Interestingly, while slowly growing LAB (*L. paracasei*) were adapted to grow at μ_{max} over a wide range of temperature (30–37 °C) or pH (5.5–6) values, fast growing LAB, like *Streptococcus thermophilus*, showed a narrow range of temperature (44 °C) and pH (6.6) for growth at μ_{max} [57].

Auxoaccelerostats have also been used to study stress responses of yeast in detail. Characterizing the toxic effects of aliphatic monocarboxylic acids (e.g., acetic, formic, propionic acid) on *S. cerevisiae* growth using CO_2-auxoaccelerostats revealed immediate decline of μ and the growth yield with slowly increasing acid concentrations [58]. Furthermore, they showed that the dose effect curves (IC_{50}; acid concentration causing a 50% decrease in either μ or growth yield) are five-times higher when the inhibitory acid is added smoothly to the growth environment (as in auxoaccelerostat) compared to rapidly (by pulse). Two studies using auxoaccelerostats and metabolic modeling for studying the effect of smoothly changing stress conditions (increasing temperature, ethanol, salt, or organic acid concentration) on a laboratory and a recombinant *S. cerevisiae* strain determined similar threshold levels of stress responses causing the decrease of μ_{max} [51,59]. Interestingly, these threshold values determined under nutrient excess and μ_{max} are very similar to those of glucose-limited D-stats at $D = 0.09/$h [51]. In conclusion, the latter studies highlight that high-resolution auxoaccelerostats can successfully be used for determination of the quantitative effects of environmental conditions on growth characteristics.

9.3.8 Adaptastat

To further complement the array of changestat methods, adaptastat was developed for studying cells near μ_{max} under substrate-limitation in aerobic cultures a decade ago in National Institute of Chemical Physics and Biophysics in Tallinn, Estonia by Dr. Kalju Vanatalu. In adaptastat, μ of cells is raised stepwise until near μ_{max} through increasing D according to activation of oxygen consumption by the microorganism [60]. More specifically, D is controlled through a feedback loop as follows. After attaining steady-state in chemostat, feeding (D) is abruptly increased followed by stop (or reduction) of nutrient inflow. Next, the time needed to exhaust the residual substrate is measured, indicated by a rapid rise in the dissolved oxygen concentration. The shorter the time, the faster the growth. Subsequently, D is increased or decreased depending on the ratio of feeding pulse and substrate exhaustion durations. Adaptastat is an attractive method since it enables to study cell metabolism near μ_{max} under substrate-limitation and also the first to produce nutrient-limited growth on two substrates. The method should be efficient for isotopic labeling of bacterial cultures and their components produced abundantly at fast growth, for example, ribosomes, RNA.

9.4 SSGSA Using Parallel-Sequential Cultivations

The fact that multiple reactors operating in series provide high conversion rates and productivities is well established in chemical engineering [80]. The concept has been successfully applied in bioprocesses as well, and is termed parallel-sequential fermentation (PSF). PSF has been used widely throughout the years starting from 1960s. It is very popular also in the industry for production of different target products. A fully automated, impressively executed and technologically advanced example of production optimization in multireactor environment was carried out recently by Luttmann et al. in case of malaria vaccine production using seven bioreactors [81].

In Ref. [7] the parallel-sequential set-up was used with the novel focus on maintaining the cell culture in the fermenters as well as during the transfer between the bioreactors in physiological steady-state throughout the whole experiment. The scheme described was termed "multiplying the physiological steady-state" and in the heart of the method lies the concept of maintaining controlled steady physiological states (quasi steady-states as in A-stat etc. – QSS) in various multiplexed multireactor set-ups (7). The application for the method was analysis of the effects of environmental variables to microbial physiology in QSS with minimal experiment time, or to speed up production processes.

The bioreactor system employed in testing the QSS parallel-sequential scheme consists of a seeder and receiver reactor(s), connected with a transfer line. The procedure was started as a typical chemostat cultivation in a single reactor, but after stabilization the culture volume was increased in fed-batch cultivation at constant growth rate equal to that in chemostat. An aliquot (corresponding to the working volume of receiver reactors) is transferred to the preset empty reactors. The reactor from which culture is transferred is termed the seeder reactor, and the target reactors are termed the receiver reactors (Figure 9.4). It is emphasized here that in order to keep the physiological state unaltered the environmental parameters (such as dilution rate set point, temperature, pH, dO_2) in the receiver units must match these of the seeder unit already at the moment of transfer – and the receiver units must be empty from any liquid. Following the transfer, the QSS is validated, and the environment inside the receiver units could be changed as planned. The experiments could be designed such that during the analysis in the receiver units the culture volume in the seeder unit is increased anew, and the next culture transfer is conducted at the moment the receiver reactors are prepared.

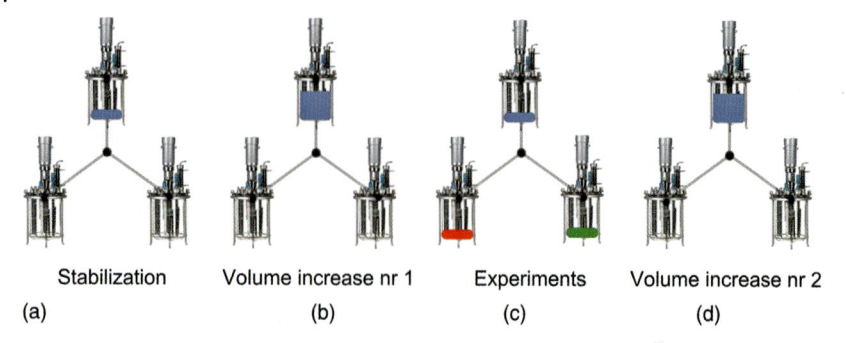

Stabilization	Volume increase nr 1	Experiments	Volume increase nr 2
(a)	(b)	(c)	(d)

Figure 9.4 Generic process of parallel-sequential experiments. The "mother" culture in the topmost reactor, in the seeder reactor, is stabilized in a required physiological steady-state in chemostat determined by D (a). Volume of the culture in the seeder reactor is increased in quasi steady-state (QSS) in fed-batch with the $\mu = D$ (b). This phase is completed with culture transfer into the receiver reactors, and planned experiments are conducted in the receiver reactors (c). After obtaining the information desired, the receiver reactors are cleaned and reset for the next transfer. Meanwhile, the culture volume in the seeder reactor is restored (d). The system is ready for the next round of experiments.

The processes conducted in the PSF-scheme vary in complexity as basically any QSS fermentation can take place in the seeder unit. Similarly, the nature of processes run in the receiver unit is not limited, provided that they are started with QSS of the seeder unit. This essentially means that the PSF allows extend possibilities for SSGSA (Figure 9.3). Completely new opportunities could be developed using different schemes of multiplexed microscale continuous culture platforms [32–34].

9.5 Modeling in Steady-State Growth Space Analysis

Currently, constraint-based metabolic network models are generally used for the quantitative analysis of essentially steady-state cell metabolism [82,83]. They have proved their usefulness [9] but at the same time their shortcomings have become also apparent. These models enable calculation and analysis of intracellular steady-state flux patterns of metabolic networks (central components of the models are fluxes of intracellular compounds) based on different constraints. The models referred have been studied very extensively during the last 30 years [84]. However, despite all the developments these models are describing growth and functioning of cell-less biomass excluding from the analysis a number of important physiological mechanisms of cells (cell cycle, cell geometry, growth of individual cells, overflow metabolism, suboptimal metabolism etc.). Therefore, it is not possible to calculate related to the parameters indicated or growth boundaries associated with mentioned physiological mechanisms in the framework of these models [85–87]. In addition, usually the biomass composition must be determined to calculate fluxes [88], and it is not possible to estimate/calculate both – flux patterns and biomass composition. Biomass synthesis requirements are determined from the cellular biomass composition and are thus essentially independent of genome sequence. Finally, as mentioned, metabolic network models generally do not characterize the growth of cells but biomass (gdw) with an average composition.

Some notable examples of evolution from regular network models toward whole-cell models can be listed. First attempt to create formalism for describing translation and transcription processes in detail was made already 10 years ago [89,90]. The main new feature of the model was the possibility to calculate metabolic demands associated with sequence-based polymer composition data (RNA, protein) resulting from RNA and protein polymerization. Initial polymerization network modeling framework has been developed over time evolving toward whole-cell description and applied also for genome-scale metabolic models [91]. Recently, metabolic and polymerization networks were integrated and the resulting model has been used to demonstrate changes in cellular composition and gene expression at different growth rates [92]. However, despite of the improvements such models still do not involve cell cycle, geometry, and molecular characteristics of cell components (enzymes, other macromolecules etc).

Very promising models applying whole-cell concept are so-called single cell models (SCM) that take into account all the main cell components, ratio of cell surface and volume, coordination between DNA replication initiation and cell growth, coordinated synthesis of monomers and macromolecules and spatial organization [93]. SCM have been used to predict cell size, shape, and volume as a function of growth rate and limiting substrate concentration [94], changes in growth rate and metabolism in response to switching from aerobic growth to anaerobic growth [95]. Effects of different lac promoter mutations [96] and molecular mechanisms of cell cycle [97,98] have been studied. In addition, SCM framework developed by Shuler and coworkers has been applied to model synthesis of membrane components [99] and nucleotides [100] in theoretical minimal cells [101]. Main distinctive features of SCM are whole- (involving all main cell components and processes including regulation) and single-cell concepts (accurate representation of cell processes), lack of various simplifying assumptions, and ability to describe dynamical changes (including during cell cycle).

A very interesting genome-scale, whole-cell level SCM was developed by Covert and coworkers to describe growth of *M. genitalium* [87]. This was possible due to extensive experimental studies that have produced enough data for model building and validation processes. A novel submodule concept was developed. Assuming that different modules function independently during short timescales, it was possible to construct hybrid model exploiting very different mathematics for sub-models that were solved separately and integrated together. The model was validated and tested with broad range of independent datasets and used to predict dynamics of DNA-binding proteins, to explain variability of cell cycle, to point out details of energy balance and so on. This model has been intensively developed further during the recent years [102–104].

Despite the principal power of the whole-cell models they have not "conquered" the world. It was summarized by Shuler in Ref. [93]. Too complicated structure of the model cells, myriad of numbers in the equations, and need, at least apparent, of extensive experimental data not available currently [104] make use of the models inconvenient and not transparent.

Summarizing, there is a need for the development of the modeling framework that would allow to analyze physiology of cells from different angles. It would mean that compatible with each other models of different complexity addressing different "layers" of cell metabolism and different regulatory levels should be developed. And as argued in the present review, models for the analysis of steady-state metabolism, essentially stoichiometric models will be, should be in the foundation of this system of models. Needless to say that these stoichiometric models should include cell cycle, geometry, and so on, currently by and large neglected in the generally used constraint-based flux models. This system of models should make possible also *in silico ab initio* "first time right" cell design. An attempt to develop this type of system of models has been undertaken, and the first results obtained by our group can be found in Ref. [105] (additional publications and software in preparation).

References

1 Hoskisson, P.A. and Hobbs, G. (2005) Continuous culture–making a comeback? *Microbiology*, **151**, 3153–3159.
2 Adamberg, K., Valgepea, K., and Vilu, R. (2015) Advanced continuous cultivation methods for systems microbiology. *Microbiology*, **161**, 1707–1719.
3 Novick, A. and Szilard, L. (1950) Description of the chemostat. *Science*, **112**, 715–716.
4 Monod, J. (1950) La technique de culture continue, theorie et applications. *Ann. Inst. Pasteur*, **79**, 390–410.
5 Paalme, T., Kahru, A., Elken, R., Vanatalu, K., Tiisma, K., and Vilu, R. (1995) The computer-controlled continuous culture of *Escherichia coli* with smooth change of dilution rate (A-stat). *J. Microbiol. Methods*, **24**, 145–153.
6 Kasemets, K., Drews, M., Nisamedtinov, I., Paalme, T., and Adamberg, K. (2003) Modification of A-stat for the characterization of microorganisms. *J. Microbiol. Methods*, **55**, 187–200.
7 Erm, S., Adamberg, K., and Vilu, R. (2014) Multiplying steady-state culture in multi-reactor system. *Bioproc. Biosyst. Eng.*, **37**, 2361–2370.
8 Bordbar, A., Monk, J.M., King, Z., and Palsson, B.Ø. (2014) Constraint-based models predict metabolic and associated cellular functions. *Nat. Rev. Genet.*, **15**, 107–120.
9 Lee, S.Y. and Kim, H.U. (2015) Systems strategies for developing industrial microbial strains. *Nat. Biotechnol.*, **33**, 1061–1072.
10 Becker, J., Zelder, O., Häfner, S., Schröder, H., and Wittmann, C. (2011) From zero to hero–design-based systems metabolic engineering of *Corynebacterium glutamicum* for L-lysine production. *Metab. Eng.*, **13**, 159–168.
11 Van Dien, S.J. (2013) From the first drop to the first truckload: commercialization of microbial processes for renewable chemicals. *Curr. Opin. Biotechnol.*, **24**, 1061–1068.

12 Huang, C.-J., Lin, H., and Yang, X. (2012) Industrial production of recombinant therapeutics in *Escherichia coli* and its recent advancements. *J. Ind. Microbiol. Biotechnol.*, **39**, 383–399.

13 Jantama, K., Zhang, X., Moore, J.C., Shanmugam, K.T., Svoronos, S.A., and Ingram, L.O. (2008) Eliminating side products and increasing succinate yields in engineered strains of *Escherichia coli* C. *Biotechnol. Bioeng.*, **101**, 881–893.

14 Lee, J.W., Na, D., Park, J.M., Lee, J., Choi, S., and Lee, S.Y. (2012) Systems metabolic engineering of microorganisms for natural and non-natural chemicals. *Nat. Chem. Biol.*, **8**, 536–546.

15 Nakamura, C.E. and Whited, G.M. (2003) Metabolic engineering for the microbial production of 1, 3-propanediol. *Curr. Opin. Biotechnol.*, **14**, 454–459.

16 Paddon, C.J., Westfall, P.J., Pitera, D.J., Benjamin, K., Fisher, K., McPhee, D., Leavell, M.D., Tai, A., Main, A., Eng, D., Polichuk, D.R., Teoh, K.H., Reed, D.W., Treynor, T., Lenihan, J., Fleck, M., Bajad, S., Dang, G., Dengrove, D., Diola, D., Dorin, G., Ellens, K.W., Fickes, S., Galazzo, J., Gaucher, S.P., Geistlinger, T., Henry, R., Hepp, M., Horning, T., Iqbal, T., Jiang, H., Kizer, L., Lieu, B., Melis, D., Moss, N., Regentin, R., Secrest, S., Tsuruta, H., Vazquez, R., Westblade, L.F., Xu, L., Yu, M., Zhang, Y., Zhao, L., Lievense, J., Covello, P.S., Keasling, J.D., Reiling, K.K., Renninger, N.S., and Newman, J.D. (2013) High-level semi-synthetic production of the potent antimalarial artemisinin. *Nature*, **496** 528–532.

17 Sun, J. and Alper, H.S. (2015) Metabolic engineering of strains: from industrial-scale to lab-scale chemical production. *J. Ind. Microbiol. Biotechnol.*, **42**, 423–436.

18 Yim, H., Haselbeck, R., Niu, W., Pujol-Baxley, C., Burgard, A.P., Boldt, J., Khandurina, J., Trawick, J.D., Osterhout, R.E., Stephen, R., Estadilla, J., Teisan, S., Schreyer, H.B., Andrae, S., Yang, T.H., Lee, S.Y., Burk, M.J., and Van Dien, S. (2011) Metabolic engineering of *Escherichia coli* for direct production of 1, 4-butanediol. *Nat. Chem. Biol.*, **7**, 445–452.

19 Málek, I. (1958) The physiological state of microorganisms during continuous culture. A Symp in Continuous Cultivation of Microorganisms, Publishing House ASCR, Prague, p. 21.

20 Konstantinov, K.B. (1996) Monitoring and control of the physiological state of cell cultures. *Biotechnol. Bioeng.*, **52**, 271–289.

21 Stephanopoulos, G., Misra, J., Hwang, D., Schmitt, W., Alevizos, I., Silva, S., and Gill, R. (2002) Defining biological states and related genes, proteins and patterns. US Patent US20,020,169,562 A1.

22 Le Marc, Y., Baranyi, J., and Pin, C. (2005) Methods to determine the growth domain in a multidimensional environmental space. *Int. J. Food Microbiol.*, **100**, 3–12.

23 Koseki, S. (2009) Microbial responses viewer (MRV): a new ComBase-derived database of microbial responses to food environments. *Int. J. Food Microbiol.*, **134**, 75–82.

24 Arike, L., Valgepea, K., Peil, L., Nahku, R., Adamberg, K., and Vilu, R. (2012) Comparison and applications of label-free absolute proteome quantification methods on *Escherichia coli*. *J. Proteom.*, **75**, 5437–5448.

25 Esquerré, T., Laguerre, S., Turlan, C., Carpousis, A.J., Girbal, L., and Cocaign-Bousquet, M. (2013) Dual role of transcription and transcript stability in the regulation of gene expression in *Escherichia coli* cells cultured on glucose at different growth rates. *Nucleic Acids Res.*, **42**, 2460–2472.

26 Ishii, N., Nakahigashi, K., Baba, T., Robert, M., Soga, T., Kanai, A., Hirasawa, T., Naba, M., and Hirai, K. other authors. (2007) Multiple high-throughput analyses monitor the response of *E. coli* to perturbations (SOM). *Science*, **316**, 593–597.

27 Ferenci, T. (2008) Bacterial physiology, regulation and mutational adaptation in a chemostat environment. *Adv. Microb. Physiol.*, **53**, 169–229.

28 Bull, A.T. (2010) The renaissance of continuous culture in the post-genomics age. *J. Ind. Microbiol. Biotechnol.*, **37**, 993–1021.

29 Harder, W. and Kuenen, J.G. (1977) Microbial selection in continuous culture. *J. Appl. Bacteriol.*, **43**, 1–24.

30 Helling, R.B., Vargas, C.N., and Adams, J. (1987) Evolution of *Escherichia coli* during growth in a constant environment. *Genetics*, **116**, 349–358.

31 Gresham, D. and Hong, J. (2014) The functional basis of adaptive evolution in chemostats. *FEMS Microbiol. Rev.*, **39**, 2–16.

32 Moffitt, J.R., Lee, J.B., and Cluzel, P. (2012) The single-cell chemostat: an agarose-based, microfluidic device for high-throughput, single-cell studies of bacteria and bacterial communities. *Lab. Chip.*, **12**, 1487–1494.

33 Dénervaud, N., Becker, J., Delgado-Gonzalo, R., Damay, P., Rajkumar, A.S., Unser, M., Shore, D., Naef, F., and Maerkl, S.J. (2013) A chemostat array enables the spatio-temporal analysis of the yeast proteome. *Proc. Natl. Acad. Sci. USA*, **110**, 15842–15847.

34 Long, Z., Nugent, E., Javer, A., Cicuta, P., Sclavi, B., Cosentino Lagomarsino, M., and Dorfman, K.D. (2013) Microfluidic chemostat for measuring single cell dynamics in bacteria. *Lab. Chip.*, **13**, 947–954.

35 Valgepea, K., Adamberg, K., Nahku, R., Lahtvee, P.-J., Arike, L., and Vilu, R. (2010) Systems biology approach reveals that overflow metabolism of acetate in *Escherichia coli* is triggered by carbon catabolite repression of acetyl-CoA synthetase. *BMC Syst. Biol.*, **4**, 166.

36 Peebo, K., Valgepea, K., Nahku, R., Riis, G., Õun, M., Adamberg, K., and Vilu, R. (2014) Coordinated activation of PTA-ACS and TCA cycles strongly reduces overflow metabolism of acetate in *Escherichia coli*. *Appl. Microbiol. Biotechnol.*, **98**, 5131–5143.

37 Valgepea, K., Adamberg, K., Seiman, A., and Vilu, R. (2013) *Escherichia coli* achieves faster growth by increasing catalytic and translation rates of proteins. *Mol. Biosyst.*, **9**, 2344–2358.

38 Van Dijken, J.P., Bauer, J., Brambilla, L., Duboc, P., Francois, J., Gancedo, C., Giuseppin, M., Heijnen, J.J., Hoare, M. *et al.* (2000) An interlaboratory comparison of physiological and genetic properties of four *Saccharomyces cerevisiae* strains. *Enzyme Microb. Technol.*, **26**, 706–714.

39 Klein, C.J.L., Rasmussen, J.J., Rønnow, B., Olsson, L., and Nielsen, J. (1999) Investigation of the impact of MIG1 and MIG2 on the physiology of *Saccharomyces cerevisiae*. *J. Biotechnol.*, **68**, 197–212.

40 Marc, J., Feria-Gervasio, D., Mouret, J.-R., and Guillouet, S.E. (2013) Impact of oleic acid as co-substrate of glucose on 'short' and 'long-term' Crabtree effect in *Saccharomyces cerevisiae. Microb. Cell Fact.*, **12**, 83.

41 Albergaria, H., Torrao, A., Hogg, T., and Girio, F. (2003) Physiological behaviour of *Hanseniaspora guilliermondii* in aerobic glucose-limited continuous cultures. *FEMS Yeast Res.*, **3**, 211–216.

42 Van der Sluis, C., Westerink, B.H., Dijkstal, M.M., Castelein, S.J., van Boxtel, A.J., Giuseppin, M.L., Tramper, J., and Wijffels, R.H. (2001) Estimation of steady-state culture characteristics during acceleration-stats with yeasts. *Biotechnol. Bioeng.*, **75**, 267–275.

43 Lahtvee, P.-J., Adamberg, K., Arike, L., Nahku, R., Aller, K., and Vilu, R. (2011) Multi-omics approach to study the growth efficiency and amino acid metabolism in *Lactococcus lactis* at various specific growth rates. *Microb. Cell Fact.*, **10**, 12.

44 Adamberg, K., Seiman, A., and Vilu, R. (2012) Increased biomass yield of *Lactococcus lactis* by reduced overconsumption of amino acids and increased catalytic activities of enzymes. *PLoS One*, **7**, e48223.

45 Girbal, L., Rols, J.L., and Lindley, N.D. (2000) Growth rate influences reductive biodegradation of the organophosphorus pesticide demeton by *Corynebacterium glutamicum. Biodegradation*, **11**, 371–376.

46 Barbosa, M.J., Hoogakker, J., and Wijffels, R.H. (2003) Optimisation of cultivation parameters in photobioreactors for microalgae cultivation using the A-stat technique. *Biomol. Eng.*, **20**, 115–123.

47 Barbosa, M.J., Zijffers, J.-W.F., Nisworo, A., Vaes, W., Wijffels, R.H., and van Schoonhoven, J. (2005) Optimization of biomass, vitamins, and carotenoid yield on light energy in a flat-panel reactor using the A-stat technique. *Biotechnol. Bioeng.*, **89**, 233–242.

48 Paalme, T., Elken, R., Kahru, A., Vanatalu, K., and Vilu, R. (1997) The growth rate control in *Escherichia coli* at near to maximum growth rates: the A-stat approach. *Antonie Van Leeuwenhoek*, **71**, 217–230.

49 Teich, A., Meyer, S., Lin, H.Y., Andersson, L., Enfors, S.-O., and Neubauer, P. (1999) Growth rate related concentration changes of the starvation response regulators sigmaS and ppGpp in glucose-limited fed-batch and continuous cultures of *Escherichia coli. Biotechnol. Prog.*, **15**, 123–129.

50 Hoekema, S., Douma, R.D., Janssen, M., Tramper, J., and Wijffels, R.H. (2006) Controlling light-use by *Rhodobacter capsulatus* continuous cultures in a flat-panel photobioreactor. *Biotechnol. Bioeng.*, **95**, 614–626.

51 Nisamedtinov, I., Lindsey, G.G., Karreman, R., Orumets, K., Koplimaa, M., Kevvai, K., and Paalme, T. (2008) The response of the yeast *Saccharomyces cerevisiae* to sudden vs. gradual changes in environmental stress monitored by expression of the stress response protein Hsp12p. *FEMS Yeast Res.*, **8**, 829–838.

52 Hoekema, S., Rinzema, A., Tramper, J., Wijffels, R.H., and Janssen, M. (2014) Deceleration-stats save much time during phototrophic culture optimization. *Biotechnol. Bioeng.*, **111**, 792–802.

53 Van Der Sluis, C., Rahardjo, Y.S.P., Smit, B.A., Kroon, P.J., Hartmans, S., Ter Schure, E.G., Tramper, J., and Wijffels, R. (2002) Concomitant extracellular

accumulation of alpha-keto acids and higher alcohols by *Zygosaccharomyces rouxii*. *J. Biosci. Bioeng.*, **93**, 117–124.

54 Adamberg, K., Adamberg, S., Laht, T.-M., Ardö, Y., and Paalme, T. (2006) Study of cheese associated lactic acid bacteria under carbohydrate-limited conditions using D-stat cultivation. *Food Biotechnol.*, **20**, 143–160.

55 Ochoa-Estopier, A. and Guillouet, S.E. (2014) D-stat culture for studying the metabolic shifts from oxidative metabolism to lipid accumulation and citric acid production in *Yarrowia lipolytica*. *J. Biotechnol.*, **170**, 35–41.

56 Laht, T.-M., Kask, S., Elias, P., Adamberg, K., and Paalme, T. (2002) Role of arginine in the development of secondary microflora in Swiss-type cheese. *Int. Dairy J.*, **12**, 831–840.

57 Adamberg, K., Kask, S., Paalme, T., and Laht, T.-M. (2003) The effect of temperature and pH on the growth of lactic acid bacteria: a pH-auxostat study. *Int. J. Food. Microbiol.*, **85**, 171–183.

58 Kasemets, K., Kahru, A., Laht, T.-M., and Paalme, T. (2006) Study of the toxic effect of short- and medium-chain monocarboxylic acids on the growth of *Saccharomyces cerevisiae* using the CO_2-auxo-accelerostat fermentation system. *Int. J. Food Microbiol.*, **111**, 206–215.

59 Kasemets, K., Nisamedtinov, I., Laht, T.-M., Abner, K., and Paalme, T. (2007) Growth characteristics of *Saccharomyces cerevisiae* S288C in changing environmental conditions: auxo-accelerostat study. *Antonie Van Leeuwenhoek*, **92**, 109–128.

60 Tomson, K., Vanatalu, K., and Barber, J. (2006) Adaptastat--a new method for optimising of bacterial growth conditions in continuous culture: Interactive substrate limitation based on dissolved oxygen measurement. *J. Microbiol. Methods*, **64**, 380–390.

61 Drews, M., Paalme, T., and Vilu, R. (1995) The growth and nutrient utilization of the insect cell line Spodoptera frugiperda Sf9 in batch and continuous culture. *J. Biotechnol.*, **40**, 187–198.

62 Paalme, T., Elken, R., Vilu, R., and Korhola, M. (1997) Growth efficiency of *Saccharomyces cerevisiae* on glucose/ethanol media with a smooth change in the dilution rate (A-stat). *Enzyme Microb. Technol.*, **20**, 174–181.

63 Adamberg, K., Lahtvee, P.-J., Valgepea, K., Abner, K., and Vilu, R. (2009) Quasi steady state growth of *Lactococcus lactis* in glucose-limited acceleration stat (A-stat) cultures. *Antonie Van Leeuwenhoek*, **95**, 219–226.

64 Kask, S., Laht, T.-M., Pall, T., and Paalme, T. (1999) A study on growth characteristics and nutrient consumption of *Lactobacillus plantarum* in A-stat culture. *Antonie Van Leeuwenhoek*, **75**, 309–320.

65 Nahku, R., Peebo, K., Valgepea, K., Barrick, J.E., Adamberg, K., and Vilu, R. (2011) Stock culture heterogeneity rather than new mutational variation complicates short-term cell physiology studies of *Escherichia coli* K-12 MG1655 in continuous culture. *Microbiology*, **157**, 2604–2610.

66 Albergaria, H., Duarte, L.C., Amaral-Collaço, M.T., and Gírio, F.M. (2000) Study of *Saccharomyces uvarum* CCMI 885 physiology under fed-batch, chemostat and accelerostat cultivation techniques. *Food Technol. Biotechnol.*, **38**, 33–38.

67 Valgepea, K., Adamberg, K., and Vilu, R. (2011) Decrease of energy spilling in *Escherichia coli* continuous cultures with rising specific growth rate and carbon wasting. *BMC Syst. Biol.*, **5**, 106.

68 Nahku, R. (2012) Validation of critical factors for the quantitative characterization of bacterial physiology in accelerostat cultures. Tallinn University of Technology.

69 Nahku, R., Valgepea, K., Lahtvee, P.-J., Erm, S., Abner, K., Adamberg, K., and Vilu, R. (2010) Specific growth rate dependent transcriptome profiling of *Escherichia coli* K12 MG1655 in accelerostat cultures. *J. Biotechnol.*, **145**, 60–65.

70 De Deken, R.H. (1966) The Crabtree effect: a regulatory system in yeast. *J. Gen. Microbiol.*, **44**, 149–156.

71 Herwig, C., Marison, I., and von Stockar, U. (2001) On-line stoichiometry and identification of metabolic state under dynamic process conditions. *Biotechnol. Bioeng.*, **75**, 345–354.

72 Postma, E., Verduyn, C., Scheffers, W.A., and Van Dijken, J.P. (1989) Enzymic analysis of the crabtree effect in glucose-limited chemostat cultures of *Saccharomyces cerevisiae*. *Appl. Environ. Microbiol.*, **55**, 468–477.

73 Stouthamer, A.H. (1973) A theoretical study on the amount of ATP required for synthesis of microbial cell material. *Antonie Van Leeuwenhoek*, **39**, 545–565.

74 Zijffers, J.-W.F., Schippers, K.J., Zheng, K., Janssen, M., Tramper, J., and Wijffels, R.H. (2010) Maximum photosynthetic yield of green microalgae in photobioreactors. *Mar. Biotechnol.*, **12**, 708–718.

75 Lahtvee, P.-J., Valgepea, K., Adamberg, K., Nahku, R., Abner, K., and Vilu, R. (2009) Steady state growth space study of *Lactococcus lactis* in D-stat cultures. *Antonie Van Leeuwenhoek*, **96**, 487–496.

76 Bryson, V. and Szybalski, W. (1952) Microbial selection. *Science*, **116**, 45–51.

77 Martin, G.A. and Hempfling, W.P. (1976) A method for the regulation of microbial population density during continuous culture at high growth rates. *Arch. Microbiol.*, **197**, 41–47.

78 Watson, T.G. (1969) Steady state operation of continuous culture at maximum growth rate by control of carbon dioxide production. *J. Gen. Microbiol.*, **59**, 83–89.

79 Markx, G.H., Davey, C.L., and Kell, D.B. (1991) The permittistat: a novel type of turbidostat. *J. Gen. Microbiol.*, **137**, 735–743.

80 Coker, A.K. (ed.) (2001) Industrial and laboratory reactors. in *Modeling of Chemical Kinetics and Reactor Design*, Gulf Professional Publishing, Boston, pp 218–259.

81 Luttmann, R., Borchert, S.-O., Mueller, C., Loegering, K., Aupert, F., Weyand, S., Kober, C., Faber, B., and Cornelissen, G. (2015) Sequantial/parallel production of potential malaria vaccines–a direct way from single batch to quai-continuous integrated production. *J. Biotechnol.*, **213**, 83–96.

82 O'Brien, E.J., Monk, J.M., and Palsson, B.O. (2015) Using genome-scale models to predict biological capabilities. *Cell*, **161**, 971–987.

83 McCloskey, D., Palsson, B.Ø., and Feist, A.M. (2013) Basic and applied uses of genome-scale metabolic network reconstructions of *Escherichia coli*. *Mol. Syst. Biol.*, **9**, 661.

84 Edwards, J.S. and Palsson, B.O. (1998) How will bioinformatics influence metabolic engineering? *Biotechnol. Bioeng.*, **58**, 162–169.

85 Goel, A., Wortel, M.T., Molenaar, D., and Teusink, B. (2012) Metabolic shifts: a fitness perspective for microbial cell factories. *Biotechnol. Lett.*, **34**, 2147–2160.

86 Molenaar, D., van Berlo, R., de Ridder, D., and Teusink, B. (2009) Shifts in growth strategies reflect tradeoffs in cellular economics. *Mol. Syst. Biol.*, **5**, 323.

87 Karr, J.R., Sanghvi, J.C., Macklin, D.N., Gutschow, M.V., Jacobs, J.M., Bolival, B., Assad-Garcia, N., Glass, J.I., and Covert, M.W. (2012) A whole-cell computational model predicts phenotype from genotype. *Cell*, **150**, 389–401.

88 Feist, A.M. and Palsson, B.O. (2010) The biomass objective function. *Curr. Opin. Microbiol.*, **13**, 344–349.

89 Allen, T.E. and Palsson, B.Ø. (2003) Sequence-based analysis of metabolic demands for protein synthesis in prokaryotes. *J. Theor. Biol.*, **220**, 1–18.

90 Allen, T.E., Herrgård, M.J., Liu, M., Qiu, Y., Glasner, J.D., Blattner, F.R., and Palsson, B.Ø. (2003) Genome-scale analysis of the uses of the Escherichia coli genome: model-driven analysis of heterogeneous data sets. *J. Bacteriol.*, **185**, 6392–6399.

91 Thiele, I., Jamshidi, N., Fleming, R.M., and Palsson, B.Ø. (2009) Genome-scale reconstruction of *Escherichia coli*'s transcriptional and translational machinery: a knowledge base, its mathematical formulation, and its functional characterization. *PLoS Comput. Biol.*, **5**, e1000312.

92 Lerman, J.A., Hyduke, D.R., Latif, H., Portnoy, V.A., Lewis, N.E., Orth, J.D., Schrimpe-Rutledge, A.C., Smith, R.D., Adkins, J.N., Zengler, K., and Palsson, B.O. (2012) *In silico* method for modelling metabolism and gene product expression at genome scale. *Nat. Commun.*, **3**, 929.

93 Shuler, M.L. (1999) Single-cell models: promise and limitations. *J. Biotechnol.*, **71**, 225–228.

94 Domach, M.M., Leung, S.K., Cahn, R.E., Cocks, G.G., and Shuler, M.L. (1984) Computer model for glucose-limited growth of a single cell of *Escherichia coli* B/r-A. *Biotechnol. Bioeng.*, **26**, 203–216.

95 Ataai, M.M. and Shuler, M.L. (1985) Simulation of the growth pattern of a single cell of *Escherichia coli* under anaerobic conditions. *Biotechnol. Bioeng.*, **27**, 1027–1035.

96 Laffend, L. and Shuler, M.L. (1994) Structured model of genetic control via the lac promoter in *Escherichia coli*. *Biotechnol. Bioeng.*, **43**, 399–410.

97 Browning, S.T., Castellanos, M., and Shuler, M.L. (2004) Robust control of initiation of prokaryotic chromosome replication: essential considerations for a minimal cell. *Biotechnol. Bioeng.*, **88**, 575–584.

98 Atlas, J.C., Nikolaev, E.V., Browning, S.T., and Shuler, M.L. (2008) Incorporating genome-wide DNA sequence information into a dynamic whole-cell model of *Escherichia coli*: application to DNA replication. *IET Syst. Biol.*, **2**, 369–382.

99 Castellanos, M., Kushiro, K., Lai, S.K., and Shuler, M.L. (2007) A genomically/chemically complete module for synthesis of lipid membrane in a minimal cell. *Biotechnol. Bioeng.*, **97**, 397–409.

100 Castellanos, M., Wilson, D.B., and Shuler, M.L. (2004) A modular minimal cell model: purine and pyrimidine transport and metabolism. *Proc. Natl. Acad. Sci. USA*, **101**, 6681–6686.

101 Shuler, M.L., Foley, P., and Atlas, J. (2012) Modeling a minimal cell. *Methods Mol. Biol.*, **881**, 573–610.

102 Karr, J.R., Sanghvi, J.C., Macklin, D.N., Arora, A., and Covert, M.W. (2013) WholeCellKB: model organism databases for comprehensive whole-cell models. *Nucleic Acids Res.*, **41**, D787–D792.

103 Purcell, O., Jain, B., Karr, J.R., Covert, M.W., and Lu, T.K. (2013) Towards a whole-cell modeling approach for synthetic biology. *Chaos*, **23**, 025112.

104 Karr, J.R., Williams, A.H., Zucker, J.D., Raue, A., Steiert, B., Timmer, J., Kreutz, C. DREAM8 Parameter Estimation Challenge Consortium, Wilkinson, S., Allgood, B.A., Bot, B.M., Hoff, B.R., Kellen, M.R., Covert, M.W., Stolovitzky, G.A., and Meyer, P. (2015) Summary of the DREAM8 parameter estimation challenge: toward parameter identification for whole-cell models. *PLoS Comput. Biol.*, **11**, e1004096.

105 Abner, K., Aaviksaar, T., Adamberg, K., and Vilu, R. (2014) Single-cell model of prokaryotic cell cycle. *J. Theor. Biol.*, **341**, 78–87.

Part Five

Continuous Downstream Bioprocessing

10

Continuous Downstream Processing for Production of Biotech Therapeutics

Anurag S. Rathore, Nikhil Kateja, and Harshit Agarwal

Indian Institute of Technology, Department of Chemical Engineering, Hauz Khas, 110016 New Delhi, India

10.1 Introduction

Continuous processing involves running unit operations of a process at steady state such that production can continue for a considerable amount of time (typically in months). This is different from batch production, where the material moves stage-by-stage and different batches of product are produced and production time of each batch is typically in weeks. Historically, many industries have successfully migrated from batch processing to continuous processing, including the petroleum, steel, automobile, and fast moving consumer goods (FMCGs) industry. Despite the significant differences in the underlying complexity of the production processes used in these industries, this migration highlights the strong drivers for continuous processing. These include high productivity (reduced cycle time) that can also be translated to the need for a significantly smaller facility (and thereby reduced capital cost), improved facility utilization, streamlined process flow, improved process control, and more consistent product quality [1].

The pharmaceutical industry, and in particular the biotherapeutic industry, has been somewhat of a laggard in this respect and only in the past decade the industry has started to show interest in adopting continuous processing. The interest perhaps stems from the ever increasing pressure on the biotech industry to develop innovative solutions that facilitate flexible and cost-effective manufacturing [1]. Recent years have seen efforts from both the regulators and the industry to implement continuous processing for manufacturing of biopharmaceuticals. The U.S. FDA and the European Medicines Agency (EMA) have adapted the pertinent regulatory guidelines to allow for definition of a batch through quantity of material rather than the mode of manufacturing [2]. An example of industry–academia collaboration is the Novartis-MIT Center for Continuous Manufacturing that is focused at design of an integrated, continuous pharmaceutical manufacturing process [3,4].

The efforts toward implementation of continuous processing in the biotech industry have so far focused primarily on upstream processing, commonly called perfusion. Use of perfusion reactors for high cell density cell cultures has been demonstrated extensively over the last two decades [1,5]. In comparison, efforts in

Continuous Biomanufacturing: Innovative Technologies and Methods, First Edition.
Edited by Ganapathy Subramanian.
© 2018 Wiley-VCH Verlag GmbH & Co. KGaA. Published 2018 by Wiley-VCH Verlag GmbH & Co. KGaA.

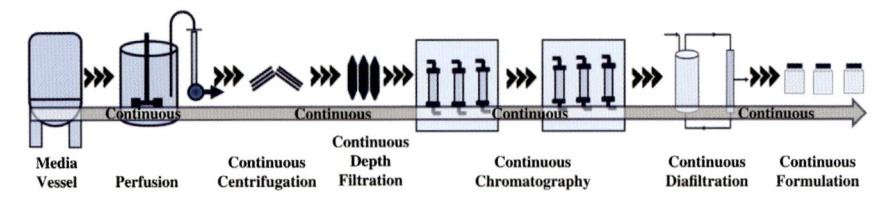

Figure 10.1 Schematic of a new integrated continuous manufacturing platform. (Reproduced with permission from Ref. [8]. Copyright 2015, Elsevier.)

making downstream processing continuous are more recent. Researchers have focused on performing key downstream unit operations such as process chromatography, refolding, precipitation, cell lysis, filtration, and aqueous two-phase extraction in a continuous fashion [6,7]. However, only a handful of references address the important topic of integration of the various unit operations of the process [1,6,8,9]. Figure 10.1 illustrates the flow scheme in a typical fully integrated continuous manufacturing platform that has been recently proposed for manufacturing of recombinant monoclonal antibodies [8]. An important topic that is yet to receive its due attention is that of control of continuous processes. We believe that this will be quite challenging though an essential component if continuous processing is to be actually used in industrial production.

In this chapter, we review various technologies that can enable continuous downstream processing for production of biotech therapeutics.

10.2 Continuous Manufacturing Technologies for Downstream Processing

In a typical process used for production of biotherapeutics, the upstream process is focused on expression of the product while the downstream process focuses on purifying the product from the myriad of host cell-related impurities (host cell proteins, host cell DNA, endotoxins, etc.), process-related impurities (antifoam, media components, protein A leachate, etc.), product-related impurities (aggregation, fragmentation, etc.), and product-related variants (acidic variants, basic variants, etc.). A typical downstream process consists of a mixture of orthogonal unit operations that may include cell lysis, centrifugation, refolding, precipitation, chromatography, and filtration. In this section, we will review advancements that have been made to enable performing these unit operations in a continuous fashion.

10.2.1 Continuous Cell Lysis

Cell lysis is typically the first step of downstream processing when the product is being expressed intracellularly. The step involves rupture of the cell membrane to release the intracellular components via chemical, thermal, enzymatic, or mechanical means.

High-pressure homogenizer is the method that is most widely used, due to its high efficiency and robustness. Cell lysis occurs when using a piston pump and a slit cavitation is generated at the valve and immediately behind the valve. When the pressure is suddenly released, cell breakage occurs. Modern high-pressure homogenizers can be operated in a continuous mode such that the cells can be disrupted in a single pass. Researchers have also demonstrated how multiple homogenization valves can be connected in series to achieve cell disruption in a continuous mode [10–12].

Microfluidic devices have also been recently explored as an alternative solution for performing continuous cell lysis [13]. The fermentation broth is pumped through a microfluidic channel continuously and cell lysis is achieved via chemical, mechanical, electroporation, or thermal means. In chemical lysis, the lysis buffer contains surfactants that solubilize the lipids and proteins in the cell membranes, making it porous and eventually resulting in lysis [14–16]. The lysis efficiency during chemical lysis was found to be dependent on both cell concentration and cell flow rate [17]. Researchers have also demonstrated lysis by concentrating and amplifying friction forces and shearing stresses using a microchannel with nanostructures (nanoknives) [18] and ultrasharp nanoblades [19]. These devices have been found to be simple and effective, with no harm to the proteins. Besides these, laser lysis in the microchannel has also been explored by researchers that involves utilization of the fluid motion produced by a focused laser to break the cell membrane [20,21]. When the laser pulse is focused at the buffer interface of a cell solution, it results in formation of a localized cavitation bubble and it is the site of this bubble that determines the start time of lysis. While use of microfluidic-based lysis systems has been demonstrated to offer several advantages, including high throughput and better control of shear stress (to prevent damage to shear sensitive biomolecules), presently, these devices are available only for laboratory-scale operation and large-scale solutions are yet to arrive [13].

10.2.2 Continuous Centrifugation

Centrifugation is a commonly used unit operation in bioprocessing for solid–liquid separation, including biomass recovery, cell homogenate clarification, and recovery and washing of inclusion bodies. The three most commonly used designs in industry today are the tubular centrifuge, chamber centrifuge, and disk stack centrifuge. Of these, the disk stack centrifuge has been successfully used in a pseudocontinuous manner as the design allows for transient removal of the solid particles from the centrifuge, and thus the suspensions can be fed and the clarified liquid can be removed continuously. Two different designs of the disc stack centrifuge are commonly used, namely, the split bowl and the disc nozzle. The type of feed affects the choice of design with the split bowl used for high solid content feeds and the disc nozzle for low solid content feeds [7,22]. Various other designs have also been proposed for continuous centrifugation, including the hybrid centrifuge rotor [23] and the continuous tubular bowl centrifuge [24]. Scalability of continuous centrifugation, however, remains a significant challenge. The smallest commercially available continuous disk stack centrifuge is suitable only for pilot plant operation. Therefore, a gap remains between the lab-scale devices and the commercial-scale continuous centrifuges, handicapping any efforts to perform lab-scale optimization. Thus, it can be concluded that continuous centrifugation is feasible and has been in practice over the past decade.

10.2.3 Continuous Refolding

Recombinant proteins produced in some of the microbial hosts such as the *Escherichia coli* are often over-expressed and result in formation of insoluble, biologically inactive inclusion bodies (IB) [25–27]. Protein in these IB is present in misfolded form and has to be refolded to its native state for the product to be biologically active [28]. Typically, refolding is carried out by solubilizing the IB followed by dilution into a refolding buffer having a composition that is favorable for restoration of the native structure. The degree of dilution in such methods is quite high (usually 10×), resulting in significant dilution of protein concentration (<100 μl/min) and requirement of large tanks to handle the required process volumes (typically 1000–10000 l) [7]. Another major concern with protein refolding is the step recovery, which decreases with the number of disulphide bonds and can be anywhere from 20–90%.

Mixing has been known to significantly impact efficiency of protein refolding. Typically, mixing in protein refolding can be characterized into two components – initial mixing (during the addition of the denatured proteins to the refolding vessel) and process mixing (sustained mixing throughout the refolding process) [29]. Poor initial mixing can result in higher aggregation due to the high concentration of intermediates. Similarly, adequate process mixing is necessary for maintaining homogeneity of the reactants and stability of dispersion. Insufficient mixing can also result in precipitation of proteins, which can be avoided by using dynamic inline mixing. In a batch process, achieving this can be nontrivial with the complexities increasing with scale [29].

A number of studies have targeted implementation of continuous refolding (Figure 10.2). These can be broadly classified into two classes of approaches –

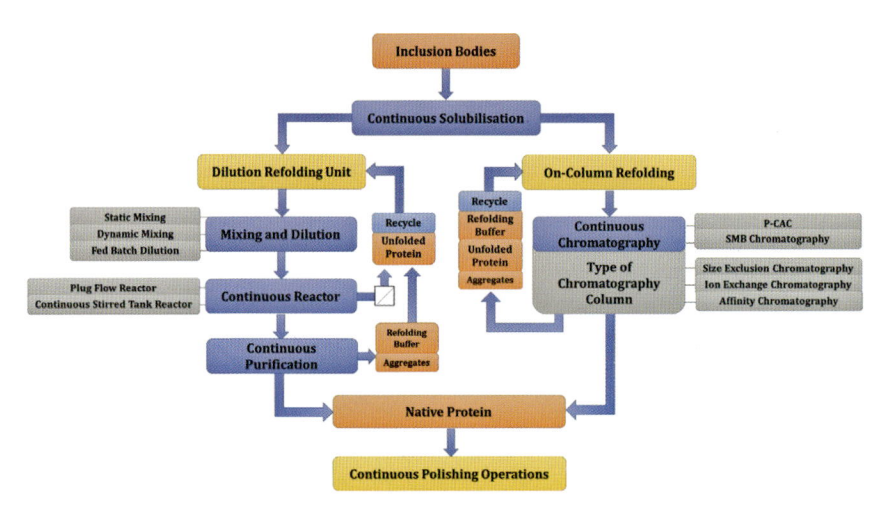

Figure 10.2 Illustration of the different process schemes that can enable continuous protein refolding. (Reproduced with permission from Ref. [6]. Copyright 2015, Taylor & Francis.)

continuous on-column refolding and continuous flow reactor refolding. Use of in-line static mixing units for "fast mixing" and improving refolding performance has been reported about a decade ago [2]. A packed column plug flow reactor, supplemented by an initial mixing unit, has been successfully used with the resulting product shown to match with the batch product with respect to quality [30]. The proposed configuration has been shown to offer additional flexibility of fine-tuning the various process variables (temperature, reagent addition, etc.) to their optimal values at specific stages of refolding, thus resulting in a narrower distribution of the degree of refolding at the outlet. Fed-batch addition of denatured protein has also been employed using a ceramic membrane tube to facilitate gradual addition and mixing of the denatured protein into the flow reactor [31]. In this case, the refolding efficiencies were found to be higher than the batch dilution method as the fed-batch addition significantly reduced the tendency of aggregation. In another configuration, a continuous stirred tank reactor (CSTR) along with a diafiltration circuit has been used to lower the denaturant levels during protein refolding and has been found to be advantageous over batch refolding when the required residence times are high [32]. It has also been suggested that the unfolded protein in the outlet can be recycled to improve refold efficiency. This configuration resulted in better cumulative yield of 62% at recycling rate of 0.75 compared to refolding yield of 40% in a batch reactor. In an improved version, the IB have been solubilized in a continuous manner before dilution of the protein solution with refolding buffer [33]. Continuous refolding has also been performed in a tubular reactor with higher overall productivity than the batch reactor without any requirement of emptying and refilling of reactor in continuous mode [34]. The performance of the reactor was found to be better than direct batch refolding for pulse and temperature leap strategies in continuous mode, demonstrating the flexibility of the proposed configuration. In an improved version, the tubular reactor has been used in tandem with continuous dissolution, refolding and precipitation, further exemplifying the use of the configuration in an integrated assembly [35]. In a very recent development, a novel coiled flow inverter (CFI)-based plug flow reactor has been used for continuous refolding [36,37]. Solubilized inclusion bodies were continuously diluted with refolding buffer using a dynamic in-line mixer followed by refolding in a CFI based tubular reactor. The configuration consisted of helical coils bent at equidistant right angles to cause flow inversion resulting in a sharper residence time distribution along with good cross-sectional mixing, better emulating a plug flow than a simple straight tube or helix. The proposed configuration was found to result in better productivity compared with batch refolding process along with faster kinetics and higher purity.

On-column refolding has also been attempted by many researchers in the past decade with the promise of simultaneous refolding and purification in a single step at high protein concentrations and yields [38,39]. Innovative designs of chromatographic equipment that allow for continuous refolding include preparative continuous annular chromatography (P-CAC) and simulated moving bed (SMB) chromatography. P-CAC involves continuous feeding of the denatured protein along with the refolding buffer onto a rotating annular column. The refolded target protein, protein aggregates, and other impurities elute at different angles at the bottom of the annular column. The elution profiles for size-exclusion-based P-CAC have been shown to be similar to that from batch operation [40]. Since the aggregates are continuously separated from the product and are recycled back to the feed stream, the refolding yield increased to 45% at a recycling rate of 0.51 compared to 27% without recycling and 10% in batch refolding [41,42]. A size-exclusion-based four-zone SMB chromatography has also been configured to enable on-column refolding of the protein continuously with high efficiency and with low consumption of refolding buffer [43]. Continuous refolding has recently been performed in an assembly that uses a pipe reactor and an expanded bed adsorption column [44]. The denatured stream was diluted in a pipe reactor, after which the nascent refolded protein was sent onto the adsorption column, through which the concentrated and purified product was eluted continuously.

Exploiting the abovementioned continuous chromatography systems; there are four principally different chromatography modalities that have been used for on-column refolding. Solvent-exchange size-exclusion chromatography (SEC) is the oldest on-column protein refolding technique [45–50]. The general protocol involves injecting a feed pulse, composed of denaturant, reducing agent, and denatured and reduced (D&R) protein into the SEC column, pre-equilibrated with refolding buffer. Due to differences in distribution coefficient, the concentration waves of denaturant, reducing agent, and the denatured and reduced protein separate as they migrate through the column [39,51,52]. SEC offers a couple of unique advantages over other chromatographic refolding techniques, including elimination of steric hindrance due to nonadsorptive protein matrix interaction and inhibition of aggregation due to restriction of diffusion of the various intermediate forms, thereby suppressing nonspecific interactions between them. The major drawback of SEC is low process productivity and product concentration. Ion exchange chromatography (IEC) involves adsorption of the denatured protein onto the surface of IEC resin followed by refolding on or off the resin surface depending on the protein and the environment. This adsorption results in reduced interaction between protein molecules, and thereby reduced aggregation [53–56]. Hydrophobic interaction chromatography (HIC) has also been used for on-column refolding of several biotherapeutics, including human interferon and pro-insulin [57,58]. This requires that strength of hydrophobic interactions should not be so much that it prevents protein folding. Hence, a proper combination of additives (binding strength modifying agents) is required to influence both stability and solubility of native, denatured, and intermediates states [59,60]. Immobilized metal ion affinity chromatography (IMAC) has also been demonstrated to be useful for on-column refolding of proteins equipped with engineered poly-histidine tags. These tags form high-affinity complexes with the immobilized divalent metal ions even in the presence of high concentrations of chaotropic agents, thereby allowing isolation and refolding of the tagged protein [61].

In summary, further development is required to make on-column refolding feasible. On the other hand, some of the other configurations already proposed (especially the CFIR) can be readily used for performing continuous refolding.

10.2.4 Continuous Precipitation

Precipitation is among the simplest and least expensive fractionation methods as it involves modest changes in solution conditions such as addition or removal of salt and organic solvents or changes in temperature and pH [62].

Continuous precipitation has been demonstrated in a variety of configurations such as stirred tank reactor, straight tube, and coiled flow inverter reactor. Crystallization of copper sulfate, nickel ammonium sulfate, and soy protein in continuous mixed suspension and mixed product removal reactors (MSMPR) has been reported [63]. Likewise, MSMPR precipitator was also used for isoelectric precipitation of sunflower protein [64]. Another configuration that has been widely explored by researchers is straight tube configuration. A 20 m long, 6 mm diameter glass tubular precipitator has been successfully used for isoelectric precipitation of sunflower protein [65]. In another finding, a two-stage batch precipitation process consisting of mineral salt ($CaCl_2$) and an organic solvent (ethanol) has been performed in a continuous mode using continuous tubular reactors, and stable operation has been demonstrated over several hours at steady state without the need of manual intervention, delivering antibody at a constant yield and purity [66]. A more complex procedure, centrifugal precipitation chromatography, has also been used for continuous precipitation and purification. The method was found capable of high yield separation of target species of proteins such as human serum proteins, plasmid DNA, polysaccharide, and PEG-protein conjugates [67]. More recently, a novel coiled flow inverter reactor has been used for precipitation of process impurities from a Chinese hamster ovary (CHO) cell harvest [36].

In summary, continuous precipitation is feasible and offers a simple, robust, and economical method to efficiently remove host cell impurities [68].

10.2.5 Continuous Chromatography

Traditional packed column chromatography is inherently a batch process, with the column loaded, washed, and eluted in a sequential process. Continuous chromatography has been pursued by many researchers due to the abovementioned advantages of higher productivity, higher purity, lesser operating cost, and lesser footprint. The simplest way to perform chromatography in a continuous mode is to run several columns in parallel [69]. While one column is loaded, others are washed, eluted, regenerated, and finally re-equilibrated. Many other methods to achieve continuous chromatographic separation (although not always with true steady-state operation) have been suggested and are briefly described in this section.

Continuous annular chromatography (CAC) and carrousel chromatography can be considered closest to true continuous chromatography. CAC (Figure 10.3a) consists of an annular adsorbent bed packed between two concentric cylinders. The feed is continuously introduced from a fixed point in the annular space between two rotating concentric cylinders. Product is recovered from the fixed outlet ports at the bottom. Individual helical bands of each component are formed depending on adsorption of the components and the elution velocity coupled with bed rotation. These helical bends extend to the bottom of the bed and separated components are

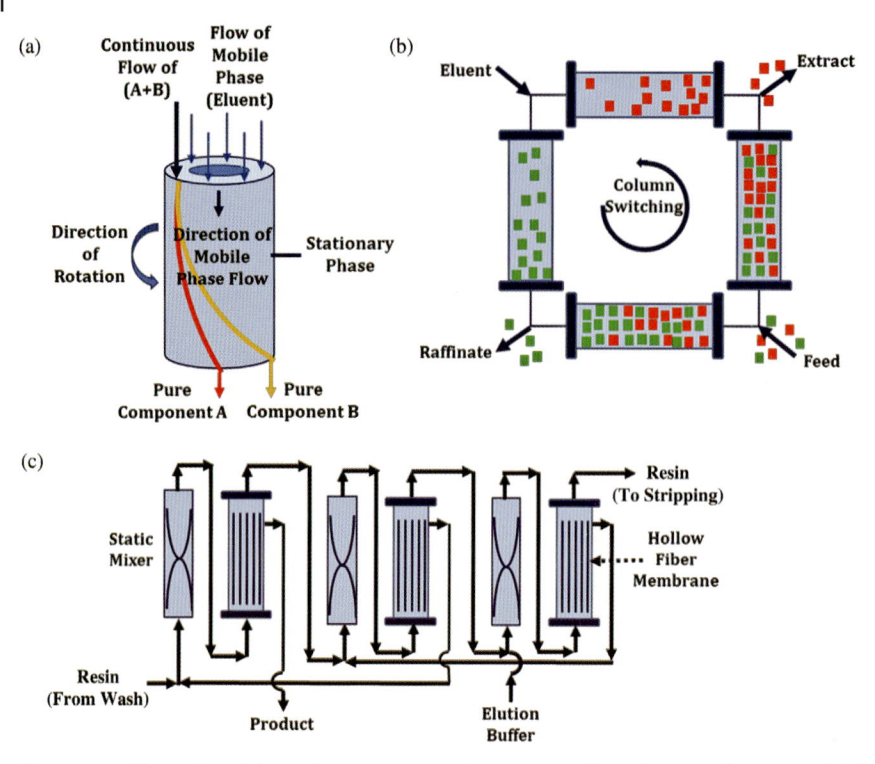

Figure 10.3 Illustration of the various continuous chromatography techniques that are available today. (a) Continuous annular chromatography. (b) Simulated moving bed chromatography. (c) Counter current tangential chromatography.

collected at different angular positions from the fed point. CAC is not limited to isocratic operation, and hence allows implementation of gradient-based elution strategies. Also, CAC provides a major advantage in the form of straightforward transfer of step-elution protocol for batch chromatography to a CAC column [70]. Some significant recent advances include implementation of stepwise elution (gradient elution when multiple concentrations are used) and operation in displacement mode [71]. Annular chromatography has also been used for removal of protein aggregates from an intravenous (serum) immunoglobulin preparation by size exclusion [72,73] and for continuous purification of recombinant factor VIII produced in a perfusion bioreactor using a weak anion exchange resin [74]. While feasible, achieving uniform column packing and flow distribution remain challenging and the technology is yet to be implemented for large-scale protein purification. A relatively more obscure approach to perform continuous chromatography is that of carrousel chromatography, where the columns are mounted on the carrousel with interconnecting valves and can be operated in concurrent or in countercurrent mode [75].

SMB chromatography has been used to provide separation of binary mixtures, for example, removal of product variants or oligomers [76–78]. SMB uses a multiple chromatographic columns connected in series (Figure 10.3b). The position of the sample inlet and the collection points are moved continuously to provide a simulation of a true moving bed chromatography with countercurrent flow (but without any real motion of the solid phase). Hence, SMB operates cyclically, with the concentration of the recovered product decaying from its maximum value at the start of each cycle to its minimum right before switching. SMB can be used to provide high-resolution separation of compounds that otherwise have minor differences in their retention times. This is achieved via cumulative addition of these differences throughout the bed. Compared to theoretical plate-dependent separation, the cost associated with maintaining, operating, and obtaining larger single columns is significantly reduced [79]. SMB has been successfully implemented in large-scale manufacturing for separation of various components [80]. Several SMB systems have also been developed for protein purification, including the Semba Octave Chromatography System and Tarpon Biosystems BioSMB. A recent development has led to a design of single-column SMB system with six containers and two pumps to collect and distribute the feed, product, and impurities [81].

Countercurrent tangential (CCT) chromatography combines the advantages of SMB and EBC formats (Figure 10.3c). The resin is in the form of slurry and flows through a series of static mixers and hollow fiber membrane modules [82–84]. During binding, the protein binds to the resin retained in the membrane while impurities flow-through the membrane and are removed as waste. Since the buffers used in the binding, washing, and elution steps flow countercurrently to the resin through the multiple hollow fiber membranes, high-resolution separations can be achieved while increasing the product yield and reducing the amount of buffer needed for protein purification. In contrast to periodic behavior of multicolumn countercurrent chromatography systems, CCTC truly provides a continuous steady-state operation where the product obtained in the elution step is at constant concentration and quality. Capability of using CCTC for initial capture and purification of two commercial mAbs has been successfully demonstrated confirming the significantly higher production and comparable purity with conventional packed (batch) bed [85].

Periodic countercurrent chromatography (PCC) utilizes multiple columns to run the different chromatographic steps (equilibration, regeneration, loading, and washing) discretely and continuously in a cyclic fashion (Figure 10.4). Since the columns are in series, the flow-through and wash from one column is captured by the second column. This unique feature allows for loading the resin close to its static binding capacity rather than dynamic binding capacity (as in batch mode chromatography). Integration of high-density perfusion cell culture with a directly coupled cell capture step through a four-column periodic countercurrent chromatography has been recently demonstrated [1]. The PCC system was directly integrated and run continuously for a period of 30 days. In a similar study, the continuous capture step has been integrated with continuous intermediate and polishing steps using two different PCC systems with protein A and cation exchange resins, respectively [8]. In yet another development, a novel end-to-end continuous monoclonal antibody purification template comprising of flocculation-based clarification integrated with PCC has been reported [9].

Expanded bed chromatography (EBC) permits expansion of the adsorbent bed due to upward movement of column fluid, thus allowing the passage of crude raw

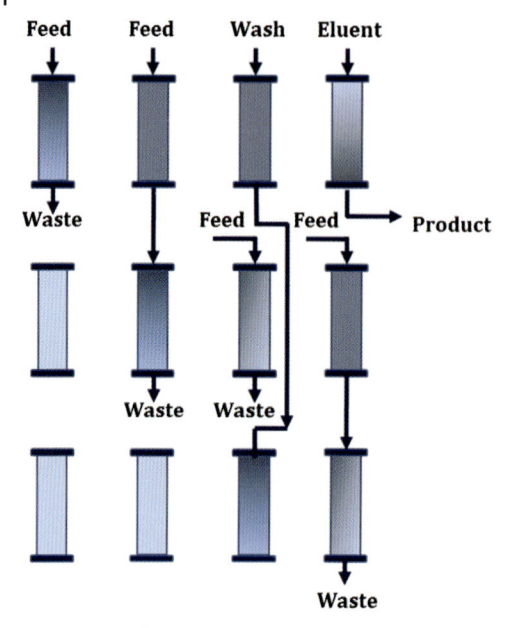

Figure 10.4 Illustration of the periodic countercurrent chromatography (PCC) cyclic operation.

materials, such as fermentation broth, without clogging the column [86]. Affinity chromatography has been successfully used in this format for isolation of proteins from cell broths [87]. An expanded bed chromatography can be run in a continuous manner by using a series of contactors (countercurrent or cocurrent) in a loop for continuous loading, washing, elution, and re-equilibration of the transported adsorbent. Continuous countercurrent expanded bed adsorption has been used successfully for direct purification of lysozyme [88].

Multicolumn countercurrent solvent gradient purification chromatography (MCSGP) is a continuous multicolumn chromatographic approach for performing high-resolution separation using linear, nonlinear, or segmented gradients [89–91]. The system is split into multiple columns and each column performs the task analogous to batch chromatography. However, the countercurrent movement of the impure fractions through the column provides high yield and high purity, simultaneously [91]. MCSGP can also be used for more than just binary separations and a four-column MCSGP system has been successfully used for initial capture of an IgG2 mAb from clarified cell culture fluid using cation exchange chromatography with gradient elution [89]. The use of MCSGP has also been demonstrated for complex separations such as for separation of charge variants using cation exchange chromatography [90].

Radial flow chromatography is similar to annular chromatography except that the flow is in the radial direction. The feed, wash buffer, and elution buffer are introduced at different and fixed angular points and the product and impurities are collected at different angular positions. The flow can be in either direction, from the inner cylinder to the outer cylinder or vice versa. Continuous operation can be achieved by rotating the annulus past the fixed positions of the feed and buffer inlets and product collection ports [92]. It is also possible to use radial flow in membrane chromatography configuration and such a DEAE membrane radial column has been successfully used for purification of human prothrombin [93]. Although various applications of radial flow chromatography have been exploited [94], commercial continuous radial flow chromatography system does not exist today.

Besides the hardware developments already mentioned, another attribute that is essential for continuous processing is that of novel resin matrices and ligands so as to reduce process time by overcoming mass transfer limitations originating from diffusion, and handle the throughput demands for foreseeable future [95]. Multimodal resins are one such example and offer a combination of multiple interaction types such as hydrophobic and ionic interactions resulting in significantly higher selectivity and fewer purification steps. Multimodal chromatography has been successfully used for a single-step purification of granulocyte colony stimulating factor (GCSF) with recovery >85% and purity >99% [96]. Multimodal cation exchange chromatography has also been successfully used for industrial scale capture of human growth factor [97]. The continuing popularity of multimodal chromatography has resulted in greater academic interest in understanding of the effects of salt, ligand homogeneity, and ligand density in HCIC systems [98,99]. The ability of multimodal chromatography to reduce the number of chromatography steps required in a process facilitates ease of implementation of continuous processing.

Monoliths have emerged as another facilitator of continuous processing due to their offer of relatively low-pressure drops and connective mass transport resulting in higher flow rates and higher productivity. They have been successfully used for both flow-through and capture of various molecules like viruses, nucleic acids, and proteins [100]. Similar advantages are also offered by membrane chromatography. Although this technology has been in existence for more than a decade, binding capacity limitations have limited the use of these materials to flow-through operations. Recent advancements have made membrane chromatography more attractive for bind/elute operations, particularly for large biomolecules [101]. Salt-tolerant membrane adsorbers have been developed and used for viral clearance. Membrane chromatography has also been employed for the purification of PEGylated proteins [102]. Recently introduced HIC membranes have comparable dynamic binding capacities to that of the conventional HIC resins. These materials have been recently successfully employed for a bind/elute purification step for a recombinant mAb [103]. An industrial scale continuous platform purification process has been achieved with rapid cycling of membrane chromatography units, processing more than 1 000 000 l of cell culture supernatant till date [104].

10.2.6 Continuous Extraction

Liquid–liquid extraction (LLE), such as aqueous two-phase extraction (ATPE), is a promising alternative to process chromatography and has been extensively examined in the last few decades [105]. Product separation in ATPS is based on differential partitioning between the two aqueous phases, for example, the two phases formed from mixtures of immiscible polymers like polyethylene glycol

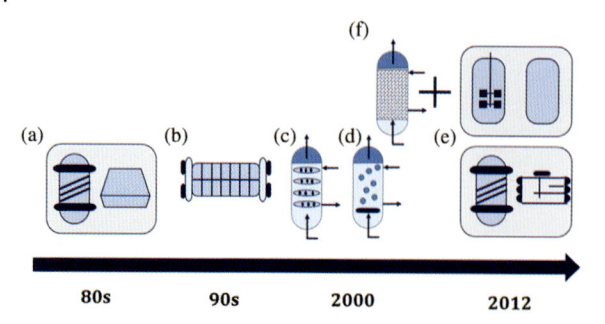

Figure 10.5 Illustration of evolution of continuous ATPS devices. (a) Mixer-settler devices (static mixer/centrifuge). (b) Graesser contactors. (c) Spray columns. (d) PRDCs. (e) Novel settler/separators. (f) Column allies. (Reproduced with permission from Ref. [110]. Copyright 2014, Elsevier.)

(PEG) and dextran or from phase-separating mixtures of PEG and appropriate salts. In the last few years, several continuous ATPS methods have been developed for antibodies [106,107], therapeutic proteins [108], and enzymes [109]. Mixer-settler units were one of the first devices employed for continuous ATPS, either in static or dynamic modes of operation consisting of a mixing stage in tanks and columns, coupled to a series of settling and separation units (Figure 10.5). Column contractors, such as spray columns, perforated rotating disk contractors (PRDCs), pulsed cap columns, and other columns (packed, sieve plate, and vane agitated columns) have been the common choice for achieving continuous ATPS using polymer-salt systems for proteins and enzymes extractions (Figure 10.5). However, the lack of versatility and limited mass transfer and separation in these column contractors have made the mixer-settler to re-appear in the stage [110].

Although continuous ATPE may be economically viable and its scalability has been demonstrated in the pharmaceutical industry for other classes of products, its adoption in the biotherapeutic industry has been slow due to concerns related to scalability as well as impact on product stability.

10.2.7 Continuous Filtration

Filtration is the simplest and also the most ubiquitous unit operation in bioprocessing. It is typically used to achieve solid–liquid separation such as for clarification of fermentation broth (microfiltration and depth filtration), concentration of proteins (ultrafiltration), and buffer exchange of process solutions (diafiltration).

For initial clarification, continuous filtration has been attempted both inside and outside the fermenter. Spin filters have been placed directly in the stirred tank reactor, typically mounted as an annulus around the impeller shaft at the center of the bioreactor, enabling continuous removal of cell-free culture harvest from inside the rotating filter. A number of case studies have examined the effects of membrane pore size, filter geometry, and rotation rate on cell retention and fouling [111,112]. CFD analysis has also been performed to demonstrate the presence of radial flow in this operation [113]. Despite all these developments, spin filters pose serious scale-up challenges in terms of availability of sufficient filter area to handle large volumes and this limits their industrial use. In a recent development, an *in situ* membrane stirrer has been developed as an alternative to the classical spin filter, wherein a micromesh membrane is incorporated directly in the impeller blades used to provide agitation to the bioreactor [114]. Although excellent performance was reported with yeast cells, the effectiveness of this technology with mammalian cell lines is yet to be demonstrated. Submerged membrane filtration is another technology for clarification of cell culture harvest using membranes that are placed directly in the bioreactor. A variety of geometries have been explored, including cylindrical rods and suspended hollow fibers [115]. These membranes can also be periodically backflushed to remove foulants from the membrane surface, which when performed together with scouring of external membrane surface through aeration, further reduces cell adhesion to the membrane [116]. An alternative approach for initial clarification is to perform separation outside bioreactor and recycle the retentate back into the bioreactor.

Alternating tangential flow filtration (ATFF) is presently the most established initial clarification technology that has been successfully demonstrated for industrial use, particularly with perfusion bioreactor. It uses standard hollow fiber membranes to separate cells and product with the flow driven through the module using a very low shear diaphragm pump that cyclically withdraws and returns the cell suspension to the bioreactor [117]. ATFF has been shown to offer near-linear scalability, low shear, and self-cleaning through backflushing.

Continuous filtration in other unit operations such as formulation and concentration has been demonstrated by using single-pass tangential flow filtration (SPTFF) and diafiltration. SPTFF is a continuous process (Figure 10.6a) where

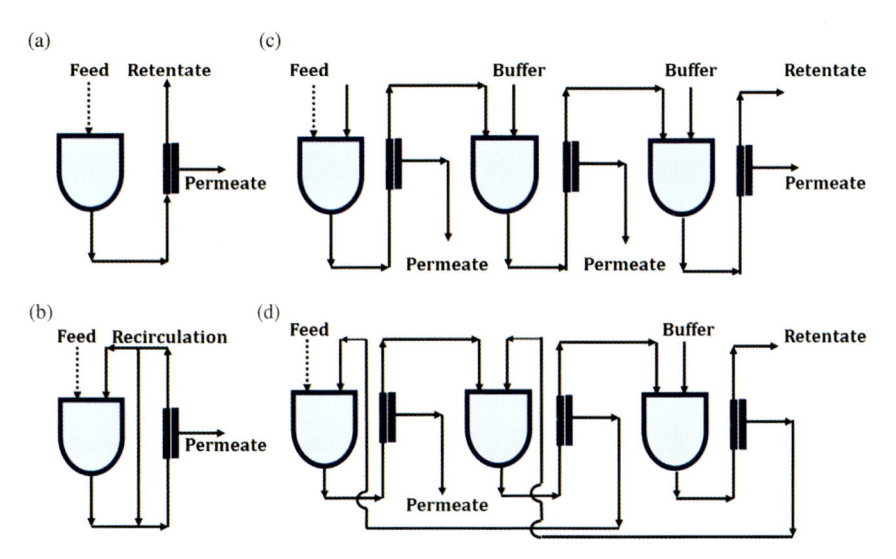

Figure 10.6 Illustration of the various modes of continuous filtration. (a) Single-pass tangential flow filtration. (b) Batch-topped off filtration. (c) Continuous cocurrent diafiltration. (d) Continuous counter current diafiltration.

product must reach the desired concentration in a single pass through the membrane unit [118]. Various modifications of SPTFF are possible, including using multiple membrane cascades in series, partial recycling of retentate, and combining SPTFF with batch topped off filtration. In batch topped off filtration (Figure 10.6b), the feed addition rate in the tank is kept same as the permeate flow rate through the membrane. The advantage of the batch topped off filtration is that the retentate is recycled back, thereby increasing the residence time of the product in the system. Owing to the longer residence time, the filter area may be saved, and hence this design can accommodate larger volumes than conventional batch processing [7].

Continuous diafiltration has been accomplished by applying continuous countercurrent and concurrent diafiltration (Figure 10.6c and d). In concurrent diafiltration, a minimum of three filtration units are connected in series. At the first stage, feed and buffer are added at a rate equal to the permeate flux and at all other subsequent stages only buffer is added and permeate is withdrawn. Rate of buffer addition is kept to be equal to the permeate flux. Countercurrent diafiltration enables buffer exchange even more efficiently [119]. The fresh feed is continuously added to the final diafiltration stage. In all other stages, permeate is added. Countercurrent diafiltration efficiently utilizes buffers and has been proven to be the most efficient method for buffer exchange.

10.3 Continuous Process Development

While significant advancements have been made in our ability to perform the various unit operations in a continuous fashion, much more needs to be done toward successful integration of these unit operations into a complete process. One of the reasons is the added complexity that is brought in by the equipment and process compared to a conventional batch system. For example, the hardware required to perform continuous chromatography is significantly more complex and expensive than what is needed to perform the traditional batch column chromatography. Hence, for successful development and implementation of a continuous downstream process a systematic stepwise approach has to be followed (Figure 10.7). We describe such an approach in this section.

First, a batch downstream process has to be developed keeping in mind the overall motivation for continuous processing. Majority of the case studies available till date involve use of a continuous process that has been created by conversion of an existing batch processing protocol. Once the batch protocol has already been developed and validated, the process has to be adapted for continuous processing. For example, in the case of chromatography, an existing batch protocol can be appropriately adopted and operated in continuous fashion using the abovementioned strategies such as periodic countercurrent chromatography (PCC) and SMB. However, if one is working on developing a continuous process without any existing batch process platform, minimization of the intermediate sample conditioning should be achieved to make the implementation in continuous mode more feasible. Integrated optimization of two or more unit operations should be performed to achieve optimal alignment [120]. A process having less

10.3 Continuous Process Development | 275

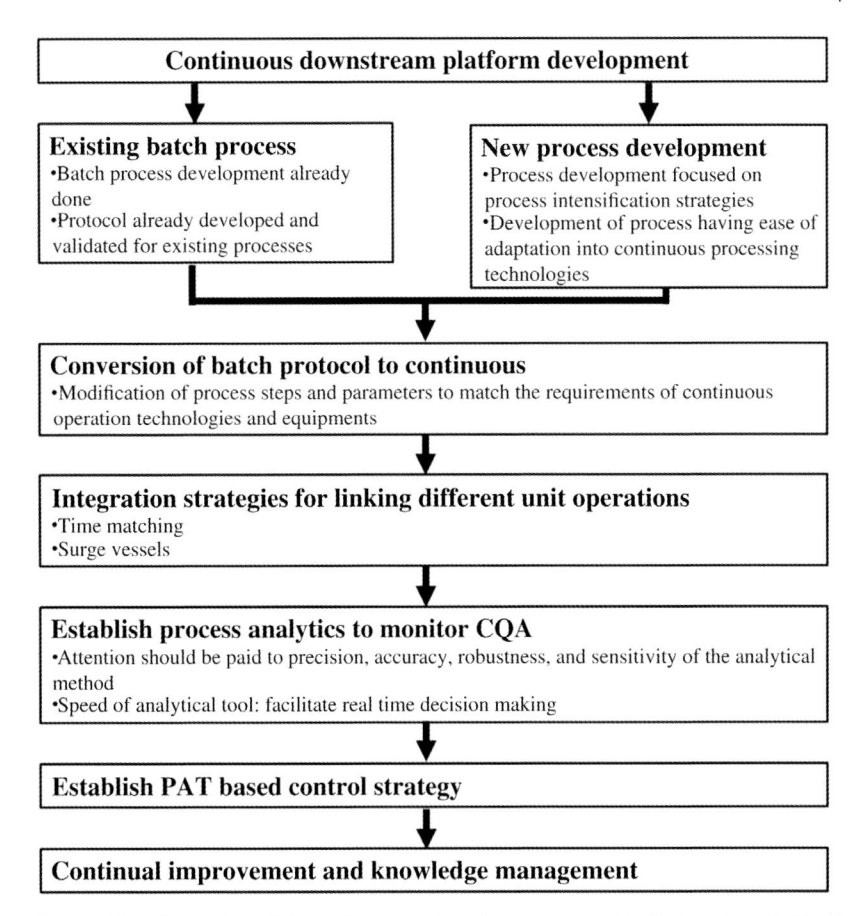

```
┌─────────────────────────────────────────────────────┐
│      Continuous downstream platform development     │
└─────────────────────────────────────────────────────┘
           │                          │
           ▼                          ▼
┌──────────────────────────┐  ┌──────────────────────────────┐
│ Existing batch process   │  │ New process development      │
│ •Batch process           │  │ •Process development focused │
│  development already done│  │  on process intensification  │
│ •Protocol already        │  │  strategies                  │
│  developed and validated │  │ •Development of process      │
│  for existing processes  │  │  having ease of adaptation   │
│                          │  │  into continuous processing  │
│                          │  │  technologies                │
└──────────────────────────┘  └──────────────────────────────┘
           │                          │
           └──────────────┬───────────┘
                          ▼
┌─────────────────────────────────────────────────────┐
│ Conversion of batch protocol to continuous          │
│ •Modification of process steps and parameters to    │
│  match the requirements of continuous operation     │
│  technologies and equipments                        │
└─────────────────────────────────────────────────────┘
                          ▼
┌─────────────────────────────────────────────────────┐
│ Integration strategies for linking different unit   │
│ operations                                          │
│ •Time matching                                      │
│ •Surge vessels                                      │
└─────────────────────────────────────────────────────┘
                          ▼
┌─────────────────────────────────────────────────────┐
│ Establish process analytics to monitor CQA          │
│ •Attention should be paid to precision, accuracy,   │
│  robustness, and sensitivity of the analytical      │
│  method                                             │
│ •Speed of analytical tool: facilitate real time     │
│  decision making                                    │
└─────────────────────────────────────────────────────┘
                          ▼
┌─────────────────────────────────────────────────────┐
│ Establish PAT based control strategy                │
└─────────────────────────────────────────────────────┘
                          ▼
┌─────────────────────────────────────────────────────┐
│ Continual improvement and knowledge management      │
└─────────────────────────────────────────────────────┘
```

Figure 10.7 Illustration of the various considerations that need to be kept in mind for development of a continuous downstream process: stepwise approach and strategies.

number of chromatographic steps and with a continuous flow of sample between steps (no pH, conductivity adjustment, or both) is more desirable for operation in continuous processing. Evolution of multimodal resins, membrane adsorbers, and single-use technologies can also become key facilitators in this transition.

Second, the batch protocol has to be converted to continuous process. Reactors like CSTR and tubular reactor have been utilized for continuous refolding and precipitation. Likewise, SMB and PCC have been widely exploited for performing continuous chromatography. In either case, the optimized batch condition is unlikely the optimum for continuous processing. Modification of the existing process parameters may be required to achieve consistent product quality. Awareness of these considerations during early stages of process development can minimize major modifications later.

Third, the different unit operations have to be integrated into one process. Evolution of new technologies facilitates the integration between different unit operations. For instance, alternative tangential flow filtration technology facilitates integration of a perfusion upstream culture with the downstream process [121]. Technologies such as EBC, SMB, and PCC can also play a facilitating role in integration of upstream and the downstream processes. Moreover, utilization of continuous flow reactors for various downstream unit operations such as CFIR for refolding [37] can also make integration with the downstream process easy and efficient.

Successful integration between the unit operations helps in achieving a pseudo-steady state. This pseudo-steady state refers not only to the flows across each unit operation but also for gradients of concentration, pH, and conductivity that are expected in the effluent of several unit operations. These gradients can be significantly decreased by selecting an appropriate volume of surge tank to act as a buffer going into the next unit operation [122]. In addition, the length of different unit operations has to be matched to ensure continuous operation. Hence, incorporation of surge vessels and balancing the flow rates between different steps helps in ensuring effective integration.

Fourth, appropriate process analytics and a PAT-based control strategy has to be established [123–127]. Attention should not only be paid to precision, accuracy, robustness, and sensitivity of the analytical method but also on the speed of the analytical method. To serve the purpose of continuous processing, it is necessary that the analysis be carried out in a time frame that facilitates real-time monitoring and control. This part is quite important because unlike batch manufacturing where any deviations can be investigated and perhaps mitigated before moving to the next step, continuous processing requires a much faster response to avoid process failure. Based on literature, the analytical tools could be rapid forms of HPLC, UPLC, fluorescence spectroscopy, UV-vis spectroscopy (A_{280}, A_{550}), multiwavelength absorbance, and so on. Based on the capability of the installed controls and the expected deviations, appropriate amount of surge capacity should be put in place for storing solutions during downtime of a process section together with strategies for orderly process line startup and shut down.

Finally, demonstration runs of the proposed continuous manufacturing platform need to be performed to verify the capability of the proposed controls as well as the efficacy of the integration.

10.4 Case Studies Related to Continuous Manufacturing

In this section we wish to showcase a few of the major, recent developments that have occurred in the field of continuous manufacturing.

Case study 1: Continuous refolding in a novel coiled flow inverter reactor (CFIR).
   In a recent study, a novel use of CFI-based reactor has been proposed for refolding of GCSF (Figure 10.8) [36,37]. In this study, an existing batch refolding process is effectively implemented in a continuous process using a dynamic in-line mixer connected to a CFI-based reactor.
   To develop the continuous process, a full factorial design of experiments (DOE) consisting of two levels of pH [9,10], three levels of degree of dilution (5, 7.5, and 10 times), and three levels to DTT ratio (1.3, 1.8, and 2.3, in the final solution) with three centerpoints was performed. Product purity and

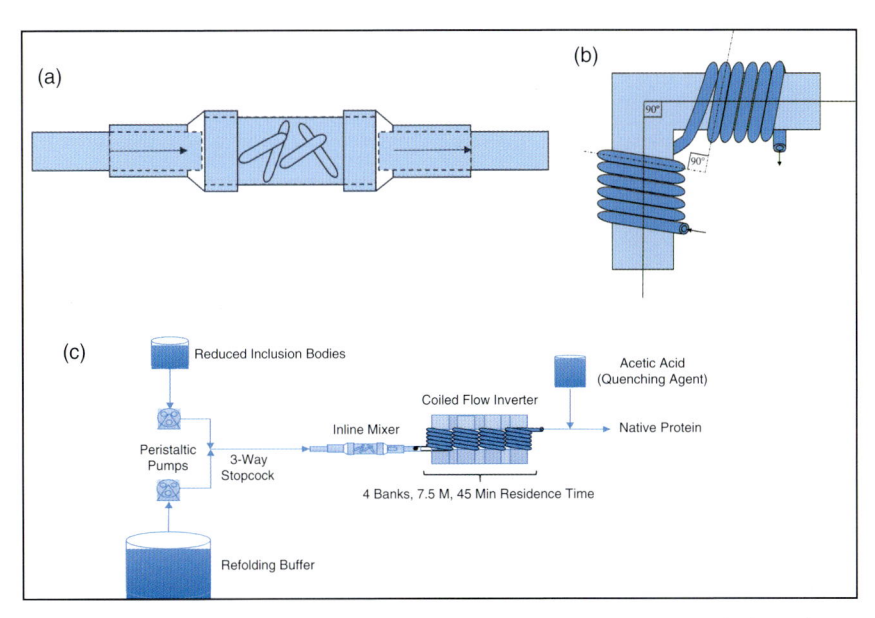

Figure 10.8 (a) Illustration of the dynamic mixing unit. Connections were made through Tygon tubing of higher diameter and reinforced through cable ties. (b) Illustration of a bend for CFI. (c) Process flow diagram for the continuous refolding process. A bank is a collection of four branches, each with five turns of helix. (Reproduced with permission from Ref. [37]. Copyright 2016, Elsevier.)

productivity were measured with respect to time both for continuous refolding and the corresponding batch refolding. From the DOE it was evident that, to obtain maximum purity and productivity, the reactor should be operated for 50–60 min with refold buffer pH 9 and dilution of reduced inclusion bodies 5× in refold buffer.

The developed continuous process was found to be at par with the industrial batch process in terms of critical process parameters such as percentage of native protein (84.2%), oxidized protein (10–12%), reduced protein (~0.1%), and aggregates (<0.5%) (Table 10.1). More importantly, the performance of the developed process was found to be 15 times better in terms of reactor-specific productivity as compared to the batch process due to elimination of shutdown, cleaning, and filling steps. This is also because the proposed continuous

Table 10.1 Comparison of batch and continuous protein refolding processes.

Aspect for comparison	Batch process	Continuous process
Dilution	10×	5×
Time to reach purity > 90%	90 min	Less than 60 min
Shutdown time	Required	Not required
Overall productivity	Y	$15Y$

process, due to its mixing features, allowed for operation at lower dilutions, and thus at higher protein concentrations (0.38 mg/ml as compared to 0.19 mg/ml of batch process). This resulted in a significant decrease in the cost of purification as well as requirement of the size of refolding vessel due to the reduction in process volume and faster refolding. The proposed unit can seamlessly be placed in a continuous bioprocessing train, taking input from an inclusion body solubilizing unit and providing the output to a suitable continuous purification step.

The proposed configuration can also be used for intensification of other biotech unit operations such as precipitation and inclusion body solubilization as well. Effectively, the configuration contributes toward enabling the development of an integrated continuous bioprocessing platform.

Case study 2: End-to-end integrated fully continuous production of recombinant monoclonal antibodies.

In a recent publication, researchers have demonstrated integration of high-density perfusion cell culture with a capture step through a four-column periodic countercurrent chromatography [1]. The high-density CHO cells were operated at a quasi-steady stage for more than 60 days. The PCC system was directly integrated into the upstream process and the resulting process was successfully performed continuously for 30 days. Product quality was found to be comparable with the batch column operation. Advantage of this assembly was realized in the form of elimination of several nonvalue-added unit operations, such as clarification and intermediate hold steps. Another recent publication has further extended this work by further integrating the capture step with continuous intermediate and polishing steps [8]. Running in an automated and continuous manner, bioreactor-PCC1-PCC2 system performed the entire upstream/downstream operation. Also, a continuous low-pH hold was included to perform continuous viral inactivation. The whole assembly had the capacity of end-to-end continuous production from bioreactor to the polished product (Figure 10.9).

The integrated system was operated continuously for 31 days. The rate of production of mAb in the bioreactor was ~10 g/day. The performance of each unit operation was monitored by analyzing in-process protein concentration, host cell protein impurity clearance (such as HCP), and aggregation. The process

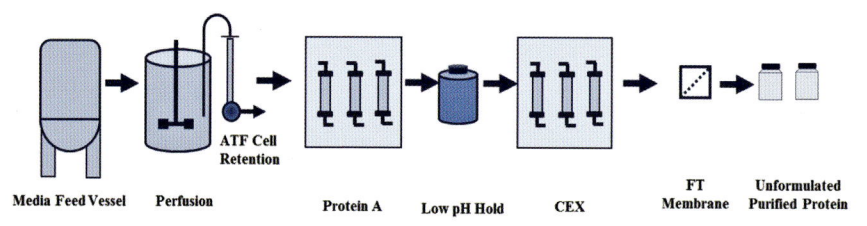

Figure 10.9 Process flow diagram of the proposed end-to-end continuous bioprocessing platform for production of a monoclonal antibody therapeutic. The platform includes a perfusion bioreactor with ATF as cell retention device for upstream processing. The downstream process included two 3-Column PCC systems. (Reproduced with permission from Ref. [8]. Copyright 2015, Elsevier.)

performance was demonstrated to be comparable to the batch mode resulting in HCP clearance of 4 logs, removal of residual protein A to 1.0 ppm, and <5% aggregation. Further, the critical attributes did not appear to exhibit any changes in performance over time and the resulting product was found to successfully meet the release specifications of the traditional batch process. Moreover, the volumetric productivity for both the upstream and downstream was found to increase many fold. Chromatography media capacity utilization was increased by 25%, buffer usage was reduced by 20%, and individual column size was reduced at least 20-fold, as compared to batch mode. The combination of a dramatic decrease in cycle time, increased equipment utilization, and increased buffer and resin utilization was shown to result in significant economic benefits.

The key distinguishing features of the proposed facility were uninterrupted and fully automated purification, steady-state operation, and consistent product quality throughout operation. In addition, increased process throughput, decreased equipment footprint, and elimination of several hold steps have been further demonstrated and, indeed, the expected benefits of continuous processing can be realized in biotherapeutic manufacturing.

10.5 Summary

Continuous processing is gradually being recognized as the future of biotherapeutic manufacturing. The drivers for this optimism are numerous. Significant advantages that continuous processing offers over batch manufacturing, and the fact that these have been realized in other industries. Other unique advantages for biotherapeutic production include better scalability, improved process controls, and a more consistent product quality. There is still a lot that needs to be done before continuous processing reaches widespread implementation in biotherapeutic manufacturing. Major concerns include the high risk associated with implementation of new technologies that are yet to be proven on commercial scale. Also, the time and cost associated with continuous process development (given that time to launch a new biotherapeutics is very critical) can be a deterrent [128,129]. However, recent reports from biotech majors seem to be encouraging [130]. Major technology suppliers are also bringing new solutions to market that will ease the implementation. A major concern that still needs to be effectively addressed is that of appropriate monitoring and control of continuous processes [130]. We are confident that this topic will continue to be an interesting one both for academia and industry in the years to come.

References

1 Warikoo, V., Godawat, R., Brower, K., Jain, S., Cummings, D., Simons, E. *et al.* (2012) Integrated continuous production of recombinant therapeutic proteins. *Biotechnol. Bioeng.*, **109** (12), 3018–3029.
2 U.S. Food and Drug Administration, (2015) 21 CFR 210.3, Code of Federal Regulations, Title 21, Volume 4.

3 Schaber, S.D., Gerogiorgis, D.I., Ramachandran, R., Evans, J.M.B., Barton, P.I., and Trout, B.L. (2011) Economic analysis of integrated continuous and batch pharmaceutical manufacturing: a case study. *Ind. Eng. Chem. Res.*, **50** (17), 10083–10092.

4 Bisson, W. (2008) Continuous manufacturing –- the ultra lean way of manufacturing, in ISPE Innovations in Process Technology for Manufacture of APIs BPCs, Copenhagen, 7 (11)

5 Langer, E.S. (2011) Trends in perfusion bioreactors. *Bioprocess Int.*, **9** (10), 18–22.

6 Rathore, A.S., Agarwal, H., Sharma, A.K., Pathak, M., and Muthukumar, S. (2015) Continuous processing for production of biopharmaceuticals. *Prep. Biochem. Biotechnol.*, **45** (8), 836–849.

7 Jungbauer, A. (2013) Continuous downstream processing of biopharmaceuticals. *Trends Biotechnol.*, **31** (8), 479–492.

8 Godawat, R., Konstantinov, K., Rohani, M., and Warikoo, V. (2015) End-to-end integrated fully continuous production of recombinant monoclonal antibodies. *J. Biotechnol.*, **213**, 13–19.

9 Xenopoulos, A. (2015) A new, integrated, continuous purification process template for monoclonal antibodies: process modeling and cost of goods studies. *J. Biotechnol.*, **213**, 42–53.

10 Sauer, T., Robinson, C.W., and Glick, B.R. (1989) Disruption of native and recombinant *Escherichia coli* in a high-pressure homogenizer. *Biotechnol. Bioeng.*, **33** (10), 1330–1342.

11 Barazzone, G.C., Carvalho, R., Kraschowetz, S., Horta, A.L., Sargo, C.R., Silva, A.J. *et al.* (2011) Production and purification of recombinant fragment of pneumococcal surface protein A (PspA) in *Escherichia coli*. *Procedia Vaccinol.*, **4**, 27–35.

12 Saboya, L.V., Maillard, M.-B., and Lortal, S. (2003) Efficient mechanical disruption of *Lactobacillus helveticus*, *Lactococcus lactis* and *Propionibacterium freudenreichii* by a new high-pressure homogenizer and recovery of intracellular aminotransferase activity. *J. Ind. Microbiol. Biotechnol.*, **30** (1), 1–5.

13 Nan, L., Jiang, Z., and Wei, X. (2014) Emerging microfluidic devices for cell lysis: a review. *Lab Chip*, **14** (6), 1060–1073.

14 Marcus, J.S., Anderson, W.F., and Quake, S.R. (2006) Microfluidic single-cell mRNA isolation and analysis. *Anal. Chem.*, **78** (9), 3084–3089.

15 Kotlowski, R., Martin, A., Ablordey, A., Chemlal, K., Fonteyne, P.-A., and Portaels, F. (2004) One-tube cell lysis and DNA extraction procedure for PCR-based detection of *Mycobacterium ulcerans* in aquatic insects, molluscs and fish. *J. Med. Microbiol.*, **53** (9), 927–933.

16 Cichová, M., Prokšová, M., Tóthová, L., Santha, H., and Mayer, V. (2012) On-line cell lysis of bacteria and its spores using a microfluidic biochip. *Open Life Sci.*, **7** (2), 230–240.

17 Chen, X., Cui, D., Liu, C., and Cai, H. (2006) Microfluidic biochip for blood cell lysis. *Chin. J. Anal. Chem.*, **34** (11), 1656–1660.

18 Di Carlo, D., Jeong, K.-H., and Lee, L.P. (2003) Reagentless mechanical cell lysis by nanoscale barbs in microchannels for sample preparation. *Lab Chip*, **3** (4), 287–291.

19 Yun, S.-S., Yoon, S.Y., Song, M.-K., Im, S.-H., Kim, S., Lee, J.-H. *et al.* (2010) Handheld mechanical cell lysis chip with ultra-sharp silicon nano-blade arrays for rapid intracellular protein extraction. *Lab Chip*, **10** (11), 1442–1446.
20 Vogel, A., Noack, J., Nahen, K., Theisen, D., Busch, S., Parlitz, U. *et al.* (1999) Energy balance of optical breakdown in water at nanosecond to femtosecond time scales. *Appl. Phys. B Lasers Opt.* **68** (2), 271–280.
21 Vogel, A., Busch, S., Jungnickel, K., and Birngruber, R. (1994) Mechanisms of intraocular photodisruption with picosecond and nanosecond laser pulses. *Lasers Surg. Med.*, **15** (1), 32–43.
22 Roush, D.J. and Lu, Y. (2008) Advances in primary recovery: centrifugation and membrane technology. *Biotechnol. Prog.*, **24** (3), 488–495.
23 Ivory, C.F., Gilmartin, M., Gobie, W.A., McDonald, C.A., and Zollars, R.L. (1995) A hybrid centrifuge rotor for continuous bioprocessing. *Biotechnol. Prog.*, **11** (1), 21–32.
24 Lander, R., Daniels, C., and Meacle, F. (2005) Efficient, scalable clarification of diverse bioprocess streams. *Bioprocess Int.*, **11**, 32–40.
25 Rathore, A.S., Bade, P., Joshi, V., Pathak, M., and Pattanayek, S.K. (2013) Refolding of biotech therapeutic proteins expressed in bacteria: review. *J. Chem. Technol. Biotechnol.*, **88** (10), 1794–1806.
26 Middelberg, A.P.J. (1996) The influence of protein refolding strategy on cost for competing reactions. *Chem. Eng. J. Biochem. Eng. J.*, **61** (1), 41–52.
27 Bade, P.D., Kotu, S.P., and Rathore, A.S. (2012) Optimization of a refolding step for a therapeutic fusion protein in the quality by design (QbD) paradigm. *J. Sep. Sci.*, **35** (22), 3160–3169.
28 Pathak, M., Rathore, A.S., Dixit, S., and Muthukumar, S. (2016) Analytical characterization of *in vitro* refolding in the quality by design paradigm: refolding of recombinant human granulocyte colony stimulating factor. *J. Pharm. Biomed. Anal*, **126**, 124–131.
29 Mannall, G.J., Titchener-Hooker, N.J., Chase, H.A., and Dalby, P.A. (2006) A critical assessment of the impact of mixing on dilution refolding. *Biotechnol. Bioeng.*, **93** (5), 955–963.
30 Terashima, M., Suzuki, K., and Katoh, S. (1996) Effective refolding of fully reduced lysozyme with a flow-type reactor. *Process Biochem.*, **31** (4), 341–345.
31 Katoh, S. and Katoh, Y. (2000) Continuous refolding of lysozyme with fed-batch addition of denatured protein solution. *Process Biochem.*, **35** (10), 1119–1124.
32 Schlegl, R., Tscheliessnig, A., Necina, R., Wandl, R., and Jungbauer, A. (2005) Refolding of proteins in a CSTR. *Chem. Eng. Sci.*, **60** (21), 5770–5780.
33 Schlegl, R. (2007) Method for refolding a protein. U.S. Patent; 2,007,023. Available at http IR://www.google.com.tr/patents/US7651848 .
34 Pan, S., Zelger, M., Hahn, R., and Jungbauer, A. (2014) Continuous protein refolding in a tubular reactor. *Chem. Eng. Sci.*, **116**, 763–772.
35 Pan, S., Zelger, M., Jungbauer, A., and Hahn, R. (2014) Integrated continuous dissolution, refolding and tag removal of fusion proteins from inclusion bodies in a tubular reactor. *J. Biotechnol.*, **185**, 39–50.
36 Rathore, A.S., Nigam, K.D.P., Pathak, M., Kateja, N., Hebbi, V., Sharma, A.K. *et al.* (2017) Coiled flow inverter as a reactor for biotech unit operations.

Indian Patent Application No.185/DEL/2015, PCT Patent Application Number PCT/IN2016/000022.

37 Sharma, A.K., Agarwal, H., Pathak, M., Nigam, K.D.P., and Rathore, A.S. (2016) Continuous refolding of a biotech therapeutic in a novel coiled flow inverter reactor. *Chem. Eng. Sci.*, **140**, 153–160.

38 Jungbauer, A., Kaar, W., and Schlegl, R. (2004) Folding and refolding of proteins in chromatographic beds. *Curr. Opin. Biotechnol.*, **15** (5), 487–494.

39 Batas, B. and Chaudhuri, J.B. (1996) Protein refolding at high concentration using size-exclusion chromatography. *Biotechnol. Bioeng.*, **50** (1), 16–23.

40 Lanckriet, H. and Middelberg, A.P.J. (2004) Continuous chromatographic protein refolding. *J. Chromatogr. A*, **1022** (1), 103–113.

41 Schlegl, R., Necina, R., and Jungbauer, A. (2005) Continuous matrix-assisted refolding of inclusion-body proteins: effect of recycling. *Chem. Eng. Technol.*, **28** (11), 1375–1386.

42 Schlegl, R., Iberer, G., Machold, C., Necina, R., and Jungbauer, A. (2003) Continuous matrix-assisted refolding of proteins. *J. Chromatogr. A*, **1009** (1), 119–132.

43 Wellhoefer, M., Sprinzl, W., Hahn, R., and Jungbauer, A. (2014) Continuous processing of recombinant proteins: integration of refolding and purification using simulated moving bed size-exclusion chromatography with buffer recycling. *J. Chromatogr. A*, **1337**, 48–56.

44 Ferré, H., Ruffet, E., Nielsen, L.-L.B., Nissen, M.H., Hobley, T.J., Thomas, O.R.T. et al. (2005) A novel system for continuous protein refolding and on-line capture by expanded bed adsorption. *Protein Sci.*, **14** (8), 2141–2153.

45 Batas, B. and Chaudhuri, J.B. (1999) Considerations of sample application and elution during size-exclusion chromatography-based protein refolding. *J. Chromatogr. A*, **864** (2), 229–236.

46 Luo, M., Guan, Y.-X., and Yao, S.-J. (2011) On-column refolding of denatured lysozyme by the conjoint chromatography composed of {SEC} and immobilized recombinant DsbA. *J. Chromatogr. B*, **879** (28), 2971–2977.

47 Chen, Y. and Leong, S.S.J. (2010) High productivity refolding of an inclusion body protein using pulsed-fed size exclusion chromatography. *Process Biochem.*, **45** (9), 1570–1576.

48 Park, B.-J., Lee, C.-H., and Koo, Y.-M. (2005) Development of novel protein refolding using simulated moving bed chromatography. *Korean J. Chem. Eng.*, **22** (3), 425–432.

49 Middelberg, A.P.J. (2002) Preparative protein refolding. *Trends Biotechnol.*, **20** (10), 437–443.

50 Freydell, E.J., van der Wielen, L.A.M., Eppink, M.H.M., and Ottens, M. (2010) Size-exclusion chromatographic protein refolding: fundamentals, modeling and operation. *J. Chromatogr. A*, **1217** (49), 7723–7737.

51 Gu, Z., Su, Z., and Janson, J.-C. (2001) Urea gradient size-exclusion chromatography enhanced the yield of lysozyme refolding. *J. Chromatogr. A*, **918** (2), 311–318.

52 Saremirad, P., Wood, J.A., Zhang, Y., and Ray, A.K. (2015) Oxidative protein refolding on size exclusion chromatography: from batch single-column to multi-column counter-current continuous processing. *Chem. Eng. Sci.*, **138**, 375–384.

53 Freydell, E.J., van der Wielen, L., Eppink, M., and Ottens, M. (2010) Ion-exchange chromatographic protein refolding. *J. Chromatogr. A*, **1217** (46), 7265–7274.

54 Li, M., Zhang, G., and Su, Z. (2002) Dual gradient ion-exchange chromatography improved refolding yield of lysozyme. *J. Chromatogr. A*, **959** (1), 113–120.

55 Jin, T., Guan, Y.-X., Fei, Z.-Z., Yao, S.-J., and Cho, M.-G. (2005) A combined refolding technique for recombinant human interferon-γ inclusion bodies by ion-exchange chromatography with a urea gradient. *World J. Microbiol. Biotechnol.*, **21** (6–7), 797–802.

56 Wang, C., Zhang, Q., Cheng, Y., and Wang, L. (2010) Refolding of denatured/reduced lysozyme at high concentrations by artificial molecular chaperone-ion exchange chromatography. *Biotechnol. Prog.*, **26** (4), 1073–1079.

57 Geng, X., Bai, Q., Zhang, Y., Li, X., and Wu, D. (2004) Refolding and purification of interferon-gamma in industry by hydrophobic interaction chromatography. *J. Biotechnol.*, **113** (1), 137–149.

58 Bai, Q., Kong, Y., and Geng, X. (2003) Studies on renaturation with simultaneous purification of recombinant human proinsulin from *E. coli* with high performance hydrophobic interaction chromatography. *J. Liq. Chromatogr. Relat. Technol.*, **26** (5), 683–695.

59 Hwang, S.-M., Kang, H.-J., Bae, S.-W., Chang, W.-J., and Koo, Y.-M. (2010) Refolding of lysozyme in hydrophobic interaction chromatography: effects of hydrophobicity of adsorbent and salt concentration in mobile phase. *Biotechnol. Bioprocess Eng.*, **15** (2), 213–219.

60 Li, J.-J., Liu, Y.-D., Wang, F.-W., Ma, G.-H., and Su, Z.-G. (2004) Hydrophobic interaction chromatography correctly refolding proteins assisted by glycerol and urea gradients. *J. Chromatogr. A*, **1061** (2), 193–199.

61 Glynou, K., Ioannou, P.C., and Christopoulos, T.K. (2003) One-step purification and refolding of recombinant photoprotein aequorin by immobilized metal-ion affinity chromatography. *Protein Expr. Purif.*, **27** (2), 384–390.

62 Kumar, A., Galaev, I.Y., and Mattiasson, B. (2003) Precipitation of proteins: non specific and specific. *Isolation and Purification of Proteins*, CRC Press Boca Raton, FL 225–276.

63 Tavare, N.S. and Patwardhan, V. (1992) Agglomeration in a continuous MSMPR crystallizer. *AIChE J.*, **38** (3), 377–384.

64 Raphael, M. and Rohani, S. (1996) Isoelectric precipitation of sunflower protein in an MSMPR precipitator: modelling of PSD with aggregation. *Chem. Eng. Sci.*, **51** (19), 4379–4384.

65 Raphael, M. and Rohani, S. (1999) Sunflower protein precipitation in a tubular precipitator. *Can. J. Chem. Eng.*, **77** (3), 540–554.

66 Hammerschmidt, N., Hintersteiner, B., Lingg, N., and Jungbauer, A. (2015) Continuous precipitation of IgG from CHO cell culture supernatant in a tubular reactor. *Biotechnol. J.*, **10**, 1196–1205.

67 Ito, Y. and Qi, L. (2010) Centrifugal precipitation chromatography. *J. Chromatogr. B*, **878** (2), 154–164.

68 Hammerschmidt, N., Tscheliessnig, A., Sommer, R., Helk, B., and Jungbauer, A. (2014) Economics of recombinant antibody production processes at various scales: industry-standard compared to continuous precipitation. *Biotechnol. J.*, **9** (6), 766–775.

69 Jungbauer, A. (1993) Preparative chromatography of biomolecules. *J. Chromatogr. A*, **639** (1), 3–16.

70 Giovannini, R. and Freitag, R. (2001) Isolation of a recombinant antibody from cell culture supernatant: continuous annular versus batch and expanded-bed chromatography. *Biotechnol. Bioeng.*, **73** (6), 522–529.

71 Bloomingburg, G.F., Bauer, J.S., Carta, G., and Byers, C.H. (1991) Continuous separation of proteins by annular chromatography. *Ind. Eng. Chem. Res.*, **30** (5), 1061–1067.

72 Vogel, J.H., Nguyen, H., Pritschet, M., Van Wegen, R., and Konstantinov, K. (2002) Continuous annular chromatography: general characterization and application for the isolation of recombinant protein drugs. *Biotechnol. Bioeng.*, **80** (5), 559–568.

73 Buchacher, A., Iberer, G., Jungbauer, A., Schwinn, H., and Josic, D. (2001) Continuous removal of protein aggregates by annular chromatography. *Biotechnol. Prog.*, **17** (1), 140–149.

74 Iberer, G., Schwinn, H., Josic, D., Jungbauer, A., and Buchacher, A. (2002) Continuous purification of a clotting factor IX concentrate and continuous regeneration by preparative annular chromatography. *J. Chromatogr. A*, **972** (1), 115–129.

75 Chin, C.Y. and Wang, N.-H.L. (2004) Simulated moving bed equipment designs. *Sep. Purif. Rev.*, **33** (2), 77–155.

76 Mun, S., Xie, Y., Kim, J.-H., and Wang, N.-H.L. (2003) Optimal design of a size-exclusion tandem simulated moving bed for insulin purification. *Ind. Eng. Chem. Res.*, **42** (9), 1977–1993.

77 Buhlert, K., Lehr, M., and Jungbauer, A. (2009) Construction and development of a new single-column simulated moving bed system on the laboratory scale. *J. Chromatogr. A*, **1216** (50), 8778–8786.

78 Aniceto, J.P.S. and Silva, C.M. (2015) Simulated moving bed strategies and designs: from established systems to the latest developments. *Sep. Purif. Rev.*, **44** (1), 41–73.

79 Grabski, A. and Mierendorf, R. (2009) Simulated moving bed chromatography. *Genet. Eng. Biotechnol. News.*, **29** (18), 54–55.

80 Juza, M., Mazzotti, M., and Morbidelli, M. (2000) Simulated moving-bed chromatography and its application to chirotechnology. *Trends Biotechnol.*, **18** (3), 108–118.

81 Zobel, S., Helling, C., Ditz, R., and Strube, J. (2014) Design and operation of continuous countercurrent chromatography in biotechnological production. *Ind. Eng. Chem. Res.*, **53** (22), 9169–9185.

82 Shinkazh, O., Kanani, D., Barth, M., Long, M., Hussain, D., and Zydney, A.L. (2011) Countercurrent tangential chromatography for large-scale protein purification. *Biotechnol. Bioeng.*, **108** (3), 582–591.

83 Zydney, A.L. (2016) Continuous downstream processing for high value biological products: a review. *Biotechnol. Bioeng.*, **113**, 465–475.

84 Napadensky, B., Shinkazh, O., Teella, A., and Zydney, A.L. (2013) Continuous countercurrent tangential chromatography for monoclonal antibody purification. *Sep. Sci. Technol.*, **48** (9), 1289–1297.

85 Dutta, A.K., Tran, T., Napadensky, B., Teella, A., Brookhart, G., Ropp, P.A. *et al.* (2015) Purification of monoclonal antibodies from clarified cell culture fluid using Protein A capture continuous countercurrent tangential chromatography. *J. Biotechnol.*, **213**, 54–64.

86 Kalyanpur, M. (2002) Downstream processing in the biotechnology industry. *Mol. Biotechnol.*, **22** (1), 87–98.

87 Lali, A. (2002) Expanded bed affinity chromatography, *Methods for Affinity-Based Separations of Enzymes and Proteins*, Springer, pp. 29–64.

88 Owen, R.O. and Chase, H.A. (1997) Direct purification of lysozyme using continuous counter-current expanded bed adsorption. *J. Chromatogr. A*, **757** (1), 41–49.

89 Müller-Späth, T., Aumann, L., Ströhlein, G., Kornmann, H., Valax, P., Delegrange, L. *et al.* (2010) Two step capture and purification of IgG2 using multicolumn countercurrent solvent gradient purification (MCSGP). *Biotechnol. Bioeng* **107** (6), 974–984.

90 Müller-Späth, T., Krättli, M., Aumann, L., Ströhlein, G., and Morbidelli, M. (2010) Increasing the activity of monoclonal antibody therapeutics by continuous chromatography (MCSGP). *Biotechnol. Bioeng.*, **107** (4), 652–662.

91 Aumann, L. and Morbidelli, M. (2007) A continuous multicolumn countercurrent solvent gradient purification (MCSGP) process. *Biotechnol. Bioeng.* **98** (5), 1043–1055.

92 Lay, M.C., Fee, C.J., and Swan, J.E. (2006) Continuous radial flow chromatography of proteins. *Food Bioprod. Process.*, **84** (1), 78–83.

93 Sun, T., Chen, G., Liu, Y., Bu, F., and Wen, M. (2000) Purification of human prothrombin from Nitschmann fraction III using DEAE membrane radial flow chromatography. *J. Chromatogr. B Biomed. Sci. Appl.*, **742** (1), 109–114.

94 Singh, S.M., Sharma, A., and Panda, A.K. (2009) High throughput purification of recombinant human growth hormone using radial flow chromatography. *Protein Expr. Purif.*, **68** (1), 54–59.

95 Cramer, S.M. and Holstein, M.A. (2011) Downstream bioprocessing: recent advances and future promise. *Curr. Opin. Chem. Eng.*, **1** (1), 27–37.

96 Bhambure, R., Gupta, D., and Rathore, A.S. (2013) A novel multimodal chromatography based single step purification process for efficient manufacturing of an *E. coli* based biotherapeutic protein product. *J. Chromatogr. A*, **1314**, 188–198.

97 Kaleas, K.A., Schmelzer, C.H., and Pizarro, S.A. (2010) Industrial case study: evaluation of a mixed-mode resin for selective capture of a human growth factor recombinantly expressed in *E. coli*. *J. Chromatogr. A*, **1217** (2), 235–242.

98 Xia, H.-F., Lin, D.-Q., Chen, Z.-M., and Yao, S.-J. (2010) Salt-promoted adsorption of an antibody onto hydrophobic charge-induction adsorbents. *J. Chem. Eng. Data*, **55** (12), 5751–5758.

99 Zhang, L., Bai, S., and Sun, Y. (2010) Molecular dynamics simulation of the effect of ligand homogeneity on protein behavior in hydrophobic charge induction chromatography. *J. Mol. Graph. Model.*, **28** (8), 863–869.

100 Rajamanickam, V., Herwig, C., and Spadiut, O. (2015) Monoliths in bioprocess technology. *Chromatography*, **2** (2), 195–212.

101 Fraud, N., Kuczewski, M., and Hirai, M. (2009) Hydrophobic membrane adsorbers for large-scale downstream processing. Advanstar Communications Inc.

102 Yu, D. and Ghosh, R. (2010) Purification of PEGylated protein using membrane chromatography. *J. Pharm. Sci.*, **99** (8), 3326–3333.

103 Zhou, J.X. and Tressel, T. (2006) Basic concepts in Q membrane chromatography for large-scale antibody production. *Biotechnol. Prog.*, **22** (2), 341–349.

104 Vogel, J.H., Nguyen, H., Giovannini, R., Ignowski, J., Garger, S., Salgotra, A. et al. (2012) A new large-scale manufacturing platform for complex biopharmaceuticals. *Biotechnol. Bioeng.*, **109** (12), 3049–3058.

105 Rosa, P.A.J., Ferreira, I.F., Azevedo, A.M., and Aires-Barros, M.R. (2010) Aqueous two-phase systems: a viable platform in the manufacturing of biopharmaceuticals. *J. Chromatogr. A*, **1217** (16), 2296–2305.

106 Rosa, P.A.J., Azevedo, A.M., Sommerfeld, S., Mutter, M., Bäcker, W., and Aires-Barros, M.R. (2013) Continuous purification of antibodies from cell culture supernatant with aqueous two-phase systems: from concept to process. *Biotechnol. J.*, **8** (3), 352–362.

107 Rosa, P.A.J., Azevedo, A.M., Sommerfeld, S., Bäcker, W., and Aires-Barros, M.R. (2012) Continuous aqueous two-phase extraction of human antibodies using a packed column. *J. Chromatogr. B*, **880**, 148–156.

108 Bhambure, R., Sharma, R., Gupta, D., and Rathore, A.S. (2013) A novel aqueous two phase assisted platform for efficient removal of process related impurities associated with *E. coli* based biotherapeutic protein products. *J. Chromatogr. A*, **1307**, 49–57.

109 Vázquez-Villegas, P., Aguilar, O., and Rito-Palomares, M. (2015) Continuous enzyme aqueous two-phase extraction using a novel tubular mixer-settler in multi-step counter-current arrangement. *Sep. Purif. Technol.*, **141**, 263–268.

110 Espitia-Saloma, E., Vázquez-Villegas, P., Aguilar, O., and Rito-Palomares, M. (2014) Continuous aqueous two-phase systems devices for the recovery of biological products. *Food Bioprod. Process.*, **92** (2), 101–112.

111 Himmelfarb, P., Thayer, P.S., and Martin, H.E. (1969) Spin filter culture: the propagation of mammalian cells in suspension. *Science*, **164** (3879), 555–557.

112 Deo, Y.M., Mahadevan, M.D., and Fuchs, R. (1996) Practical considerations in operation and scale-up of spin-filter based bioreactors for monoclonal antibody production. *Biotechnol. Prog.*, **12** (1), 57–64.

113 Figueredo-Cardero, A., Chico, E., Castilho, L.R., and Medronho, R.A. (2009) CFD simulation of an internal spin-filter: evidence of lateral migration and exchange flow through the mesh. *Cytotechnology.*, **61** (1–2), 55–64.

114 Femmer, T., Carstensen, F., and Wessling, M. (2015) A membrane stirrer for product recovery and substrate feeding. *Biotechnol. Bioeng.*, **112** (2), 331–338.

115 Carstensen, F., Apel, A., and Wessling, M. (2012) In situ product recovery: submerged membranes vs. external loop membranes. *J. Memb. Sci.*, **394**, 1–36.

116 Meng, F., Chae, S.-R., Shin, H.-S., Yang, F., and Zhou, Z. (2012) Recent advances in membrane bioreactors: configuration development, pollutant elimination, and sludge reduction. *Environ. Eng. Sci.*, **29** (3), 139–160.

117 Bonham-Carter, J. and Shevitz, J. (2011) A brief history of perfusion biomanufacturing. *BioProcess Int.*, **9** (9), 24–30.
118 Alford, J.R., Kendrick, B.S., Carpenter, J.F., and Randolph, T.W. (2008) High concentration formulations of recombinant human interleukin-1 receptor antagonist: II. aggregation kinetics. *J. Pharm. Sci.*, **97** (8), 3005–3021.
119 Kim, W.-S., Hirasawa, I., and Kim, W.-S. (2002) Aging characteristics of protein precipitates produced by polyelectrolyte precipitation in turbulently agitated reactor. *Chem. Eng. Sci.*, **57** (19), 4077–4085.
120 Rathore, A.S., Pathak, M., and Godara, A. (2015) Process development in the QbD paradigm: role of process integration in process optimization for production of biotherapeutics. *Biotechnol. Prog.*, **32**, 355–362.
121 Voisard, D., Meuwly, F., Ruffieux, P.-A., Baer, G., and Kadouri, A. (2003) Potential of cell retention techniques for large-scale high-density perfusion culture of suspended mammalian cells. *Biotechnol. Bioeng.*, **82** (7), 751–765.
122 Brower, M., Hou, Y., and Pollard, D. (2014) Monoclonal antibody continuous processing enabled by single use. *Continous Processing in Pharmaceutical Manufacturing*, Wiley-VCH Verlag GmbH, Weinheim 255–296.
123 Rathore, A.S. (2009) Roadmap for implementation of quality by design (QbD) for biotechnology products. *Trends Biotechnol.*, **27** (9), 546–553.
124 Read, E.K., Park, J.T., Shah, R.B., Riley, B.S., Brorson, K.A., and Rathore, A.S. (2010) Process analytical technology (PAT) for biopharmaceutical products: part I. Concepts and applications. *Biotechnol. Bioeng.*, **105** (2), 276–284.
125 Read, E.K., Shah, R.B., Riley, B.S., Park, J.T., Brorson, K.A., and Rathore, A.S. (2010) Process analytical technology (PAT) for biopharmaceutical products: Part II. Concepts and applications. *Biotechnol. Bioeng.*, **105** (2), 285–295.
126 Rathore, A.S., Bhambure, R., and Ghare, V. (2010) Process analytical technology (PAT) for biopharmaceutical products. *Anal. Bioanal. Chem.*, **398** (1), 137–154.
127 Rathore, A.S. and Winkle, H. (2009) Quality by design for biopharmaceuticals. *Nat. Biotechnol.*, **27** (1), 26–34.
128 Zydney, A.L. (2015) Perspectives on integrated continuous bioprocessing – opportunities and challenges. *Curr. Opin. Chem. Eng.*, **10**, 8–13.
129 Croughan, M.S., Konstantinov, K.B., and Cooney, C. (2015) The future of industrial bioprocessing: batch or continuous? *Biotechnol. Bioeng.*, **112** (4), 648–651.
130 Konstantinov, K.B. and Cooney, C.L. (2015) White paper on continuous bioprocessing. May 20–21 2014 continuous manufacturing symposium. *J. Pharm. Sci.*, **104** (3), 813–820.

11

Evolving Needs For Viral Safety Strategies in Continuous Monoclonal Antibody Bioproduction

Andrew Clutterbuck,[1] Michael A. Cunningham,[2] Cedric Geyer,[1] Paul Genest,[2] Mathilde Bourguignat,[1] and Helge Berg[1]

[1]Technology Management, Millipore SAS, 39 Route Industrielle de la Hardt, 67124 Molsheim, France
[2]Technology Management, EMD Millipore Corporation, 290 Concord Road, Billerica, MA 01821, USA

11.1 Introduction

The viral safety of biologics is a well-known but still evolving subject for pharmaceutical manufacturers and regulators alike. A long history of contamination events has led regulatory authorities to put a standard framework in place to limit the likelihood of a viral contamination leading to an adverse event in a patient administered a biotherapeutic medication. Despite significant advances in process knowledge and technological advancements over the preceding decades, clinical manufacturers of biological entities focus not if a viral contamination will happen, but rather when and where it may occur. In this context, the techniques developed to limit viral contamination and propagation risk in classical batch mAb production processes and how they can be applied to the new and evolving continuous production processes will be reviewed in this chapter.

It is well known and documented that a variety of biological contaminants can be present or introduced in to manufacturing processes. They include bacteria, mycoplasma, transmissible spongiform encephalopathies (TSE), and viruses; all of which present unique and challenging problems in the manufacturing of biologics. The topic of viral safety is broad; this chapter will focus on virus mitigation and removal strategies for monoclonal antibody protein manufacturing processes with a strong emphasis on continuous biomanufacturing. While plasma derivatives or vaccines are also subject to similar kinds of risk mitigation concerns, they will not specifically be discussed in this chapter.

Monoclonal antibodies for human use are produced mainly by animal cell culture. These typical production processes can generate, or be contaminated with, potential pathogenic viruses capable of causing disease in humans. Viruses are obligate intracellular parasites composed of nucleic acid and proteins, and require a living host for their replication and propagation.

Continuous Biomanufacturing: Innovative Technologies and Methods, First Edition.
Edited by Ganapathy Subramanian.
© 2018 Wiley-VCH Verlag GmbH & Co. KGaA. Published 2018 by Wiley-VCH Verlag GmbH & Co. KGaA.

Viruses can be classified in the following two categories, based on their source:

- *Exogenous/adventitious virus:* Unintentionally introduced contaminant viruses that may originate through the addition of contaminated raw materials or through extraneous contamination, for example, parvovirus.
- *Endogenous virus:* Virus whose genome is part of the germ line of the species of origin of the cell line and is covalently integrated into the genome of the animal from which the parental cell line was derived, for example, Retrovirus-like particles (RVLP), X-MuLV.

If a purification process does not effectively eliminate or inactivate viral particles present in the product, they can be transmitted to humans and potentially have severe clinical manifestations. Past contaminations of biologic drugs have resulted in product recalls, plant shutdowns, and adverse events in patients. A virus contamination, in most cases, will cause production interruptions in order to ensure production plant is well decontaminated, which represents a significant loss for the company financially, and in terms of public image/perceptions, as well as increased regulatory scrutiny. For all these reasons, regulatory authorities such as the Food & Drug Administration (FDA), as well as consortiums such as the International Conference on Harmonisation (ICH), which involves members from both regulatory and industry, formulated guidelines to lower the risk of potential viral transmission.

As stated in the FDA code of federal regulations – 21 CFR 610.13 document

> "Products shall be free of extraneous material except that which is unavoidable in the manufacturing process described in the approved biologics license application".

Nowadays, to ensure, to the greatest extent possible, that pharmaceutical products are free of adventitious agents, most countries follow the same guidelines, with minor differences. The International Conference on Harmonisation, ICH Q5A R1 document gives a good overview of the established rules.

This layered viral safety strategy is composed of three principle complementary approaches, which has evolved to control the potential of viral contamination of biotechnology products[1]:

- Selecting and testing cell lines and other raw materials, including media components, for the absence of undesirable viruses that may be infectious and/or pathogenic for humans.
- Assessing the capacity of the production process to clear infectious viruses.
- Testing the product at appropriate steps of production for absence of contaminating infectious viruses.

The selection of raw materials is a first step in the process. Many past contamination events arose from animal-derived raw materials. Fetal bovine serum (FBS), for instance, was commonly used in cell culture media formulations in production processes until the late 1980s, with the emergence of bovine

---

1 ICH Q5A R1 September 1999 and European pharmacopoeia Ph EUR 5.1.7.

spongiform encephalopathy (BSE). For this reason, the use of bovine serum has been regulated, and the products classified based on their country of origin. Biomanufacturers then started to look for serum replacements, but this is still a challenge today with certain cell lines, where cell growth and/or productivity can be challenging in the absence of serum. Individual proteins of animal origin have been used with a certain success, but need to be replaced by recombinant equivalents, as they have also been involved in viral contamination events. An additional safety layer, meant to mitigate pathogen safety risk of using contaminated media components during the cell culture step is the introduction of virus barriers into the upstream manufacturing process. Virus barrier technologies can also be useful tools for screening animal free recombinant media additives, as they have also been identified as potential sources of viral contamination events [1]. This will be discussed in more detail in Section 11.2.2.

An example of a recent serious event was a contamination of rotavirus vaccine[2] with a porcine circovirus (PCV-1) that was, after investigation, found to be present in the master cell bank (MCB). A root cause investigation showed that virus-contaminated porcine trypsin was used to establish the MCB. Interestingly, this contamination was discovered in 2010 by an academic research team from the University of California testing the massive parallel sequencing (MSP) metagenomic technology on eight live attenuated vaccines [2]. They discovered DNA sequences from PCV-1 in two batches of Rotarix® (Glaxo Smith Kline's (GSK) live attenuated rotavirus vaccine). Tests performed after this discovery confirmed presence of PCV-1 nearly full-length DNA in the master and working cell banks as well as in process intermediates and final drug product.

This contamination was found to have originated in a master cell bank generated during early process development work, and carried through to production lots used during successful clinical trials. Luckily, even if PCV-1 had been able to replicate in cell culture media, this virus is not known to cause disease in humans, and thus the 70 million doses produced and injected into patients prior to detection of the contamination did not cause adverse reactions in patients. GSK continued producing the vaccine but has since started working on a new master cell bank to manufacture PCV-1 free Rotarix [3].

The second aspect of an effective pathogen safety risk mitigation strategy is the use of analytical methods to effectively test raw materials. However, virus detection methods are not without challenges. Viruses can be highly host-specific parasites, which explains why *in vitro* detection techniques using living cells to detect viruses will only detect the organisms capable of replicating under the specific/conditions of the test. These tests can also be very time consuming, which often prevents their use as *online* analytics. Newer detection techniques have been developed such as quantitative polymerase chain reaction (qPCR) or transcription-mediated amplification (TMA) methodologies. These techniques are able to deliver quicker results in comparison to more conventional cell-based assays, and are able to detect more virus types when appropriate primers are available.

---

2 http://www.fda.gov/BiologicsBloodVaccines/Vaccines/ApprovedProducts/ucm205539.htm.

These techniques can also show inherent limitations of sampling and testing in that, as the sample volume is low, a low viral particle load can easily not be detected in the *bulk* material. Another common issue is the prevalence of false positive results (noninfectious viral genome parts), as viral inactivation methods, such as low pH or solvent detergent treatment, can generate genomic fragments that can be detected by the testing method. In other words, the test is not able to differentiate between active and inactive viruses or viral DNA. Improvement of virus detection analytical methods continues to be a technical area of focus in the biotechnology industry in order to improve legitimate virus detection, and reduce pathogen safety risk. Advances in bioinformatics should help improve detection methods, and wider industry adoption will likely occur as assay and detection methods become cheaper, more effective, and more accessible. Currently, however, it is recognized that no universal test exists to detect every virus.

Inherent limitations of sampling and testing methods also exist. Because of this, validation of virus removal capabilities of manufacturing processes is an integral part of a robust and comprehensive viral safety strategy. For example, in the European pharmacopoeia Ph EUR 5.1.7, which provides general requirements concerning the viral safety of medicinal product whose manufacture has involved the use of materials of human and animal origin, it is advised that after a risk assessment is carried out, "one or more validated procedures for removal or inactivation of viruses are applied." This document also recommends referring to the Committee for Medicinal Products for Human Use document CPMP/BWP/268/95 of the CPMP and the ICHQ5A guideline for more details on how to perform these virus removal validation studies.

Biomanufacturers are required to show that their production processes are able to robustly remove significant amounts of virus. While an exact objective value is not given, the following guideline is provided in this regulatory document:

> "It is important to show that not only is the virus eliminated or inactivated, but that there is excess capacity for viral clearance built into the purification process to assure an appropriate level of safety for the final product".

A similar example given in Appendix 5 of the ICH Q5 A is

> "An excess clearance of 6 log, meaning that statistically one viral particle could be found in 1 million produced doses".

Pharmaceutical manufacturers usually target a log reduction value of 12–18 over an entire purification process, through elimination or inactivation steps, which will be described in greater detail in subsequent sections of this chapter.

It is accepted that a layered approach to viral safety has certain limitations, and that it is not possible to guarantee an absolute absence of contamination. However, this risk mitigation approach has been successfully used for many years to minimize the virus risk of batch processes with a high degree of success. As manufacturers are contemplating transitioning from batch processes toward continuous processes, where similar virus safety risk mitigation strategies will be applied using the same guiding principles.

11.1.1 Current Regulations and Practices

Viral contamination may arise from several sources, such as raw materials used in the production process (cell banks, cell culture media, chemicals, etc.), environmental sources (water, HVAC, personnel), or inappropriate/ineffective manufacturing controls. In addition, retrovirus-like particles can be generated by the host cell line during production, especially under stressed conditions. As discussed earlier, a viral safety strategy is put in place by biomanufacturers to secure the outcome of their production processes, and it ensures that viral contaminations are minimized. Mitigation of pathogen safety risk is generally achieved by incorporating a variety of risk reduction strategies into the facility design and production process, including specific viral inactivation and reduction steps within the production process itself.

A viral inactivation/reduction step itself is considered effective if it is capable of achieving a minimum of a 4 log reduction value (LRV) in viral load[3]. The design of experiments to characterize LRV capability is very important; process parameters that could potentially impact the effectiveness of virus removal or inactivation must be investigated. A quality by design experimental approach may help to define these critical process parameters impacting viral clearance. For chromatographic steps, for instance, process parameters such as column diameter, process flow rate, or resin lifetime should be studied, and appropriate limits should be set and tested in viral validation studies. In addition, viral clearance steps incorporated into the manufacturing process must be orthogonal, meaning that they inactivate or remove viruses by different mechanisms of action (e.g., chemical inactivation, size exclusion, or adsorption).

The aim of a viral validation study is twofold:

- To provide evidence that the process will effectively inactivate/remove viruses that are known to either contaminate the starting material or which could conceivably do so.
- To provide indirect evidence that the process might inactivate/remove novel or unpredictable virus contamination (CPMP/BWP/268/95 from EAEM).

Virus removal validation studies are generally performed using scale-down models of the unit operation in question, and the spiking of an appropriate model virus to assess the capacity of the step to remove or inactivate this model virus. The scale-down model must to be an accurate representation of the full-scale production process. If differences between this small-scale model and the manufacturing scale operation are unavoidable, it can be considered an acceptable model as long as these differences are documented and explained.

The choice of model viruses for the validation study is mainly dependent on the cell line used to produce the drug. Model viruses should be chosen for their similarity to pathogenic viruses that might contaminate the product, and must also represent a wide range of physicochemical properties.

---

3 CPMP (1996) Note for Guidance on Virus Validation Studies: The Design, Contribution and Interpretation of Studies Validating the Inactivation and Removal of Viruses. CPMP/BWP/268/95.

As GMP practices do not allow the introduction of viruses in production plants, the virus removal validation study should be conducted in a separate laboratory equipped specifically for virology work:

> "An effective virus removal step should give *reproducible* reduction of virus load shown by at least two independent studies.... Performed by staff with virological expertise *in conjunction with the production bioengineers*"[1,3]

During validation studies, the quantity of viruses added to the product should be as high as possible to ascertain the effective capacity of the step to inactivate or clear viruses. However, the amount of virus used in validation studies should be such that the addition of viruses does not alter the composition of the product or significantly alter the unit operation performance seen with representative feed and no virus spike present. Once the experiment is performed, the quantitative infectivity assays must be performed according to the principles of GLP, and the method ($TCID_{50}$ assays, plaque formation, detection of other cytopathic effects, and so on) should have a good sensitivity and reproducibility.

11.1.2 Evolving Needs: Process versus Regulatory

In the context of a continuous biomanufacturing process, the challenges of viral clearance/safety mimic those of conventional batch bioprocesses. There are no specific regulations concerning virus safety approaches in continuous processes, which is the reason why many of the viral safety considerations used in traditional batch processes are also incorporated into continuous production processes, and validated to show similar effectiveness. In batch processes, each unit operation is separate and well defined. In continuous processes, however, one unit operation is directly fed by the preceding one, potentially creating challenges with downstream processing. For instance, quality attributes of the recombinant product vary with time. As an example, if a bind and elute chromatographic step is directly linked to the following processing step, the concentration of the column eluate will vary, and this could have an impact on the performance of this following step. Therefore, it is critical to define the term *batch* in continuous processing in the context of a viral safety strategy. According to the regulatory documents, specifically 21 CFR 210.3, the definition of a batch is

> "A specific quantity of a drug or other material that is intended to have uniform character and quality, within specified limits, and is produced according to a single manufacturing order during the same cycle of manufacture".

In summary, a batch refers to the quantity of material, and does not specify the mode of manufacture.

The following are additional ways to define a batch:

- According to the production time period (24 h, 2 days, etc.)
- All material processed in a specified time period

- A specific mass of protein
- According to the product variation (e.g., different lots of feedstock)
- According to equipment cycling capability (for instance, a batch will be the quantity of product going through a simulated moving bed (SMB) process for 20 cycles)

Once the term *batch* has been defined, methods must be implemented to prove that the continuous production process or steps within a continuous process are able to remove or inactivate viruses. In certain cases, alternative spiking techniques (continuous, inline spiking, for instance) will replace the more typical *spiked run* technique. Inline spiking will allow viral reduction quantitation of a specific step without introducing a disruption in a process where one does not exist. Virus spiking approaches will be described in more detail later in this chapter.

11.1.3 Current Technology Landscape

With the previous descriptions of variations in *continuous* processing strategies, potential implications to virus safety relative to the traditional batch processing method can be considered. As illustrated in Figure 11.1, the specific processing details of continuous/connected processing vary from manufacturer to manufacturer. Some of these differences in processing details include method of cell culture, method of cell separation for perfusion, concentration of protein A column load material, method of chromatography (bind/elute, flow-through modes), choice of chromatography resin, order of chromatography polishing steps, and utilization of single-use technologies (i.e., bags, mixers, etc.). Conversely, there are some common processing logistics among the different versions

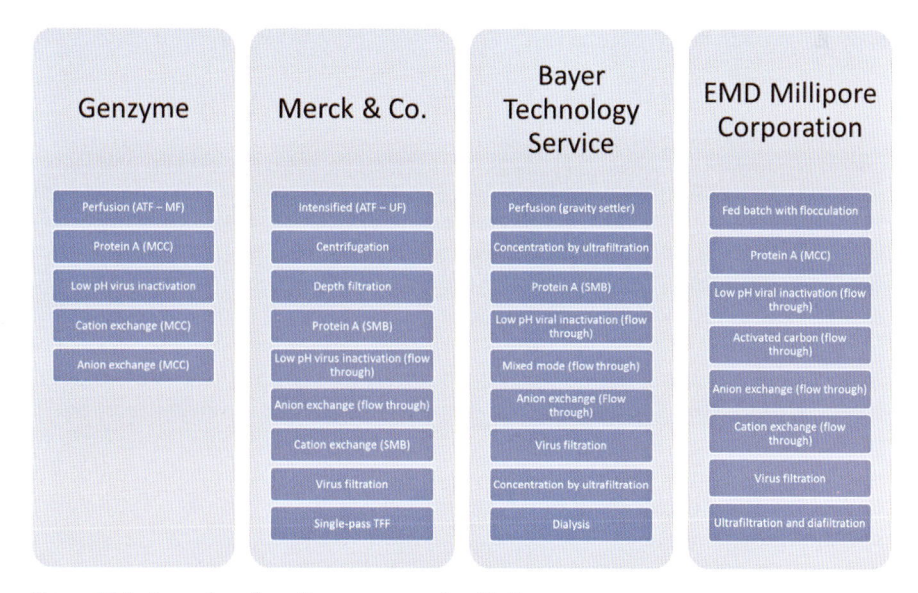

Genzyme	Merck & Co.	Bayer Technology Service	EMD Millipore Corporation
Perfusion (ATF – MF)	Intensified (ATF – UF)	Perfusion (gravity settler)	Fed batch with flocculation
Protein A (MCC)	Centrifugation	Concentration by ultrafiltration	Protein A (MCC)
Low pH virus inactivation	Depth filtration	Protein A (SMB)	Low pH viral inactivation (flow through)
Cation exchange (MCC)	Protein A (SMB)	Low pH viral inactivation (flow through)	Activated carbon (flow through)
Anion exchange (MCC)	Low pH virus inactivation (flow through)	Mixed mode (flow through)	Anion exchange (flow through)
	Anion exchange (flow through)	Anion exchange (Flow through)	Cation exchange (flow through)
	Cation exchange (SMB)	Virus filtration	Virus filtration
	Virus filtration	Concentration by ultrafiltration	Ultrafiltration and diafiltration
	Single-pass TFF	Dialysis	

Figure 11.1 Examples of continuous processing [4–8].

of continuous processing, and include the use of multicolumn or simulated moving bed systems, utilization of protein A for affinity capture, and the use of small intermediate tanks or surge bags to balance flow in between the various unit operations for processing flexibility.

A current industry survey gives a brief snapshot of the current landscape as it pertains to continuous processing and an insight in to what continuous means to various companies [4–8]. The survey includes a description of the unit operations used by each company in their version of continuous processing. The different approaches are compared and assessments made with regard to what these different approaches mean related to virus safety, relative to the traditional batch processes.

Based on a given manufacturer's history and preferences, *continuous/connected* processing can mean different things. As a baseline, a typical traditional batch process might include: a fed-batch bioreactor production, centrifugation and/or depth filtration, sterile filtration, protein A affinity chromatography, low pH virus inactivation, cation exchange chromatography run in bind and elute mode, anion exchange flow-through polishing, virus filtration, ultrafiltration/diafiltration concentration and formulation, and a final terminal sterile filtration step. The process train may include a number of process hold points.

It is worth noting that among the various biopharmaceutical manufacturers there are some differences with respect to the number, types, and the order of the chromatography steps. This may also include the modes of chromatography operation (e.g., weak partitioning).

A number of manufacturers have published and/or patented various descriptions of continuous processing. Figure 11.1 shows the general steps in continuous processing utilized by a number of industry practitioners.

All of the given examples include a protein A affinity capture chromatography step using some form of multicolumn or SMB approach for capture. These systems include a number of smaller chromatography columns used over a larger number of repeated cycles (equilibration, load, wash, regenerate, back to equilibration again). The columns are typically loaded to 100% saturation, avoiding drug product loss by running multiple columns in series and switching to a different column in sequence at a specified point (i.e., when product breakthrough is detected during column loading). The columns may also be run at faster flow rates in multicolumn/SMB configurations. These faster flowing, more highly loaded, and more frequently used columns are more productive than traditional bind and elute batch protein A capture processes, which use large columns, slower flow rates, and lower product loadings (typically 80% of the dynamic binding capacity).

An additional difference between the traditional batch processes described by Merck & Co [6] and Bayer [7] that may impact viral safety is the switch from batch to flow-through mode of operation for low pH virus inactivation. In contrast to batch mode, the flow mode of operation postprotein A column elution must rely on a mechanism by which all of the eluate is exposed to a low-pH solution for a specified period of time in order to ensure virus inactivation efficacy. The logistics of accomplishing this inactivation step would rely either on utilizing a product hold tank to conduct the low-pH hold, or on devising another mechanism such as a flow cell in which protein A column eluate is acidified and held for the required length of time while traveling through the flow cell.

Following this low-pH hold step, the different versions of continuous processing use either another SMB bind and elute cation exchange chromatography (Genzyme) or flow-through steps (anion exchange, mixed mode, or activated carbon – Merck & Co. [6], Bayer [7], and EMD Millipore Corporation, respectively). Merck & Co. follows its flow-through anion exchange step with another SMB bind and elute cation exchange step [6]. Following completion of the adsorptive purification steps, all manufacturers use a virus filter in order to conduct a second dedicated virus removal step. Product is then concentrated using single-pass tangential flow filtration (SPTFF) (Merck & Co. [6]), traditional ultrafiltration and diafiltration (EMD Millipore Corporation), or ultrafiltration concentration followed by dialysis (Bayer [7]).

11.2 Batch versus Continuous: Potential Impacts on Virus Safety

As shown in Section 11.1.3, the nomenclature *continuous* can be applied to a multitude of processing permutations, but the basic premise is that feed from one unit operation feeds into a linked unit operation and transformed product is produced, at the same time, and under steady-state conditions where by product is being produced at in a continuous, constant manner. This is in contrast to the batch method where feed goes in, some period of time passes, and then transformed product is removed or collected. Fed-batch bioreactors, bind and elute chromatography, and ultrafiltration/diafiltration are examples of unit operations that have traditionally been run in batch mode. However, these processes can be and are being modified to run in a continuous mode of operation.

Small volume bioreactors and chromatography columns run continuously, more frequently, and with less down time can be equally or more productive than larger batch processes. Utilization of small volume bioreactors and chromatography columns is facilitated by perfusion cell culture and multicolumn or SMB chromatography systems. These smaller systems are less expensive and take up less plant floor space. Implementation of smaller bioreactors and column sizes also enables manufacturers to take advantage of recent advances in single-use disposable technology (bags, purification devices, connected flow paths/valves, etc.), which can further reduce costs, support facility flexibility, and may facilitate closed processing where closed processing is defined as a process step or system that utilizes processing equipment in which the product is not exposed to the immediate room environment[4].

The use of perfusion production techniques can be an intensive process, particularly from a media perspective, utilizing anywhere from 1 to 2 bioreactor volumes per day [9]. If perfusion is used and the media consumption is much greater than that used in batch mode, there is the possibility of an increased risk of

---

4 www.ispe.org/glossary?term=Closed+Process.

virus contamination. Not only are larger media volumes used in perfusion, but there have also been past virus contamination events that were shown to have been due to virus-contaminated media. These facts may provide a stronger driver to use high-temperature short-time (HTST) treatment or virus barrier filters during media(s) reconstitution to reduce risk whenever possible.

If perfusion is utilized, is the retrovirus-like particle's load greater compared to fed-batch and how does it change over the longer culture time? It is not a simple one and may be stated that it really depends on the host cell line and should be evaluated on a case-by-case basis. With either batch/fed-batch or perfusion modes of bioreactor operation, it is important to know how much endogenous virus impurity is present to ensure that the downstream process is appropriately designed to handle the inactivation and or removal of potential virus to levels required by relevant regulatory agencies. Furthermore, when using a perfusion bioreactor and utilizing minimal clarification (no depth filters) might lead to fewer steps in the process that can contribute to virus reduction [4,5,7].

Strategies are being investigated to more efficiently *connect* and *close* the continuous processes to further reduce the overall facility footprint (more cost savings) and compress the overall production process time (more progress toward greater efficiency). The modifications to connected processes, in conjunction with continuous processing, include trying to closely match process flow rates from adjacent unit operations and reducing the size of, or eliminating, intermediate break tanks needed, compared to the large tanks used with conventional batch processes. Currently, utilized connection methods utilize large tanks that hold feeds still for relatively long period of time (as much as 4 h). One example where continuous processing can reduce the need for intermediate tanks is the changing out of a batch ultrafiltration step that requires a large feed tank to a process with an in-line SPTFF system. In this example, no tank is needed for SPTFF. The feed enters from one end and concentrated product exits at the other end. Lower fluxes and more filter area are needed, but there are cost savings associated with capital expenditure and floor space requirements (no *system*: no feed/retentate tank, no recirculation lines, etc.).

In chromatography operations, either smaller tanks or disposable surge bags can be used together with smaller columns to accommodate any minor changes in the continuous unit operation (feed or impurity concentration, pH, conductivity, etc.). These smaller bags or tanks can also be used to make changes, when needed, in pH or conductivity before the feed is moved onto the next continuously run unit operation. The increased or complete use of disposables further reduces cost by decreasing the need for clean-in-place (CIP) cycles (water, chemicals, time/labor) and it adds flexibility by reducing the turnaround time in between batches of the same drug product or switchover to new drug products. The reduction in cost and increases in efficiency and flexibility from *continuous* and *connected* processing, allow the manufacturer to better compete in the evolving biopharma marketplace.

The fact that the drug product may move more rapidly downstream in a continuous process might present a greater pathogen safety risk when virus is not detected early in bioprocessing (i.e., bioreactor operation). New, more reliable, and faster virus detection methods would aid in lowering this risk of detection. In addition, the use of a fully disposable and closed continuous processes would help reduce decontamination efforts if a virus contamination did occur and was not detected in the bioreactor. One aspect of continuous processing that may mitigate virus contamination risk is the high level of automation used for continuous operations. Increased automation reduces the personnel interaction with the process and thus, reducing the likelihood of virus exposure to the bioprocess contributed through the operator.

Viral clearance validation of the continuous (multicolumn or SMB) chromatographic steps will show the impact, if any, to the log virus reduction value, relative to a similar batch process. In addition, and as practical, as many aspects of the continuous process such as slower flow rates that may potentially be used for AEX flow-through or virus filtration, should be mimicked in the scale-down model validation as well to cover any impact to the virus reduction capabilities.

A further risk reduction consideration would be to answer the following question:

> "What would happen if the virus filter failed a post-use integrity test after the batch was already concentrated by single pass TFF?"

In batch processes, integrity testing of the virus reduction filters can be completed before subsequent product processing steps. Confirmation of virus filter integrity postuse reduces pathogen safety risk and also facilitates the termination of processing should a filter failure be detected. A continuous process may not accommodate completion of a postuse virus filter integrity test prior to subsequent product processing, which may result in processing of material with virus safety risk should the integrity test fail, or in unnecessary operations expense caused by completion of subsequent process steps prior to integrity test results being available as a test failure likely would result in the disposal of processed product.

In a continuous process, the virus filter integrity could be tested prior to use, and care would need to be taken in operation not to damage the filter. This would help avoid a failing integrity test, post use. Preuse integrity test and care in operation are even more important for continuous processing. In fact, it could be foreseen to placing some kind of hold at this step, before further downstream processing, to verify the virus safety risk level.

11.2.1 Raw Material Safety/Testing

Raw material safety and testing is necessary in all commercial manufacturing of biologics regardless of whether conventional batch or continuous manufacturing procedures are employed. Raw materials that are utilized in typical bioprocesses include, but are not limited to: cell culture media, supplements, single-use assemblies/disposable bags, bioreactors and sampling kits, resins, and other chemicals used in buffer preparation and cleaning solutions. Pathogen safety risk mitigation strategies have been commonplace in downstream manufacturing processes, but they have also become significant areas of focus in upstream processes, driven at least in part by recent biomanufacturing contamination reports attributable to tainted raw materials [10].

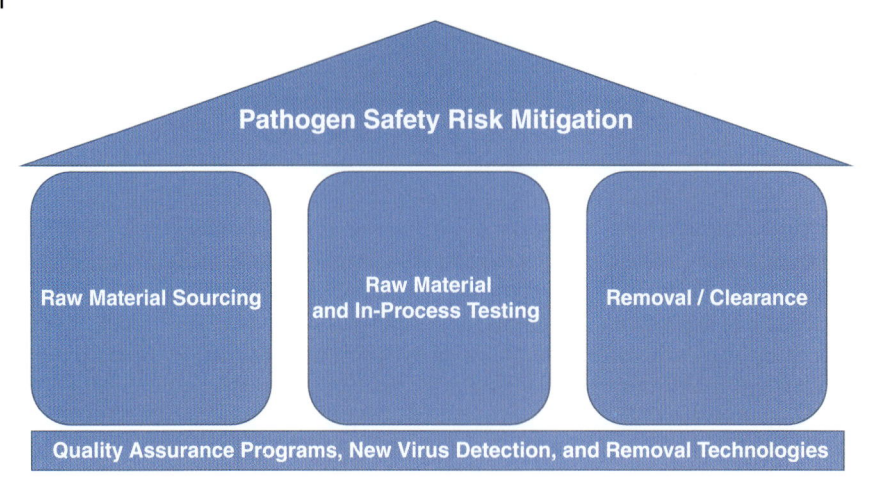

Figure 11.2 Common integrated approaches to virus safety risk mitigation strategies. Raw material sourcing and testing contribute to assurance of the absence of virus, while removal or clearance processes contribute additional risk reduction. Success of this strategy is dependent on the implementation of effective quality assurance programs and virus detection technologies.

Upstream pathogen safety risk mitigation strategies typically rely on three main areas of focus (Figure 11.2). They include raw material sourcing, raw material and in-process testing, and virus reduction or removal tool utilization. Raw material sourcing initiatives focus on selecting low-risk raw materials where animal origin free substances are used where feasible. Historically, recombinant production of biologics relied on the use of cell culture media formulations that contained animal-sourced components, including serum, albumin, and transferrin (among others). Because these animal-derived media components are potential sources of adventitious agents that can cause human disease (such as transmissible spongiform encephalopathies), they are being progressively replaced by nonanimal origin-sourced alternatives where possible. Plant-derived hydrolysates have been successfully used as effective serum substitutes, although their variable composition and performance are drawbacks to their widespread use. In processes where serum cannot be substituted by a nonanimal-sourced alternative, serum and serum-derived products can sometimes be sourced from animals located in areas of low pathogen safety risk (designated geographically based risk 1, or GBR-1 sources) [11]. Some serum-derived components, such as transferrin and insulin, are also manufactured as recombinant molecules, which lower the risk that these raw materials can contribute to introducing adventitious agents to a biomanufacturing process.

More recently, the development and optimization of chemically defined cell culture media of known composition have shown to be effective alternatives to traditional serum-containing formulations. An additional benefit to the known composition of chemically defined cell culture media is that, like disposable bags and sampling assemblies, they can potentially be sourced from suppliers who can provide regulatory-focused dossiers describing raw material specifications, manufacturing process, purity and stability information, and other quality-related information. This traceability can potentially mitigate pathogen safety risk attributable to raw material sourcing.

In addition to effective sourcing, raw materials are often tested prior to, and/or during use in biomanufacturing. Effective testing relies both on the sensitivity of the detecting assay system, as well as on the sample sizes that are evaluated. In practice, only a fraction of the total amount of raw materials used in a given bioprocess are tested. One additional challenge with conducting virus testing on raw materials of animal (particularly bovine) origin is that inhibitory antibodies of target viruses may interfere with virus detection [12]. It is, therefore, crucial to utilize highly sensitive virus (and other adventitious agents) detection assays in order to minimize pathogen safety risk. In addition, virus testing of sufficient raw material sample sizes is an important consideration in any pathogen safety risk mitigation program. Moreover, the amount of virus introduced to a bioprocess that could ultimately result in a contaminated batch (i.e., where virus clearance methods were insufficient to minimize pathogen safety risk to an acceptable level) may be very low.

In addition to minimizing virus introduction to bioprocesses, virus removal steps are commonly incorporated into bioprocessing manufacturing procedures in order to provide additional reduction in pathogen safety risk. Virus reduction steps are commonly used in downstream bioprocessing activities, and include at least two orthogonal methods by which viruses are removed from the product being purified. These downstream virus reduction methods include low-pH incubation, exposure to solvent and/or detergent solutions, and mechanical removal by virus filtration or chromatography. In upstream bioprocessing, incorporating a dedicated virus removal step has recently been a technical area of focus. Several technologies, including exposure of cell culture media to HTST technique, ultraviolet radiation type-C (UV-C), and gamma irradiation (these technologies are described in greater detail elsewhere in this chapter), have been evaluated to reduce virus levels in upstream bioproduction. Evaluation of the impact of these technologies on both virus reduction effectiveness as well as residual cell culture media performance continues to be characterized. More recently, virus barrier technologies (media filtration) have been evaluated in cell culture media filtration in the upstream space to reduce virus exposure risk. The efficiency of virus removal is typically inversely proportional to filter throughput, which makes one challenge the ability to filter virus in large volumes of cell culture media in processing time- and cost-effective manner.

In summary, raw material safety and testing is a multifaceted approach to minimize bioprocessing exposure to pathogens. In addition to maximizing raw material quality assurance, virus reduction methods are employed to maximize the reduction of pathogen exposure risk.

11.2.2 Upstream and Bioreactor Safety

There has been a rapid increase in the demand for recombinant biologics produced from animal cell culture bioprocesses [10]. Monoclonal antibodies (mAbs) have been particularly effective biologically based medications and their use at relatively high concentrations in an ever-increasing number of indications has resulted in the development of more efficient bioprocesses. Recent upstream bioprocessing development activities have focused on utilizing continuous processing to increase productivity. A universal definition of a continuous bioprocess (particularly an upstream one) has remained elusive, but the premise of this biomanufacturing model is that recombinant products generated in a mammalian bioprocess are collected from the production bioreactor (separately from the cells producing the product) as soon after generation as feasible, and subsequently purified using a modified downstream mAb processing template (as described in Figure 11.1).

Batch and fed-batch bioprocesses are based on inoculation of cell culture media with cells capable of expressing the mAb product of interest. The bioreactor contents are then harvested at a predetermined time point, where cells and insoluble debris are separated from the soluble extracellular milieu that contains the monoclonal antibody of interest. In other words, the bioreactor is harvested *en masse* as a batch. In an upstream continuous process, cells are grown in a cell culture media-containing bioreactor, where conditioned media is separated from and fresh media is replenished into the bioreactor, usually at equivalent rates (Figure 11.3). Removal of conditioned media (media exposed to cells and containing the mAb product) is usually accomplished by using a cell retention device, where cell-free conditioned media (perfusate) is removed from cells that are recirculated back into the bioreactor to continue to secrete recombinant product. The perfusate can then be either collected into a surge tank and processed as a single batch once the bioreactor process has been terminated or alternatively, may be processed through the purification train as it becomes available.

Operation of a continuous upstream bioprocess creates additional challenges for implementing effective virus safety risk mitigation strategies. In conventional batch or fed-batch processes, media and feeds (when relevant) are generally

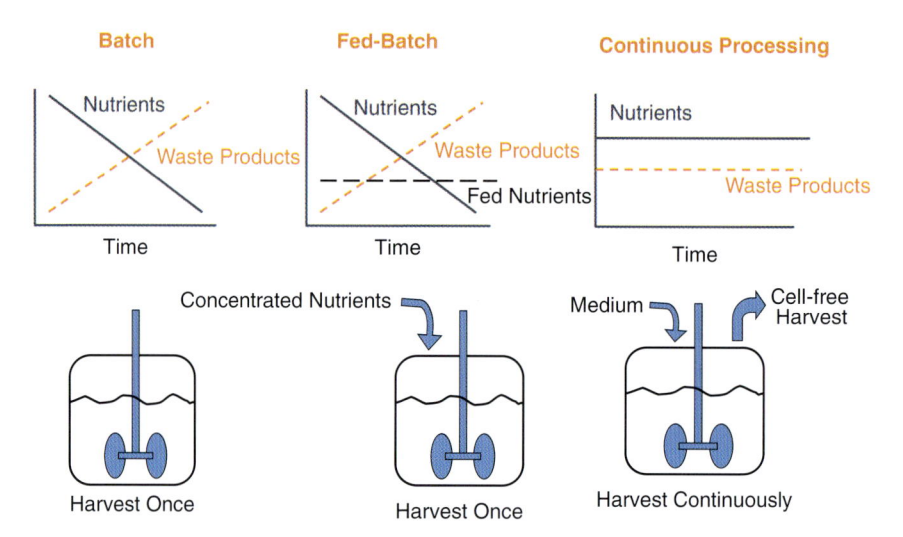

Figure 11.3 A schematic comparing typical batch, fed-batch, and continuous processing modes of operation.

reconstituted and sterile filtered close to the point of use. Moreover, the volume of feeds and/or media requiring sterile filtration is approximately equivalent to the bioreactor volume at which the upstream bioprocess is executed. In contrast, media requirements for a continuous upstream bioprocess, where media exchange rates range between 1 and 2 bioreactor vessel volumes per day, can exceed 20–200 times the working bioreactor volume [11]. From a virus safety risk perspective, these larger media volumes can increase the possibility that a virus is introduced to the process. Historically, viral clearance has been quantified, from a regulatory perspective, in units of log reduction value (LRV). This concept relies on the reduction in the probability of virus exposure, not on the actual elimination of all viruses from a process. Based on this LRV concept, the potentially large increase in media volume requirements in a continuous process increases the risk of virus introduction to the upstream process.

One challenge with implementing an effective virus safety risk reduction strategy in upstream continuous bioprocessing is that the continuous nature of bioprocessing may make virus detection problematic while the process is still in progress. Specifically, conventional virus detection assays require processing times that prevent results from being available prior to subsequent processing of the harvest material, which either requires the process to be continued at risk or that the harvest material be stored prior to subsequent processing until after virus testing results are obtained [11]. Several technologies have been evaluated to enhance upstream virus safety risk reduction and include HTST media exposure, ultraviolet-C or gamma irradiation exposure, and virus barrier filtration. The details of these technologies are described in greater detail in this chapter, but the focus of these technologies has been on reducing process risk, and not on developing additional virus reduction claims. Because continuous processing relies on the utilization of relatively large volumes of cell culture media, upstream processing will likely require *just in time* media generation due to facility constraints in large media volume storage. Technologies used to reduce virus safety risk will not only have to be effective in doing so, but also will need to accommodate prescribed processing time requirements as well as be cost-effective.

An additional consideration in developing an effective virus safety risk reduction strategy in continuous upstream bioprocessing is related to the bioreactor conditions under which recombinant products are produced. Cell densities in conventional fed-batch bioprocesses are typically observed in the $15–30 \times 10^6$ cells/ml range, whereas cell densities greater than 80×10^6 cells/ml are common in some continuous upstream processing applications. CHO cells have previously been characterized demonstrating that they express endogenous retroviruses [12]. Some known viruses, such as *murine leukemia virus* (MLV) as well as other noninfectious virus particles, have been detected by electron microscopy in mammalian cells used for recombinant product generation (including CHO cell and hybridomas). Despite a lack of evidence connecting murine retroviruses and human disease, the potential for oncogenic agents to contaminate a biotherapeutic preparation raises regulatory concerns in biomanufacturing. As such, virus safety risk strategies are important steps in bioprocessing. In addition, given the higher cell densities typically utilized in continuous bioprocessing, a higher retroviral load should be considered when viral safety risk reduction schemes are being developed.

Table 11.1 Common viral reduction and inactivation unit operations and their respective effectiveness in batch downstream purification of monoclonal antibodies [13].

Technology	Primary clearance mechanism	Typical capabilities	Comment
Filtration	Size exclusion	3–6 LRV (Parvo) +6 LRV (MuLV)	Robust and easy to implement
Chemical: • Low pH • Solvent/Detergent	Inactivation	+6 LRV (MuLV)	Simple, robust but can denature proteins
Chromatography: • Packed bed • Membrane adsorption	Adsorption	0–6 LRV	Process specific
Heat (Pasteurization)	Inactivation	No data	High-temperature short-time (HTST) technique used for cell culture media processing. Mandated for albumin or plasma processes
Ultraviolet (UV-C)	Inactivation	2–6 LRV	Masking due to protein, virus dependant can denature proteins
Gamma irradiation	Inactivation	2–6 LRV	Batch process. Can denature proteins

11.2.3 Downstream Virus Removal Strategies

As stated in previous sections, a purification strategy must contain a number of orthogonal viral reduction and inactivation steps to ensure both patient safety and regulatory compliance. The information provided in Table 11.1 represents *typical* values obtained during the validating cycle of downstream unit operations used within the production of therapeutic monoclonal antibodies. The efficiency of the viral clearance capacity of these steps when run in a connected and continuous fashion remains to be fully characterized; in this section, this will be discussed in more detail.

The effectiveness of a process step for viral clearance is quantified in terms of its log removal value (LRV). This is defined as

$$\text{LRV} = \log\left(\frac{C_{\text{feed}}}{C_{\text{perm}}}\right),$$

where C_{feed} and C_{perm} are the inlet and outlet virus concentrations, respectively. Several commercial virus clearance technologies and typical virus LRVs are summarized in Table 11.1.

11.2.3.1 Viral Reduction by Normal Flow Filtration (NFF)

Normal flow filtration (NFF, or sometimes called dead-end filtration) operates on the principle of size exclusion, and is a physical and mechanical separation of

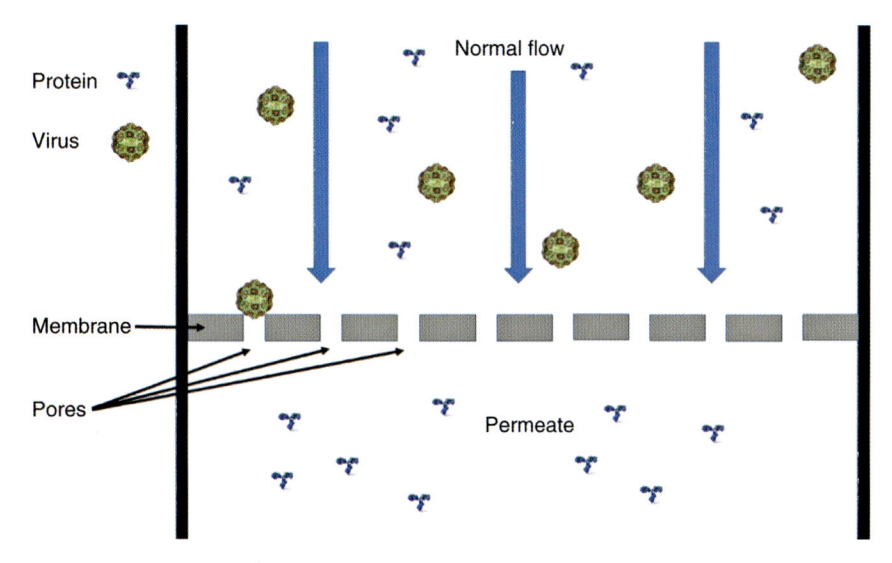

Figure 11.4 Principles of membrane-based normal flow filtration.

particles from fluid by passage through a permeable medium where the fluid flow is perpendicular to the membrane surface (Figure 11.4). In the case of virus reduction filtration, a tight exclusion limit is required from the membrane (typically in the region of 50 to 20 nm) in order to effectively remove viruses or virus-like particle from the processing stream and allow the passage of proteins.

During process development, selecting, implementing, and properly optimizing a virus filtration process, as well as establishing process robustness, are important considerations. Processing variables that impact virus retention (LRV), product recovery, and product throughput also should be evaluated. This is the typical approach for a batch process and would also be applicable when developing a viral filtration step for a continuous process.

When developing a viral reduction filtration strategy, in either continuous or batch platforms, there are a number of locations within the purification scheme where a virus reduction filtration step could be implemented. Several factors may influence the decision where to place the virus reduction filtration step(s) within the production process. These include the following:

- *Protein concentration:* Lower protein concentration of the feed material used at the filtration step can reduce the fouling of the membranes, resulting in reduced membrane surface area requirements and ultimately lower cost of production.
- *Buffer compatibility:* pH and conductivity can both directly influence the efficiency of a viral reduction filtration step.
- *Facility fit:* A significant proportion of manufacturing facilities have segregated pre- and postviral reduction production areas. The facility size and logistics of facility operations may influence the placement of the viral filtration step(s).
- *Incoming raw materials:* The further down the purification process, the less additional raw materials will be introduced into the production process, thus reducing the risk of an unforeseen external contamination.

Filters are broadly classified into two categories: filters that provide >4 or >6 $\log_{10}$ removal of large viruses (typically 80–100 nm endogenous retroviruses), and filters that provide >4 $\log_{10}$ removal of small and large viruses (larger than 18–24 nm parvoviruses)[5]. A well designed viral reduction filtration step in a typical monoclonal antibody production process should be validated to minimum of a 4 $\log_{10}$ removal of parvovirus, which is considered a robust removal step and generally accepted from a viral reduction filtration step for a classical mAb purification process. In addition, a virus filtration unit operation should enable the following:

- *Highly reproducible effectiveness in virus clearance that can be validated:* A log reduction >4 is a requirement for a robust viral reduction filtration step.
- *High protein mass capacity:* Virus reduction filters are expensive. A high mass capacity enabling the processing of a large mass of protein with the minimum of filter area is required to reduce operational costs.
- *Scalable processing times (less than 4 h):* Efficient throughput and a robustly scaled process are key requirements for any purification unit operation.
- *Disposable flow path at all scales:* Reducing the requirement for cleaning validation and decreasing processing times.
- *Effective cleaning using caustic solutions:* To allow efficient and effective cleaning and depyrogenation of the membranes.
- *Ease of installation, use, and execution of integrity test:* A common cause of failure is due to operator error. Simplifying the unit operation and reducing complexity will reduce the chance of operator error such as incorrectly installed membranes leading to failed integrity tests.

Viral reduction filters are available from a number of suppliers with a variety of membrane materials of construction in a variety of formats and sizes from small-scale validation kits to commercial production scale production filters and housings.

There are a number of additional considerations when applying a viral reduction filtration step to a continuous process. Filters have a finite lifespan meaning only a given mass of protein can be processed on a specific membrane surface area. The filter should not be pushed to or beyond this capacity as a postuse integrity test is still required after completion of the unit operation. It is generally considered as good practice by filter suppliers to not overly foul the filters and that as the mass a filter can process is product/process specific, the filter should be sized accordingly. These points should be considered when implementing virus filtration into a continuous process.

Worth noting is the fact that all of the continuous processing examples in Figure 11.1 show the placement of the virus reduction filter at the end of the process, post polishing/prior to formulation. It does not appear that the feed concentration in these processes would vary with the virus filter, so validation would be the same as the batch processes. If one wanted to include the virus filter post CEX bind and elute, and surge bags did not smooth out the eluate conditions, the virus filter would see varying concentration to it. Virus filters, in general, can be limited by high feed concentration that slows the flux due to polarization and can foul the filter more,

---

5 https://www.emdmillipore.com/US/en/ps-learning-centers/virus-safety-learning-center/assessment-of-needs/KJub.qB._tAAAAFAF1IENH5P,nav.

due to higher presence of aggregates [14]. Again, if the continuous process were designed this way, in-line virus injection could be used to assess the virus filter efficiency or LRV, even with the changing eluate concentration coming to it. Process development and pilot-scale data using a connected CEX bind and elute and a virus filter would provide proof of concept and proof of robustness for this kind of process, and validation using in-line virus injection would show how the LRV might be impacted, compared to the straight batch approach.

Another technical consideration for virus filtration in continuous processing is that viral reduction filters are generally run at constant pressures between 1 and 4 bar while allowing the flow to decay over time, and not at a constant flow where the pressure increases over time. However, it is possible to run virus filters in a constant flow configuration, this should not pose any technical difficulties or affect the viral reduction capacity [15]. However, if slower filter flow rates or process interruptions are utilized, it should be noticed that these can have an impact on the LRV (slower flows, stops, and starts can show lower LRVs for some virus filters) [16], and should be taken into consideration when devising virus filter validation plans.

Given that the filters have a finite lifespan, how the filters are run in continuous mode without introducing process interruptions should be considered. Oversizing the filters for the duration of the continuous process is one option that may not be cost effective nor sustainable. The simplest option would be to manifold a number of filters in parallel and switch to a secondary filter at a predefined point in the production process. This would allow for the filters to be prepared offline of the process (CIP, flush, and preuse integrity test) prior to usage and without process interruptions. This would also allow for the used filters to be postuse integrity tested offline as well, and reduce the processing risk associated with an unknown virus filter postuse integrity status. An example of a filter manifold system is shown in Figure 11.5.

In this example, the point of the process cycle in which each filter is at will change over time, for example, equilibration, process, and wash, but the premise is to have one filter running, a second filter is pre-prepared and ready to take over the process when a predefined set point is achieved on the current *"running"* filter, and a third filter is used in either pre- or post-use preparation (i.e., CIP, flush, and integrity test). It would also be possible and perhaps preferential under some circumstances to have the pre-use filter preparation and post-use integrity test performed offline, even in a different production area. This proposed filter configuration could accommodate these processing requirements. Consideration also needs to be given to the start-up and shutdown phases of these filters, and if the product and flushes processed during these phases can be recovered or discarded or if only material processed at *"steady state"* constituted material is taken forward through the next processing stage.

Viral filters are also susceptible to performance fluctuations due to changes in process conditions. Therefore, it is generally regarded as good practice to maintain a consistent load material throughout the duration of the filtration step that may not be the case if the filtration step is preceded by a bind and elute cation exchange chromatography step with no break tank to collect the elution peak and ensure homogeneity of the product. Specifically, protein concentration, pH, and conductivity should be kept as consistent as possible. It has also been shown that with some virus reduction filter, process interruptions may reduce viral clearance efficiency.

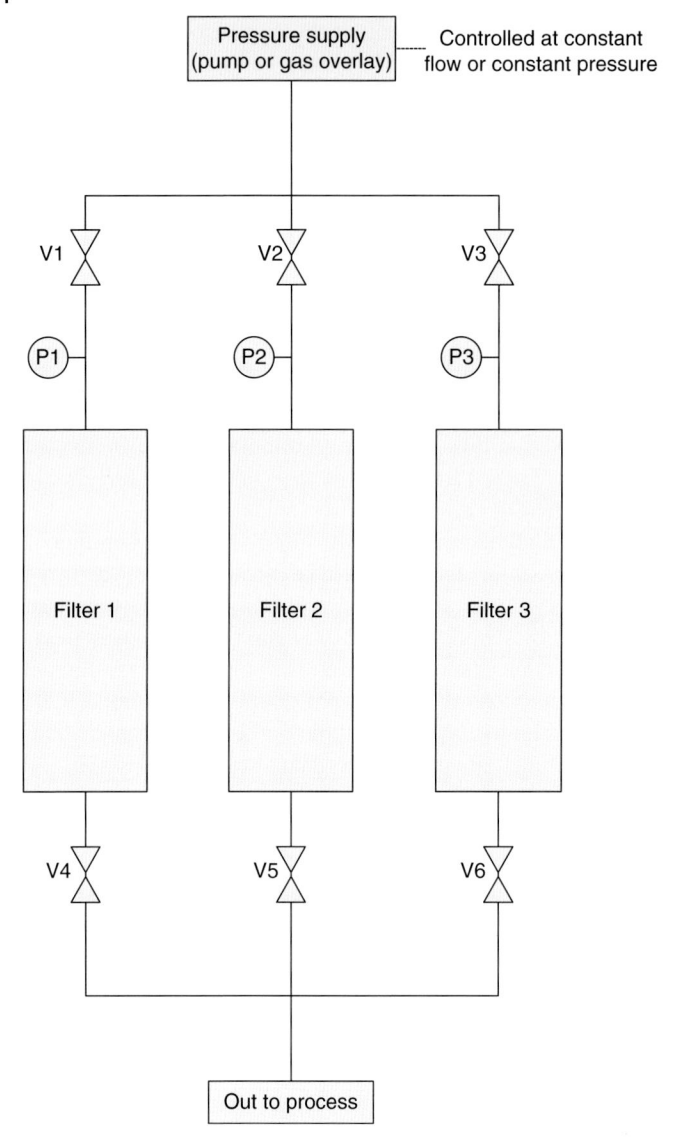

Figure 11.5 General, simplified example of a virus filtration manifold for continuous processing.

These requirements need to be considered when connecting an upstream unit operation (such as a chromatography step) to a viral reduction filtration step.

11.2.3.2 Chemical Inactivation (Low pH or Solvent Detergent)

Low-pH Inactivation
A low-pH hold step is a common viral inactivation methodology employed in the majority of mAb production processes. Typically, the mAb protein is captured from the clarified bioreactor supernatant using protein A affinity chromatography. To

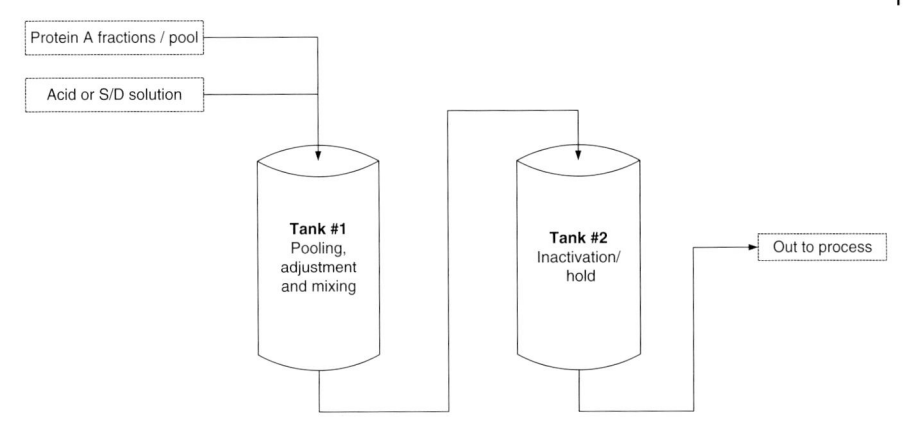

Figure 11.6 Simplified low pH or S/D viral inactivation manufacturing set-up for batch manufacturing.

elute the mAb from the column, a low-pH buffer (citrate or glycine) is commonly used. This means that the mAb pool eluted from the column is typically at a pH between 3.5 and 4.5 already, so with minor adjustment, lowering the pH to between 3.0 and 4.0 required for effective inactivation is a relatively simple, cheap, and effective way of inactivating enveloped retroviruses.

The protein A column eluate is collected directly from the column into a mixing vessel. The pH is then checked and lowered to the desired range via the addition of an acid (typically acetic acid). Once the eluate is within the desired pH inactivation range (typically between 3.0 and 4.0, more commonly 3.75), the process solution is mixed for a given length of time to ensure homogeneity. The process solution is then transferred to a second mixing vessel for the timed inactivation period. The purpose of liquid transfer to a second vessel is to ensure complete inactivation of the process fluid (i.e., complete homogeneity where no splashes or droplets not in contact with the mixed bulk escape the inactivation process (Figure 11.6)).

The ease of implementation of this step, requiring only appropriate mixing vessels, calibrated pH measuring equipment, and an acid/buffer dosing system, makes it an extremely attractive option for virus inactivation. However, there can be drawbacks with this application. Some proteins may not be stable for prolonged period of low pH leading to degradation or aggregation of the product. Precipitation may also occur when the pH is raised postinactivation, leading to loss of product. Host cell proteins (HCPs) may also precipitate from solution, which may reduce the HCP content of the process solution, but will also likely require an additional clarification step in the process in order to protect the unit operations further downstream.

When developing and evaluating the effectiveness of low pH on virus inactivation step, an accurate and predictable scale-down model should be employed. A number of variables can have an effect on inactivation kinetics, such as shear force, temperature, inactivation time, and pH, so critical parameters, such as vessel geometry, mixing time, mixer type (overhead/bottom mounted impeller versus magnetic stirrer and beaker), should be maintained. The consistency of the process stream is also important, and the pH and inactivation time must be monitored and controlled throughout the unit operation with suitably calibrated equipment.

Solvent Detergent (S/D) Inactivation

As stated above in Section "Low-pH Inactivation", low-pH viral inactivation may not be possible for all applications due to protein instability. There is, however, an alternative; developed in the 1980s for the plasma industry, chemical virus inactivation with solvent detergent (S/D) treatment. This technique works by disrupting the lipid membrane surrounding enveloped viruses, thus rendering the viruses noninfective. Common solvents used for the solvent detergent inactivation step include tri-n-Butyl phosphate (TnBP), Tween® 80 (Polysorbate), or Triton® X-100 with a typical final inactivating solution being a (final) mixture of 1% TNBP and 1% Triton X-100. It is important to note that as with low-pH inactivation, this method is not effective for nonenveloped viruses (such as parvoviruses), and there may be loss of product since additional steps of removing S/D must be introduced in the process once the virus inactivation step has been completed.

In practice, the S/D inactivation procedure is similar to that already described for the low-pH inactivation, differing in that the pH of the protein A eluate may be increased by the addition of a basic buffer/solution to increase protein stability. A solvent detergent mixture is then added, mixed, transferred to the inactivation vessel, and held for the appropriate duration to ensure complete viral inactivation. As with the low-pH inactivation step, a suitable scale-down model is required for process development, where critical control parameters are monitored and maintained.

Chemical Inactivation in Continuous Processing

Chemical viral inactivation is well suited to batch processing. Generally, the eluate from the protein A column is collected into a pooling tank. Either the pH is lowered to the desired range or a solvent detergent solution is added and then mixed for a predetermined time. Once the solution is homogeneous, the process solution is transferred to the inactivation tank and held at a set mixing rate and temperature for the predefined time (typically 60 min) to ensure complete inactivation of the process solution eliminating risks of splashed liquid on the tank walls/head plate, inadequate mixing in tubing and piping, dead zones in the process tank, and so on.

There are a number of potential technologies that could be applied for a continuous chemical viral inactivation strategy. The preferred option is a flow cell with in-line pH/chemical modification capabilities, turbulent (high shear) flow, and a defined residence time. There is some speculation that, dependant on design, this type of flow-through device may, in fact, be more effective at inactivating viruses due to shear effects within an efficiently designed flow cell. A flow cell may be as simple as a section of tubing of a defined length run at a specified flow rate to give a predetermined residence time of the process solution. Static mixers may also be used in a continuous process, similar to batch processes.

Another potential option would be to have multiple process streams aliquoted into smaller pooling and viral inactivation vessels. While one is filling, a second tank could be inactivating. The use of single-use mix technology would mitigate the requirement for CIP/SIP of tanks in between fill/inactivation cycles. With either approach, and as with batch processes, consideration should be given to the development of an accurate and representative scale-down model for validation studies.

11.2.3.3 Chromatography

While an appropriate LRV may be achieved using an orthogonal approach of filtration and chemical inactivation for early clinical phases of the drug development life cycle, biologicals moved toward clinical phase III and commercial production are required to increase the safety factor by validating the chromatography steps for their viral clearance capacity.

There are several iterations of the way in which chromatography techniques may be applied in continuous processing. These include, but are not limited to, the following:

- *Multicolumn*: Using a defined sequence between 2 and 4 bind and elute chromatography columns, where columns are switched between the loading step/nonloading steps with column cycling through equilibration, loading, wash, elution, and clean-in-place (CIP) and/or regeneration sequences.
- *Simulated moving bed*: Also has a multitude of permutations but the basic premise is that a standard SMB unit operation consists of four or more bind and elute chromatographic columns connected in a semiclosed-loop fashion, with two inlet streams for desorbent and feed and two outlet streams for waste and purified protein, respectively [17].
- *Flow-through*: Process-related impurities are bound and the molecule of interest, in this case a monoclonal antibody, flows through the chromatography matrix unimpeded.

Designed and developed primarily for the removal of closely related product impurities such as host cell protein and DNA, both enveloped and nonenveloped viruses also can be separated from the protein of interest by chromatography techniques with varying degrees of effectiveness. The viral clearance capacity of the chromatographic step varies, and is dependent on the physicochemical and biochemical properties of individual viruses, and has to be assessed on a case-by-case basis.

A typical monoclonal antibody purification scheme consists of three chromatographic purification steps. The capture step, usually protein A affinity, an intermediate purification step, and a final polishing step. Typically the intermediate and final polishing steps are ion exchange (cation or anion), hydrophobic interaction (HIC), or a mixed mode chromatography step.

There are a number of iterations but the simplest and most widely adopted purification strategy for monoclonal antibodies is the following:

Protein A → Cation exchange (bind and elute) → Anion exchange (flow through).

Each of the chromatographic steps has varying degrees of viral clearance capacities dependant on separation chemistry and buffer compositions used in the separation process.

There are some differences how protein A capture columns could be run in continuous processes, namely, in multicolumn or SMB modes. In both multicolumn and SMB chromatography, columns are typically smaller, run at higher loading (to breakthrough), run faster, and cycled rapidly. For ease of use, it is preferable to run the CEX and AEX polishing steps in flow-through mode, as either packed bed volumes or disposable membrane adsorption devices, where practical. New and novel separation techniques, such as activated carbon flow-through, are becoming more widespread. All of these should be taken into consideration from a viral clearance capabilities perspective.

It is worth noting that the use of disposable membrane technology in a flow-through format offers many benefits, such as the elimination and the requirement for sanitization and resin reuse studies. Prepacked columns available from most vendors should also be considered to reduce plant processing time and remove column packing requirements from the production facility.

With the exception of operating columns or membranes in flow-through mode, there is currently very little information or literature available with regards to the impact on viral safety of the bioproduction process when moving from classical batch chromatography to the more novel chromatography techniques, and the methodologies currently being employed in continuous processes. As the techniques mature, become more defined and more widely adopted in the industry, more information on viral removal efficiency of these chromatography steps reported from both suppliers and biologic manufacturers should become available.

11.2.3.4 Other Techniques

Inactivation by Ultraviolet Radiation (UV-C)
Ultraviolet-C (UV-C) radiation produced by a low-pressure mercury vapor lamp at a wavelength of 254 nm and a reaction chamber are used to generate photochemical reactions that alter molecular components essential to cell function (i.e., damaging nucleic acids) to inactivate viruses while maintaining structural and functional integrity of the protein of interest. Protein irradiation level expressed as $mJ/cm^2/min$ is dependent on the optical density of the test sample and the average irradiance emitted by the lamps. The residence time of the protein in the reaction chamber, therefore, must be optimized for each protein/buffer system.

This technique has been shown to be effective against both retro and parvoviruses. However, UV-C radiation has been shown to denature proteins through photolysis of disulphide bonds or the generation of reactive species resulting in protein/thiol oxidation, and therefore, is not widely used for the commercial production of mAbs.

There are commercially available UV-C viral inactivation devices. However, in practice, this technique has not gained much traction within the therapeutic protein industry, but researchers are exploring the application of this technology for treating cell culture media used for recombinant protein production as a means for mitigating viral contamination risk (see Section 11.2.1). Should UV-C treatment become a viable virus reduction option in bioprocessing, its implementation into continuous processing modes of operation will need to be elucidated.

Heat Treatment/Pasteurization

Although the application of heat can be a simple and effective method of viral inactivation, it is not a common technique used in the bioproduction of mAbs. This is due to the fact that a significant proportion of proteins are denatured when heated to >60 °C in liquid form [18]. For example, the general guidelines state that albumin solutions must be heated as a liquid at 60 ± 0.5 °C for 10–11 h continuously [18]. Protein stabilizers can also be used during pasteurization process. These may include amino acids such as N-acetyl-DL-tryptophan and glycine, sugars such as sucrose, citrate, sodium caprylate, or caprylic acid (octanoic acid). However, it is important to consider the effect of the stabilizer on the virus too as some excipients may also stabilize viruses.

In monoclonal antibody production, heat treatment in the form of HTST techniques are most commonly applied upstream as a cell culture media treatment as part of an overall viral risk mitigation strategy for the process (Section 11.2.2).

Gamma Irradiation (20–40 kGy)

Gamma irradiation is not commonly used in bioproduction, but it is worth mentioning, as the increased reliance on enabling technologies such as single-use systems, bags, and assemblies in continuous processes require a high degree of control with respect to raw materials entering the process. Gamma irradiation of various raw materials reduces the risk of both microbial and viral contamination. Typically, single-use components are sterilized by irradiation using γ-rays. Gamma irradiation of raw materials requires the items to be sent offsite to specialized companies, as the process must be strictly controlled to ensure consistent, reproducible, effective, and efficient virus inactivation.

11.3 Validation of Viral Reduction Steps in Continuous Manufacturing Processes

Due to the multiple configurations of continuous processes that are currently being investigated by biopharmaceutical companies, it is difficult to comprehensively define a prototypical process, along with its associated raw materials and process parameters. However, it appears that it may not strictly be necessary to run a continuous connected process where material flows from one unit operation to the next with absolutely no collection nor process interruptions or holds. Regardless of perfectly balanced flows in between every unit operation, the drug product would see much less hold time (standing still) in the continuous process. Continuous processing may impact the viral inactivation and removal capabilities because of these higher flow rates. Another question that arises in considering a virus safety strategy in continuous processing is how to best mimic a potential virus contamination when the downstream process is under constant flow.

The process feed from a continuous process may differ from that generated in a batch process in that it may not have constant characteristic properties over time. Indeed, the concentration of the virus filtration feed, for instance, may change over time, especially if filtered immediately after becoming available from elution of an upstream chromatography step, where multiple columns will likely be utilized. In this scenario, alternative virus validation methods could be used, like in-line virus spiking for instance: the virus spike is added continuously to the eluate in an in-line mixer as it is eluting from the bind and elute step, then this spiked feed goes into the next unit operation (flow-through chromatography or virus filter). This will be described more in detail in Section 11.3.4.

To mimic as closely as possible the manufacturing process, the validation may have to be performed with the previous step in-line. For instance, to validate the AEX step, the CEX bind and elute may have to be performed, so that the product that enters the AEX is representative of large-scale operation.

Another way to perform the virus validation could be to validate the worst cases of the process. For instance, if protein concentration is considered a variable, then the step will have to be validated using the highest concentration or worst case/most challenging conditions that might be reached during the step. If this step is able to remove viruses in these worst conditions, then it will be able to remove viruses in all conditions. However, this approach involves defining and demonstrating what is considered worst case at each process step, and how these impact the steps and would require a high concentration feed stream for the validation studies that may not be feasible/practical.

11.3.1 Protein A Capture Chromatography

The first unit operation that is validated for virus removal is the protein A step. As stated in Section 11.2.3.3, the main difference between a batch and a continuous process is the way in which the column is run (80% of dynamic binding capacity and lower flow rates for the batch mode versus saturation of the columns and faster loading flow rates for the continuous mode).

There may be the following two options to run the virus validation for this step:

- *Option 1:* Perform the validation with a single column running it multiple times using the same process conditions as utilized at large scale, for example, the same buffers and the same flow rates, and check if the efficiency of the resin to remove the viruses is adversely impacted. It should not be necessary to validate all columns in the system, however, it is worth noting that but when running the chromatography in continuous mode, each column in the system is first loaded with the breakthrough from the previous column, which would not be the case here and it would remain to be proven that this option is representative of what happens at manufacturing scale.
- *Option 2:* Perform a scale-down model of the continuous chromatography process running it with a representative product using the same sequences and operating conditions as that used at large scale. Analysis of the flow-through, washes, and the elution pool would be required to calculate the virus mass balance.

With both options, it will be necessary to prove that the resin cleaning steps are efficient and the column performance does not determinate over time. The resin used for the validation studies can be a new, used, or both. Defining the number of cycles for the resin lifetime is important but the considerations regarding column lifetime would still be the same as with batch column operation: As with batch processes, resin lifetime studies would show when the column resin would need to be replaced.

11.3.2 Chemical Inactivation (Low pH/Solvent Detergent)

The second unit operation that is validated for virus reduction is the low-pH (3.75) hold intended to inactivate enveloped virus. Depending on how the continuous processor performs this step (static in a bag or a tank, similar to batch method versus flow in a flow cell with the proper residence time), a representative scale-down model, especially for the more uncommon flow process would need to be qualified. This would include sampling points along the flow path length to assess the consistency of virus concentration, and confirm that all the parameters that can impact virus inactivation are constant (stable pH, stable temperature, etc.). If the dynamic method is chosen, the virus inactivation could be faster since there would be an increase in inactivation kinetics due to better mixing. Shear force may also decrease the virus inactivation time.

11.3.3 Intermediate and Polishing Chromatography

As described in Section 11.3.1, the same discussion and questions for SMB or multicolumn protein A bind and elute would also apply to SMB or multicolumn CEX bind and elute.

The AEX flow-through validation would be the next unit operation to consider. The main question is what the difference in operation (feed to the unit operation) between the strictly batch processing and these continuous variants would be. For the scenario with CEX bind and elute into flow-through AEX, the concentration and the buffer conditions may change throughout the process. In this case, the use of surge bags and planned staggering of pump stops and starts could smooth out any processing changes. These general questions are applicable to the unit operations after the two bind and elute steps (protein A and CEX if used as bind and elute).

If the continuous process had a CEX elution peak with changing concentration entering directly into an AEX flow-through (with no CEX elution pooling), an in-line method could be used, meaning that the product is spiked continuously into the eluate as it is coming off of the bind and elute step, then passed through a mixer to make it homogeneous, and loaded into the AEX. A coupled virus spiking study as already described would accurately determine the LRV, and if it were different from the batch method (elute into a collection tank first, then homogeneous feed run on AEX flow-through).

As an example, Merck & Co. published their method explaining that small surge bags can be effectively used to smooth out changing conditions during column elutions and the subsequent processing steps [6]. Over an extended period of time (20 h), the cyclical changes in pH coming from the multiple protein A elutions are dampened in the surge bag. This means that the feed in this surge bag is approximately homogenous prior to the next *continuous* step (acid addition and residence time for virus inactivation). This method should also work with a CEX bind and elute SMB system eluting multiple times into a surge bag. The concentration and solution conditions in the surge bag would be approximately homogeneous and ready to be processed on the AEX flow-through. This would make the validation process as similarly straightforward as the traditional batch process. Running coupled validations could also be an option, one step (eluting CEX) right into the other (AEX flow-through). This is an acceptable technique as long as the validation mimics the exact conditions used in the continuous manufacturing process. If surge bags are effective at smoothing process transitions and inconsistencies, coupled validation of two or more unit operations run together may not be necessary.

11.3.4 Viral Reduction Filtration

One of the last steps in the process that impacts virus safety is virus filtration. All of the continuous processing examples show its placement at the end, postpolishing and prior to formulation (see Figure 11.1).

For the validation of a batch virus reduction filtration step, the virus filter is run in decoupled mode, as it is considered to be worst case. The feed is passed through the adsorptive filter first and then the virus is spiked into the pool and run on the virus filter. This is because the adsorptive prefilter could remove some viruses and only the virus filter has to be validated. However, sometimes the product to be filtered can be very sensitive; for example, aggregates may form, foul the virus filter, and make it nonrepresentative of the larger scale manufacturing process.

As an example and as shown in a collaborative paper from EMD Millipore Corporation and Amgen, a validation strategy was devised that couples an adsorptive filter used for aggregate removal and a parvovirus filter [19]. In initial validation attempts, the feed was passed through the adsorptive filter first in batch mode, then the virus was spiked into the pool and this was run on the virus filter as already described. But aggregates formed and fouled the virus filter. The solution that was used to overcome this situation was to run the adsorptive filter and the virus filter in-line and inject virus continuously into the stream leaving the adsorptive filter (Figure 11.7). Then this spiked feed went through an in-line mixer, and then a sample was drawn out prior to the virus filter to measure the actual virus challenge to the virus filter.

This arrangement allowed for the required aggregate removal to protect the virus filter, the needed virus addition, and the needed sample prior to the virus filter so that the virus LRV could be assessed. And this was done in an in-line mode with a similar fouling profile as the large scale manufacturing process.

This is an important part of the validation process, the scaled down model used in validation should match as closely as possible the manufacturing scale. This same method could also be used if the continuous process had a CEX elution peak with changing concentration entering directly into an AEX flow-through with no break. A coupled virus spiking study as described would accurately determine the LRV and if it were different from the old batch method (elute into a collection tank first, then homogeneous feed run on AEX flow-through).

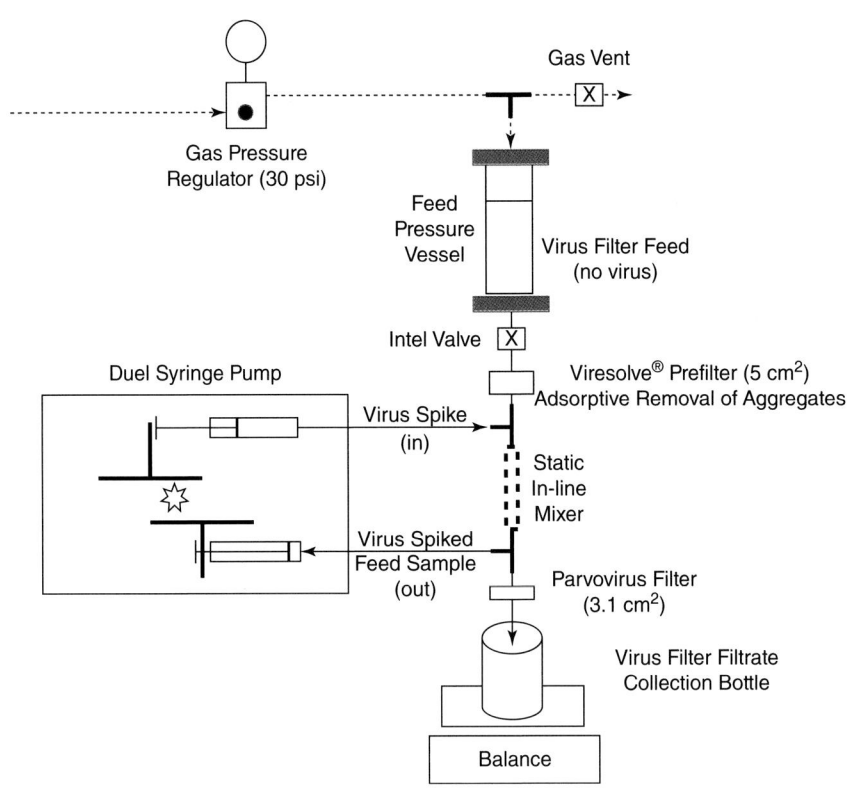

Figure 11.7 Schematic for in-line virus injection validation method for coupled unit operations. For the decoupled virus spiked runs ∼ 310 ml were spiked with 0.1% of virus to a concentration of ∼ 2×10^6 MVM/ml. So 0.31 ml of MVM virus preparation at 2×10^9 MVM/ml ($0.31 \times 2 \times 10^9 = 310 \times 2 \times 10^6$). For the in-line virus injection method, the same total virus was added to ∼ 16 ml of buffer and loaded into a syringe (0.31 ml $\times 2 \times 10^9$ MVM/ml $= 6.2 \times 10^8$ MVM into ∼ 16.31 ml $= 3.8 \times 10^7$ MVM/ml in the syringe). The starting syringe flow rate was 8 ml/h or 0.13 ml/min. The staring flow out of the VPF was ∼ 2.5 ml/min, so this will dilute the injected virus by ∼ 19×, (6.2×10^8 MVM/ml/19 $= 2 \times 10^6$ MVM/ml). As the flow out of the VPF dropped the rate of injection was adjusted proportionately to keep the ∼ 19× dilution (1.25 ml/min VPF matched with 0.065 ml/min or 3.9 ml/h injection, for example, and 0.625 ml/min VPF matched with 0.0325 ml/min or 2 ml/h, for example, and so on). So after 174 min, 300 ml of feed was processed through the VPF in-line with the Vpro and >14 ml of the syringe load was added. 14 ml $\times 3.8 \times 10^7$ MVM/ml $= 5.32 \times 10^8$ MVM/300 ml $= 1.8 \times 10^6$ ∼ 2×10^6 MVM/ml.

If the feed concentration in these processes does not vary with the virus filter, then the validation should be performed similar to that done for batch processes. As stated in Section 11.2.3.1, if one wanted to include the virus filter post CEX bind and elute, and surge bags did not smooth out the eluate conditions, the virus filter would see varying conditions. Virus filters in general can be limited by high-feed concentration that slows the flux due to polarization and can foul the filter more due to higher presence of aggregates (see Figure 11.8). Again, if the continuous processes were designed this way, in-line virus injection could be used to assess the virus filter LRV even when exposed to variable eluate concentrations. Process development and pilot-scale data using a connected CEX bind and elute and a virus

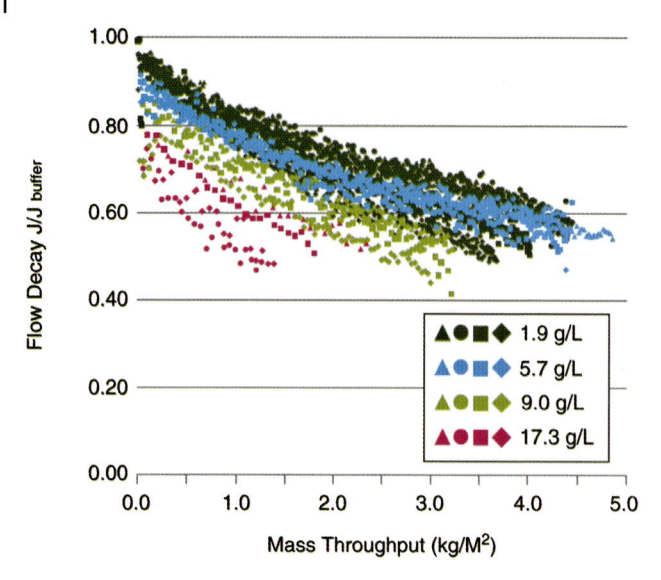

Figure 11.8 Virus filter flow decay versus mass throughput at various feed concentrations.

filter would provide proof of concept and proof of robustness for this kind of process. Validation using in-line virus injection would show how the LRV might be impacted, compared to the conventional batch approach.

11.4 Conclusion

In summary, the specific parameters of continuous/connected processing vary from manufacturer to manufacturer. Several process parameters, including the method of cell culture, method of cell separation for perfusion, concentration of pre-protein A drug product, the type, the mode (bind and elute or flow-through) and order of the polishing steps, and the implementation of fully disposable infrastructure, can differ not only from batch processing, but also between various continuous processes. These differences in processing require developers of biomanufacturing processes to be mindful of the impact of these process modifications on virus safety risk mitigation activities

If a perfusion process is used, where media consumption is typically greater than batch, there may be an increased risk of virus contamination. Past contaminations have shown virus introduction with the media. If this is true, there may be a stronger driver to use virus barrier filters wherever possible. The fact that the drug product moves more quickly downstream might present a greater risk when virus is not detected early in the upstream. New, more reliable, and faster virus detection methods should be developed to reduce risk.

The use of a fully disposable and closed continuous process could also help reduce decontamination efforts if a virus contamination did occur and was not caught early in the bioreactor. One aspect that may reduce the risk of virus contamination is the introduction of process automation. Continuous processing may have lower risk to virus contamination because there is less personnel interaction in the process.

If surge bags are used, especially after bind and elute chromatography steps, coupled in-line virus injection validation may not be needed. However, if feed concentrations are changing when entering to a specific unit operation, the validation could utilize the in-line virus injection method. The general principles of validation are still fundamentally valid: A qualified representative scale-down model for the unit operation should be used. Representative feeds should be used and all operating sequences and operating conditions used in the full-scale process should be mimicked during validation. Validations with the higher chromatography flow rates and higher loadings to saturation should demonstrate the impact, if any, to the achieved LRV relative to a batch process.

To further reduce virus exposure risk, virus filter integrity could be preuse tested and care should be taken in operation not to damage the filter. This would help to avoid a failing integrity test, postuse. This is even more important for continuous processing. In fact, manufacturers could place some kind of pause or process interruption, before further downstream processing, to verify an acceptable virus safety risk level.

The fundaments of virus safety and validation remain the same regardless of bioprocessing approaches (batch versus continuous). Continuous processing includes some processing modifications that could have impact on virus safety risk, and should be considered. Ultimately, the industry will continue in their efforts to maintain the highest level of virus safety possible as they continue to evaluate more efficient bioprocessing manufacturing schemes.

References

1 Moody, M., Alves, W., Varghese, J., and Khan, F. (2011) Mouse minute virus (MMV) contamination: a case study – detection, root cause determination, and corrective actions. *PDA J. Pharm. Sci. Technol.*, **65** (6), 580–588.
2 Victoria, J.G., Wang, C., Jones, M.S., Jaing, C., McLoughlin, K., Gardner, S., and Delwart, E.L. (2010) Viral nucleic acids in live-attenuated vaccines: detection of minority variants and an adventitious virus. *J. Virol.*, **84**, 6033–6040.
3 http://www.fda.gov/downloads/AdvisoryCommittees/CommitteesMeeting Materials/BloodVaccinesandOtherBiologics/VaccinesandRelatedBiological ProductsAdvisoryCommittee/UCM212015.pptx.
4 Warikoo, V., Godawat, R., Brower, K., Jain, S., Cummings, D., Simons, E., Johnson, T., Walther, J., Yu, M., Wright, B., McLarty, J., Karey, K.P., Hwang, C., Zhou, W., Riske, F., and Konstantinov, K. (2012) Integrated continuous production of recombinant therapeutic proteins. *Biotechnol. Bioeng.*, **109** (12), 3018–3029.
5 Konstantinov, K., Godawat, R., Warikoo, V., and Jain, S. (2014) Integrated continuous manufacturing of therapeutic protein drug substances. US 2014/ 0,255,994 A1.
6 Brower, M., Hou, Y., and Pollard, D. (2014) Monoclonal antibody continuous processing enabled by single use, *Continuous Processing in Pharmaceutical Manufacturing*, Wiley-VCH Verlag GmbH, Weinheim, 255–294.

7 Klutz, S., Magnus, J., Lobedann, M., Schwan, P., Maiser, B., Niklas, J., Temming, M., and Schembecker, G. (2015) Developing the biofacility of the future based on continuous processing and single-use technology. *J. Biotechnol.*, **213**, 120–130.

8 Pollock, J., Ho, S.V., and Farid, S.S. (2013) Fed-batch and perfusion culture processes: economic, environmental, and operational feasibility under uncertainty. *Biotechnol. Bioeng.*, **110** (1), 206–219.

9 LaCasse, D., Genest, P., Pizzelli, K., Greenhalgh, P., Mullin, L., and Slocum, A. (2013) Impact of process interruption on virus retention of small-virus filters. *Bioprocess Int. Bioprocess Tech.*, **11** (10), 34–44.

10 Qiu, Y., Jones, N., Busch, M., Pan, P., Keegan, J., Zhou, W., Plavsic, M., Hayes, M., McPherson, J.M., Edmunds, T., Zhang, K., and Mattaliano, R.J. (2013) Identification and quantitation of Vesivirus 2117 particles in bioreactor fluids from infected Chinese hamster ovary cell cultures. *Biotechnol. Bioeng.*, **110** (5), 1342–1353.

11 Butler, M. (2005) Animal cell cultures: recent achievements and perspectives in the production of biopharmaceuticals. *Appl. Microbiol. Biotechnol.*, **68**, 283–291.

12 Pollock, J., Ho, SV., and Farid, SS. (2013) Fed-batch and perfusion culture processes: economic, environmental, and operational feasibility under uncertainty. *Biotechnol. Bioeng.*, **110** (1), 206–219.

13 Phillips, M.W., Bolton, G., Krishnan, M., Lewnard, J.J., and Raghunath, B. (2006) Virus filtration design and implementation, in *Process Scale Bioseparations for the Biopharmaceutical Industry* (eds A.A. Shukla, M.R. Etzel, and S. Gadam), CRC Press, pp. 334–364.

14 Viresolve® Pro Solution Performance Guide, Merck KGaA, Darmstadt, Germany, 2013.

15 Genest, Paul, Ruppach, Horst, Geyer, Cedric, Asper, Marcel, Parrella, Joseph, Evans, Bill, and Slocum, Ashley (2013) Artefacts of virus filter validation. *Bioprocess Int., Bioprocess Tech.*, **11** (5), 54–61.

16 LaCasse, D., Genest, P., Pizzelli, K., Greenhalgh, P., Mullin, L., and Slocum, A. (2013) Impact of process interruption on virus retention of small-virus filters. *Bioprocess. Int.*, **11** (10), 34–44.

17 Sreedhar, B. and Kawajiri, Y. (2014) Multi-column chromatographic process development using simulated moving bed superstructure and simultaneous optimization – model correction framework. *Chem. Eng. Sci.*, **116**, 428–441.

18 Sofer, G., Lister, D.C., and Boose, J.A. (2003) Part 6, Inactivation methods grouped by virus. BioPharm International, April.

19 Lutz, H., Chang, W., Blandl, T., Ramsey, G., Parella, J., Fisher, J., and Gefroh, E. (2010) Qualification of a novel inline spiking method for virus filter validation. *Biotechnol. Progress*, **27** (1), 121–128.

Part Six

Continuous Chromatography

12

Multicolumn Continuous Chromatography: Understanding this Enabling Technology

Kathleen Mihlbachler

Lewa Process Technologies, Inc, Separations Development, 8 Charlestown Street, Devens, MA 01434, USA

12.1 Introduction

In recent years, the pharmaceutical industry has invested enormous resources in the development of continuous downstream processes for the manufacturing of biomolecules[1] [1–19]. These efforts are fueled by improvements in upstream processing, such as increased titer and implementation of continuous fermentations. Downstream processes became the "bottleneck," especially for the production of monoclonal antibodies. Additionally, the introduction of biosimilar/biobetter increased pressure on the industry to improve process productivities and sustainability, in particular, with the first approval of Remsima in 2012 by Celltrion[1]. Innovative downstream approaches are required to purify in cost-effective ways while retaining the biomolecule's characteristics. Approaches such as integrated continuous downstream process (as displayed in Figure 12.1) show promising results with increased productivity and reduced capital and operational expenses, especially when using multicolumn continuous chromatography. In this process scheme, the single units are linked by surge bags and their operations are synchronized. The first production scale of these DSP schemes was recently reported by Amgen [8]; however, only conventional batch chromatographic steps were implemented.

With "Innovation and Continuous Improvement in Pharmaceutical manufacturing, Pharmaceutical CGMPs for the 21st Century," the FDA created their first guidelines in 2004 [20] when they urged the industry to invest in new technologies. Together with the guidance on process analytical tools (PAT), the foundation for continuous manufacturing approaches was generated [21]. In addition, the ICH guidance 8–11 [22–25] provide the framework for continuous manufacturing. Other professional organizations and regulatory bodies also provide guidance, such as the ISPE on risk-based manufacturing [26] and the ASTM standard on continuous processing [27]. However, the urgent call for investment in new technologies came from Dr. J. Woodcock directly. At recent meetings, she was outspoken in her support for integrated continuous downstream processing – at

---

1 www.biosimilarnews.com/worlds-first-biosimilar-antibody-is-approved-in-korea.

Continuous Biomanufacturing: Innovative Technologies and Methods, First Edition.
Edited by Ganapathy Subramanian.
© 2018 Wiley-VCH Verlag GmbH & Co. KGaA. Published 2018 by Wiley-VCH Verlag GmbH & Co. KGaA.

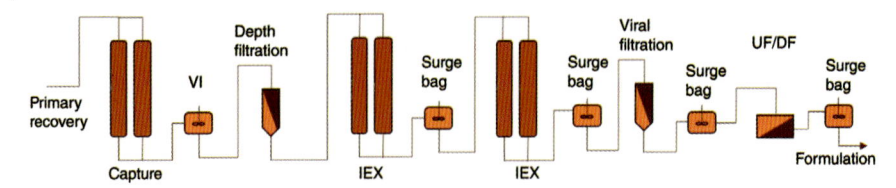

Figure 12.1 Generic integrated continuous downstream process.

the International Symposium on Continuous Manufacturing of Pharmaceuticals at MIT in 2014 and at the Integrated Continuous Biomanufacturing in Berkley in 2015 [28,29], to mention only two.

Many researchers [1–7,9–13], including the author [14–19], have called for investing in these new technologies – so, what hinders implementing the technologies in downstream processing? What are the real and perceived technical or regulatory hurdles? To answer these questions, it is helpful to look briefly at the history of the technology. Since its introduction, process chromatography has been well established in the industry [30]. A typical DSP scheme includes chromatographic capture and purification and polishing steps using different chromatographic interaction mechanisms (more details in Section 12.3). The chromatographic steps are traditionally executed by batch loading onto large LPLC columns, which require large buffer tank systems. Between chromatographic steps, there are filters and membranes for viral clearance or buffer exchange, as well as large hold-up tanks that are also used for viral inactivation. Membranes and filters are operated continuously by installing parallel units.

Although the continuous operation of process chromatography has been discussed [31–36] from the beginning, it has not yet been implemented in the manufacturing of biomolecules. In the 1960s, continuous chromatography was successfully applied to the production of hydrocarbons [37–39], in particular simulated moving bed chromatography, invented by Broughton [37]. It is a billion dollar business [40] today. Since the 1980s, high fructose corn syrup has been enriched by the technology [41,42]. Since the 1990s, chiral compounds have been separated by SMB chromatography in the pharmaceutical industry [43–61], which includes synthetic amino acids. Due to its high productivity and yield as well as reduced solvent consumption, the SMB technology became an economically but also ecologically attractive tool for the production of enantiopure compounds [62,63] as an alternative to traditional enantiospecific synthesis or crystallization. Several commercial products such Citalopram (H.Lundbeck), Sertraline (Pfizer), and Kepra (UCB) apply the technology for enantioseparation [60].

After successful implementation of the SMB technology for API manufacturing 15 years ago, research on implementing continuous chromatography into the DSP [64–82] continued. Significant research was conducted at ETH Zurich, University of Magdeburg, and Purdue University [66–68]. However, it took another 5 years after the first industrial investigations were published [68,69] for the industry as a whole to pick up the technology. A development that was not only supported by the previously mentioned business and market demands of the pharmaceutical industry but also by an initiative to convert traditional stainless steel production facilities to facilities using single-use technologies. Because multicolumn continuous chromatography reduces the size of process equipment, when incorporated into integrated continuous downstream processes it enables single-use technologies (Figure 12.1) [1,2,16–19,82,83]. It creates functional closed systems with limited or no hold points. Missing tanks reduce the footprint of the entire DSP scheme. All these efforts together simplify cleaning procedures and process validations, and, therefore, shorten the turnaround time between batches. This is especially important for CMOs.

Replacing batch chromatography with multicolumn continuous chromatography as shown in Figure 12.1, improves process efficiency and robustness of the chromatographic steps and also reduces buffer consumption. At the bench-top scale, increased throughput and packing utilization were already demonstrated [3–7]. Now, the full potential of the integrated continuous DSP platform has to be proven at the pilot and production scale. The first prototypes were introduced such as by GE, Pall, and LEWA Process Technologies (Figure 12.2).

Previously, the author had presented technical and regulatory hurdles and provided guidance on how to overcome them [14–19]; however, feedback from the audience showed that there is still reluctance. This reluctance might be caused

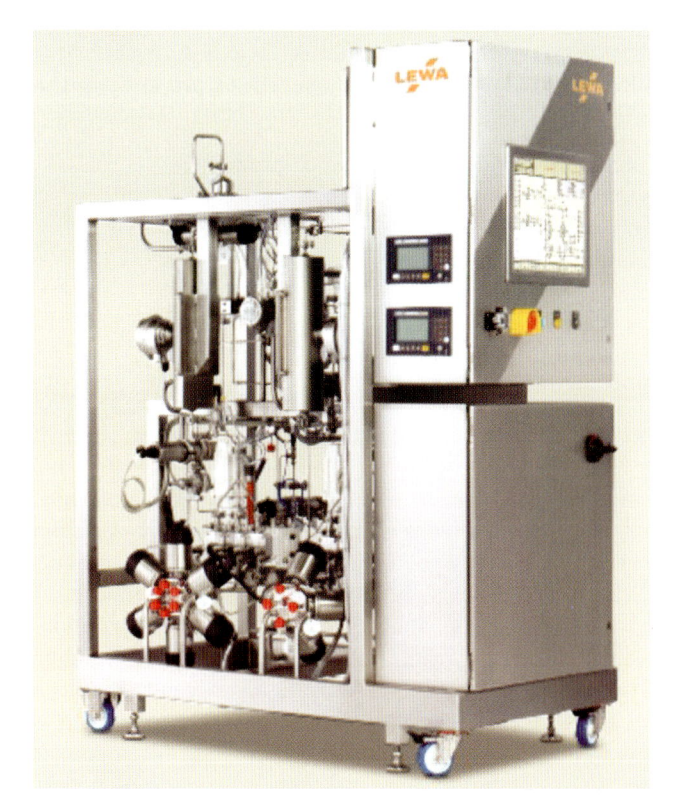

Figure 12.2 EcoPrime Twin 100 prototype. (Reproduced with permission from LEWA Process Technologies.)

by something as simple as ambiguous terminology used to describe the chromatographic technology. The principles of continuous chromatography are not foreign to the field; nevertheless, there are misunderstandings about what continuous chromatography is. Thus, the main focus of this chapter is on common definitions in DSP and chromatography such as batch. Developing a correct and shared understanding of these definitions can help overcome the remaining hurdles, in particular when evaluating the multicolumn continuous chromatographic processes under QC/QA and regulatory aspects. Understanding this terminology will give readers a deeper knowledge of continuous chromatography, answer remaining technical questions about process development and equipment design, and provide insights into process evaluation and equipment assessment.

12.2 Modes of Chromatography

Chromatographic separation/purification processes are classified in multiple ways. Each of them describes the process from a different viewpoint regarding how molecules interact with packing materials and liquid phases such as aqueous buffers and organic solvents. The two major classifications are modes of operation and chromatographic interaction mechanisms. These mechanisms are not the primary focus of this chapter but are briefly summarized in Section 12.3. Eight different kinds of modes of operations are specified as follows:

a) Purpose of operation: analytical and preparative or productive
b) Pressure rating: low, medium, and high pressure
c) Mode of elution: isocratic and linear or step gradient
d) Mode of binding: bind-elute or flow-through
e) Flow direction: cocurrent and countercurrent
f) Number of columns: single and multicolumn
g) Mode of operation: batch or continuous
h) Type of load: batch or continuous

When going through this list, some of the criteria are not always clearly distinguishable and are used interchangeably. For example, the term "batch" has multiple meanings when used in chromatographic applications for the pharmaceutical industry. Batch chromatography is a traditional term that describes distinct injection/load and elution steps yielding one or more distinct product peaks (Figure 12.3a). When applied in analytical and preparative (production scale) chromatography, the focus is on analysis of samples and separation/purification of products, respectively. During the analysis, compounds are identified, qualified, and/or quantified; thus, baseline separation between the peaks is essential. In production, peak shapes and their resolution are not as important as productivity and yield as well as precision and reproducibility. In the following sections, the focus is on the production scales and their process development. However, analytical chromatography remains a crucial PAT in monitoring and controlling the process separation.

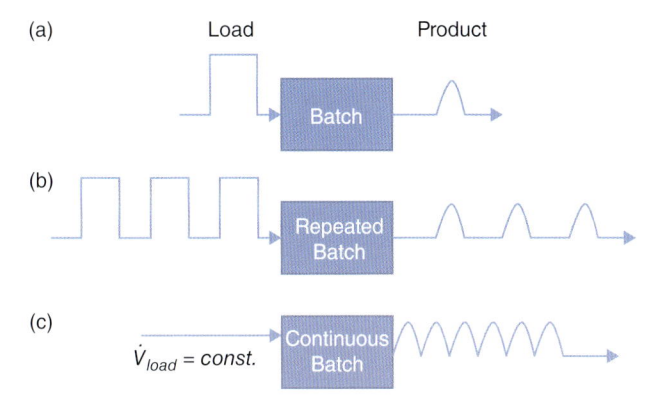

Figure 12.3 Schematic batch chromatography. (a) Batch chromatograph. (b) Repeated loading. (c) Continuous loading (multiple columns).

Another meaning of "batch" is adopted from chemical engineering. Here, batch processes are defined as discontinuous processes where process parameters and product characteristics change with time and within spatial dimensions. Translated to the pharmaceutical industry, critical process parameters (CPPs) and critical quality attributes (CQAs) become time dependent. Examples of CPPs of chromatographic processes include flow rate, process pressure, and buffer compositions with their pH and conductivity values. CQAs are feed and product concentrations, and their compositions. All these attributes are time dependent, in particular the latter ones that change during the elution of the product peak (Figure 12.3).

Both definitions of "batch" are used interchangeably in chromatographic applications, which is not entirely correct. The traditional batch process can be run continuously by loading repeatedly onto the same single-column or a multicolumn system until entire feed batch is purified (Figure 12.3b and c). In the case of single-column systems, the feed stream is discontinuous. As explained later in this chapter, a chromatographic system with multiple columns can be fed continuously (Figure 12.3c); nevertheless, the process steps executed on one of the columns are the same steps as in traditional batch chromatography (more details in Section 12.5). Alternately, multicolumn systems can operate continuously, countercurrent processes such as simulated moving bed (SMB) chromatographic processes (Section 12.6). In all of these processes, CPPs and CQAs are time dependent even if the overall process is fed continuously.

The term "batch," when used in a regulatory context, has a slightly different meaning. It defines a specific amount of product (mass or volume), with specific characteristics [28] that were obtained during a specific time interval. This definition does not account for how, when, and where the product is generated. It is independent of the chromatographic mode. A batch in the regulatory sense can be a single injection/load in a traditional, single-column batch process, repeated loads of continuous batch chromatography using single or multicolumn systems, or multicolumn continuous countercurrent chromatography. The term "batch" is interchangeable with "lot" or "charge." In addition, operation sometimes defines "batch," which can be a single or multiple loads run batchwise or continuously, as campaign.

Because the term "batch" has traditional meanings that were adopted from different fields, these meanings cannot be easily substituted. The descriptions presented in the following section clarify and distinguish the term by making readers aware of differences in the processes and by providing needed background information. In this sense, it is also important to understand the counterpart expression "continuous," especially its use in chromatography. This term is also not straightforward, as will be explained in detail later. At this point, the author only wants to point out one aspect of continuous operation that is associated with chemical engineering: reaching "steady state." This state is defined by constant, nonchanging characteristics. All process parameters and quality attributes are time independent. Obviously, this is the desired operation state that shows control of the process. However, process parameters are actually variables with variabilities within control ranges. It becomes a state of control[2].

Additionally, devices used to monitor and control processes have accuracy ranges, not absolute values. For instance, if a flow meter has an accuracy of 0.1% across its range of measurement, then measurements have only accuracy within +/− range. One cannot specify more precise fluid delivery than this range. Feed and buffer pumps do not deliver constant fluid streams. Depending on the pump type, they pulse more or less. Some have pulse frequencies within milliseconds, others in minutes. If measurement intervals get small enough, more variability might become visible. In contrast, the flow of pumps with high frequencies appears smooth if measurement intervals are sufficiently large and control loops control or adjust pulsations.

Pressure ratings of chromatographic systems depend on the molecules to be separated and the type of packing material defined by the interaction mechanisms. The following section provides a brief overview of chromatographic interactions. More details can be found elsewhere [79]. Polymeric resins have lower pressure ratings compared to silica-based packing; thus, polymeric resins are primarily applied in low-pressure liquid chromatography (LPLC), and only rarely in medium and high-pressure liquid chromatography (HPLC). The mechanical and chemical stability of the packing are key for chromatographic separations. One can apply almost any chromatographic mode as long as the flow rates are chosen within the pressure limits of the columns with their packings.

12.3 Interaction Mechanisms Used in Chromatographic Systems

Although interaction mechanisms are not the focus, they are important to understand how different modes of operation work best and what their advantages are. The following section including Table 12.1 provides the needed background information on different interaction mechanisms.

---

2 Discussion at AIChE-FDA Workshop.

Table 12.1 Type of chromatographic interaction.

Mode	Interaction	Elution	DSP application
Normal phase	Polar	Isocratic or positive polar organic gradient	NA
Reversed phase	Hydrophobic	Isocratic or positive organic gradient	Polishing
Chiral NP or RP	Specific + polar or hydrophobic	Isocratic or positive organic gradient	NA
HIC	Hydrophobic	Negative salt gradient	Polishing
Size exclusion	Steric	Isocratic	Capture, polishing
Affinity	Specific	Solvent modification	Capture
Ion exchange	Electrostatic	Positive salt gradient	Purification

Due to the use of organic solvents, normal phase (NP) and reverse phase (RP) chromatography are typically not implemented in downstream purification. Biomolecules are not chemically stable enough to withstand the harsher solvent environment, and denature during the chromatographic processes. Exceptions can be found when smaller biomolecules such as synthetic peptides or proteins, such as insulin, are purified.

As alternative to RP chromatography, hydrophobic interaction chromatography (HIC) is implemented, which utilizes the hydrophobic characteristics of the biomolecules by adjusting the buffer strength.

Chiral separations are primarily applied for racemic mixtures of small molecules [30,59], but also smaller biomolecules like amino acids. Due to the characteristics of these molecules, alternative separation approaches are usually more resource intensive. This binary separation is perfectly suited for multicolumn, continuous, countercurrent chromatography, as explained later in Section 12.6 about SMB chromatography.

The major mechanisms in downstream processing are affinity, ion exchange, hydrophobic interaction, and size-exclusion chromatography (SEC). A generic downstream process is outlined in Figure 12.1. Affinity chromatography is the essential capture step in the scheme. In case of monoclonal antibody purification, this step is known as the protein A capture. Due to the very tight binding of the mAb molecules onto the protein A, the process runs in an on–off bind–elute mode. The purification and polishing steps use cation and/or anion exchange chromatography. Depending on the pH and conductivity of the buffers, charged sites of the biomolecules can bind to the resin or these interactions can be broken. Impurities with different charge sites flow through during the load or are strongly bound. IEX chromatography is also run in bind–elute or flow-through mode. These steps are sometimes substituted by HIC or SEC steps. In the latter case, one takes advantage of the size differences of the biomolecules such as native molecules and their fragments or aggregates.

12.4 Batch Chromatography

Batch chromatography is the traditional mode of chromatography where the feed mixture is injected or loaded onto the column batchwise (Figure 12.3a). As described previously, traditional batch chromatography can also be run continuously (more details in Section 12.5). Let us first look more closely at this chromatographic meaning and then examine its implementations.

Figure 12.4 outlines a typical batch chromatographic system with its main components. Two pumps deliver buffers and can generate gradient formations. Feed can also be delivered by one of the pumps or by an injection system. UV detectors (with single or multiple wavelengths), and sensors for pH, conductivity, temperature, and pressure monitor the process parameters. The operation of the system is controlled by a PLC. It also collects and stores process data and exports them to MES systems. Generally, the separation takes place on a single column. If needed, multiple columns can be connected in series to increase the column length/volume. Process outlet streams are fractionated for collecting products and by-products.

Similar components are used in analytical or production scale systems of batch chromatography.

Batch chromatography has the following five major process steps (Figure 12.5):

1) *Equilibration*: Preparation of the column
2) *Load*: Injections of feed solution onto the column
3) *Wash*: Removal of loosely bound impurities
4) *Elution*: Release of products from the resin and their collections
5) *Regeneration*: Removal of very strongly bound impurities from the resin and its cleaning

In isocratic chromatography, the buffer compositions stay constant during all process steps. Differences in characteristics of the molecules to be separated and, therefore, in their binding strengths are minimal. Adjustments in the buffer strength are not needed. CIP steps are usually not required. Injection/load are only a small percentage of the column volumes. This simple operation is usually applied in chiral batch separations.

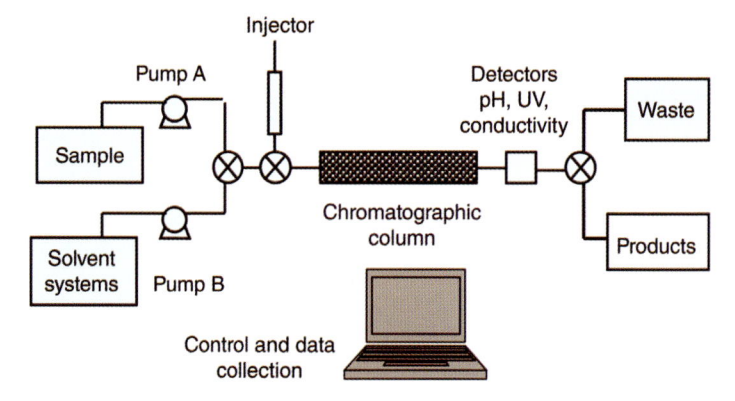

Figure 12.4 Batch chromatographic system.

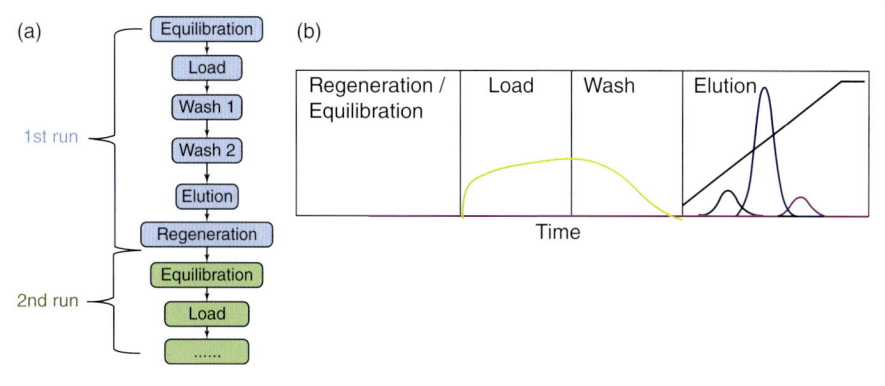

Figure 12.5 Process steps for batch chromatography with linear elution gradient (a) and corresponding chromatogram (b).

For gradient chromatography, buffer compositions are modified to separate feed mixtures with very different molecule characteristics. Some are very closely related to the product. Others are very different; thus, their binding strengths are very different. Implementing gradients and additional *wash* steps prior to the *elution* step guarantees the isolation of the products. Due to low feed concentrations and/or due to operating in a bind/elute mechanism, load volumes are generally multiple column volumes (CVs). In Figure 12.5, the process steps are summarized for batch separation using a linear elution gradient: equilibration, load, wash(es), elution, and regeneration. The corresponding chromatogram is shown on the right.

12.5 Semicontinuous and Continuous Batch Chromatography

Without changing the process steps as already defined (Section 12.4), batch chromatography can be operated semicontinuously (semibatch) or continuously if a single injection/load step does not separate the complete feed batch. As shown in the flowcharts of this unit operation (Figure 12.3), traditional batch chromatography with only one load step (Figure 12.3a) can be completed by repeated load steps (Figure 12.3b), and continuous feed stream (Figure 12.3c). In chromatographic sense, the batch recipe can be repeated on a single-column configuration (middle scheme) or multicolumn system (lower scheme). This allows defining batch sizes based on feed batch being completely purified, or requirements such as size of product tanks or shift durations. Re-equilibration steps are needed to bring the column back into the starting conditions prior to repeated injections.

12.5.1 Single Column

If only one small column is available, the injection/load steps are simply repeated onto the same column. In Figure 12.6, multiple runs are displayed. Each of the product peaks is collected into a single product batch. Of course, each of the

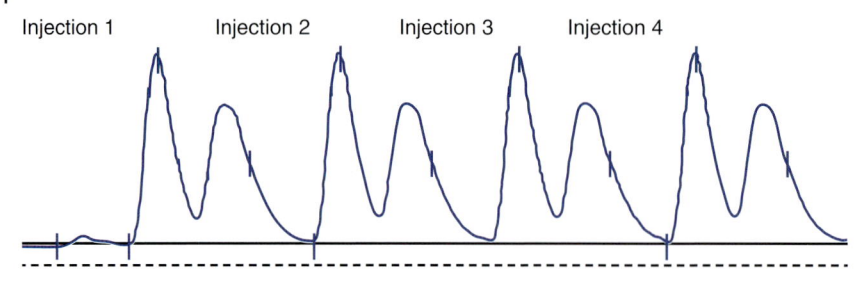

Figure 12.6 Stack injection.

product peaks need to meet the purity requirements. Here, the front parts of the first peak and the back part of the second one meet the purity requirements.

For isocratic chromatography, process efficiency can be improved by stacking the injections. Process time is reduced by injecting the load when the previous load is eluting. Additionally, buffer consumptions are reduced. To ensure that no strong binding impurities elute with one of the next product peaks, regeneration steps are added in-between to remove all the strong binding impurities at once.

This simple stacking is not possible during gradient processes; however, intelligent gradient schemes can reduce the time needed to regenerate and re-equilibrate the column. As Mihlbachler *et al.* [84] have shown previously, implementing a reverse gradient during elution and combining regeneration and re-equilibration reduces process time and buffer consumption significantly. As shown in Figure 12.7, the front site impurities elute earlier while the separation between the two main peaks is slowed down, allowing higher loads. When the separation is completed, the buffer strength is stepped up, allowing the fast elution of strongly retained compounds plus the re-equilibration of the column at the same time. According to the authors, in one application the load of a fermentation product was increased by 76%, and in another case the productivity increased by 33%, plus the solvent consumption was reduced by 50%.

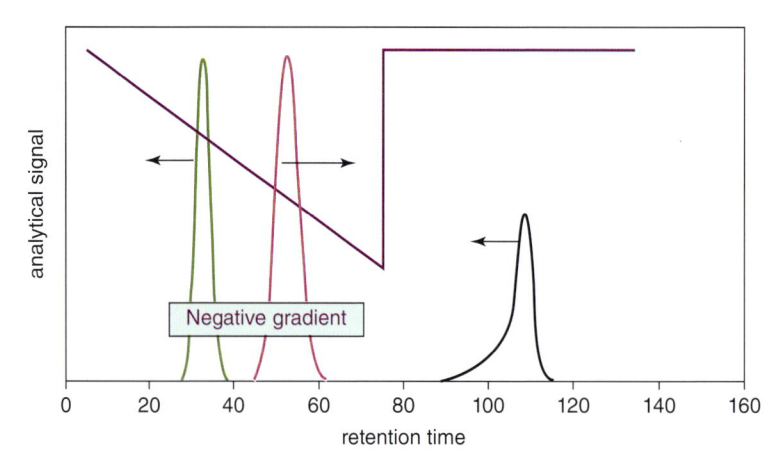

Figure 12.7 Innovative negative elution gradient.

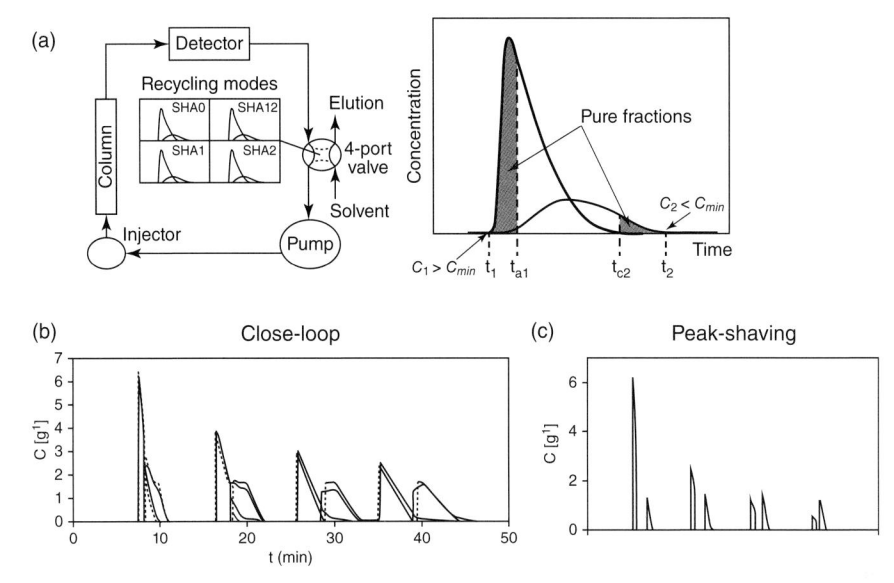

Figure 12.8 Single-column recycling schemes. (a) System setup (b) Chromatograms when close-loop recycling.(c) Chromatograms when peak-shaving. (Reproduced with permission from Ref. [85]. Copyright 1995, Elsevier.)

Another approach to improve the process yield and throughput is to inject high loads onto a single column, while recycling the entire peak (close-loop) or the overlapping fractions (peak-shaving) back onto the column as shown in Figure 12.8 [85]. The first approach essentially simulates a longer column. These approaches are best applied during isocratic chromatography. The approaches are common practice, particularly in the early process development phase.

Another recycling process is the steady-state recycling (SSR) process [86,87], which has been commercialized as CycloJet by NovaSep, Inc[3]. In comparison to the previous peak shaving process, the next load is introduced while the first one elutes in SSR. The injection is timed in such a way that it falls between the two product peaks. "Shaving" both pure products off the front and back of the peaks begins after a few cycles when the internal profile is sufficiently high concentrated. This process is mainly applied to chiral separation during the PR&D phase. In case of gradient chromatography, buffer compositions have to be readjusted before recycling is possible. These processes were proposed by different academic teams [88,89] but have not been implemented into the downstream purification of biomolecules (Figure 12.9).

12.5.2 Multicolumn Parallel Operation

Because smaller columns are packed more efficiently (higher number of theoretical plates) and are more mechanically stable, using multiple smaller columns packed with the same material increases the process productivity just by itself.

---

3 www.novasep.com.

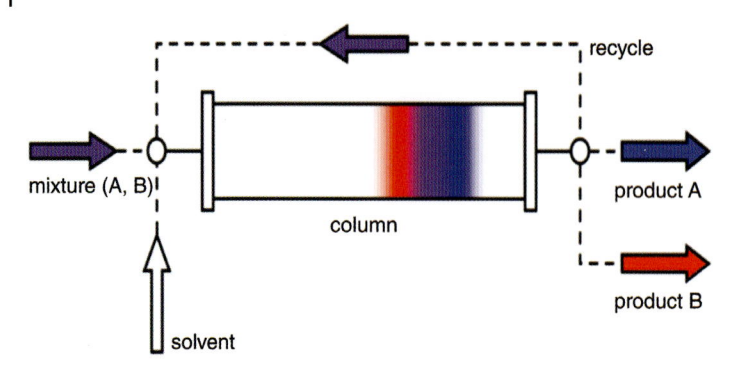

Figure 12.9 Flow diagram of steady-state recycling process.

Transferring the process steps: equilibration, load, wash(es), elution, and regeneration, onto parallel columns as shown in Figure 12.10 allows continuous loading. Coordinating these process steps eliminates wait times between loads.

Figure 12.11 shows how this process works. The same batch process steps run parallel on each column but are one step off on each column. While column 1 is regenerated and equilibrated, column 2 is loaded, column is 3 washed, and column 4 is eluted. Splitting the column into separate, smaller column allows the time-to-space transition of the process steps. The columns cycle through the four steps by a set switching time until column 1 is back in the regeneration/equilibration position. Feed is loaded without interruption onto Col2, Col1, Col4, and Col3 before going back to Col2. The cycle repeats until the complete feed batch is separated.

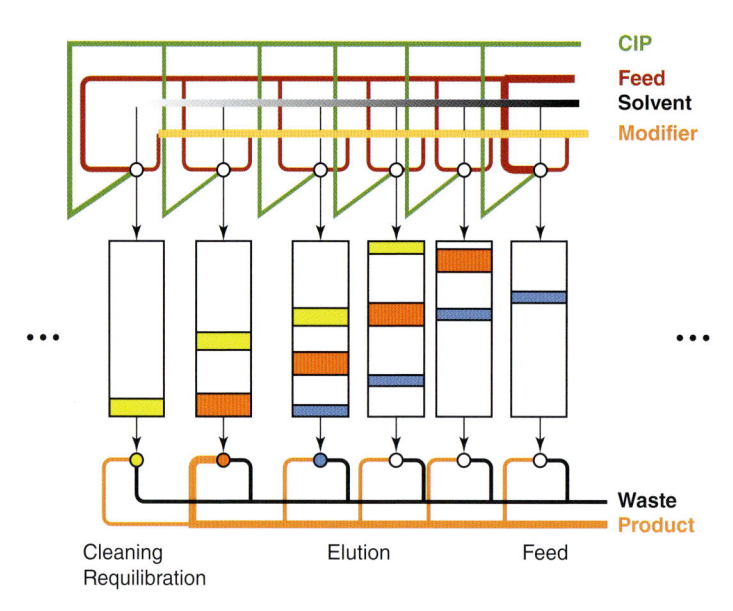

Figure 12.10 Parallel operation of multiple columns: right column – load, middle columns – wash(es) and elution, and left columns – regeneration and re-equilibration.

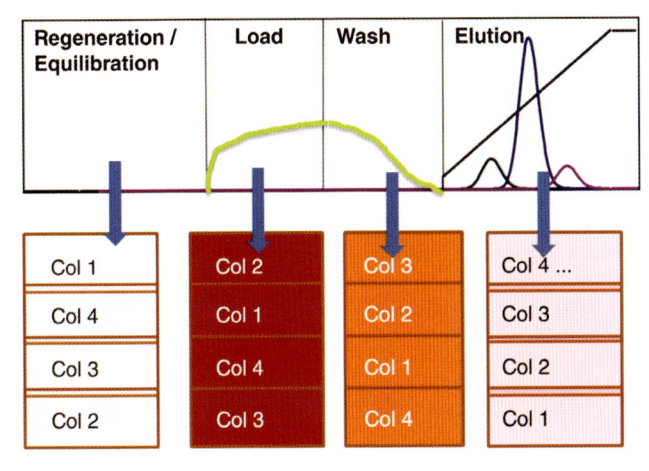

12.5 Semicontinuous and Continuous Batch Chromatography | 335

Figure 12.11 Transformation of batch chromatography from single to multiple column process.

The number of required columns can be reduced by combining the process steps, such as wash(es), elution, regeneration, and equilibration onto one column. Their flow rates are optimized to guarantee their completion before the column switch. The switching time is now defined as sum of these process steps:

$$t_{switch} = t_{washes} + t_{elution} + t_{regeneration} + t_{equilibration}. \quad (12.1)$$

The flow rates are not only dependent on the pressure limitations and residence time requirements [90], but also on whether they need to fit into the overall separation time scheme.

Due to the slow mass transfer of biomolecules during chromatographic processes (Figure 12.12a) [90], the loading capacity of single columns is limited

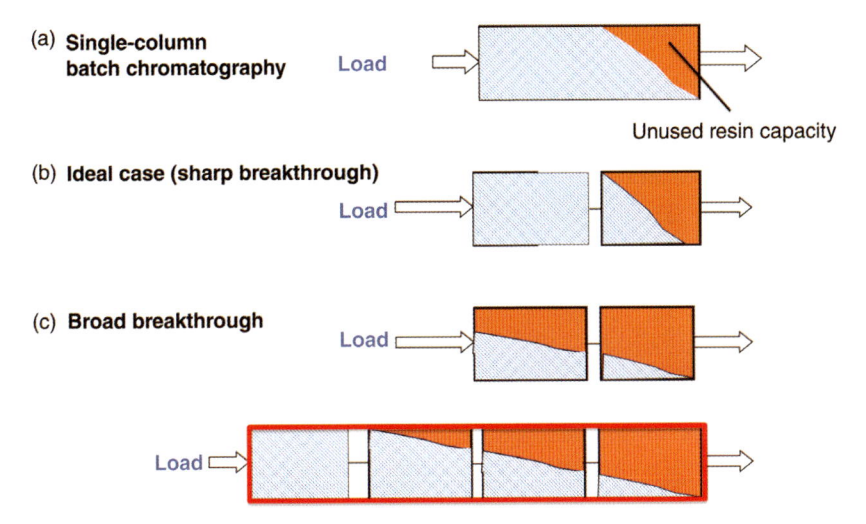

Figure 12.12 Loading to column capacity. (Reproduced with permission from Ref. [16]. Copyright 2014, LEWA Process Technologies.)

in batch chromatography. Products are lost when the breakthrough occurs. Usually, loading is stopped at 5 or 10% breakthrough of the dynamic breakthrough (DBC) (Figure 12.31). At this point, the resin is not fully utilized. One way to improve the utilization is the reduction of the flow rate and, therefore, the increase in residence time on the column, which sharpens the breakthrough. Sufficient time is now available for the molecule to travel to binding sites within resins. However, the operating time is increased significantly (Section 12.8).

To improve productivity, the original column length is split into two columns (Figure 12.12b). In an ideal case, the first column (here on the left side) is loaded to its capacity while the breakthrough flows onto the connected second column. The columns are then disconnected. The left one is washed, eluted, and regenerated in a parallel column configuration, while the loading of the other column is continuous. Due to the splitting of the column, the linear velocities/flow rates can increase when columns are operated in parallel, yet still remain within the pressure limits.

In case of broader breakthrough curves, two columns in the loading zone might not be sufficient (Figure 12.12c) to completely saturate the left column. As shown in the lower part of Figure 12.12c, more than two columns might be used. In this case, the other three columns are only loaded partially. The loaded column is removed from the loading zone and washed, eluted, and regenerated in the parallel operation, while the other three are continuously loaded (Figure 12.13a). Of course, when operating three and four columns sequentially, flow rates have to be reduced due to pressure limits of columns and their packing.

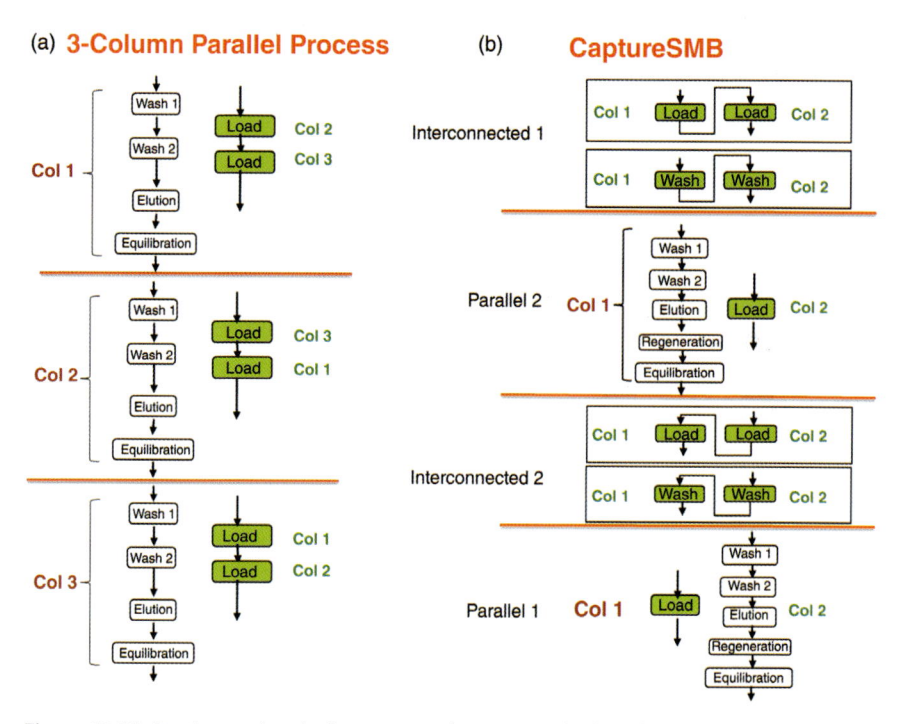

Figure 12.13 Continuous batch chromatography using multiple columns. (a) Three columns with two always in load. (b) Two-column system (CaptureSMB process by ChromaCon).

When applying the switching schedule for the continuous batch process as shown in Figure 12.13a, one operates essentially in parallel. While one column is always in recovery, the other two are connected in the loading position. Due to the connection, the first column can be higher loaded and, therefore, more products can be eluted in the next elution step from this column and the buffer consumption is reduced. The elution can be either a step or linear gradient. However, during the loading a significant part of the packing in the second column (Figure 12.12c) is not utilized. When calculating the production rate of the process, this underutilized packing has to be accounted for as described in more detail in Section 12.8.

12.5.3 Multicolumn Parallel and Interconnected Operation

To overcome the underutilization of the packing of the previous configuration and, therefore, improve the production rate, an innovative approach was introduced by ChromaCon [7]. Interconnected steps are added between parallel operations, as shown in Figure 12.13b, for the CaptureSMB process[4]. Here, only two columns are operated in either parallel or interconnected configurations. This process is best employed for continuous affinity chromatography, such as the protein A capture step in the DSP scheme of monoclonal antibodies. This process can become the first chromatographic step in an integrated continuous downstream purification concept.

This CaptureSMB process performs the same process steps as already described for batch processes, however, using two unique process configurations. During the parallel configurations, one column is in the load position while the other is in wash(es), elution, regeneration, and equilibration. The elution of product usually occurs due to a step gradient, however, linear gradients are also possible. During the interconnected configurations, both columns are connected in series. Depending on when the interconnection steps occur, either column 1 or column 2 is in first position. While loading the columns, the breakthrough of the first column is retained on the second column. Prior to disconnecting the columns, an additional wash step can be implemented that captures all unbounded molecules from the first column onto the second column.

The process steps are repeated until the entire feed batch is purified. To improve process performance, a startup and shutdown sequence can be implemented. During the initial load step, the load time is extended to simulate a preload of the first column. Due to this preload, "steady-state" conditions are reached faster and operation is more robust from the start up on. During the final cycle, the feed is replaced by the equilibration buffer, which allows the recovery of the remaining product from the columns. Without these steps, the still adsorbed product would be lost. This shutdown sequence starts the cleaning of the column in preparation of the next batch or for storage.

A twin-column system, such as the EcoPrime Twin from LEWA Process Technologies, Inc. (Figures 12.2 and 12.14), allows the continuous loading and purifying of the feed batch by alternating the position and functionalities of the two identical columns. When looking at the overall operation of this system, the

---

4 www.ChromaCon.com.

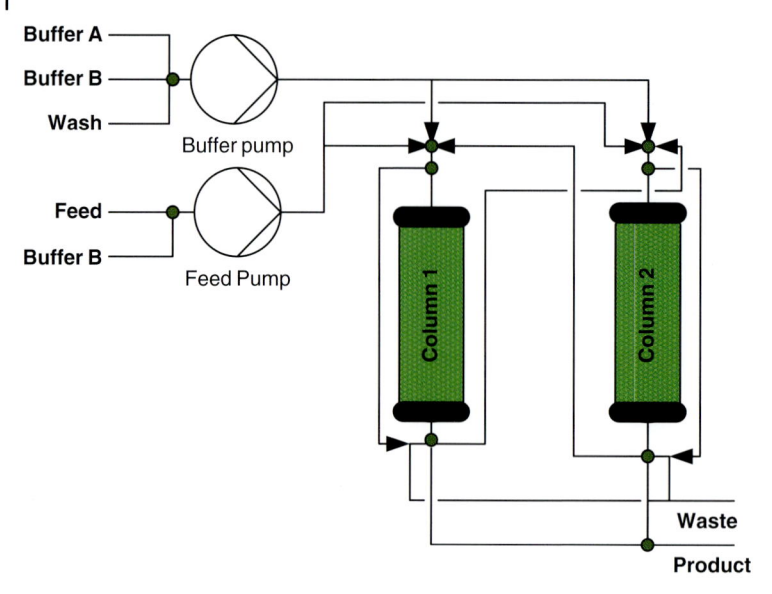

Figure 12.14 Flow diagram of EcoPrime Twin LPLC system.

process has a continuous loading, and the product peaks are collected continuously but alternately from columns 1 and 2, as shown in the lower graph of Figure 12.3c. This is continuous batch operation.

Section 12.8 provides more details about optimizing continuous batch processes for capture. There different configurations are compared that operate in parallel configuration only or in parallel and sequent configurations.

If more closely related molecules are to be separated by the twin-column configuration, the process is modified to enable the internal recycling of the front and back site portions of the product peak. This process is based on the patented multicolumn solvent gradient purification (MCSGP) process by ChromaCon [7]. In Figure 12.15, the process steps for continuous MCSGP are outlined. A complete process cycle has eight switches between parallel and interconnected process steps. The process returns then to the initial column configuration. The process cycle can be repeated until the feed batch is purified.

In the startup phase, the same process steps are executed as in traditional batch chromatography with buffer gradient. After equilibration of both columns, the first column is loaded with a predefined amount of feed (not shown in Figure 12.15). The column Col1 is then flushed with one or more different wash buffers (first parallel step). During continuous operation, column Col2 is regenerated and re-equilibrated, which is not required during startup.

When the front side of the main peaks starts eluting (first red line in Figure 12.16) from column Col1, its outlet is now connected to the second column (first interconnected step). To guarantee binding of the front side fraction onto the receiving column, the process stream is diluted to the equilibration buffer strength. Then both columns are disconnected, the product eluting from the first column is collected (first green line in Figure 12.16) while the second column is now loaded

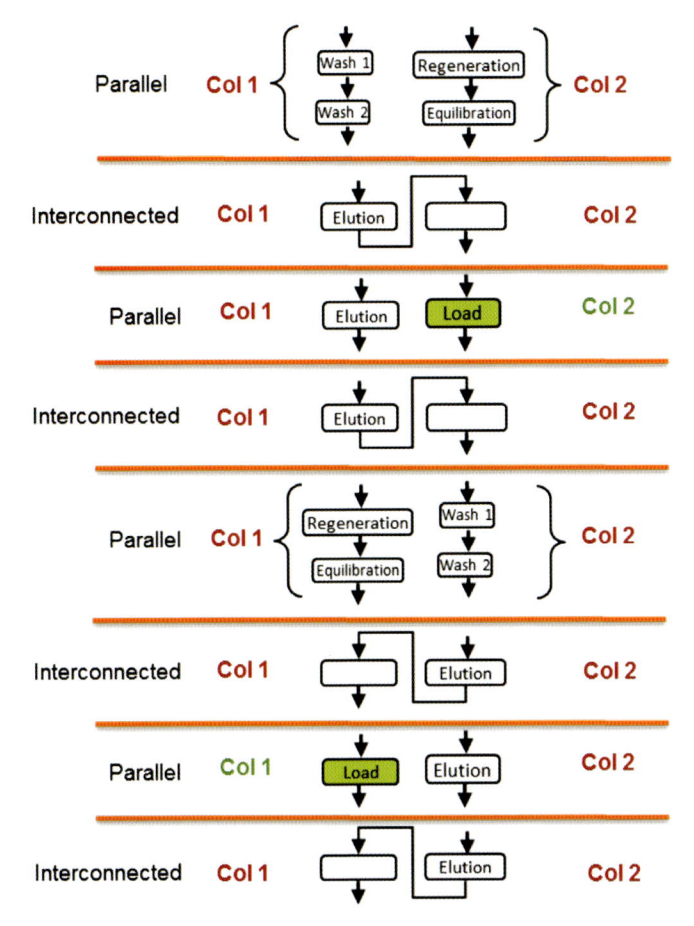

Figure 12.15 Standard process cycle of MCSGP with four alternating parallel and interconnected process steps.

(the first loading step during the continuous cycle). After completing the elution of column Col1 and loading of column Col2, both are reconnected to retain the backside of the product peak onto column Col2 (first magenta line in Figure 12.16). Again, the column outlet stream is diluted to the equilibration buffer strength, which guarantees the retention of the molecules on column Col2. The first half of the cycle is now completed and both columns are disconnected.

To recover the product from column Col2, the column goes through the wash steps before elution starts. In the meantime, column Col1 is regenerated and re-equilibrated in parallel operation. After equilibration of column Col1 is completed, the outlet of column Col2 is connected to the inlet of column Col1 to retain the front side of the product peak eluting from column Col2 (second red line in Figure 12.16). Then, in parallel operation, the product is eluted from column Col2 (second green line in Figure 12.16) while the first column is loaded. Both columns are connected again, to retain the backside of the product peak (second magenta line in Figure 12.16) on column Col1. While column Col2 is

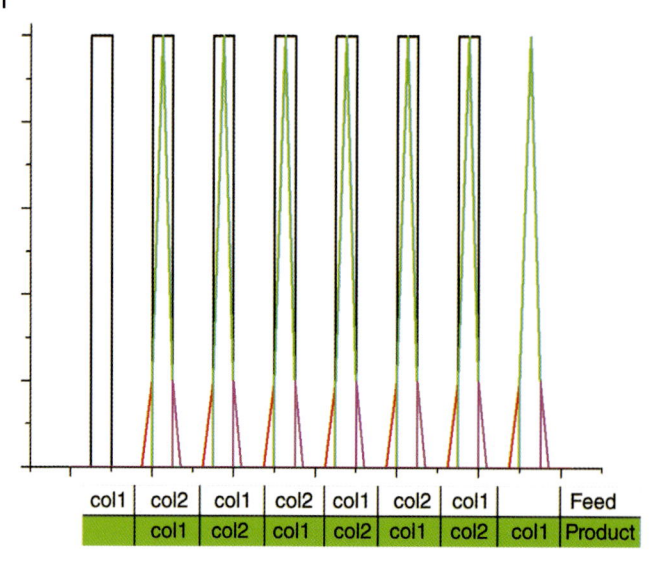

Figure 12.16 MCSGP process with three consecutive cycles: Black line – Load Col1 and Col2 alternating, green line – product alternating from Col1 and Col2, red and magenta line – front and back side of product peak alternating from Col1 and Col2 going to Col2 and Col1, respectively.

regenerated and re-equilibrated, the first column goes through the wash steps. From this point all steps are repeated until the feed batch is completely purified. During the last cycle no new feed is loaded, but it is substituted by the equilibration buffer. Due to the recovering of this last product, the process yield can be increased. The corresponding chromatograms of the complete process (here three cycles) are shown in Figure 12.16.

Similar as described for the steady-state recycling process in Section 12.5.1, the MCSGP process internally recycles product fractions to enhance process yield and productivity. The fractions from the front and back side of the product peak are usually discarded from the product peak in traditional batch chromatography. Due to this unique switching algorithm, these parts of the product peak can also be recovered by maintaining the purity requirements. The partial recycling of product fractions causes an increase of internal product concentration compared to traditional batch chromatography. Due to the concentration increase, the amount of buffers needed for purification is also reduced.

The previous description of the MCSGP process is for a typical IEX process. The process steps can be adjusted to HIC and RP processes, which also use gradient during product elution.

12.6 Multicolumn, Countercurrent, Continuous Chromatography

Up to this point, most of the separation processes run essentially the same process steps as described in Section 12.4 for traditional batch chromatography. They allow the continuous but batchwise loading of multiple columns. Exceptions are the partial internal recycling of the product fractions during steady-state recycling and MCSGP processes. The loading in these cases is periodical.

Combining the continuous loading onto multiple columns with the internal recycling of product fractions becomes only possible with a multicolumn, countercurrent, continuous chromatographic process, in particular SMB chromatography.

12.6.1 Implementing Traditional SMB Technology

The SMB technology is based on the true moving bed (TMB) principles, where two different phases (here solid and liquid phase) move in opposite directions. Depending on the adsorption behavior, the more and less strongly adsorbed compounds move in the direction of the solid and liquid phase, respectively. Two pure product fractions are collected: extract and raffinate, respectively.

Since solid flow raises numerous technical issues, the solid phase is instead packed into columns that are connected in a ring formation (Figure 12.17a) with inlet and outlet valve blocks in between. The countercurrent movement is simulated by periodical valve switching in the direction of the liquid phase at predefined switching times. The ring formation is divided into four zones, each of which has one or more columns. These zones move with the valve switching in the liquid phase directions; however, the physical columns remain in place. Due to binary separation, this process is most suitable for chiral separations.

The major tasks of zones 1 and 4 (Figure 12.17) are the cleaning of the solid and liquid phases, respectively. Fresh solvent continuously enters the system at the inlet of zone 1, which is then combined with the outlet stream of zone 4. At this point, the internally recycled liquid stream from zone 4 should be in the component-free state. The solid phase packed in the first column of zone 1 is also clean at this point; thus breakthrough of the components can be avoided between zones 1 and 4.

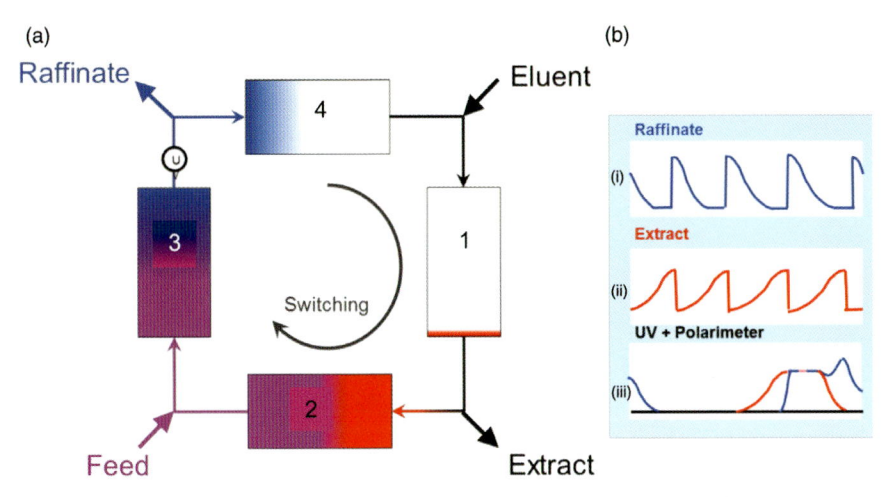

Figure 12.17 Simulated Moving Bed (SMB) Chromatography. (a) Flowchart of 4-zone system. (b) Data recording: (i) UV signal on raffinate outlet, (ii) UV signal on extract outlet, and (iii) inline combined UV and polarimeter signal.

The continuous feed stream is introduced between zones 2 and 3. The primary task of these zones is separation. Combining UV and polarimeter signals [91] allows the recording of both enantiomer concentrations separately in the chromatogram (Figure 12.17b(iii)). The overlapping portion of the concentration profile should remain in zones 2 and 3. It should not reach the product ports. Due to this internal recycling of the mixture, the residence time of single molecules might be very long. Wang and coworkers conducted statistical studies [92].

The more retained compound moves with the solid phase and is collected between zones 1 and 2, the extract port. The less retained compound moves in the direction of the liquid phase. It is collected at the outlet of zone 3, the raffinate port. None of the more retained compounds should reach this port. During each of switching periods, the same amount of extract and raffinate are collected. A recording of these product profiles is shown in Figure 12.17b. These "identical" product peaks repeat continuously in a sawtooth formation. During the startup period the product peaks are still building up; therefore, no product is usually collected. The author previously evaluated how variability of the column characteristics influences separation performance (for more details, review Ref. [91]).

The flow rate of each zone as well as the inlet and outlet streams, and the corresponding switching t_{switch}, are defined based on thermodynamic and kinetic principles [93] allowing under steady-state conditions both product streams meet the purity requirement. Flow rates in each section can be expressed as a non-dimensional number m_j (section $j = 1,2,3,$ and 4):

$$m_j = \frac{Q_I}{Q_S} - \frac{1}{F}, \quad \text{with} \quad Q_S = (1-\varepsilon)Au_S = (1-\varepsilon)AL_C/t^*, \tag{12.2}$$

where F is the phase ratio, Q_S and Q_I are the solid- and mobile-phase flow rate, respectively, ε is the porosity, u_s is the velocity of the solid phase, and A and L_c are the cross section and the length of the column, respectively. For a diluted operating condition, the following relationship must be fulfilled to completely separate a binary mixture:

$$0 < m_4 < a_1 < m_2 < m_3 < a_2 < m_1, \tag{12.3}$$

where a_1 and a_2 are the equilibrium constants of the two compounds.

In case of Langmuirian-type isotherm, complex relationships are determined from the analytical solution of the mass balance equations, which are explained in more detail by Storti et al. [94,95]. The flow rate ratios m_2 and m_3 are usually displayed by a triangular area where complete separation is theoretically achieved within its boundaries (Figure 12.18). The so-called triangle theory was developed in the late 1980s by Storti et al. [94,95]. Process productivities are the highest when operating conditions are at the apex of the triangle. When choosing operating conditions outside the triangle, either pure extract or raffinate streams are obtained as shown in Figure 12.18. To optimize robust SMB separations, mass transfer effects and column characteristics have to be included in the design process [49,53,68,91].

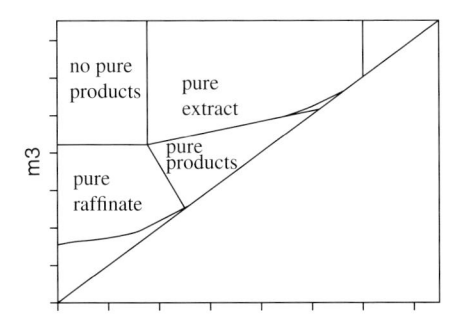

Figure 12.18 Region of complete separation (pure products) under nonideal separation conditions (competitive Langmuirian adsorption isotherm at high concentration) based on the so-called "triangle theory."

The following equations calculate the production rates of both process streams, that is, the kilogram produced per day using the total amount of column packing (kg/(day l_{pack})):

$$\text{Prod}_{Ex} = \frac{Q_{Ex} \int_0^t c_{Ex,2}(t)\,dt}{n(1-\varepsilon)V_c}, \quad \text{Prod}_{Ra} = \frac{Q_{Ra} \int_0^t c_{Ra,1}(t)\,dt}{n(1-\varepsilon)V_c}, \quad (12.4)$$

where n is the number of columns within the SMB unit and $(1-\varepsilon)V_c$ is the total volume of the packing material. The solvent consumption is calculated by

$$\text{Consumption} = \frac{Q_D + Q_F}{Q_F c_{Feed}} \rho_D, \quad (12.5)$$

where c_{Feed} and ρ_D are the total feed concentration and the solvent density, respectively. Values for enrichment and recovery can be used to compare the performance of alternative process steps, which can be derived from the purity and productivity of the process.

12.6.2 SMB Technology for Biomolecules

A well-defined process, such as SMB chromatography, is not easily implemented into the DSP due to the nature of the biomolecules. Process modifications are needed, although the same fundamental thermodynamic and kinetic principles apply as for chiral compounds. Chemical and mechanical concerns specific to biomolecules need to be addressed. First of all, separations are usually not binary. Complex mixtures with many related impurities have to be purified to ppm levels. Thus, molecules are grouped together to reduce the process to a binary or ternary separation. The following sections explain some of these modifications and limitations in more detail.

Downstream processing of biopharmaceuticals consists of more than one step for capturing, purification, and polishing of biomolecules (Section 12.3). To accomplish the required purity levels of sometimes very closely related substances, different separation mechanisms are necessary, such as size exclusion, ion exchange, reverse phase, hydrophobic interaction, or affinity chromatography. Due to the characteristics of biomolecules, retention mechanisms might result from multimodal interactions with additional nonspecific interactions. Often new impurities are created during processing, due to the sensitivity of biomolecules to their chromatographic environment.

The simplest mode of chromatography, SEC (Section 12.3), has been utilized for multicolumn SMB configurations. Due to the isocratic operation of SEC, the transformation from batch to SMB is straightforward compared to other modes. For the process design, the simple linear "triangle" model is applied (Eq. (12.2)). Generally, SMB skids have the traditional four-zone configuration, with one or more columns per zone.

Hashimoto et al. [96] were the first to use an SMB for desalting proteins. In their study, bovine serum albumin (BSA) was separated from ammonium sulfate with yields between 70 and 80%. At the beginning of the 1990s, Roper and Lightfoot [97] used the binary mixture of BSA and ovalbumin as a model system to study the feasibility of size-exclusion SMB. Their theoretical comparison to batch SEC showed that an SMB system reduced solvent consumption by 76% and decreased SEC resin volume by almost four orders of magnitude.

Multicomponent separations are performed using multiple conventional SMB systems in tandem as described by Hritzko et al. [98] and Nicolaos et al. [99] for a ternary system. Later researchers evaluated different numbers of SMB zones for ternary separations, including a nine-zone system as shown in Figure 12.19.

Nicolaos et al. [99] also evaluated different tandem configurations with complete and incomplete intermediate separation of the target component (Figure 12.20). Due to the complex equipment and process design required for nine-zone SMB, tandem configurations using conventional SMBs are preferred. Depending on the separation requirement – which is the key component – one configuration will

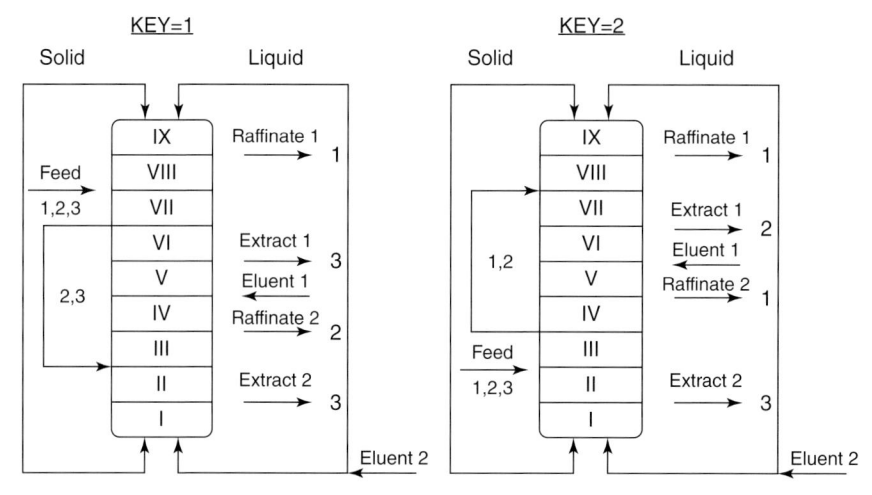

Figure 12.19 9-zone SMB designed ternary separations. (a) Key component 1. (b) Key component 2. (Reproduced with permission from Ref. [96]. Copyright 1988, Taylor & Francis.)

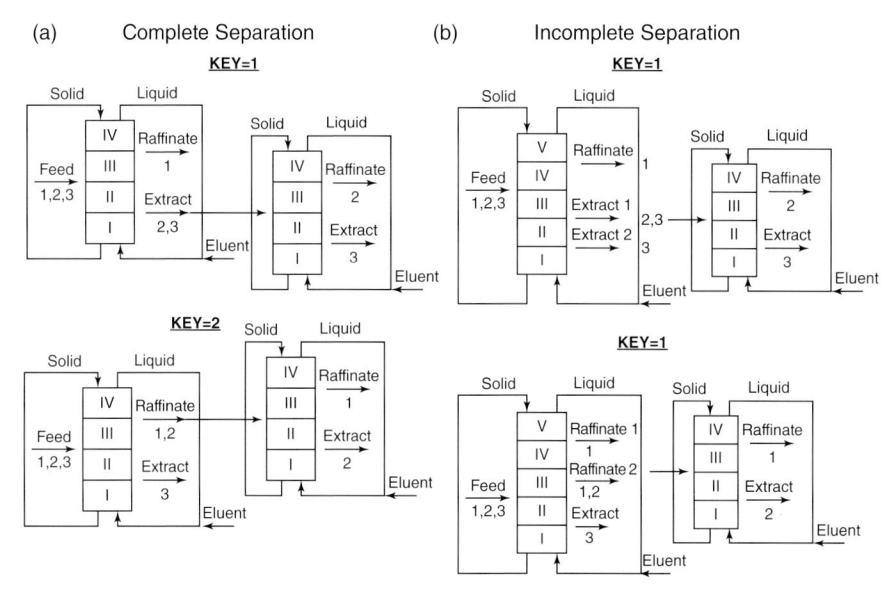

Figure 12.20 Tandem SMB configurations. (a) Complete separations. (b) Incomplete separations. (Reproduced with permission from Ref. [96]. Copyright 1988, Taylor & Francis.)

always be preferred over the others. The design can be simplified if only one component is desired.

In their earlier work, the research group of Prof. Wang proposed a nine-zone SMB for the recovery of glucose and xylose [100] that is essentially a combination of two systems. This tandem approach was also applied to an SEC–SMB system for insulin purification [68,69,92,101,102]. In this case, SEC–SMB was investigated for purification of insulin from high molecular weight proteins and zinc ions (Figure 12.21). Their experimental data proved a fourfold increase in productivity, 10% increase in yield, and two-thirds reduction in solvent consumption [68]. Due to the SEC applications, the separation was isocratic; thus, no re-equilibration was required and a conventional SMB was implemented. Several concerns, such as design optimization, process robustness, residence time, startup and shutdown, and column CIP, were addressed as well [69–73].

These SMB systems were university-built and are not commercially available; therefore, all required validations of the equipment are not provided. However, Prof. Wang's group published an FMEA analysis of their pump system [101], which provides essential information about the performance of the multicolumn system in case of different pump failures.

In comparison to SEC processes, the implementation of traditional SMB technology is not as straightforward for other modes of chromatography (Section 12.4). Usually, batch processes using the chromatographic modes with salt or pH gradients are scaled up in a more empirical or statistical approach. After high-throughput screening different solvent compositions, gradients, and packing materials, the chromatographic system is optimized at lab scale to achieve highest purity, yield, and robustness using resolution at different loads. Figure 12.5 shows

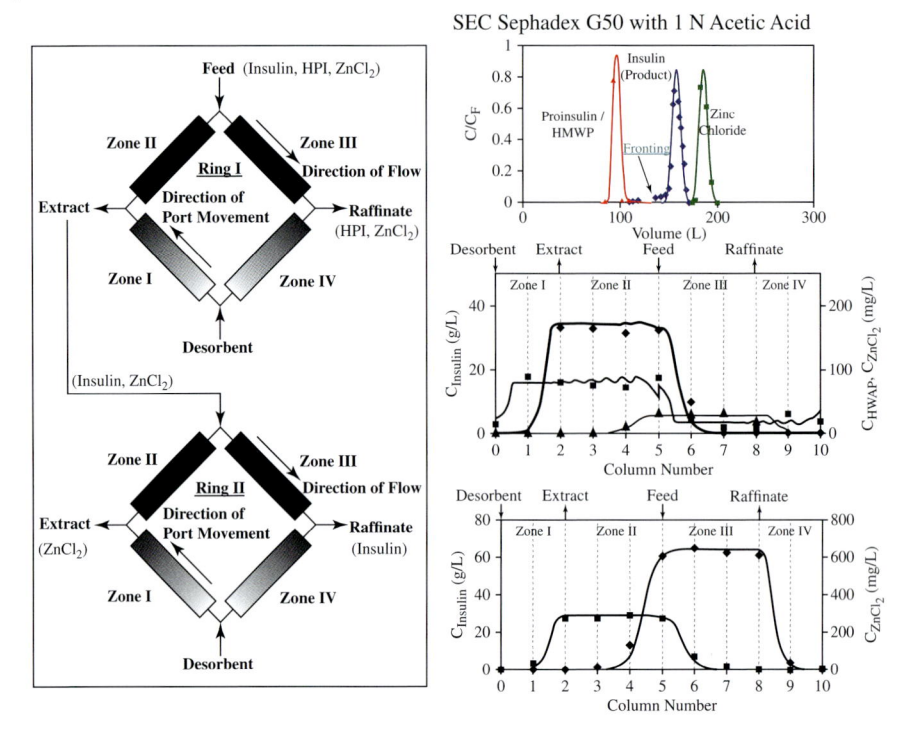

Figure 12.21 Tandem SMB for SEC purification of insulin. (Reproduced with permission from Ref. [68]. Copyright 2002, John Wiley & Sons, Inc., New Jersey.)

a typical chromatogram for a multicomponent mixture and the traditional process steps. Usually, columns are regenerated and rinsed by CIP solutions to ensure complete cleansing of the packing from late-eluting compounds and from denatured proteins.

The configuration in Figure 12.22 utilizes the partial countercurrent movement of the liquid and solid phases, as explained in Section 12.6.1, for the separation of a binary mixture. Due to the implementation of a solvent gradient and the CIP section, the system is now an open-loop SMB system with four zones. The less retained proteins elute as raffinate, and more retained proteins elute as extract. The process is suitable for capture and polishing steps.

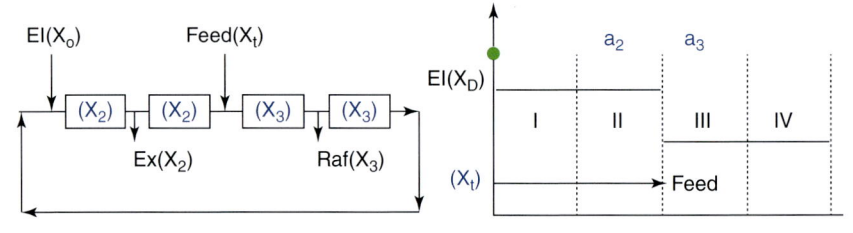

Figure 12.22 Binary SMB system with buffer gradient (eluent and feed have different buffer strength). (a) Schematic and (b) gradient configuration in different zones. (Reproduced with permission from Ref. [74]. Copyright 2002, Elsevier.)

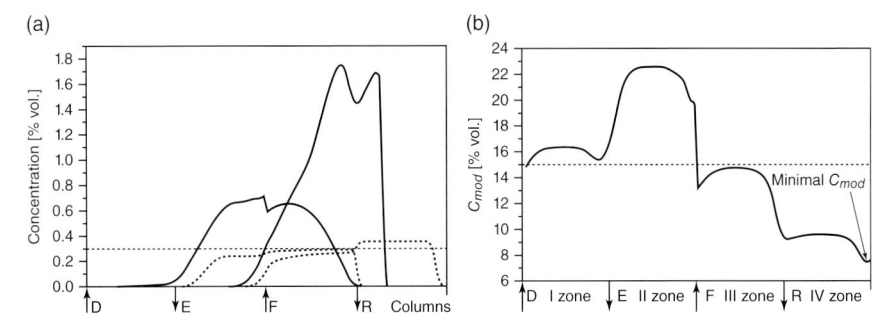

Figure 12.23 Binary SMB chromatography using step gradients. (a) Simulated internal concentration profiles of the extract and raffinate. (b) Simulated internal modifier concentration. (Reproduced with permission from Ref. [67]. Copyright 2001, Elsevier.)

To guarantee the required separation, the switching time t_{switch} cannot be too long to prevent the more retained proteins from reaching the raffinate outlet. However, the proteins should be able to leave through the extract port during one switching period. Buffer compositions are not as predictable as shown in Figure 12.22. Columns are not re-equilibrated when changing from zone 3 to 2 and from zone 1 to 4. During the transitions, the compositions change with time and over the column length. However, these SMB configurations reach "steady-state" conditions when the modifier compositions are constant at a particular time and position in the switching cycle. As shown in Figure 12.23b, Antos et al. [67] determined the corresponding modifier concentrations (buffer compositions) across the SMB system.

With the help of these buffer compositions, the researchers were able to determine the internal concentrations of the more and less retained proteins (Figure 12.23a). By applying the "triangle theory" to the different buffer compositions, they were able to determine the corresponding flow rates in each zone, as shown in Figure 12.24.

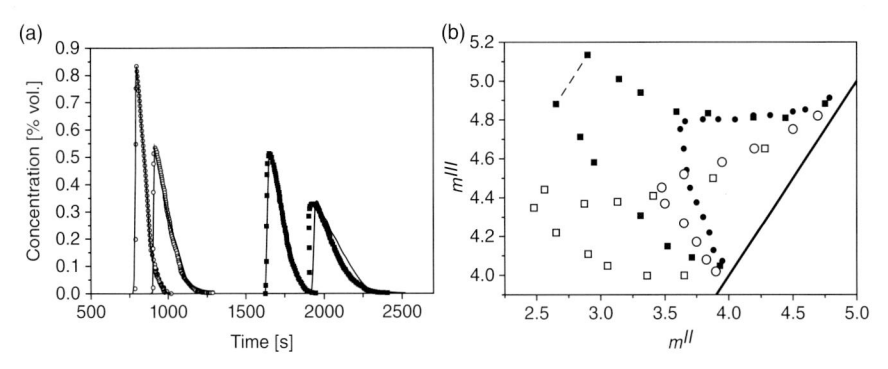

Figure 12.24 Flow rate determination for different SMB zones when using step gradient for elution. (a) Retention of binary mixtures at the two modifier concentration. (b) Separation triangles for both modifier concentrations based on the "triangle theory". Open symbols – strong modifier and closed symbols – weak modifier. (Reproduced with permission from Ref. [67]. Copyright 2001, Elsevier.)

The correlations derived for calculating the flow rates for each zone are listed as following:

$$H_1^{I,II} = f(C_{mod}^{I,II}) < H_1^{III,IV} = f(C_{mod}^{III,IV}),$$
$$H_2^{I,II} = g(C_{mod}^{I,II}) < H_2^{III,IV} = g(C_{mod}^{III,IV}).$$
$$H_2^{I,II} < m^I,$$
$$H_1^{I,II} < m^{II} < H_2^{I,II},$$
$$H_1^{III,IV} < m^{III} < H_2^{III,IV},$$
$$m^{IV} < H_1^{III,IV}.$$
(12.6)

Based on the "triangle theory" (Eq. (12.3)), the flow rate ratios in the different zones depend directly on the linear retention factors (Henry coefficients) of both components at the different modifier concentrations.

As already mentioned, protein mixtures are difficult to group into two fractions. Impurities are in the front and at the back of the protein peak of interest. To simplify the process to a ternary separation [14, 115], the front and back side impurities are grouped together into "Imp1" (less retained impurities) and "Imp2" (more retained impurities), respectively. Column configurations are modified by adding an additional zone that allows a ternary separation and one that regenerates the columns before moving from zone 1 to 4. Three different solvent strengths are introduced as indicated by "modifier 2" and "modifier 1" as well as the solvent composition of the feed stream. Modifier 1 generates the step gradient. Modifier 2 removes very strongly adsorbed impurities. If needed, a CIP step can be incorporated. The columns are re-equilibrated by the weak buffer composition before re-entering zone 4. In this configuration, the main product is collected at the extract port. Front and back side impurities can also be collected at the raffinate port and the column outlet after introducing modifier 2, respectively.

In Eq. (12.7,) the relationships between the switching time t_{switch} and the retention times of the compounds in different zones of a ternary separation are shown. The switching time has to be long enough to let the less retained impurity and the protein of interest move out the raffinate and extract outlet, respectively. Linear adsorption models (Eq. (12.3)) and mass transfer correlations provide the corresponding flow rates in the different zones. Prof. Wang's research group developed appropriate design tools in 2002 [68].

An additional design restriction is that the more retained impurity needs to be removed from the column by the second modifier, and that the column in the cleaning zone needs to be regenerated during the switching time. If the allowed time for a particular zone is not sufficient, additional columns are implemented in that zone:

$$t_{r\,raf,imp\,1} < t_{switch} < t_{r\,raf,protein},$$
$$t_{r\,ex,protein} < t_{switch} < t_{r\,ex,imp\,2},$$
$$t_{r\,regen,imp\,2}, t_{equal/regen} < t_{switch}.$$
(12.7)

Incorporating linear solvent gradients into a system significantly complicates the design of a robust separation due to nonlinear mixing of different solvents, the

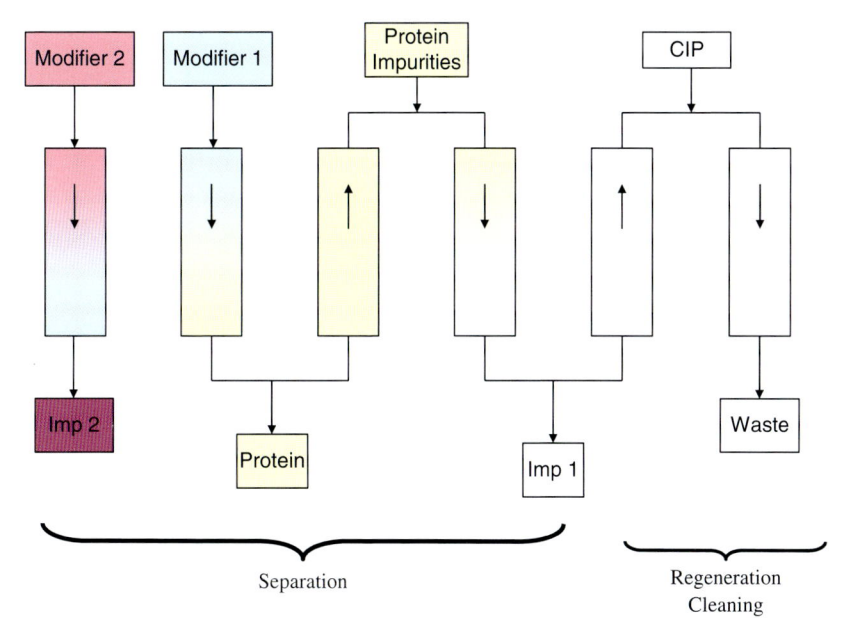

Figure 12.25 Schematic design of multicomponent SMB system: Imp 1 – less retained impurities, and Imp 2 – more retained impurities; red – highest solvent strength, blue – medium strength, yellow – protein, and white – lowest solvent strength.

continuously changing retention behavior of the molecules, and the semicyclic operating conditions of the system. Therefore, step gradients should be applied to reduce variability in the solvent composition and provide more robust operating conditions. If the process cannot be operated in isocratic mode, processes with step gradients are developed at the bench-top, and then scaled up to the manufacturing SMB system.

12.6.3 Additional Examples of SMB Purifications

For many years Carbon Calgon has used the ISEP and CSEP systems for large-scale separations[5]. The former one is based on the principle shown in Figure 12.10 that exhibits a more simulated annular chromatographic system. The columns are operated in parallel mode, and it is possible to change the inlet and outlet stream according to the process requirements.

The CSEP process displayed in Figure 12.26 takes advantage of countercurrent flow to improve the separation performance as already described (Figure 12.25). The flowchart in Figure 12.26 shows a binary separation. Both processes use a single multiport valve. A bench-top and semi-prep systems are distributed by Knauer[6] (Figure 12.27a). In Figure 12.27b, an industrial scale ion-exchange skid is displayed. Kessler and Seidel-Morgenstern describe how to implement the multiport valve for the continuous separation of a multicolumn mixture and the CIP of the system [103].

---

5 www.carboncalgon.com.
6 http://www.knauer.net/e/produkte/e_produkte.htm.

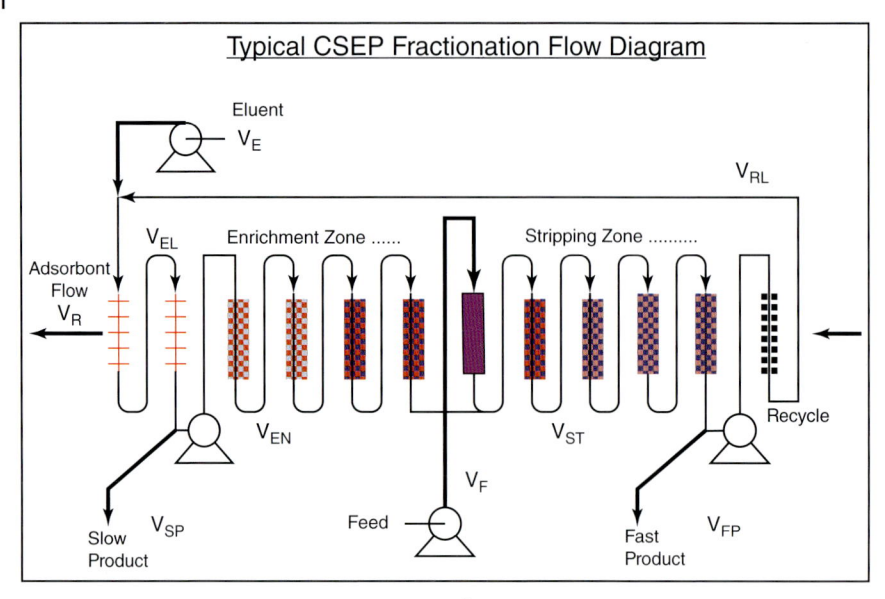

Figure 12.26 CSEP process from Carbon Calgon[5].

Conventional skid manufacturers for the pharmaceutical industry, such as GE Healthcare and NovaSep, modified their batch process equipment and traditional SMB equipment, respectively, into SMB systems for the purification of biomolecules. Examples [75] were presented using a lab-scale SMB based on the Äkta explorer system. Initial work was performed in collaboration with the research groups of Morbidelli and Mazzotti. By modifying the process control system and

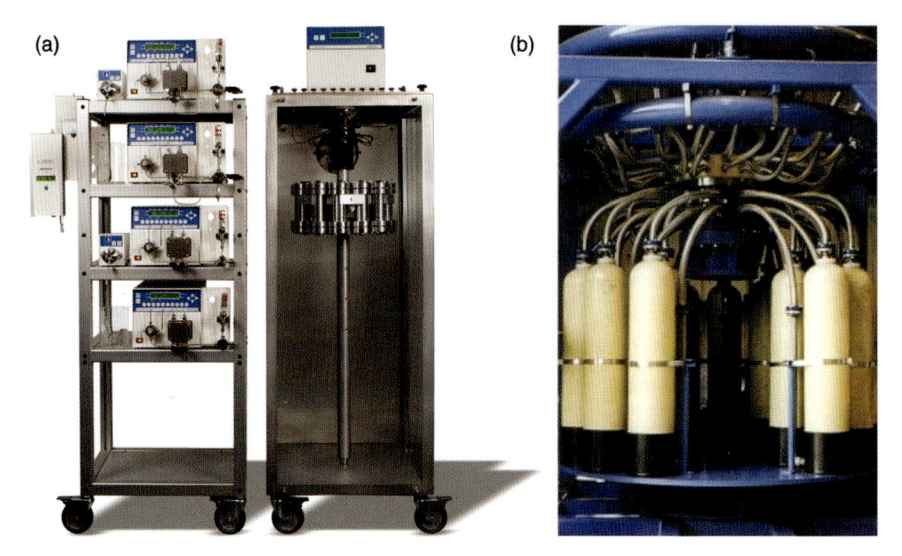

Figure 12.27 (a) CSEP SMB system (with kind permission from Knauer) and (b) Calgon Carbon Corporation's patented Adrian MN ISEP® system are engineered to meet specific application requirements. (Reproduced with permission from Ref. [15]. Copyright 1999, Elsevier.)

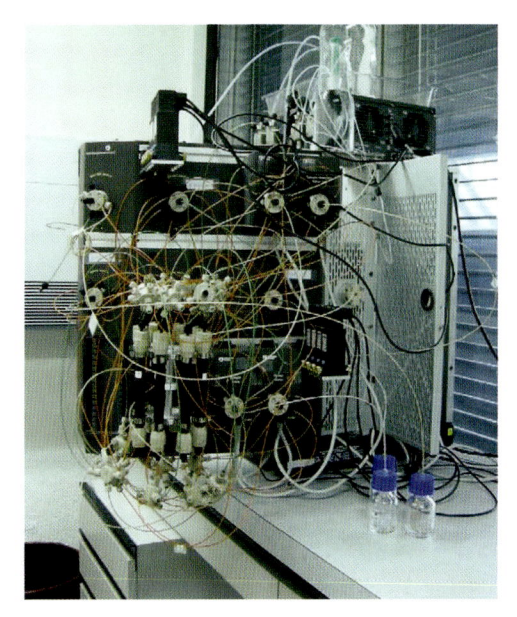

Figure 12.28 Picture of the modified batch chromatography system AKTAexplorer used as an SMB. (Reproduced with permission from Ref. [75]. Copyright 2004, Elsevier.)

incorporating an extra pump, check-valves, four 8-port valves, and manifolds, a conventional SMB system can be set up (Figure 12.28). The system has eight columns with manifolds between the columns to direct flow. Three fraction separations and coupled CIP, with cocurrent and countercurrent flow, can be realized.

Using the equipment setup already mentioned, Stroehlein *et al.* [76] published a design for continuous ternary separations with a modifier gradient (Figure 12.29). The system is divided into five zones, each having different modifier strengths on its inlet to allow the elution of pure products from zones 1, 3, and 5 during one switching period. As introduced in Figure 12.25, the liquid flow is not always continuous but interrupted for the three product outlets and modifier inlets. Zone 2

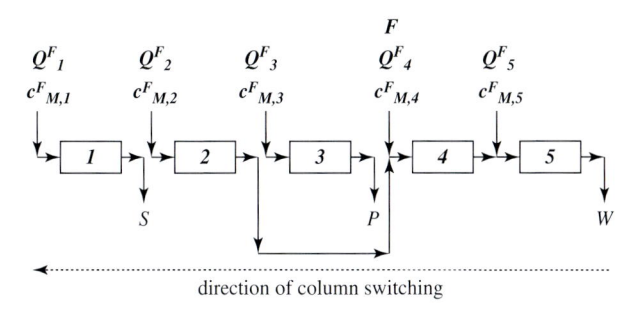

Figure 12.29 Five-zone continuous process for ternary separations using modifier step-gradients and feed pulse injections. P – product, S – strongly adsorbed impurity, and W – weakly adsorbed impurity. (Reproduced with permission from Ref. [76] (Figure 3). Copyright 2006, Elsevier.)

outlet stream bypasses zone 3 and directly re-enters the inlet of zone 4. Using similar separation conditions to those shown in Figure 12.24, the complete separation of the three compounds is guaranteed and has been proven experimentally.

In 2004, Paredes et al. [104] presented their results for the separation of a binary mixture of nucleosides with a strongly retained impurity. Their theoretical results were verified using the SMB modification from GE Healthcare for three fractions (Figure 12.28). The researchers also published their work on plasmid DNA [66]. With their experimental set up, a three-fraction separation with CIP was accomplished using SEC for the primary purification step. Additionally, Paredes and Mazzotti optimized the batch and continuous separation [105] where the latter has double productivity while still having similar solvent consumption.

Univalid from Leiden, the Netherlands, reported work on an IgG capture step using SMB technology[7]. The work was performed for IDEC Pharmaceuticals. To reduce the large amount of solvent required for the large-scale batch process, and to efficiently use an expensive packing material, 20 columns packed with ProSep rA (Millipore) were run in a CSEP-SMB.

Merck published a two-step process purifying therapeutic oligonucleotides that applies step SMB chromatography after the initial synthesis [106]. A single SMB step achieved the required purity values with higher throughput. With this platform, not only antisense oligonucleotide but also siRNAs, aptamers, and CpG nucleotides could be produced at lower costs. Merck and Bayer Technology Services GmbH collaborated in the area of column packing and SMB technology [106].

Houwing et al. [107] studied the continuous separation of BSA and myoglobin using an SEC–SMB process. The purity of BSA was as high as 99%. The authors redesigned the separation using ion exchange chromatography [108] based on charge differences between the two molecules. By implementing a salt gradient, the compounds were further concentrated. With the help of the "triangle theory," operating conditions were found for the complete separation of the two molecules using a reverse gradient. The theoretical results were verified experimentally. In their later work [109], the authors improved the salt gradient position within the SMB unit by using an ion-exclusion isotherm model for the salt ions.

At the 2005 Prep Symposium, Auman et al. [110] presented experimental results of the separation of calcitonin by continuous multicolumn chromatography. At lab-scale they used a modified Akta purifier system (Figure 12.28), which allowed the separation of a feed mixture with 55% purity into three streams where the middle stream was the compound of interest. Optimal operating conditions reached purities and yields greater than 85 and 95%, respectively, than in corresponding batch process.

At the Prep 2006, Seidel-Morgenstern's group presented their work on the isolation of proteins and antibodies using gradient SMB processes [111] as a continuation of their previous theoretical work [67]. A two-step gradient reduces solvent consumption and increases product concentration compared to the isocratic SMB process. Two model systems were described: Bovine IgG from

---

7 http://www.actip.org/manuals/filesmarburg/univalid1-16.pdf.

lysozyme and/or BSA, and the active dimeric and monomeric form of bone morphogenic protein-2 using a three- or four-zone SMB.

12.7 Risk Assessment of Continuous Chromatography

Previous sections provided detailed insights into different classifications, terminologies, and process configurations used in chromatography, which allows now the risk assessment of different chromatographic processes and their associated technologies. Although SMB technology has been successfully implemented in the manufacturing of chiral compounds, and researchers have provided proof of concepts for continuous DSP [1–7,9], real and/or perceived hurdles still exist that hinder the implementation of continuous chromatography into the manufacturing of biomolecules. Technical and regulatory hurdles have been presented by the author and guidance was provided on how to overcome them [14–19], which will be summarized here.

The major technical challenges for multicolumn continuous chromatography [14–19] are the complex skid designs with additional valves and chromatographic columns, as well as the implementation of PAT. These challenges are weighted with the high initial capital investment of the skids, multiple pumps, and columns, as well as more complex process design and control when working in the regulated environment. Comprehensive risk assessment for each of the chromatographic steps is required. Some of the inherent risks of the traditional batch chromatography are reduced due to the continuous operation. For instance, the entire batch is not lost in case of a failure, as happens during single, one-time injections into large columns. Other risks are enhanced, but can be mitigated by implementing control strategies [3,80,81], like using PAT to guarantee stability of the biomolecules during long-term operation.

These complexities increase the failure risk of processes and their equipment designs. These risks can be mitigated by conducting risk assessments and implementing the appropriate control strategies (Table 12.2). Due to the improved productivities, smaller columns and skids with smaller pumps, valves, and piping are required; thus, capital investment of complete chromatographic systems (for both skids and columns) are reduced. Smaller equipment scales improve the process robustness and, consequently, lower the process risks. Smaller scales also reduce the footprint of chromatographic skids and the entire DSP. Using modular design allows parallel operation of multiple chromatographic skids, or entire manufacturing schemes, without any additional scale-up efforts. Modular design provides more flexibility when demands change, without adjusting the equipment scale. For instance, one of the CaptureSMB units can be added or removed if batch sizes change due to schedule modification.

As shown by the complex piping design of traditional SMB processes (Figure 12.26), there is an obvious benefit to minimize the number of columns. Lab-scale systems with three or more columns have been proposed [3,9]. The CaptureSMB and MCSGP [7] technology by ChromaCon, and now the scale-up EcoPrime Twin system (Figure 12.2), only have two columns. This simple skid design (far less complex valve arrangements) with less complex process

Table 12.2 Parts of generic risk assessment for EcoPrime twin process.

Description	Probability	Severity	Impact (GMP, GAMP5...)	Detectable	Comments complexity, novelty... detectable	Risk control measures
General risks capture SMB						
Process: batch versus continuous	Medium	Medium	Medium	Yes	Same process steps only feed continuously using multiple column, possible long-term operation (24 h to 6 weeks), perception that different process	Monitoring using PAT, process, and cleanability verification on bench-top scale, adjusted automation
Skid: batch versus continuous	Medium	Medium	Medium	Yes	Very similar design that is capable to run two column parallel or sequential	Verification of design (see below) no dead legs or back mixing
Mechanical and chemical stability of biomolecules	Medium	Medium	High	Yes	Novel continuous process, long-term stability data needed under this operating conditions	Long-term feasibility studies, control strategies, PAT implementation, equipment cleanablity studies
Mechanical and chemical stability of resin	Low	Medium	Medium	Yes	Novel continuous process, long-term stability data needed under this operating conditions	Long-term feasibility studies, control strategies, PAT implementation, equipment cleanablity studies
Skid design						
Complexity	High	Medium	Medium	Yes	More complex design with additional parts, need for more complex automation and control strategy	Rigorous design to avoid any dead volumes, monitoring CPP, implementing cleaning procedure
Valves						
Multiple port valves	Medium	High	High	Yes	When one fails more potential negative effects	Double valves, feedback from valves
Single on–off valves	High	Medium	High	Yes	Large number of valves but the effect of one failing is not as	Double valves on important points, valve feedback
Columns						
Twin	Medium	Medium	Medium	Yes	Two column but smaller design, more robust and efficient	Testing of columns, pressure monitoring, cleaning of skid and column according to strategy

configurations and controls has a lower risk of failure. Unique pre and post-column valve block arrangements (Figure 12.14) allow operating two columns. The skid design is similar to the traditional batch skids. The two lines connecting both columns enable the interconnected operation in both configurations: going from the outlet of column 1 to column 2 and from outlet of column 2 to column 1 (Figure 12.14). To ensure proper operation, the piping design between the valve blocks must be symmetrical. All skid piping must be minimized to reduce hold-up volumes. The ratio between these external volumes to the column volumes should remain constant while scaling up. This simple design for twin-column process enables a robust operation by reaching comparable throughputs and high productivities, as described in Section 12.8.

Skid piping must be designed to secure accurate liquid flow through the skids without any cavitation. Components such as sensors, filters, and valves should be chosen accordingly. Pressure regulators can be installed to adjust the pressure distribution. Computational fluid dynamic (CFD) tools are helpful to estimate not only the pressure distributions within the skid, its piping, and components, but also to determine the correct fluid and cleaning flow design.

The product's materials of construction (MoC) must be compliant with FDA requirements. Characteristics such as inertness (no impact on the mechanical and chemical stability, or the biocomparability of the biomolecules) and cleanability are important. This is in particular important for valve and pump design. In the past, only high-grade stainless steels were implemented. Recently, polymers such as polystyrene, polysulfone, polyethylene, polyamide, and ethylene vinyl acetate have gained popularity [1,2,82,83]. Their advantages are low costs and flexibility. The cost for skid and column cleaning and their validation are minimized. However, single-use components might not have the same chemical and mechanical stability as steel components. The components are more affected by their operational environment, for example, temperature, light, oxygen levels, pressure, and sterilization irradiation. Their lifetime is limited. Stricter monitoring of extractables is also required [2,112].

CIP steps can be a direct part of the process step sequence, ensuring that the feed streams are segregated from the CIP lines and that no CIP solution remains in hidden dead legs. Valve configurations are, therefore, important.

In critical components, such as pumps and valves, hold-up volumes should be avoided. Any hold-up volumes result in stagnant liquids and allow cross-contamination. It is best to source pumps with hygienic designs.

Valves are important to control the flow in the multicolumn continuous chromatography skid. Are configurations containing simple two-way valves preferable to complex multiport valves, if a large number of simple valves are needed? Or, is this a matter of scale? At the bench-top scale, there might be other valve types available than at the pilot or large manufacturing scale. Additionally, FDA requirements are different when working in PRD or GMP manufacturing environments.

Having reproducible column packing is particularly important in multicolumn processes. Any variability in the column characteristics affects the process performance [91]. By implementing control strategies, like setting tight efficiency limits, column variability can be controlled, yet not avoided. Knowing the challenge in packing "identical" columns, one should include variability during the process design phase. As required in the QbD directive of the FDA, this will allow more robust yet flexible processing.

Due to the scale down of the columns in an MCC process, prepacked single-use columns can be implemented into manufacturing processes. These columns are factory-packed and might already be qualified, performing therefore, more accurately and reproducibly. This is particularly attractive for CMOs due to reduced demands on cleaning and precampaign qualification, leading to shorter turnaround times between campaigns.

During the continuous operation of the columns, the mechanical and chemical stability of the packing material, and the characteristics thereof, must be guaranteed. Switching valves changes not only the flow directions in the MCC process, but also influences the pressure in the system and columns. The packed bed should not shrink or expand when going through the process steps, as this negatively affects the column performance and lifetime.

Detailed risk assessments of equipment components and process aspects of multicolumn, continuous chromatography should be conducted prior to building the chromatographic systems, and implementing such processes [23,24,26]. In each phase, a group of subject matter experts with backgrounds in quality control, regulatory affairs, automation, skid design, pharmaceutical PRD, and manufacturing should come together. Using formal risk assessment tools, they are able to provide quantitative evaluations of the processes or systems [26]. Parts of a simple, generic assessment and analysis are shown for the capture step in Table 12.2.

Besides evaluating throughput and capacity during design of the capture process, which is displayed in Figure 12.14, a cost–risk assessment has to be conducted. The EcoPrime Twin system (Figures 12.2 and 12.14) includes only two pumps: one for loading feed and one for delivering buffers for the washes, elution gradient, CIP, regeneration, and equilibration. The assessment should also include performance improvements of small columns that have higher efficiency and, therefore, better separation performance. Nevertheless, less complicated systems with fewer valves, pumps, and columns will have lower performance risks.

General risks as well as risks of skid components are listed in Table 12.2. Probability and severity of failure or risk are evaluated when considering patient safety. Additional comments evaluate impact on GMP and GAMP5 directions, complexity, novelty of risk, and if it is detectable. Within the scope of product life cycle management, these assessments are revised, as new knowledge about the process and skid becomes available.

Each of the subject matter experts has different sets of experiences, and, thus, different understanding of the risks and the requisite mitigation. For instance, research scientists have data about the long-term chemical and mechanical stability of molecules, can conduct the required experiments to characterize molecules, and know how to monitor these characteristics. The design engineer implements the recommendations for risk control and mitigation into revised P&ID and functional specifications.

As already shown, for the continuous capture process and its equipment implementation, similar assessments have to be executed for each of the process units of integrated DSP. When assessing the entire scheme, one needs to be aware of the complexity and interconnections of each step, and potential risk mitigations. Process changes in the upstream affect the performance of the DSP units. As mentioned earlier, variability in the feed concentration and composition can be designed into processes, by adjusting the loading time for instance. Appropriate control strategies can be implemented while monitoring the concentration online [80,81]. Inaccurate or nonreproducible buffer delivery during elution causes variability in the composition of the product stream. Changes in the feed compositions have to be investigated thoroughly to ensure that subsequent process units handle higher impurity without negatively affecting the CQAs of the drug product.

12.8 Process Design of Continuous Capture Step

Implementing continuous chromatography into the DSP scheme is straightforward for the capture step. In affinity chromatography, the molecules of interest are captured in the bind–elute or flow-through mode. Figure 12.5 shows a typical recipe and chromatogram of a batch chromatographic process [17] that includes regeneration, equilibration, load, wash(es), and elution steps. Columns are typically loaded up to 10% breakthrough (DBT) of their dynamic binding capacity of the resin (Figure 12.31). Thus, only this loading capacity is utilized to minimize the yield losses of the capture step.

During elution, the buffer compositions are adjusted by step or linear gradients. Their adsorption behaviors are modified from the on- and off-mechanisms. Depending on the CIP strategy, the column is cleaned and re-equilibrated. Due to the limited load onto the column, the buffer consumption is relatively high in batch processes. Additionally, the exposure of the column to the CIP solutions is high that has significant impact on the lifetime of the protein A.

To increase the loading onto the column, residence times of biomolecules are increased to overcome the slower mass transfer [90]. The breakthrough curve of the slower loading velocity (blue line in Figure 12.31) is steeper than for faster loading velocities; thus, more feed can be loaded onto the column. However, due to the slower feed rate, processing time increases significantly and, therefore, the throughput and productivity of the process are reduced.

Generally, high resolutions and high yields are obtained due to the specific binding behavior in affinity chromatography. Process optimization now becomes a question between column capacity and throughput, and therefore, buffer consumption, as illustrated in Figure 12.30.

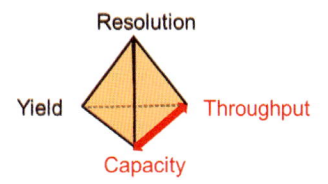

Figure 12.30 Process design objective for affinity chromatography.

Due to the high cost of the resin (i.e., protein A), the focus has been on optimizing the resin utilization. One simple approach is dividing a long batch column into multiple shorter columns (Figure 12.12), which are then operated in parallel or/and sequentially. Without any modifications of the process steps, the capture step can be run as a multicolumn, continuous batch chromatograph. Detailed process descriptions are in Section 12.5.

Productivity is an excellent parameter when optimizing configurations and evaluating their performance:

$$\text{PR} = \frac{m_{\text{mAb}}}{V_{\text{resin}} \cdot t} \tag{12.8}$$

being the amount purified per total resin volume per time (kg/(l_{resin} day)). The total resin volume here refers to the resin volume of all columns, not only the column that is in the recovery phase. The production rate is directly correlated to the feed flow rate. If the feed is put into the column faster, the bind–elute cycles can be repeated more frequently, and, therefore, more product per time period is recovered. Elution is generally less dependent on the flow rate during capture, and, therefore, can be performed at the highest possible flow rate. The wash, regeneration, and equilibration steps are usually defined by column volumes, and can also be performed at the highest possible flow rate.

In traditional batch chromatography, the highest possible flow rate is limited by the pressure rating of the column and its packing. The pressure drop Δp across the column depends on column length, linear velocity, and resin porosity, and can be calculated by the Carman–Kozeny equation. In multicolumn processes, columns operate not only in parallel but also sequentially (Figure 12.13). The pressure drop depends now on the total length of connected columns; thus the flow rates might have to be reduced during the loading. Consequently, the load time and processing time increase, and, therefore, the production rate reduces. During recovery, usually only one column is segregated, thus, higher flow rates can be applied.

Optimal loading times are determined from the dynamic breakthrough curves at different linear velocities (Figure 12.31). Of course, when determining the loading time for the sequential connected column, the preload has to be taken into account.

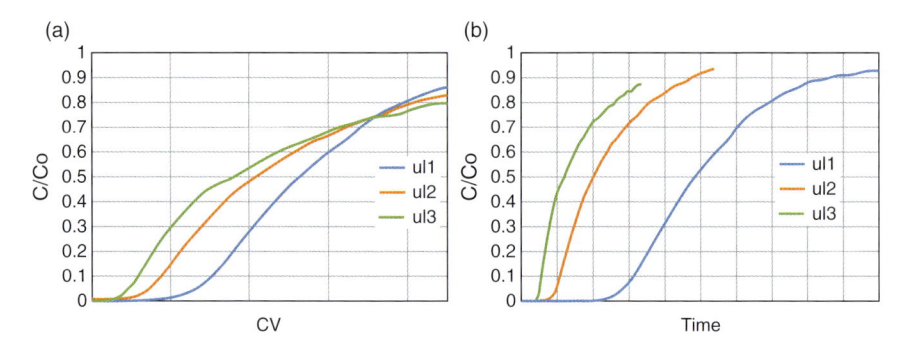

Figure 12.31 Dynamic breakthrough curves. (a) Based on CV and (b) based on time. Three different linear velocities: $ul3 > ul2 > ul1$. Gray area indicates 10% DB.

Table 12.3 Theoretical comparison of different column configurations.

No. of col	L sin CV	L con CV	uL sing (cm/h)	uL con (cm/h)	Cycle (min)	Load per cycle (g)	L per col (kg/l$_{res}$)	L per h (g/h)	Prod (kg/(l$_{res}$ day))	Buffer (l/gprod)
1	15	0	200	0	120	1.84	0.038	0.92	0.45	0.67
2	15	5	200	200	60	2.45	0.050	2.45	**0.60**	0.50
3	15	20	200	200	105	4.30	0.029	2.45	0.40	0.29
3	25	0	200	0	75	3.07	0.021	2.45	0.40	0.40
4	25	18.75	200	150	150	5.37	0.027	2.15	0.26	**0.23**

A capture process example [16] is transferred from a single column to a continuous batch process using multiple columns. The feed titer is 2.5 g/l of protein. The recipe of the original batch process (Figure 12.5) is defined as follows:

Equilibration: 5 CVs
Load: 5 CVs
Wash1: 5 CVs low salt
Wash2: 5 CVs high salt
Elution: step gradient with 5 CVs

Applying mass balances at different column configurations (Table 12.3): single-column and continuous batch processes, production rates and buffer consumptions are calculated based on their feed flow rates [16]. The feed flow rates were determined with the help of the corresponding breakthrough curves at three linear velocities. Figure 12.31 shows the breakthroughs based on CV (Figure 12.31a) and time (Figure 12.31b). When evaluating resin utilization, the slowest linear velocity has the largest number of CVs loaded onto the column. At this condition, mAbs have sufficient residence time to overcome mass transfer limitations when binding onto the resin. However, when evaluating productivity, longer loading times significantly reduce the throughput per time period. When increasing the linear velocity from $u3$ to $u1$, the load at 0% breakthrough doubles while the loading time increases four times. For a single-column batch process, both process conditions might have the same productivity (now depending on the other process steps, such as the number of wash steps and their length). However, in case of continuous batch processes running shorter columns parallel, the faster process velocity leads to significant improvements in productivity. The following calculation uses the medium velocity to avoid possible high-pressure limitations.

For the single-column batch process, the estimated production rate and buffer consumption are 0.45 kg/(l$_{resin}$ day) and 0.67 l/g$_{prod}$, respectively (Table 12.3). For the calculations, loads are repeated for all processes. Using two columns in the CaptureSMB configuration (Figure 12.13b) allows an increase in loading of 5 CV while interconnected. The production rate rises to its highest value of 0.6 kg/(l$_{resin}$ day) (values given in bold in Table 12.3), while reducing the buffer consumption to 0.5 l/g$_{prod}$.

Adding more columns into the loading zone does not lead to higher productivities, even if the process is operated in the CaptureSMB configuration. In case of three-column and four-column configurations, two or three columns are in the loading zone and one in regeneration, respectively, during parallel operation (Figure 12.13b). Between the parallel operations, interconnected process steps are introduced during which additional 20 and 18.75 CVs are loaded for a three-column and four-column configurations, respectively. Although, the highest load per column (per elution step) was obtained in the latter case; this configuration has still the lowest productivity. In addition, the linear velocity of the feed stream was reduced to 150 cm/h due to pressure limitations of the three columns that are connected sequentially.

In contrast, buffer consumptions decline with increasing column numbers. The lowest buffer consumption of 0.23 l/g_{prod} was reached (values in bold in Table 12.3) when using four columns in the CaptureSMB configuration.

The three-column configuration without interconnected steps has a slight reduction in the buffer consumption by a significantly lower productivity.

In summary, configurations with fewer columns can generally run faster, improving the productivity. However, buffer consumptions decline when more columns are implemented. Due to shorter loading periods and faster regeneration and elution cycles, faster processing using fewer columns is generally more favorable.

12.9 Conclusion

Implementing continuous chromatographic processes into the downstream purification of biologics, in particular mAbs, is not only driven by the pharmaceutical industry to improve process efficiency and robustness but also by regulatory agencies. In recent years, there has been a tremendous increase of activities in this field. All major biopharmaceutical companies and manufacturers of resin, sensors, and equipment have invested significant resources. New results and developments are presented to the community constantly. Initial investigations are scaled up into the GMP environment of manufacturing facilities. Biochemists and engineers work closely together to make the transition from the bench-top to the manufacturing scale possible.

In this chapter the author provided insight into different modes of chromatography, in particular continuous chromatography by clarifying terminology, such as the term "batch" with its different meanings. Batch in its traditional chromatographic meaning describes a distinct load onto columns that is operated batchwise. Nevertheless, this batch process can also operate continuously in a single-column but also multiple-column configuration. In the latter case, the chromatographic skid can even be loaded continuously (Figure 12.3); however, this process is *not a traditional SMB chromatography* due to missing internal recycling. It still uses the same process steps as applied in traditional batch chromatography. There are no changes to the binding characteristics and product peak composition and concentration. None of these "batch" terms reflects exactly the regulatory meaning of "batch" where it describes an amount of product with defined characteristics obtained during a specific time interval.

Understanding the differences in chromatographic, engineering, and regulatory definitions describing the technologies will overcome still existing real and perceived hurdles in implementing continuous chromatography, in particular under QA/QC and regulatory aspects. It will support cost–risks assessments when evaluating the different modes of operation by highlighting the advantages of continuous operation, such as segregation of out-of-spec product without losing the entire feed batch. In the case of continuous batch chromatography, the residence time of molecules remains defined; thus molecules can be easily followed throughout the process.

Insight into different chromatographic technologies starting from batch to semibatch/semicontinuous to continuous operation on single and multiple column to multicolumn, continuous, countercurrent chromatography provided technical and process know-how to identify the proper technology for the separation task at hand. Guidelines on how to overcome remaining hurdles were summarized, in particular when simplifying complex process and equipment designs and still taking advantage of continuous operation.

From this information, it can be concluded that multicolumn continuous technology reduces not only the footprint of manufacturing facilities but also improves process performance and robustness. It enables the implementation of single-use technologies due to the size reduction. It also allows modular design of process trains that can be easily multiplied during process scale-up and that provide flexibility in multiproduct facilities with different batch sizes (e.g., when producing personalized medicine, higher titer products, orphan drugs, and so on). Therefore, fast turnaround times are not only important for CMOs but for all biopharmaceutical companies.

Design and optimization of the capture step using multicolumn continuous chromatography was described in detail. The effects of the number of columns on the production rate (amount purified per time per total packing amount) and buffer consumption were evaluated. The highest production rate was obtained for a twin-column system in the example presented. The lowest amount of buffer was consumed on a four-column system.

From this information, it can be concluded that savings in resin costs, especially protein A resins, have significant impact on the capital portion of the production costs of monoclonal antibodies. The twin-column system is, therefore, attractive for the early development phase to production scale. Its fast turnaround times based on higher production rates are desirable. Buffer consumption is not as critical due to the small scale of this development phase.

Compared to the traditional SMB process, the simpler skid design of the twin-column system is preferred based on the risk assessments; in particular at the manufacturing scale. There are fewer risks of failure of equipment components. Less complexity in the process enables the separation of more complex feed mixtures with their variability common in DSP processes.

References

1 Subramanian, G. (ed.) (2015) *Continuous Processing in Pharmaceutical Manufacturing*, Wiley-VCH Verlag GmbH, Weinheim, Germany.

2 Subramanian, G. (ed.) (2012) *Biopharmaceutical Production Technology*, Wiley-VCH Verlag GmbH, Weinheim, Germany.
3 Godawat, R., Brower, K., Jain, S., Konstantinov, K., Riske, F., and Warikoo, V. (2012) Periodic counter-current chromatography – design and operational considerations for integrated and continuous purification of proteins. *Biotechnol. J.*, **7** (12), 1496–1508.
4 Warikoo, V., Godawat, R., Brower, K., Jain, S., Cummings, D., Simons, E., Johnson, T., Walther, J., Yu, M., Wright, B., McLarty, J., Karey, K., Hwang, C., Zhou, W., Riske, F., and Konstantinov, K. (2012.) Integrated continuous production of recombinant therapeutic proteins. *Biotechnol. Bioeng.*, **109** (12), 3018–3029.
5 Jungbauer, A. (2013) Continuous downstream processing of biopharmaceuticals. *Trends Biotechnol.*, **31** (8), 479–492.
6 Pollock, J., Bolton, G., Coffman, J., Ho, S.V., Bracewell, D., and Farid., S. (2013) Optimising the design and operation of semi-continuous affinity chromatography for clinical and commercial manufacture. *J. Chromatogr. A*, **1284**, 17–27.
7 Auman, L., Mueller-Spaeth, T., Stroehlein, G., and Morbidelli, M. (2007) Multicolumn countercurrent solvent gradient purification of biomolecules (MCSGP), in *Advances in Large-Scale Biopharmaceutical Manufacturing and Scale-Up Production*, 2nd edn, ASM Press and BioPlan Associates, Inc., p. 907.
8 BioPharm International (2014) Amgen Opens Single-Use Manufacturing Plant in Singapore, Available at www.biopharminternational.com/amgen-opens-single-use-manufacturing-plant-singapore, November 20.
9 Holzer, M., Osuna-Sanchez, H., and David, L. (2008) Multicolumn chromatography – a new approach to relieving capacity bottlenecks for downstream processing efficiency. Bioprocess International, September, p. 74.
10 MIT (2014) International Symposium on Continuous Manufacturing of Pharmaceuticals, Boston, MA, May 20–21.
11 Goudar, C.T., Titcherner-Hooker, N., and Konstantinov, K. (2015) Integrated continuous biomanufacturing: a new paradigm for biopharmaceutical production. *J. Biotechnol.*, **213**, doi: 10.1016/j.jbiotec.2015.08.015.
12 Goudar, C.T., Farid, S., Hwang, C., Lacki, K., Titcherner-Hooker, N., Konstantinov, K, Betenbagh, M., and Buckland, B. (2015) Integrated Continuous Biomanufacturing II. An ECI Conference Series, Berkeley, CA.
13 Morbidelli, M., Mueller-Spaeth, T., and Mihlbachler, K. (2015) Webinar, Part 1: Twin-Column Chromatography for Purification of Biomolecules, March 3; and Part 2: Comparison of Multi-Column Continuous Chromatography Processes October 22.
14 Mihlbachler, K. (2005) Obstacle of SMB implementation into the production of bio-molecules, PREP Symposium.
15 Mihlbachler, K. (2007) Simulated moving bed chromatography, in *Advances in Large-Scale Biopharmaceutical Manufacturing and Scale-Up Production*, 2nd edn, ASM Press and BioPlan Associates, Inc., p. 817.
16 Mihlbachler, K. (2014) Implementing Continuous Chromatography into DSP of Bio-Molecules, PREP Symposium in Boston, MA.

17 Mihlbachler, K. and Morbidelli, M. (2015) Continuous chromatography for biomolecules purification, Workshop at 2015 PREP Symposium in Philadelphia, PA.
18 Mihlbachler, K. (2014) Implementing integrated continuous downstream processing, SPICA 2014, Basel, Switzerland.
19 Mihlbachler, K. (2015) White Paper Integrated Continuous Downstream Processing – An Enabling Manufacturing Approach.
20 FDA-Guidance (2004) Innovation and Continuous Improvement in Pharmaceutical Manufacturing Pharmaceutical CGMPs for the 21st Century, Available at www.fda.gov.
21 FDA-Guidance (2004) Guidance for Industry PAT — A Framework for Innovative Pharmaceutical Development, Manufacturing, and Quality Assurance, Available at www.fda.gov, September.
22 ICH (2009) ICH Harmonised Tripartite Guideline, Pharmaceutical Development Q8(R2), Current Step 4, August.
23 ICH (2005) ICH Harmonised Tripartite Guideline, Quality Risk Management Q9, Current Step 4, November.
24 ICH (2008) ICH Harmonised Tripartite Guideline, Pharmaceutical Quality System Q10, Current Step 4, June .
25 ICH (2012) Harmonised Tripartite Guideline, Development and Manufacture of Drug Substances (Chemical Entities and Biotechnological/Biological Entities) Q11, Current Step 4, May.
26 ISPE (2010) Risk-Based Manufacture of Pharmaceutical Products: A Guide to Managing Risks Associated with Cross-Contamination, vol. 7, Tampa, FL.
27 ASTM, Standard E2968-14 (2014) Standard Guide for Application of Continuous Processing in the Pharmaceutical Industry, revision WK51471.
28 Woodcock, J. (2014) Modernizing pharmaceutical manufacturing – continuous manufacturing as a key enabler, at the International Symposium on Continuous Manufacturing of Pharmaceuticals, MIT, Boston, MA, May 20–21.
29 Woodcock, J. (2015) Introducing new technology in pharmaceutical at the Integrated Continuous Biomanufacturing II conference, Berkley, CA, November 1–5.
30 Subramanian, G. (ed.) (1995) *Process Scale Liquid Chromatography*, Wiley-VCH Verlag GmbH, Weinheim, Germany.
31 Neretnieks, I. (1975) A simplified theoretical comparison of periodic and countercurrent adsorption. *Chem. Ing. Technik*, **47**, 773.
32 Svedberg, U.G. (1976) Numerical solution of multicolumn adsorption processes under periodic countercurrent operation. *Chem. Eng. Sci.*, **31**, 345.
33 Liapis, A.I. and Rippin, D.W.T. (1979) The simulation of binary adsorption in continuous countercurrent operation and a comparison with other operating modes. *AIChE J.*, **25**, 455.
34 Carta, G. and Pigford, R.L. (1986) Periodic countercurrent operation of sorption processes applied to water desalination with thermally re-generable ion-exchange resins. *Ind. Eng. Chem. Fundam.*, **25**, 677.
35 Liu, P.D. and Pigford, R.L. (1987) Preparative separation of proteins by periodic countercurrent sorption. Paper presented at the AIChE Spring National Meeting, Houston, TX.

36 Arve, B.H. and Liapis, A.I. (1988) Biospecific adsorption in fixed and periodic countercurrent beds. *Biotechnol. Bioeng.*, **32**, 616.
37 Broughton, D.B. and Gerhold, C.G. (1961) U.S. Patent No. 2,985,589.
38 Broughton, D.B. (1968) Molex: case history of a process. *Chem. Eng. Prog.*, **64**, 60.
39 Broughton, D.B., Neuzil, R.W., Pharis, J.M., and Brearley, C.S. (1970) The parex process for recovering paraxylene. *Chem. Eng. Prog.*, **66**, 70.
40 Cheng, L. (2014) Advancement of UOP petrochemical technology. Annual AIChE Meeting, Atlanta GA, 35d.
41 Broughton, D.B. (1991) Production-scale adsorptive separation of liquid mixture by simulated moving bed technology. *Sep. Sci. Technol.*, **19**, 723.
42 Barker, P.E. and Joshi, K. (1991) The recovery of fructose from inverted sugar beet molasses using continuous chromatography. *J. Chem. Technol. Biotechnol.*, **52**, 93.
43 Ching, C.B., Ruthven, D.M., and Hidajat, K. (1985) Experimental study of a simulated counter-current adsorption system – III. Sorbex operation. *Chem. Eng. Sci.*, **40**, 1411.
44 Nicoud, R.M., Fuchs, G., Adam, P., Bailly, M., Kusters, E., Antia, F.D., Reuille, F., and Schmid, E. (1993) Preparative scale enantioseparation of a chiral epoxide: comparison of liquid chromatography and simulated moving bed adsorption technology. *Chirality*, **5**, 267.
45 Strube, J., Altenhöner, U., Meurer, M., Schmidt-Traub, H., and Schulte, M. (1997) Dynamic simulation of simulated moving-bed chromatographic processes for the optimization of chiral separations. *J. Chromatogr. A*, **769** (1), 81.
46 Xin Wang, X. and Ching, C.B. (2004) Chiral separation and modeling of the three-chiral-center β-blocker drug nadolol by simulated moving bed chromatography. *J. Chromatogr. A*, **1035** (2), 167.
47 Pais, L.S., Loureiro, J.M., and Rodrigues, A.E. (1997) Separation of 1,1′-bi-2-naphthol enantiomers by continuous chromatography in simulated moving bed. *Chem. Eng. Sci.*, **52**, 245.
48 Seidel-Morgenstern, A., Blümel, C., and Kniep, H. (1998) Efficient design of the SMB process based on a perturbation method to measure adsorption isotherms and on a rapid solution of the dispersion model, in *Fundamentals of Adsorption 6* (ed. F. Meunier), Elsevier, Amsterdam, p. 303.
49 Azevedo, D., Pais, L.S., and Rodrigues, A.E. (1999) Enantiomers separation by simulated moving bed chromatography. Non-instaneous equilibrium at the solid-fluid interface. *J. Chromatogr. A*, **865** (1/2), 187.
50 Miller, L., Orihuela, C., Fronek, R., Honda, D., and Dapremont, O. (1999) Chromatographic resolution of the enantiomers of a pharmaceutical intermediate from the milligram to the kilogram scale. *J. Chromatogr. A*, **849** (2), 309.
51 Pedeferri, M., Zenoni, G., Mazzotti, M., and Morbidelli, M. (1999) Experimental analysis of a chiral separation through simulated moving bed chromatography. *Chem. Eng. Sci.*, **54**, 3735.
52 Lehoucq, S., Verheve, D., Wouwer, A.V., and Cavoy, E. (2000) SMB enantioseparation: process development, modeling, and operating conditions. *AIChE J.*, **46**, 247.

53 Biressi, G., Ludemann-Hombourger, O., Mazzotti, M., Nicoud, R.-M., and Morbidelli, M. (2000) Design and optimisation of a simulated moving bed unit: role of diviations from equilibrium theory. *J. Chromatogr. A*, **876**, 3.

54 Juza, M., Mazzotti, M., and Morbidelli, M. (2000) Simulated moving-bed chromatography and its application to chirotechnology. *Trends Biotechnol.*, **18** (3), 108.

55 Khattabi, S., Cherrak, D., Mihlbachler, K., and Guiochon, G. (2000) Enantiomerseparation of 1-phenyl-1-propanol by simulated moving bed under linear and nonlinear conditions. *J. Chromatogr. A*, **893**, 307.

56 Subramanian, G. (2001) *Techniques in Preparative Chiral Separations* (ed. G. Subranamian), Wiley-VCH Verlag GmbH, New York.

57 Xie, Y, Hritzko, B., Chin, C., and Wang, N.-H.L. (2003) Separation of FTC-ester enantiomers using a simulated moving bed. *Ind. Eng. Chem. Res.*, **42**, 4055.

58 Mihlbachler, K., Seidel-Morgenstern, A., and Guiochon, G. (2004) Detailed study of Tröger's base separation by SMB process. *AIChE J.*, **50**, 611.

59 Schmidt-Traub, H. (2005) *Preparative Chromatography for fine Chemicals and Pharmaceutical Agents*, Wiley-VCH Verlag GmbH, Weinheim.

60 Cox, G. (2005) *Preparative Enantioselective Chromatography*, John Wiley & Sons, Inc., Chichester.

61 Juza, M. (2004) SMB technology: high-quality enantioseparations. Available at www.sp2.uk.com.

62 Dunn, P., Wells, A., and Williams, M. (2010) *Green Chemistry in Pharmaceutical Industry*, Wiley-VCH Verlag GmbH, Weinheim.

63 Mihlbachler, K. and Dapremont, O. (2012) Preparative chromatography, in *Green Techniques for Organic Synthesis and Medicinal Chemistry* (eds W. Zhang and B. Cue), Wiley-VCH Verlag GmbH, Weinheim.

64 Nicoud, R.M. (2000) *Handbook of Bioseparations (Separation Science and Technology)* (ed. S. Ahuja), Academic Press, San Diego, CA, p. 475.

65 Imagogly, S. (2002) Advances in biochemical engineering/biotechnology. *Simulated Moving Bed Chromatography (SMB) for Application in Bioseparation*, vol. 76, Springer, Berlin.

66 Paredes, G., Mazotti, M., Stadler, J., Makart, S., and Morbidelli, M. (2005) SMB operation for three-fraction separations: purification of plasmid DNA. *Adsorption*, **11** (1), 841–845.

67 Antos, D. and Seidel-Morgenstern, A. (2001) Application of gradients in the simulated moving bed process. *Chem. Eng. Sci.*, **56**, 6667–6682.

68 Xie, Y., Mun, S., Kim, J., and Wang, L. (2002) Standing wave design and experimental validation of a tandem simulated moving bed process for insulin purification. *Biotechnol. Prog.*, **18**, 1332–1344.

69 Mun, S., Xie, Y, and Wang, N.H.L. (2003) Optimal design of a size exclusion simulated moving bed for insulin purification. *Ind. Eng. Chem. Res.*, **42**, 1977.

70 Xie, Y., Mun, S., and Wang, N.-H.L. (2003) Startup and shutdown strategies of simulated moving bed for insulin purification. *Ind. Eng. Chem. Res.*, **42**, 1414.

71 Mun, S., Xie, Y., and Wang, N.-H.L. (2003) Residence time distribution in a size-exclusion SMB for insulin purification. *AIChE J.*, **49**, 2039.

72 Xie, Y., Mun, S., Chin, C., and Wang, N.-H.L. (2003) Simulated moving bed technologies for producing high purity biochemicals and pharmaceuticals, in

New Frontiers in Biomedical Engineering (ed. N.H.-C. Hwang), Kluwer Academic Publishers, New York.

73 Mun, S, Xie, Y., and Wang, N.-H.L. (2003) Robust pinched wave design of a size-exclusion simulated moving bed process for insulin purification. *Ind. Eng. Chem. Res.*, **42**, 3129.

74 Abel, S., Mazzotti, M., and Morbidelli, M. (2002) Solvent gradient operation of simulated moving beds: I. Linear isotherm. *J. Chromatogr. A*, **944**, 225.

75 Abel, S. et al. (2004) Two-fraction and three-fraction continuous simulated moving bed separation of nucleosides. *J. Chromatogr. A*, **1043**, 201.

76 Stroehlein, G., Aumann, L., Mazzotti, M., and Morbidelli, M. (2006) A continuous, counter-current multi-column chromatographic process incorporating modifier gradients for ternary separations. *J. Chromatogr. A*, **1126** (1–2), 338–346.

77 Andersson, J. and Mattiasson, B. (2006) Simulated moving bed technology with a simplified approach for protein purification: separation of lactoperoxidase and lactoferrin from whey protein concentrate. *J. Chromatogr. A*, **1107** (1–2), 88–95.

78 Keßler, L.C., Gueorguieva, L., Rinas, U., and Seidel-Morgenstern, A. (2007) Step gradients in 3-zone simulated moving bed chromatography: application to the purification of antibodies and bone morphogenetic protein-2. *J. Chromatogr. A*, **1176** (1–2), 69–78.

79 Kastner, M. (2000) *Protein Liquid Chromatography*, Journal of Chromatography Library, vol. 61, Elsevier, Amsterdam.

80 Mueller-Spaeth, T., Ulmer, N., Aumann, L., and Bavand, M. (2015 Control and optimization of a twin-column counter-current chromatography process for affinity capture of biopharmaceuticals. Talk (388), 2015 ACS Annual Meeting, Denver, March.

81 Brestrich, N., Briskot, T., Osberghaus, A., and Hubbuch, J. (2017) A tool for selective inline quantification of co-eluting proteins in chromatography using spectral analysis and partial least squares regression. *Biotechnol. Bioeng*, **111**, 1365–1373.

82 Pollard, D. (2015) mAb for the masses: automated continuous processes enabled by single-use. PepTalk 2015 in San Diego.

83 Eibl, R. and Eibl, D. (2011) *Single-Use Technology in Biopharmaceutical Manufacture*, John Wiley & Sons, Inc., New Jersey.

84 Mihlbachler, K., Quiroz, F., and Chen, D. (2010) Negative solvent gradient in RP chromatography for purifying pharmaceutical compounds, 2010 Annual AIChE Meeting. Salt Lake City, 470b.

85 Heuer, C. and Seidel-Morgenstern, A. (1995) Experimental investigation and modelling of closed-loop recycling in preparative chromatography. *Chem. Eng. Sci.*, **50** (7), 1115–1127.

86 Grill, C.M. (1998) Closed-loop recycling with periodic intra-profile injection: a new binary preparative chromatographic technique. *J. Chromatogr. A*, **796** (1), 101–113.

87 Grill, C.M., Miller, L., and Yan, T.Q. (2004) Resolution of a racemic pharmaceutical intermediate: a comparison of preparative HPLC, steady state recycling, and simulated moving bed. *J. Chromatogr. A*, **1026** (1–2), 101.

88 Abunasser, N. and Wankat, P. (2005) Ternary separations with one-column analogs to SMB. *Sep. Sci. Technol.*, **40** (16), 3239–3259.

89 Zobel, S., Helling, C., Ditz, R., and Strube, J. (2014) Design and operation of continuous countercurrent chromatography in biotechnological production. *Ind. Eng. Chem. Res.*, **53** (22), 9169–9185.

90 Carta, Giorgio and Jungbauer, Alois (2010) *Protein Chromatography: Process Development and Scale-Up*, Wiley-VCH Verlag GmbH, Weinheim.

91 Mihlbachler, K. (2002) *Enantioseparation via SMB Chromatography: A study of Tröger's Base Unique Adsorption Behavior and the Influence of Heterogeneity of the Column Set on the Performance of the SMB Process*, Logos-Verlag, Berlin.

92 Xie, Y., Mun, S., and Wang, N.-H.L. (2003) Startup and shutdown strategies of simulated moving bed for insulin purification. *Ind. Eng. Chem. Res.*, **42**, 1414.

93 Guiochon, G., Felinger, A., Shirazi, D.G., and Katti, A. (2006) *Fundamental of Preparative and Nonlinear Chromatography*, 2nd edn, Elsevier.

94 Storti, G., Masi, M., Carra, S., and Morbidelli, M. (1989) Optimal design of multicomponent countercurrent adsorption separation processes involving nonlinear equilibria. *Chem. Eng. Sci.*, **44**, 1329.

95 Mazotti, M., Storti, G., and Morbidelli, M. (1997) Optimal operation of simulated moving bed units for nonlinear chromatographic separations. *J. Chromatogr. A*, **769**, 3.

96 Hashimoto, K., Adachi, S., and Shirai, Y. (1988) Continuous desalting of proteins with a simulated moving-bed adsorber. *Agric. Biol. Chem.*, **52**, 2161.

97 Roper, D.K. and Lightfoot, E.N. (1993) Comparing steady counterflow separation with differential chromatography. *J. Chromatogr. A*, **654**, 1.

98 Hritzko, B.J., Xie, Y., Wooley, R.J., and Wang, N.-H.L. (2002) Standing-wave design of tandem SMB for linear multicomponent systems. *AIChE J.*, **48** (12), 2769–2787.

99 Nicolaos, A., Muhr, L., Gotteland, P., Nicoud, R.-M., and Bailly, M. (2001) Application of equilibrium theory to ternary moving bed configurations (four+four, five+four, eight and nine zones): I. Linear case. *J. Chromatogr. A*, **908**, 71–86.

100 Wooley, R., Ma, Z., and Wang, N.-H.L. (1998) A nine-zone simulating moving bed for the recovery of glucose and xylose from biomass hydrolyzate. *Ind. Eng. Chem. Res.*, **37** (9), 3699–3709.

101 Chin, C., Xie, X., Alford, J., and Wang, N.-H.L. (2006) Analysis of zone and pump configurations in simulated moving bed purification of insulin. *AIChE J.*, **52**, 2447.

102 Linda Wang, N.-H. and Chin, C.Y. (2006) Versatile simulated moving bed systems. U.S. Patent US7,141,172 B2. Published November 28, 2006.

103 Keßler, L.C. and Seidel-Morgenstern, A. (2006) Theoretical study of multicomponent continuous countercurrent chromatography based on connected 4-zone units. *J. Chromatogr. A*, **1126** (1–2), 323–337.

104 Paredes, G., Abel, S., Mazzotti, M., Morbidelli, M., and Stadler, J. (2004) Analysis of a simulated moving bed operation for three-fraction separations (3F-SMB). *Ind. Eng. Chem. Res.*, **43** (19), 6157–6167.

105 Paredes, G. and Mazzotti, M. (2007) Optimization of simulated moving bed and column chromatography for a plasmid DNA purification step and for a chiral separation. *J. Chromatogr. A*, **1142** (1), 56–68.

106 Voigt, U., Hempel, R., Kinkel, J., and Nicoud, R.-M. (1997) Chromatographic process for obtaining very pure cyclosporin a and related cyclosporins. WO1997034918 A1.

107 Houwing, J., Billiet, H.A.H., and van der Wielen, L.A.M. (2003) Mass-transfer effects during separation of proteins in SMB by size exclusion. *AIChE J.*, **49** (5), 1158.

108 Houwing, J., Hateren, S.H.van., Billiet, H.A.H., and van der Wielen, L.A.M. (2002) Effect of salt gradient on the separation of diluted mixtures of proteins by ion-exchanges in simulated moving beds. *J. Chromatogr. A*, **952**, 85–98.

109 Houwing, J., Jensen, T.B., Hateren, S.H.van., Billiet, H.A.H., and van der Wielen, L.A.M. (2003) Positioning of salt gradients in ion-exchange SMB. *AIChE J.*, **49** (3), 665.

110 Aumann, L., Stroehlein, G., Grimm, S., Tarafder, A., Mazzotti, M., and Morbidelli, M. (2005) Chromatographic purification of proteins. Prep Symposium, L201.

111 Gueorguieva, L., Keßler, L.C., and Seidel-Morgenstern, A. (2006) Applying gradient SMB processes for the isolation of proteins and antibodies. PREP 2006 – 19th International Symposium, Exhibit and Workshops on Preparative/Process Chromatography, Ion Exchange, Adsorption/Desorption, Processes & Related Separation Techniques, May 14–17, 2006, Baltimore, MD.

112 Ding, W., Madsen, G., Mahajan, E., O'Connor, S., and Wong., K. (2014) Standardized extractable testing protocol for single-use systems in biomanufacturing. *Pharm. Eng.*, **34** (6), 74.

13

Continuous Chromatography as a Fully Integrated Process in Continuous Biomanufacturing

Steffen Zobel-Roos, Holger Thiess, Petra Gronemeyer, Reinhard Ditz, and Jochen Strube

Clausthal University of Technology, Institute for Separation and Process Technology, Leibnizstr 15, 38678 Clausthal-Zellerfeld, Germany

13.1 Introduction

Chromatography is the main working horse in the separation of biomolecules. As wide as the field of biotechnology is, as vast is the diversity in terms of chromatographic unit operations, resins, and operation modes.

Nevertheless, compared to other industrial fields, the biotechnology and especially the biochromatography used to be very batch oriented. There is an ongoing trend for continuous manufacturing [1].

For the upstream part, there are several solutions to achieve continuous operation [2]. This makes continuous downstream an urgent need. On the other hand, this is a good opportunity to break with old structures and outdated approaches.

Many downstream problems could be solved by an integrated upstream and downstream approach. The upstream side should not only go for high titers at all costs. This normally leads to a huge amount of side components and, therefore, to problems on the downstream side (Chapters 3 and 17).

Beside continuous chromatography, there are several other unit operations that can be very beneficial, especially in continuous downstream. This is, for example, the case for aqueous two-phase liquid extraction in the downstream of monoclonal antibodies (mAb). Hence, all unit operations should be taken into account when developing a new biotechnological process (Chapters 3 and 17).

Nevertheless, a chromatography free process is most likely not possible. Therefore, continuous chromatography processes are necessary as well. In general, there are four different approaches.

The fully continuous simulated moving bed (SMB) is capable of continuous loading and continuous product elution. The advantages are numerous with only few drawbacks. Gradient elution is hardly possible and the demands on hardware are rather large [3,4].

A semicontinuous alternative with a quite reduced demand on hardware is the one-column SMB [5,6].

Continuous Biomanufacturing: Innovative Technologies and Methods, First Edition.
Edited by Ganapathy Subramanian.
© 2018 Wiley-VCH Verlag GmbH & Co. KGaA. Published 2018 by Wiley-VCH Verlag GmbH & Co. KGaA.

Continuous loading with batch like elution is achieved with sequential chromatography. The main benefit for this interconnection lies in the better resin utilization due to fully loading one column and collecting the overflow in another one [7,8].

Continuous loading and product elution by gradient is solely achieved by multicolumn countercurrent solvent gradient purification (MCSGP) in the six column setup. Nevertheless, a concentration increase is reached compared to the corresponding batch chromatography. The alternatives with fewer columns are semicontinuous. Loading from one column to another, however, calls for extensive buffer adjustments [9,10].

The integrated countercurrent chromatography (MCSGP) is also a semicontinuous chromatography process. Here, an increase in concentration is achieved as well. Despite the MCSGP, no or little buffer adjustments are necessary due to the utilization of two different resin types [6].

In general, the chromatographic operations suitable for the biotechnological sector are mostly semicontinuous. Nevertheless, the cycle times of these operations are not comparable with batch chromatography. To fit into a fully continuous downstream, only small tanks are needed to buffer the incoming and outgoing streams.

13.2 Continuous Chromatography

13.2.1 SMB

Classical single-column chromatography, either isocratic or gradient elution, is always operated in batch mode. As Figure 13.1 indicates, there is a difference in the velocity of elution for each component. Hence, a second load step must be scheduled so that the faster flowing components of the second loading do not mix with the strong binding components of the previous loading.

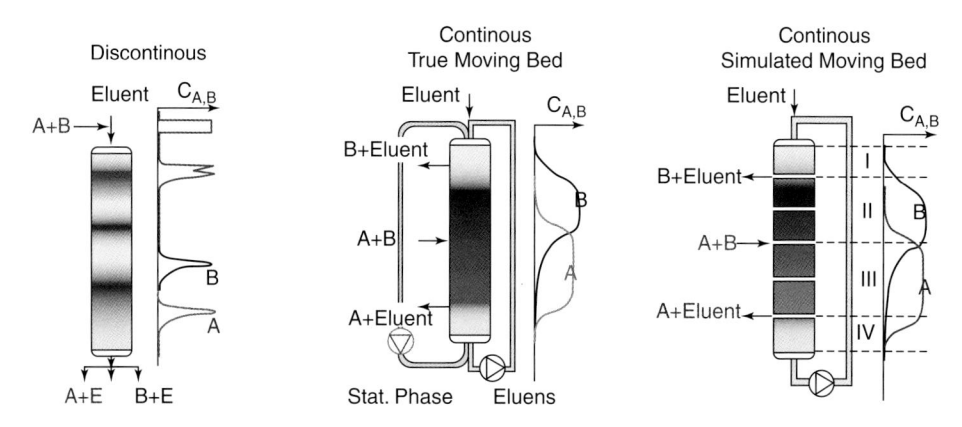

Figure 13.1 Comparison between batch chromatography, true moving bed, and simulated moving bed. (Reproduced with permission from Ref. [11]. Copyright 1994, John Wiley & Sons.)

As known from other different unit operations, this issue might be overcome with countercurrent flow. For a two-component system, the linear velocity of the solid phase should exceed the relative velocity of the strong binding component through the solid phase, thus transporting the strong binding component against its flow direction. As shown in Figure 13.1, loading the column in the middle leads to a binary split. The weak binding component follows the eluent flow to the bottom end of the column. The strong binding component tags along the countercurrent solid phase and elutes at the top. This setup is called true moving bed. Although this setup is very promising, it is hard to achieve due to serious hardware issues. Mainly because pumping the abrasive solid phase would lead to massive hardware wear out. Besides, achieving a well-packed bed is barely possible [11].

The SMB solves the problems mentioned above by applying an intermittent countercurrent. Figure 13.1 shows the basic idea. Instead of a continuous solid flow, the column is separated into several smaller columns. These columns are moved in countercurrent direction simultaneously after a specific amount of time. For an infinite number of columns with infinitesimal column height and switch time, this intermittent flow would resemble the true moving bed [12]. The countercurrent movement is achieved by switching the inlet and outlet streams in the same direction as the solvent flow. This leads to an opposite movement of the columns and, therefore, the solid phase as indicated in Figure 13.2. Despite the intermittent countercurrent, the inlet and outlet streams are fully continuous [3].

Figure 13.2 shows the concentration profile above the column lengths of all columns involved in one SMB cycle shortly before switching. The SMB can be separated into the following four zones:

- Zone I has the highest elution strength due to fresh desorbent applied at the column inlet. Strong binding component gets eluted at the end of zone I and is captured as extract.
- Zone II is the separation zone for the binary split. The strong binding components are retained and later switched to zone I for elution. The weak binding component has to be eluted completely for high purity.

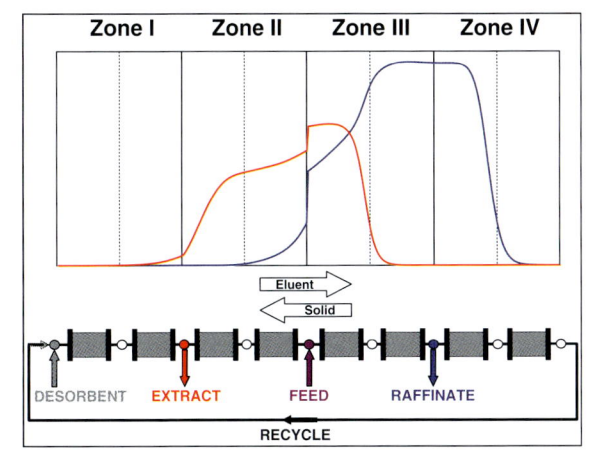

Figure 13.2 SMB in column concentration profiles.

- Zone III is loaded with feed and pure raffinate is eluted.
- Zone IV captures the low binding components and, therefore, preloads the column(s) before feed is applied in zone III. Thus, a displacement effect is achieved [13].

Besides being fully continuous, the simulated moving bed chromatography provides several major advantages.

Figure 13.2 indicates that there is no need for a baseline separation. In theory, separation factors close to 1 should be sufficient for a binary split with 100% purity and yield. Only the ascending or descending desorption fronts are collected of the raffinate or extract, respectively. The overlapping areas stay within zone II and III and, therefore, are completely recycled.

This internal recycling leads to an increased concentration in the columns as well as the outlet streams. Therefore, the product concentrations are much higher for the SMB compared to the corresponding isocratic batch chromatography.

Although it might look complicated, the basic SMB design is relatively easy and straightforward. A short introduction is given here. The detailed design criteria are well known and elaborated in detail elsewhere [12,14–19].

Besides the column dimensions, for a steady-state SMB operation point, the following parameters have to be matched:

- Recycle stream
- Feed stream
- Desorbent stream
- Extract stream
- Raffinate stream
- Switch time

As indicated in Figure 13.2, for a pure extract, the desorption front of the low-binding component has to leave zone II completely before the inlets are switched. In contrast, for a pure raffinate, the adsorption front of the strong binding component has to remain within zone III before the switch. Also, the concentration profile of the strong binding component has to leave zone I and the profile of the weak binding component has to remain in zone IV.

Therefore, the relative velocities of each zone have to be matched with respect to every other zone. To do so, the isotherm parameters have to be known. For linear isotherms, or the linear range of Langmuir or other isotherm types, the following considerations are valid. For nonlinear isotherms, there are some changes but the basic idea remains the same [18].

According to Eq. (13.1,) the theoretical retention time t_{th} depends on the total porosity ε and the Henry coefficient H [20]:

$$t_{th} = t_0 \cdot \left(1 + \frac{1 - \varepsilon_t}{\varepsilon_t} \cdot \frac{dq_i}{dc_i}\bigg|_{\bar{c}}\right), \tag{13.1}$$

$$H = \frac{dq_i}{dc_i}. \tag{13.2}$$

Since there is only one switching time possible that accounts for every zone, the velocities have to be calculated in respect to the other zones and the intermittent solid flow. Hence, a mass flow ratio m_j can be defined for each zone j [14]:

$$m_j = \frac{\dot{V}_{L,j} \cdot t_s - V_{Column} \cdot \varepsilon_t}{V_{Column} \cdot (1 - \varepsilon_t)}. \tag{13.3}$$

Rewriting Eq. (13.1) for the Henry coefficient H shows that Eqs (13.1) and (13.3) are identical for $t_{th} = t_s$. Considering the restrictions already made a set of inequalities (Eqs. (13.4)–(13.8)) can be obtained [16]:

$$H_A < m_1 < \infty, \tag{13.4}$$

$$H_B < m_2 < H_A, \tag{13.5}$$

$$H_B < m_3 < H_A, \tag{13.6}$$

$$\frac{-\varepsilon_p}{(1-\varepsilon_p)} < m_4 < H_B, \tag{13.7}$$

$$H_B < m_2 < m_3 < H_A. \tag{13.8}$$

These inequalities can be illustrated in a simple diagram. The relative flow in zone III m_3 is plotted against m_2. For linear isotherms, this gives a triangle with (H_A/H_A), (H_B/H_B), and (H_B/H_A) as corner coordinates [19]. For nonlinear isotherms, the triangle loses its right-angled shape, dependent on the feed concentration. The higher the feed concentration gets, the higher becomes the deviation, as shown in Figure 13.3. For pure raffinate and extract, the working point has to be within the triangle [19].

The upper left corner would be the best point in terms of eluent consumption and dilution. In respect to, for example, pump inaccuracies, a safety margin should be included, so that the operation point would be within the triangle rather than on the edges. This design approach should only be used as a starting point for further optimizations, like process simulations [3].

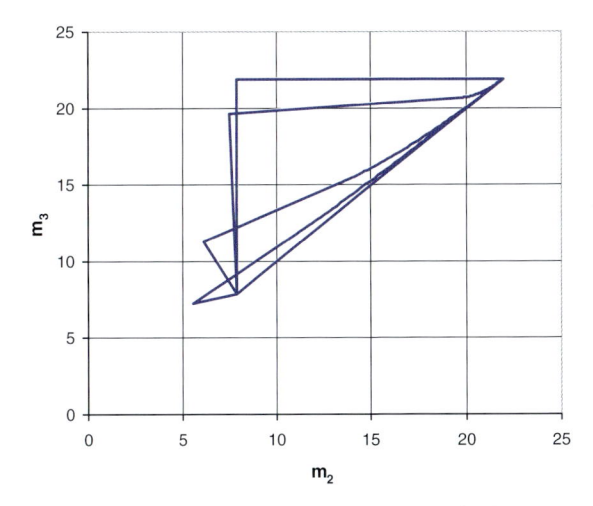

Figure 13.3 Operation diagram for increasing feed concentration.

Comparing Figure 13.3 with the inequalities shown before indicates that the concentration increase already mentioned cannot exceed the feed concentration when operating in the linear range of the isotherms. Equations (13.6) and (13.8) show that the net flow rate of zone II must exceed zone IV. Therefore, the raffinate stream must be higher than the feed stream. In steady state, the mass of the weak binding component provided with the feed stream must leave the system with the raffinate stream. Hence, the concentration in the feed must be higher than the raffinate concentration. This applies for the mean outlet concentrations of one cycle. Temporary concentrations, however, might be much higher. Nevertheless, for nonlinear isotherms, an increase above the feed concentration might be possible.

Despite the many advantages, SMB chromatography has the drawback of being fairly rigid in its operating regimes and hardware intensive.

At least one column has to be present in every zone summing up to at least four columns. Very often, particular zones are utilized with more than one column to enhance the separation [3]. Although at least five pumps are needed, and there is a high demand in valves. These valves have to provide very short closing and opening times. There are several different approaches for the setup and design of these port switching valves [21]. At last, the classical SMB does only work with isocratic elution [22] or step elution with high elution strength in zones I and II and the step to low elution strength at the feed inlet [23].

To overcome some of the problems or to enhance the performance, several SMB-like unit operations have been created [4,24–30]. In addition, there are more than four zone concepts to obtain a multicomponent rather than a binary split [31].

Three examples of an improved SMB shall be shown here. The first one is the Varicol concept. This concept does not work with one fixed switch time. Instead, the inlet and outlet ports are switched individually that allows for a better utilization of every zone, as shown in Figure 13.4 [4,32,33].

Another very popular idea is the intermittent SMB (iSMB) [35,36]. As shown in Figure 13.5, the iSMB process can be separated into two steps. The first step is equivalent to an open four-zone SMB. There is no flow in zone IV and, therefore, no recycling. The recycling is done in the second step, where all inlet and outlet valves are closed and the separation takes place until the concentration fronts are

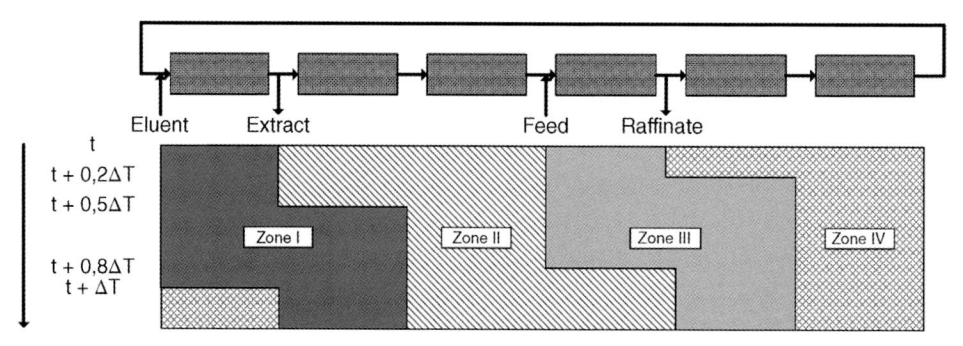

Figure 13.4 Schematic visualization of Varicol-SMB switching times. (Reproduced with permission from Ref. [34]. Copyright 2012, John Wiley & Sons.)

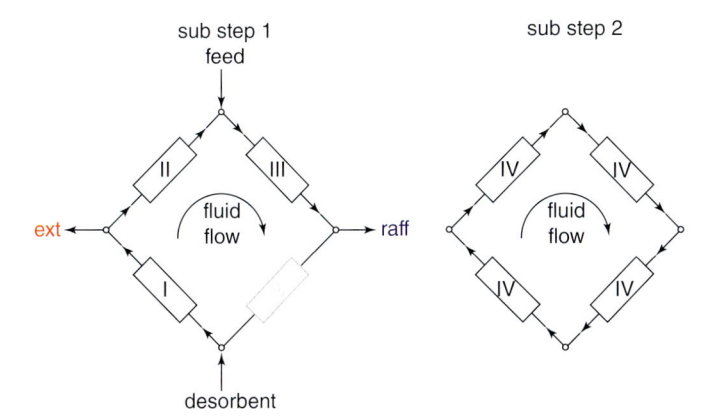

Figure 13.5 iSMB process scheme. (Reproduced with permission from Ref. [35]. Copyright 2014, Elsevier.)

separated well enough [35]. This setup reduces the amount of columns needed at the expense of the continuous operation. The iSMB is, therefore, a semicontinuous operation.

One very simple, semicontinuous SMB-like unit operation is the 1-SMB [5,6,37]. Utilizing only one column, this SMB clone has the least hardware demand with very similar separation behavior as the four zones SMB. Moreover, it provides more degrees of freedom for tuning the separation process.

Figure 13.6 shows the basic idea. During a conventional SMB process, each column is switched in countercurrent mode through the four zones [6]:

- The column is preloaded in zone IV, switched to zone III where the feed loading takes place.
- After that, the column is switched to zone II where the separation takes place.
- The column is then switched to zone I for elution of the strong binding component before being preloaded in zone IV.

Unlike the normal SMB, all these steps do not require to have the same switch time. Advantages similar to those off the Varicol process can be obtained.

There are several possibilities for the 1-SMB setup. The one in Figure 13.6 is the maximum setup using eight tanks and two pumps, whereas the four tanks for feed, eluent, extract, and raffinate are the same needed in the conventional SMB as well. The other four tanks are exchangeable.

Figure 13.6 indicates the possibility of removing all four zone tanks. In this case, the streams for zone tank I–III are recycled in the extract, raffinate, or feed tank, respectively. This leads to a very simple setup of only one column, one pump, and four tanks at the expense of slight losses in separation performance.

To obtain the exact same separation as achieved in the four-zone SMB, the zone tanks I–III must be designed as delay coils. Otherwise, the concentration profiles generated in the column are destroyed in the stirred tanks. In the next cycle, this leads to a loading of the column with a mean concentration instead of a profile, as shown in Figure 13.6.

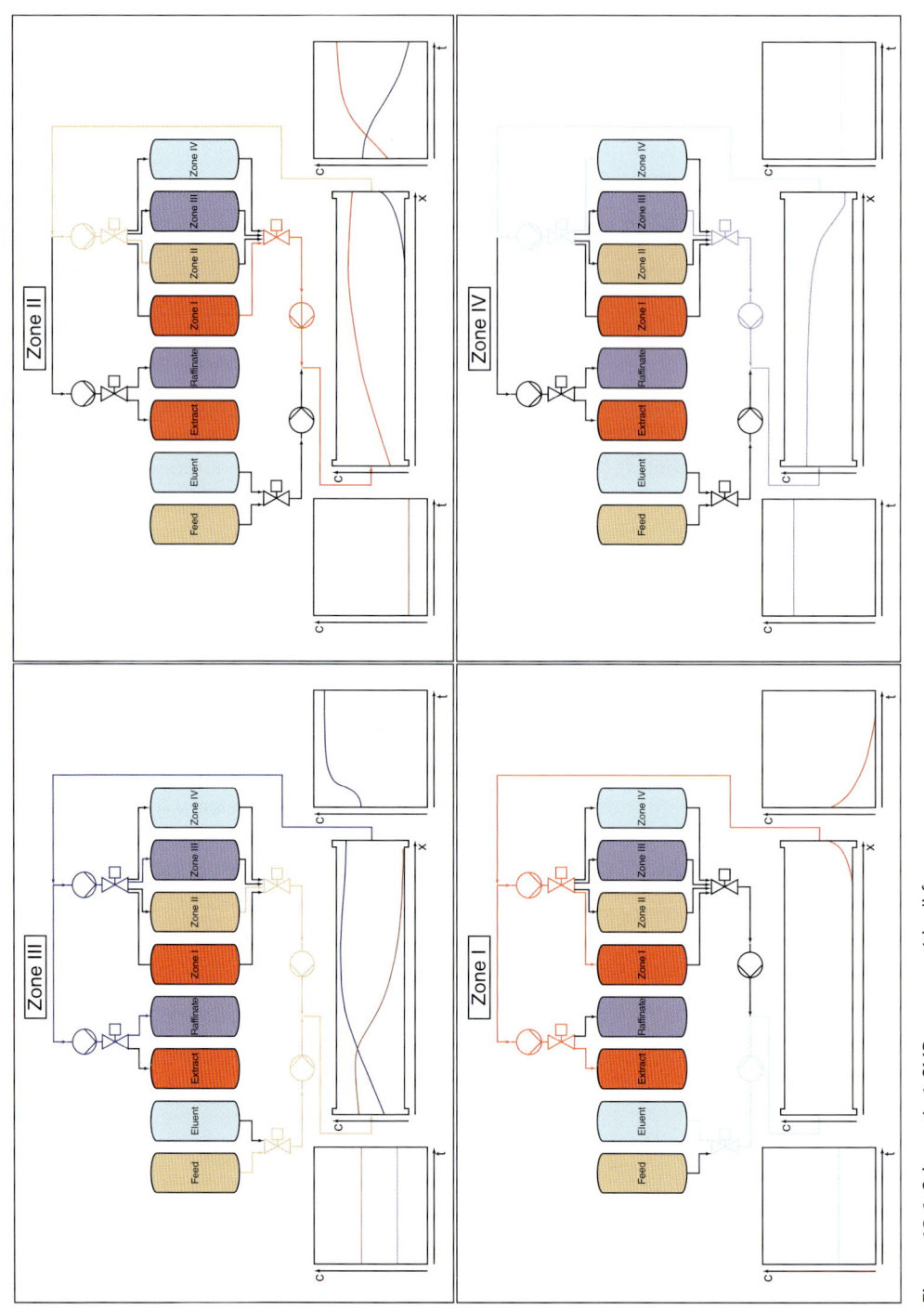

Figure 13.6 Schematic 1-SMB process with all four zones.

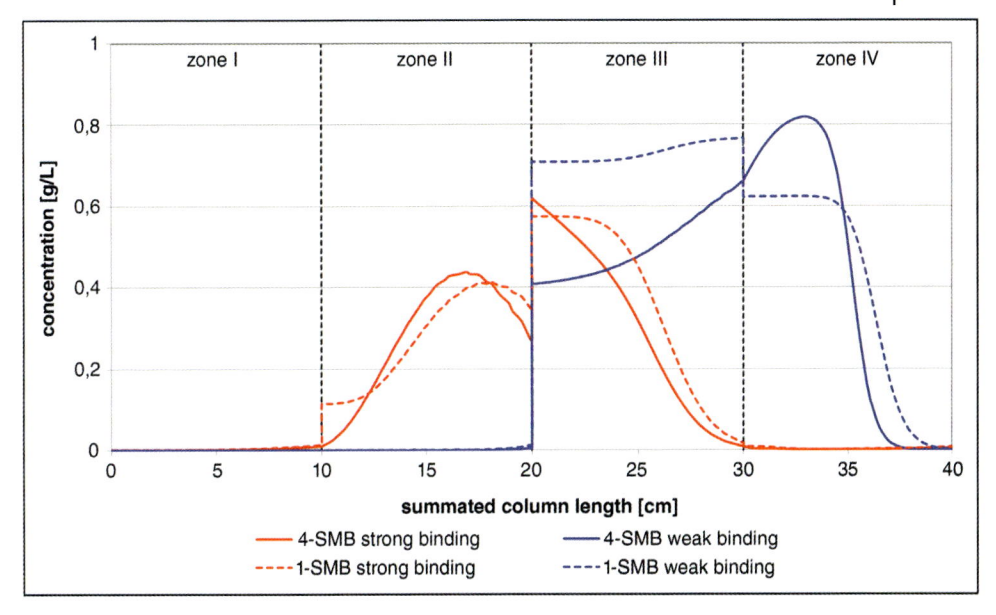

Figure 13.7 Comparison of classical SMB (4-SMB) with 1-SMB with 7 stirred tanks.

Nevertheless, this loss of separation performance is only problematic for very difficult separations.

Figure 13.7 shows a simulated steady-state chromatogram of a 1-SMB with four product and three zone tanks (zone IV excluded), implemented as stirred tanks, compared to a classical SMB. This one is implemented with four columns in the 1:1:1:1 setup. The simulations are carried out with a distributed plug flow model [6,38,39].

The basic separation results for both processes are very similar. For the 1-SMB, the raffinate has 98.6% purity and 99.9% yield compared to 99.3% purity and 100% yield for the 4-SMB.

The extract purity in the 1-SMB is 99.9% with 98.7% yield. In the 4-SMB, 99.9% purity and 99.3% yield is achieved.

The operating point for both simulations is the same. As shown in Figure 13.7, the main difference between the simulated curves for the 1-SMB (dashed lines) and the 4-SMB (solid lines) are the loading conditions. The solid lines show a loading with a concentration profile whereas the dashed lines show that the columns are loaded with a constant concentration. This is due to the storage in stirred tanks.

With delay coils instead of stirred tanks for zones I, II, and III, the results would have been the same and the dashed lines would match the solid lines.

Nevertheless, one advantage is achieved as well. The mean concentration of each product stream for this setup is higher for the 1-SMB compared to the 4-SMB [6].

13.2.2 Serial Multicolumn Continuous Chromatography

A very straightforward idea to achieve continuous feed loading is the serial connection of several columns. As shown in Figure 13.8, the basic setup combines

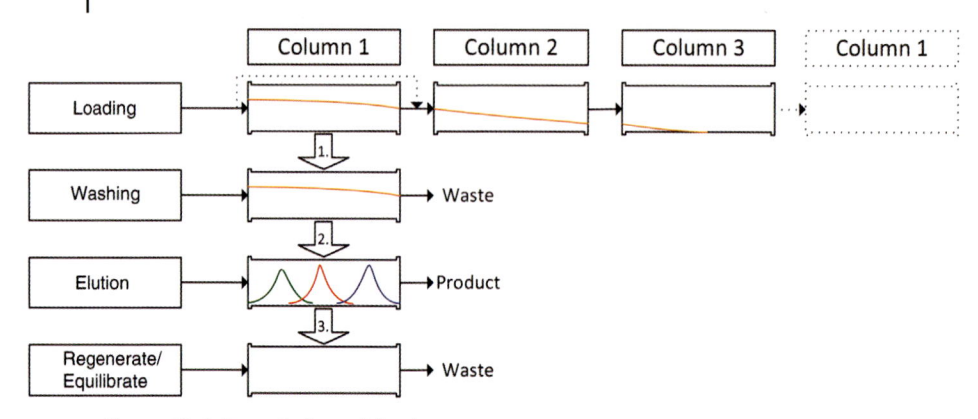

Figure 13.8 Example for serial column setup.

several columns into one chain. The loading starts with the first column but, despite batch chromatography, is not stopped when the breakthrough of the first column occurs. Instead, the column is loaded further. The breakthrough is captured by the second column. Hence, the first column can be loaded to a higher degree, leading to superior capacity utilization. The number of columns in the loading step depends on several variables such as the dynamic binding capacity, the loading velocity, and the desired degree of loading for each column.

When the first column is loaded completely, it is disconnected from the line and runs through the normal batch operation modes: washing, elution, regeneration, and equilibration. After that, it is switched back to the loading step at the end of the line.

In the meantime, the second column is loaded and the third column, if existing, catches the breakthrough. Although countercurrent behavior is aspired, it is hardly achieved.

There are several different versions of this concept available. They differ in the amount of columns as well as the elution behavior. Although the loading is continuous, most of the concepts do not have continuous product streams. The loaded column is handled as a batch column after being disconnected from the loading line. This is the case for the periodic countercurrent chromatography (PCC) of GE [2,8,40] or the sequential multicolumn chromatography (SMCC) concept from NovaSep [7]. Both utilize three or four columns, respectively. Also, semicontinuous in terms of product outlet, but with only two columns is the capture SMB from ChromaCon [41]. In contrast to the other ideas, the BioSMB from Tarpon is fully continuous. Due to a huge amount of columns, there are always several columns in the loading and at least one column in every other step. Therefore, the loading and the product outlet are continuous [42,43].

13.2.3 Continuous Countercurrent Multicolumn Gradient Chromatography

In general, the application of a gradient elution in combination with countercurrent behavior is difficult to implement [22]. The target or side components, once eluted by a solvent with the required elution strength, are not going to bind again unless the elution strength is lowered. This is the concept of MCSGP [44].

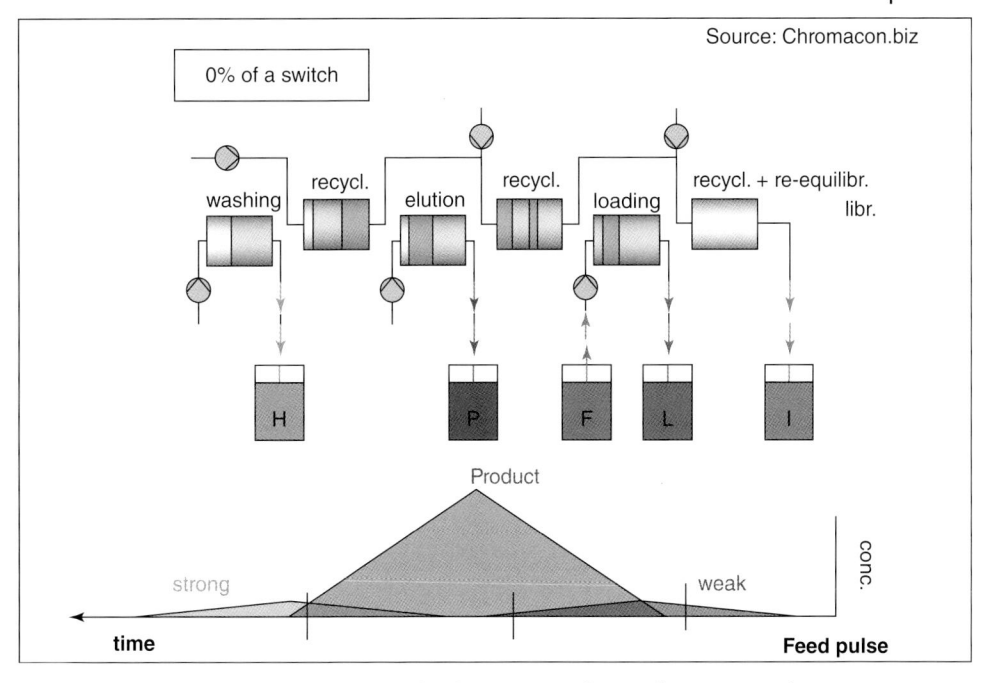

Figure 13.9 6-column MCSGP. (Reproduced with permission from Ref. [10]. Copyright 2013, Elsevier.)

In the full continuous setup, at least six columns are needed. As pictured in Figure 13.9, three of them are connected whereas three are operated individually. The individual ones are for column loading, product elution, and column cleaning. The interconnected ones are for the separation. In Figure 13.9, the columns are shifted from right to left. The elution strength in the line with the connected columns is adjusted in front of every column. Starting with the highest concentration on the left, the elution strength is lowered by dilution [9,10,45].

To reduce the amount of columns, a three-column system can be used at the cost of the continuous mode. Then the three columns are switched from the upper, interconnected positions to the lower, individual operated ones simultaneously. The separation stays the same [46].

Although the idea is very promising, the hardware and process control demand is very high. Therefore, a twin-column setup was introduced [10] (Chapter 14).

13.2.4 Integrated Countercurrent Chromatography

As already mentioned, fully continuous countercurrent chromatography with gradient is very hardware intensive. To simplify the process, a tradeoff between the following parameters has to be found:

- Full continuous operation versus semicontinuous
- Countercurrent versus batch-like separation
- Gradient elution versus isocratic elution

In terms of the operation mode, a semicontinuous mode with quasi-steady state behavior is acceptable. To implement a semicontinuous unit operation into a fully continuous site requires short cycle times and buffer tanks before and after the unit to harmonize the volume streams.

For good separations, as found in protein A chromatography, countercurrent behavior is not advisable. In this case, the focus should lie on high-resin utilization instead of separation improvement. Nevertheless, for most of the other chromatography types, gains in separation factor can be achieved by countercurrent operations.

In general, gradient elution is superior in terms of peak capacity, resolution, eluent consumption, and dilution [47]. Hence, for most of the bioseparations, a semicontinuous, countercurrent, gradient elution chromatography is necessary instead of elution chromatography. This should favor the twin-column MCSGP. Nevertheless, MCSGP has some drawbacks in itself. In addition, there is a need for in-line dilution while loading from one column to the other [10]. Besides, there are usually up to three different chromatographic steps in one biopharmaceutical downstream process [48]. This would lead to a total of six chromatographic columns.

The basic idea of the iCCC [6] is to reduce the overall amount of columns to just two columns. Avoiding the dilution resulting from loading directly from one column to another column with the same resin type, two different separation mechanisms are chosen. As shown in previous work, it is possible to run an ion exchange chromatography and a hydrophobic interaction chromatography with the same buffer system [49]. This leads to a multimodal, twin-column, semicontinuous countercurrent chromatography as shown in Figure 13.12. The basic idea is similar to those of the twin-column MCSGP.

There is one ion exchange and one hydrophobic interaction column. Both are operated with the same buffer system but with opposite gradient behavior. The IEX is loaded with low salt concentration and eluted with a linear gradient to a high salt concentration. The HIC is operated vice versa.

The basic idea of the iCCC is to use narrow product fractions to achieve a high purity but to reload all fractions containing product and side component to achieve a high yield.

The following considerations refer to an iCCC process setup with the ion exchange column in front of the hydrophobic interaction column. This depends on the classical batch process with protein A, ion exchange, and hydrophobic interaction chromatography [48]. Nevertheless, there are iCCC setups with a different order or with reversed phase and normal phase chromatography as the complementary resin types.

In the batch operation mode, IEX fractions would be collected with relatively wide cut points to obtain the maximum of purity and yield (one has to be selected in favor over the other). In Figure 13.10, this is indicated by the red, dashed lines. In the iCCC mode, a first fraction (black, dashed lines) would be collected, containing the overlapping region of the weak binding and target component. This fraction is low in salt concentration and can, therefore, be used for reloading in the next cycle. In the meantime, this fraction is stored in a stirred tank. In case of a linear gradient, the stirred tank mixes up the linear gradient. Besides

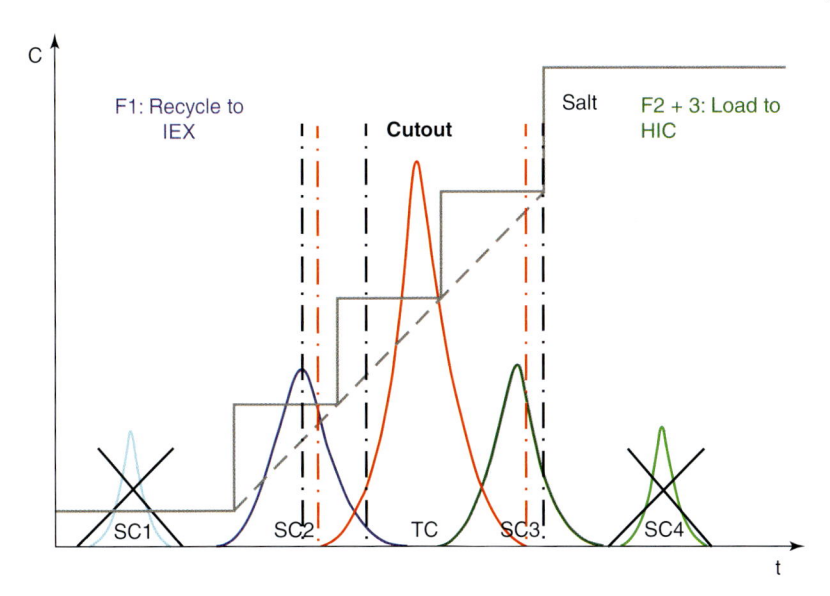

Figure 13.10 IEX chromatogram showing the cut points for batch (red) and iCCC (black) operation.

collecting target product to achieve a high yield, the main aim is to obtain a fraction with an ideal mean salt concentration. In this case, the salt concentration should be low enough to bind the target in the next cycle, but high enough not to bind the side components.

Following the logical order, there would be two more fractions to collect, the target fraction and the overlap of target and strong binding side component. Since both contain a relatively high amount of salt, they are both loaded to the hydrophobic interaction column. Hence, there is only one fraction.

The general considerations for the IEX apply for the HIC as well. As shown in Figure 13.11, there are three different fractions (black, dashed lines). The target fraction in the middle can be cut out with very narrow cut points. This leads to very high purity. The first fraction contains a high amount of salt and is, therefore, recycled to the HIC in the next step. The last fraction has a low salt content and can be loaded to the IEX column.

The recycling operations allow for very high product purities with very high yields at the same time. This is not only due to the very narrow product cut points but the recycling also leads to a concentration increase and, therefore, to displacement effects. Furthermore, the columns are preloaded before the feed injection, leading to typical separation enhancements known for countercurrent operations.

For an ideal case there is no need for an in-line dilution due to the reloading to different resin types.

As shown in Figure 13.12, the iCCC setup contains two columns with different resin types. In this case, an ion exchange and a hydrophobic interaction resin. Besides, five tanks are also required. Two of them contain the feed and the product. The other three collect the side component and target product mixture fractions.

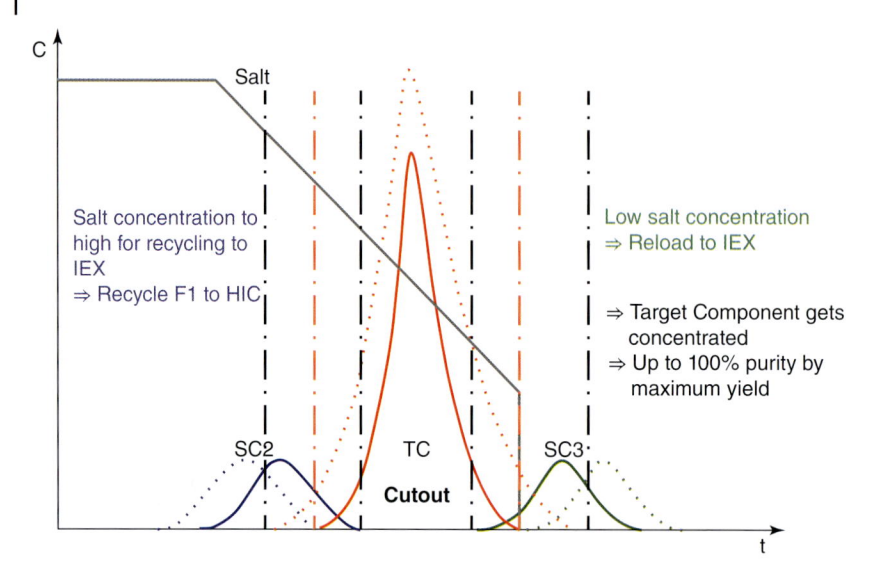

Figure 13.11 HIC chromatogram showing the cut points for batch (red) and iCCC (black) operation.

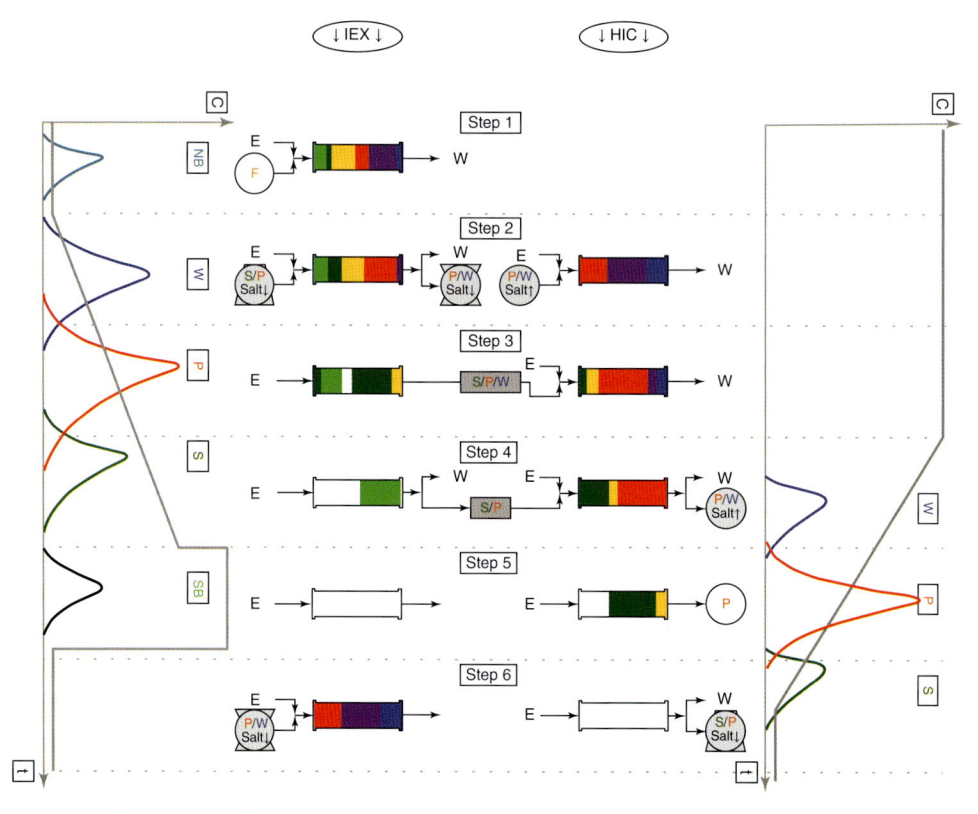

Figure 13.12 iCCC process scheme.

Figure 13.12 shows the six steps of one complete iCCC cycle.

Step 1: The IEX column got preloaded in the previous cycle. It is now loaded with feed.
Step 2: The gradient starts for the IEX column. At the same time, the overlapping region of strong binding side component and target product from HIC step 6 is loaded. Meanwhile, the HIC gets preloaded with the high salt fraction eluted in HIC step 4.
Step 3: Product is eluted from the IEX and directly loaded to the HIC.
Step 4: The high salt mixture from the IEX is loaded to the HIC for which the gradient starts. The HIC elutes the high salt fraction, which is stored for reloading.
Step 5: Pure product is eluted from the HIC. The IEX gets cleaned and preconditioned.
Step 6: The IEX gets preloaded with the low salt IEX fraction from step 2. The HIC gets cleaned and conditioned.

Although it might look complicated, the iCCC design is relatively straightforward. As a starting point, the ion exchange and hydrophobic interaction chromatography get tuned in batch operation mode separately. The focus is on the more difficult separation. This has to be tuned for a sufficient separation. The other column is adapted to fit the same time scale.

The choice of the cut points seems easy, but there are two problems. At first, in most of the cases you cannot see the real concentration progress due to overlapping peaks. Therefore, the start and end points of the overlapping regions are hard to find.

Furthermore, the increase of concentration has to be foreseen. Higher concentrations lead to displacement effects and, therefore, to a shift in the retention times.

Both problems can be solved simultaneously by rigorous process simulation. For the following process simulations, a distributed plug flow model with linear driving force is used to describe the behavior of the liquid phase. The solid phase is considered as a porous system. Hence, pore diffusion is included. The equilibrium is described by multicomponent competitive Langmuir isotherms with respect to the modifier concentration, in this case the salt concentration. For more detailed information, see Refs [3,6,38,50–52].

Figure 13.13 shows 15 cycles of the iCCC. It can be seen that the low binding component is displaced to the very beginning of the chromatogram in the IEX column. This is due to the right amount of salt captured in the first IEX fraction. Although the gradient was high enough to elute some target component, the salt concentration is mixed in the stirred fraction tank. Therefore, the target component in this fraction binds with the next loading whereas the low binding component doesn't.

The concentration of the target component increases in the IEX and the HIC as well, although the side component concentrations remain the same. In the hydrophobic interaction chromatography, this concentration increase forces the target peak to shorter retention times.

Figure 13.14 shows the experimental validations. These experiments were carried out using fermentation broth containing IgG with approximately 1 g/l and the 10-fold of side components. The fermentation broth was not processed with protein A

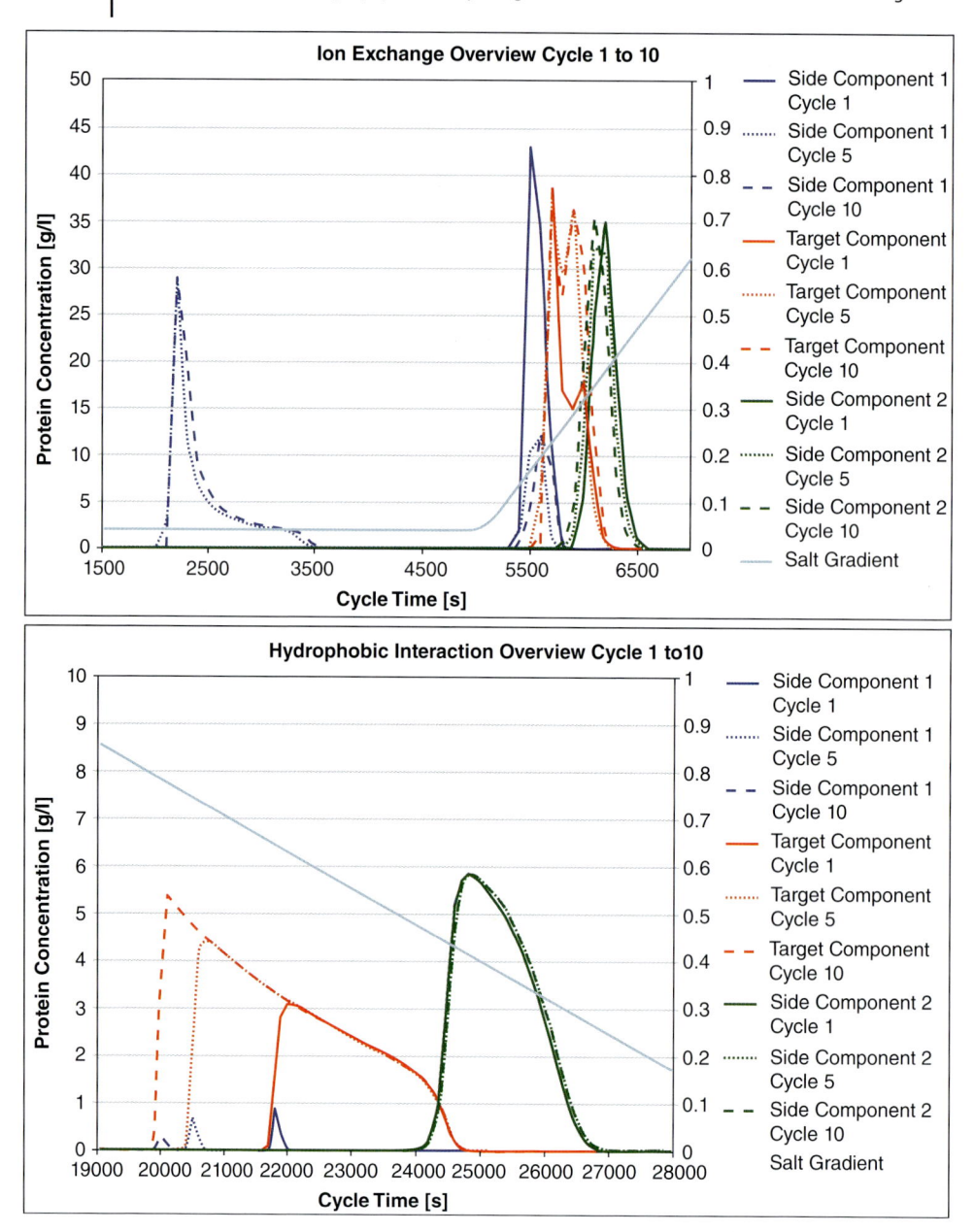

Figure 13.13 Simulation results for 15 iCCC cycles.

chromatography or other separation operations before this experiment except for a diafiltration to a 0.02 molar sodium phosphate buffer at pH 6.0. This buffer is used as the IEX loading and HIC elution buffer as well. In respect to previous work, the IEX elution and HIC binding was done with the same buffer containing 1 molar ammonium sulfate [49]. The diafiltration step might become necessary in respect

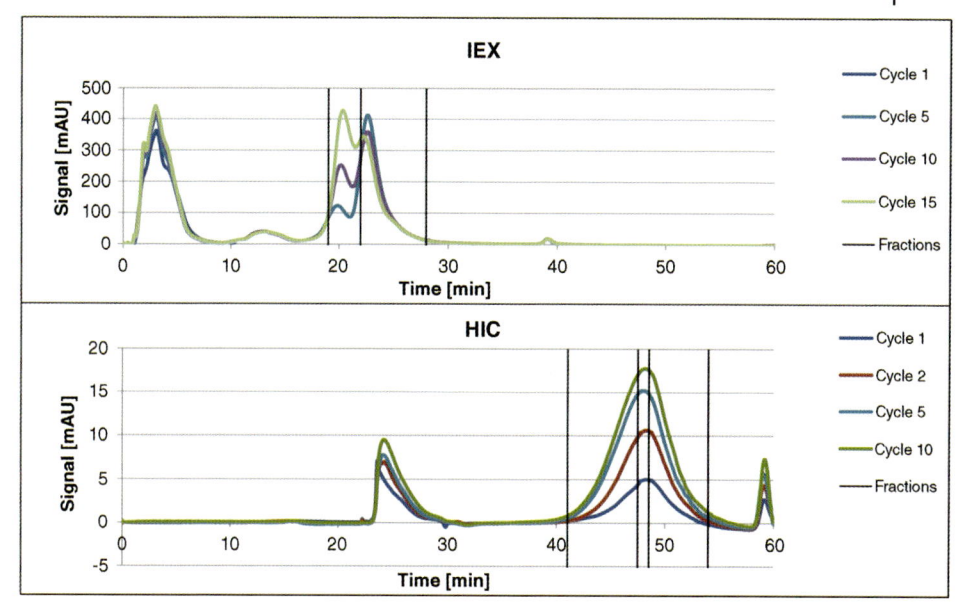

Figure 13.14 Experimental validation.

to the high amount of salts used in the fermentation process. Alternatively, the iCCC sequence might be changed, so that the hydrophobic interaction column is the first for loading. This was not done in this case because the ion exchange binding capacity is higher and the first column has to load the largest amount of side components.

The cut points are shown as black lines in Figure 13.14. For the hydrophobic interaction chromatography and, thereby the final product cut, the target component cut points were set very narrow. This lead to an analytical purity of 100%, analyzed by analytical protein A and SEC chromatography.

Although a pure experimental design is possible, Figure 13.14 shows the correlated problems. Especially, for the HIC column in this example, there is only one peak visible, although it is a multicomponent system. This shows the high need of offline, or better online analytics.

Besides the continuous chromatography, overlapping peaks are very common phenomena for all chromatographic processes. The classical detection methods are not suitable to solve this problem. Nevertheless, the well-known diode array detector (DAD) is capable of performing an online peak deconvolution.

Because of its diode array, the DAD does not only record several chromatograms but also provides the corresponding spectral data. The chromatogram plots the signal intensity over time for one given wavelength. A spectrum, on the other hand, is the signal intensity as a function of the wavelength for a given time.

Although most proteins have very similar spectra, for example, an absorption maximum at 280 nm, there are always some differences. According to the Beer–Lambert law, the spectrum measured for a mixture of components equals the sum of the single-component spectra in respect to their concentration.

This can be used to identify the single component peaks out of overlapping peaks. This idea was first shown to work by Hubbuch and coworkers for test

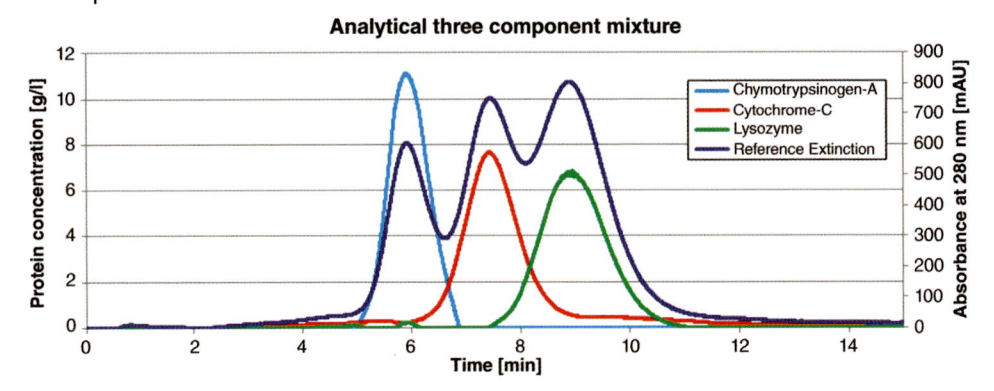

Figure 13.15 Online peak deconvolution of three proteins.

mixtures as well as real separation problems based on mAb purification on a cation exchange column [53,54].

In general, there are two different approaches to the DAD peak deconvolution idea.

The first and easier idea is to measure the extinction coefficients for each component present in the mixture as a single-component spectrum. In this case, the online peak prediction is relatively easy. For a three-component system of chymotrypsinogen-α, cytochrome c, and lysozyme, this is shown in Figure 13.15.

The second approach is to predict the extinction coefficients online as well. This can be done sequentially and might come with some time offset due to more extensive calculations.

Both ideas can be enhanced by mathematical peak identification methods as well as model assumptions.

In terms of biomanufacturing, a combination of both ideas might apply. Normally, the target component should be known as well as some major impurities. Therefore, the extinction coefficients for these components could be measured beforehand. The peak deconvolution is then carried out with known extinction coefficients for these components. The missing coefficients for other side components are then measured according to the second method.

All in all, online peak deconvolution has the potential to shift the pooling decisions from empirical to data-based decisions and should, therefore, be taken into account not only for continuous but also for all chromatographic processes.

13.3 Conclusion and Outlook

The field of continuous chromatography is relatively wide. A brief overview and comparison can be found in Table 13.1. Although the best process has to be chosen for every case individually, some general considerations apply.

SMB is a very useful process. Unfortunately, it is not well suited for the resins typically used in biotechnological processing. Especially, when a solvent gradient is needed, the SMB is often not the operation of choice.

13.3 Conclusion and Outlook

Table 13.1 Comparison of continuous chromatography processes.

	Gradient	Continuous	Countercurrent	Multimodal	Number of columns	Invest costs	Complexity	Eluent consumption	Dilution	Resin utilization
Batch	Yes	No	No	No	1	Low	Low	0	0	0
SMB	Step	Full	Yes	No	>4	High	High	+++	++	+++
1-SMB	Step	Semi	Yes	No	1	Low	Medium	+++	++	+++
iCCC	Yes	Semi	Yes	Yes	2	Low	Medium	++	+++	+++
SMCC	Yes	Semi	No	No	>3	Medium	High	+	+	+
6-MCSGP	Yes	Full	Yes	No	6	High	Extreme	++	++	+++
2-MCSGP	Yes	Semi	Yes	No	2	Low	Medium	+	+	++

0 as reference, + as better, ++ as much better, +++ as way better.

For continuous loading and better resin utilization, the sequential column setups such as SMCC, PCC, or BioSMB are a good choice. Nevertheless, these concepts are pretty conservative in terms of elution. A better separation than in batch mode is not to be expected.

The more innovative approach is the twin-column MCSGP or iCCC. These not only achieve better resin utilization but also enhance the whole separation in terms of purity and yield.

For downstream processes, where several different resin types are needed, the iCCC is especially beneficial due to the combination of two different resin types in one unit operation. This also reduces the in-line dilution that is mandatory for the twin-column MCSGP.

For chromatographic processes in general and continuous ones in particular, there is a high demand for online analytics. For a better process control, online peak deconvolution should be implemented when overlapping peaks are involved.

The knowledge of every component concentration instead of the sum signal of the mixture shifts the process control from predefined parameters setup in the process development to data-based decisions. This reduces the batch failure and is especially beneficial for semicontinuous recycling strategies such as MCSGP and iCCC.

Nevertheless, continuous biomanufacturing cannot be achieved by continuous chromatography by itself. There are several challenges for the total process that has to be taken into account as well (Chapters 3 and 17). In addition, there is a high demand for an integrated approach for upstream and downstream (Chapters 3 and 17).

Symbols

Symbols	Name	Dimension
c	Concentration	(g/l)
H	Henry coefficient	
m	Mass flow ratio	–
t	Time	(s), (min)
q	Loading	(g/l)
V	Volume	(l), (ml)
Greek		
α	Separation factor	–
ε	Porosity	–
Indices		
1	Zone I	
2	Zone II	
3	Zone III	
4	Zone IV	
A	Component A	

(Continued)

Symbols	Name	Dimension
B	Component B	
Column	Column	
L	Liquid	
p	Particle	
s	Switch	
t	Total	
th	Theoretical	

References

1 Konstantinov, K. (2011) Continuous bioprocessing: an interview with Konstantin Konstantinov from Genzyme. Interviewed by Prof. Alois Jungbauer and Dr. Judy Peng. *Biotechnol. J.*, **6** (12), 1431–1433.

2 Warikoo, V., Godawat, R., Brower, K., Jain, S., Cummings, D., Simons, E., Johnson, T., Walther, J., Yu, M., Wright, B., McLarty, J., Karey, K.P., Hwang, C., Zhou, W., Riske, F., and Konstantinov, K. (2012) Integrated continuous production of recombinant therapeutic proteins. *Biotechnol. Bioeng.*, **109** (12), 3018–3029.

3 Strube, J. (2000) *Technische Chromatographie: Auslegung, Optimierung, Betrieb und Wirtschaftlichkeit*, Shaker, Aachen.

4 Rodrigues, A. (2015) *Simulated Moving Bed Technology: Principles, Design and Process Applications*, Elsevier Science, Burlington.

5 Mota, J.P.B. and Araújo, J.M.M. (2005) Single-column simulated-moving-bed process with recycle lag. *AIChE J.*, **51** (6), 1641–1653.

6 Zobel, S., Helling, C., Ditz, R., and Strube, J. (2014) Design and operation of continuous countercurrent chromatography in biotechnological production. *Ind. Eng. Chem. Res.*, **53** (22), 9169–9185.

7 Holzer, M., Osuna-Sanchez, H., and David, L. (2008) Multicolumn chromatography: a new approach to relieving capacity bottlenecks for downstream processing efficiency. *BioProcess Int.*, **6** (8), 74–82.

8 Godawat, R., Brower, K., Jain, S., Konstantinov, K., Riske, F., and Warikoo, V. (2012) Periodic counter-current chromatography – design and operational considerations for integrated and continuous purification of proteins. *Biotechnol. J.*, **7** (12), 1496–1508.

9 Aumann, L. and Morbidelli, M. (2007) A continuous multicolumn countercurrent solvent gradient purification (MCSGP) process. *Biotechnol. Bioeng.*, **98** (5), 1043–1055.

10 Krättli, M., Steinebach, F., and Morbidelli, M. (2013) Online control of the twin-column countercurrent solvent gradient process for biochromatography. *J. Chromatogr. A*, **1293**, 51–59.

11 Deckert, P. and Arlt, W. (1994) Simulierte gegenstromchromatographie. *Chem. Ing. Tech.*, **66** (10), 1334–1340.

12 Storti, G., Baciocchi, R., Mazzotti, M., and Morbidelli, M. (1995) Design of optimal operating conditions of simulated moving bed adsorptive separation units. *Ind. Eng. Chem. Res.*, **34**, 288–301.

13 Seidel-Morgenstern, A. (1995) *Mathematische Modellierung der präparativen Flüssigchromatographie*, DUV: Naturwissenschaft, Deutscher Universitätsverlag Wiesbaden.

14 Mazzotti, M., Storti, G., and Morbidelli, M. (1994) Robust design of countercurrent adsorption separation processes: 2. Multicomponent systems. *AIChE J.*, **40** (11), 1825–1842.

15 Mazzotti, M., Storti, G., and Morbidelli, M. (1996) Robust design of countercurrent adsorption separation: 3. Nonstoichiometric systems. *AIChE J.*, **42** (10), 2784–2796.

16 Mazzotti, M., Storti, G., and Morbidelli, M. (1997) Optimal operation of simulated moving bed units for nonlinear chromatographic separations. *J. Chromatogr. A*, **769**, 3–24.

17 Mazzotti, M., Storti, G., and Morbidelli, M. (1997) Robust design of countercurrent adsorption separation processes: 4. Desorbent in the feed. *AIChE J.*, **43** (1), 64–72.

18 Storti, G., Mazzotti, M., Morbidelli, M., and Carrà, S. (1993) Robust design of binary countercurrent adsorption separation processes. *AIChE J.*, **39** (3), 471–492.

19 Migliorini, C., Mazzotti, M., and Morbidelli, M. (2000) Robust design of countercurrent adsorption separation processes: 5. Nonconstant selectivity. *AIChE J.*, **46** (7), 1384–1399.

20 Guiochon, G., Felinger, A., Shirazi, D.G., and Katti, A.M. (2006) *Fundamentals of Preparative and Nonlinear Chromatography*, 2nd edn, Elsevier Academic Press.

21 Faria, R.P.V. and Rodrigues, A.E. (2015) Instrumental aspects of simulated moving bed chromatography. *J. Chromatogr. A*, **1421**, 82–102.

22 Belcheva, D. (2004) Theoretische und experimentelle Studie der Gradienten-Gegenstromchromatographie unter linearen Bedingungen. Dissertation, Otto-von-Guericke-Universität. .

23 Li, P., Xiu, G., and Rodrigues, A.E. (2007) Proteins separation and purification by salt gradient ion-exchange SMB. *AIChE J.*, **53** (9), 2419–2431.

24 Meurer, M., Altenhöner, U., Strube, J., Untiedt, A., and Schmidt-Traub, H. (1996) Dynamic simulation of a simulated-moving-bed chromatographic reactor for the inversion of sucrose. *Starch/Stärke*, **48** (11–12), 452–457.

25 Xie, Y., Mun, S., Kim, J., and Wang, N.-H.L. (2002) Standing wave design and experimental validation of a tandem simulated moving bed process for insulin purification. *Biotechnol. Progress*, **18** (6), 1332–1344.

26 Zhang, Z., Mazzotti, M., and Morbidelli, M. (2003) PowerFeed operation of simulated moving bed units: changing flow-rates during the switching interval. *J. Chromatogr. A*, **1006**, 87–99.

27 Zhang, Z., Mazzotti, M., and Morbidelli, M. (2004) Continuous chromatographic processes with a small number of columns: comparison of simulated moving bed with Varicol, PowerFeed, and ModiCon. *Korean J. Chem. Eng.*, **21** (2), 454–464.

28 Pötschacher, P. (2005) Optimization of SMB technology working as a single column system. *Chem. Eng. Technol.*, **28** (11), 1426–1434.
29 Lübke, R., Seidel-Morgenstern, A., and Tobiska, L. (2007) Numerical method for accelerated calculation of cyclic steady state of ModiCon–SMB-processes. *Comput. Chem. Eng.*, **31** (4), 258–267.
30 Küpper, A. and Engell, S. (2009) Optimierungsbasierte Regelung des Hashimoto-SMB-Prozesses. *at - Automatisierungstechnik*, **57** (7), 360–370.
31 Wooley, R., Ma, Z., and Wang, N.-H.L. (1998) A nine-zone simulating moving bed for the recovery of glucose and xylose from biomass hydrolyzate. *Ind. Eng. Chem. Res.*, **37**, 3699–3709.
32 Blehaut, J. and Nicoud, R.-M. (1998) Recent aspects in simulated moving bed. *Analusis Mag.*, **26** (7), 60–70.
33 Ludemann-Hombourger, O., Nicoud, R.-M., and Bailly, M. (2000) The "VARICOL" process: a new multicolumn continuous chromatographic process. *Sep. Sci. Technol.*, **35** (12), 1829–1862.
34 Da Silva, A.C., Salles, A.G., Perna, R.F., Correia, C.R.D., and Santana, C.C. (2012) Chromatographic separation and purification of mitotane racemate in a Varicol multicolumn continuous process. *Chem. Eng. Technol.*, **35** (1), 83–90.
35 Jermann, S. and Mazzotti, M. (2014) Three column intermittent simulated moving bed chromatography: 1. Process description and comparative assessment. *J. Chromatogr. A*, **1361**, 125–138.
36 Katsuo, S. and Mazzotti, M. (2010) Intermittent simulated moving bed chromatography: 2. Separation of Tröger's base enantiomers. *J. Chromatogr. A*, **1217** (18), 3067–3075.
37 Araújo, J.M.M., Rodrigues, R.C.R., Ricardo, J.S.S., and Mota, J.P.B. (2007) Single-column simulated moving-bed process with recycle lag: analysis and applications. *Adsorption Sci. Technol.*, **25** (9), 647–659.
38 Borrmann, C., Helling, C., Lohrmann, M., Sommerfeld, S., and Strube, J. (2011) Phenomena and modeling of hydrophobic interaction chromatography. *Sep. Sci. Technol.*, **46** (8), 1289–1305.
39 Schmidt-Traub, H., Strube, J., Paul, H.-I., and Michel, S. (1995) Dynamische Simulation des kontinuierlichen SMB (Simulated Moving Bed)-Chromatographie-Prozesses. *Chem. Ing. Tech.*, **67** (3), 323–326.
40 Mahajan, E., George, A., and Wolk, B. (2012) Improving affinity chromatography resin efficiency using semi-continuous chromatography. *J. Chromatogr. A*, **1227**, 154–162.
41 Angarita, M., Müller-Späth, T., Baur, D., Lievrouw, R., Lissens, G., and Morbidelli, M. (2015) Twin-column CaptureSMB: a novel cyclic process for protein A affinity chromatography. *J. Chromatogr. A*, **1389**, 85–95.
42 Whitford, W.G. (2010) Single-use systems as principal components in bioproduction. *BioProcess Int.*, **8** (11), 34–44.
43 Gjoka, X., Rogler, K., Martino, R.A., Gantier, R., and Schofield, M. (2015) A straightforward methodology for designing continuous monoclonal antibody capture multi-column chromatography processes. *J. Chromatogr. A*, **1416**, 38–46.
44 Ströhlein, G., Aumann, L., Mazzotti, M., and Morbidelli, M. (2006) A continuous, counter-current multi-column chromatographic process

incorporating modifier gradients for ternary separations. *J. Chromatogr. A*, **1126** (1–2), 338–346.

45 Aumann, L., Stroehlein, G., and Morbidelli, M. (2007) Parametric study of a 6-column countercurrent solvent gradient purification (MCSGP) unit. *Biotechnol. Bioeng.*, **98** (5), 1029–1042.

46 Aumann, L. and Morbidelli, M. (2008) A semicontinuous 3-column countercurrent solvent gradient purification (MCSGP) process. *Biotechnol. Bioeng.*, **99** (3), 728–733.

47 Unger, K.K. and Weber, E. (1995) *Handbuch der HPLC: Teil 1 - Leitfaden für Anfänger und Praktiker*, 2nd edn, GIT Verlag GmbH, Darmstadt.

48 Sommerfeld, S. and Strube, J. (2005) Challenges in biotechnology production – generic processes and process optimization for monoclonal antibodies. *Chem. Eng. Process.*, **44** (10), 1123–1137.

49 Helling, C., Borrmann, C., and Strube, J. (2012) Optimal integration of directly combined hydrophobic interaction and ion exchange chromatography purification processes. *Chem. Eng. Technol.*, **35** (10), 1786–1796.

50 Wiesel, A., Schmidt-Traub, H., Lenz, J., and Strube, J. (2003) Modelling gradient elution of bioactive multicomponent systems in non-linear ion-exchange chromatography. *J. Chromatogr. A*, **1006** (1–2), 101–120.

51 Helling, C. and Strube, J. (2012) Modeling and experimental model parameter determination with quality by design for bioprocesses, in *Biopharmaceutical Production Technology*, 1st edn (ed. G. Subramanian), WILEY-VCH, pp. 409–443.

52 Altenhöner, U., Meurer, M., Strube, J., and Schmidt-Traub, H. (1997) Parameter estimation for the simulation of liquid chromatography. *J. Chromatogr. A*, **769** (1), 59–69.

53 Brestrich, N., Briskot, T., Osberghaus, A., and Hubbuch, J. (2014) A tool for selective inline quantification of co-eluting proteins in chromatography using spectral analysis and partial least squares regression. *Biotechnol. Bioeng.*, **111** (7), 1365–1373.

54 Brestrich, N., Sanden, A., Kraft, A., McCann, K., Bertolini, J., and Hubbuch, J. (2015) Advances in inline quantification of co-eluting proteins in chromatography: process-data-based model calibration and application towards real-life separation issues. *Biotechnol. Bioeng.*, **112** (7), 1406–1416.

14

Continuous Chromatography in Biomanufacturing[*]

Thomas Müller-Späth[1,2] and Massimo Morbidelli[2]

[1]ChromaCon AG, Process Development, Technoparkstrasse 1, 8005 Zurich, Switzerland
[2]ETH Zurich, Institute for Chemical and Bioengineering, Department of Chemistry and Applied Biosciences, Vladimir-Prelog-Weg 1, 8093 Zürich, Switzerland

14.1 Introduction to Continuous Chromatography

Chromatography is an integral part in the manufacturing of biologics due to its capability to purify the target compounds with high yields, high purity, and under conditions that preserve their structure.

The impurities to be removed by chromatography include process- and product-related impurities. The most prominent process-related impurities are host cell proteins (HCPs) and host cell DNA, while the most common product-related impurities are aggregates and target compound fragments.

Most modern downstream purification processes of biopharmaceuticals include at least two single-column chromatography steps.

The first chromatography step is generally referred to as "capture" step while additional chromatography steps are called "polishing" steps.

Single-column, or batch chromatography, is either run in bind/elute mode with the product adsorbing on the stationary phase (resin) and the impurities passing through the chromatographic bed or in flow through mode with the impurities adsorbing on the stationary phase and the product passing through. A concentration increase of the product, which is desirable in downstream processing to reduce volume streams and equipment size, is only possible with bind/elute chromatography.

In downstream processing, the main purpose of the capture step is increasing the concentration of the product (bind/elute mode) and the removal of a large part of the process-related impurities. Polishing steps are generally used for removal of product-related impurities and are run either in bind/elute mode or in flow-through mode.

In this chapter, the term "chromatography" is used to describe processes, which uses a solid stationary phase, which is coherent with the original description of

---

[*]Part of this work has been previously published in *Continuous Processing in Pharmaceutical Manufacturing*, Wiley-VCH Verlag GmbH.

Continuous Biomanufacturing: Innovative Technologies and Methods, First Edition.
Edited by Ganapathy Subramanian.
© 2018 Wiley-VCH Verlag GmbH & Co. KGaA. Published 2018 by Wiley-VCH Verlag GmbH & Co. KGaA.

chromatography. Although this appears obvious, in the literature sometimes liquid–liquid extraction processes are termed "chromatography" or "continuous chromatography."

Despite its great separation power, traditional batch chromatography does have disadvantages. These include mainly high buffer consumption (usually several liters of buffer/solvent are required to produce 1 g of pure product), low productivity (usually only a few grams of product can be produced per liter of packed bed per hour), and high stationary phase costs (typically a couple of 1000 US$ per liter). Stationary phase costs are driven up further by underutilization of the resin in terms of capacity or lifetime.

Continuous chromatography can address these challenges and improve the situation, when operated in a countercurrent manner.

Numerous definitions of the term "continuous chromatography" exist. In a strict definition, "continuous chromatography" implies that the inlet and the outlet streams of the process are not interrupted over time. Less strict definitions allow short interruptions of the streams for valve switching. Even less strict definitions require only the entering or the feed stream to be uninterrupted or interrupted only for valve switching.

Some definitions refer to the process of being "continuous" when the feed stream into the process is not interrupted [1] or interrupted only for valve switching and of constant flow rate.

Moreover, in this chapter the term "continuous chromatography" is being used more generally to describe chromatography processes that are designed to run in a cyclic manner and include a start-up phase, a cyclic steady-state phase, and a shut down phase. In the cyclic steady-state phase, product concentration and quality are identical from cycle to cycle.

All types of continuous chromatography that are in use in biopharmaceutical industry today use at least two columns of the same type.

Historically, continuous annular chromatography (pCAC) has been evaluated as single column continuous chromatography process. It fulfills the strict definition of "continuous" because both its inlet and outlet streams are uninterrupted. However, the technology was not developed further due to difficulties related to packing of the annular bed and due to questions about scalability [2], since this technique requires specialized equipment.

In countercurrent chromatography, the mobile phase (buffer) and the stationary phase (resin) move in opposite directions. Different compounds traveling with the mobile phase are traveling through the stationary phase at different velocities due to their different adsorptive properties. By adjusting the relative speed of the mobile and the stationary phase, even components with very similar adsorptive properties can be separated.

This concept can be explained by a simple allegory: a turtle and a cat, representing two compounds in a mixture, are dropped on a conveyor belt (stationary) and immediately start to run to the right (migration with the mobile phase) (Figure 14.1). The cat runs faster than the turtle (it is less adsorbing). The conveyor belt moves to the left, opposite to the direction in which the animals are running (countercurrent movement). If the speed of the conveyor belt is adjusted to a certain value, the slow turtle (strongly adsorbing) will be transported in the

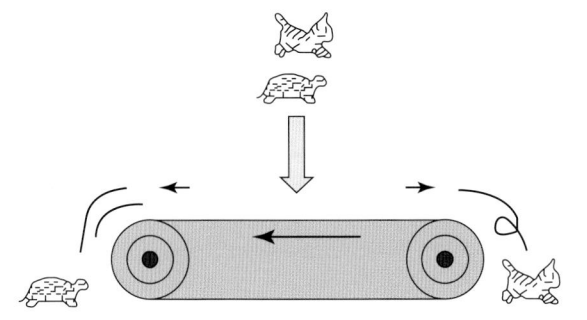

Figure 14.1 Countercurrent process analogy.

direction of the conveyor belt movement to the left while the cat still reaches the right-hand side of the belt. Thus, the animals are separated successfully, even if they are dropped continuously on the conveyor belt.

In countercurrent chromatography, by tuning the relative speed of the stationary and the mobile phase, even components with very similar adsorptive properties can be eventually separated with high yield and purity.

Since a physically moving stationary phase, a so-called true moving bed (TMB), is hard to realize from an engineering point of view and could lead to mechanical stress on the resin particles, the simulated moving bed (SMB) technology has been developed. In SMB, the resin "conveyor belt" is approximated by a circle of interconnected single columns (Figure 14.2). The movement can be simulated by valve switching, such that the inlet and outlet ports of the streams entering and leaving the circle are advancing from one column to the other over time. The columns maintain their position for a certain period of time before moving to the next position. Physical movement of the columns can be realized for instance by mounting the columns in a revolver as demonstrated in cSEP and iSEP processes [3]. However, in most cases, SMB with fixed columns is used because it avoids the use of special hardware and allows for greater flexibility with regards to individual switching times than a revolver-type setup. Four-zone SMB has successfully been scaled-up to large-

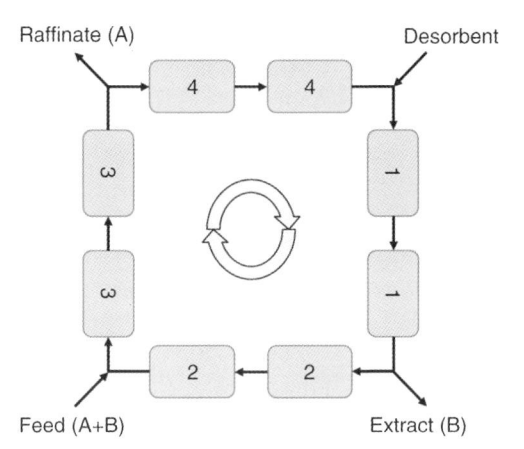

Figure 14.2 Schematic of an eight-column four-zone SMB. Each zone comprises two columns. The circular arrow indicates the direction of port switching and the remaining arrows represent liquid flows.

scale production and is widely used in the purification of chiral small molecule compounds.

As in the conveyor belt analogy a traditional four-zone SMB is capable of separating only two different compounds and it cannot perform linear gradients, which represents a major limitation in purification of biologics, and explains why alternative countercurrent processes are being used here.

It is clear that at least two columns are required in countercurrent chromatography. With regard to the fluid path, the columns can be either in interconnected mode where a mobile phase stream from one column is directed to another column or the columns can be operated in batch mode with independent supply of mobile phase. In case where multicolumn countercurrent chromatography is used with more than two columns, interconnected and batch states may be present at the same time.

14.2 Introduction to Manufacturing Aspects of Chromatography

Multicolumn countercurrent chromatography is capable of improving the performance of many chromatographic batch separations. Different countercurrent processes are applicable for capture and polishing processes. In capture applications, improvements can be expected if the product exhibits a broad, diffuse breakthrough curve; in polishing applications, improvements can be achieved for "difficult" separations, where product and impurities partially overlap in the corresponding chromatogram.

In downstream processing, the chromatographic steps are subject to certain manufacturing constraints. The most tangible ones are listed as follows:

1) The produced product needs to be in specification.
2) The number of cycles n multiplied with the duration of one cycle must not exceed the transit time, that is, the total time that is permitted for this downstream processing step

    $n \times t_{cycle} \leq t_{transit}.$

3) The linear flow rates used in the chromatography step must not exceed the maximum linear flow rate permitted for the used resin.

    $u \leq u_{max}.$

4) The time needed to load the column with feed material in each cycle must not exceed the cycle time

    $t_{feed} \leq t_{cycle}.$

    Other manufacturing constraints such as available floor space, equipment, and so on may apply and strongly depend on the manufacturing situation.

The abovementioned parameters and constraints allow calculation of the required column inner diameter.

From these constraints, the required column diameter d can be estimated for the manufacturing scenario. t_{cycle}, t_{feed}, and u_{feed} can be determined using lab-scale chromatography equipment.

$$d \geq \sqrt{\frac{V_{feed}}{t_{transit}} \times \frac{n}{u_{feed}} \times \frac{t_{cycle}}{t_{feed}} \times \frac{4}{\pi}}. \tag{14.1}$$

The productivity of a chromatographic process is defined as

$$\text{Prod} = \frac{L \times Y}{t_{cycle}} = \frac{u_{feed} \times (\pi \, d^2/4) \times c_{feed} \times t_{feed} \times Y}{V_{col} \times t_{cycle}}. \tag{14.2}$$

Here

V_{feed} (l): volume of feed to be processed by downstream operation
$t_{transit}$ (h): transit time. Time available to process entire feed volume
u_{feed} (cm/h): linear feed flow rate (average feed flow rate)
d (cm): column inner diameter
t_{feed} (h): time during which the feed is loaded onto the column
t_{cycle} (h): duration of the chromatographic run (cycle)
n (–): number of cycles run within transit time
Prod (g/l/h): productivity in gram of product produced per liter of packed bed per hour.
L (g/L): load in gram of product per liter of packed bed
c_{feed} (g/l): feed concentration
Y (%): yield

From Eq. (14.1,) following important aspects can be derived by changing one of the parameters and leaving the others constant:

- t_{cycle}, difficulty of separation: In polishing applications, the separation of product and impurities can be generally improved by running a shallower gradient. Prolonging the gradient may turn a "difficult" separation with overlapping product and product-related impurity peaks into a "less difficult" one with better separation. However, increasing the separation time (t_{cycle}) leads to an increased column diameter d and increased resin demand. There are manufacturing constraints with respect to maximum column diameters, which are related to liquid flow distribution. Today in industry, maximum column diameters in biopharmaceutical manufacturing are 100–200 cm.
- u_{feed}: Increasing the feed flow rate decreases the required column diameter d. The flow rate constraint (u_{max}) due to pressure drop needs to be taken into account.
- V_{feed}: A larger feed amount requires a larger column diameter d.
- $t_{transit}$: Increasing the transit time reduces the required column diameter d. This aspect is particularly relevant for capture processes that are used in conjunction with continuous fermentation.
- When referring to the right part of Eq. (14.1) it can be seen that the column diameter d can be kept constant if the product of productivity and column volume is kept constant.

The productivity (Eq. (14.2)) is the most important measure to describe the production capacity of a chromatographic process. It expresses how much product is produced per resin volume and per time. The definition of the productivity elucidates further aspects that are relevant to chromatography processes. The following statements can be made when assuming that one parameter is changed while the others are kept constant:

- Increasing the load L increases the productivity. This is exploited when optimizing capture processes. It is worth noting that increasing the load by increasing the residence time may not significantly increase the productivity, since it requires a prolongation of the cycle time. Furthermore, in polishing chromatography (nonaffinity) increasing the load generally leads to a worse separation. In this case, a summarizing the above considerations, a trade-off between productivity, and separation or purity exist.
- Decreasing the cycle time t_{cycle} increases the productivity. A decrease in cycle time can be achieved by optimizing the chromatographic protocol equilibration, washing, recovery, and regeneration steps, both in terms of duration by maximizing the flow rate u and by optimizing the number of column volumes of buffer needed for the nonloading tasks of the process.
- Increasing the yield increases the productivity. This is obvious as more material is produced per time and resin volume if the yield is increased. This context is exploited in multicolumn countercurrent polishing processes that make use of internal recycling.

In the context of multicolumn processes, the increase in column number increases the resin volume. Thus, a process with more columns of the same dimensions that has the same or even increased product output per time (in grams per hour) has a lower productivity than a process with fewer columns due to the increased resin volume in the system.

As an example, a four-column process producing 100% more material (in gram per hour) than a two-column process has exactly the same productivity as the two-column process. Essentially, in order to produce the same amount as the four-columns process, the column dimensions of the 2-column process could be increased so that the total resin volume is the same or a 2nd 2-column process of the same type could be installed. Generally, the former option is preferable as it keeps equipment complexity to the minimum.

Every single-column batch process can be transformed into a process with continuous feed flow rate (except for valve switching) by parallelizing single column operation and scheduling the start of the runs such that the loading of a new column starts as soon as the loading of the previous column is finished. The number of columns n required for this kind of process is given by the following equation:

$$n = \frac{t_{\text{cycle}}}{t_{\text{feed}}}. \tag{14.3}$$

In this equation n needs to be rounded up to the next natural number.

A multicolumn continuous process of this type (parallel batch) has exactly the same performance (productivity, buffer consumption, product concentration, resin costs) as a single-column batch process. The advantage of this process is the continuous and uniform feed flow rate. The disadvantage is the increased equipment complexity and cost.

Performance improvements of multicolumn continuous processes can only be achieved if the processes make use of countercurrent principles.

14.3 Trade-Offs in Batch Chromatography

In capture chromatography, there is a trade-off between capacity utilization and the productivity. In batch capture the column is loaded with feed until a certain value in relation to the dynamic capacity is reached. Exceeding the dynamic capacity leads to breakthrough of the product at the column outlet and to yield losses. The dynamic capacity is dictated by the sharpness of the breakthrough curve (caused by mass transfer effects). Decreased loading flow rates (increased residence times) lead to sharper breakthrough curves and improved resin capacity utilization. On the other hand, decreased loading flow rates lead to decreased productivities. This means that larger columns have to be used to process the feed within a certain transit time.

Increased loading flow rates increase productivity; however, they lead to shallower breakthrough curves, lowering the dynamic capacity and the capacity utilization.

Taking into account the aforementioned productivity considerations and manufacturing constraints, in practice many batch separations have a shallow breakthrough curve. For these processes, multicolumn countercurrent chromatography represents the optimal solution.

In polishing chromatography, for "difficult" separations, there is a trade-off between yield and purity as in batch chromatography. In difficult separations, even after reasonable optimization effort, impurities overlap with the product in the front or in the tail of the product compound peak or on both sides. This means that a broad product fraction will be of low purity because it includes a large amount of impurities. On the other hand, when narrowing down the product fraction, the purity is increased but an increasing amount of product ends up in the side (waste) fractions, which lowers the yield.

Figure 14.3 shows a guideline for the selection of the optimal chromatography process (in terms of productivity) for the most common chromatography challenges that are depicted by small schematic chromatograms.

In the case of the capture step, sharp or diffuse breakthrough curves may occur. In case of sharp breakthrough curves, batch chromatography is the method of choice. In case of diffuse breakthrough curves, multicolumn sequential loading processes are optimal.

In binary and ternary separation polishing applications, where baseline separation is present, batch chromatography is the optimal process and delivers product with maximum yield and purity. In difficult ternary separations, where the product of interest is overlapping with closely eluting impurities, the MCSGP (Multicolumn Countercurrent Solvent Gradient Purification) process is the optimal separation process, mitigating the yield-purity trade-off of batch chromatography.

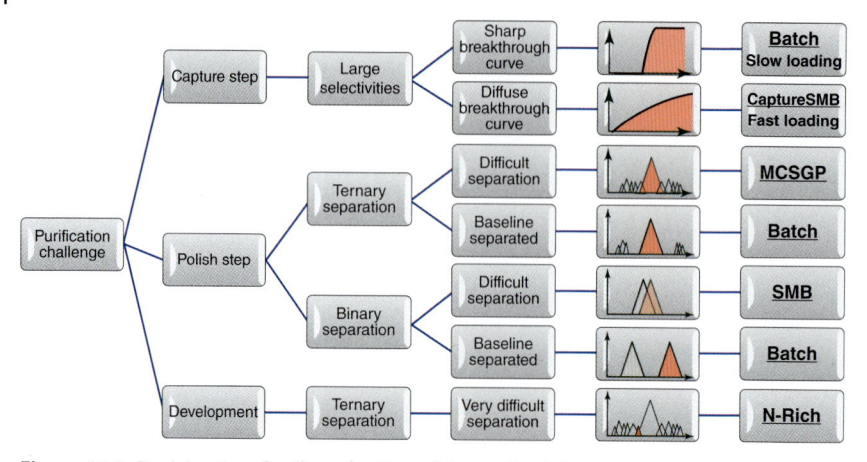

Figure 14.3 Decision tree for the selection of the optimal downstream process.

If only relatively small amounts of product are required for characterization purposes and the product recovery is not of primary importance, the N-Rich process presents a better option for product isolation than the MCSGP process. However, the N-Rich process is not ideal for manufacturing purposes since it does not reach a cyclic steady state, in contrast to MCSGP.

In difficult binary separations where the product of interest is overlapping with closely eluting impurities in the front or in the tail and the separation can be run under isocratic conditions, the four-zone or three-zone SMB process is the optimal separation process. Four-zone or three-zone SMB is also applicable for pseudoternary separations where the product is overlapping with impurities in the front or in the tail and the remaining impurities are baseline-separated on the respective other side of the product peak. However, in separations of biopharmaceuticals, mostly ternary, center-cut separations are required, which cannot be addressed by four-zone SMB.

Thus, summarizing, multicolumn countercurrent processes are beneficial in the capture step in case of a diffuse breakthrough curve and in polishing steps when an overlap between the product and the impurities is present, which includes most separation challenges in biologics downstream processing.

A selection of twin-column countercurrent processes will be presented in the following sections.

14.4 Capture Applications

14.4.1 Introduction

The purpose of the capture step in the downstream processing of biopharmaceuticals is the removal of nonproduct-related impurities and the concentration of the product. Typically, the process has to deal with large volume streams from the upstream fermentation. Therefore, typically stationary phases with high capacities for the target molecule and large particle diameters are used. Large particle diameters exhibit low backpressure at high flow rate allowing high throughput. On the other hand, large particles lead to increased mass transfer phenomena translating into fronting and tailing of the internal product profiles in the columns. For chromatographic breakthrough (BT) curves, increased mass transfer effects/increased feed flow rates/smaller residence times lead to shallower BT curves [4].

It is important to distinguish between the static binding capacity (SBC), corresponding to the maximum binding capacity of the stationary phase and the dynamic binding capacity (DBC), which corresponds to the capacity under flow conditions. Shallower breakthrough curves lead to lower DBC values. DBC values are frequently used as measure to compare resin capacities. Different DBC standards are in use, for example, 1 or 10% DBC values. These values refer to the load L when the value of 1 or 10%, respectively, of the feed titer has been reached at the column outlet. Exceeding the load corresponding to 1% DBC is generally not desired in order to minimize product losses. In fact, in order to account for resin degradation over time it is common practice to load the column only up to 80 or 90% of the load that corresponds to 1% DBC (10–20% load safety margin).

For the sake of completeness it is mentioned here that for affinity resins the static capacity and the equilibrium binding capacity are the same for representative titers. Thus, an affinity stationary phase can be loaded to its static capacity when full equilibrium is established.

The ratio between load and static capacity is termed "capacity utilization" (CU):

$$\mathrm{CU} = \frac{L}{\mathrm{SBC}} \quad (\%). \tag{14.4}$$

In batch capture typical values of capacity utilization are 40–60%.

To improve capacity utilization in batch chromatography, the loading flow velocity has to be decreased. This leads to steeper breakthrough curves, increases the DBC values and the possible load.

However, lower loading flow rates also lower the productivity, increase buffer consumption and product concentration. Thus, in affinity capture processes there is a trade-off between productivity (buffer consumption, product concentration) and capacity utilization.

Using multicolumn countercurrent chromatography, the capacity utilization can be drastically increased due to sequential loading, which is described in the following:

In multicolumn capture processes at least two columns are loaded in series beyond the dynamic capacity of the first column. Thereby, the product breaking through from the first column is captured in a second column. The first column is loaded up to 70% breakthrough (70% DBC) [5] or beyond. Consequently, the capacity of the first column is used to a large extent and values of 90% or larger of the static capacity, which represents the maximum available capacity, have been reported [6]. The larger capacity utilization decreases resin costs and increases productivity. In the case of monoclonal antibody (mAb) capture using protein A affinity, 40% resin cost reduction and up to 40% productivity increase have been reported [7,8].

As described in Ref. [9], Figure 14.4 schematically illustrates the concept of increasing stationary phase capacity utilization by sequential capture processes.

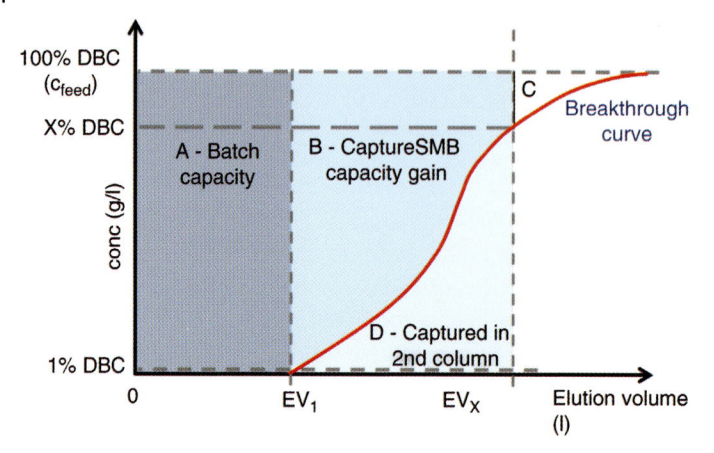

Figure 14.4 Schematic illustration of a breakthrough curve. EV_1, EV_X: elution volumes corresponding to 1% feed concentration (1% DBC), and X% feed concentration, respectively (X% DBC). The area A corresponds to the resin capacity utilized in batch chromatography, when loading to 1% DBC without safety margin.

In the following, we assume that the x-axis shows all elution volumes after subtraction of the dead volume. In Figure 14.4, the area A represents the mass that can be loaded on a single column before reaching the 1% DBC value. On the right-hand side, the area is limited by the volume of 90% of EV1, whereby EV1 is the elution volume corresponding to 1% DBC. As said, the 90% are a loading safety factor accounting for column aging. When loading two columns in series until reaching a desired breakthrough value of X% of the feed concentration, the upstream column contains the additional mass corresponding to area B and the downstream column contains the mass corresponding to area D in Figure 14.4. The total area $A + B + C$ corresponds to the static capacity. Thus, the capacity utilization of batch capture chromatography is $A/(A + B + C)$ and the capacity utilization of the twin-column process is $(A + B)/(A + B + C)$ according to Figure 14.4. An important consequence of these findings is that in single column chromatography, the DBC determines the effective column loading.

In contrast, in the sequential loading process the capacity utilization can be extended closer to the static capacity and therefore, the static capacity is more important.

The increased capacity utilization of the sequential loading process leads to significant resin cost savings, which is particularly relevant in the case of capture with expensive affinity materials with low resin lifetime. In this regard, protein A affinity chromatography plays a special role since the stationary phases have been optimized within the last decades for high capacity and caustic stability, allowing for running protein A columns for hundreds of cycles without significant loss of capacity. In contrast, most other affinity materials that are available today have much lower capacity and caustic stability making capacity utilization improvement even more important.

In sequential loading processes the number of columns required to accommodate the internal product concentration profile (from zero close to the feed

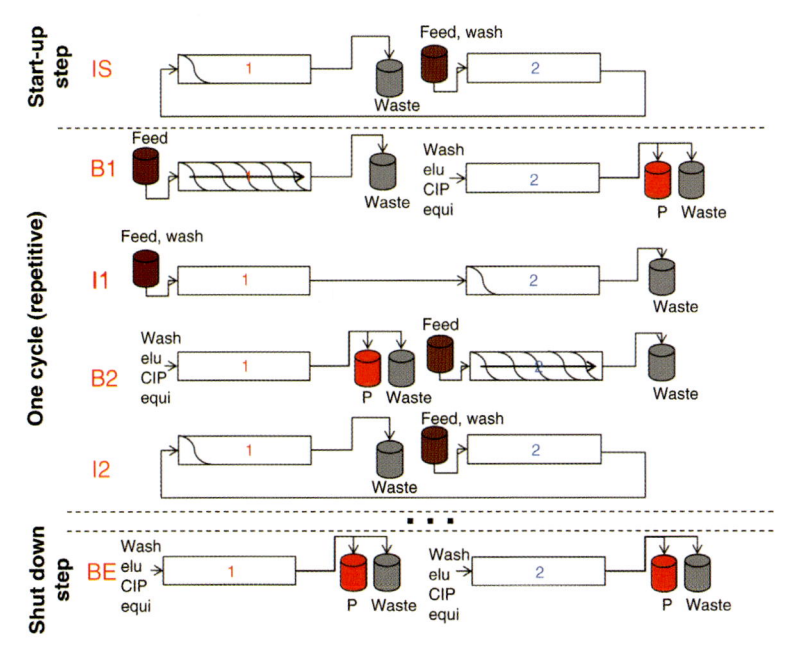

Figure 14.5 Schematic illustration of the twin-column CaptureSMB process. The letters indicate the following: IS, interconnected start-up phase, B1, first batch phase, I1, first interconnected phase, B2, second batch phase, I2, second interconnected phase, BE, final batch elution phase, wash, washing buffer, elu, elution buffer, CIP, CIP buffer, equi, equilibration buffer, P, product.

concentration) is dependent on the stationary phase, the product to be captured, the bed height, and the flow velocity. In the case of monoclonal antibody capture using protein A chromatography, two sequentially loaded columns are sufficient [8,10], while for other cases like the capture of an enzyme using hydrophobic interaction chromatography (HIC), the use of three sequentially loaded columns has been reported [6].

14.4.2 Process Principle

In this section the sequential loading process is explained for a twin column sequential loading process (CaptureSMB) (Figure 14.5) [8].

The process comprises interconnected phases I1, I2, where the columns are loaded and washed sequentially and batch phases B1, B2, where the formerly upstream column is washed, eluted, and regenerated, and the formerly downstream column is continued to be loaded. The elution may be achieved through a step gradient but linear gradients are also possible.

In the subsequent interconnected phase the regenerated column is placed in the downstream position and the previously loaded column is placed in the upstream position.

The process applies a dual loading flow rate strategy to optimize the overall process performance, as described in Ref. [9]. In the interconnected phase the columns are operated at maximum possible feed flow rate while in the batch phase the column that previously was in the downstream position is continued to be loaded at a lower flow rate. The lower feed flow rate ensures that no breakthrough of feed from the single column occurs while the other column performs the tasks of washing, recovery, and regeneration (wash, elute, CIP, re-equilibration) in Figure 14.5. Moreover, the lower feed flow rate leads to an improved dynamic capacity for the batch feeding step. In twin column Capture-SMB the duration of the batch phases B1 and B2 is determined by the overall duration of the washing, recovery, and regeneration tasks. An increased duration of these tasks may lead to a very low loading flow rate in the batch phase. However, the advantages of the process versus batch chromatography in terms of capacity utilization persist, since these are due to the sequential loading in the interconnected phase where the upstream column is loaded to $X\%$ DBC (X is typically in the range of 60–90). In the interconnected step, following the loading, the columns are washed sequentially to adsorb unbound protein on the downstream column. The washing, recovery, and regeneration tasks are identical in batch and in CaptureSMB chromatography except for the first washing step. In CaptureSMB the first washing step is split into two parts with the first part of the washing step taking place as interconnected wash (typically three column volumes) during the phases I1 and I2 following the load and the second part taking place in the subsequent batch phase. The stationary phase and buffers are the same as in batch capture chromatography.

Processes with more columns are based on the same principle that is illustrated in Figure 14.5 and include a mix of interconnected and batch states with two columns loaded in series at the most. Three ways of running a three-column capture process (3C-PCC) are reported in the literature [5–7]. It was found that in sequential loading processes a washing step in the interconnected phase following the interconnected loading phase is important to ensure a high process yield. The interconnected washing step is typically three to five single column volumes. The washing step is important to wash unbound biomolecules and molecules desorbing due to isotherm effects from the interstitial void volume of the upstream column into the downstream column for readsorption.

It is worth mentioning that theoretical optimizations of multicolumn countercurrent capture processes may lead to optimal column bed heights below 7–10 cm. These bed heights are difficult to pack in large scale, and it is recommended to include minimal bed height constraints of 10 cm bed height or larger in optimization considerations.

Multicolumn capture processes with two, three, four, and more columns have been presented. A simulation study for 2.5 and 5.0 g/l feed titer using an agarose-based protein A resin has been carried out in order to compare the process in terms of capacity utilization and productivity [11]. The simulation study showed that with all multicolumn processes it was possible to achieve an equally high capacity utilization of >95% that was significantly higher than the value for batch chromatography (Figure 14.6). In addition, the study showed that the two-column process was superior over processes with three and more columns in terms of productivity while operating at the same low buffer consumption. From a perspective of process and equipment complexity, the two-column process is preferable, too, as complexity (number of pumps, valves, detectors, and piping)

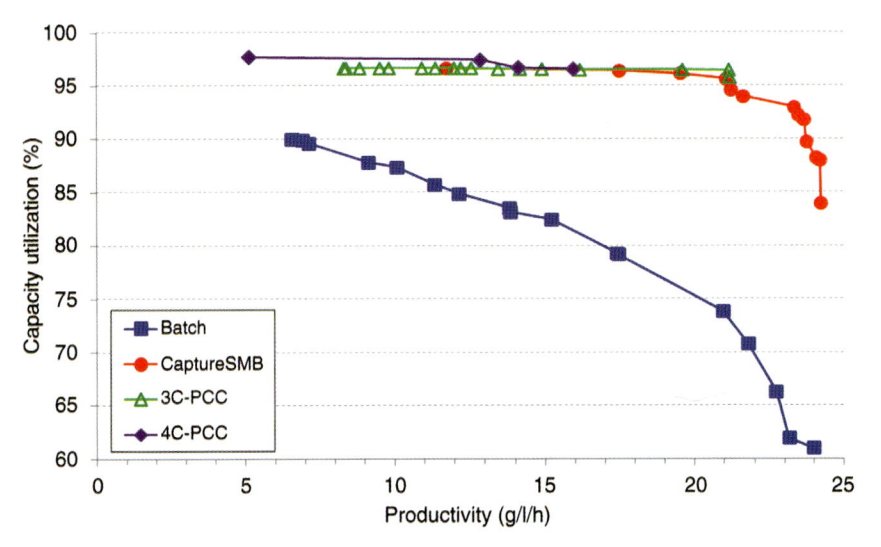

Figure 14.6 Capture comparison for 1–4 column processes in terms of capacity utilization and productivity for a feed titer concentration of 2.5 g/l.

increases with the number of columns. Equipment for countercurrent chromatography capture processes is available from a number of suppliers. Two-column lab equipment is available from ChromaCon (CaptureSMB process), three-column equipment from GE (3C-PCC) process and equipment with more columns from Pall, Novasep, and Semba. A GMP twin column pilot/production scale system is available from LEWA Process Technologies.

14.4.3 Application Examples

For production of mAbs, purification platforms have been developed based on protein A affinity capture followed by one or two additional polishing steps. With sequential loading processes, the performance of the protein A capture step can be improved without changing the stationary phases or buffers used in the downstream process.

The application of multicolumn sequential loading processes for monoclonal antibody (mAb) capture has been evaluated by Genentech and Genzyme/Sanofi. In both cases a three-column setup was used (with at most two columns loaded in series) [5,7]. Genzyme/Sanofi has also evaluated a four-column setup for the capture of enzymes using pseudoaffinity and hydrophobic interaction (HIC) stationary phases [6]. Godawat *et al.* carried out a comparison between a batch process and the multicolumn process showing significant advantages of the multicolumn process in terms of column size reduction (factor 23–35), resin capacity utilization (1.3–3.3 fold), and buffer savings (20–70%) [6]. This analysis included process integration with continuous fermentation instead of batch fermentation, which contributed to a large extent to the savings.

Mahajan *et al.* have evaluated the improvements of the single processing step alone [7]. They concluded that the multicolumn chromatography and modified batch processes have the potential to save approximately 40% on the cost of resin, buffer, and processing time.

A twin-column sequential loading process, CaptureSMB, was evaluated by Angarita *et al.* for mAb capture using protein A affinity chromatography and similar improvements were found in terms of resin capacity utilization and productivity (up to 30–40%) [8]. The advantages were larger for increased interconnected state feed flow rates, which is expected as the dynamic binding capacity of the stationary phase decreases with increasing flow rate. This affects the batch chromatography load to a much greater extent than the CaptureSMB load.

The capture of mAb fragments using twin-column CaptureSMB was shown by Ulmer *et al.* [12] indicating resin cost savings of 40% in comparison to single column batch chromatography ($ 2 million for 100 kg annual production).

All authors have reported comparable product quality of sequential loading processes and batch processes.

A detailed cost modeling analysis for a 3C-PCC process was carried out by Pollock *et al.* [13] showing savings of 30% in the proof of concept scenario, while savings in phase III and commercial manufacturing were lower. This was attributed to the fact that in commercial scale the full lifetime of the stationary phase is exploited while the resin is discarded after a few cycles in the proof of concept scenario, leading to high relative resin costs ($ resin cost/g product) so that resin cost savings by sequential loading processes have a larger impact. However, in terms of absolute numbers, according to Pollock *et al.* [13], the annual costs savings through multicolumn capture processes could exceed 1 million of US$ since the estimated cost savings per 10 kg batch are in the range of 30–45 kUS$.

14.5 Polishing Applications

14.5.1 Introduction

One of the major tasks in chromatographic polishing applications is the clearance of product related impurities such as protein fragments, isoforms, or aggregates. A typical ternary polishing separation challenge is illustrated in Figure 14.7 (simulation data). The figure shows peaks of the weakly adsorbing impurities W, the product P, and the strongly adsorbing impurities S and the cumulative UV signal, that is, the signal that would be recorded by the UV detector in a preparative batch chromatography run. In practice, the W, P, S concentrations could be obtained by fractionation and offline analysis. The overlap between W/P in the front and P/S in the tail of the profile indicate that this is a "difficult" separation. The dashed vertical lines in Figure 14.7 confine a narrow product fraction with high purity but low yield. The dotted lines confine a much wider fraction, including the entire product, thus corresponding to maximum yield but very low purity. Fractions of intermediate width and intermediate purity and yield could be generated. This situation where high purity is possible only at the cost of yield and vice versa is called yield/purity trade-off. As outlined above, it is detrimental to productivity to prolong the gradient to improve the separation.

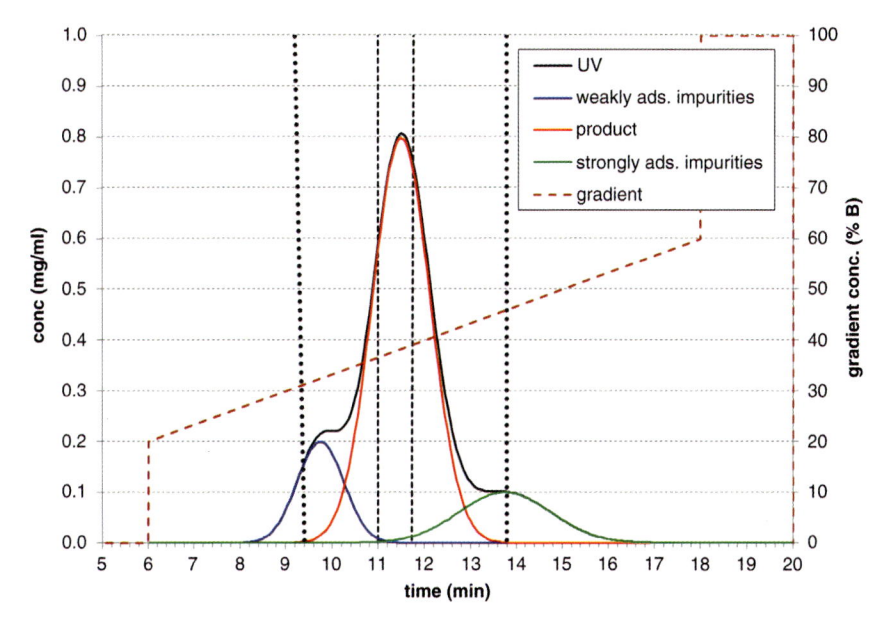

Figure 14.7 Schematic illustration of a typical bioseparation problem where the product (P) is flanked by weakly (W) and strongly (S) adsorbing impurities with similar adsorptive properties. The overlapping regions of product and impurities are clearly visible. The dashed vertical lines confine an interval where the product has high purity while the dotted vertical lines confine an interval that includes the entire product.

Also it is detrimental to productivity to decrease the load to improve separation. The overlap of product and impurities that persists after reasonable batch chromatography optimization effort inevitably leads to a yield/purity trade-off.

Multicolumn countercurrent chromatography in the form of MCSGP is the only process capable of resolving the yield/purity trade-off, performing a ternary separation and preserving the gradient that was used in the reference batch run. The idea behind the MCSGP process is the internal recycling of the overlapping side fractions to recover the product contained therein.

14.5.2 MCSGP (Multicolumn Countercurrent Solvent Gradient Purification) Principle

The process principle of twin-column MCSGP has been outlined in Ref. [9] and is shown in Figure 14.8. The schematic chromatogram at the bottom of Figure 14.8 represents a batch chromatogram that has been divided into different sections (vertical dashed lines) according to the tasks that are carried out in the batch chromatography run (equilibration in zone 1, feeding in zone 2, washing in zone 3, elution in zones 4-7, cleaning and re-equilibration in zone 8). The elution phase is subdivided into additional zones according to the elution order of W, P, and S in the chromatogram (elution of weakly adsorbing impurities W in zone 4, elution of the overlapping part W/P in zone 5, elution of pure P in zone 6, elution of the overlapping part of P/S in zone 7). In the twin-column MCSGP process, these individual tasks of zones 1–8 are carried out as in batch chromatography, with the

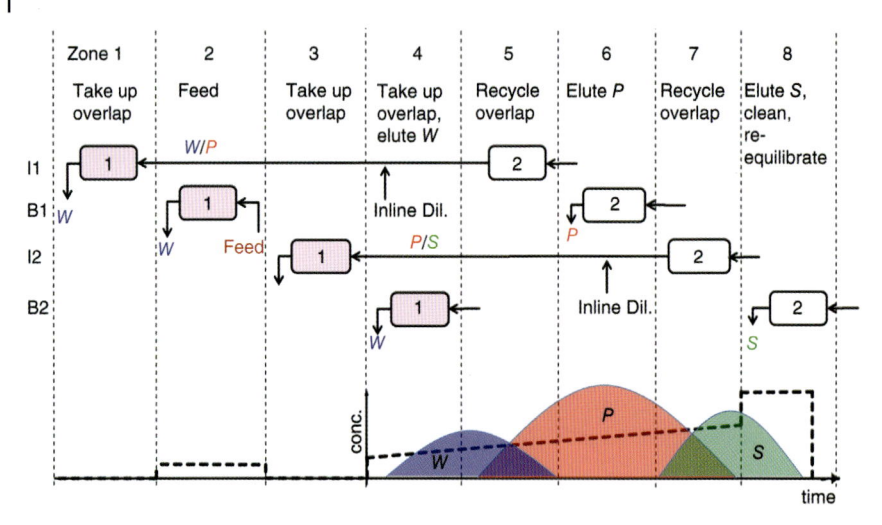

Figure 14.8 Schematic illustration of the twin-column MCSGP process principle (1st switch). The dashed vertical lines separate the different MCSGP process tasks corresponding to the zones of the schematic batch chromatogram shown in the lower part of the figure. Phases I1, B1, I2, and B2 are carried out sequentially.

decisive difference that the W/P and the P/S eluate are directed to a second column for recovery of P (zones 5 and 7). Thus, the process tasks of the single column batch process and the MCSGP process are analogous and it is possible to derive the operating parameters for MCSGP from the batch operating parameters and the corresponding chromatogram. The MCSGP process design procedure is outlined in the next chapter.

A complete cycle of a twin-column MCSGP process comprises two "switches" with four pairs of tasks each (I1, B1, I2, B2) as illustrated in Figure 14.8. The phases in each switch are identical; the difference is only in the column position: In the first switch, column 1 is downstream of column 2 while in the second switch (not shown in Figure 14.8) column 2 is downstream of column 1. The four phases include the following tasks:

- *Phase I1*: The overlapping part W/P is eluted from the upstream column (zone 5), and internally recycled into the downstream column (zone 1). In between the columns, the stream is diluted in-line with buffer/solvent to readsorb P (and overlapping W) in the downstream column. At the end of phase I1, pure product is ready for elution at the outlet of the upstream column (zone 5).
- *Phase B1*: Pure P is eluted and collected from the column in zone 6 (column 2 in Figure 14.8), keeping the overlapping part P/S and S in the column. At the same time, fresh feed is injected into the column in zone 2.
- *Phase I2*: The overlapping part P/S is eluted from the upstream column (zones 7), and internally recycled into the downstream column (zone 3). In between the columns, the stream is diluted in-line with buffer/solvent to readsorb P in the downstream column. At the end of the step, all remaining P has been eluted from the upstream column and only S is left in the upstream column.

- *Phase B2*: The column in zone 8 (column 2 in Figure 14.8) is cleaned to remove S and re-equilibrated. At the same time, W is eluted from the other column in zone 4.

After having completed these tasks, the columns switch positions and in the next phase I1 (not shown in Figure 14.8), column 2 is in the downstream position (zone 1) and column 1 is in the upstream position (zone 5). At the beginning of this I1 phase, column 2 is cleaned and re-equilibrated and ready for uptake of the W/P fraction from column 1. After having completed B1, I2, and B2 for the second time, the columns are returning to their original positions and one cycle has been completed. Column 1 is now clean and ready for uptake of W/P from column 2 in phase I1 (as shown in Figure 14.8).

As in other countercurrent chromatographic processes, in practice in MCSGP the column movement is simulated by connecting and disconnecting column inlets and outlets through valve switching and not by physical movement of the columns.

The process is run in a cyclic manner and reaches a cyclic steady state, in which the amount of product withdrawn in each cycle is equal to the amount of product fed and the product quality is the same from cycle to cycle. Figure 14.9 shows start-up and achievement of cyclic steady state, for example, of lysozyme purification without a dedicated start-up method. Through loading of an increased amount of feed in the first cycle, the start-up time can be significantly reduced such that the process is in cyclic steady state from the second cycle onwards. In cyclic steady state, the UV signals of subsequent cycles match exactly and the process delivers product of constant quality. By overlay of the UV signals of subsequent cycles, it can be determined if the process has reached cyclic steady state. For typical applications of MCSGP, differences in bed height or packing quality can be tolerated within certain limits (approximately, up to 10% bed height difference). However, it has to be ensured that the product quality is analyzed as average from both columns, that is, product collected over one complete cycle (corresponding to one product elution per column) or multiple complete cycles should be analyzed.

The average residence time of the product in the MCSGP process is slightly longer than in batch chromatography due to the internal recycling. No detrimental effect on product quality has been observed or reported in the literature so far. A simulation analysis using typical MCSGP operating conditions has shown that a population of tracer product molecules is reduced to 0.01%, approximately, within 5 cycles, that is, the maximum residence time in MCSGP is fivefold larger than the maximum residence time of batch chromatography.

14.5.3 MCSGP (Multicolumn Countercurrent Solvent Gradient Purification) Process Design

As stated above, the MCSGP process operating parameters can be determined from a single column batch chromatogram (Figure 14.7). The regions of pure product P and the overlapping regions with weakly adsorbing impurities W/P, (phase 5 in Figure 14.8) and strongly adsorbing impurities P/S, (phase 7

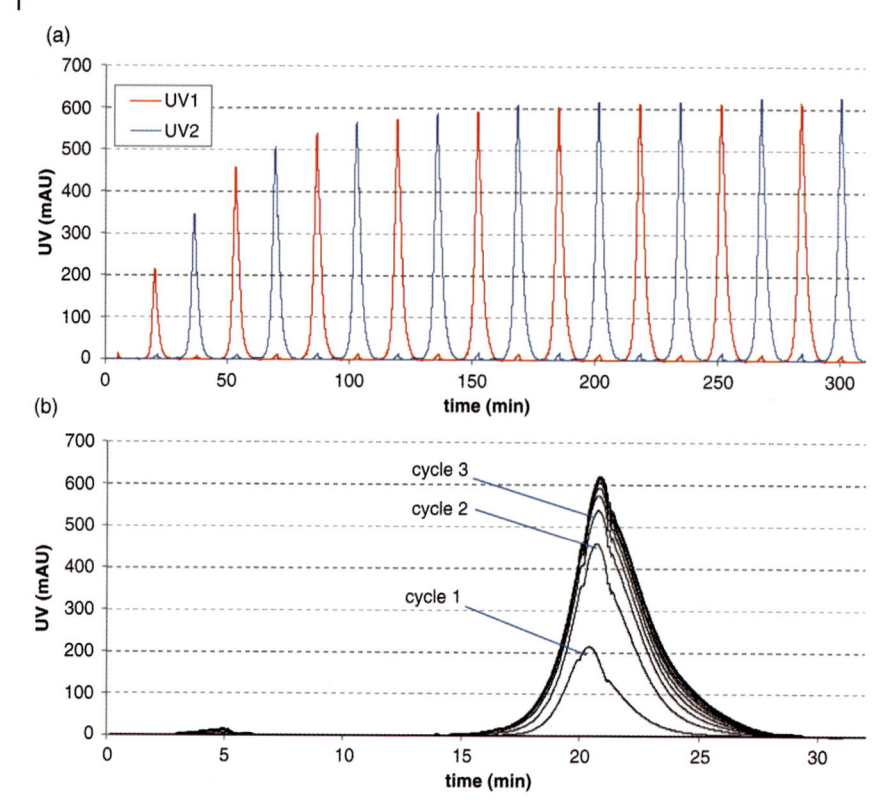

Figure 14.9 Start-up and steady state of MCSGP for lysozyme purification. (a) Internal chromatograms measured by the UV detectors located at the outlet of each column. Each peak corresponds to a product elution. (b) Overlay of the internal chromatograms of UV1. The arrow indicates the increase in concentration from cycle to cycle until a cyclic steady state is reached where peak height and shape remain constant from cycle to cycle.

Figure 14.8) have to be known approximately, for instance by offline analysis. The "design" chromatogram, from which the MCSGP operating parameters are derived, should fulfill the following specifications (Figure 14.7):

- Product elution in between minimum 10% B and maximum 70% B with a linear gradient
- Product elution in between minimum 2 and maximum 15 column volumes after elution start
- Product elution complete at least 0.5 CV before wash/CIP step starts
- The load for the batch run should be approximately 25–50% of the maximum load of the 1% DBC value of the protein mixture on the resin. Typical values in protein purification are 5–30 mg protein per ml of packed bed.
- At least part of the product (20–50%) isolated in the fractions of the batch run fulfills the purity specifications.

If one of the above conditions is not met, a new "design" batch run with improved conditions should be recorded. From the analyzed batch process, MCSGP operating parameters can be determined based on the schematic shown in Figure 14.10.

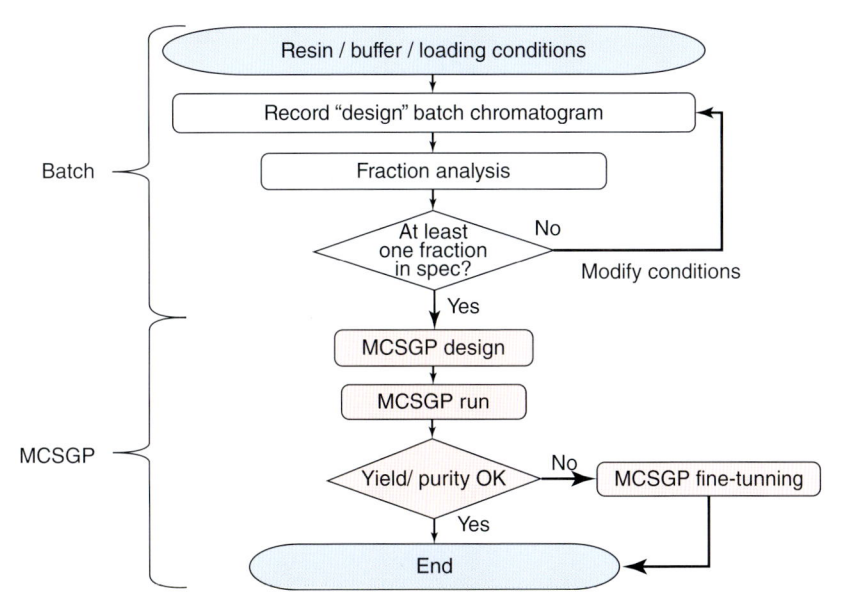

Figure 14.10 MCSGP process design schematic.

The MCSGP design is done based on the previously evaluated "design" batch chromatogram along the following lines:

- The width of the zone intervals is known since the time span and the flow rate of the batch chromatogram is known. From this, the volume to be delivered by the pumps to the upstream column (column 2 in Figure 14.8, phases 5, 6, and 7) is known. The volume corresponding to phase 4 is defined by the start of the gradient and the start of phase 5. Phase 8 parameters (cleaning and re-equilibration) can be chosen the same as in batch chromatography.
- The gradient concentrations (%B) at the start and the end of each zone are known. Therefore, the gradient to be delivered by the pump to the upstream column (column 2 in Figure 14.8, phases 5, 6, and 7) is known.
- The magnitude of the in-line dilution streams in phases 1 and 3 is chosen such that the maximum concentration occurring in W/P and P/S, respectively, is diluted to the starting concentration of the gradient or to conditions corresponding to the adsorptive strength of the feed, whichever is higher.
- The amount of feed added in phase 3 corresponds to the amount of feed that is removed via the product elution window. This determines the feed volume to be added each cycle.

Thus, summarizing the process design, the procedure aims at reproducing the underlying batch chromatogram within the twin column system with the key difference that the impure side fractions are internally recycled in MCSGP whereas, they are collected and discarded in batch chromatography.

The operating parameter determination procedure for MCSGP design has been automated in a software tool ("MCSGP wizard," ChromIQ software, ChromaCon AG, Switzerland).

The initial operating point is designed to deliver the product with the same productivity as the underlying "design" batch chromatography run. For process optimization, further MCSGP runs are carried out in order to improve productivity for instance by increasing the load or running a steeper gradient. In batch chromatography these actions would lead to an increased overlapping of product and impurities, causing a decrease in yield for the desired purity. In contrast, the MCSGP process can afford having increased overlaps between product and impurities due to its internal recycling capabilities. Since the MCSGP can transform low-yield batch processes into high-yield process it can help speeding up process development timelines. Required product and process quality thresholds (e.g., in terms of purity, yield, productivity) can be reached much earlier using MCSGP than by optimizing the batch process. This factor can be very important in an industrial setting for both originator and biosimilar producers.

14.5.4 MCSGP (Multicolumn Countercurrent Solvent Gradient Purification) Case Study

MCSGP has been successfully applied to protein and peptide purifications using a variety of stationary and mobile phases including ion exchangers, hydrophobic interaction and reverse phase chromatography using aqueous buffers (proteins) and reverse phase solvents (peptides), respectively. Application examples include the purification of mAbs, of mAb isoforms, bispecific antibodies, antibody conjugates, fusion proteins, PEGylated proteins, and peptide hormones.

Historically, the first MCSGP processes were operated with six columns. In the following years the process was simplified to a two-column setup. Aumann et al. have shown the purification of calcitonin using reverse phase chromatography using six- and three-column setups [14–16].

The isolation of mAb isoforms was shown by Müller-Späth et al. [17,18]. Apart from removing unwanted isoforms, antibody isoform separation is of interest for characterization purposes or potentially for production of biobetters. MCSGP can also be useful to straighten out product isoform patterns resulting from variations in upstream fermentation. This capability can be potentially of high interest for biosimilar manufacturers. An illustration of mAb isoform straightening is shown in Figure 14.11 [18] for the case of Herceptin®, a mAb manufactured by Roche for cancer treatment. Figure 14.11a shows the isoform patterns obtained by analytical cation exchange chromatograms of two different feed materials: Herceptin obtained from the pharmacy and Herceptin spiked with weakly adsorbing isoforms. Figure 14.11b shows the analytical chromatograms of the product pools obtained using cation-exchange MCSGP running exactly the same operating parameters. It is obvious that the product isoform pattern is identical despite the large variations in the feed pattern.

The application of MCSGP to PEGylated protein purification has been presented in detail elsewhere [9]. Briefly, the purification challenge was the purification of monoPEGylated α-lactalbumin from a mixture of higher PEGylated species. It was shown that the same product quality (93% purity) could be obtained using batch and MCSGP chromatography. However, the yield of the batch run was only 56% while the yield for MCSGP was 83%. A batch product pool

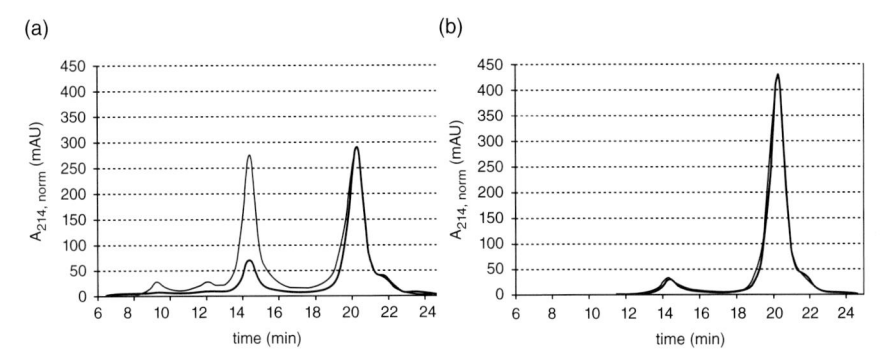

Figure 14.11 mAb isoform separation capabilities of MCSGP visualized for the case of Herceptin® using analytical cation exchange chromatograms. (a) Overlay of original Herceptin® (blue) and Herceptin® spiked with weakly adsorbing isoforms (red) chromatograms. (b) Overlay of product pools obtained using the same MCSGP operating conditions for the two feed materials.

with the same yield as MCSGP would have a purity of only 82% and include more than the double amount of impurities. Furthermore, the buffer consumption in this case was reduced 50% using MCSGP.

The results of the comparison are summarized in Figure 14.12, where the yield/purity trade-off of batch chromatography and the simultaneous achievement of high yield and purity become evident.

The purification of bispecific antibodies using cation-exchange MCSGP has been shown by Müller-Späth *et al.* [19]. The purification of a bispecific antibody represents a classical ternary difficult separation since the target

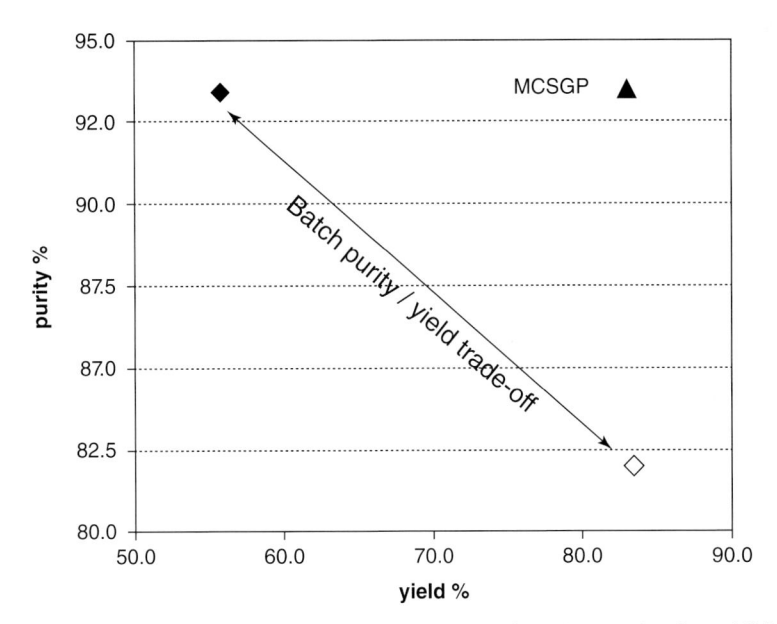

Figure 14.12 Yield/purity chart showing the results of a high purity batch pool (full diamond), a high yield batch pool (empty diamond), and an MCSGP pool (triangle).

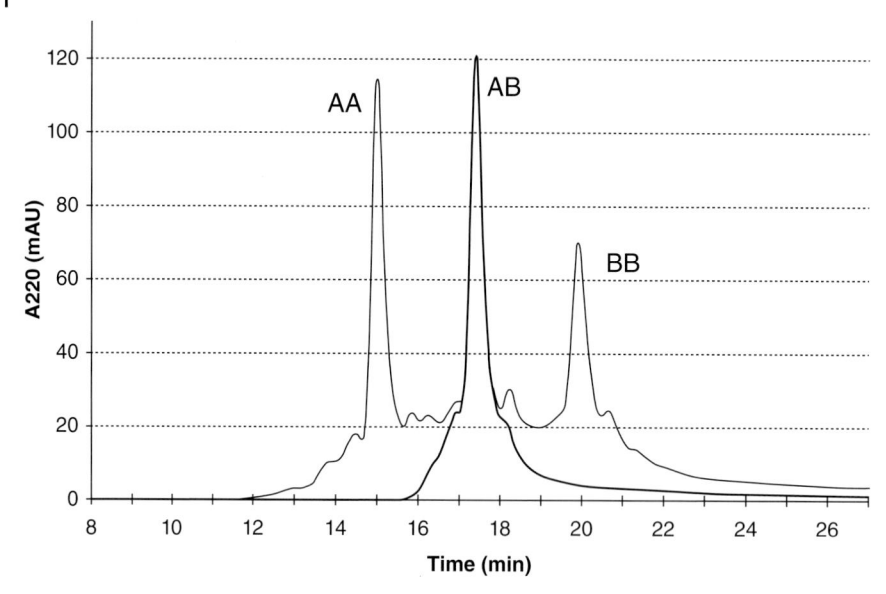

Figure 14.13 Analytical cation exchange chromatogram of a bispecific antibody mixture and the product AB (bold line) produced using the MCSGP process.

bispecific antibody AB is accompanied by a number of undesired antibody forms AA and BB and their charged isoforms that are coexpressed by the host cells. Analytical results of a weak cation exchange analysis of the feed material and MCSGP-purified product obtained in this study are shown in Figure 14.13.

In another case study, MCSGP was used to isolate a fraction of an antibody–drug conjugate (ADC) model system with a narrow drug–antibody ratio (DAR) distribution. A fluorescent dye was coupled to Trastuzumab through the antibodies' lysine residues leading to a broad DAR profile due to the high number of available lysine residues.

Antibody with DAR = 2 was selected as product and could be isolated with a purity of 59% and a yield of 34% using batch chromatography. Using MCSGP, a purity of 70% and a yield of 61% was achieved for the same productivity Figure 14.14. Consistent product quality was obtained over several cycles as confirmed by mass spectrometry analysis. This procedure could be a very attractive option for producers of first-generation ADCs that use a nonspecific coupling chemistry and confirms that even under strict purity constraints, MCSGP can deliver good yields.

14.6 Discovery and Development applications

Multicolumn countercurrent chromatography is also applicable in the case of drug identification and drug development for the isolation of minor compounds in a complex mixture. Similarly as MCSGP, the N-Rich process uses the concept of internal recycling to recycle a portion of the chromatogram

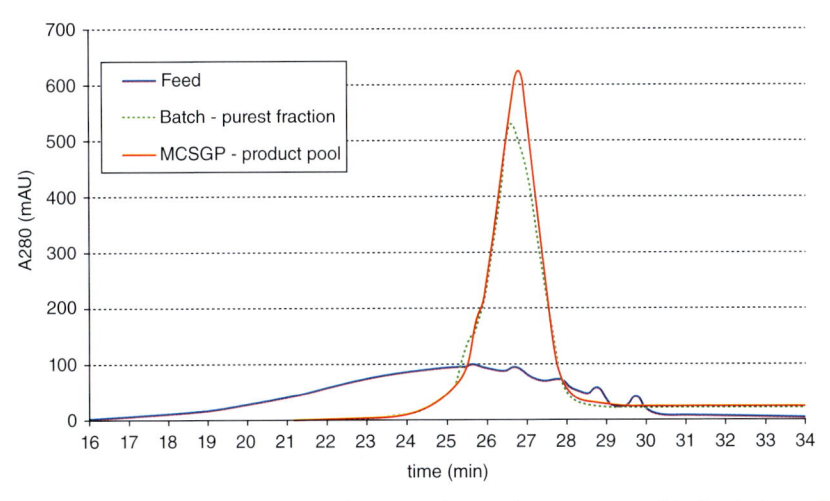

Figure 14.14 Overlay of an analytical cation exchange chromatogram of the feed mixture (blue), the purest batch fraction (green) and the MCSGP product pool (red) corresponding to DAR = 2.

from one column to the other. However, the fundamental difference to MCSGP is that all fractions except the product fraction (phase 6 in Figure 14.10) are discarded and only the product fraction is internally recycled (Figure 14.15). Since the amount of product that is removed per cycle (through the discarded W/P and P/S fractions) is smaller than the amount of product that is fed, the product P accumulates within the system while the other components are removed. The duration of one cycle is similar as the duration of a batch chromatography run. Using the N-Rich process principle, enrichments of up to 1000-fold can be achieved. The enrichment comprises a simultaneous increase of the concentration and improvement of the purity. Once the process has been operated over a number of cycles and sufficient enrichment has been achieved, the product of interest is recovered through a very shallow isocratic or gradient elution with fine fractionation. The fractionation ensures the recovery of the compound of interest and potentially also neighboring compounds. The N-Rich process amplifies an entire region of the underlying "design" chromatogram; therefore, the exact position of the compound in the chromatogram does not need to be known. Since the process does not operate with maximum yield and does not reach a cyclic steady state it is not suited to continuously or periodically manufacture product. In fact the process is used for the one-time production of compounds of interest. Possible applications include the identification of potential drug candidates or the isolation of product-related impurities in milligram to gram amounts for characterization purposes (e.g., product isoforms). Through the N-Rich process, the effort of repeated HPLC runs, fractionation, manual handling, and pooling can be significantly reduced and replaced by a fully automated process.

The N-Rich process has been demonstrated for the isolation of product-related impurities of fibrinopeptide A, produced by chemical synthesis [20]. Enrichment factors of 600-fold were achieved by improving the purity from 0.13 to over 80%. The concentration of the target compound was increased 10-fold.

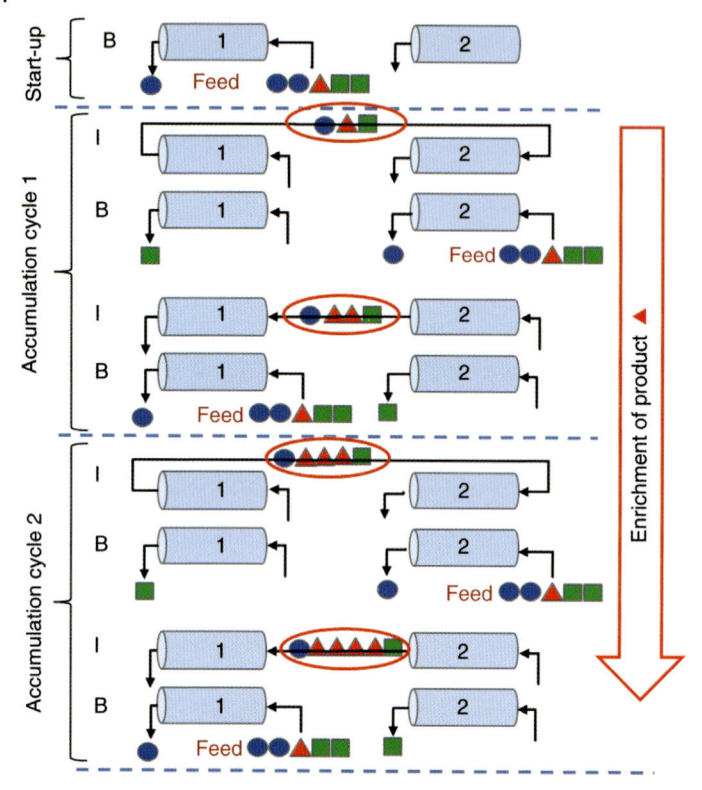

Figure 14.15 N-Rich process principle. The compound of interest (red triangles) is accumulated over multiple cycles by internal recycling from one column to the other. A portion of the weakly adsorbing impurities (circles) and the strongly adsorbing impurities (squares) are removed every cycle. The readsorption of the product in the downstream column is ensured by in-line dilution (not shown).

14.7 Scale-Up of Multicolumn Countercurrent Chromatography Processes

As in single column chromatography, the scale-up of multicolumn countercurrent chromatography is usually done by increasing the column diameter and keeping the bed height constant. The minimum bed height considered "packable" in large scale is 10 cm. This bed height should also be used during small-scale process development.

In most scale-up cases no specialized hardware components are required that could pose a risk. Existing pump, valve, and detector technology is suited for large-scale multicolumn countercurrent chromatography. The ratio of dead-volumes to column volume decreases when scaling-up from the lab scale and a negative impact is not expected in that regard. Skids capable of GMP manufacturing of biopharmaceuticals are in development by a handful of equipment manufacturers (as of 2015). A twin-column skid is available from LEWA Process Technologies.

The scale-up of multicolumn countercurrent processes in pharmaceuticals production has been successfully demonstrated for four-zone SMB processes [21].

An MCSGP scale-up cost calculation has been carried out by Takizawa in collaboration with Sandoz for a biosimilar manufacturing process showing significant cost savings of the multicolumn countercurrent process and a net present value of several million US$ for a large biopharmaceutical product [22].

Due to its capability of improving yields, it is easier to develop economically viable scenarios for difficult separations when MCSGP is included.

Moreover, for the production of biosimilars or biobetters the capabilities of MCSGP to adjust the isoform pattern can be very attractive.

For sequential loading processes (CaptureSMB, 3C-PCC, 4C-PCC, SMCC), the resin costs savings and buffer volume savings are in the range of 40–60% compared to batch affinity chromatography, as mentioned above.

The economic advantages of multicolumn countercurrent processes come at the cost of increasing equipment complexity. When comparing multicolumn processes, the advantage of using fewer columns is obvious from the perspective of equipment complexity, equipment downtime risk, and capital expenditures. While for the capture process a number of multicolumn alternatives exist (CaptureSMB, 3C-PCC, 4C-PCC, SMCC, etc.), the MCSGP process is the only multicolumn chromatography option for bind/elute polishing of biopharmaceuticals. The state of the art MCSGP process uses in a twin-column configuration.

14.8 Multicolumn Countercurrent Chromatography as Replacement for Batch Chromatography Unit Operations

As outlined in the previous chapters, multicolumn countercurrent processes are suited for all chromatographic unit operations. Therefore, in existing or planned chromatographic downstream processes, they can be used as replacement for any batch chromatography step, thereby significantly improving productivity and process economics.

Reflecting the productivity (Eq. (14.2), an elevated productivity allows the following production options:

1) More product can be produced within the same timeframe and the same bed volume
2) The same amount of product can be produced within the same timeframe using smaller columns
3) The same amount of product can be produced using the same bed volume within a shorter time frame.

If resin costs are a key economic parameter option (2) would be very attractive; in the case of purification of an instable product option (3) would be most interesting.

When replacing batch steps by multicolumn countercurrent processes, the economic advantages of multicolumn countercurrent processes are cumulative

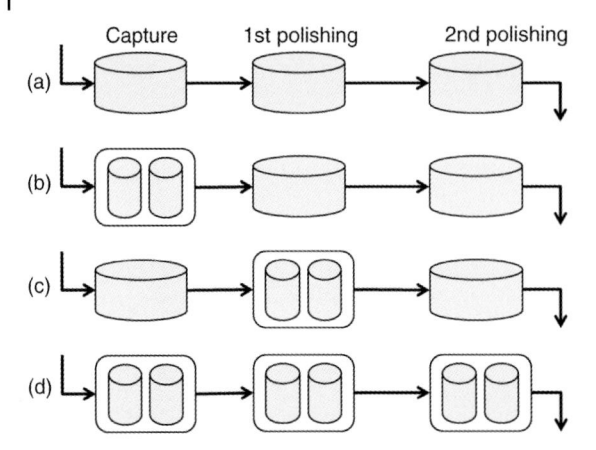

Figure 14.16 Examples for downstream processes with varying degree of multicolumn countercurrent process implementation. (a) Traditional downstream process with three batch steps. (b) Hybrid process with multicolumn countercurrent capture step and batch polishing step. (c) Hybrid process with batch capture step, one multicolumn countercurrent polishing step, and one batch polishing step. (d) Downstream process with three multicolumn countercurrent steps.

and improve with the number of batch steps replaced. Also hybrid processes are possible. For instance, a multicolumn countercurrent capture step can be used instead of a batch capture step, while the polishing steps remain in batch mode. Independent of the capture step, a multicolumn countercurrent step can be used for polishing (e.g., MCSGP in ion exchange or HIC mode). Four of the eight possible combinations of batch and multicolumn countercurrent processes for a three-chromatography step downstream process are shown in Figure 14.16. Intermediate nonchromatographic steps and hold steps are omitted in the figure for the sake of simplicity.

Independent of the type of process used in the single chromatographic unit operation (batch or multicolumn countercurrent), the downstream process may be optimized by process integration. In traditional downstream processing product from a first unit operation is collected and pooled and then reconditioned before being supplied as feed for the following chromatography step. Depending on the batch definition, the later unit operation has to wait for a certain number of runs of the earlier unit operation to be completed before starting, causing a high degree of inactivity. By process integration, the scheduling of the chromatographic steps is improved so that the later chromatographic unit operation starts as soon as enough material is available from the earlier unit operation. Ideally, the product eluting from the first unit operation is directly loaded into the second unit operation without intermediate hold step and with in-line dilution for conditioning.

Multicolumn countercurrent processes are very suited for process integration because they produce product in smaller intervals than batch chromatography that allows for further improvement of the scheduling of the chromatography steps that are coupled (integrated).

14.9 Multicolumn Countercurrent Chromatography and Continuous Upstream

Continuous chromatography has intensively been discussed in conjunction with continuous fermentation. When coupled to continuous fermentation and operating over a prolonged time period in the range of weeks or months, the sterile coupling of the equipment needs to be warranted. Since all currently available lab-scale systems are not suited for sterile operation, at least the backgrowth of microbes through the harvest line needs to be avoided for instance by means of sterile filters. On larger scale, sterile equipment is needed with a CIP or SIP (steam-in-place) option. In the presented cases a surge bag is used to derisk the downstream process by avoiding a chain reaction of failures in case one of the unit operations in the downstream or upstream process experience a failure [5,6].

To ensure continuous operation, the net inflow and net outflow to/from the surge bag need to be identical. In continuous manufacturing the columns can be significantly downsized as the product from the bioreactor is withdrawn continuously at a low flow rate and purified as the fermenter is running. Thus, the transit time includes the fermentation time, while in fed-batch fermentation the transit time is mostly <48 h. This means that equipment with relatively small column diameter can be sufficient for production purposes (Eq. (14.1)).

Current cell specific perfusion rate values in the range of 0.025 nl/cell/day have been reported. Assuming a cell density of 150 million cells/ml the flow rate exiting a 200 l fermenter would be 31 l/h or 520 ml/min that could be handled by equipment considered "pilot scale" in existing processes [1]. Assuming a mAb titer of just 1.0 g/l a fermenter of this type could produce around 750 g mAb per day and about 150 kg mAb per year, which is a typical demand for a mAb product [23]. In this case twin-column pilot scale equipment with 10–15 cm inner diameter columns would be capable of capturing the volume flow from continuous upstream and to supply the market as indicated above.

Most industrial processes for mAb and Fc fusion protein purification rely on protein A chromatography for capture. Even in multicolumn capture processes, the elution from protein A is discontinuous, which enables volume reduction. The protein A eluates are collected and pooled before being processed further. As described previously, also the polishing steps can be operated in multicolumn mode, for example, as MCGSP process.

14.10 Regulatory Aspects and Control of Multicolumn Countercurrent Processes

Regulatory authorities support the introduction of continuous manufacturing in pharmaceuticals production as they promise to lead to drug quality advantages [24]. Continuous manufacturing is in-line with the FDA quality initiatives (QbD).

No specific FDA guidance exists about continuous manufacturing apart from the definition of a "lot." Nothing in FDA guidance prohibits the use of continuous manufacturing.

Both definitions of "batch" and "lot" are applicable to continuous manufacturing. According to 21 CFR 210.3 "Lot means a batch, or a specific identified portion of a batch, having uniform character and quality within specified limits; or, in the case of a drug product produced by continuous process, it is a specific identified amount produced in a unit of time or quantity in a manner that assures its having uniform character and quality within specified limits."

Moreover, a batch is defined as follows: "Batch means a specific quantity of a drug or other material that is intended to have uniform character and quality, within specified limits, and is produced according to a single manufacturing order during the same cycle of manufacture."

Thus, the regulations provide great flexibility in terms of lot definition while not even specifying the mode of manufacturing in the batch definition. Mass-based, time-based, or raw materials-based definitions are possible.

Continuous chromatography processes can be controlled to a stronger extent than batch processes. Online information on product quality can be obtained more regularly and over longer periods of time compared to batch or fed-batch-based processes that are never in steady state. Cyclically continuous processes produce a steady output of small batches that can be subject to state-of-the-art analytical tools.

Reported online analytics for multicolumn processes for biopharmaceuticals have been mainly employing UV-based monitoring and control strategies. In cyclic steady state, the UV signals of subsequent cycles should match when superimposed as product concentration and quality remain the same from cycle to cycle. UV and other spectral data can be used to extract purity information [25] or at least to determine a steady-state "fingerprint." Besides spectral data, conductivity and pH are measured to monitor cycle-to-cycle buffer and gradient consistency.

Online information (e.g., UV) can be used for feedback control and to correct deviations from cyclic steady state, which may occur for instance due to column aging.

UV-based feedback control concepts have been demonstrated for both capture and polishing processes.

A first control concept is based on monitoring the breakthrough of the target compound while loading two columns in series and comparing the breakthrough UV value with the feed UV value determined by a second detector and using this information to determine the point of column switching [5]. A second control concept uses the comparison of the breakthrough values and the elution peak areas from cycle to cycle for process control [26] and was validated for the twin-column CaptureSMB process.

In polishing processes, the determination of the product purity can be realized effectively using at line HPLC and analyzing the product from each cycle, however it is elaborate in terms of equipment setup and integration. This concept was applied to a twin-column MCSGP process by Krättli *et al.* [27].

References

1 Konstantinov, K. (2013) The promise of continuous bioprocessing. Castelldefells, Spain, October 20–24.
2 Vogel, J.H. et al. (2002) Continuous annular chromatography: general characterization and application for the isolation of recombinant protein drugs. *Biotechnol. Bioeng.*, **80** (5), 559–568.
3 Kaspereit, M. et al. (2012) Process concepts. *Preparative Chromatography*, Wiley-VCH Verlag GmbH, pp. 273–320.
4 Carta, G. and Jungbauer, A. (2010) Effects of dispersion and adsorption kinetics on column performance. *Protein Chromatography*, Wiley-VCH Verlag GmbH, pp. 237–276.
5 Warikoo, V. et al. (2012) Integrated continuous production of recombinant therapeutic proteins. *Biotechnol. Bioeng.*, **109** (12), 3018–3029.
6 Godawat, R. et al. (2012) Periodic counter-current chromatography – design and operational considerations for integrated and continuous purification of proteins. *Biotechnol. J.*, 7 (12), 1496–1508.
7 Mahajan, E., George, A., and Wolk, B. (2012) Improving affinity chromatography resin efficiency using semi-continuous chromatography. *J. Chromatogr. A*, **1227**, 154–162.
8 Angarita, M. et al. (2015) Twin-column CaptureSMB: a novel cyclic process for protein a affinity chromatography. *J. Chromatogr. A*, **1389**, 85–95.
9 Müller-Späth, T. and Morbidelli, M. (2014) Multicolumn countercurrent gradient chromatography for the purification of biopharmaceuticals, in *Continuous Processing in Pharmaceutical Manufacturing*, Wiley-VCH Verlag GmbH, pp. 227–254.
10 Gjoka, X. et al. (2015) A straightforward methodology for designing continuous monoclonal antibody capture multi-column chromatography processes. *J. Chromatogr. A*, **1416**, 38–46.
11 Baur, D. (2016) Comparison of batch and continuous multi-column protein A capture processes. *J. Biotechnol.*, **11**, 920–931.
12 Ulmer, N., Muller-Spath, T., Neunstoecklin, B., Aumann, L., Bavand, M., and Morbidelli, M. (2015) *Affinity Capture of F(ab')2 Fragments: Using Twin-Column Countercurrent Chromatography*, Bioprocess International.
13 Pollock, J. et al. (2013) Optimising the design and operation of semi-continuous affinity chromatography for clinical and commercial manufacture. *J. Chromatogr. A*, **1284**, 17–27.
14 Aumann, L. and Morbidelli, M. (2007) A continuous multicolumn countercurrent solvent gradient purification (MCSGP) process. *Biotechnol. Bioeng.*, **98** (5), 1043–1055.
15 Aumann, L. and Morbidelli, M. (2008) A semicontinuous 3-column countercurrent solvent gradient purification (MCSGP) process. *Biotechnol. Bioeng.*, **99** (3), 728–733.
16 Aumann, L., Stroehlein, G., and Morbidelli, M. (2007) Parametric study of a 6-column countercurrent solvent gradient purification (MCSGP) unit. *Biotechnol. Bioeng.*, **98** (5), 1029–1042.

17 Müller-Späth, T. *et al.* (2008) Chromatographic separation of three monoclonal antibody variants using multicolumn countercurrent solvent gradient purification (MCSGP). *Biotechnol. Bioeng.*, **100** (6), 1166–1177.
18 Müller-Späth, T. *et al.* (2010) Increasing the activity of monoclonal antibody therapeutics by continuous chromatography (MCSGP). *Biotechnol. Bioeng.*, **107** (4), 652–662.
19 Th. Müller-Späth, N.U., Aumann, L., Ströhlein, G., Bavand, M., Hendriks, L.J.A., Kruif, J.de., Throsby, M., and Bakker, A.B.H. (2013) Purifying common light-chain bispecific antibodies: a twin-column, countercurrent chromatography platform process. *Bioprocess Int.*, **11** (5), 36–45.
20 Müller-Späth, T., Ströhlein, U.N., Bavand, G., and N-Rich, M. (2013) A novel automated enrichment process for the isolation of product-related impurities from active pharmaceutical ingredients., in PREP Symposium 2013, Boston MA.
21 Hamende, M. (2007) Case study in production-scale multicolumn continuous chromatography, *Preparative Enantioselective Chromatography*, Blackwell Publishing Ltd., pp. 253–276.
22 Takizawa, B.T. (2011) Evaluation of the financial impact of continuous chromatography in the production of biologics. Available at http://hdl.handle.net/1721.1/66045.
23 Kelley, B. (2009) Industrialization of mAb production technology: the bioprocessing industry at a crossroads. *mAbs*, **1** (5), 443–452.
24 Janet, W. (2014) Modernizing pharmaceutical manufacturing – continuous manufacturing as a key enabler. International Symposium on Continuous Manufacturing of Pharmaceuticals. 2014: Cambridge MA.
25 Brestrich, N. *et al.* (2015) Advances in inline quantification of co-eluting proteins in chromatography: process-data-based model calibration and application towards real-life separation issues. *Biotechnol. Bioeng.*, **112** (7), 1406–1416.
26 Müller-Späth, T. (2014) Automated process development and control of a twin-column counter-current process (CaptureSMB) for affinity capture. PREP Conference, Boston.
27 Krättli, M., Steinebach, F., and Morbidelli, M. (2013) Online control of the twin-column countercurrent solvent gradient process for biochromatography. *J. Chromatogr. A*, **1293**, 51–59.

15

Single-Pass Tangential Flow Filtration (SPTFF) in Continuous Biomanufacturing

Andrew Clutterbuck, Paul Beckett, Renato Lorenzi, Frederic Sengler, Torsten Bisschop, and Josselyn Haas

Millipore SAS, Process Solution Technologies, 39 Route Industrielle de la Hardt, 67124 Molsheim, France

15.1 Introduction

The phenomenon of osmosis, which is characterized as the transport of water or solvent through a semipermeable membrane, has been known and understood since 1748. This phenomenon was observed by Abbel Nollet when he noted that water diffuses from a dilute solution to a more concentrated one when separated by a semipermeable membrane [1]. Then, in 1907, the first reference to the term "ultrafiltration" was used in literature by Bechold [2]. His experiment was to drive solutions at high pressures (up to several bar) through a membrane prepared by impregnating filter paper with acetic acid and acid collodion [2]. Then, in the early 1950s, Samuel Yuster, of the University of California, Los Angeles, predicted that it should be possible to produce fresh water from brine using this membrane separation methodology [3]. However, it took until the 1960s to see the first ultrafiltration (UF) and reverse osmosis (RO) continuous molecular separation processes that did not involve a phase change or interphase mass transfer, that is, the transport of mass within a single phase depends directly on the concentration gradient as appose to mass may also transport from one phase to another such as the separation of a liquid and gas at a defined boundary.

In both microfiltration and ultrafiltration processes, the basic premise is that a pressure gradient across a membrane drives solvents and salts to pass through a membrane of a defined molecular weight cutoff, where the larger molecules or particles are retained and concentrated in the retentate side, while smaller molecules, below the molecular size exclusion limit of the membrane, pass through to the permeate side. These ultrafiltration type processes have evolved significantly with the development of membrane technology, devices, and different module formats such as hollow fiber, spiral wound, and flat sheet devices.

Ultrafiltration and diafiltration, often abbreviated as UF/DF, is commonly used in bioproduction to concentrate and buffer exchange biopharmaceuticals at various points within the production process and is commonplace in a significant majority of today's downstream bioprocesses templates. Ultrafiltration is commonly performed by tangential flow filtration (TFF) where the liquid flow is parallel to the surface of the membrane to minimize fouling of the filter membrane during processing. Ultrafiltration and tangential flow filtration are not mutually exclusive but the terminologies are used interchangeably within the industry.

In the biopharmaceutical industry TFF is a well-established technology [4–6]. As a key bioprocess unit operation, ultrafiltration is applied often multiple times during downstream processing. The main use is for concentration, buffer exchange purification, and the final formulation of biological molecules.

The wide range of TFF applications includes the following:

- Concentration and desalting of proteins, peptides, and nucleic acids (i.e., DNA, RNA, oligonucleotides).
- Recovery and purification of antibodies or recombinant proteins from cell culture media.
- Plasmid DNA from cell lysates.
- Chromosomal DNA from whole blood.
- Fractionation of dilute protein mixtures.
- Clarification of cell lysates or tissue homogenates.
- Harvest of whole cells or viruses.
- Removal of viruses.
- Depyrogenation (endotoxin removal) from water, buffers, and media solutions.

From a manufacturing perspective, biomanufacturing process developers must support the trend of higher yield cell cultures, higher protein concentrations, and new buffer formulations, while ensuring process safety and reproducibility from one batch to another. Thus, continued development of innovative products requires the following:

- Access to a robust portfolio of processing products.
- Versatile single-use platform solutions to support multiproducts manufacturing and continuous manufacturing.
- Robust and scalable unit operations, from bench to commercial scale.
- Template-able process solutions to speed up time to market.
- Innovative solutions that balance speed to market with cost containment [3].

As the biopharmaceutical industry evolves the pressure to improve, intensify, and streamline biomanufacturing processes increases. Biomanufacturers have investigated many different solutions although most of the efforts have focused on the following three areas:

- Improving existing process templates by eliminating unit operations.
- Connected processing by combining unit operations.
- The introduction of technologies that enables continuous processing.

Single-pass tangential flow filtration (SPTFF) is an enabling technology that significantly aids in the implementation of a fully continuous bioprocess. The many benefits include the fact that SPTFF runs at constant operating conditions throughout the process, simplifies hardware requirements by reducing volumes, and higher product recovery, reduces shear damage to sensitive products, and can be easily implemented in-line before or after a chromatography step.

15.1 Introduction

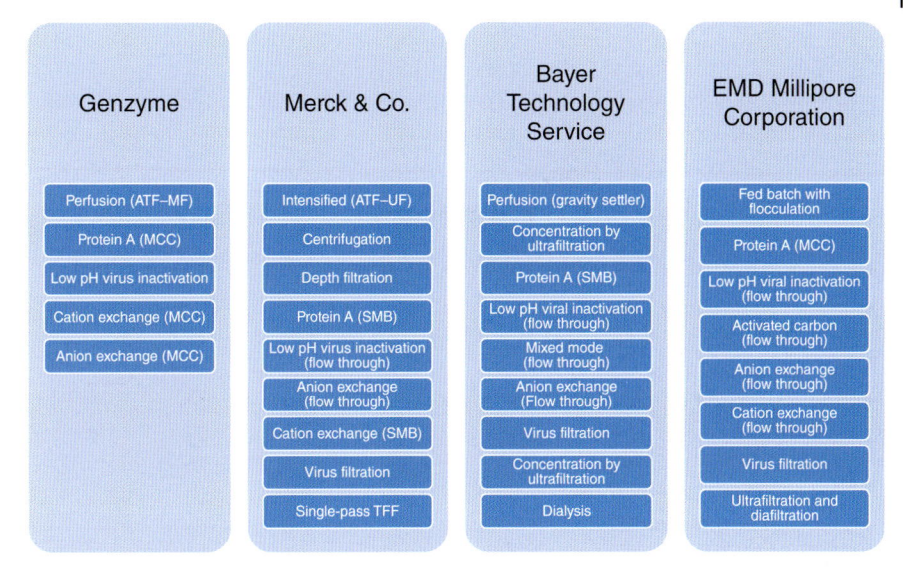

Figure 15.1 Examples of continuous processing in the literature [7–11].

Genzyme	Merck & Co.	Bayer Technology Service	EMD Millipore Corporation
Perfusion (ATF–MF)	Intensified (ATF–UF)	Perfusion (gravity settler)	Fed batch with flocculation
Protein A (MCC)	Centrifugation	Concentration by ultrafiltration	Protein A (MCC)
Low pH virus inactivation	Depth filtration	Protein A (SMB)	Low pH viral inactivation (flow through)
Cation exchange (MCC)	Protein A (SMB)	Low pH viral inactivation (flow through)	Activated carbon (flow through)
Anion exchange (MCC)	Low pH virus inactivation (flow through)	Mixed mode (flow through)	Anion exchange (flow through)
	Anion exchange (flow through)	Anion exchange (Flow through)	Cation exchange (flow through)
	Cation exchange (SMB)	Virus filtration	Virus filtration
	Virus filtration	Concentration by ultrafiltration	Ultrafiltration and diafiltration
	Single-pass TFF	Dialysis	

Continuous processing can mean different things. As a baseline, a typical traditional batch process might include: a fed-batch bioreactor production, centrifugation and/or depth filtration, sterile filtration, protein A affinity chromatography, low-pH virus inactivation, cation exchange chromatography run in bind and elute mode, anion exchange flow-through polishing, virus filtration, ultrafiltration/diafiltration concentration and formulation, and a final terminal sterile filtration step before filling.

It is worth noting that among the various biopharmaceutical manufacturers there can be some differences with respect to the number, types, and the order of the processing steps. A number of manufacturers have published and/or patented various descriptions of continuous processing. Figure 15.1 shows the general steps in continuous processing utilized by a number of industry practitioners.

In general, each template needs to be designed to meet the new robust process requirements taking into account the process constraints. The following different constraints need to be taken into account:

- Product stability, efficacy, and high yield.
- Consistent quality from batch to batch.
- Limitations on chemical consumption.
- Process duration.
- Smaller ecological footprint.
- Lower capital costs.
- Reduction of handling to enable increased operator safety, that is, using toxic compounds such as when processing antibody drug conjugates (ADCs) or when processing toxoid vaccines.

In order to meet all these objectives, ultrafiltration has to be optimized. The identification of the critical parameters is crucial. That is why a new trend tends to develop new approaches like Quality by Design (QbD) where Design of Experiments (DOE) could help achieving these goals.

In the next section, we will focus on the current existing technologies and methodologies, to understand better how TFF technology could help to move from batch to continuous processing.

15.2 Tangential Flow Filtration in Bioproduction

15.2.1 Batch versus Single-Pass Tangential Flow Filtration

Tangential flow filtration is generally operated in two modes within downstream unit operations: Batch whereby the process solution is recirculated or passed over the surface of the membrane multiple times or in single-pass mode. Batch TFF is commonplace and has been widely used in the biopharmaceutical industry for downstream processing applications for many years. Typical batch TFF steps concentrate product through volume reduction, and buffer exchange through diafiltration, to achieve the final targeted concentration and/or buffer formulation.

Traditional TFF operates in batch mode, where the feed/retentate is recirculated through the filter assembly (Figure 15.2a). Typically, TFF cassettes operate in parallel, with multiple passes through membranes required to achieve the desired concentration. Single-pass TFF is a different application of an existing technology (Figure 15.2b). The basic underlying principle of SPTFF is that increased residence time in the feed channel results in increased conversion. Increased residence time can be accomplished by reducing flow rate or increasing path length in a serial configuration. Cassettes in series have a higher mass transfer compared to parallel configurations at equivalent residence times (Figure 15.2c and d).

Next, we will focus specifically on single-pass TFF and its implementation as part of a continuous biomanufacturing strategy.

15.2.2 Membrane Type and Format for TFF Applications

Membranes can be put in several type of configurations or format such as cassettes, spiral wound, or hollow fiber, each with its own advantages and drawbacks.

- Spiral wounds are considered as an economical option for very large process volumes. These devices incorporate membrane and screen separators wrapped around a central tube. However, the fact that these devices are difficult to scale in a linear fashion has to be taken into consideration.
- Hollow fibers consist of bundles of membrane tubes. The advantage of hollow fiber is that it has a gentle flow and generates little shear. However, they are less efficient than cassettes for processing protein-rich feeds, and at large scale it needs a relatively high feed flow rate to generate an acceptable flux. The retention and yield are also lower compared to the use of flat sheet membranes.

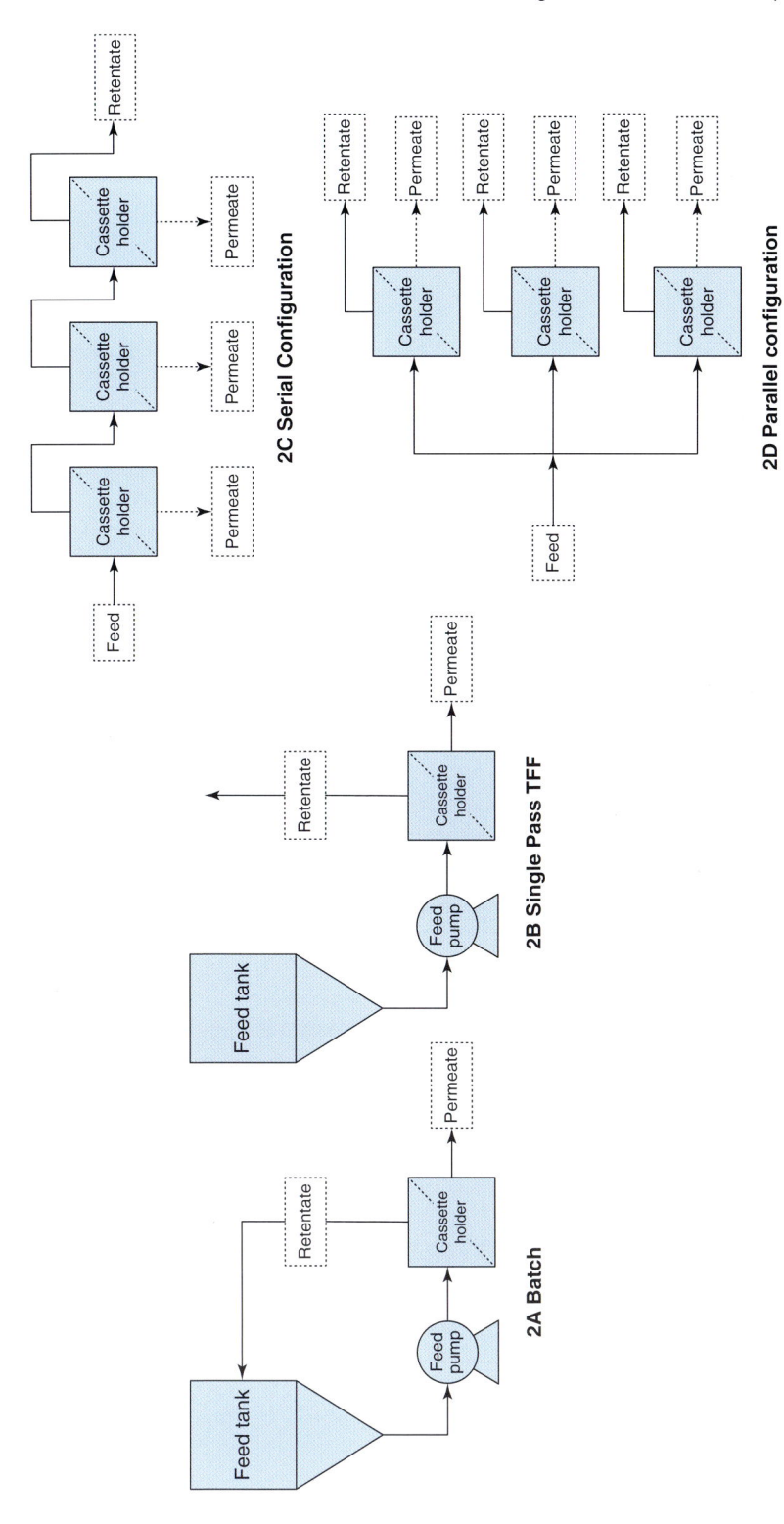

Figure 15.2 TFF modes of operation [12]. (Courtesy of EMD Millipore Corporation.)

- Flat sheet cassettes are the most widely used UF/DF device in biomolecule production processes. In this configuration, there are alternate layers of membranes and screens that are stacked together and then sealed. The screens are used to increase turbulence in the channels and to improve back transport of retained solutes away from the membrane surface. The turbulence-promoted channels have higher mass transfer coefficients at lower cross-flow rates, meaning that higher fluxes are achieved with lower pumping requirements. Turbulence-promoted feed channels are, therefore, more efficient than open channels.

Determining the ideal UF membrane configuration depends on many different criteria, including but not limited to: the character of the feed stream, floor space, scalability, and cost, but the predominant membrane type currently used for single-pass TFF applications are flat sheet cassettes due to their ease of implementation, scalability, and widespread adoption in the bioprocessing industry.

15.2.3 Single-Pass Tangential Flow Filtration (SPTFF)

A properly designed continuous process should eliminate or minimize the requirement for full process volume hold tanks and reduce the requirement for intermediate break tanks to a minimum. Therefore, to fully realize the key advantages of continuous bioprocessing, including increased efficiency, reduced footprint, and increased use of capacity, it is critical to control and reduce process volumes.

There are a number of stages within a typical bioprocess where the process volume is increased substantially. This is typically found in any stage that requires load pH/conductivity adjustment, such as post-virus inactivation in a monoclonal antibody process or around ion exchange chromatography steps, where desalting is required to condition the feed stream for optimal binding capacity. Batch or in-line dilution is typically utilized for this conditioning, but as the dilution can be up to a factor of 3 this requires very large tanks or piping, which do not readily fit into a continuous process design paradigm.

Historically, and indeed in some existing processes, batch TFF is used for feed conditioning, both to concentrate the feed stream for easier handling and to diafilter away salts that will raise the conductivity. However, batch TFF is challenging to implement into a continuous bioprocess midway through the downstream process, as it does not produce a continuous homogeneous product stream. Single-pass TFF, on the other hand, is an extremely effective TFF method that allows limited diafiltration and full concentration and adapts well to a continuously fed product stream. Single-pass TFF can also be used to reduce process stream volume at many points during a continuous bioprocess (Figure 15.3).

The applications for single-pass TFF also include [12] the following:

- *Product concentration and volume reduction*: Single-pass TFF can be used to reduce intermediate pool volumes. In turn, this can debottleneck a process limited by tank volumes and/or reduce chromatography column cycle time.
- *In-line dilutions/desalting*: Single-pass TFF can be used for in-line desalting before ion exchange chromatography steps, or virus prefilters without expanding the pool volume by dilution.

15.2 Tangential Flow Filtration in Bioproduction

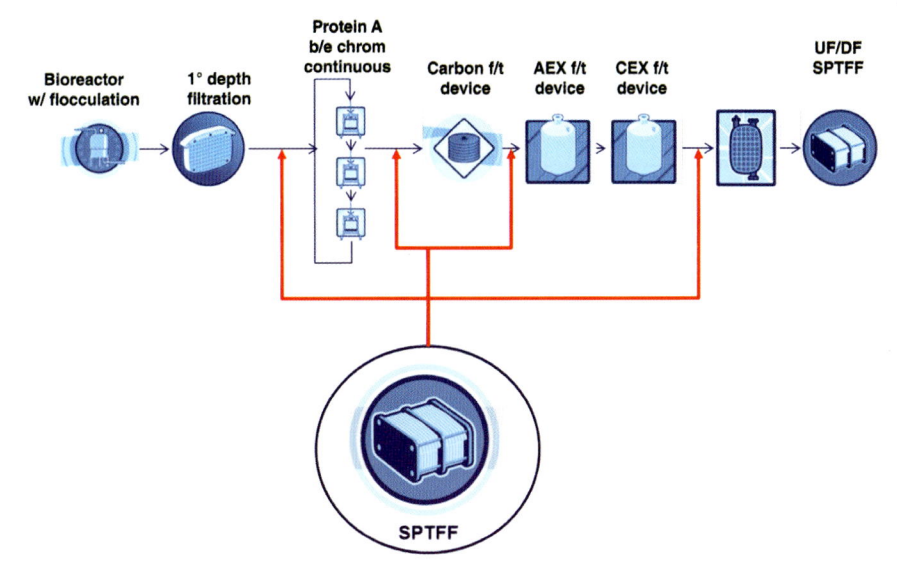

Figure 15.3 One possible permutation of a continuous mAb template including single-pass TFF [13]. (Courtesy of EMD Millipore Corporation.)

- *Final formulation/concentration (postbatch ultrafiltration/diafiltration)*: Single-pass TFF reduces working volume limitations compared with traditional TFF, allowing the process to achieve higher final concentrations by minimizing product dilution during recovery from the TFF system.

As the name suggests, in single-pass TFF, the product stream goes through the TFF cassettes only once, compared to the dozens of times with a standard recirculating batch TFF step. The conversion is kept high by increasing the residence time with a slow flow rate and also by serializing the flow-through up to three sections of cassettes, extending the flow path length. This has a number of key advantages:

- The footprint for the operation is much smaller, as there is no tank and no recirculation loop. In certain situations the pump can also be dispensed with, leaving only the holder and cassettes.
- The output of a single-pass TFF operation is both continuous and homogenous.
- Certain products, such as viral vaccines or vectors, are very shear sensitive. A large yield loss can be avoided by having only a single pass through the pump and valve apparatus.
- As the pipework and operation is linear and compact, it makes recovery of product, particularly those that are viscous or highly concentrated, much more effective. Product recoveries of 100% have been reported.

Single-pass TFF, however, is not a panacea to replace all batch TFF within the process, as it has a number of disadvantages.

- Large numbers of dia-volumes are not currently possible, limiting the scope for the technology for final formulation or mid-downstream contaminant removal.

- More membrane area is required to produce the same volumetric concentration factor (VCF). This typically ranges between 1.2 and 3×.
- The flow rate has to be very slow to achieve the required conversion rates. This may be greatly different from the preceding step or may be limited by pump turn down constraints.

15.2.4 Process Design

Depending on the membrane cutoff, the TFF applications are defined as microfiltration (MF) with a membrane cutoff >0.22 µm and ultrafiltration with a range of 300 kDa to 1 MDa and from 300 kDa down to 1 kDa for open UF and UF, respectively . TFF below 1 kDa is referred to as nanofiltration (NF).

There are several approaches for the execution of a TFF operation. The most used and known approach is batch or a modified version of it called fed-batch TFF where a feed solution is recirculated over a semipermeable membrane using a pump and a recirculation vessel (Figure 15.4a).

In wastewater treatment or in the dairy whey industry where continuous process streams are handled, also single-pass TFF (SPTFF) (Figure 15.4b) is used, where the TFF operation is performed during a single pass through the TFF device. With the current trend to continuous bioprocesses or high concentration protein solution, SPTFF gained a lot of interest also in the biopharmaceutical industry due to its specific properties. It offers a simple flow path that allows easy recovery at high yield and can operate in a continuous way. In addition to that low feed flow rates require a smaller pump compared with batch TFF and enable low shear processing of sensitive biomolecules.

Figure 15.5 is illustrating a concentration performed in a batch and in single-pass mode. In the batch process, the feed is recirculated by a pump over the membrane that retains the product of interest. During every passage, a small portion of the feed is passing the membrane as filtrate. The remaining process fluid exits the TFF device at a slightly higher concentration as retentate. This

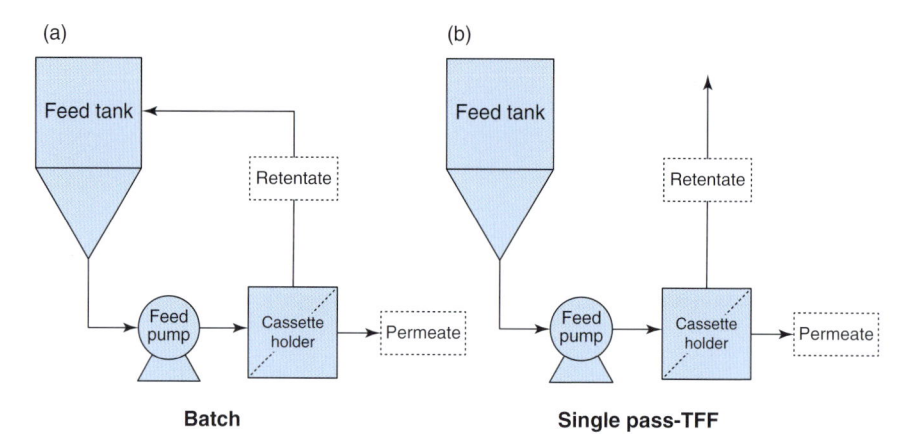

Figure 15.4 System setup for (a) batch TFF and (b) SPTFF [12]. (Courtesy of EMD Millipore Corporation.)

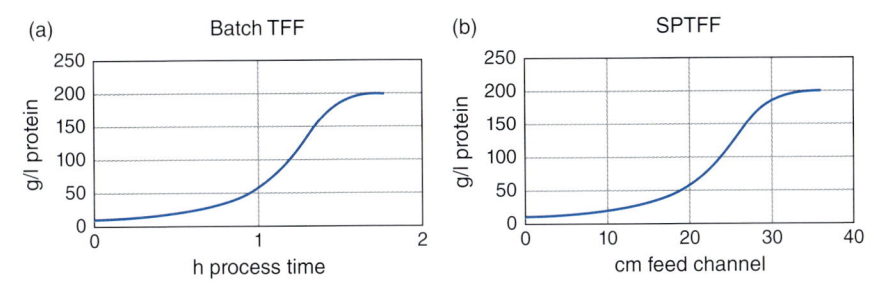

Figure 15.5 Concentration using batch and SPTFF [14]. (Courtesy of EMD Millipore Corporation.)

recirculation continues until the final targeted concentration is reached in the recirculation tank.

In a SPTFF process, the same concentration of the feed is achieved at a single passage through the TFF device. The feed is pumped at a much lower velocity through the TFF device to increase the residence time under pressure in the device and allow the conversion of feed to filtrate. While it travels though the TFF device, more and more feed is converted to filtrate and the product concentration rises. When the remaining liquid exits the TFF device as retentate, the target concentration is reached in this one pass.

To achieve the final concentration in just one single pass, the residence time in the TFF device is increased. Next to lowering the feed flow, this can be done using a parallel or serial configuration of multiple TFF devices. In a parallel configuration as shown in Figure 15.6, the feed flow is split by three devices that increases the residence time also by factor of 3 compared with single device. The other possibility is to increase residence time by serialization of the same three devices. The retentate of the first device is then the feed of the succeeding one and so on. By installing the devices in series the residence time is also increased by the same factor of 3.

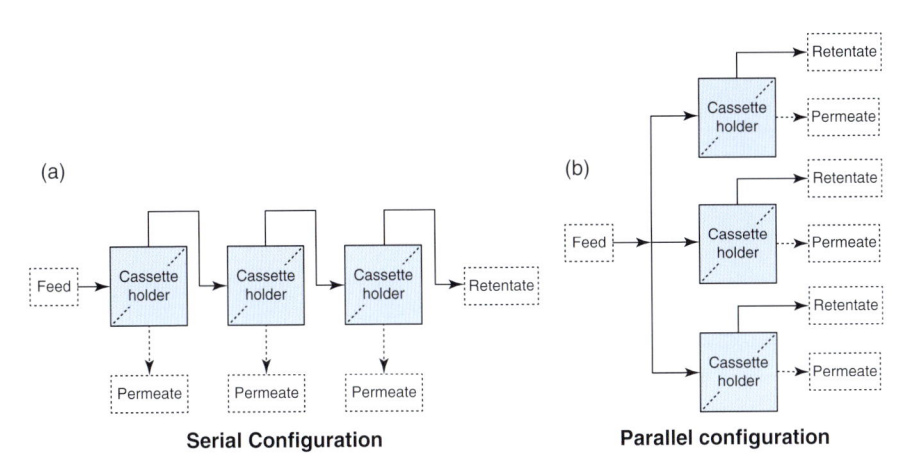

Figure 15.6 Serial and parallel configuration – two way of increasing the residence time in TFF devices [12]. (Courtesy of EMD Millipore Corporation.)

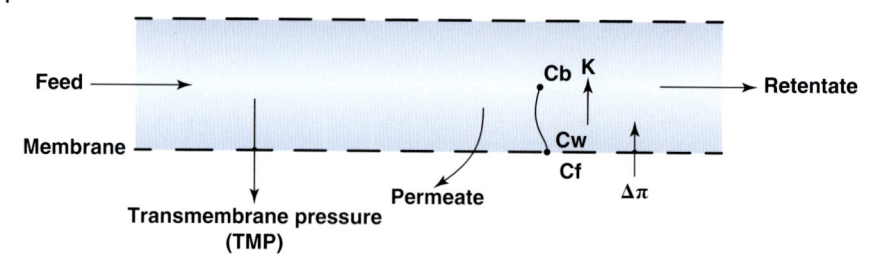

Figure 15.7 Inside a feed channel [15]. (Courtesy of EMD Millipore Corporation.)

See Figure 15.7 for a feed channel, where TMP is the transmembrane pressure (bar), $\Delta\pi$ is the osmotic pressure (bar), K is the mass transfer coefficient (LMH), C_b is the bulk concentration (g/l), C_w is the wall concentration (g/l), and C_f is the filtrate concentration (g/l).

In each TFF process, independent from its mode of operation (batch, fed-batch, or SPTFF), the transmembrane pressure is providing the driving force for the feed to pass the membrane as filtrate. The liquid removed from the feed increases the concentration of the retained solutes at the membrane surface and forms the polarization layer. The feed flow in the feed channels is tangential to the membrane surface and provides a sweeping action for the retained solutes that influences the thickness of this polarization layer ($C_w > C_b$). Together with a turbulence promotor (screen) built into the feed channels of TFF devices it supports the back transport (mass transfer) of retained solutes from the membrane surface. For steady operating conditions, a balance between TMP and feed flow needs be determined and maintained for optimal filtrate performance with a controlled polarization layer. Excessive TMP can lead to excessive polarization or even membrane fouling and filtrate flux decay. Membrane fouling can only be removed by cleaning while the polarization layer buildup can be reversed by lower TMP or increased feed flow.

In Figure 15.8, data from a typical experiment are shown, flux versus TMP, performed during the development of a batch TFF operation in total recirculation mode. It shows the filtrate flux dependency from TMP for different feed flow rates at a set feed concentration. The graph would look similar when two different feed concentrations are used at identical feed flows.

At low TMP, the filtrate flux increases linearly with the TMP in the "pressure-dependent" region, with flux values dictated by the fouled membrane resistance, up to a point where flux is not a function of applied pressure (TMP) anymore. This is called the "pressure (or TMP)-independent region."

The pressure-independent region showed the maximum achievable or limiting flux at each feed flow, which minimizes the required membrane area when processing time is fixed. Under similar conditions, it is important to mention that operating at the pressure-dependent region would require more membrane area due to the lower fluxes.

In batch TFF mode, the feed flow that produces the highest flux is often selected, and TMP is chosen at a value on the knee point of the curve, in close proximity to the "pressure-independent" region (Figure 15.8).

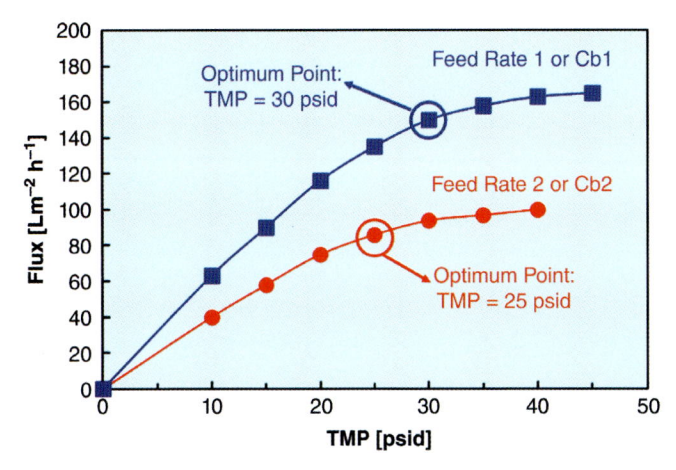

Figure 15.8 Flux excursion batch TFF – Filtrate flux versus TMP [15]. (Courtesy of EMD Millipore Corporation.)

The design and the working parameter optimization of the ultrafiltration unit operation has a direct impact on the overall process yield and product quality.

A characteristic of the batch TFF UF process is the low conversion per pass (where conversion is defined as ratio of feed flow that move into the permeate or filtrate flow); to reach the target final concentration, feedstock had to pass through the membrane multiple times [4]. That means a recirculation loop, a large feed pump, and a relatively large feed tank are needed. The protein solution is forced to flow tangentially over the membrane surface by a pump and a portion of the liquid passes through the membrane while the product of interest is retained. The product concentration increases over time in the recirculation loop and feed vessel.

Mixing and foaming during the process can also be problematic and may reduce the product yields and quality via aggregation or precipitation. Moreover, because of multiple passes of feed material through the pump, shear-sensitive and fragile biomolecules can be damaged.

Due to the large recirculation tanks, conventional TFF systems have large minimum working volumes, which limits the maximum achievable concentration factor (CF) for a given batch and complicates the product recovery too.

When final concentration is reached, concentrated solution recovery can be achieved by a sequence of buffer recirculation and/or buffer flush steps post system emptying by drainage. This means that the concentrate will be diluted by the recovery steps and compromise the final product concentration. Typically, this is addressed by running an overconcentration before recovery. However, there are constraints to that approach, such as limits on process stream viscosity or system minimum working volumes.

Batch and fed-batch TFF applications are the best design approach when purification, formulation, and buffer exchange are needed. However, by nature they are batch operations and could hardly satisfy process requirements whose target should be the "continuous mode."

The SPTFF design leverages the idea that longest residence time, achieved by reduced flow velocities and/or the use of much longer flow paths, produce high conversion rates.

Conversion rate is defined as ratio of feed flow to cumulated permeate or filtrate flow:

$$\varphi = \frac{Q_f}{Q_F}$$

or

$$\varphi = \frac{JA}{Q_F},$$

where Q_f is the filtrate flow rate (l/h), J is the normalized filtrate flow rate or flux (l/m²/h or LMH), A is the filtration area (m²), and Q_F is the feed flow rate (l/h).

The conversion defines the resulting volumetric concentration factor (Figure 15.9), the two parameters being linked by the formula

$$VCF = \frac{1}{1 - \varphi}$$

or

$$VCF = \frac{1}{1 - (JA/Q_F)}.$$

In order to increase the conversion rate, the filtrate flow rate should increase (this can be achieved by increasing the surface area and keep the same flux) while keeping feed flow rate constant, or feed flow rate has to be decreased at a higher ratio than permeate flux (Figure 15.10).

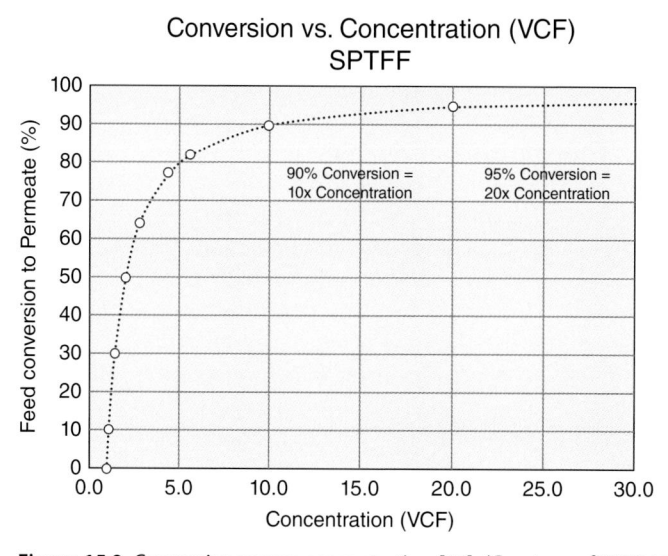

Figure 15.9 Conversion versus concentration [14]. (Courtesy of EMD Millipore Corporation.)

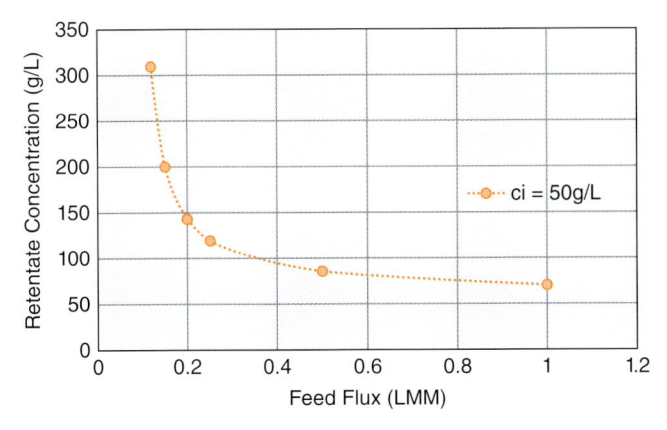

Figure 15.10 Retentate concentration vs. feed flux [14]. (Courtesy of EMD Millipore Corporation.)

In SPTFF applications, the effect of pressure (TMP, or retentate pressure at low Δp) on the performance of SPTFF is not very significant [16]. Unlike traditional TFF, where the protein is concentrated gradually after several passes, with SPTFF the target protein concentration is immediately reached after one pass. Therefore, SPTFF reaches the TMP-independent region (mass transfer limited) very quickly and any increase in TMP achieved through an increase in retentate pressure has no significant effect on permeate flux and VCF. Only a minimum of pressure needs to be applied on retentate side to overcome the osmotic pressure reached at the retentate exit with the highest product concentration (Figure 15.11).

For the determination of the size and configuration of a SPTFF operation, a feed flux versus conversion graph is generated, where the cumulative conversion achieved after each section is plotted against the area normalized feed flux (Figure 15.12).

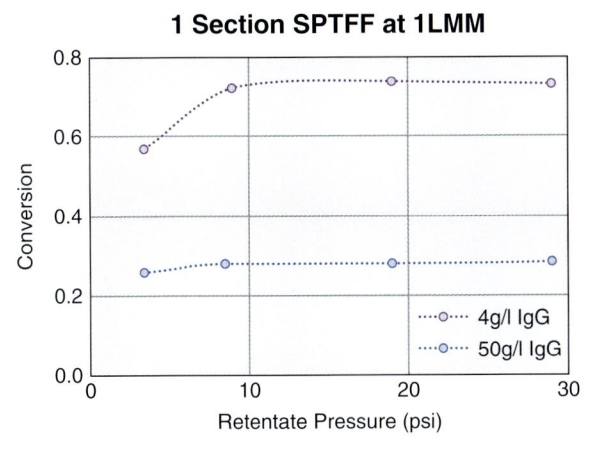

Figure 15.11 Conversion dependency on retentate pressure [14]. (Courtesy of EMD Millipore Corporation.)

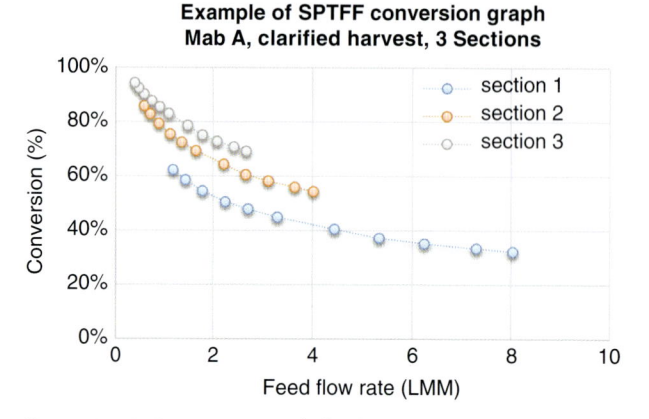

Example of SPTFF conversion graph Mab A, clarified harvest, 3 Sections

Figure 15.12 Conversion graph for three section configuration.

One model to describe UF process in the pressure-independent region is the stagnant film model defined as

$$J = k \ln\left(\frac{C_w}{C_i}\right),$$

where J is filtrate flow rate (or flux), C_w is the protein concentration at the membrane wall, C_i is the bulk protein concentration in the feed, and k is the mass transfer coefficient.

The mass transfer coefficient is a function of feed flow rate given by an empirical correlation as [17]

$$K = K_0 \left(\frac{Q_F}{Q_{F0}}\right)^a,$$

where K is determined as a function of feed flow rate Q by a constant exponent a and mass transfer coefficient K_0 at one feed flow rate Q_{F0}.

So,

$$J = K_0 \left(\frac{Q_F}{Q_{F0}}\right) \ln\left(\frac{C_w}{C_i}\right).$$

And conversion can be expressed as

$$\varphi = \frac{JA}{Q_F} = AkQ_F^{(a-1)} \ln\left(\frac{C_w}{C_i}\right).$$

In the last equation, conversion is expressed as a function of filtration area A. Conversion can be predicted when, once the filtration area is fixed, mass transfer K and C_w have been empirically determined (with dedicated experiments) [17]. By rearranging the equation (and sentence), flux can be determined by fixing the target conversion at predetermined filtration area A, C_w, and K parameters.

Data from literature report that typical value for index a should be approximately 0.80–0.87 in turbulent flow [17].

One of the major benefits in single-pass mode is that feedstock remains at a constant protein concentration in each section of the module assembly, which allows the system to come to steady-state equilibrium [16]. Therefore, with a constant feed flow rate (Q_{feed} or Q_F), retentate flow rate ($Q_{retentate}$ or Q_r), filtrate flow rate ($Q_{filtrate}$ or Q_f), relatively uniform TMP profile, and constant concentration factor during the entire process, SPTFF represents the best candidate to be used as a unit operation in a continuous process [18].

The operation mode of SPTFF, because characterized by constant operating conditions, clearly shows that this step could have been easily coupled with the other bioprocess unit operation, such as virus removal and chromatography, for higher throughput and immediate use without the need for accumulating a batch.

SPTFF ultrafiltration step could utilize the same pump from the coupled unit operation. The outlet of the previous unit operation is connected to the inlet on the UF cassette holder and the retentate sent to the pool tank while the filtrate is sent to drain. Minimal additional equipment is required in that case for SPTFF: no recycle tank is needed, only a UF cassette holder with adapted diverter plates but no additional instrumentation or control system. The system design is simplified and hardware cost is limited.

Considering the typical purification process for a monoclonal antibody (mAb), we can identify potential places where SPTFF UF could be used in a hypothetical continuous process (Figure 15.2): upstream of the post clarification chromatography Protein A capture step, in-between ion exchange chromatography purification steps for volume reduction, or in-line viral filtration step. The development of the SPTFF process is simpler in the latter case because the protein concentration will be relatively constant throughout the entire virus filtration operation.

In the other cases, SPTFF performances would be difficult to predict because the concentration of protein entering the UF module would change as a function of time during the clarification or elution phase. In such cases, temporarily pooling the product in an intermediate vessel could help to solve the risk of an unsteady state.

When SPTFF is positioned after or between chromatography steps, we also need to take into account the back pressure generated by the UF modules placed in-line. Back pressure could increase the absolute pressure in the chromatography column and the pressure ratings on conventional columns are relatively low.

Another possibility is the use of single-pass TFF in the final concentration step of a protein manufacturing process where high protein concentrations are required [4]. The reduced hold up volume/working volume in combination with a plug flow buffer flush allows higher final concentrations and a high yield/recoveries at minimal concentrate dilution.

This technique is also indicated for concentration of shear sensitive proteins. In a single-pass operation, the feed solution is passed only once over a device or a series of devices containing the semipermeable membrane retaining the protein and critically only once through the pump, a traditional source of protein shear damage. Lower shear stress exposure for sensitive proteins provide higher final product quality.

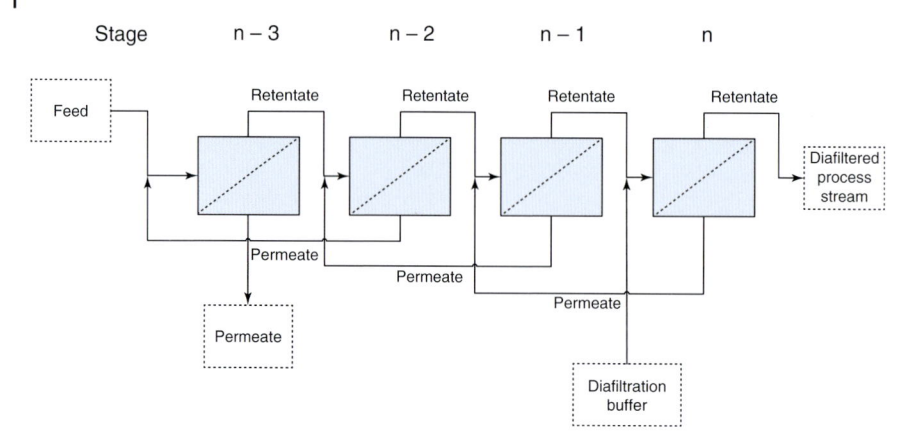

Figure 15.13 Countercurrent single-pass diafiltration. (Reproduced with permission from Ref. [4]. Copyright 2015, Elsevier.)

SPTFF technique has some limitations. Compared to traditional batch operations, SPTFF is more expensive because of higher filtration area requirements (in most cases from 1.2 to 3 times more filter area than a traditional TFF configuration, application dependent). As long residence times are required within the cassettes to achieve the necessary conversion, feed flow rates are very low. This limits the mass transfer advantages from the fluid flow sweeping the membrane and reducing the polarization layer and, therefore, more membrane surface area is required. These additional costs can be partially mitigated by reusing the SPTFF cassettes.

SPTFF has also limited use for buffer exchange, purification (protein separation), or final formulation where batch TFF systems are more efficient and are, therefore, preferred for such applications. Single-pass diafiltration is more difficult to manage and involves a series of concentrations and in-line dilutions. "Buffer passes into the permeate stream during the concentration step and volume is restored in the inline dilution" [4].

A way to run diafiltration in single-pass configuration is the so-called *countercurrent diafiltration* (Figure 15.13). Diafiltration buffer is added before the last section, with the last section permeate used as the diafiltrate to the previous stage. All flows are the same for each section and fully passing salt concentrations decrease from section to section, but are the same within each section [4].

15.2.5 Laboratory-Scale Process Development Example

One of the most important objectives of process development in SPTFF is to determine the membrane area or feed flow required to give the desired volume reduction.

In TFF applications, changing membrane area at fixed feed flow rate changes the mass transfer coefficient, which means the volume reduction will be a nonlinear function of the membrane area. Qualitatively, increasing the UF membrane area at fixed total flow rate will lead to a lower flow rate per membrane area and a

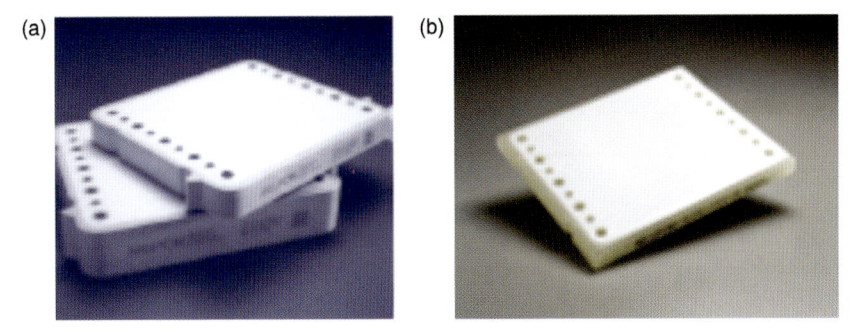

Figure 15.14 EMD Millipore's Pellicon® 3 (a) and Pellicon® 2 (b) cassette devices [3].

decreased mass transfer coefficient again. In SPTFF process development activity, it is "experimentally more convenient to keep the membrane area constant and vary feed flow rate to determine the effect of feed flow rate on mass transfer and conversion rate" [17].

For process development at small scale (laboratory) scale, the use of existing scalable and cleanable validated UF cassette modules installed in standard holders could be a practical solution [3,19]. An example is shown in Figure 15.14.

Serial configuration of sections to increase length of feed channel is done by the use of dedicated diverter plates placed between each module. Diverter plates have been installed to allow the retentate of the previous device to enter into the feed port of the succeeding one. This configuration uses equal areas per section, and because of the use of standard materials the design can be easily implemented at pilot or manufacturing scale.

A serial flow-through in all installed devices can, therefore, be established while the permeate of each device can be collected individually. Figure 15.15 shows a typical setup for process development containing three sections. Each TFF device represents one section.

The feed pump provides the feed flow through the system. Pressure gauges located before each section indicate the pressure profile over the setup. A

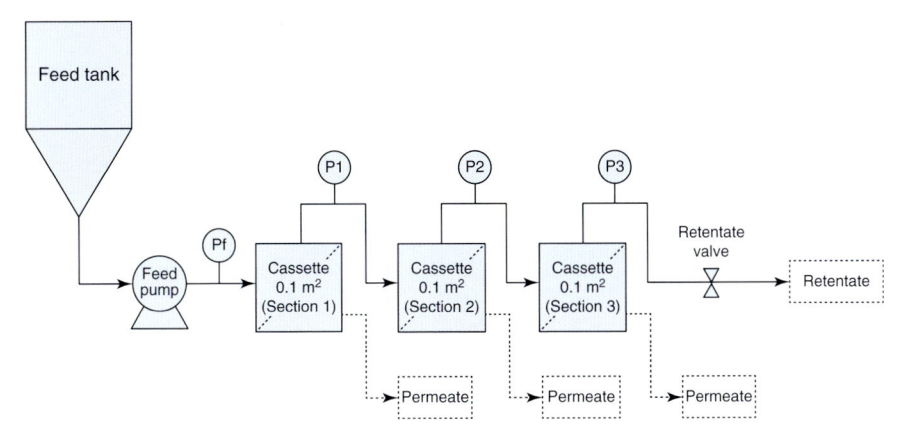

Figure 15.15 SPTFF experimental setup [12]. (Courtesy of EMD Millipore Corporation.)

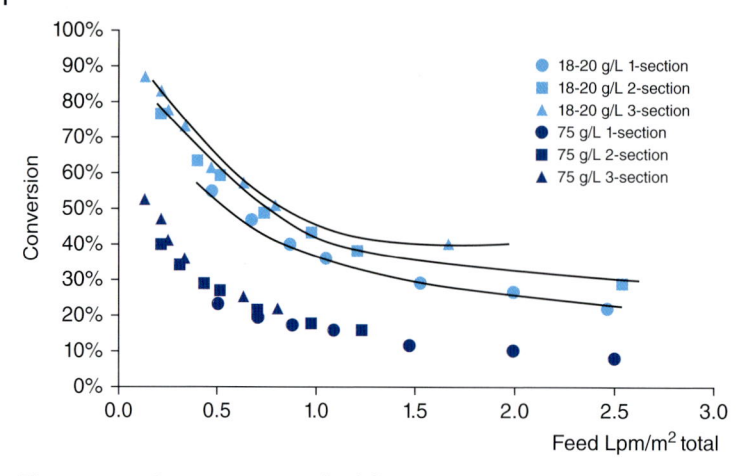

Figure 15.16 Conversion versus feed flow [12]. (Courtesy of EMD Millipore Corporation.)

retentate valve after the final section is used for retentate pressure adjustment. All permeate lines are open to atmosphere and pressures are assumed to be zero. The filtrate can be collected for each section. This setup allows calculation of the corresponding TMP and cumulated conversion for each section individually. Feed flow flux and conversion data for each section can be utilized for process scale-up.

For sizing the flux dependence on feed flow, pressure (retentate valve) needs to be tested.

First, feed flow excursion experiments should be performed (Figure 15.16) where the conversion/concentration will be evaluated against the feed flow after each of the three sections of the single-pass process. The different feed flows (the range will depend on the conversion/concentration target) will be tested from the fastest to the slowest in order to increase progressively during the trials the residence time and so the concentration of the product. For each feed flow tested, the retentate flow and three permeate flows will be measured as well as the concentration in the feed retentate and permeate and the feed and retentate pressures.

Second, the retentate pressure impact (retentate pressure excursion) can be tested by doing the same experiments and setting a higher retentate pressure. The retentate pressure should start between 5 and 10 psi during the first set of experiments (to overcome the osmotic pressure effects) and should be gradually increased until the conversion remains stable.

Higher retentate pressures should be useful to increase conversion for dilute products but might not have added value with more concentrated feed stream.

All of these experiments can be performed by collecting permeates and retentate separately (real single pass) or by directing permeate and retentate lines back into the feed container (total recirculation). If sufficient feed material is available, single-pass operation remains the best choice. But otherwise, the total recirculation option is a good solution while keeping a good mixing in the feed container and checking the concentration of the feed material at different step of the optimization.

Calculation of cumulated conversion for sections is done by the following formula:

$$Y_{\text{Section }n}(\%) = \frac{\sum_{i=1}^{n} Q_i}{Q_{\text{Feed}}} \times 100,$$

where Y_n is the cumulated conversion up to section n (%), Q_i is the permeate flow of section (l/min), and Q_{Feed} is the feed flow (l/min).

The VCF can then be calculated:

$$\text{VCF}_n = \frac{100}{100 - Y_n}.$$

The CF depends on the retention of the product (R), which typically needs to be very high for concentration applications ($R = 0.995$–0.9999); CF can then be determined with

$$\text{CF} = \text{VCF}^R.$$

The final retentate concentrations can be measured via absorbance at 280 nm (A_{280} nm) in order to confirm the conversion versus concentration factor relationship.

When conversion is plotted against feed flux for the individual sections, the resulting graph can be used for a scale up calculations and for the definition of the module configuration or number of sections to be placed in series.

In Figures 15.16 and 15.17, we report an example where three sections in series have been chosen for experiments. By simple interpolation of the curve with the required conversion factor, we can select the feed flow, the number of sections, and the required filtration area needed for each section able to reach the target conversion rate.

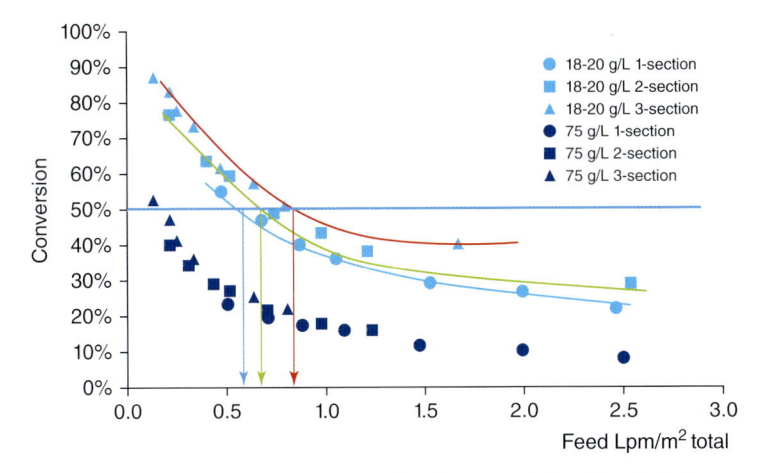

Figure 15.17 Use of conversion versus feed flow graph for sizing [12]. (Courtesy of EMD Millipore Corporation.)

Table 15.1 SPTFF sizing example.

Number of sections	Feed flux (LMM)	Total area (m^2)	Proposed configuration (m^2)
1	0.60	17	1×17
2	0.70	14	2×7
3	0.85	12	3×4

For batch operation, where batch volume and process time are predetermined, the required area is calculated by

required area = scaled up feed flow rate/required feed flux.

Example: 2400 l in 4 h (required feed flux = 10 l/min) with data that generated graph in Figure 15.17 results in the choice of one of the sizing as shown in Table 15.1.

All of the proposed scenarios fit with the process target. The choice comes from several considerations about costs, ease of work, and product recovery practices able to guarantee highest product yields. The cheapest solution is in most cases the preferred choice. In this specific case, the solution with three sections placed in series is able to provide the lowest required surface area that means lowest installation cost.

In case of continuous operation, the feed flow rate is very often a given parameter. Time is not a useful parameter anymore and filtration area is calculated based on the pre-established conversion target.

Taking as an example of SPTFF application in post-virus filtration, where a constant feed flow of 10 l/min is provided and 50% conversion is required, calculations reveal that we will have the same sizing approach as in the previous example. What potentially changes would be the selection of the number of sections, although the pressure drop at each of the different section configurations would have to be considered. The configuration having the lowest impact on working parameters of the previous or following step has to be chosen.

As an example, virus filter maximum operating pressure should be considered the primary constraint for the choice of the best SPTFF configuration in this application. In spite of the fact that single section configuration requires more filter area and higher cost, such configuration is able to guarantee the simpler installation/operation and lower pressure drop at fixed feed flow.

15.2.6 Consideration on Equipment Configuration and Requirements

Different possibilities exist for SPTFF implementation at pilot or manufacturing scale when standard cassette modules are used. Unlike TFF batch operation where cassettes modules are installed in parallel in a standard TFF UF holder, SPTFF can be configured in different ways.

Single-section installations (Figure 15.18) with all the modules fed in parallel, as previously described, is the simplest configuration but in most cases it represents the higher installation costs because of the need of higher surface area.

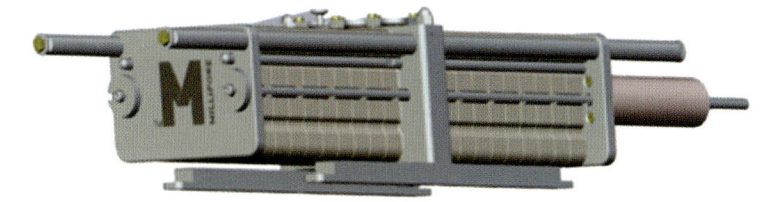

Figure 15.18 Single section configuration.

Multiple SPTFF sections can be assembled by placing the single modules in series inside a single holder (Figure 15.18) or placing in series multiple holders with modules configured in parallel (Figures 15.19 and 15.20).

In the first case, dedicate diverter plates have to be placed between each module in order to connect the outlet port of the first module with the inlet port of the succeeding one [6].

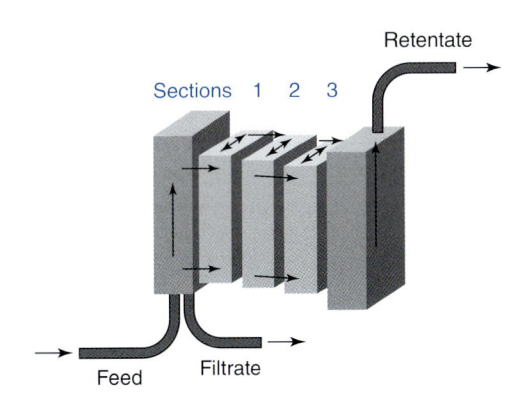

Figure 15.19 Single holder configuration (three modules in series in one holder) [19]. (Courtesy of EMD Millipore Corporation.)

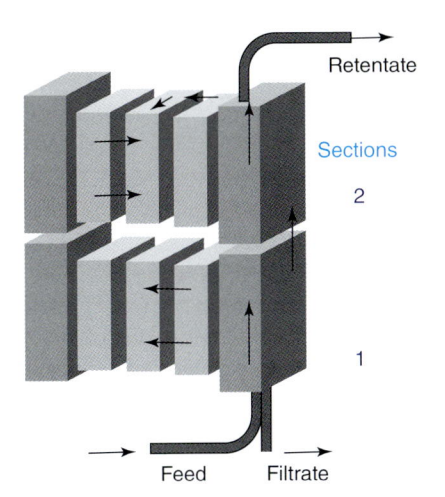

Figure 15.20 Multilevel sections configuration (two holder in series) [19]. (Courtesy of EMD Millipore Corporation.)

For all the described configurations, feed pump is designed and regulated in order to provide the predefined residence time by setting the required feed flow:

$$Q_F = QA,$$

where Q is the normalized feed flow (l/min/m^2 or LMM) and A is the total installed surface area. SPTFF operations require as minimum that the feed flow and retentate pressure be controllable parameters. Sensors to measure flows (feed, retentate, and/or permeate), pressures (feed, retentate, permeate), and product composition in the retentate and permeate (UV absorbance as an indicator of protein content, pH, conductivity as desired) are in most cases installed on board of a SPTFF unit. Process temperature can also be monitored with dedicated instruments, but SPTFF does not typically need a temperature control management and a heat exchanger.

Sensor readings may be displayed on a PC screen. Working parameters can also be monitored and logged to trigger alarms, and used to trigger a subsequent step in a process. Abnormal pressure excursions or fluxes, high permeate UV absorbance, and high temperatures generate alarms and stop/place on hold the process. Process data are typically logged as per GMP requirements, for use in trending and diagnosis of any unusual excursions [4].

In Figure 15.21, an example of unit installation is shown. A pump, whose speed is regulated by a flowmeter control, feeds the modules assembly. Note that the pump could be a piece of equipment from the previous unit operation (virus filtration as an example, in the case that virus filtration is run at constant flow). Pressure sensors are placed in the feed, retentate, and permeate side to monitor the pressure drop and any backpressure in the retentate and permeate side. Retentate pressure sensors (PR) could also regulate the retentate backpressure valve and set the outlet pressure based on pre-established working parameters from process development work. Feed pressure sensors (PF) should work as an alarm switch when inlet pressure exceeds the maximum specification of the module.

Several in-line controls could be installed on the retentate side. For protein applications, UV detector is the most used on board instrument. When adequately calibrated and validated, UV detectors can monitor the actual protein

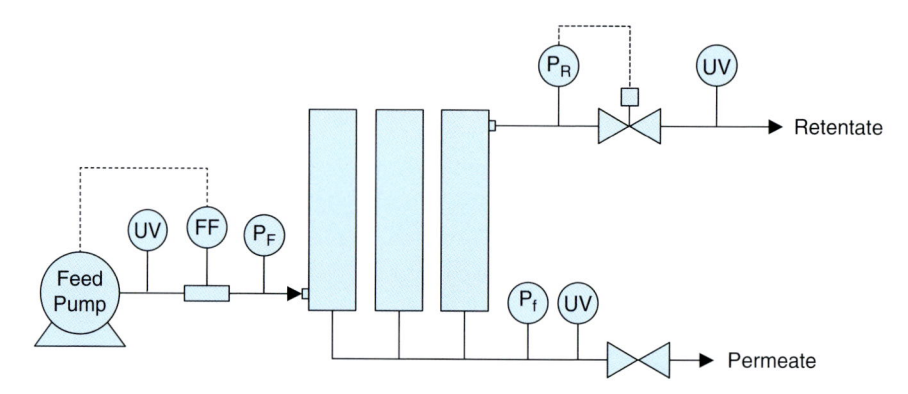

Figure 15.21 Example of SPTFF unit installation.

concentration of the retentate; it should also be used as regulation tool by interacting with pump speed (or feed flow) in order to set the right value able to reach the desired final concentration. It should be said that feed protein concentration has to be controlled as well, in order to be able to verify any variability along the process and to establish if the target conversion factor can be reached.

UV detectors should also be placed on the permeate side as an alarm for high protein passage, revealing potential failures in the assembly or in the membrane integrity.

15.3 Validation

15.3.1 Key Validation Considerations between Batch and Continuous Processing

Validation[1] [20] of a bioprocess to the stringent regulatory and GMP standards is an enormous task, requiring substantial quantities of time, material, and labor to perform. It is much easier to break up the bioprocess into individual unit operations and validate them separately, with each unit operation then further broken up into suboperations, such as cleaning, processing, or flushing. The level of knowledge required is such that validation engineers often specialize in specific parts of the process such as upstream or final fill.

The question has to be asked as to whether the discrete unit operation paradigm of validation for a bioprocess is appropriate for a continuous system. As an example, secondary clarification depth filter processes are typically followed straight away by membrane filtration, a typical case where a section of a templated monoclonal antibody bioprocess is run continuously as standard now. It is a common practice to validate and process develop these two operations together as one rather than separately, as they are so interdependent. The argument, therefore, follows that the whole continuous bioprocess should be validated as one unit, although it is doubtful as to whether this would be feasible in reality.

Even in batch bioprocesses, there is a great deal of interdependence between the unit operations growing as you proceed down the downstream process. Marginal lot-to-lot variability is expected within bioprocesses due to the biological nature of the production, but these small changes are magnified as the product proceeds downstream and each unit operation knocks on. This is typically managed by stringent specifications, in-process controls, risk-assessed control strategies, and liberal use of large safety factors, which can be as high as 3 in safety critical operations. However, in continuous processing, the whole process (except perhaps the final formulation step which is usually more batch like in nature due to operational constraints) is in essence a single-unit operation and really should be validated as such.

These leads to a number of interesting challenges as regards validation of a process that is being run continuously, such as how you define a batch, how you

---

1 ICH Q5 A-E, ICH Q7, ICH Q8R2, ICH Q9, ICH Q10.

validate a control strategy, how you scale down the process for development or validation, and how quality by design can be leveraged.

As previously mentioned, the validation of a full bioprocess is a mammoth undertaking with specialists required at each point. Validating the entire continuous process in one go would, therefore, be very challenging, so it is necessary to find some logical segmentation to section the work.

There are two obvious *breaks* in the continuous process, those based on a change of unit operation technology and those based on break tanks if they exist. If the bioprocess is hybrid continuous/batch, then these natural segments appear on their own, but with a full continuous process hold tanks are eliminated and break tanks are undesirable.

The change in technology is a logical segmentation. This is relatively straightforward with the upstream perfusion cell culture that can be validated independently, primarily as it is the first step and is not dependent on feed stream characteristics from a previous unit operation. This can be done further down the process too, but potential variabilities in the feed stream, particularly if the feed has been held before validation work, needs to be risk assessed and tested.

For the rest of the downstream process, it makes sense to validate it in the order that the process occurs and *build* the process to provide feedstock for the development and validation of the next segment. Logical breaks could be defined by time in each stage or perhaps by the complexity in the validation, for example, after bind/elute chromatography steps.

A continuous process could last many weeks and it is a regulatory requirement to have batches of product that can be released separately. The FDA defines a batch as

> "a specific quantity of a drug or other material that is intended to have uniform character and quality, within specified limits, and is produced according to a single manufacturing order during the same cycle of manufacture".[2]

One of the key advantages of a continuous process is that the product stream has excellent consistency in terms of product quality. In theory, on the basis of this guidance, you could consider the entirety of the campaign (which could last months) as a single huge batch. This, however, is not recommended for the following reasons.

First, it pays to prepare for the worst. In the case of a manufacturing deviation, or worse an out of conformance, a batch may have to undergo additional testing or indeed even be disposed of. If the whole campaign is classed as one large batch, this puts a great dollar value risk on the process that would be unacceptable. Having batches within the campaign, however they are defined, will allow excision of a section of the campaign for further testing or disposal, should the need arise.

Secondly, batch documentation is very lengthy since, from a purely practical perspective, manufacturing cycles are long and hence the record itself would is very large and hard to reference.

---

2 Code of Federal Regulations Title 21 (Food and Drugs). Section 210.3.

Third, a requirement for a batch is to be "uniform in character." In a batch process, the whole bulk drug substance is pooled before final filtration and filling, leading to a uniform character within the batch by definition. However, a continuous process is continually filling, and it could be argued that to ensure uniform character the specifications would have be narrower, in case of small changes over time. Breaking the campaign into batches will increase the chance of uniformity and potentially allow transfer of drug substance into pools before filling, ensuring the uniformity of the batch. If a pool is utilized before filling, the effect of worst- case hold time on product quality, safety, and efficacy will need to be investigated.

The definition of the batch can be implemented by various means in a continuous process. The simplest would be on a basis of time/volume (output of drug substance is continuous at a fixed flow rate, so time and volume are directly proportional). The exact time/volume chosen is dependent on how long the hold can be validated if a final pool is used and the analytical and quality resources available to correctly release the batch. Batch definition on equipment cycling is also a possibility, as it will tie processing equipment lot numbers to product batch numbers. Parametric release reduces the analytical requirements substantially, if it can be implemented appropriately [21].

It is widely expected that continuous process verification (CPV) [22], utilizing process analytical technology (PAT) [23], can greatly support validation activities and will be essential in the control of a continuous process. CPV typically monitors the process in real time and, therefore, generates a great deal of data. As such electronic batch records are necessary to fully record the process.

Validation of the control strategy has to be approached with the view that actual access to the product stream is likely to be very challenging as regards taking samples. This is further complicated in that removing a reasonable sample volume from a continuous stream will impact on the process, as flow rate is critical to both chromatography and single-pass TFF steps in particular.

There are points where a sample could be drawn, such as break tanks if they are implemented, but under normal circumstances reliance will be on the in-line and online sensors to provide feedback on process control. This is very challenging regarding those analytical tests that cannot be done online, such as bioburden or endotoxin testing. Validating the process interruption of taking a sample is a potential solution to collecting this data, either for the validation process or for standard in-process control.

For chromatography columns and to a certain extent TFF cassettes, it is expected the same resin/device will be in use for the entire campaign, being cleaned and cycled as required. However, some parts of the process, in particular viral clearance filters and depth filters, are inherently single-use devices and this will not be possible.

This adds another layer of complexity to the process, as at some point these filters will need to be switched out for new ones. This is very challenging in a continuous process where by definition you cannot stop it.

The need for change out will require creative use of hardware and manifolds, in much the same way as multicolumn chromatography is handled. With the right manifolds, the feed stream can be redirected to the new, flushed filters while the used ones are removed and replaced. Most of these filters have entirely single-use flow paths that negate the need for cleaning of hardware.

A key advantage of continuous processing is, however, the very small footprint that it takes up; by requiring two hardware skids each for the depth filtration and viral clearance steps, the footprint is increased substantially. The system and process design will also have to be sensitive to the complex connections and tubing assemblies required to allow processing, flushing, and cleaning of two separate hardware skids for a single-unit operation, and to potential operator error in its installation and control.

The most complex part of change out is ensuring a spike of inconsistent product is not introduced into the continuous process, via a slug of buffer or even air, at the point of change out. From an engineering perspective this would be nontrivial to implement, as it will be very difficult to prime a filter or column manifold without disrupting the flow to the rest of the process. A potential solution to this would be to dispose of a volume of product that corresponds to the window when the change out occurred, the challenge being the calculation of where and when you need to dispose of the product without wasting too much yield. Introducing a tracer into the process should help during validation to give a timing for product progress and this will have to be tagged product to correctly evaluate passage through a bind/elute chromatography step. Another possibility is to schedule all of the change outs throughout the process at once, then run a whole process volume to drain before recollected bulk drug substance. This does rely on different unit operations not having greatly different change out times, which is unlikely in practice.

Parametric release (or real-time release) of biological products is considered to be infeasible due to the level of characterization required. However, as previously mentioned, the CPV, PAT, QbD, and detailed process control required for continuous processing forces the collection of much more high-quality data about the process and the product, potentially enough data to allow a parametric release. At the time of writing, this was yet to be fully tested in the industry, but it is theoretically possible.

There are some tests required for batch release that cannot be performed in process, such as bioburden. Bioburden testing, however, usually takes 14 days, which makes it a poor candidate for CPV anyway. The limiting of postprocess release testing does have great value, however, particularly as regards those products with limited shelf lives and those with complex distribution requirements.

It is clear that continuous processes are difficult to validate fully with a quality by inspection approach. The additional multivariate complexity is difficult to test with discrete validation activities; therefore, quality by design (QbD) is an excellent fit.

In existing quality by design on a single unit operation, the feed stream from the previous processing step is considering a critical material attribute (CMA) for the current one. Continuous processes are more vulnerable to "knock on" effects upstream primarily from lack of detection windows; therefore, this CMA analysis would need to be considered in series for each of the preceding steps. The level of in-process data necessary for a continuous process makes a quality by design approach much easier, as a great deal of the data is generated within the process itself rather than as a specialized validation project.

QbD should allow, once the validation is complete for the individual process segments, for the continuous process to be considered as a whole, with the ramifications of any change upstream known for the downstream product quality. Considering the process in its entirety rather than the sum of its parts is a profound paradigm shift, but should be feasible even with currently available technology.

15.3.2 Validation of Single-Pass TFF

Single-pass TFF can have multiple roles within the bioprocess, from reducing process volume, desalting, and to improving recoveries for final formulation. Each of these applications will have a different take on the validation requirements.

Fundamentally, however, single-pass TFF will be easier to validate than batch TFF. The reason for this is that the process stream does not change during the process, the output is totally consistent if it has been developed correctly. This is in contrast to batch TFF, which has to deal with a steadily concentrating solution with non-Newtonian rheological properties and gradual changes into a variety of potential diafiltration buffers, all those that need to be controlled to prevent excessive process and product variability.

Fundamentally, the key validation and control parameters for a single-pass TFF step relate to the consistency of the input feed stream, the flow rate control (which dictates conversion and, therefore, concentration factor), and the length of time the SPTFF can operate with a continuous process stream before a buildup of residues on the membrane starts impacting concentration factors, pressures, and other critical product quality attributes such as aggregation. Although TFF is considered a steady-state process it will not keep processing forever and even the least binding of proteins will buildup on surfaces over time. This process intermediate endpoint needs to be identified, characterized, risk assessed, and finally validated to provide a change out point, with suitable safety factor included. The change out point should be developed on a set of cassettes on the point of replacement as a worst-case scenario if the cassettes are reused, that is, when the normalized water permeability reaches specification, usually around 70% of the reference NWP (see the discussion on reuse).

It is fully possible that the change out point may be beyond the length of the campaign, which is an ideal scenario as no change out procedure would need to be validated. For particularly fouling products or very long processes, however, there could be multiple change outs during a single campaign.

The flushing operation and its validation, to ensure that storage and preservative solutions are removed before processing, will be performed on a separate skid that will need to be qualified according to GMP. The validation of the flush volumes and temperatures is identical to the validation performed on a batch TFF system.

The sanitization step is highly recommended for a continuous process, due to the potential impact of high bioburden on product quality in a process that could last weeks. This should be performed on a separate skid after flushing and it is logical to validate them together. It is not typically necessary to test bioburden reduction with a microbial spike unless an outstanding contamination is being addressed on site but the flush out of the sanitization fluids needs to be validated fully afterward and bioburden/endotoxin should be checked at this point. Both retentate and permeate sides of the membrane should be tested, typically by dynamic flush residual.

The methods of validating sanitization agent (or indeed cleaning agent or preservative) flush out depend on the chemicals used. pH or conductivity is highly effective for sodium hydroxide or acid-based agents, while oxidizing agents such as sodium hypochlorite have commercially available assays.

The installation procedure will need to be validated and the operators appropriately trained. Single-use technology is highly recommended for continuous processing as it cuts down the large validation burden. As previously mentioned, the flow paths can be highly complex in continuous processing, therefore, predesigned assemblies with single-use connectors are desirable.

It has been demonstrated that TFF membranes used for single-pass operations are cleanable with existing industrial CIP regimes [4]. As such, the cleaning procedure for a single-pass TFF system is not drastically different to a batch TFF other than its simpler as there are no tanks or retentate lines, but this is offset by the need to test three separate permeate lines in a three section TFF system.

Cleaning can only be performed on change out or campaign end, so will require an independent qualified (IQ/OQ) skid for this operation. It is possible that the manifold used to manage change out during the process could be used to feed in WFI and cleaning solutions to the cassettes, but this would increase an already complex suite of flow paths and tubing around the process area.

It is a common practice in all TFF systems for parenteral production that TOC flush residuals are tested periodically (especially during the validation) and NWP is tested after every use to monitor cleaning effectiveness. For the TOC flush residuals, both permeate and retentate lines should be tested [24,25].

The validation of the effectiveness of the cleaning regime has greater ramification for single-pass TFF in a continuous process, however. With conventional batch TFF membrane, cleaning effectiveness is monitored by normalized water permeability measurements (NWP) and TOC flush residuals. NWP, in particular, is used as guidance to the number of reuses the cassettes is capable of in that specific process and typically the cassettes are discarded when the NWP falls below 70% or so of the reference value, although this value varies from process to process. While a reuse validation (discussed further) can be performed concurrently with production it does leave the last batch at risk of failure upon release, as it will need to be quarantined until the extra validation testing is complete. This risk of failure occurring with a long, expensive, continuous process is most likely unacceptable.

Another possibility is to simply single use the cassettes and not attempt to validate reuse, which simplifies matters greatly and is in keeping with the drive toward disposable technology [26]. The manifold lines may be reused; however, the cleaning of these will need to be validated if the intention is to change out the cassettes again before the campaign end.

If the cassettes are intended to be reused, then the storage procedure and duration will need to be validated. Storage durations could be very short if this is simply storage offline while waiting for another change out (two sets of cassettes swapped back and forth), but the limit of storage duration where it is deemed unnecessary to validate depends on the risk assessments performed.

If diverter plates are used and storage time exceeds a few days, then it is recommended that the installation be dismantled before storage. The additional permeate lines and long flow path would make it more challenging to ensure the whole product contact surface of the cassettes and lines were adequately exposed to the storage solution if storage was attempted in the holder.

The validation of storage of the cassettes and diverter plates will need to include bioburden, endotoxin, normalized water permeability, and air diffusion integrity tests.

Reuse of the TFF cassettes is greatly complicated with continuous processing. Typically the full-scale validation of reuse is done concurrently with production, as it is uneconomical in the extreme to run large numbers of engineering batches just to validate a large number of cassette reuses.

This, however, presents a risk problem with continuous processing. It is critical to know at which point the cassettes need to be changed out, usually way before the fouling can impact conversion or product quality. This change out point is likely to change considerably depending on how fouled the cassette was when installed into the process, which will correlate to how many reuses the cassettes have.

There is little choice here but to test the reuses on a scale-down model. If you consider that a continuous process could last a month with multiple change outs, this would require substantial resources to collect the data and validate. Once a limit of reuses have been reached, determined via NWP measurement, then these worse case cassette sets can be used to develop the change out point in the process and run the validation from there.

As previously noted, this will be very complicated and time consuming to implement. It is, therefore, recommended that TFF cassettes are either used only once or are only used for a single campaign with limited change outs before being discarded.

Continuous process verification and implementation of process analytical technology will be critical to adequately control and monitor a closed-continuous process. As taking samples is highly problematic, most of the analytics will need to be done in-line or at worst online.

In the case of single-pass TFF, various analytical technologies can be used to support the validation and monitor the process going forward to manufacture.

Pressure sensors, which can be obtained in single-use variants, will need to be installed on the feed line and the retentate line at a minimum. Each retentate connection between each section requires its own pressure gauge during process development (see process development considerations) and potential validation but this is not necessary for manufacturing.

The key control aspect that needs to be validated for single-pass TFF is the conversion rate to permeate, as this controls the CQA product concentration at the output. Measuring the feed flow is possible with a magnetic flow meter on the feed line and with either a load cell or flow meter on the permeate lines, with a mass balance then providing the retentate flow rate. If there is a pump before the SPTFF step (i.e., the flow is not continuous using the pump power from a previous step) and this pump is fully qualified and validated, the feed line flow meter can be dispensed with. There are dead volume and pressure drop concerns with excessive amounts of instrumentation in the flow path, so it is worthwhile limiting the sensors if possible.

It is recommended for both batch and single-pass TFF processes that a UV flow cell is installed on the permeate lines, usually set to 220/280 nm. This will provide assurance that the membrane is integral and retention is as expected throughout the whole process. As the permeate line is not product contact, this UV cell can be transferred during change out to the new cassette set.

For validating flush outs, be it of cleaning, sanitizing, or storage agents, pH or conductivity is effective. TOC is the standard technique, as referred to in USP Chapter <643>, for validating product residue carryovers and for extractables validation (see further).

TFF cassettes are linearly scalable by design, which allows partial validation at small scale. This is particularly useful for validating the parts of the SPTFF process that are feedstock and time/labor intensive such as cassette reuse.

Continuous processes are typified by the small size of the unit operations involved, as productivity comes from the constant flow of product over time rather than large volume handling. For this reason, even for substantial commercial processes, the SPTFF system is unlikely to exceed pilot scale relative to a standard batch TFF process. This makes a transfer of development and validation from a scale down system under 500 cm^2 to manufacturing scale more acceptable. However, as these are biologics, it is not acceptable to directly jump from scale down to manufacturing; the validation has to be performed at least once (preferably three times or as determined by risk assessment) at manufacturing scale to confirm the validation. It is possible to use these batches for clinical or even commercial supply, however, with the appropriate testing.

Extractables risk assessments, at least, and further testing are expected for parenteral products, with regulatory and operational guidance freely available [27]. A continuous bioprocess is no exception to this.

Extractables analysis and validation is primarily based on worse case models. It is known that worst-case conditions for extractables occur in periods of extended hold or when large contact surface areas are present, such as in final fill assemblies or in container closure. In this respect, the continuous process has an advantage as the process stream is always moving, greatly limiting the chance of extractables finding their way into the vial in any detectable concentration.

It is, however, still necessary to do the risk assessments and potentially the testing for those high-risk operations. Potentially, with a continuous bioprocess, more of the unit operations will fall into the low risk category as based on FMEA, although this is unlikely for final concentration SPTFF as there are no further purification steps before the vial.

If the extractables testing performed by soak, the standard first port of call with extractables testing, fails then it is justifiable to flush and take a dynamic flush residual at the flow rate specified in the process.

The permeate side of the TFF cassettes is usually discounted during extractables analysis due to it not being product contact. However, the Starling flow phenomenon [28], where fluid passes backward from the permeate to the retentate side, can occur when the polarization layer is very concentrated and the cross-flow/transmembrane pressure are low. These conditions are highly probable in single-pass TFF, particularly if the final formulation is specified for production concentrations above 100 g/l.

15.4 Conclusion

Process intensification through moving from batch to a continuous manufacturing process has already been achieved in a number of industries such as steel casting, petrochemicals, or glass manufacturing. As the biopharmaceutical industry evolves under the same pressure to intensify and streamline production, classical batch mAb production templates are being redesigned to achieve fully integrated end-to-end continuous production [29].

Tangential flow filtration, run as either batch or single pass, is now widely implemented within most monoclonal antibody production templates for a variety of applications, as discussed previously in this chapter. Single-pass TFF used in both batch and continuous processes is an effective solution to debottlenecking downstream capacity, especially when the pool tank size is a limiting factor. Its elegance lies in that it is a new way of operating an existing technology, using proven membrane cassettes, processing hardware and equipment. Where biomanufacturing process developers are looking into alternatives to reduce volume within a process or to desalt and buffer exchange their product, they are now seeing the advantages of using single-pass TFF as a viable solution for these applications, in that single-pass TFF enables higher final formulations and allows facilities to meet the demands of higher titer processes without major investments in new equipment [4].

The specific parameters of continuous/connected processing varies from manufacturer to manufacturer, but there is a fundamental need in all continuous processes for process volume reduction and buffer exchange. Single-pass TFF is not only a practical, but is also an enabling solution for many continuous manufacturing applications allowing for the in-line concentration and buffer exchange of biomolecule intermediates and bulk drug substances allowing for continuous, closed, and connected processing while addressing the volumetric challenges, such as tank limitations of high titer processes, to improve process flexibility.

References

1 Langer, E.S. (2011) Perfusion bioreactors are making a comeback, but industry misperceptions persist. *BioProcess J.*, **9**, 49–52.
2 Voisard, D. (2003) Potential of cell retention techniques for large-scale high-density perfusion culture of suspended mammalian cells. *Biotechnol. Bioeng.*, **82**, 751–765.
3 Cheryan, M. (1998) *Ultrafiltration and Microfiltration Handbook*, 2nd edn, CRC Press.

4 Lutz, H. (2015) *Ultrafiltration for Bioprocessing*, Woodhead Publishing Series in Biomedicine: Number 29, Woodhead Publishing, Cambridge.
5 Jungbauer, A. (2013) Continuous downstream processing of biopharmaceuticals. *Trends Biotechnol.*, **31** (8), 479–492.
6 Rathore, A.S. and Shirke, A. (2011) Recent developments in membrane-based separations in biotechnology processes: review. *Prep Biochem Biotechnol*, **41** (4), 398–421.
7 Warikoo, V., Godawat, R., Brower, K., Jain, S., Cummings, D., Simons, E., Johnson, T., Walther, J., Yu, M., Wright, B., McLarty, J., Karey, K.P., Hwang, C., Zhou, W., Riske, F., and Konstantinov, K. (2012) Integrated continuous production of recombinant therapeutic proteins. *Biotechnol. Bioeng.*, **109** (12), 3018–3029.
8 Konstantinov, K., Godawat, R., Warikoo, V., and Jain, S. (2014) Integrated Continuous Manufacturing of Therapeutic Protein Drug Substances. United States Patent Application US2014/0255994 A1.
9 Brower, M., Hou, Y., and Pollard, D. (2014) *Continuous Processing in Pharmaceutical Manufacturing*, **11**, John Wiley & Sons, Inc., New York, 255–294.
10 Klutz, S., Magnus, J., Lobedann, M., Schwan, P., Maiser, B., Niklas, J., Temming, M., and Schembecker, G. (2015) Developing the biofacility of the future based on continuous processing and single-use technology. *J. Biotechnol.*, **213**, 120–130.
11 Pollock, J., Ho, S.V., and Farid, S.S. (2013) Fed-batch and perfusion culture processes: economic, environmental, and operational feasibility under uncertainty. *Biotechnol. Bioeng.*, **110** (1), 206–219.
12 EMD Millipore Corporation (2014 Application Note AN5572EN00_EM: Single-pass tangential flow filtration.
13 EMD Millipore Corporation (2014) Biopharm Application Guide.
14 Bisschop, T. (2015) Process Development and Scale Up of SPTFF, Dechema Himmelfahrtstagung, EMD Milipore Corporation.
15 EMD Millipore Corporation (2013) TFF Technical Brief Protein Concentration and Diafiltration by Tangential Flow Filtration.
16 Dizon-Maspat, J., Bourret, J., and D'Agostini, A. (2012) Feng Li1 - single pass tangential flow filtration to debottleneck downstream processing for therapeutic antibody production. *Biotechnol. Bioeng.*, **109** (4), 962–970.
17 Teske, C.A. and Lebreton, B. (2010) Robert van reis - inline ultrafiltration. *Biotechnol. Prog.*, **26** (4), 1068–1072.
18 Subramanian, G. (2015) *Continuous Processing in Pharmaceutical Manufacturing*, Wiley-VCH Verlag GmbH, Weinheim.
19 Steen, J. et al. (2011) Single pass tangential flow filtration, ACS meeting poster, Anaheim, CA, March.
20 FDA Guidelines (2011) Process Validation: General Principles and Practices.
21 EMA (2012) Guideline on Real Time Release Testing, March 29th, 2012, EMA/CHMP/QWP/811210/2009-Rev1.
22 Kettlewell, R. et al. (2011) Continuous verification – providing an alternative approach to process validation. *Pharm. Eng.*, **31** (1), 18–26.

23 Rathore, A.S. *et al.* (2010) Process analytical technology (PAT) for biopharmaceutical products. *Anal. Bioanal. Chem.*, **398** (1), 137–154.
24 Parenteral Drug Association (2009) Technical report 15, Validation of Tangential Flow Filtration in Biopharmaceutical Applications.
25 Parenteral Drug Associate (2010) Technical report 49, Points to Consider for Biotechnology Cleaning Validation.
26 Shukla, A. (2013) Single use disposable technologies for biopharmaceutical manufacturing. *Trends Biotechnol.*, **31** (3), 147–154.
27 Ding, W. *et al.* (2014) Standardized extractables testing protocol for single-use systems in biomanufacturing. *Pharm. Eng.*, **34** (8), 74–83.
28 Starling, E.H. (1896) On the adsorption of fluid from the connective tissue space. *J. Physiol.*, **19**, 312–326.
29 Konstantinov, K. (2014) White paper on continuous bioprocessing.

Part Seven

Integration of Upstream and Downstream

16

Design of Integrated Continuous Processes for High-Quality Biotherapeutics

Fabian Steinebach, Daniel Karst*, and Massimo Morbidelli*

ETH Zurich, Institute for Chemical and Bioengineering, Department of Chemistry and Applied Biosciences, Vladimir-Prelog-Weg 1, 8093 Zurich, Switzerland

16.1 Introduction

Therapeutic proteins represent one of the newest drug families considered by the pharmaceutical industry to contrast a large variety of diseases. This family is largely dominated by monoclonal antibodies whose market grew from about $10 billion in 2000 to more than $140 billion in 2014 [1,2]. These involve relevant production facilities, both in size and value, which justify considering the adaption of continuous manufacturing. Currently, the mAb production in the biopharma industry is based on a rather well established platform relying on batch operation, rigorously divided into up and downstream sections. In particular, this is constituted by a fed-batch bioreactor, typically applying CHO cell recombinant technology, followed by the typical downstream cascade, consisting of a protein A capture step and a couple of polishing steps based on ion exchange or similar chromatography, combined with filtration and virus inactivation steps. Indeed, the possibility of operating each one of these steps in the continuous mode has been considered and investigated [3]. A detailed account of these efforts is reported in various chapters of this monograph. In this chapter we want to discuss the possibility of integrating all such steps in a single process where all steps are designed, optimized, and operated together like a single conceptual unit.

The continuous integrated version of the typical batch wise mAb production platform mentioned above is schematically illustrated in Figure 16.1. The first unit is now a perfusion reactor, which is connected directly, through a small buffering tank, to the purification train: capture, polishing, and virus inactivation, where all operations are conducted in the continuous mode.

Indeed, perfusion reactors are well known in biotechnology since many years and are currently used in several industrial productions, such as in the case of factor VIII (whose high instability does not allow for the long operating times of fed-batch reactors) [4]. Recently, significant improvements in the operation of

---

*Fabian Steinebach and Daniel Karst have equally contributed to the chapter.

Continuous Biomanufacturing: Innovative Technologies and Methods, First Edition.
Edited by Ganapathy Subramanian.
© 2018 Wiley-VCH Verlag GmbH & Co. KGaA. Published 2018 by Wiley-VCH Verlag GmbH & Co. KGaA.

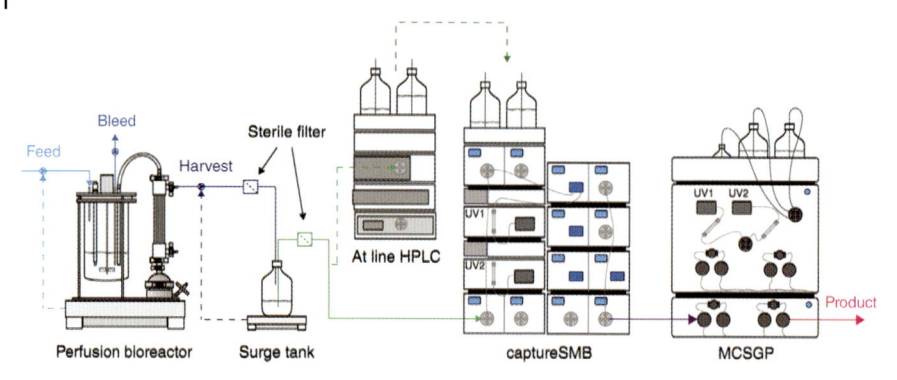

Figure 16.1 Schematic representation of a continuous integrated process consisting of a perfusion bioreactor, a continuous capture, and subsequent polishing steps.

such reactors have been achieved, particular with respect to the devices able to retain the cells in the reactor while harvesting the product. This, coupled with the recent advances in the development of continuous chromatographic purification units, has made it possible to directly connect the units while retaining the reliability degree needed by the biopharma industry. It is in fact clear that the cost and quality requirements of the end products do not allow this industry to tolerate failures nor disturbances to propagate along the production process. For this reason, batch operation has been preferred over the year. Each batch can in fact be controlled independently, thus, leading to an apparently safer production. However, the current technology in the various up and downstream units, as mentioned above, is now in the position to guarantee reliable operation. This has therefore opened up the possibility also for the biopharma industry to initiate the transition from batch to continuous operation that over the last century has characterized a number of industrial sectors. Eventually, this always led to better quality products and more reliable and less expensive productions.

The objective of this chapter is to review the current status of development of continuous integrated mAb manufacture processes. In particular, we focus on the technological aspects, which made the realization of such process at the industrial scale possible.

As mentioned above, each step of the process shown in Figure 16.1 is performed continuously, thus exhibiting advantages and disadvantages with respect to the corresponding batch operation. We have already observed that batch operation allows confining failures/disturbances to a single batch and this is indeed valuable when the understanding and control of the process is poor. However, once one can master what happens in the process better, it becomes convenient to enjoy the advantages of continuous operation. First, once at steady state, continuous processes produce continuously and consistently high-quality end products. With appropriate integrated monitoring and control strategies, disturbances can be rejected and the operating conditions maintained at the optimal level. Second, continuous operation implies lower residence times in the production line, which implies, for the same productivity, smaller volumes. This typically reduces the footprint of the production units by about 10 times, which particularly in a GMP environment is quite significant. Before proceeding with examples to illustrate the points above, let us discuss a bit further the differences between batch and continuous operation.

The perfusion reactor offers the possibility of producing better quality proteins than the corresponding fed-batch reactor. This is due to the much shorter residence time, which reduces unfavorable post-translational modifications (e.g., deamidation, fragmentation, and others) and aggregate formation. May be even more importantly, due to the uniform composition inside the reactor at steady state, all cells experience the same environment and therefore, contrary to a fed-batch reactor, express always very similar proteins. This results in a much decreased heterogeneity of the produced proteins, for example, in terms of glycoforms and charge variant distributions. The consequence is a lighter purification process and eventually and more importantly safer drugs for the patients. Another important advantage of the perfusion reactor is that it allows controlling the composition inside the cell culture in terms of both nutrients and metabolic byproducts, including those generated within the culture such as ammonia and lactic acid. This is not possible with reactors not having a continuous outlet stream. This becomes a strong limitation when one has to control the cell culture composition so as to control the protein quality profile, as it is typically done when aiming at biosimilars. On the other hand the presence of this outlet stream is also responsible for an important disadvantage of perfusion reactors, which is related to the larger consumption of media. This leads to costs of goods typically higher for perfusion than for fed-batch reactors. Another delicate aspect in the operation of perfusion reactors is the bleed flowrate used to control the viable cell density in the reactor, which actually represents a loss not only of media but also of the desired product. This aspect may negatively affect the overall product yield and therefore has to be considered carefully in the design of the reactor operation. Finally, it is worth mentioning that in the last 10–20 years fed-batch reactors have undergone a significant optimization, mainly with respect to media composition and cell line characteristics. This has, for example, produced an increase in titer by more than one order of magnitude. It is clear that this same "optimal" choices do not necessarily hold true for perfusion operation, and therefore such optimization process will have to be repeated hopefully leading again to significant process performance improvements.

When considering the chromatographic purification units, their operation in the continuous mode is realized through the multicolumn technology, which may involve the simultaneous use of two or more chromatographic columns. When compared to batch operation, this technology exhibits advantages, which are different depending on whether we are considering the capture or one of the polishing steps. In the first one the batch processes suffer for a productivity versus resin utilization trade-off. High flow rates lead in fact to shallow break through curves, which force early switch to column regeneration, which in turn implies low-resin utilization. The addition of a second column significantly alleviates this problem since the outlet stream from the first column can be fed to the next column after breakthrough so as to allow for saturation of the first column and then a higher degree of resin utilization. In the case of batch polishing, in order to achieve high purity one has to keep the pooling window rather narrow so as to avoid the collection of too many impurities. But this of course leads to important losses of the desired product, and then low yields. Enlarging the pooling window would indeed increase the yield but of course at the expenses purity. Also in this case, the use of the multicolumn technology in the form of the MCSGP allows strongly reducing such trade-off by allowing higher values of both yield and purity. It is worth noting that such a performance improvement in MCSGP is so strong to make it possible to integrate this unit in the purification train in order to modify, for example, the charge variant profile of the end product, which would be unthinkable with a simple batch operation. This appears particularly promising when considering the problem of biosimilar production or modern high quality therapeutic proteins.

In the case of an integrated process, all the above steps are connected to form a single process, which has to be designed and operated as a single unit. Thereby, all the advantages discussed above arising from the continuous operation of the individual units are merged. In addition, we have to take advantage from the integration of the different units, by operating each of them depending on the current performance of the others. This requires a higher level monitoring and control of the process, where the signals coming from the single units are all collected and made available for a hierarchical control and operation of the process. Here an upper layer of control would be established where all data collected locally from the single units are simultaneously available and can be used. All of this is currently open to technology development where mathematical models will indeed play an important role. We can think to mechanistic models to reject disturbances, for example, changing the operating conditions of a purification unit to compensate for a malfunctioning in the perfusion reactor or a decrease in cell productivity. Another aspect is the use of multivariate analysis of the collected data in order to identify the behavior of the process, based on the statistical analysis of the data collected in the history of the plant operation. One could also think of combining the two approaches through hybrid models. Indeed, the variety of possibilities that we have to improve the performance of the process, particularly in terms of the quality of the end product, is very rich and indeed we will see a lot of progress in this direction in the near future.

Recent contributions appeared in the literature reporting applications of the integrated manufacturing concept to various degrees. It is worth mentioning several contributions by Konstantinov and coworkers where the integration of the perfusion reactor with a continuous capture step for the production of a mAb and a fusion protein [5–7] is shown. The same group, as well as Schembecker and coworkers, reported the concept in an end-to-end continuous manner by including the polishing step [8,9].

In the following we analyze, with reference to a specific mAb of industrial relevance, the continuous operation of both the culture and the purification steps integrated in a single process [10]. We discuss in particular the relevant technological issues and compare the process performance with that of the classical batch platform. Particular emphasis is placed on the quality of the end product, which is indeed the key aspect driving technological innovation since this has the strongest impact on the patient.

16.2 Perfusion Cell Culture Development

16.2.1 Objectives and Requirements

Mammalian cell perfusion cultures are well established in the production of labile therapeutic proteins [11]. However, the ease in operation and control in combination with the significant potential to increase product yields of fed-batch cultivations has prevented a more general application of perfusion processes. Recently, technical advances and the call for increasing manufacturing flexibility have refueled the interest in perfusion mode. In particular, the stable operation at high viable cell density and the short product residence time is beneficial to increase the overall productivity while minimizing heterogeneities of critical quality attributes [8].

Moving from classical fed-batch to integrated continuous operation requires changes in the bioreactor setup and control. The achievement of high viable cell density and the harvest of cell free supernatant are coupled to the incorporation of a cell retention device. Among multiple retention principles, such as sedimentation [12–14], centrifugation [15,16], or ultrasound [17,18], only external filtration devices offer the total exclusion of cells [19]. In particular, the application of cross flow filtration using hollow fiber modules in tangential or alternating tangential flow mode have been shown to limit excessive filter blockage and increase filter life-time. In combination with the control of the in and outlet flow rate to maintain the perfusive flow, the additional complexity introduced by the cell retention device needs to be considered in the bioreactor setup design.

High cell densities are inevitably linked to increased overall production and consumption rates, which make the supply of sufficient nutrients and oxygen as well as the removal of accumulating carbon dioxide challenging [20]. In addition, inadequate homogenization of the culture broth, can lead to spatial gradients in metabolite and gaseous composition, harming the attainable number of cells [21]. Therefore, a thorough physical characterization of the system in terms of hydrodynamics, mixing effectiveness, gas–liquid mass transfer and maximum hydrodynamic stress is required. In close alignment with set culture objectives, suitable operating ranges of key process parameters can be defined. This integrated approach of developing perfusion cultures has to be complemented with a precise control of key operating parameters, such as volumetric flow rates (harvest, bleed, and feed rate) and cell density to eventually achieve a stable operation.

In the pursuit of an integrated manufacturing process for the production of monoclonal antibodies, in this study two identical stirred tank perfusion bioreactor setups with either the commercial alternating tangential flow (ATF) or the self-built tangential flow filtration (TFF) device were characterized and compared.

16.2.2 Bioreactor Setup

Two identical stirred tank bioreactor perfusion setups only deviating in their retention device and have been designed based on a DASGIP 2.5 L bioreactor system (Eppendorf AG, Switzerland), as shown in Figure 16.2. The first one was

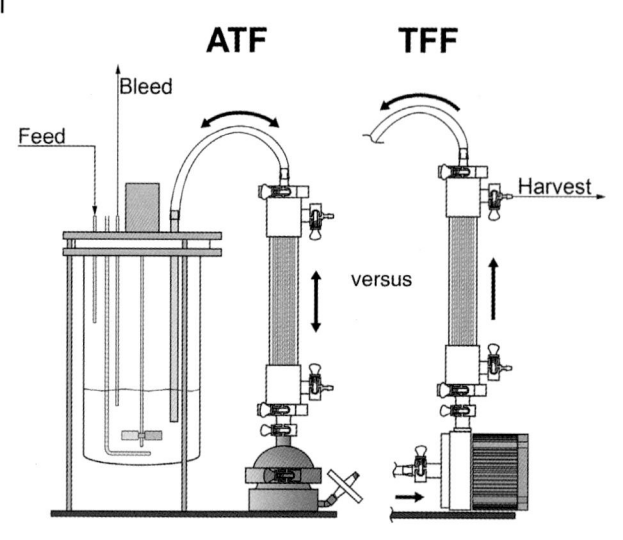

Figure 16.2 Stirred tank perfusion bioreactor setup equipped with either ATF or TFF system for cell retention. The TFF hollow fiber was driven by a bearingless centrifugal pump withdrawing the cell culture broth from a bottom outlet of the glass vessel.

equipped with the commercial ATF2 (Repligen, USA) device. Whereas, the second was a tangential flow filtration (TFF) system, consisting of a bearingless centrifugal pump (Levitronix AG, Switzerland) with a hollow fiber modulus mounted on top. Two types of hollow fibers have been evaluated, here referred to as long (length: 60 cm, area: 1300 cm^2, pore size: 0.5 μm, material: PES, fiber diameter: 1 mm) and short (length: 25 cm, area: 1570 cm^2, pore size: 0.5 μm, material: PES, fiber diameter: 1 mm) hollow fiber. A single six-blade Rushton turbine impeller (diameter: 4.5 cm) and a four-whole open pipe sparger were used for mixing and aeration of the bioreactor. A more detailed description is available in Ref. [22].

16.2.3 Physical Bioreactor Characterization

As described in Section 2.1 high cell density perfusion cultures require enhanced nutritional supply, gas liquid mass transfer, and mixing effectiveness. An accurate *a priori* physical characterization of the bioreactor system including the cell retention device, allows the identification of suitable operating ranges for the key process parameters. In addition, it reduces the necessity of laborious and excessive screening cell culture experiments. To guarantee a comparable operation of the ATF and TFF setup, the hydrodynamics of both retention devices were investigated (Figure 16.3a/b). Using an ultrasound clamp-on flow rate sensor the flow rate in the external loop of the ATF and TFF system was monitored. In the ATF the diaphragm pump creates a periodic change in flow direction to the reactor. Given the constant pump displacement volume, the cycle time determines the effective flow rate through the hollow fiber. For example, when changing the flow rate from 1.0 to 1.5 l/min, the cycle time is adjusted from 11 to 8 s. In the TFF system a linear increase in the flow rate with higher rotational

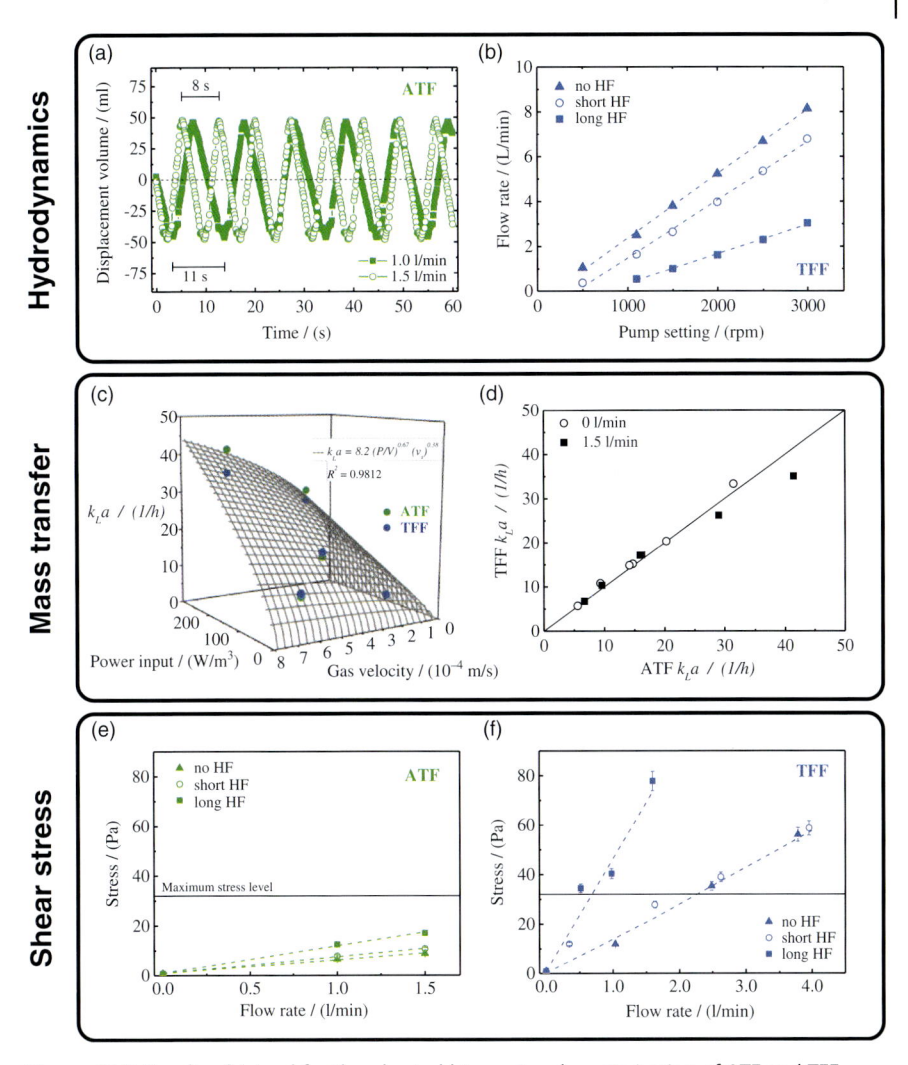

Figure 16.3 Results obtained for the physical bioreactor characterization of ATF and TFF, including hydrodynamics of the retention device (a/b), gas–liquid mass transfer (c/d), and hydrodynamic shear stress characterization (e/f).

speeds of the centrifugal pump was observed as shown in Figure 16.3b. In contrast to the short hollow fiber, the long hollow fiber provided considerably lower flow rates at similar rotational speeds. The increased length and smaller effective cross-section area represented an enhanced hydrodynamic resistance for the nonvolumetric centrifugal pump.

The knowledge of the bioreactor hydrodynamics is essential to predict the gas–liquid mass transfer and mixing effectiveness in the perfusion culture. The homogenous supply of sufficient oxygen and removal of carbon dioxide is essential not to limit the attainable number of cells. The oxygen mass transfer coefficient was determined at various Rushton turbine impeller stirring speeds and gas flow rates. Independent of the retention device and its incorporation, $k_L a$

values ranged from 5 to 40 h^{-1}, increasing for higher stirring and sparging rates as shown in Figure 16.3c/d. The observed values are well in the range proposed to support high cell density cultures in the order of 10^8 cells/ml [21]. The good dispersion of bubbles and mixing times below 6 s at all tested conditions, indicate no limitations arising from spatial gradients in the nutritional supply of cells. To prevent the accumulating carbon dioxide and given the fast saturation of gas bubbles with CO_2, high volumetric gas flow rates are favorable [23,24]. In practice, the choice of impeller speeds and sparging rates have to be balanced with vortex formation, foam built-up, cavity formation, and hydrodynamic stress that cells are exposed to.

To identify safe operating ranges, the maximum effective hydrodynamic stress has been determined using a shear sensitive PMMA particulate system, as described in Ref. [25]. A maximum tolerable stress value of approximately 32 Pa for the cell line under investigation was determined in a previous study [26]. Maximum operating ranges to prevent cell damage of impeller speed and volumetric gas flow rate were determined without attaching a cell retention device. Stress values increased as function of the Re_{imp} (in a range from 30 to 500 rpm) with a maximum of 37 Pa at 500 rpm. The presence of a 0.22 vvm gas flow rate only led to higher stress values at low Re_{imp} numbers, but were dominated by the increase in stirring speed thereafter [22]. The operation of the cell retention device can introduce additional stress to the cells. Therefore, the impact of the external loop flow rate in the ATF and TFF system with either long or short hollow fiber mounted was investigated and is presented in Figure 16.3e/f. In the ATF, although increasing with higher external loop flow rates, stress values did not exceed the threshold of 32 Pa independent of the hollow fiber type. In the TFF, stress levels were found significantly higher than compared to the ATF. In particular, the long hollow fiber created much higher maximum hydrodynamic stress values, compared to the short and no hollow fiber, since increased pump speeds of the centrifugal pump had to be set for similar flow rates. To remain below the maximum stress level, the short hollow fiber and an external loop flow rate of 1.5 l/min for both ATF and TFF was chosen. In order to ensure good mixing and sufficient oxygen supply an impeller speed of 400 rpm and 0.22 vvm were selected.

16.3 Continuous Capture Development

16.3.1 Objectives and Requirements

Continuous capture steps allow the direct processing of cell free harvest from perfusion cultures. The integration of these two continuous processes excludes nonproductive storage and filtration steps that are necessary in typical batch-wise operation, thus leading to a leaner process [6]. The high product affinity of the chromatographic capture process facilitates the separation of most product harming proteins (e.g., proteases) while achieving high concentration of the target protein in a predefined buffer system. This on one hand leads to a significant reduction of the handling volume and also prevents undesired modifications of the protein product.

A unique feature of the continuous capture step operation is the possibility to adsorb the target compound on a cascade of several columns. While the first column of this "adsorption train" can be completely saturated, the breakthrough product is captured on the subsequent column(s). Compared to batch-wise capture, higher loading flow rates can be used leading to enhanced productivities [27,28]. In addition, the complete saturation of a single column results in increased capacity utilization and is thus linked to savings in resin and buffer usage [28]. Simulations have shown that multicolumn capture processes based on this concept are especially superior to batch capture in cases of titer concentrations lower than 2.5 g/l [29], like often observed for perfusion cultures.

Several technical setups for the continuous capture operation are commercially available. The flow distribution to the different columns can either be performed in a central valve block, like in case of the BioSMB™ (Pall) and OCTAVE™ SMB (Tarpon) systems, or with distributed valves as implemented in the ÄKTA pcc (GE Healthcare), Contichrom® CUBE (Chromacon), Contichrom® Lab-10 (Knauer), and BioSC® (Novasep) setups. All mentioned setups are suitable for integrated processing since they allow the independent column switching with multiple in and outlet valves.

Two key aspects have to be considered in the technical implementation of the integration between the perfusion bioreactor and the continuous capture step. First, since chromatography itself is not a sterile process, the implementation of a sterile barrier in between the two unit operations or sterilization of the capture unit and its columns is a necessity. Second, low suction power chromatographic pumps, especially on bench-scale, are not able to actively draw the supernatant directly from the hollow fiber module. Therefore, in most cases a surge tank in between the harvest pump of the bioreactor and the feed pump to the chromatographic device is installed. This low volume reservoir allows the decoupling of the bioreactor and capture operation, beneficial for robustness of the entire process.

16.3.2 Continuous Two-Column Capture Process

In this study a two column capture step for the continuous processing of the cell free harvest from the perfusion culture was developed [30]. The operation of only two columns, reduces the equipment footprint and operational complexity [28]. As shown in Figure 16.4, the process consisted of three subsequent steps, with the two columns either interconnected (step A and B) or operated in batch mode (step C). During step A, column 1 is loaded with fresh feed beyond its dynamic binding capacity, while any target protein in the breakthrough is adsorbed onto column 2. In the subsequent step B, column 1 is washed and any unbound target protein is mixed in-line with additional feed and adsorbed on column 2. In the third step, the saturated and washed column 1 is regenerated, while column 2 is further loaded.

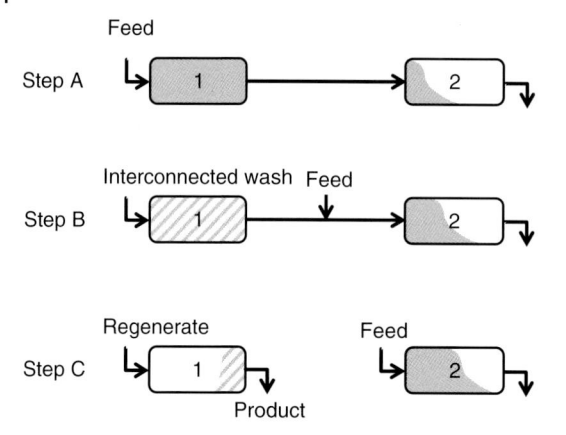

Figure 16.4 Flowchart of the two-column capture process. After three consecutive steps, the two columns switch their positions. During the first two steps, the two columns are interconnected while they are operated batch-wise in the third step.

The designed capture step is capable to continuously process a cell free harvest stream from the perfusion bioreactor, while eluting product in a discrete manner during the regeneration step. A complete saturation of the first column, while avoiding product breakthrough from the second one, can be achieved by properly scaling the column size and defining the length of each step.

16.3.3 Process Performance

Parameters to assess the process performance of the capture operation include the product yield (the difference between the total load amount and the amount of product lost in the flow through divided by the total load amount), the time-based productivity (the amount of product purified per unit time and column volume) and the capacity utilization (the fraction of resin capacity saturated by the product in one cycle).

The performance of the continuous capture process was evaluated in comparison to the batch-wise processing, using simulations with a mechanistic model [30]. Assumptions included an equal total column volume and regeneration protocol, a continuous constant inlet flow and equal loading as well as regeneration durations of the two column batch-wise capture. When operated at the same productivity and yield (99.9%), the continuous process exhibits a 2.5 higher capacity utilization compared to the batch-wise process. This is in agreement with other simulations [29] and experimental results [28,31,32] for mAb capture on Protein A column reporting a 40% improvement in capacity utilization using continuous operation. The resulting much higher product concentrations, decreases the volume handling in further downstream steps and increases the amount of product purified during the column lifetime.

16.3.4 Process Control

The process design and its practical implementation has to be aligned with a rigid process control. In particular, the adaptation of the capture step to variations in the feed concentration and binding capacity with column age need to be considered. The product concentration may not only show fluctuations depending on the bioreactor operation but also due to a reduced cell-specific productivity with culture duration. The reduction of column binding capacity is a well-known effect in affinity capture, resulting from the loss of functional groups during the cleaning-in-place step. In order to adapt the process to these changes, suitable detectors measuring the target protein concentration, such as online UV [6,33], at line HPLC [34], or Raman spectroscopy are necessary. Depending on this feedback, the cycle length of the capture operation is adapted to retain defined process performance targets. In particular, the length of the interconnected step A can be used to adjust the process, since the duration of step B and C are given by the preset wash volume (2–5 column volumes) and the regeneration protocol (obtained from batch experiments). In Figure 16.5, the process performance parameters, such as yield, resin utilization, and buffer consumption are shown as function of the feed volume per cycle for a given feed concentration and flow rate. The increase of the interconnected time and hence the feed volume per cycle results in a better capacity utilization of the resin and reduction of buffer consumption. In contrast, the yield remains constant until breakthrough of product occurs in the second column and decreases thereafter. The optimal operation is achieved when selecting the largest interconnected time still fulfilling the specified product yield target. A look-up table of the interconnected times at defined yield targets for varying feed concentrations and flow rates was generated by simulations, and used to maintain stable and optimal operation of the capture section while operating the unit in the integrated mode as discussed in Section 16.4.

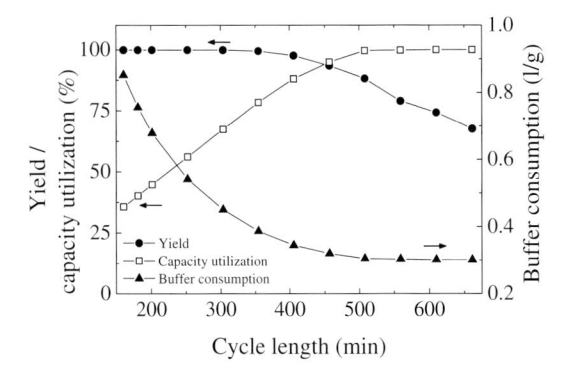

Figure 16.5 Simulated process performance parameters yield, capacity utilization, and buffer consumption as function of the cycle length, which is varied by changing the interconnected feed time. Optimal operation performance is achieved by maximizing the cycle length up to the point where yield decreases below an acceptable level. (Reproduced with permission from Ref. [30]. Copyright 2016, Elsevier.)

16.4 Operation of the Continuous Integrated Process

16.4.1 Bioreactor Operation

Based on the physical characterization described in Section 2.3, suitable operating parameters for the ATF and TFF systems were determined. An external loop flow rate of 1.5 l/min, a gas flow rate of 0.22 vvm and a Rushton impeller stirring speed of 400 rpm were found suitable to support higher demands in gas–liquid mass transfer without harming cellular integrity. Their applicability was investigated in a series of viable cell density set points at 20, 60, and 40×10^6 cells/ml, each kept for one week. In order to achieve a constant culture at these operating points, a precise control strategy and suitable media design had to be defined. Since a low perfusion rate is beneficial to increase process productivity [35], a harvest rate of one reactor volume per day (RV/day) was fixed throughout the culture. In contrast, the bleed rate was continuously adjusted according to an online cell density measurement (Aber Instruments Ltd., UK) to remove exceeding biomass from the culture. The perfusion bioreactor was placed on a balance, regulating the inlet feed flow rate to keep the overall weight constant. Depending on the cell density set point, the fresh nutrient media was composed of a distinct ratio between a base and an enriched media (VCD set point 20, 40, and 60×10^6 cells/ml – Fraction base/enriched media 100/0, 65/35, 30/70%, respectively). The physiological environment of the cells was maintained at a pH of 7.1 and 50% dissolved oxygen tension, varying the fraction of carbon dioxide and oxygen in the inlet gas stream. The temperature was fixed at 36.5 °C.

16.4.2 Cell Growth

After thawing, the CHO cell line secreting a human IgG 1 isotype monoclonal antibody was cultured in a shaking incubator for one week in a chemical defined medium. Further expansion was carried out in a perfusion seed bioreactor using a TFF cell retention device during another week. This procedure allowed a rapid increase of cell number and the inoculation close to the first cell density set point in the production bioreactor, not limiting filter life time. Irrespective of the retention system, stable operation was achieved at the three viable cell density set points (20, 40, and 60×10^6 cells/ml), as shown in Figure 16.6a. Although cell viability remained high throughout the culture, higher cell densities resulted in a decrease of growth rate und thus lower bleed rates. In addition with a fixed harvest of 1 RV/day the overall perfusion rate was simultaneously decreased. The reduction of the cell specific perfusion rate was thus compensated by the change in medium composition depending on the cell density set point. Therefore, resulting metabolite levels in the bioreactor changed between these operating points but remained constant in time after an initial transient. Steady-state operation in terms of metabolite consumption and production rates was achieved [22]. In general, the cellular growth behavior in both reactor systems with ATF or TFF showed comparable trends. The careful definition of operating parameters based on the physical characterization enabled the reliable long-term culture of cells at different viable cell density set points.

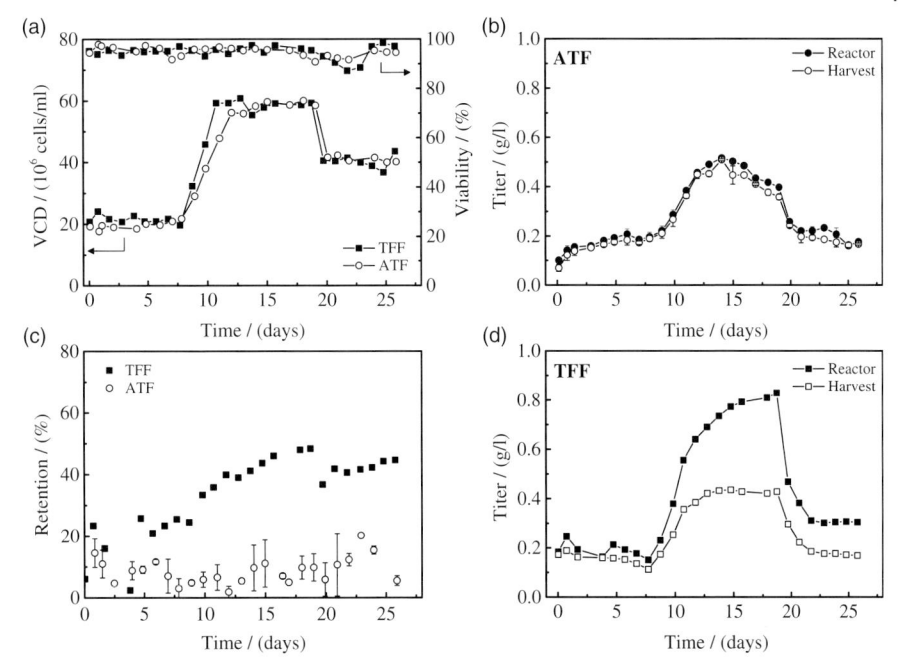

Figure 16.6 (a) Steady-state operation was achieved at three viable cell density set points during 25 days in both ATF and TFF. Time evolution of mAb concentration in the ATF (b) and TFF (d) in the reactor (closed) and harvest stream (open). (c) Retention of mAb in the hollow fiber module comparing both retention devices.

16.4.3 Monoclonal Antibody Production

Although the viable cell densities in the two systems evolved very similarly in time through the three distinct set points (Figure 16.6a), a significant difference in the harvest concentration between ATF and TFF was observed (Figure 16.6c). Whereas, an average retention of 10% in the ATF hollow fiber was present, up to 50% of the product was retained in the TFF. The higher retention in the TFF system is most probably due to the difference in flow pattern between the two retention devices affecting membrane fouling. Compared to the TFF, the change in flow direction in the ATF creates a backflush through the membrane pores from the permeate site, thus limiting the extend of cellular debris deposition. A higher degree of fouling, given the unidirectional flow behavior of the TFF was confirmed by the steady increase of the centrifugal pump speed to maintain a flow rate of 1.5 l/min. This results in differences in the titer inside the reactor and in the harvest flow that are observed with TFF system (Figure 16.6d) but not with the ATF one (Figure 16.6b). Similar results for the TFF have been observed in other studies [36]. However, care must be taken to generalize these findings since the product retention is inevitably influenced by multiple parameters including the dimensions, material, pore size, permeate flux, and flow pattern of the hollow fiber [19]. As previously reported [37], in both systems a reduction of specific antibody productivity during the culture is observed. The stable expression of protein product in long-term perfusion cultures can be compromised by unfavorable media composition or cell line instability [38–40].

16.4.4 Monoclonal Antibody Capture

A small surge tank with a working volume of around 5% of the reactor volume was placed in between the perfusion bioreactor and the continuous capture process (Figure 16.1). The decoupling of the peristaltic harvest pump on the bioreactor system and the piston feed pump of the capture process allowed the balancing of small fluctuations in flow rates, which increased the overall process robustness. In addition, sterile filters were placed before and after the surge bottle to prevent the contamination of the bioreactor from the nonsterile capture operation [10].

Three control loops were implemented to coordinate the flow rates of the integrated process. To ensure a constant inlet flow rate to the capture unit, the harvest rate from the surge tank was set to one reactor volume per day. The peristaltic pump controlling the harvest rate from the bioreactor was adjusted to keep the liquid volume in the surge tank constant. Gravimetric control of the bioreactor volume was used as input to regulate the feed pump at the reactor inlet. As mentioned above, the viable cell density was automatically controlled adjusting the bleed rate according to an online cell density measurement (Aber Instruments, UK).

The capture process was implemented on a Contichrom® Lab-10 (ChromaCon AG, Switzerland) system using two 2 mL MabSelect SuRe protein A columns (GE Healthcare, USA). Each product elution was followed by a cleaning-in-place step. An analytical HPLC (Agilent) was used at line to monitor the concentration of the target protein in the feed stream. For the analysis, 3 ml of cell-free and filtrated sample were automatically transferred to the skid to determine the product concentration on a strong cation exchange column (TOSOH TSKgel CM-STAT, Tosoh, Japan).

In combination with the generated look-up table of interconnected times for varying feed concentrations discussed in the previous section, the product titer was used to adapt the cycle length of the capture process. As shown in Figure 16.7, depending on the harvest concentration at different cell density set points in the bioreactor the cycle length was adapted so as to maintain optimal operation in terms of resin utilization and buffer consumption while maintinaing the desired yield. It is seen that in order to maintain a yield target of 99% the duration of the interconnected load (Step A in Figure 16.4) was decreased for higher feed concentrations [10]. It is worth mentioning that the mechanistic model was able to account for the decrease of functional groups during the process operation (dotted lines), notable due to the reduction of cycle length at constant feed concentration. All operating points of the capture process in the integrated operation are shown in Figure 16.7, where every point corresponds to the process parameters for one cycle. An average yield of 99.2% was achieved during the entire process.

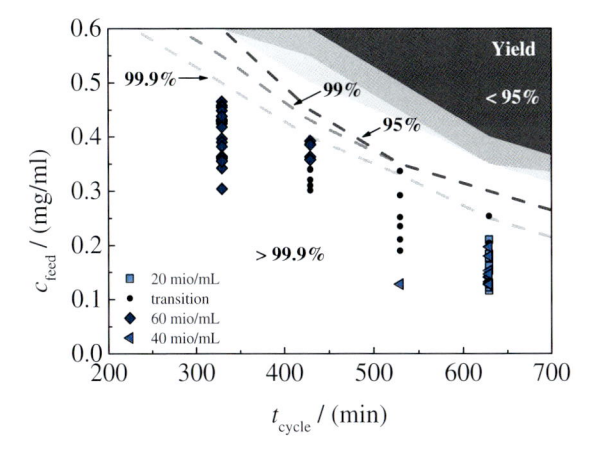

Figure 16.7 The yield of the capture process was simulated for different cycle durations (t_{cycle}) and feed concentrations (c_{feed}), as shown in the background. Each point represents the operation conditions that were used during one cycle. Here the cycle duration in the continuous capture process was adapted according to the feed concentration. At low feed concentrations, the cycle length was increased thereby operating the process under close to optimal conditions for every steady state in the bioreactor, while maintaining the yield at the desired value, 99%.

16.4.5 Process Performance

The integrated process was operated for a total of 26 days, while maintaining the sterility of the perfusion cell culture and the surge tank. Steady-state operation at three consecutive viable cell density set points was achieved. Although none of these has been optimized, it is useful to compare their performance in terms of mAb productivity, media and buffer consumption, and product yield as shown in Figure 16.8. The reduction of cellular growth at higher cell concentrations resulted in a decrease of the bleed rate, thus reducing the loss of noncaptured product. Given the 99.2% yield of the optimized capture step, this resulted in a higher overall yield of the integrated process as shown in Figure 16.8. Since the harvest rate was fixed at 1 reactor volume per day, the decrease of the bleed rate

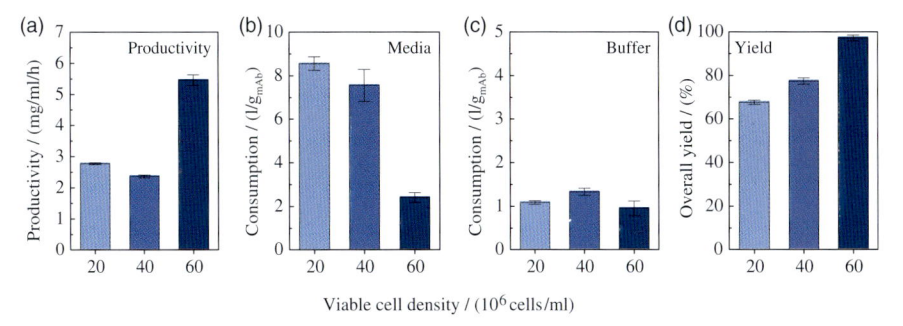

Figure 16.8 Productivity, media consumption, buffer consumption, and overall yield for the three different steady states differing in the viable cell density.

resulted in a reduction of perfusion rate. The simultaneous increase of mAb concentration from 20 to 60×10^6 cells/ml (Figure 16.6), significantly increased the overall process productivity and reduced the amount of media supplied per gram of mAb. However, productivity dropped at 40×10^6 cells/ml due to the reduction of the cell-specific productivity with culture duration. In general the operation at high viable cell density, low cell specific perfusion rate, and low bleed rate is favorable to optimize the operation of the integrated process in terms of both productivity and yield. The online optimization of the capture process not only resulted in a high yield but also significantly reduced its buffer consumption. Figure 16.8 shows in fact that independent of the viable cell density the buffer consumption remained constant, as the cycle length was adapted to changes in feed concentration.

Possible limitations arise from long-term operation, including cell-line stability, filter blockage, or the chromatography column lifetime. However, filters and columns can be substituted during such a process and should therefore not limit the total duration.

16.4.6 Product Quality

Several quality attributes of the monoclonal antibody, including N-linked glycosylation, charge isoforms, aggregates, and fragments were quantified in the post capture pool. Aggregate and fragment content was determined by size exclusion chromatography, charge isoforms by weak cation exchange chromatography and glycan patterns by capillary gel electrophoresis with laser-induced fluorescence detection. Their evolution in time during the entire operation of the unit (left column) and their average values corresponding to each of the three steady states are shown in Figure 16.9. Irrespective of the quality attribute and steady state, a transient phase of six days was observed until product quality remained constant [10].

Five charge isoforms were detected by weak cation exchange chromatography and displayed in Figure 16.9a/b. The main form made up to 80% of the whole distribution, but reduced with higher viable cell densities, leading to the relative occurrence of more basic isoforms. As no impact of the capture process on the charge isoform pattern is expected, these variations are to be attributed to a change of culture conditions. Possible explanations include the difference in metabolite composition or the change of product residence time in the bioreactor, given the variation of perfusion rate with the cell density steady state.

The formation of recombinant antibody aggregates and fragments can be affected by several conditions, including temperature, pH, ionic strength, shear forces, concentration, and released proteases [41]. Within this study, aggregate contents between 0.5 and 4% were observed, whereas total fragment levels remained below 5% (Figure 16.9c/d). While more aggregates were formed at higher viable cell densities, no clear influence on the fragment content of the different steady states was observed. The increased amount of aggregates can be attributed to the elevated antibody concentrations at higher cell densities. In general, the short product residence time in the continuous bioreactor and the direct capture of the mAb reduces the impact of undesired product degradation and released proteases.

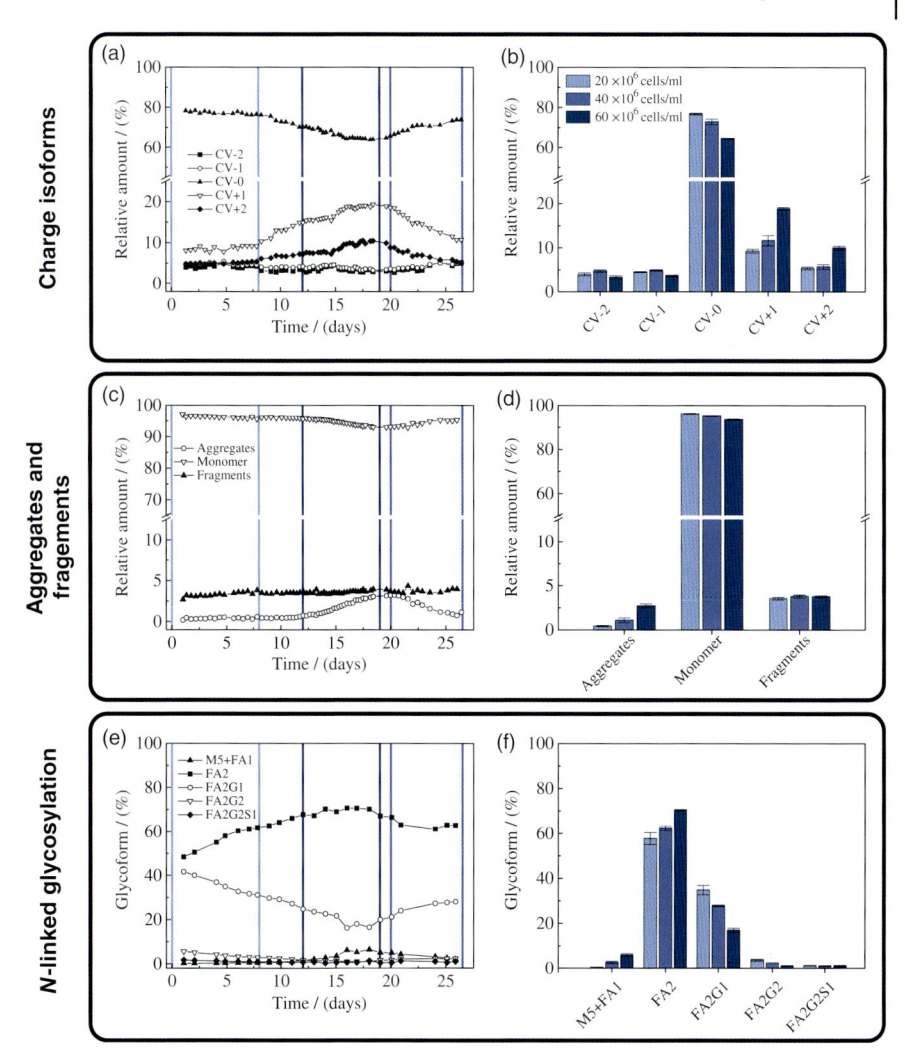

Figure 16.9 Product quality in terms of charge isoform profile, aggregate and fragment content, and glycoform profile for the post Protein A product obtained at the three different steady states.

The *N*-linked glycosylation pattern is one of the most crucial quality attributes of monoclonal antibodies. As for the charge isoforms, the constant environment at steady state operation favors the consistent glycosylation of the product. With increasing cell density, a higher fraction of less complex species (e.g., HM, FA2) is formed. This trend can be due to the change of metabolite composition in the reactor depending on the viable cell density set point. In particular, the presence of increasing ammonia levels with higher cell densities, has been reported to decrease the extend of galactosylation. The good mechanistic knowledge of the glycosylation machinery and the numerous reports on the effect of media supplements (trace elements, nucleotide sugar precursors, and enzymatic inhibitors) as well as bioreactor operating conditions (pH, temperature, osmolality, and dissolved oxygen) make the modulation of a desired glycosylation pattern possible [42,43].

In addition to product-related quality attributes, process-related impurities were evaluated after each unit operation. While in the perfusion bioreactor, approximately 10^4 ng/ml of DNA and 10^5 ng/ml of HCP were measured [10], only a small decrease in the HCP concentration and a similar level of DNA was determined in the harvest. However, the continuous capture operation significantly decreased both concentrations by two orders of magnitude.

When comparing the perfusion culture to the traditional fed-batch and batch operation, the enhanced product quality from continuous processing is one of the most important benefits. The short residence time and uniform conditions in the bioreactor reduce the variability of critical product quality attributes, which is shown here for the three cell density steady states. It is worth noting that for these steady states both the cell specific perfusion rate and the media consumption is changed, which has both an impact on the product quality as well as on the economics of the entire process.

16.5 Conclusion

With adequate equipment and process design, a long-term operation of a continuous mAb production process was achieved. The sterile barrier between the perfusion bioreactor and the two-column continuous countercurrent capture process was provided by suitable filters. Online control of the process was achieved by gravimetric control loops for the liquid handling and at line monitoring of the harvest concentration for the capture operation. Thereby, stable operation at three different set points was achieved [10]. It is worth pointing out that in order to fully exploit the potential of continuous integrated processing we need not only to properly connect the liquid streams but also to connect and integrate control loops and model optimization.

Product quality attributes such as charge isoforms, aggregates, and glycolsylation profile remain constant during each steady state. Due to changes in the media composition, different product quality in between the steady states was obtained. Compared to classical batch operation, the continuous operation with process integration reduces the residence time and thereby the product heterogeneity.

It is to be noted that the online model-based adaptation of the operating conditions of the capture unit allowed to maintain a high yield, and not to affect product quality and amount. Indeed, the reduced residence time in the reactor and integration without storage steps results in an improved product quality. In addition, the integrated capture step reduces the liquid handling, as the volume is decreased by a factor 10 to 20, thus leading to higher product concentration compared to classical batch operation. This is beneficial for the performance and economics of the subsequent purification steps in the downstream cascade.

Acknowledgment

The authors would like to acknowledge Merck-Serono SA for providing the cell line, nutrition media and carrying out the analysis of N-linked glycans as well as Levitronix AG for providing the centrifugal pump and the ultrasonic flow sensor. This work was financially supported by the SNF Grant 206021_150744/1.

References

1 Walsh, G. (2003) Biopharmaceutical benchmarks-2003. *Nat. Biotechnol.*, **21**, 865–870.
2 Walsh, G. (2014). Biopharmaceutical benchmarks 2014. *Nat. Biotechnol.*, **32**, 992–1000.
3 Jungbauer, A. (2013). Continuous downstream processing of biopharmaceuticals. *Trends Biotechnol.*, **31**, 479–492.
4 Bödeker, B.G.D., Newcomb, R., Yuan, P., Braufman, A., and Kelsey, W. (1994). Production of recombinant factor viii from perfusion cultures: i. large-scale fermentation, in: *Animal Cell Technology*, Elsevier, pp. 580–583.
5 Konstantinov, K.B. and Cooney, C.L. (2015). White paper on continuous bioprocessing. May 20–21, 2014 continuous manufacturing symposium. *J. Pharm. Sci.*, **104**, 813–820.
6 Warikoo, V., Godawat, R., Brower, K., Jain, S., Cummings, D., Simons, E. et al. (2012). Integrated continuous production of recombinant therapeutic proteins. *Biotechnol. Bioeng.*, **109**, 3018–3029.
7 Walther, J., Godawat, R., Hwang, C., Abe, Y., Sinclair, A., and Konstantinov, K. (2015). The business impact of an integrated continuous biomanufacturing platform for recombinant protein production. *J. Biotechnol.*, **213**, 3–12.
8 Godawat, R., Konstantinov, K., Rohani, M., and Warikoo, V. (2015). End-to-end integrated fully continuous production of recombinant monoclonal antibodies. *J. Biotechnol.*, **213**, 13–19.
9 Klutz, S., Magnus, J., Lobedann, M., Schwan, P., Maiser, B., Niklas, J. et al. (2015). Developing the biofacility of the future based on continuous processing and single-use technology. *J. Biotechnol.* doi: 10.1016/j.jbiotec.2015.06.388
10 Karst, D.J., Steinebach, F., Soos, M., and Morbidelli, M. (2016). Process performance and product quality in an integrated continuous antibody production process. *Biotechnol. Bioeng.*, **114**, 298–307.
11 Boedeker, B.G.D. (2001). Production processes of licensed recombinant factor VIII preparations. *Semin. Thromb. Hemost.*, **27**, 985–994.
12 Pohlscheidt, M., Jacobs, M., Wolf, S., Thiele, J., Jockwer, A., Gabelsberger, J. et al. (2013). Optimizing capacity utilization by large scale 3000L perfusion in seed train bioreactors. *Biotechnol. Prog.*, **29**, 222–229.
13 Lipscomb, M.L., Mowry, M.C., and Kompala, D.S. (2004). Production of a secreted glycoprotein from an inducible promoter system in a perfusion bioreactor. *Biotechnol. Prog.*, **20**, 1402–1407.

14 Choo, C., Tian, Y., Kim, W., Blatter, E., Conary, J., and Brady, C.P. (2007). High-level production of a monoclonal antibody in murine myeloma cells by perfusion culture using a gravity settler. *Biotechnol. Prog.*, **23**, 225–231.

15 Johnson, M., Lanthier, S., Massie, B., Lefebvre, G., and Kamen, A. (1996). Use of the centritech lab centrifuge for perfusion culture of hybridoma cells in protein-free medium. *Biotechnol. Prog.*, **12**, 855–864.

16 Kim, B.J., Oh, D.J., and Chang, H.N. (2008). Limited use of entritech Lab II centrifuge in perfusion culture of rCHO cells for the production of recombinant antibody. *Biotechnol. Prog.*, **24**, 166–174.

17 Shirgaonkar, I.Z., Lanthier, S., and Kamen, A. (2004). Acoustic cell filter: a proven cell retention technology for perfusion of animal cell cultures. *Biotechnol. Adv.*, **22**, 433–444.

18 Gorenflo, V.M., Smith, L., Dedinsky, B., Persson, B., and Piret, J.M. (2002). Scale-up and optimization of an acoustic filter for 200L/day perfusion of a CHO cell culture. *Biotechnol. Bioeng.*, **80**, 438–444.

19 Woodside, S.M., Bowen, B.D., and Piret, J.M. (1998). Mammalian cell retention devices for stirred perfusion bioreactors. *Cytotechnology.*, **28**, 163–175.

20 Goudar, C.T., Piret, J.M., and Konstantinov, K.B. (2011). Estimating cell specific oxygen uptake and carbon dioxide production rates for mammalian cells in perfusion culture. *Biotechnol. Prog.*, **27**, 1347–1357.

21 Ozturk, S.S. (1996). Engineering challenges in high density cell culture systems. *Cytotechnology.*, **22**, 3–16.

22 Karst, D.J., Serra, E., Villiger, T.K., Soos, M., and Massimo, M. (2015). Characterization and comparison of ATF and TFF in stirred bioreactors for the continuous production of therapeutic proteins. *Biochem. Eng. J.*, **110**, 17–26.

23 Goudar, C.T., Matanguihan, R., Long, E., Cruz, C., Zhang, C., Piret, J.M. et al. (2007). Decreased pCO_2 accumulation by eliminating bicarbonate addition to high cell-density cultures. *Biotechnol. Bioeng.*, **96**, 1107–1117.

24 Gray, D.R., Chen, S., Howarth, W., Inlow, D., and Maiorella, B.L. (1996). CO2 in large-scale and high-density CHO cell perfusion culture. *Cytotechnology.*, **22**, 65–78.

25 Villiger, T.K., Morbidelli, M., and Soos, M. (2015). Experimental determination of maximum effective hydrodynamic stress in multiphase flow using shear sensitive aggregates. *AIChE J.*, **00**, 1–10.

26 Neunstoecklin, B., Stettler, M., Solacroup, T., Broly, H., Morbidelli, M., and Soos, M. (2014). Determination of the maximum operating range of hydrodynamic stress in mammalian cell culture. *J. Biotechnol.*, **194**, 100–109.

27 Carta, G. and Perez-Almodovar, E.X. (2010). Productivity considerations and design charts for biomolecule capture with periodic countercurrent adsorption systems. *Sep. Sci. Technol.*, **45**, 149–154.

28 Angarita, M., Müller-Späth, T., Baur, D., Lievrouw, R., Lissens, G., and Morbidelli, M. (2015). Twin-column CaptureSMB: a novel cyclic process for protein A affinity chromatography. *J. Chromatogr. A.*, **1389**, 85–95.

29 Baur, D., Angarita, M., Müller-Späth, T., and Morbidelli, M. (2015). Optimal model-based design of the twin-column CaptureSMB process improves capacity utilization and productivity in protein A affinity capture. *Biotechnol. J.* **11** (1), 135–145.

30 Steinebach, F., Angarita, M., Karst, D.J., Müller-späth, T., and Morbidelli, M. (2016). Model based adaptive control of a continuous capture process for mAb production. *J. Chromatogr. A.*, **1444**, 50–56.

31 Ng, C.K.S., Rousset, F., Valery, E., Bracewell, D.G., and Sorensen, E. (2014). Design of high productivity sequential multi-column chromatography for antibody capture. *Food Bioprod. Process.*, **92**, 233–241.

32 Mahajan, E., George, A., and Wolk, B. (2012). Improving affinity chromatography resin efficiency using semi-continuous chromatography. *J. Chromatogr. A.*, **1227**, 154–162.

33 Krättli, M., Ströhlein, G., Aumann, L., Müller-Späth, T., and Morbidelli, M. (2011). Closed loop control of the multi-column solvent gradient purification process. *J. Chromatogr. A.*, **1218**, 9028–9036.

34 Krättli, M., Steinebach, F., and Morbidelli, M. (2013). Online control of the twin-column countercurrent solvent gradient process for biochromatography. *J. Chromatogr. A.*, **1293**, 51–59.

35 Konstantinov, K., Maria, G., Renato, N., Thrift, J., Chuppa, S., Matanguihan, C. et al. (2006). The "push-to-low" approach for optimization of high-density perfusion cultures of animal cells operating point. *Adv. Biochem. Eng. Biotechnol.*, **101**, 75–98.

36 Clincke, M.-F., Mölleryd, C., Samani, P.K., Lindskog, E., Fäldt, E., Walsh, K. et al. (2013). Very high density of CHO cells in perfusionby ATF or TFF in WAVE bioreactor™ - Part II: applications for antibody production and cryopreservation. *Biotechnol. Prog.*, **29**, 768–777.

37 Frame, K.K. and Hu, W.S. (1990). The loss of antibody productivity in continuous culture of hybridoma cells. *Biotechnol. Bioeng.*, **35**, 469–476.

38 Bailey, L.a., Hatton, D., Field, R., and Dickson, A.J. (2012). Determination of Chinese hamster ovary cell line stability and recombinant antibody expression during long-term culture. *Biotechnol. Bioeng.*, **109**, 2093–2103.

39 Mercille, S., Johnson, M., Lanthier, S., Kamen, A., and Massie, B. (2000). Understanding factors that limit the productivity of suspension-based perfusion cultures operated at high medium renewal rates. *Biotechnol. Bioeng.*, **67**, 435–450.

40 Ozturk, S.S. (1990). Loss of antibody productivity during long-term cultivation of a hybridoma cell line in low serum and serum-free media. *Hybridoma*, **9**, 167–175.

41 Wang, W. (2005). Protein aggregation and its inhibition in biopharmaceutics. *Int. J. Pharm.*, **289**, 1–30.

42 Ivarsson, M., Villiger, T.K., Morbidelli, M., and Soos, M. (2014). Evaluating the impact of cell culture process parameters on monoclonal antibody N-glycosylation. *J. Biotechnol.*, **188**, 88–96.

43 Brühlmann, D., Jordan, M., Hemberger, J., Sauer, M., Stettler, M., and Broly, H. (2015). Tailoring recombinant protein quality by rational media design. *Biotechnol. Prog.*, **31**, 615–629.

17

Integration of Upstream and Downstream in Continuous Biomanufacturing

Petra Gronemeyer, Holger Thiess, Steffen Zobel-Roos, Reinhard Ditz, and Jochen Strube

Clausthal University of Technology, Institute for Separation and Process Technology, Leibnizstr 15, 38678 Clausthal-Zellerfeld, Germany

17.1 Introduction

Biologically active components, such as monoclonal antibodies, offer a broad range of applications. They are established in diagnostics and therapies, including cancer, multiple sclerosis, or immunological diseases [1]. Treatment costs for patients are exceeding the financial capabilities of most healthcare systems around the world. Due to their global use and currently high product costs, the demand for antibodies at reasonable prices is constantly increasing resulting in the necessity for improvements and innovation of established manufacturing processes [2]. Biosimilars are another aspect in technology development and process optimization. The first generic antibodies have entered the market, impacting the cost/profit structure of companies. Their introduction to market has raised the pressure of developing cost effective and faster manufacturing processes in order to stay competitive [3]. Stratified medicine approaches require specific products for smaller patient populations, resulting in more products at smaller scale [1–5]. Increasing safety requirements of regulatory agencies increase the pressure on existing manufacturing processes in addition to increasing economic and political pressure as well [6,7]. However, it is not only necessary to improve process development and implement new unit operations [3,5], but also quality improvements of antibodies need to go along [3] with developing more efficient and economic processes [2].

In order to address this challenge, several improvements in USP and DSP have been pursued. In addition to shortened development timelines [7], platform technologies [1,8–12], high-throughput methods [13,14], statistically planned experimental designs (DoE) [15–17], and Quality by Design (QbD) approaches [18–21] have been implemented for process optimization in USP and DSP [6].

In USP, process efficiency and product concentrations have been increased significantly over the last two decades [1,2,5,22], resulting in antibody concentrations in the 10–13 g/l range by fed-batch processes [5,23,24] or even up to 25 g/

Continuous Biomanufacturing: Innovative Technologies and Methods, First Edition.
Edited by Ganapathy Subramanian.
© 2018 Wiley-VCH Verlag GmbH & Co. KGaA. Published 2018 by Wiley-VCH Verlag GmbH & Co. KGaA.

l via modified perfusion processes [22,25]. The focus of USP development lies on high product titer, high productivity, and defined quality [5,23,26]. Not sufficiently recognized and subsequently addressed in the discussed optimization concepts is the role of a nonlinear increase of product impurities, and even worse, the increasing bio/chemical similarity of impurities with the target compounds, which result in much more complex and expensive downstream operations.

In comparison, DSP development has focused on yield, productivity, purity, and process capacities [6]. The optimization procedure concentrates on an increase of separation capacity by expanding existing facilities, optimizing existing processes, and developing new technologies [1,6]. New methods applied to process development include the integration of modeling and simulation and the use of mini-plant facilities, in addition to the methods mentioned before [1,13,27]. Innovative technologies applied to antibody purification include the use of aqueous two-phase extraction (ATPE) [28–33], precipitation [34–36], and membrane-based adsorption [37–39]. Other notable unit operations are crystallization [40], flocculation, and magnetic separations.

A closer view on USP optimization shows that product titers can be increased without additional costs. The optimized USP manufacturing takes place in the same facilities that were used for "lower titer" processes before. The resulting product batches enter downstream processing containing 15–100 kg mAb/batch instead of the earlier 5–10 kg mAb/batch [23,24]. This increase in productivity does, however, not translate directly into a similar productivity increase in downstream processing. The usage of chromatography resins in particular is limited by the mass of product to be purified, not by its volume. In order to handle higher concentrated product volumes, the processing time increases as well as material consumption and costs. As a consequence, this limitation of technology and equipment affects the cost of goods as well as facility costs due to a necessary increase of space and volumes for downstream (DSP) unit operations and periphery [41].

Changed compositions and concentrations of impurities, especially of host cell proteins (HCP) present another problem generated by USP development processes [42–45]. These changes are created by varying cultivation conditions. The HCP spectrum becomes more similar to the product regarding characteristic properties like pI, molecular weight or hydrophobicity, increasing process time, material consumption, and costs due to a more difficult separation process. In particular, the increase in processing time is critical, because at the higher titers the degradation of target molecules by proteases becomes critical and obviously counterproductive. New methods, technologies, and alternative as well as faster processes are required. They include, for example, the optimization in chromatographic separations by introducing new separation media [3,46], continuous [47] or integrated chromatographic processes for complex mixtures [48]. Furthermore, optimization trends in downstream processing address the development of new nonchromatographic methods, including liquid–liquid extraction [28], precipitation [49], and crystallization [40]. These trends aim at partially reducing or even completely eliminating chromatographic separation steps. Another option for process optimization consists in designing new facilities that are more flexible, set up in smaller scale [50], and operated continuously [51].

Development strategies for platform processes based on continuous operations have already been published [12,52]. Trends concentrate on new cell retention devices and modifications of operation modes in USP, the implementation of single-use technologies, and new continuous operation modes for conventional and innovative technologies in DSP. One example of device development in USP consists in the alternating tangential flow technology (ATF) in modified perfusion mode or as concentrated fed-batch process [53–55].

17.2 Background on Upstream Development in Continuous Manufacturing

In USP, continuous processes are used for unstable products, in particular, but they are also applied in high-density seed bioreactors and in cell bank manufacturing [6,53,54]. The main advantage consists in the short retention time of the product in the bioreactor while the cultivation time is extended and the cell number is high as well as the productivity [53,54,56–58]. As consequence, the production time increases and the product quality is preserved [6]. On the other hand, large harvest volumes require a high consumption of medium during the cultivation process and need to be purified continuously [53,54,57].

The process performance of perfusion processes is determined by the cell retention devices used [53–56,58,59]. Cell retention devices are filtration or acceleration based and need to be robust [54,56] and scalable [55]. Devices, which are easy to scale up, are mostly based on filtration, gravity settling, and centrifugation [6]. Gravity-based cell settlers, spin-filters, centrifuges, alternating tangential flow filters, vortex flow filters, acoustic settlers, and hydrocyclones are commonly used in small-scale applications [6,55,56]. As an example for the scalability of a device: internal spin-filters are used in production volumes of up to 500 l, and external ones up to 1000 l. On a larger scale, mostly gravity settlers are used [6,58]. An overview of industrially applied retention devices in continuous antibody manufacturing is published in Ref. [58] and another overview on principles of sedimentation, centrifugation, and filtration as cell retention devices is provided in Ref. [57].

One of the newest developments on cell retention devices is the ATF System (Refine Technology) [22,53,55,58] that is based on tangential flow filtration with an alternating direction of the flow of cultivation broth. A diaphragm pump pumps the broth alternately out of the bioreactor and back inside without additional shear stress. The back flush reduces possible fouling effects [6,53,54]. This system is supposed to be scalable and can be used as single-use device [54,58]. The main advantage consists in the generation of cell densities up to 2.14×10^8 cells/ml [54] as well as product titer of up to 25 g/l by concentrated fed-batch processes [6,22]. Further studies of generated impurities have not been published up till now.

Product quality and stability, existing facilities, and experiences most often determine the choice of operation mode in biomanufacturing processes. Process development and optimization procedures focus – both, fed-batch and continuous process development – on defining optimal operation parameters, including temperature shifts, gas exchange, shear stress, feeding strategy, duration of the cultivation, and, if necessary, perfusion rate [1,9,57,60–63]. USP development aims for high cell numbers, defined product quality in the bioreactor, high titer, and an extension of fermentation duration [9,62,63]. Development activities are mostly performed according to statistically designed experimental plans [6].

These development and optimization activities led to an increase in monoclonal antibody concentration from 50 mg/l in 1986 to 5–20 g/l today [22,24,64]. During the development procedures, the type and concentration of impurities as well as their impact on DSP is often neglected [1,6,43]. Impurities are considered if they are highly toxic or if changes in their metabolic routes result in a higher production of the desired product [6].

17.3 Background on Downstream Development in Continuous mAb Manufacturing

The focus of DSP development lies on yield, productivity, purity, and process capacity. Expansion of existing facilities and optimization of established unit operations result in increasing process efficiency of the respective unit operation [1,6]. The traditional manufacturing process of monoclonal antibodies is shown in Figure 17.1. New technologies are developed and optimized to present an alternative to established unit operations in order to increase process efficiency, reduce costs, and/or maintain product quality [6]. New methods for process development include the establishment of the following:

- Platform technologies [1]
- High-throughput methods with approaches based on QbD [13]
- DoE-based experimental optimization [27]
- Integration of modeling and simulation of unit operations [6]
- Use of mini-plant facilities [6].

New separation technologies need to be significantly more efficient and less expensive because their implementation on an industrial scale will require high investments in process development, scale up, and validation [6,23]. In DSP manufacturing, flexible facilities and faster turnaround times allow companies to manage an evolving portfolio of ever changing product numbers, volumes, and types [6,52]. Single-use technologies [1,66], continuous process strategies [51,67], and decentralized manufacturing concepts [50] are worked on in order to support these changing requirements [6]. Continuous unit operations, for example, continuous chromatography, are discussed in detail in Chapter 13.

Despite limited experience with continuous protein purification at manufacturing scale [68], many companies go from batch to continuous processes to improve their flexibility and cost of goods without losing their operational excellence [52]. Other advantages of this transition are standardization, easy scale up, and a more consistent product quality [52]. Until recently, necessary unit

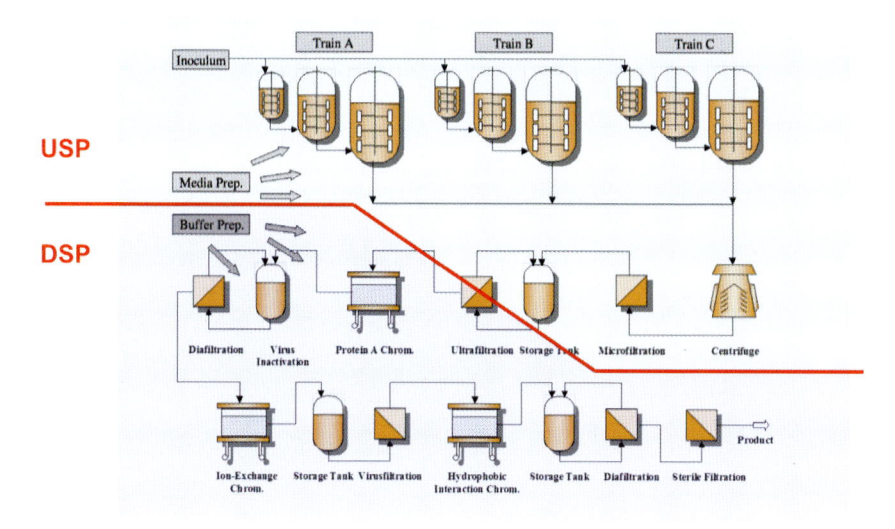

Figure 17.1 Two worlds of a generic manufacturing process for monoclonal antibodies. (Reproduced with permission from Ref. [65]. Copyright 2005, Elsevier.)

operations were not available for industrial continuous manufacturing, but now continuous chromatography systems are available in small and pilot scale [68]. Because of this, their implementation will still take some time and make evaluations necessary to be accepted completely in biopharmaceutical industry. Completed transitions from batch to continuous demonstrated that this conversion required the design of new fit-for-purpose unit operations, processes, and facilities [68].

Continuous chromatography systems as well as continuous concepts of aqueous two-phase extraction or precipitation – as nonchromatographic alternatives – are already known and, in part, even implemented in industrial manufacturing processes. Other unit operations such as continuous viral inactivation and removal and filtration processes are required in continuous mode as well, but they are not available right now to be implemented in fully integrated continuous processes [68].

17.4 Challenges in Process Development

17.4.1 Impact of Changing Titers and Impurities on Cost Structures

Conventionally, process development for active pharmaceutical ingredients produced by fermentation is divided into two separate fields – upstream and downstream processing – see Figure 17.1. Both differ significantly regarding their optimization foci. In USP, the focus is on high product titer, high productivity, and a defined quality, whereas the focus of DSP development is on yield, purity, and process capacity [6]. In process development, the optimization of USP and DSP is handled separate from each other and the needs of one manufacturing part do rarely fit the requirements of the other. In consequence, these optimization foci are partly antagonistic to each other resulting in new problems and challenges. Therefore, foci of traditional process development methods need to change. The optimal process is not always the road to maximum product titer after fermentation. Instead, the optimal process distinguishes itself by the best cost to product ratio, resulting in a new focus on the downstreaming of by-products.

In times of a lower achievable antibody titer, larger facilities were constructed and are still in use. In combination with higher product titers in upstream processing, the resulting product batches enter downstream processing containing 15–100 kg mAb/batch instead of the earlier 5–10 kg mAb/batch [23]. The increase in productivity per volume in cell culture technology shifts cost distribution of the overall manufacturing process toward downstream processing (Figure 17.2) [41,69,70]. This increase in upstream productivity does not translate directly into a similar productivity increase in downstream processing. The usage of chromatography resins in particular is limited by the mass of product to be purified, not by its volume. As a consequence, this limitation of technology and equipment affects the cost of goods as well as facility costs due to a necessary increase of space and volumes for downstream (DSP) unit operations and periphery [69].

Grote and coworkers [70] presented studies of cost distribution in downstream processing depending on the product concentration. They calculated costs of DSP in industrial scale belonging to three different antibody concentrations assuming a changing spectrum of impurities. If USP development would stop at 0.1 g antibody/l, 33% of the total downstream costs per batch are caused by building, 22% by labor, and 9% by auxiliary components such as buffer, AC resin, IEX resin, HIC resin, and membranes. The rest of the costs are distributed between QA/QC, waste treatment, CIP, equipment, and engineering. If the product titer increases up to 1 g/l in USP development, a significant change can be found in cost distribution. The costs for buildings and labor are reduced to 19 and 14% per batch in DSP, whereas the percentage of auxiliary components rises up to 33%.

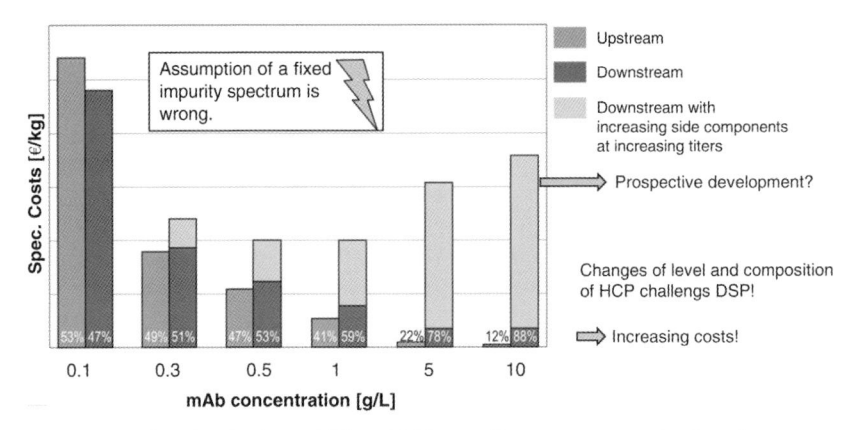

Figure 17.2 Cost distribution in USP and DSP regarding a constant and a changing spectrum of impurities and their effect on cost distribution. (According to Ref. [70].)

This increase is mainly caused by an increased use of AC resins and membranes but also by a rise in IEX and HIC resins. A scale up of product titer by a factor of 10 results in another shift of costs. Due to higher usage of different chromatographic resins, membranes, and buffers, their contribution to DSP costs rises up to 64%. During the process development, absolute costs for buildings, labor, QA/QC, equipment, and so on can be considered as constant. Consequently, an increase of costs in auxiliary components, which are needed in addition, result in an increase of the overall process costs [9,65,70].

These studies indicate the considerable shift of proportionate production costs toward DSP and a possible cost increase for the overall manufacturing process in case of higher titer in USP development of future processes. This development would not affect established processes because their fermentation conditions and, therefore, their impurity profiles, are constant.

Based on the assumption of a stable spectrum of impurities, the overall cost of goods (COG) of a manufacturing process decreases proportionally with increasing product titers. Though, downstream costs dominate the overall production costs with increasing titer. Due to the fact that the composition of impurities changes with increasing titers, this assumption has to be revised and the costs of DSP are significantly higher (see light bars in Figure 17.2) [70]. This increase in costs will be stimulated by an increased usage of auxiliary components in product purification. Considering the trend toward further increasing product titers in future processes, the trend toward increasing costs of downstream processing is likely to continue too [65,70]. Increasing titers of future manufacturing processes are going to result in a changed impurity spectrum with higher amounts of impurities combined with higher similarity to the target compound. Therefore, they are more difficult to separate from the product. A higher amount of HCP with changed property distribution will cause an increase in separation effort, and therefore, in costs.

17.4.2 Impurities as Critical Parameters in Process Development

Impurities are components of the host cell and medium, product variants or isoforms that have to be separated from the product substance itself. They consist of product and process-related components as well as contaminants [6,45,71].

Product-related components include molecular variants of the desired target molecule [71]. Such molecular variants can be precursors, degraded products [72,73], aggregates, such as dimers or multimers, or product variants by different posttranslational modification [73,74]. They are produced during manufacturing process or storage. These product variants can influence structure and function of the product molecule [45] and possess different properties compared to the product regarding activity, efficacy, or safety aspects [6].

The next class of impurities includes process-related components. These are, for example, cell components, media components, chemical additives [6,45,71,75], or leachables such as protein A in the DSP [45]. Cell components include HCP, DNA, and RNA, among others, whereas residual media or digested components are composed of carbohydrates, amino acids, vitamins, salts, and lipids, among others [6].

Contaminants are a class of impurities too, and should be strictly avoided. They include all introduced materials that are not part of the validated manufacturing process. Examples are (bio-) chemical materials or microbial species [6,71].

The most common analytical methods to identify and quantify mAb products and general impurities are listed in Table 17.1.

17.4.3 Host Cell Proteins as Main Problem in Process Development

The main sources of impurities from host cells are so-called HCP. HCP are a very complex group of proteins. Within one process, HCP distinguish themselves by a

Table 17.1 Overview of industrial relevant analytical methods for characterization of fermentation broth.

Parameter	Method	Limit of quantification[a]	References
Product			
Identification	Protein A	1 ppm	[76]
	IEF	0.005 pH	[77]
Quantification	ELISA	88 ng/ml	[78]
	Protein A	1 ppm	[76]
Biological activity	Surface plasmon resonance	200 pM–40 nM	[79]
Structural integrity	MS	15 ppm[b]	[80]
Purity	RP-HPLC	0.1 µg	[76]
	SEC	0.1 µg	[76]
Product-related impurities			
Aggregates/fragments	SEC/IEX/HIC	0.1 µg	[76]
Isoforms	IEF	0.017 mg/ml	[81]
Process-related impurities			
HCP	Gel electrophoresis	< 1 fmol, 14 kDa	[81]
	Coomassie	125 fmol	[82]
	Silver	31 fmol	[82]
	CyDyes	0.03–0.08 ng/protein	[45]
	ELISA	1 ppm	[76]
DNA	QPCR	2 pg/ml	[76]
Protein A	ELISA	1 ppm	[76]
Organic acids	AQC/HPLC	0.11 pmol/injection	[83]
Vitamins	RP-HPLC	0.1 µg/ml	[84]
Lipids	LC-MS/MS	10 fmol	[85]
Saccharides	Fluorescence	10 fmol	[86]

a) The exact limit of quantification depends on the protocol which is applied, on the sample, and on the local equipment.
b) Range of mass accuracy.

very broad variety of physical and chemical properties. They differ strongly in their structure, molecular mass, isoelectric point, and hydrophobicity [6]. Additionally, the concentration, composition, and the individual properties of HCP vary throughout the fermentation process resulting in changed impurity properties depending on the harvest time. This property variety can easily cause problems in product purification [6].

Some research groups have already taken a closer view on problematic HCP and their potential removal in DSP. The exact composition of HCP depends on multiple factors [6]:

- Host organism [44]
- Cell clone [87]
- Protein of interest [88]
- Route of metabolic expression [44,89]
- Viability [42,87]
- Stage of cell culture
- Process conditions in fermentation
- Harvest conditions [42]

It was shown that the HCP composition depends strongly on metabolic pathways. These are characteristic for the chosen cell line and clone and result in a characteristic pattern of proteins [44]. If pathways are changed to optimize the production routes of the product, production routes of secondary metabolic products, such as HCP, are changed as well [44,88,89].

However, other USP process parameters influence the production of HCP as well. Age and viability of the cell culture impact the HCP composition and concentration strongly [42]. The protein production of cells changes during the fermentation process because of changes in the bioreactor environment, metabolism, and number of cells. Therefore, other proteins in higher concentrations are produced toward the end of the fermentation process compared to its beginning [6,42]. In addition to age and viability, process parameters, such as temperature, aeration, feeding strategy, medium composition, cell culture duration, and harvest conditions, influence composition and concentration of HCP too [44]. These process parameters can result in changes of the HCP level by a factor of 0.5–7 [44]. According to these data, the time of harvest determines the composition and concentration of host cell proteins entering the DSP. A change of harvest time and conditions might be able to improve the manufacturing process because of easier to separate impurities in spite of a potential reduction of production time in USP, and maybe a slight decrease in yield [6].

In DSP development, several studies were accomplished to monitor HCP through different chromatographic separations and protein A chromatography [69,90,91]. The impact of different harvest operations on the HCP profile was described in Ref. [43]. Early DSP operations impact the HCP profile and the relative abundance of particular proteins throughout product purification [6].

The high impact of upstream and early downstream operations on concentration, composition, and property distribution of HCP presents high potential to optimize overall mAb manufacturing processes. A change of USP development focus could lead to a reduction of HCP or change existing production routes toward HCP that are easier to separate. Critical HCP could be identified and their production could be influenced [6].

17.4.4 Regulatory Aspects

In order to ensure high product quality and patient safety, risk assessment and scientific knowledge are used in process development. Control strategies have to be developed to identify critical quality attributes (CQA) that might be able to affect patient safety and to define acceptable parameter ranges. According to the guideline Q11 of the International Conference on Harmonisation (ICH), HCP are defined as critical quality attributes and have to be controlled throughout the manufacturing process [45,71].

HCP-specific tests and correlating acceptance criteria regarding the HCP concentration in drug substance and drug product are based on data from scientific experience, process development studies and documentation, pre-clinical and clinical studies, and manufacturing process history [45]. These data should be used to develop a suitable risk assessment and establish parameter ranges of HCP throughout the process. In biopharmaceutical industry, such risk assessments are applied to set limits on HCP concentrations [45,89], although there is no general guideline that defines the maximum limit of HCP [45,92].

Quality by Design process development strategies are a further method to ensure process understanding and process robustness and, therefore, product quality and patient safety. It is recommended by current regulatory agencies, the ICH, and Parenteral Drug Association (PDA) [45,71]. These process development strategies involve data from early process development up to process validation studies. Critical impurities can be identified and clearance strategies can be adapted based on these data [45,93].

17.5 Trends and Integration Approaches

Increasing demands for high-quality biologics and the trend toward stratified medicine challenge process development in USP and DSP. The total number of products will increase but the amounts of a substance to be produced will drop. Regulatory agencies press more for enhanced quality and heightened process understanding while health care systems press for lower product costs.

Possible solutions to solve rising challenges include new technologies, new materials, and methods in USP and DSP. New tools in process development and a heightened integration between these factors and different parts of process development and manufacturing are going to be established as well. However, new separation technologies need to improve process efficiency and manufacturing costs significantly in order to be established on an industrial scale. Otherwise, associated risks would be too high [24]. Development of stratified and personalized medicine will require a reduction of process volumes, higher flexibility of the facility, and faster turnaround times. This would benefit the implementation of disposables in manufacturing [1,66]. Continuous processing [51,67] and decentralized manufacturing concepts, like manufacturing in containers [50] are going to be useful as well.

An implementation of disposables leads to lower capital and operational costs and increase the production flexibility. In biopharmaceutical production, the manufacturing process could be set up rapidly for multiple products [94]. Advances in single-use technologies offer the possibility to implement complete single-use processes, including devices from storage bags, bioreactors, and chromatography units [94]. In DSP, recent advances achieved product-dedicated fixed columns and fully disposable columns [66]. Among others, membrane adsorption technologies, single-use moving bed, or countercurrent chromatography are likely to be implemented as single-use options as well [94]. In membrane science, disposable membrane cassettes present an alternative to flow-through chromatography steps [6,66]. Interactions of plastic bags and process components are still under investigation as well as the possibility of leachables and extractables [66,94]. Another disadvantage consists in the limit of scale. This affects the upstream processing and requires alternative process strategies for traditional fed-batch processes [6].

Continuous processing is one alternative process strategy. In USP, this process technology is already well established. Newest developments on cell retention devices resulted in the alternating tangential flow technology (ATF). This device allows conventional perfusion processes, which result in very high cell numbers at high viability, and concentrated fed-batch processes that concentrate the product in the bioreactor [53,54,58]. The combination of ATF, continuous DSP, and disposable technologies enable the implementation of small-scale manufacturing processes of low investments costs and risks [6].

One of the first development strategies for a platform process based on continuous operations was presented in Ref. [12]. This approach includes membrane adsorbers and operates on small scale, manufacturing biopharmaceuticals but no antibodies. The problem of a transfer to antibody purification consists in the limited capacity of membrane adsorbers [6]. Continuous ATPE [33,95,96], centrifugation [97], or chromatography such as MCSGP [98] or iCCC [47] possess higher capacity and present useful alternatives for antibody purification [6].

Individualized biologics and personalized medicines can probably be produced in small scale using single-use technologies due to small required volumes. Already established facilities at large scale (>10 000 l) will still be used, if possible, but new facilities will be added at smaller scale [6]. Modular facilities provide some advantages for this manufacturing concept. Unit operations can be housed in container-like transportable clean rooms. Doing that, the whole manufacturing process is transported, constructed, and made operational rapidly [94].

Other trends in process development concentrate on the integration of different downstream operations. Examples are the integration of ion exchange and hydrophobic interaction chromatography that permits the use of only one buffer system for both operations. This could substantially widen the selectivity window [6,48]. Further research activities regard nonchromatographic processes such as membrane adsorbers [37,99], crystallization [40,100], precipitation, or aqueous two-phase separation [28,41,95,101]. In DSP development, the application of predictive process design is already established, results in cost reduction and improves the success rate of commercial viability [102]. Equivalent optimization approaches for USP development are in progress. One example consists in the approach to model scale up and glycosylation behavior in USP of mAbs [103].

Another integration approach concerns the optimization of USP and DSP as one manufacturing process and not the optimization of single unit operations [6]. Foci of optimization procedures are product concentration, aggregates, and product quality. Other parameters can be turbidity or DNA content, depending on the individual process. Other impurities are rarely considered in process development. This is going to be a problem since process development activities to increase product titers often also result in changes of concentration and type of generated impurities. These are more similar to the product and, therefore, more difficult to separate. The new optimization approach takes these impurities into account working on the question if it would be better to produce less product in order to change the impurity profile. Such a change could be the generation of less impurities or only impurities that are easy to separate. This could result in a less expensive, yet more stable manufacturing process. The integration of upstream and downstream processing could be a first useful tool for total process optimization [6].

17.6 Methodical Approach of Integrating USP and DSP Regarding Impurity Processing

In this chapter, we present a methodical approach to integrate USP and DSP development. It is based on an iterative procedure of the process development of USP and DSP using design of experiments (DoE) in USP and physicochemical models in downstream DSP. The approach is schematically presented in Figure 17.3.

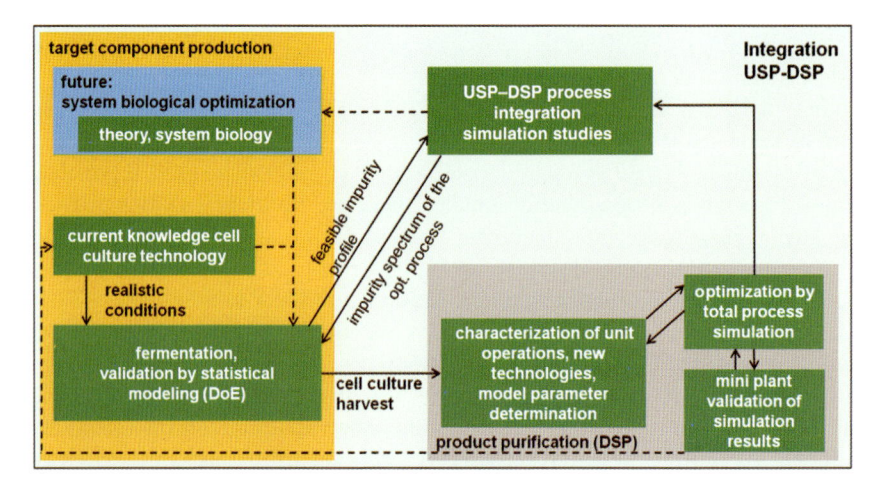

Figure 17.3 Schematic presentation of the methodical approach to integrate upstream and downstream processing. (Reproduced with permission from Ref. [104]. Copyright 2016, Elsevier.)

17.6 Methodical Approach of Integrating USP and DSP Regarding Impurity Processing

During USP development, different cultivations are carried out to optimize and select host cell and cell clone, develop media, optimize process parameter and operation mode in the bioreactor, and select the cell retention device. The different parameters are optimized by statistically experimental plans (DoE design). Foci of this development process are cell growth, product titer, product quality, and – in addition to the state-of-the-art foci – HCP concentration and composition [104].

The fermentation broths, produced in this context, do not need to be discarded. Instead they can be used for developing, investigating, and optimizing new, but also already established DSP unit operations. These unit operations include different kinds of chromatography and membrane operations as well as aqueous two-phase extraction, precipitation, flocculation, or crystallization. The evaluation of different broths, containing different compositions and concentrations of HCP, enables the identification of "good" (easy to separate) and "bad" (difficult to separate) HCP. The unit operations can be classified according to their ability of separating different kinds of HCP. For modeling and simulation of DSP operations, model parameter can be determined by these experiments as well. Such simulation studies are going to reduce experimental efforts, avoid redundant use of fermentation broth, and provide an optimized purification process. The results of these integration and simulation studies provide a feedback to USP development by indicating which HCP causes additional problems within the purification process. This can be taken into consideration regarding further development steps [104]. Figure 17.4 provides a less schematic and more experiment-oriented overview on this integration approach.

One example was carried out and is presented later in case study 1. In that case, USP development started with media development in shake flasks. Due to different media compositions, the resulting HCP concentrations and compositions are significantly different. These broths were used for a comparison of different harvest operations (centrifugation, filtration, and ATPE). Critical HCP can be followed throughout the different unit operations and provide important

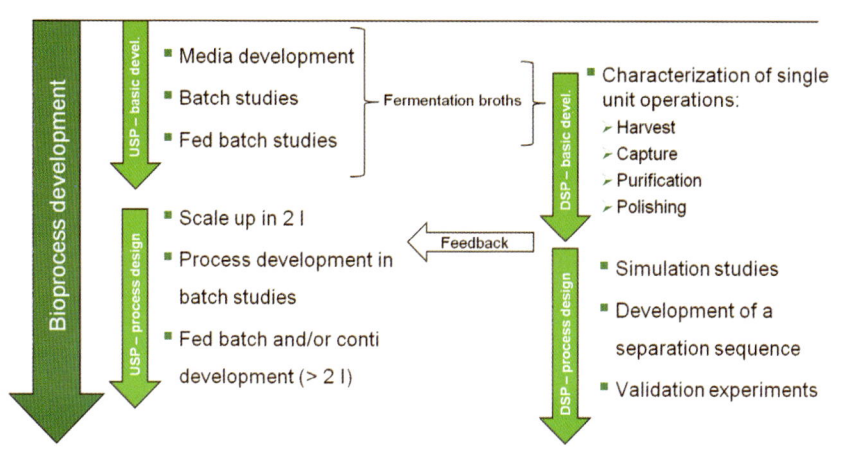

Figure 17.4 Experimental presentation of the methodical approach to integrate upstream and downstream processing.

information for USP development. This iterative approach takes place throughout the whole development process [104].

The results of these development studies allow a characterization of different impurity profiles regarding their separability from the product. Based on this information, cultivation conditions can be optimized in order to create an improved ratio of product to impurity components. Validation experiments are to be carried out in mini-plant scale. Future integration processes of USP and DSP might also include system biological approaches to optimize new host and production organisms toward an optimal manufacturing process. At the end of this development process, the manufacturing process is reliable, robust, and distinguishes itself by an optimal product-to-cost ratio [104].

17.6.1 Case Study: Influence of Media Components on Impurity Production

In this case study, the influence of single media components on the production of HCP was investigated in order to determine the possibility to control or influence this HCP production by medium adaptation. A DoE screening regarding 44 single components was carried out. Based on these results, a second fractional factorial DoE plan was carried out optimizing the most influential components. A third full factorial DoE design was used to fine-tune single concentrations [104].

The cultivations were performed in shake flasks. After cell harvest, 2D gel electrophoresis gels were run to determine the HCP profile. In Figure 17.5, a selection of the most differing HCP profiles is presented.

The general pattern of proteins was similar. The general range of molecular weight was from low to middle and a pI range from slightly acidic to neutral. Cultivations of similar IgG/protein ratio (S4, S17) produced significantly different HCP. A cultivation (S18) resulting in a high product concentration and an IgG/protein ratio of 50% contained in a high number of different HCP. These were

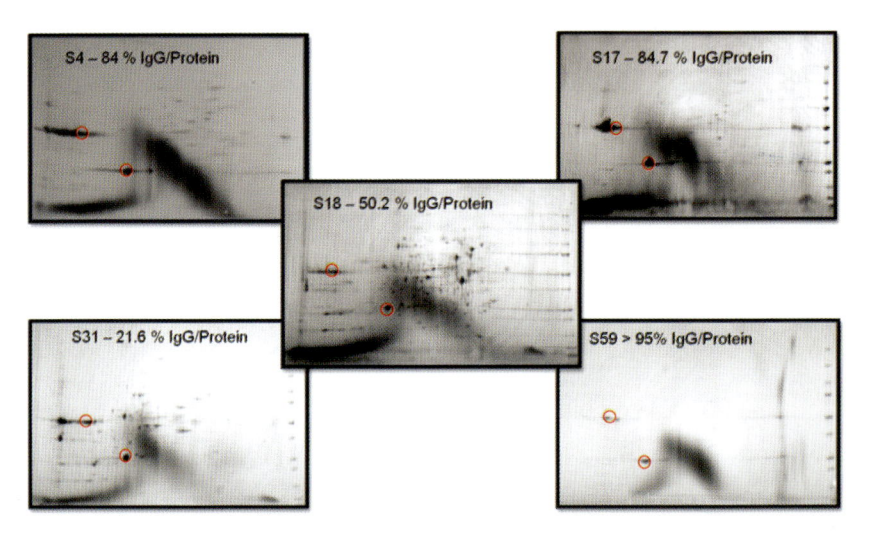

Figure 17.5 Resulting HCP profiles from cultivations in slightly different media. (Reproduced with permission from Ref. [104]. Copyright 2016, Elsevier.)

distributed over a broad pI range (4–9). The cultivation S31 also showed a good growth behavior, but a worse IgG/protein ratio. The HCP show a shift into the neutral to basic pI range. In DSP, this might be challenging because the properties of the HCP become more similar to the product's properties. Other cultivations also demonstrated a better cell growth and good IgG production, but also resulted in higher HCP concentrations of less single proteins. The HCP became more similar in their properties to the product molecules. The results of this screening demonstrated the strong influence of single-media components on the generation of HCP. Not only the component itself but also its concentration and combination with other components exhibited a certain influence [104].

In short, it is possible to generate a shift of HCP production toward less protein production without a loss in antibody concentration. In these cultivations, the HCP concentration could be reduced by up to 54%. Higher HCP improvements resulted in a significant reduction of IgG concentration by 29%. In short, composition and concentration of HCP are strongly influenced by cultivation parameters, impacting the separation efficiency of downstream processes [104].

Based on these data, 15 components out of 44 had a significant positive effect on cell growth, IgG production, HCP minimization, and an improved IgG/protein ratio. Ascorbic acid, glycine, and folic acid improved the HCP concentration. Arginine, asparagine, isoleucine, and thiamine had a positive influence on cell growth without a bad influence on HCP or IgG concentration. The product concentration was influenced in a positive way by phenylalanine, choline chloride, folic acid, and riboflavin, without negative effects on HCP. The components glucose, insulin, $MgCl_2$, $CuSO_4$, and $ZnSO_4$ improved growth and production behavior, too [104].

After two additional optimization rounds, a cultivation using the best medium composition led to an increase of product concentration by a factor of up to 2.5, of maximum cell number up to 130%, and of the ratio of IgG/proteins of at least 10%. The HCP profile was shifted significantly toward more acidic pI values, less similar to the product's properties, this way much easier to be removed [104].

17.6.2 Case Study: Influence of Harvest Operations on Impurity Production

The broths from earlier media development were used to investigate and evaluate the impact of early DSP operations on the HCP profile. In this case, centrifugation was chosen as reference method and compared with filtration (cellulose acetate, 0.45 μm) and a PEG400-phosphate-based ATPS. Four different broths were harvested using each of these unit operations. The IgG and protein concentrations in the supernatant after centrifugation were defined as 100% [104]. The results can be viewed in Figure 17.6. One broth is marked by one color. The left bar of one color represents the results of ATPE for IgG yield, the right bar the results for microfiltration.

The yield after filtration processes ranged between 86 and 96%, whereas yields of IgG within the top phase were between 92 and 99% compared to the centrifugation results. In contrast, the yield of total protein differed more between the harvest operations. The protein concentration included HCP as well as IgG. The analysis of the protein content showed that no proteins were reduced by

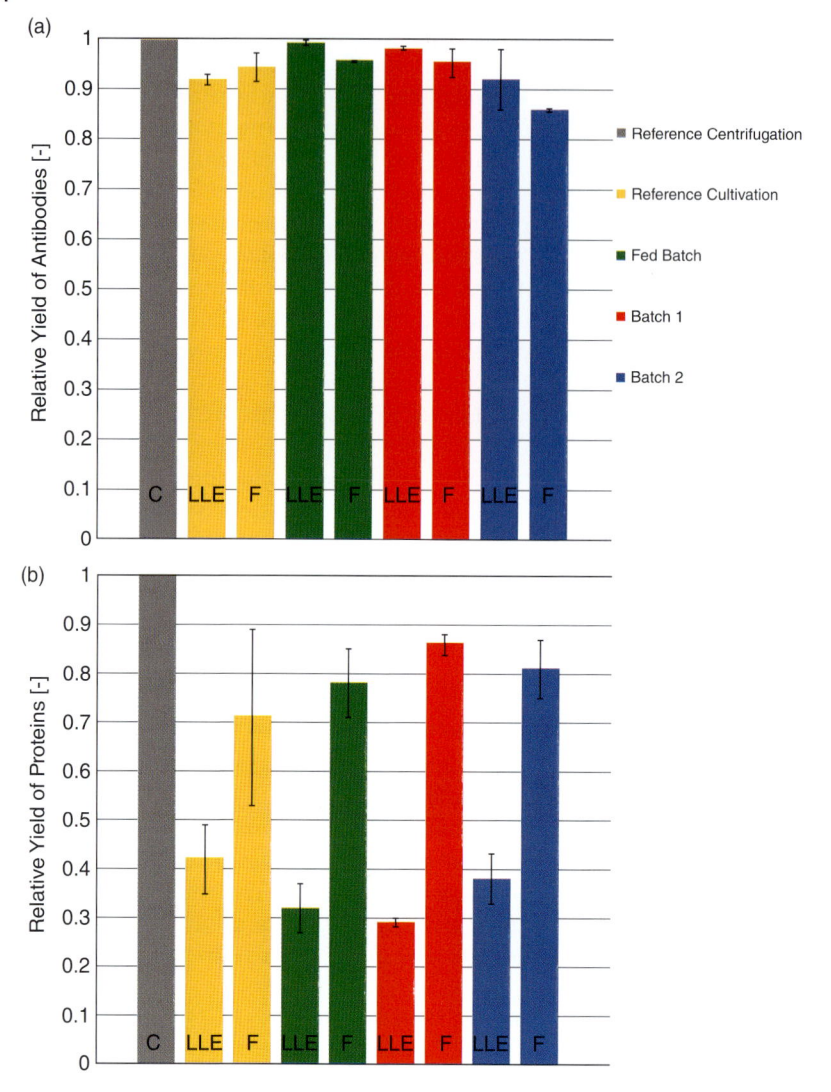

Figure 17.6 (a) Relative yield of antibodies in the filtrate (filtration) and top phase (ATPE) referred to the antibody content in the supernatant after centrifugation.; (b) Relative yield of proteins (including IgG) in the filtrate (filtration) and top phase (ATPE) referred to the antibody content in the supernatant after centrifugation. (Reproduced with permission from Ref. [104]. Copyright 2016, Elsevier.).

centrifugation and, therefore, were significantly higher than the protein contents in the filtrate and top phase. Microfiltration led to a decrease of proteins, probably by adsorption at the membrane. The most significant protein reduction was reached by ATPE for all different fermentation broths. This strong reduction was achieved without a significant loss of IgG. The top phase contained 29–41% of proteins of the centrifuged sample. ATPE always separated more proteins from IgG than filtration and centrifugation but there is a significant influence of the

17.6 Methodical Approach of Integrating USP and DSP Regarding Impurity Processing | 497

SEC diagram

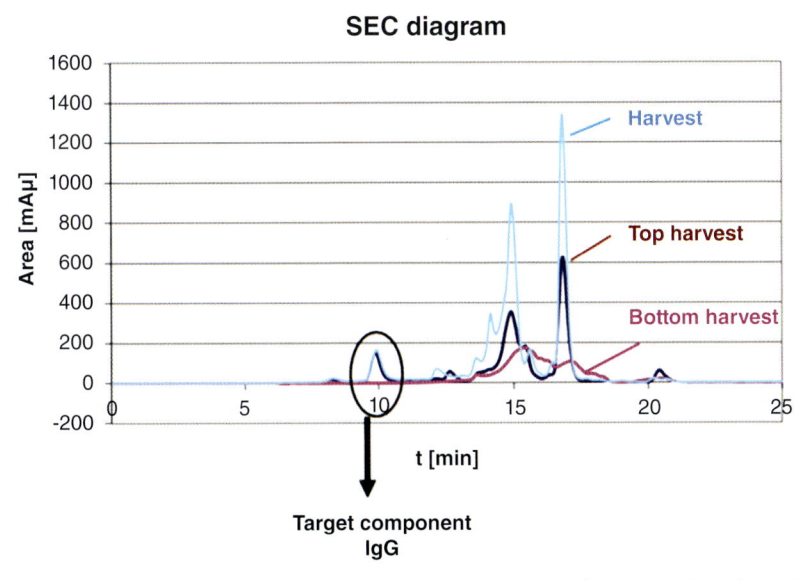

[Chart showing Area [mAµ] vs t [min], with curves labeled Harvest, Top harvest, and Bottom harvest; Target component IgG marked at ~10 min]

Figure 17.7 SEC diagram of top and bottom phase compared to a centrifuged sample [105].

HCP compositions on the degree of transferred proteins visible. The ATPE harvests resulted in a purity increase by a factor of 2.5–3.5 without any optimization – depending on the composition of broth, a 1D gel electrophoresis check supported these results. The top phases of different broths contained less low molecular weight proteins than filtrates and supernatants [104].

An example of the composition of centrifuged broth, top phase, and bottom phase, analyzed by size exclusion chromatography, can be seen in Figure 17.7.

DoE-based experiments resulted in an ATPS that achieved a yield of about 100% (up to 3 g/l), low HCP concentrations, and a cell separation efficiency higher than 95%. Purities depend on the composition of the broth and can reach higher levels than 58% [104]. The results of an ATPE-based cell harvest process are presented in Table 17.2. Three different cultivations were carried out and achieved different mAb and HCP concentrations. In the top phase of the ATPS, IgG and HCP concentrations were increased in a smaller top phase volume compared to the broth's volume. In spite of the apparent increase of HCP, the amount of HCP was significantly decreased according to the ELISA results [106].

The screening strategy above, laid out in some detail, shows clear indication of the significant potential for enhancing process selection and operating conditions over a large variety of unit operations with a direct impact on process-relevant parameters in USP and their direct impact on DSP. Whether this might become a reference for sound process assessment and decision-making for regulatory inspection is up for consideration [106].

17.6.3 Nonchromatographic Continuous DSP Operation

By taking the results of the case studies into account, the possibility of a chromatography-free/reduced downstream process of monoclonal antibodies

Table 17.2 Comparison of the purification effect of ATPE for different broth's compositions.

Parameter	Batch	Fed-batch 1	Fed-batch 2
mAb concentration (g/l)	0.2	0.3	0.4
Yield (%)	>99	>99	>99
HCP after centrifugation (g/l)	0.021	0.037	0.030
HCP after ATPE (g/l)	0.024	0.043	0.047
Reduction of HCP (%)	43.2	42.9	21.2
Increase of IgG/protein (%)	93.7	129.7	39.8

Source: According to Ref. [106].

does not seem to be impossible. A consideration of impurity production would ease the way for the implementation of alternative technologies.

In continuous DSP, there are different concepts of continuous chromatography, which are discussed in Chapters 3 and 13 in more detail. Therefore, we will take a closer look at already existing concepts of continuous ATPE/ATPS and precipitation, and consider possible scenarios to combine/integrate these unit operations.

17.6.3.1 ATPS

Aqueous two-phase systems (ATPS) present an attractive alternative for the capture of human antibodies from complex cell culture media [95]. In case of proteins, ATPS show no denaturation processes [107] due to a high water content of 70–90% (w/w) and low interfacial tension of these systems. ATPS can be used to separate small, simple, large, and complex biomolecules from cultivation broth. Examples are the purification of monoclonal antibodies (mAb) [28,32,41,49,108,109], interferon [110,111], and virus-like particles [112].

The extraction of proteins requires two immiscible solvents in which impurities and product show different partition coefficients [107]. The resulting system is spontaneously formed by the mutual incompatibility of two polymers or one polymer and salt above certain concentrations in aqueous media [101].

Advances like the application of ATPE for HTP methods, microfluidic devices [113], and molecular modeling [114–116] led to faster process development procedures.

For large-scale processes, polyethylene glycol (PEG)/salt systems have been preferred due to low cost, low viscosity, short-phase separation time, wide pH range, and possible recycling strategies of polymer and salt [101]. Optimization of phase forming components and their concentration, pH, and ionic strength results in high selectivity and yield [117]. In addition, ATPS can integrate clarification, concentration, and purification in one process step [106].

Research activities of the last few years resulted in possible applications of ATPS for purification of recombinant proteins [96], such as IgG. The establishment of a competitive ATPS-based process for antibody purification seems to be possible [41] as well as the scale-up [32] and conversion into a continuous process [95]. Its economic benefits and scalability have already been demonstrated for different products [101,107].

In the 1970s, continuous extraction processes using ATPS were developed [107]. Rosa et al. [95] evaluated the performance of a packed extraction column for a continuous countercurrent ATPE by example of monoclonal antibodies. They tested plastic, metal, and glass materials and investigated different strategies to achieve the best process performance. Wetting studies showed that the PEG-rich phase should be used as the dispersed phase using a stainless steel packed bed. A combination of the packed column with a pump mixer-settler delivered the best results for the ATPE of IgG. The use of higher flow rates and an additional mixer-settler stage benefit the IgG mass transfer. An IgG yield of 85% and a protein purity of 84% were obtained [95]. This is a significant improvement compared to batch processes in which yields of 61% and a purification factor of 1.59 were achieved [95].

17.6.3.2 Precipitation

Another simple separation operation is the precipitation. Precipitation is considered as low cost, high-yield step, and easy to scale up [118]. In general, precipitation is useful if the feedstock has a high titer and fairly high purity outperforming protein A affinity chromatography in terms of cost [34,107,118]. This method can be scaled up to the ton scale because it scales with processed solution volume only. Therefore, precipitant stock consumption is independent of titer [35,107]. However, a disadvantage consists in its selectivity that is lower compared with other methods.

Research activities in the last few years demonstrated the applicability of this technology for antibody purification [3,107]. Precipitation processes can be performed in both batch and continuous mode [35]. Continuous manufacturing offers several significant advantages such as

- higher degree of automation,
- reduction in manual work, and
- less chance for human error [35].

Its conversion into a continuous process seems to provide consistent, reproducible separation results. In addition, it produces particles with optimal size, strength, and density for efficient solid–liquid separations [107]. Until now, continuous precipitation is not used for recombinant protein production but is in use for other more labile molecules such as blood plasma proteins [35,107]. Although a continuous precipitation process would currently be available, blood plasma proteins are typically fractionated in batch mode [107].

Jungbauer and coworkers presented a complete manufacturing process based on precipitation with only one chromatography step [34]. They presented a feasibility study for the purification of a recombinant antibody from clarified cell culture supernatant using continuous precipitation. The precipitation process was a two-staged process of a calcium phosphate flocculation and a cold ethanol precipitation [35]. For these process steps, continuous manufacturing had the significant advantage of lowering the cooling requirements because smaller volumes were handled at any point in time. It was demonstrated that continuous precipitation can be a viable capture step in the manufacturing process of an IgG antibody. The batch and the continuous operating mode had similar results in yield and purity. The precipitation parameters used were determined at a small batch scale (precipitant concentration, pH, conductivity). Therefore, necessary parameters can be transferred to continuous mode with little effort.

Further advantages of precipitation as purification tool are that the precipitated antibody can be dissolved at very high concentrations in various buffer systems at a variety of pH values as well as high and low ionic strengths [35]. This will benefit subsequent unit operations. According to Ref. [35], physicochemical properties of different antibodies were not changed by several precipitation steps, storage of the precipitate, and redissolution of the antibody.

17.6.3.3 One Step Toward a Chromatography Free Purification Process

Another alternative to substitute all chromatography steps could be achieved by an integration of ATPE and precipitation as a harvest, capture, and purification step. A membrane adsorption unit could be used in polishing and presents another virus reducing step (Figure 17.8). This has not been tested until now.

Continuous ATPE has already been published and should not be a problem at all. There is work in progress on continuous precipitation as well, and this system could be implemented in two subsequent tubular reactors. The polishing step could consist in two parallel membrane adsorbers or – if necessary – continuous chromatography such as iCCC. One main problem could be the continuous filtration. Until now, there are only few continuous filtration devices available for large scale manufacturing. In general, several filtration units are used in parallel to allow changes of single filtration units during the manufacturing process.

17.7 Conclusion and Outlook

During the last decades, platform processes for manufacturing of monoclonal antibodies have been developed. They are based on a fed-batch fermentation resulting in titers of 2–5 g/l followed by protein A chromatography as capture step. In general, two subsequent chromatography steps were carried out as well as a pH virus inactivation and several filtration and diafiltration steps. The platform processes were applicable to a large number of monoclonal antibodies but the high pressure to reduce the time-to-market led to good manufacturing processes that lack the complete exploitation of the total process potential. This can only be achieved by individualized process optimization.

Necessary changes in process development and the manufacturing process itself could be simplified, if an integration of USP and DSP development regarding impurity profiles would be carried out. It could be established at reasonable expense [6] (Chapters 3 and 13).

The demand for high product quality is going to be increased in the next few years. The trend toward personalized, or better, stratified medicine will cause changes in the manufacturing processes. In consequence, lower amounts of one product will be necessary but the number of products will be increased

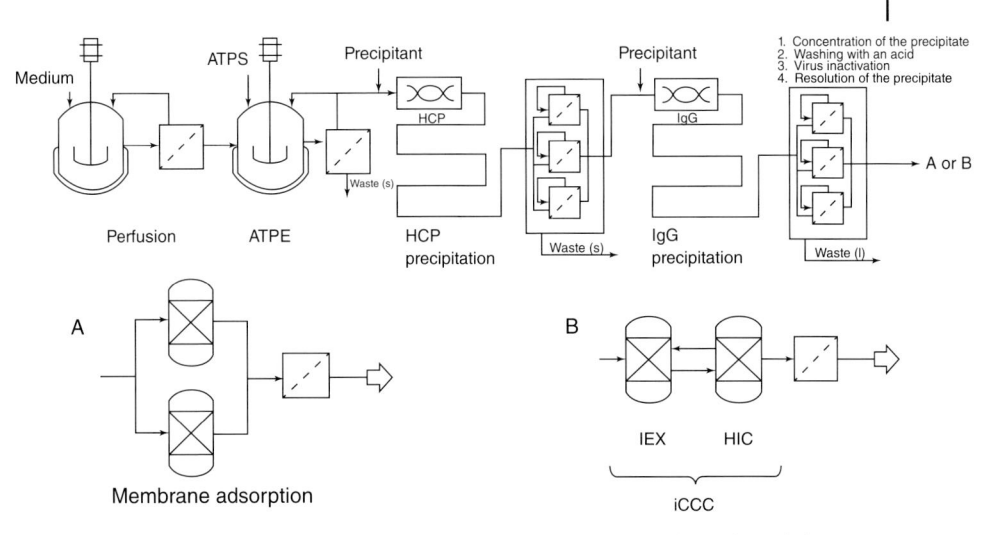

Figure 17.8 Flow sheet of a possible continuous process with a reduced number of chromatography steps. (From Ref. [47].)

significantly. Therefore, the number of blockbusters will drop significantly. In addition, the pressure from regulatory agencies regarding enhanced quality and from the health care systems toward lower costs of therapies will increase steadily. Highly flexible multipurpose facilities will be required to produce smaller volumes of a high number of different products. On the one hand, this will push the implementation of single-use equipment but, on the other hand, it will increase development activities of nonchromatographic and continuous downstream operations [6] (Chapters 3 and 13).

References

1 Shukla, A.A. and Thömmes, J. (2010) Recent advances in large-scale production of monoclonal antibodies and related proteins. *Trends Biotechnol.*, **28** (5), 253–261.
2 Jain, E. and Kumar, A. (2008) Upstream processes in antibody production: evaluation of critical parameters. *Biotechnol. Adv.*, **26** (1), 46–72.
3 Gagnon, P. (2012) Technology trends in antibody purification. *J Chromatogr. A*, **1221** (0), 57–70.
4 Elvin, J.G., Couston, R.G., and van der Walle, C.F. (2013) Therapeutic antibodies: market considerations, disease targets and bioprocessing. *Int. J. Pharm.*, **440** (1), 83–98.
5 Li, F., Vijayasankaran, N., Shen, A., Kiss, R., and Amanullah, A. (2010) Cell culture processes for monoclonal antibody production. *mAbs*, **2** (1942–0862), 466–479.
6 Gronemeyer, P., Ditz, R., and Strube, J. (2014) Trends in upstream and downstream process development for antibody manufacturing. *Bioengineering*, **1** (4), 188–212.

7 Subramanian, G. (ed.) (2015) *Continuous Processing in Pharmaceutical Manufacturing*, 1st edn, Wiley-VCH Verlag GmbH, Weinheim, Germany.

8 Liu, H.F., Ma, J., Winter, C., and Bayer, R. (2010) Recovery and purification process development for monoclonal antibody production. *mAbs*, **2** (1942–0862), 480–499.

9 Birch, J.R. and Racher, A.J. (2006) Antibody production. *Adv. Drug Deliv. Rev.*, **58** (5–6), 671–685.

10 Shukla, A.A., Hubbard, B., Tressel, T., Guhan, S., and Low, D. (2007) Downstream processing of monoclonal antibodies – application of platform approaches. *J. Chromatogr. B*, **848** (1), 28–39.

11 Nfor, B.K., Verhaert, P.D., van der Wielen, L.A., Hubbuch, J., and Ottens, M. (2009) Rational and systematic protein purification process development: the next generation. *Trends Biotechnol.*, **27** (12), 673–679.

12 Vogel, J.H., Nguyen, H., Giovanni, R., Ignowski, J., Garger, S., Salgotra, A., and Tom, J. (2012) A new large-scale manufacturing platform for complex biopharmaceuticals. *Biotechnol. Bioeng.*, **109** (12), 3049–3058.

13 Bhambure, R., Kumar, K., and Rathore, A.S. (2011) High-throughput process development for biopharmaceutical drug substances. *Trends Biotechnol.*, **29** (3), 127–135.

14 De Jesus, Maria de and Wurm, F.M. (2011) Manufacturing recombinant proteins in kg-ton quantities using animal cells in bioreactors. *Eur. J. Pharm. Biopharm.*, **78**, 184–188.

15 Jordan, M., Voisard, D., Berthoud, A., Tercier, L., Kleuser, B., Baer, G., and Broly, H. (2013) Cell culture medium improvement by rigorous shuffling of components using media blending. *Cytotechnology*, **65** (1), 31–40.

16 Zhang, H., Wang, H., Liu, M., Zhang, T., Zhang, J., Wang, X., and Xiang, W. (2013) Rational development of a serum-free medium and fed-batch process for a GS-CHO cell line expressing recombinant antibody. *Cytotechnology*, **65**, 363–378.

17 Jiang, Z., Droms, K., Geng, Z., Casnocha, S., Xiao, Z., and Gorfien, S. (2012) Fed-batch cell culture process optimization. A rationally integrated approach. *BioProcess Int.*, **10** (3), 40–45.

18 Michels, D.A., Parker, M., and Salas-Solano, O. (2012) Quantitative impurity analysis of monoclonal antibody size heterogeneity by CE-LIF: example of development and validation through a quality-by-design framework. *Electrophoresis*, **33** (5), 815–826.

19 Pathak, M., Dutta, D., and Rathore, A. (2014) Analytical QbD: development of a native gel electrophoresis method for measurement of monoclonal antibody aggregates. *Electrophoresis*, **35**, 2163–2171.

20 Jiang, C., Flansburg, L., Ghose, S., Jorjorian, P., and Shukla, A.A. (2010) Defining process design space for a hydrophobic interaction chromatography (HIC) purification step: application of quality by design (QbD) principles. *Biotechnol. Bioeng.*, **107** (6), 985–997.

21 Rathore, A.S. (2009) Roadmap for implementation of quality by design (QbD) for biotechnology products. *Trends Biotechnol.*, **27** (9), 546–553.

22 Chon, J.H. and Zarbis-Papastoitsis, G. (2011) Advances in the production and downstream processing of antibodies. *New Biotechnol.*, **28** (5), 458–463.

23 Kelley, B. (2007) Very large scale monoclonal antibody purification: the case for conventional unit operations. *Biotechnol. Prog.*, **23** (5), 995–1008.
24 Kelley, B. (2009) Industrialization of mAb production technology: the bioprocessing industry at a crossroads. *mAbs*, **1** (1942–0862), 443–452.
25 Butler, M. and Meneses-Acosta, A. (2012) Recent advances in technology supporting biopharmaceutical production from mammalian cells. *Appl. Microbiol. Biotechnol.*, **96** (4), 885–894.
26 Rita Costa, A., Elisa Rodrigues, M., Henriques, M., Azeredo, J., and Oliveira, R. (2010) Guidelines to cell engineering for monoclonal antibody production. *Eur. J. Pharm. Biopharm.*, **74** (2), 127–138.
27 del Val, IoscaniJimenez., Kontoravdi, C., and Nagy, J.M. (2010) Towards the implementation of quality by design to the production of therapeutic monoclonal antibodies with desired glycosylation patterns. *Biotechnol. Prog.*, **26** (6), 1505–1527.
28 Eggersgluess, J., Both, S., and Strube, J. (2012) Process development for extraction of biomolecules: application downstream processing of proteins in aqueous two-phase systems. *Chem. Today*, **30** (4), 32–36.
29 Eggersgluess, J.K., Richter, M., Dieterle, M., and Strube, J. (2014) Multi-stage aqueous two-phase extraction for the purification of monoclonal antibodies. *Chem. Eng. Technol.*, **37** (4), 675–682.
30 Eggersgluess, J., Wellsandt, T., and Strube, J. (2014) Integration of aqueous two-phase extraction into downstream processing. *Chem. Eng. Technol.*, **37** (10), 1686–1696.
31 Silva, M.F.F., Fernandes-Platzgummer, A., Aires-Barros, M.R., and Azevedo, A.M. (2014) Integrated purification of monoclonal antibodies directly from cell culture medium with aqueous two-phase systems. *Sep. Purif. Technol.*, **132** (0), 330–335.
32 Rosa, P.A.J., Azevedo, A.M., Sommerfeld, S., Mutter, M., Aires-Barros, M.R., and Bäcker, W. (2009) Application of aqueous two-phase systems to antibody purification: a multi-stage approach. *J. Biotechnol.*, **139** (4), 306–313.
33 Rosa, P.A.J., Azevedo, A.M., Sommerfeld, S., Mutter, M., Bäcker, W., and Aires-Barros, M.R. (2013) Continuous purification of antibodies from cell culture supernatant with aqueous two-phase systems: from concept to process. *Biotechnol. J.*, **8** (3), 352–362.
34 Hammerschmidt, N., Tscheliessnig, A., Sommer, R., Helk, B., and Jungbauer, A. (2014) Economics of recombinant antibody production processes at various scales: industry-standard compared to continuous precipitation. *Biotechnol. J.*, **9** (6), 766–775.
35 Hammerschmidt, N., Hintersteiner, B., Lingg, N., and Jungbauer, A. (2015) Continuous precipitation of IgG from CHO cell culture supernatant in a tubular reactor. *Biotechnol. J.*, **10**, 1196–1205.
36 Sommer, R., Tscheliessnig, A., Satzer, P., Schulz, H., Helk, B., and Jungbauer, A. (2015) Capture and intermediate purification of recombinant antibodies with combined precipitation methods. *Biochem. Eng. J.*, **93** (0), 200–211.
37 Fröhlich, H., Villian, L., Melzner, D., and Strube, J. (12012) Membrane technology in bioprocess science. *Chem. Ing. Tech.*, **84** (6), 905–917.
38 Chenette, H.C.S., Robinson, J.R., Hobley, E., and Husson, S.M. (2012) Development of high-productivity, strong cation-exchange adsorbers for

protein capture by graft polymerization from membranes with different pore sizes. *J. Membr. Sci. Technol.*, **423–424** (0), 43–52.

39 Weaver, J., Husson, S.M., Murphy, L., and Wickramasinghe, S.R. (2013) Anion exchange membrane adsorbers for flow-through polishing steps: part II. Virus, host cell protein, DNA clearance, and antibody recovery. *Biotechnol. Bioeng.*, **110** (2), 500–510.

40 Smejkal, B., Agrawal, N., Helk, B., Schulz, H., Giffard, M., Mechelke, M., Ortner, F., Heckmeier, P., Trout, B., and Hekmat, D. (2013) Fast and scalaable purification of a therapeutic full-length antibody based on process crystallization. *Biotechnol. Bioeng.*, **110** (9), 2452–2461.

41 Azevedo, A.M., Rosa, P.A.J., Ferreira, I.F., and Aires-Barros, M.R. (2009) Chromatography-free recovery of biopharmaceuticals through aqueous two-phase processing. *Trends Biotechnol.*, **27** (4), 240–247.

42 Tait, A.S., Hogwood, C.E.M., Smales, C.M., and Bracewell, D.G. (2012) Host cell protein dynamics in the supernatant of a mAb producing CHO cell line. *Biotechnol. Bioeng.*, **109** (4), 971–982.

43 Hogwood, C.E.M., Tait, A.S., Koloteva-Levine, N., Bracewell, D.G., and Smales, C.M. (2013) The dynamics of the CHO host cell protein profile during clarification and protein A capture in a platform antibody purification process. *Biotechnol. Bioeng.*, **110** (1), 240–251.

44 Jin, M., Szapiel, N., Zhang, J., Hickey, J., and Ghose, S. (2010) Profiling of host cell proteins by two-dimensional difference gel electrophoresis (2D-DIGE): implications for downstream process development. *Biotechnol. Bioeng.*, **105** (2), 306–316.

45 Tscheliessnig, A.L., Konrath, J., Bates, R., and Jungbauer, A. (2013) Host cell protein analysis in therapeutic protein bioprocessing – methods and applications. *Biotechnol. J.*, **8** (6), 655–670.

46 Ndocko Ndocko, E., Ditz, R., Josch, J.-P., and Strube, J. (2011) New material design strategy for chromatographic separation steps in bio-recovery and downstream processing. *Chem. Ing. Tech.*, **83** (1–2), 113–129.

47 Zobel, S., Helling, C., Ditz, R., and Strube, J. (2014) Design and operation of continuous countercurrent chromatography in biotechnological production. *Ind. Eng. Chem. Res.*, **53** (22), 9169–9185.

48 Helling, C., Borrmann, C., and Strube, J. (2012) Optimal integration of directly combined hydrophobic interaction and ion exchange chromatography purification processes. *Chem. Eng. Technol.*, **35** (10), 1786–1796.

49 Oelmeier, S.A., Ladd-Effio, C., and Hubbuch, J. (2013) Alternative separation steps for monoclonal antibody purification: combination of centrifugal partitioning chromatography and precipitation. *J. Chromatogr. A*, **1319**, 118–126.

50 Strube, J., Ditz, R., Fröhlich, H., Köster, D., Grützner, T., Koch, J., and Schütte, R. (2014) Efficient engineering and production concepts for products in regulated environments – dream or nightmare? *Chem. Ing. Tech.*, **86** (5), 687–694.

51 Warikoo, V., Godawat, R., Brower, K., Jain, S., Cummings, D., Simons, E., Johnson, T., Walther, J., Yu, M., Wright, B., McLarty, J., Karey, K.P., Hwang, C., Zhou, W., Riske, F., and Konstantinov, K. (2012) Integrated continuous

production of recombinant therapeutic proteins. *Biotechnol. Bioeng.*, **109** (12), 3018–3029.

52 Walther, J., Godawat, R., Hwang, C., Abe, Y., Sinclair, A., and Konstantinov, K. (2015) The business impact of an integrated continuous biomanufacturing platform for recombinant protein production. *J. Biotechnol.*, **213**, 3–12.

53 Clincke, M.-F., Mölleryd, C., Samani, P.K., Lindskog, E., Fäldt, E., Walsh, K., and Chotteau, V. (2013) Very high density of Chinese hamster ovary cells in perfusion by alternating tangential flow or tangential flow filtration in WAVE Bioreactor™-part II: applications for antibody production and cryopreservation. *Biotechnol. Prog.*, **29** (3), 768–777.

54 Clincke, M.-F., Mölleryd, C., Zhang, Y., Lindskog, E., Walsh, K., and Chotteau, V. (2013) Very high density of CHO cells in perfusion by ATF or TFF in WAVE bioreactor™. Part I. Effect of the cell density on the process. *Biotechnol. Prog.*, **29** (3), 754–767.

55 Bonham-Carter, J. and Shevitz, J. (2011) A brief history of perfusion biomanufacturing. *BioProcess Int.*, **9** (9) 24–32.

56 Voisard, D., Meuwly, F., Ruffieux, P.-A., Baer, G., and Kadouri, A. (2003) Potential of cell retention techniques for large-scale high-density perfusion culture of suspended mammalian cells. *Biotechnol. Bioeng.*, **82** (7), 751–765.

57 Henzler, H.-J. (2012) Kontinuierliche fermentation mit tierischen zellen. Teil 2. techniken und methoden der zellrückhaltung. *Chem. Ing. Tech.*, **84** (9), 1482–1496.

58 Pollock, J., Ho, S.V., and Farid, S.S. (2013) Fed-batch and perfusion culture processes: economic, environmental, and operational feasibility under uncertainty. *Biotechnol. Bioeng.*, **110** (1), 206–219.

59 Castilho, L.R., Anspach, F.B., and Deckwer, W.-D. (2002) An integrated process for mammalian cell perfusion cultivation and product purification using a dynamic filter. *Biotechnol. Prog.*, **18** (4), 776–781.

60 De Jesus, M.J., Girard, P., Bourgeois, M., Baumgartner, G., Jacko, B., Amstutz, H., and Wurm, F.M. (2004) TubeSpin satellites: a fast track approach for process development with animal cells using shaking technology. *Biochem. Eng. J.*, **17** (3), 217–223.

61 Gomez, N., Ouyang, J., Nguyen, M.D.H., Vinson, A.R., Lin, A.A., and Yuk, I.H. (2010) Effect of temperature, pH, dissolved oxygen, and hydrolysate on the formation of triple light chain antibodies in cell culture. *Biotechnol. Prog.*, **26** (5), 1438–1445.

62 Jing, Y., Borys, M., Nayak, S., Egan, S., Qian, Y., Pan, S.-H., and Li, Z.J. (2012) Identification of cell culture conditions to control protein aggregation of IgG fusion proteins expressed in Chinese hamster ovary cells. *Process Biochem.*, **47** (1), 69–75.

63 Ye, J., Kober, V., Tellers, M., Naji, Z., Salmon, P., and Markusen, J.F. (2009) High-level protein expression in scalable CHO transient transfection. *Biotechnol. Bioeng.*, **103** (3), 542–551.

64 Wurm, F.M. (2004) Production of recombinant protein therapeutics in cultivated mammalian cells. *Nat. Biotechnol.*, **22** (11), 1393–1398.

65 Sommerfeld, S. and Strube, J. (2005) Challenges in biotechnology production – generic processes and process optimization for monoclonal antibodies. *Chem. Eng. Process.*, **44** (10), 1123–1137.

66 Shukla, A.A. and Gottschalk, U. (2013) Single-use disposable technologies for biopharmaceutical manufacturing. *Trends Biotechnol.*, **31** (3), 147–154.

67 Konstantinov, K. (2011) Continous bioprocessing: an interview with Konstantin Konstantinov from Genzyme. Interviewed by Prof. Alois Jungbauer and Dr. Judy Peng. *Biotechnol. J.*, **6** (12), 1431–1433.

68 Konstantinov, K.B. and Cooney, C.L. (2015) White paper on continuous bioprocessing. May 20–21, 2014 continuous manufacturing symposium. *J. Pharm. Sci.*, **104** (3), 813–820.

69 Low, D., O'Leary, R., and Pujar, N.S. (2007) Future of antibody purification. *J. Chromatogr. B*, **848** (1), 48–63.

70 Strube, J., Grote, F., and Ditz, R. (2012) Bioprocess Design and Production Technology for the Future. In *Biopharmaceutical Production Technology*, G. Subramanian (ed.), Wiley-VCH Verlag GmbH, Weinheim.

71 Food and Drug Administration (1998) Federal Register/Vol. 63, No. 110/ Tuesday, June 9, 1998/Notices, 63 (110).

72 Raijada, D.K., Prasad, B., Paudel, A., Shah, R.P., and Singh, S. (2010) Characterization of degradation products of amorphous and polymorphic forms of clopidogrel bisulphate under solid state stress conditions. *J. Pharm. Biomed. Anal.*, 52 (3), pp. 332–344.

73 Pan, C., Liu, F., and Motto, M. (2011) Identification of pharmaceutical impurities in formulated dosage forms. *J. Pharm. Sci.*, **100** (4), 1228–1259.

74 Mazur, M., Seipert, R., Mahon, D., Zhou, Q., and Liu, T. (2012) A platform for characterizing therapeutic monoclonal antibody breakdown products by 2D chromatography and top-down mass spectrometry. *AAPS J.*, **14** (3), 530–541.

75 Horak, J., Ronacher, A., and Lindner, W. (2010) Quantification of immunoglobulin G and characterization of process related impurities using coupled protein A and size exclusion high performance liquid chromatography. *J. Chromatogr. A*, **1217** (31), 5092–5102.

76 Flatman, S., Alam, I., Gerard, J., and Mussa, N. (2007) Process analytics for purification of monoclonal antibodies. *J. Chromatogr. B*, **848** (1), 79–87.

77 Fonslow, B.R., Kang, S.A., Gestaut, D.R., Graczyk, B., Davis, T.N., Sabatini, D.M., and Yates J III, J.R. (2010) Native capillary isoelectric focusing for the separation of protein complex isoforms and subcomplexes. *Anal. Chem.*, **82** (15), 6643–6651.

78 Leary, B.A., Lawrence-Henderson, R., Mallozzi, C., Fernandez Ocaña, M., Duriga, N., O'Hara, D.M., Kavosi, M., Qu, Q., and Joyce, A.P. (2013) Bioanalytical platform comparison using a generic human IgG PK assay format. *J. Immunol. Methods*, **397** (1–2), 28–36.

79 Im, H., Sutherland, J.N., Maynard, J.A., and Oh, S.-H. (2012) Nanohole-based surface plasmon resonance instruments with improved spectral resolution quantify a broad range of antibody-ligand binding kinetics. *Anal. Chem.*, **84** (4), 1941–1947.

80 Zhang, Z., Pan, H., and Chen, X. (2009) Mass spectrometry for structural characterization of therapeutic antibodies. *Mass Spectrom. Rev.*, **28** (1), 147–176.

81 Timms, J.F. and Cramer, R. (2008) Difference gel electrophoresis. *Proteomics*, **8** (23–24), 4886–4897.

82 Winkler, C., Denker, K., Wortelkamp, S., and Sickmann, A. (2007) Silver- and Coomassie-staining protocols: detection limits and compatibility with ESI MS. *Electrophoresis*, **28** (12), 2095–2099.

83 Fiechter, G. and Mayer, H.K. (2011) Characterization of amino acid profiles of culture media via pre-column 6-aminoquinolyl-*N*-hydroxysuccinimidyl carbamate derivatization and ultra performance liquid chromatography. *J. Chromatogr. B*, **879** (17–18), 1353–1360.

84 Klejdus, B., Petrlová, J., Potěšil, D., Adam, V., Mikelová, R., Vacek, J., Kizek, R., and Kubáň, V. (2004) Simultaneous determination of water- and fat-soluble vitamins in pharmaceutical preparations by high-performance liquid chromatography coupled with diode array detection. *Anal. Chim. Acta*, **520** (1–2), pp. 57–67.

85 Scherer, M., Leuthäuser-Jaschinski, K., Ecker, J., Schmitz, G., and Liebisch, G. (2010) A rapid and quantitative LC-MS/MS method to profile sphingolipids. *J. Lipid Res.*, **51** (7), 2001–2011.

86 Alwael, H., Connolly, D., and Paull, B. (2011) Liquid chromatographic profiling of monosaccharide concentrations in complex cell-culture media and fermentation broths. *Anal. Methods*, **3**, 62–69.

87 Grzeskowiak, J.K., Tscheliessnig, A., Wu, M.W., Toh, P.C., Chusainow, J., Lee, Y.Y., Wong, N., and Jungbauer, A. (2010) Two-dimensional difference fluorescence gel electrophoresis to verify the scale-up of a non-affinity-based downstream process for isolation of a therapeutic recombinant antibody. *Electrophoresis*, **31** (11), 1862–1872.

88 Prieto, Y., Rojas, L., and Pérez, R. (2011) Towards the molecular characterization of the stable producer phenotype of recombinant antibody-producing NS0 myeloma cells. *Cytotechnology*, **63**, 351–362.

89 Wang, X., Hunter, A.K., and Mozier, N.M. (2009) Host cell proteins in biologics development: identification, quantitation and risk assessment. *Biotechnol. Bioeng.*, **103** (3), 446–458.

90 Pezzini, J., Joucla, G., Gantier, R., Toueille, M., Lomenech, A.-M., Le Sénéchal, C., Garbay, B., Santarelli, X., and Cabanne, C. (2011) Antibody capture by mixed-mode chromatography: a comprehensive study from determination of optimal purification conditions to identification of contaminating host cell proteins. *J. Chromatogr. A*, **1218** (45), 8197–8208.

91 Guiochon, G. and Beaver, L.A. (2011) Separation science is the key to successful biopharmaceuticals. *J. Chromatogr. A*, **1218** (49), 8836–8858.

92 Champion, K., Madden, H., Dougherty, J., and Shacter, E. (2005) Defining your product profile and maintaining control over it, part 2. *BioProcess Int.*, **3** (8), 52–57.

93 Shukla, A.A. and Hinckley, P. (2008) Host cell protein clearance during protein A chromatography: development of an improved column wash step. *Biotechnol. Prog.*, **24** (5), 1115–1121.

94 Langer, E.S. and Rader, R.A. (2014) Single-use technologies in biopharmaceutical manufacturing: a 10-year review of trends and the future. *Eng. Life Sci.*, **14** (3), 238–243.

95 Rosa, P.A.J., Azevedo, A.M., Sommerfeld, S., Bäcker, W., and Aires-Barros, M.R. (2012) Continuous aqueous two-phase extraction of human antibodies using a packed column. *J. Chromatogr. B*, **880**, 148–156.

96 Asenjo, J.A. and Andrews, B.A. (2012) Aqueous two-phase systems for protein separation: phase separation and applications. *J. Chromatogr. A*, **1238**, 1–10.

97 Chatel, A., Kumpalume, P., and Hoare, M. (2014) Ultra scale-down characterization of the impact of conditioning methods for harvested cell broths on clarification by continuous centrifugation – recovery of domain antibodies from rec *E. coli. Biotechnol. Bioeng.*, **111** (5), 913–924.

98 Müller-Späth, T., Krättli, M., Aumann, L., Ströhlein, G., and Morbidelli, M. (2010) Increasing the activity of monoclonal antibody therapeutics by continuous chromatography (MCSGP). *Biotechnol. Bioeng.*, **107** (4), 652–662.

99 Fischer-Fruhholz, S., Zhou, D., and Hirai, M. (2010) Sartobind STIC[reg] salt-tolerant membrane chromatography. *Nat. Methods*, **7** (12)

100 Zang, Y., Kammerer, B., Eisenkolb, M., Lohr, K., and Kiefer, H. (2011) Towards protein crystallization as a process step in downstream processing of therapeutic antibodies: screening and optimization at microbatch scale. *PLoS One*, **6** (9), e25282.

101 Rosa, P.A.J., Ferreira, I.F., Azevedo, A.M., and Aires-Barros, M.R. (2010) Aqueous two-phase systems: a viable platform in the manufacturing of biopharmaceuticals. *J Chromatogr. A*, **1217** (16), 2296–2305.

102 Ditz, R. (2012) Separation technologies 2030 – are 100 years of chromatography enough? *Chem. Ing. Tech.*, **84** (6), 875–879.

103 Kontoravdi, C., Samsatli, N.J., and Shah, N. (2013) Development and design of bio-pharmaceutical processes. *Curr. Opin. Chem. Eng.*, **2** (4), 435–441.

104 Gronemeyer, P., Ditz, R., and Strube, J. (2016) DoE based integration approach of upstream and downstream processing regarding HCP and ATPE as harvest operation. *Biochem. Eng. J.*, **113**, 158–166.

105 Eggersglüß, J. (2013) Entwicklung und Auslegung von Flüssig-Flüssig Bioextraktionsprozessen: Antikörperaufreinigung mit wässrigen Zweiphasensystemen. Dissertation, Technische Universität Clausthal.

106 Gronemeyer, P., Ditz, R., and Strube, J. Implementation of Aqueous two-phase extraction combined with precipitation in a monoclonal antibody manufacturing process. *Chem. Today*, 34 (3), 66–70.

107 Jungbauer, A. (2013) Continuous downstream processing of biopharmaceuticals. *Trends Biotechnol.*, **31** (8), 479–492.

108 Rosa, P.A.J., Azevedo, A.M., Ferreira, I.F., Sommerfeld, S., Bäcker, W., and Aires-Barros, M.R. (2009) Downstream processing of antibodies: single-stage versus multi-stage aqueous two-phase extraction. *J.Chromatogr. A*, 1216 (50), pp. 8741–8749.

109 Oelmeier, S.A. (2012) Aqueous two phase extraction of proteins: From molecular understanding to process development. Dissertation, Karlsruher Institut für Technologie.

110 Menge, U. and Kula, M.-R. (1984) Purification techniques for human interferons. *Enzyme Microb. Technol.*, **6** (3), 101–112.

111 Guan, Y., Lilley, T.H., Treffry, T.E., Zhou, C.-L., and Wilkinson, P.B. (1996) Use of aqueous two-phase systems in the purification of human interferon-$\alpha 1$ from recombinant *Escherichia coli. Enzyme Microb. Technol.*, **19** (6), 446–455.

112 Benavides, J., Aguilar, O., Lapizco-Encinas, B.H., and Rito-Palomares, M. (2008) Extraction and purification of bioproducts and nanoparticles using aqueous two-phase systems strategies. *Chem. Eng. Technol.*, **31** (6), 838–845.
113 Silva, D.F.C., Azevedo, A.M., Fernandes, P., Chu, V., Conde, J.P., and Aires-Barros, M.R. (2012) Design of a microfluidic platform for monoclonal antibody extraction using an aqueous two-phase system. *J. Chromatogr. A*, **1249**, 1–7.
114 Oelmeier, S.A., Ladd Effio, C., and Hubbuch, J. (2012) High throughput screening based selection of phases for aqueous two-phase system-centrifugal partitioning chromatography of monoclonal antibodies. *J. Chromatogr. A*, **1252**, 104–114.
115 Hubbuch, J. (2012) Editorial: high-throughput process development. *Biotechnol. J.*, **7** (10), 1185.
116 Wiendahl, M., Oelmeier, S.A., Dismer, F., and Hubbuch, J. (2012) High-throughput screening-based selection and scale-up of aqueous two-phase systems for pDNA purification. *J. Sep. Sci.*, **35** (22), 3197–3207.
117 Schügerl, K., (2005) Extraction of Primary and Secondary Metabolites. *Technology Transfer in Biotechnology*, Springer, Berlin, pp. 1–48.
118 Capito, F., Bauer, J., Rapp, A., Schröter, C., Kolmar, H., and Stanislawski, B. (2013) Feasibility study of semi-selective protein precipitation with salt-tolerant copolymers for industrial purification of therapeutic antibodies. *Biotechnol. Bioeng.*, **110** (11), 2915–2927.

Part Eight

Quality, Validation, and Regulatory Aspects

18

Quality Control and Regulatory Aspects for Continuous Biomanufacturing

Guillermina Forno[1,2] and Eduardo Ortí[2]

[1]*Ciudad Universitaria, Cell Culture Laboratory, UNL, FBCB, Paraje el Pozo, CC 242 Santa Fe, Argentina*
[2]*Ciudad Universitaria, R&D Zelltek S.A., UNL, FBCB, Paraje el Pozo, CC 242 Santa Fe, Argentina*

18.1 Introduction

Several key advantages have been associated with the implementation of continuous biomanufacturing, such as (i) high operational flexibility to match changing product demands, (ii) smaller facility footprint with the consequent reduction in capital expenditure and cost of goods, (iii) shorter processing times, (iv) reduced bioreactor start-up and turnaround time, (v) increased volumetric productivity, and (vi) potential to incorporate the verification of product quality and the real-time release of products [1–3].

Despite the fact its numerous advantages and that the foundational concepts are not new to commercial biotechnology, the biopharmaceutical industry has been reluctant to introduce continuous manufacturing in their production processes. At commercial scale, only a limited number of companies are currently using it (they are mainly associated with the employment of perfusion culture), or making significant investment on process development based in continuous manufacturing [4]. This, at least in part, can be explained by the general perception of a poor regulatory flexibility, unfavorable to the adoption of novel methods for biotherapeutics manufacturing. At the end, the industry hesitancy in adopting innovative, more efficient manufacturing methods makes it difficult to incorporate affordable medicines for the public health.

18.2 FDA Support for Continuous Manufacturing

The FDA recognized as a problem the lack of innovation in the manufacturing processes of the pharmaceutical industry. In an attempt to address it, the "Pharmaceutical cGMP for the 21st Century – A Risk-Based Approach" [5] initiative was launched. It constitutes a framework to enhance and modernize FDA's regulations governing pharmaceutical manufacturing and product quality for human and veterinary applications. It emphasizes the need for an agile and flexible pharmaceutical manufacturing sector that reliably produces high quality

Continuous Biomanufacturing: Innovative Technologies and Methods, First Edition.
Edited by Ganapathy Subramanian.
© 2018 Wiley-VCH Verlag GmbH & Co. KGaA. Published 2018 by Wiley-VCH Verlag GmbH & Co. KGaA.

drugs, without extensive regulatory oversight and encourages the early adoption of new technological advances by the industry, including integrated systems based on science and engineering principles and knowledge.

Moreover, continuous manufacturing has become a highly visible topic since FDA has provided a strong support to its implementation. Dr. Jane Woodcock, Director of the Center for Drug Evaluation and Research (CDER) at the FDA expressed in her oral presentation at the 25th AAPS Annual Meeting that was held in October 2011 in Washington D.C., USA: "Right now, manufacturing experts form the 1950s would easily recognize the pharmaceutical manufacturing processes of today. It is predicted that manufacturing will change in the next 25 years as current manufacturing practices are abandoned in favor of cleaner, flexible, more efficient continuous manufacturing." Therefore, FDA promotes the adoption of continuous manufacturing since it is more agile and flexible than traditional batch processes, allows a reduction in manual handling of products, and implies a better process control through real-time testing approaches, reducing the risk of drug shortages by failures at the manufacturing facilities [6]. Although there are no specific regulations for continuous manufacturing, it is understood that there are no regulatory obstacles for its implementation. However, it is accepted that some concepts will differ from the traditional approach. As an example, in-process controls and sampling consideration, validation strategy, and batch/lot definition must be carefully defined in order to achieve a proper process consistency and traceability.

18.3 PAT as a Facilitator for Continuous Manufacturing Implementation

Continuous manufacturing has a higher requirement for continual assurance of product quality during processing, as well as for monitoring and control systems to multiple unit operation with different system dynamics that work in a connected manner. Also, it is important to monitor and maintain the quality of the incoming materials that enter continuously into the manufacturing process. On the other hand, in-process monitoring and control consist of analytical measurements and, if required, the manipulation of variables and a feedback controller. It is essential to distinguish feedforward control, in which manipulations are made in response to measurements of identifiable process disturbances or deviations of the input to the process, from feedback control, in which manipulations are made in response to measurement of a process output. In consequence, a continuous process could be controlled through manipulation of downstream process parameters in response to measured upstream disturbances to maintain final product quality and in response to local real-time measurements of product quality. Using both feedforward and feedback controls to response in real time to disturbances throughout the multiple unit operation is a distinctive characteristic of continuous manufacturing. Therefore, the detection of disturbances or variables changes and a fast diagnostic of their causes are crucial for a comprehensive process understanding and control.

The improvement of process analytical technology (PAT) instrumentation and methods is an important facilitator for successful implementation of continuous manufacturing, providing greater information about product quality attributes on a real-time testing. In 2004, the FDA published the guidance: "PAT – A Framework for Innovative Pharmaceutical Development, Manufacturing and Quality Assurance" [7] that constitutes a regulatory framework for developing and implementing a manufacturing strategy based on improved production efficiency through technological innovation while maintaining or improving the current level of quality assurance of the product. That guidance document states "A desired goal of PAT framework is to design and develop well understood processes that will consistently ensure a predefined quality at the end of the manufacturing process" and this will result from, among others "facilitating continuous processing to improve efficiency and manage variability."

The PAT approach is conceived as a process control strategy based on real-time measurement of parameters that directly correlate with critical quality attributes or parameters confirming that a unit operation or piece of equipment continues to be fit for purpose [8]. It provides the elements to make go/no go decisions and to demonstrate that the process is working under a predefined operation range [9]. Its goal is to increase the understanding and control of the manufacturing processes, allowing investigating and understanding the relationship among process factors.

The advantage over the classical monitoring strategy is the higher efficiency to detect failures and sources of process variability, along their potential to generate a platform for an online/in-line active control, allowing the adjustment of the operating parameters if a variation is detected in the input materials or in the environment that could have a negative impact on the product [10]. Higher product quality can be achieved during processing, rather than testing product quality solely at the end [11], which leads to a cost reduction. Therefore, savings are associated with a cost efficient resource use, optimal machinery operation, reduced waste, and a greatly reduced risk of product recall [12].

Despite its multiple benefits, several factors had hindered the application of PAT to the practice of bioprocessing, and this can explain that the field of monitoring and control is for biotechnology manufacturing ahead of the food, agriculture, and chemical industries. This is mainly because the biotechnology industry presents significant challenges, such as (i) the sophisticated metabolism of cell-based biological systems that are highly sensitive to environmental conditions [8], (ii) the structural characteristics of proteins, substantially more complex than small molecules not only because of their size and the presence of posttranslational modifications but also for their tendency to degrade during manufacturing or storage, (iii) the difficulties for determining in a clear manner the product characteristics that are functionally/clinically significant, in terms of safety, efficacy, and stability, (iv) the substantial variability in the composition of complex raw materials utilized mainly during the fermentation step [13], and (v) the presence of mixtures of process-related and product-related impurities that can interfere with many sensitive sensor technologies. On the other hand, some factors are associated with technological restrictions such as the lack of accurate mathematical models for efficient feedback control design and the limited availability of reliable, inexpensive, *in situ* and nondestructive instrumentation [14]. Finally, it is worth mentioning the concern of some companies to find problems in their systems that would not have been observed under normal process monitoring, and the costs of restructuring the production lines, in order to adapt them for PAT implementation [12].

PAT can be offline (in a lab away from the process, utilizing a discrete sample), online (measurement where the sample is derived from the manufacturing process and not returned to the process stream), in-line (the sample is analyzed with continuous stream flow and is not removed from it), at line (where the sample is removed, isolated from, and analyzed in close proximity to the process stream), or *in situ* (in the vessel itself continuously in contact with the content). A variety of spectroscopy techniques (Fourier transform infrared, UV, X-ray, 2D fluorescence, and NIR/MIR spectroscopy) as well as chromatographic techniques (HPLC, UPLC, gas chromatography) have been developed. The advantage of spectroscopic measurement techniques is that they are noninvasive and nondestructive and provide online or at line about several variables simultaneously [14]. Analysis can be discrete (measured at periodic intervals) or continuous (measured constantly) [11,15], and its successful application demands an array of PAT tools, data acquisition and analysis tools, process analyzers, process control tools, continuous improvement and knowledge management tools, applicable to a single-unit operation or to an entire manufacturing process and its quality assurance system [8].

18.4 PAT Applications in the Pharmaceutical Industry

A growing number of PAT applications are available both for cell culture and multicolumn continuous processes, some of them listed in Tables 18.1 and 18.2. Many of them apply multivariate data analysis (MVDA), which consists in the analysis using statistical techniques of datasets composed of more than one variable or more than one type of variables, which are analyzed simultaneously to perform exploratory, regression, or classification studies [16]. Through MVDA, large datasets originating from pilot or manufacturing plants are translated into relevant and understandable information, making it possible to find cause-effect correlations useful for process control and variation. The use of MVDA as a PAT tool for biopharmaceutical processing for immediate identification of deviations in the process and to establish control strategies has greatly increased over the last years. Several examples can be found in bibliography, such as the application of surface-enhanced Raman scattering (SERS) spectroscopy for the analysis of cell culture media degradation. The use of MDVA enable the fast detection of compositional changes in a chemically defined media, demonstrating that even when media are stored at low temperature (2–8 °C) and in the dark, significant chemical changes occur, particularly in the cysteine/cystine concentration [17]. In addition, bioreactor cultivation monitoring has increased in terms of process parameters that can eventually be further correlated with CQAs. For instance, Schmidberger *et al.* established that through a performance-based modeling it was possible to forecast the product quality and quantity in two

Table 18.1 PAT applied to cell culture operations.

Direct measurement	Technology used for PAT	Samples from	Application	References
Glucose	Raman spectroscopy	Bioreactor	Controlling glucose to avoid antibody glycation	[18]
Osmolality, glucose, product titer, packed cell volume, integrated viable packed cell volume, viable cell density, integrated viable cell count	Near-infrared spectroscopy	Bioreactor	In-line monitoring of cell culture parameters for identifying batch homogeneity between lots and abnormal fermentation runs	[19]
Product concentration and affinity	Surface plasmon resonance	Bioreactor, purification train	Process control and monitoring of CQAs	[20]
Imaging of cell population	In situ microscopy	Inoculum, bioreactor	Characterize cell population	[21]
Volatile components	Electronic nose with metal oxide semiconductor field effect transistors sensors	Bioreactor head space	Early detection of microbial and viral contaminants on cell cultures	[22]
Glucose and biomass concentration	Multiwavelength fluorescence	Bioreactor	Control of specific growth rate and product quality	[23]

mammalian cell culture processes. The relationship between available information from small-scale cultivations, such as metabolite profile, and 12 critical quality attributes was studied. Good accuracy was obtained for the prediction of product concentration and some glycan isoforms, demonstrating that PAT tools can be successfully implemented to monitor the identified quality-related parameters. MDVA also has potential application for establishing control strategies for multicolumn chromatography processes, since in contrast to traditional chromatography processes that typically yield few elution peaks per batch, continuous multicolumn chromatography processes generate hundreds elution peaks per batch. This higher amount of data generated during a single batch demands sophisticated tools for identification of deviations in the process.

Additionally, another good example of PAT application is the control of glycosylation during bioreactor operation. For most therapeutic glycoproteins, glycosylation pattern constitutes a critical quality attribute, since it affects numerous biological functions. For example, for monoclonal antibodies, several alterations in N-glycosylation have clinical relevance: (i) Core fucosylation

Table 18.2 PAT applied to purification operations.

Direct measurement	Technology used for PAT	Samples from	Application	References
pH, conductivity	Online pH and conductivity	Product stream during ultrafiltration/diafiltration step	Indication of completion of diafiltration process	[24]
Product impurities	Online anion exchange HPLC	Purification process	Online discrimination of disulfide misfold species for real time pooling decisions	[25]
Charge variants	Cation exchange HPLC using nonlinear salt gradient	Purification process	Rapid analysis of charge variants of monoclonal antibodies	[26]
Product impurities	Tryptophan fluorescence	Purification process	At line measurement of misfold variants for real time pooling decisions	[27]
Product impurities	Mid-UV absorption spectra and partial least squares regression	Purification process	In-line quantification of coeluting proteins	[28]

inhibits IgG binding to FcγRIIIa, decreasing ADCC, (ii) high mannose glycans can have a negative impact on pharmacokinetics while they enhance ADCC and reduce CDC, (iii) terminal Gal has shown to play an important role in CDC activity (CDC increases with the increase in Gal Content), and (iv) higher levels of sialylation were associated with reduced activity in the ADCC assay and have shown an antiinflammatory effect of intravenous immunoglobulin G [29]. In addition, glycosylation is very sensitive to environmental conditions. The concentration of glycosylation precursors, glucose, ammonia, and dissolved oxygen, osmolality, and the culture temperature are some of the variables that have been reported to determine glycosylation patterns in different recombinant proteins [30–34]. For example, Fan et al. [33] demonstrated that the balance of glucose and amino acid concentration in the culture is important for IgG N-glycosylation. Besides, Pacis et al. [35] studied the effect of osmolality and culture duration on antibody glycosylation. Higher osmolality from NaCl addition in cell culture media affected not only cell growth and title, but also increased high mannose Man5 levels compared to the controls. The high mannose glycan levels were also sensitive to the duration of culture, since in a 22-day culture an increase in the percentage of high mannose glycans between day 7 and 21 for both control and high osmolality cultures was observed. The aforementioned findings clearly indicated that the monitoring and control of factors affecting glycosylation as well as the high-throughput methods to be employed for glycosylation analysis are essential for final product quality assurance. Important advances in method automatization have enabled for studying the product glycosylation pattern prior to commencement of purification [36–40].

Currently, PAT is pushing for easy-to-use process analyzers, mathematical integration tools for data analysis, and feedback control methods to perform process adjustments as necessary. Ultimately, the major driving force for the penetration of PAT in the industrial environment is the possibility of adjusting the operating parameters if a variation in the environment or input materials that would adversely impact the drug product quality occurs, as well as the possibility of using the continuous information collected to justify proposals for postapproval process changes, reducing the regulatory burden.

18.5 Process Validation for Continuous Manufacturing Processes

Process validation implies the demonstration that a manufacturing process is under control and can achieve the expected performance. Several key considerations for validating a continuous manufacturing process show coincidence with those applicable to validation of traditional batch processes. For example, the principles of process development as well as the guidelines, regulations, and standards governing process validation are the same for both continuous and batch processing. Some common basic concepts include the establishment of critical quality attributes, critical process parameters, system and equipment qualification, definition of acceptance criteria for products and processes, and the use of quantitative statistical methods to evaluate the validation data and intra and interbatch variation.

In a continuous manufacturing process, input raw materials are continuously fed into a process train while the processed output materials are continuously removed. The process can run over a long period of time in an integrated fully continuous process or under a hybrid approach. A hybrid approach is a combination of continuous and discontinuous steps. Usually, the upstream part is operated continuously using perfusion with a bioreactor that has been fashioned with a cell retention device, whereas downstream is performed as a succession of discontinuous chromatography steps operated in batch. On the other hand, in a fully continuous process (also known as end-to-end continuous process), different steps are sequenced together to form a continuous production. An example of this approach was described by Godawat et al. [41] for the production of a monoclonal antibody expressed in a CHO/DHFR cell line, achieving process train simplification by the elimination of several nonvalue unit operations such as harvest hold and clarification. Cells were cultivated in a bioreactor equipped with an alternating tangential flow filtration device and the clarified harvest was introduced in an uninterrupted and fully automated purification system to the drug substance over the complete process, achieving consistent product quality over around 30 days. At the end, for end-to-end or hybrid continuous manufacturing, the peculiar characteristics regarding traditional batch processes require the definition of specific considerations related to process validation.

18.6 Regulatory Documents Related to Process Validation

Several regulatory documents provide a framework for process validation. In many cases these documents exhibit overlap or are complementary. In any case, the validation of a continuous process must take into consideration its philosophy and directives to meet authorities expectations.

18.7 ICH

The guide ICH Q7: Good Manufacturing Practice Guide for Active Pharmaceutical Ingredients presents key concepts such as batch and process validation definition [42]. In addition, Q8(R2): Pharmaceutical Development, Q9: Quality Risk Management, and Q10: Pharmaceutical Quality System Guides [43–46], incorporate notions of process understanding, continual improvement, use of risk management tools, knowledge management, and efficient quality systems. They also change the focus on the product, and instead they emphasize the focus on the patient. Therefore, those guides provide tools to identify product critical quality attributes and define a design space and the process control strategy. This information is valuable to determine the best approach for process validation before and during initial commercial production batches. As an alternative to the traditional process validation, continuous process verification is introduced in ICH Q8(R2), and can be utilized in process validation protocols for the initial commercial production and for manufacturing process changes. The continuous process verification is based on the online, in-line, or at line monitoring that provides substantial information and knowledge about the process, facilitating process understanding.

18.8 FDA

The FDA *Guidance for Industry Process Validation: General Principles and Practices* [47] introduces basic concepts for the validation of manufacturing processes of human and animal drug and biological products, including active pharmaceutical ingredients (APIs or drug substances). It defines process validation as "the collection and evaluation of data, from the process design stage through commercial production, which establishes scientific evidence that a process is capable of consistently delivering quality product," and encourages a validation process approach during the entire product lifecycle, linking product and process development to the commercial manufacturing process. This guideline describes process validation activities in three stages. (i) Stage 1 – Process design: The knowledge gained through development and scale-up activities will define and support the commercial manufacturing process. Knowledge can be obtained from a retrospective review from all studies and activities, a risk assessment of the process to identify parameters that influence process performance and product quality, and the execution of experiments designed using statistical tools (Design of Experiments, DOE), (ii) Stage 2 – Process qualification: During this stage, the process design is evaluated to determine if the process is capable of reproducible commercial manufacturing, that is, a successful process qualification will confirm the process design and will demonstrate if the commercial manufacturing process is performed as expected, and (iii) Stage 3 – Continued process verification: based on maintenance, continuous verification, and process improvement, to achieve the on-going assurance that routine production process remains in a state of control by collecting and monitoring information during commercialization. This guidance describes activities typical of each stage, but in practice, some activities might occur in multiple stages, for example, the application of risk analysis tools.

18.9 EMA

The EMA Guideline on process validation for finished products – information and data to be provided in regulatory submissions [48] – states that process validation can be performed using a traditional approach regardless of the development strategy used. This guideline also returns to the concept of continuous process verification from ICH Q8 (R2). It considers that continuous process verification can be used in addition to or instead of traditional process verification if an enhanced strategy of development was used or significant process understanding was gained during manufacturing experience. Also a hybrid approach, including the traditional process validation approach and continuous process verification can be used.

18.10 ASTM

A description of continuous quality verification concept as well as principles for its application to pharmaceutical or biopharmaceutical processes, including continuous manufacturing processes or supporting utility systems, is addressed in Ref. [49]. There, continuous quality verification (CQV) is described as an approach to process validation where manufacturing process's performance is continuously monitored, evaluated, and adjusted as necessary. Using this approach the batches manufactured during development can be included in the validation exercise.

18.11 Special Considerations for Continuous Manufacturing Process Validation

Several aspects related with process validation will need to be carefully defined for continuous manufacturing and will exhibit differences regarding the traditional approach for batch processes. As for batch processes, the starting point is the definition of critical process parameters (CPP) and critical quality attributes (CQA) and the demonstration that in-process controls are capable to detect process excursions. However, as the product is formed continuously over a long period of time and its quality is expected to be susceptible to environmental conditions that potentially could change, it is mandatory to clearly define aspects such as (i) allowed start-up and shut down of the process, (ii) process run-time, (iii) operational flow rate and residence time, (iv) inter-relationship on CPPs and CQA, and (v) details of process monitoring, including sampling strategy, size of samples, and frequency of monitoring.

For example, it has been recognized that an important advantage of continuous processes is that they do not require a change of scale between clinical and commercial manufacturing, since the commercial manufacturing can be run longer in order to meet the manufacturing demands. This avoids the demonstration of process and equipment equivalency between clinical and commercial material [50]. However, this obvious convenience requires that process validation activities include the study of manufacturing campaign lengths. This includes the demonstration that the system can reach and maintain the intended process condition over the entire process.

In addition, cell line stability and product quality are aspects that require extensive characterization, including studies under the worst-case scenario (e.g., longest). Desai *et al.* [51] reported for the production process of coagulation factor BeneFIX, marketed by Pfizer, that recombinant CHO producer cells are quite stable for over eighty generations. The analysis of the cell-specific growth rate, which is a good indicator of the cell line genetic stability for that process, indicated that no obvious trend up or down was evident for 2500, 6000, and 12 000 l bioreactor scale. In addition to this kind of study, usually the limit of *in vitro* cell age study is required for validation of the maximum number of generations, and this study must be performed at full scale.

For purification operations, the maintenance over prolonged periods under strict bioburden control conditions is relevant. Although upstream processing is traditionally performed in aseptic and functionally closed systems, downstream processes are open operations designed to maintain low-bioburden state. Therefore, the downstream process components, such as skids, column hardware, and chromatography resins, currently available have been designed for low bioburden rather than aseptic operations. One of the key design elements of continuous manufacturing is the interconnection of upstream and downstream operations and the capability of the system to run for extended durations at ambient temperature, maintaining a bioburden-free state.

As it was already mentioned, the continuous processing systems suppose a challenge in the feed systems, for which not only the lot-to-lot variability must be considered but also the flow rate of material through the process must be defined and controlled, since it is essential to demonstrate material traceability during the entire production process. Therefore, a complete understanding of material flow in the system is essential to divert or recall the potentially affected material.

Regarding viral inactivation, continuous manufacturing implies a change in the way this topic is traditionally addressed in batch manufacturing. Currently, the biopharmaceutical industry is investigating the impact on viral inactivation and removal capacity of downstream process steps, low pH inactivation, or virus retentive filtration.

Viral inactivation at low pH focuses on enveloped viruses, which experiment an irreversible loss of viral activity after exposure at pH<3.9. However, variables in inactivation process include exposure time, temperature, product concentration, volumes, flow rates, and the presence or absence of contaminant proteins [52]. Therefore, the typical manufacturing process for monoclonal antibodies includes a viral inactivation step generally implemented after the protein A capture step, because the protein A eluate has a pH ~3.5, which fulfills the requirements for viral inactivation. Therefore, the eluate of a single downstream cycle of the protein A column is collected in a vessel and stored for the required inactivation time of 30–120 min, depending on the process parameters. Using this approach, the eluted product contained inside the viral inactivation vessel has the same residence time, and homogeneous viral reduction is achieved. Subsequently, process fluid is neutralized and further processed. Although the literature indicates great variation in protein aggregation behavior depending on the protein, in general it is accepted that during low pH exposure products tend to form aggregates. Moreover, it has been demonstrated for an IgG4 antibody that monomer loss over time under various pH conditions follows an exponential decay kinetic [53]. In consequence, the viral inactivation step is a critical unit operation depending on the exposure time, whereas a minimum inactivation time is needed to achieve robust viral reduction, the longer the inactivation takes, the higher the product loss due to aggregation will be.

In order to develop a low pH viral inactivation operation unit, Klutz *et al.* [50] discussed different reactor concepts regarding their applicability for continuous viral inactivation, based in the following principles: (i) narrow residence time distribution, (ii) long residence times of up to 120 min, (iii) 100% single-use technology, (iv) low investment and operation costs, (v) good scalability and wide operation window regarding flow rates, (vi) robust operation, and (vii) no active mixing to avoid fouling or product degradation inside the mixer. The coiled flow inverter was chosen as the most appropriate concept since the minimum residence time in this reactor guarantees for all molecules entering the continuous viral inactivation at least the inactivation time in the batch inactivation step. Furthermore, through theoretical calculations two different design approaches were studied. One of them was based in the maintenance of the minimum residence time inside the reactor equal to the inactivation time reached in batch inactivation, which would be the safer design method from a regulatory perspective. The second approach was based in the assumption that the same viral reduction value in continuous inactivation is reached, resulting in the preferred design method from an economic perspective. The viral reduction value and the monomer loss during inactivation were characterized for each design approach.

Another suitable approach is the viral inactivation mediated by ultraviolet-C irradiation, which has been used for successful inactivation of nonenveloped and enveloped viruses [54]. Continuous flow UVC reactor (UVivatec) provides highly efficient mixing and maximizes virus exposure to the UV light through a helical channel irradiated from the inside to the outside by rod-shaped UV light of 254 nm. These vortices generated in this helical channel would provide highly efficient mixing in the fluid stream, optimizing virus exposure to the light source. The uniform delivery of the UVC irradiation guarantee that the required residence times in the irradiation chamber are extremely short and the UVC treatment can be accurately controlled.

On the other hand, it has been reported that the ability to isolate and reject material that is out of specification if the process is no longer in a state of control is an important aspect of a continuous manufacturing strategy [6]. If the specified process controls do not provide assurance of adequate process performance and product quality, for example, during planned start-up and shut down or even during normal operation, it will be necessary to remove the affected material. The extent of material to be isolated and rejected will depend on the duration, severity of the disturbance, and the mixing patterns of the system. In addition, the availability of process monitoring systems capable to detect excursions from the target CPP or CQA values at different stages of the process and in a real-time manner as well as and the knowledge of the product residence time distribution will allow the physical separation of the non-conforming material.

Finally, for the implementation of continuous manufacturing, it is essential to define the criteria for product collection, product diversion, and rejection of an entire batch, indicating clearly when, how, and who makes such decisions.

18.12 Scale-Down for Continuous Bioprocessing

Scale-down systems are laboratory-scale models used for developing a manufacturing process that will be scaled up to pilot and eventually production scale. In addition, they can be used prospectively to imitate a large-scale process since it is known that scale-down systems are useful for characterization and process-development studies, such as the investigation of the interrelationship of process parameters and critical quality attributes, or the study of excursions from the target process parameters values, establishing process comparability as well as supporting manufacturing investigations [55]. Consequently, for licensed processes, they play an important role in supporting process changes.

From the regulatory point of view, scale-down models that accurately reflect manufacturing process steps have been specifically required in several documents. For instance, virus clearance studies in accordance with ICH-Q5A [56] are regularly performed using scale-down purification systems because it does not make sense to spike live virus into a unit located in a cGMP commercial manufacturing area. According to this document "the level of purification of the scale-down model should represent the production scale as closely as possible." In addition to viral clearance studies, scale-down models have been used to evaluate removal of host cell-derived impurities such as nucleic acids and host cell protein, evaluate removal of media additives, determine useful life of chromatographic resins, and so on. Moreover, recent guides allow the use of small-scale models to estimate process variability, since "laboratory or pilot-scale models designed to be representative of the commercial process can be used to estimate variability" [47].

Qualification of scale-down models comprises several aspects, including model design, performance, and quality. Design refers to specifications and geometry of the equipment (for instance a bioreactor) and ancillary systems. Ideally, they should be similar for the sale-down model and the commercial scale counterpart. For example, the ideal scale-down model of an upstream operation will use the identical vessel geometry and oxygen control strategy as the commercial scale. This will be reached employing the same sparger design, with the same location in the vessel and identical control set point. Performance alludes to the response to input parameters as well as the capability of accurately targeting and maintaining desired outputs, such as cell metabolism, growth, and viability of product titer. Recently, the potential of a system's biology tool for assessing the scalability of mammalian cell perfusion systems through performance comparison was reported [57]. DNA microarray-based transcriptomics was used to evaluate the dynamics of the overall macrolevel gene expression profile across long-term perfusion cultures of BHK cells, both for laboratory- and manufacturing-scale perfusion bioreactor cultures. Based on the results, it was possible to demonstrate that the laboratory-scale model provided a representative reflection of the manufacturing scale bioreactor, allowing the comparison of age-dependent behavior changes. Finally, quality includes the system's ability to make a product that meets predetermined quality attributes such as product purity and integrity, glycosylation, and potency [58,59].

To demonstrate equivalence of the scale-down model, the statistical two one-sided test (TOST) is commonly used, both for performance or quality attributes. However, it has been reported that a simpler approach may be applied comparing quality attributes of material generated from the scale-down model to the acceptance ranges and criteria established for the large-scale product.

The establishment of a suitable scale-down model is challenging because of geometric differences between scales; process performance at manufacturing scale often varies form bench scale performance, usually exhibiting differences in parameters such as cell growth, protein productivity, and/or dissolved carbon dioxide concentration. Even more complications can appear for continuous processing, because of several particular characteristics of these systems such as the difficulties to maintain long-term cultures and sterile interfaces. For microscale bioreactors, one of the barriers for its adoption is the need of efficient miniaturized systems for gas transport and fluid addition as well as efficient cell retention devices. As an example, the alternating tangential flow or ATF system is well known because it is capable of maintaining high cell density applications during long periods. They are based in the use of hollow fiber filters operated in combination with alternating fluid flow. This prolongs filters life without fouling. However, these systems are currently unable to operate at scales smaller than 1 l working volumes.

The suitability of a 12 l bioreactor scale-down model to represent the manufacturing scale (2000 l bioreactor) for a perfusion process performed at Genzyme and lasting about 60 days was demonstrated by DiCesare et al. [60]. Some differences between both models were listed, for instance, the material of the bioreactor (glass for the scale-down model and stainless steel for the manufacturing scale), the material of the device for harvest collection (plastic bags at the small scale and stainless steel vessels at the large scale), impeller configuration (dual elephant-ear impeller at small scale versus single elephant-ear impeller for the 2000 l bioreactors), and impeller to tank ratio, which was about 1.6: 1. In contrast, for model development, it was possible to keep the seed train characteristics as well as the pCO_2 and agitation level – both critical for cell culture performance.

To demonstrate equivalence between laboratory and commercial scale, the sample size (i.e., the minimum number of runs to be performed at the small scale) was calculated using the following equation:

$$N = \frac{2(Z_{1-(\alpha/2)} + Z_{1-\beta})}{k^2},$$

where N = number of runs; α = confidence level; $1-\beta$ = power; Z = inverse of cumulative standard normal distribution function; k = number of standard deviations, resulting in a minimum of 5 runs to be performed at the laboratory-scale model.

The qualification of the cell culture scale-down model was performed by evaluating the statistical equivalency, using TOST, of performance parameters such as cell culture growth, metabolism, and productivity. Besides, for product quality attributes of purified material assessment was made on the basis of the acceptance criteria for different characterization parameters such as specific activity, glycosylation profile, and charge variants.

18.13 Impact of Single-Use Systems in Process Validation

As for traditional batch processes, the implementation of single-use disposable systems can significantly reduce the efforts required for process validation, increasing significantly the flexibility in several operations. Also, if single-use equipment is used, the engineering and installation effort needed is reduced. As single-use devices are usually "ready-to-use," cleaning and sterilization efforts are made by the vendor. For example, the adoption of single-use chromatography columns or membranes is considered an attractive option for biotherapeutic manufacturing as well as the buffer storage, bioreactor, and single-use sensors.

It has been reported that by 2020 more than 50% of new commercial manufacturing facilities will be based on single-use technology [61]. Moreover, it is expected that future bioprocessing will be performed based on the principles of (i) continuous processing, (ii) 100% single-use equipment, (iii) closed processing, where the product is not exposed to the room environment, and (iv) a single cleanness class D "ballroom," including all operations from fermentation to viral filtration and even buffer and culture media preparation tanks [50]. This will allow faster, smaller, inexpensive, more flexible, and sustainable facilities.

18.14 Batch and Lot Definition

ICH Q7 defines batch (or lot) as "a specific quantity of material produced in a processes or series of processes so that it is expected to be homogeneous within specified limits. In the case of continuous production, a batch may correspond to a defined fraction of the production. The batch size can be defined either by a fixed quantity or by the amount produced in a fixed time interval." According to FDA (21 CFR 210.3 (2)) "Batch means a specific quantity of a drug or other material that is intended to have uniform character and quality, within specified limits, and is produced according to a single manufacturing order during the same cycle of manufacture" and "Lot means a batch, or a specific identified portion of a batch, having uniform character and quality within specified limits; or, in the case of a drug product produced by continuous process, it is a specific identified amount produced in a unit of time or quantity in a manner that assures its having uniform character and quality within specified limits."

Therefore, it is the responsibility of the manufacturer to define the batch in a continuous process according to an adequate scientific rationale, and it may correspond to a defined fraction of the production that is characterized by its intended homogeneity, but taking into account the availability of enough data to demonstrate material traceability. The batch size can be defined either by a fixed quantity or by the amount produced in a fixed time interval.

For example, for the manufacturing process of BeneFIX (Pfizer), the production bioreactor is inoculated at low cell density from the seed train, and as the culture reaches higher cell density, bulk of the broth is harvested and the remainder is used as seed for the next passage with the addition of fresh medium, in a process referred to refeed process. This operation is repeated several times until the validated cell generation limit is reached. For this process, a batch is defined as the product obtained from a single culture passage in a refeed process, prior to refeed. Although for other continuous process, a batch can be produced from multiple thaws, every BeneFIX batch can be traced to a single cell bank thaw. During purification, material from different bioreactors is processed separately until the second chromatography step, then the eluates are pooled and they are sequentially processed into a single eluate pool. Using this strategy, 2500 batches of BeneFIX have been produced since 1997, the year of BeneFIX approval [51].

A different approach is described in Ref. [41] for a monoclonal antibody produced with a fully continuous process. In automated and integrated system, a batch was defined as 1 day of processing, therefore, 31 independent batches can be obtained in 31 days of system operation.

Finally, it has been described that under continued state of control operation is possible to assure the homogeneity of large quantities of material, even though different lots of raw materials and different processing conditions may have been used. In those cases, it is important to clearly define the criteria that describe the state of control operation, establishing the product and process data that demonstrate continued conformance with these criteria [62].

18.15 Conclusion

Current regulatory guides and concepts, traditionally applicable to batch processes support continuous manufacturing. However, special considerations must be taken into account such as monitoring and control strategy, validation approach and batch definition. However, it is expected that multiple benefits of continuous manufacturing will trigger significant advancements both in science, engineering, and regulations to support its implementation in the production of next-generation biopharmaceuticals.

References

1 Xenopoulos, A. (2015) A new, integrated, continuous purification process template for monoclonal antibodies: process modeling and cost of goods studies. *J. Biotechnol.*, **213**, 42–53.

2 Walther, J., Godawat, R., Hwang, C., Abe, Y., and Sinclair, A. (2015) The business impact of an integrated continuous biomanufacturing platform for recombinant protein production. *J. Biotechnol.*, **213**, 3–12.

3 Konstantinov, K.B. and Cooney, C.L. (2015) White paper on continuous bioprocessing. May 20–21, 2014 continuous manufacturing symposium. *J. Pharm. Sci.*, **104** (3), 813–820

4 Palmer, E. (2015) Vertex, J&J, GSK, Novartis all working on continuous manufacturing facilities. FiercePharma Manufacturing, February 9.

5 FDA (2004) Pharmaceutical cGMPs for the 21st century. A risk-based approach. Final report. Department of Health and Human Services, Rockville, MA.

6 Lee, S.L., O'Connor, T.F., Yang, X., Cruz, C.N., Chaterjee, S., Madurawe, R.D., Moore, C.M.V., Yu, X.L., and Woodcock, J. (2015) Modernizing pharmaceutical manufacturing: from batch to continuous production. *J. Pharm. Innov.*, **10** (3), 191–199.

7 FDA (US Food and Drugs Administration) (2004) Guidance for Industry PAT – A Framework for Innovative Pharmacutical Manufacturing and Quality Insurance. US Department of Health and Human Services, Rockville, MA.

8 Read, E.K., Park, J.T., Shah, R.B., Riley, B.S., Brorson, K.A., and Rathore, A.S. (2009) Process analytical technology (PAT) for biopharmaceutical products: part I. Concepts and applications. *Biotechnol. Bioeng.*, **105** (2), 276–295.

9 Yu, L.X., Amidon, G., Khan, M.A., Hoag, S.W., Polli, J., Raju, G.K., and Woodcock, J. (2014) Understanding pharmaceutical quality by design. *AAPS J.*, **16** (4), 771–783.

10 Yu, L.X., Lionberger, R.A., Raw, A.S., D'Costa, R., Wu, H., and Hussain, A.S. (2004) Application of process analytical technology to crystallization process. *Adv. Drug Deliv. Rev.*, **56** (3), 349–369.

11 De Palma, A. (2005) Moving forward the FDA's PAT initiative. *Genet. Eng. Biotechnol. News*, **25** (15), 56–58.

12 Willis, R.C. (2004) Process analytical technology. Today's Chemist at Work, February, pp. 21–22.

13 Lopes, J.A., Costa, P.F., Alves, T.P., and Menezes, J.C. (2004) Chemometrics in bioprocess engineering: process analytical technology (PAT) applications. *Chemometr. Intell. Lab. Syst.*, **74**, 269–275.

14 Teixeira, A.P., Oliveira, R., Alves, P.M., and Carrondo, M.J.T. (2009) Advances in on-line monitoring and control of mammalian cell cultures: supporting the PAT initiative. *Biotechnol. Adv.*, **27**, 726–732.

15 Junker, B.H. and Wang, H.Y. (2006) Bioprocess monitoring and computer control: key roots of the current PAT initiative. *Biotechnol. Bioeng.*, **95** (2), 226–261.

16 Mercier, S.M., Diepenbroek, B., Wijffels, R.H., and Streefland, M. (2014) Multivariate PAT solutions for biopharmaceutical cultivation: current progress and limitations. *Trends Biotechnol.*, **32** (6), 329–336.

17 Calvet, A. and Ryder, A.G. (2014) Monitoring cell culture media degradation using surface enhanced Raman scattering (SERS) spectroscopy. *Anal. Chim. Acta*, **20** (840), 58–67.

18 Berry, B.N., Dobrowsky, T.M., Timson, R.C., Kshirsagar, R., Ryll, T., and Wiltberger, K. (2016) Quick generation of Raman spectroscopy based in-process glucose control to influence biopharmaceutical protein product quality during mammalian cell culture. *Biotechnol. Prog.*, **32** (1), 224–234.

19 Clavaud, M., Roggo, Y., Von Daeniken, R., Liebler, A., and Schwabe, J.O. (2013) Chemometrics and in-line near infrared spectroscopic monitoring of a biopharmaceutical Chinese hamster ovary cell culture: prediction of multiple cultivation variables. *Talanta*, **111**, 28–38.

20 Jacquemarte, R., Chavante, N., Durocher, Y., Hoemann, C., De Crescenzo, G., and Jolicouer, M. (2008) At-line monitoring of bioreactor protein production by surface plasmon resonance. *Biotechnol. Bioeng.*, **100** (1), 184–188.

21 Joeris, K., Frerichs, J.-G., Konstantinov, K., and Scheper, T. (2002) *In-situ* microscopy: online process monitoring of mammalian cell cultures. *Cytotechnology*, **38**, 129–134.

22 Kreij, K., Mandenius, C.F., Clemente, J.J., Cunha, A.E., Monteiro, S.M., Carrondo, M.J., Hesse, F., Los Molinas, M.M., Wagner, R., Merten, O.W., Geny. Fiamma, C., Leger, W., Wiesinger-Mayr, H., Muller, D., Katinger, H., Martensson, P., Bachinger, T., and Mitrovics, J. (2005) On-line detection of microbial contaminations in animal cell reactor cultures using an electronic nose device. *Cytotechnology*, **48**, 41–58.

23 Odman, P., Johansen, C.L., Olsson, L., Gernaey, K., and Lantz, A.E. (2009) On-line estimation of biomass, glucose and ethanol in *Saccharomyces cerevisiae* cultivations using *in-situ* multi-wavelength fluorescence and software sensors. *J. Biotechnol.*, **144**, 102–112.

24 Rathore, A.S., Sharma, A., and Chillin, A. (2006) Applying process analytical technology to biotech unit operations. *BioPharm*, **19**, 48–57.

25 Rathore, A.S., Parr, L., Dermawan, S., Lawson, K., and Lu, Y. (2010) Large scale demonstration of a process analytical technology application in bioprocessing: use of on-line high performance liquid chromatography for making real time pooling decisions for process chromatography. *Biotechnol. Prog.*, **26** (2), 448–457.

26 Joshi, V., Kumar, V., and Rathore, A.S. (2015) Rapid analysis of charge variants of monoclonal antibodies using non-linear salt gradient in cation-exchange hight performance liquid chromatography. *J. Chromatogr. A*, **1406**, 175–185.

27 Rathore, A.S., Li, X., Bartkowski, W., Sharma, A., and Lu, Y. (2009) Case study and application of process analytical technology (PAT) towards bioprocessing: use of tryptophan fluorescence as at-line tool for making pooling decisions for process chromatography. *Biotechnol. Prog.*, **25** (5), 1433–1439.

28 Brestrich, N., Sanden, A., Kraft, A., McCann, K., Bertolini, J., and Hubbuch, J. (2015) Advances in inline quantification of co-eluting proteins in chromatography: process-data-based model calibration and application towards real-life separation issues. *Biotechnol. Bioeng.*, **112** (7), 1406–1416.

29 Liu, L. (2015) Antibody glycosylation and its impact on the pharmacokinetics and pharmacodynamics of monoclonal antibodies and Fc-fusion proteins. *J. Pharm. Sci.*, **104** (6), 1866–1884.

30 Kildegaard, H.F., Fan, Y., Sen, J.W., Larsen, B., and Andersen, M.R. (2016) Glycoprofiling effects of media additives on IgG produced by CHO cells in fed-batch bioreactors. *Biotechnol. Bioeng.*, **113** (2), 359–366.

31 Ahn, W.S., Jeon, J.J., Jeong, Y.R., Lee, S.J., and Yoon, S.K. (2008) Effect of culture temperature on erythropoietin production and glycosylation in a perfusion culture of recombinant CHO cells. *Biotechnol. Bioeng.*, **101** (6), 1234–1244.

32 Hossler, P., Khattak, S.F., and Li, Z.J. (2009) Optimal and consistent protein glycosylation in mammalian cell culture. *Glycobiology*, **19** (9), 936–949.

33 Fan, Y., Jimenez Del Val, I., Müller, C., Lund, A.M., Sen, J.W., Rasmussen, S.K., Kontoravdi, C., Baycin-Hizal, D., Betenbaugh, M.J., Weilguny, D., and Andersen, M.R. (2015) A multi-pronged investigation into the effect of glucose starvation and culture duration on fed-batch CHO cell culture. *Biotechnol. Bioeng.*, **112** (10), 2172–2184.

34 Moremen, KW., Tiemeyer, M., and Nain, A.V. (2012) Vertebrate protein glycosylation: diversity, synthesis and function. *Nat. Rev. Mol. Cell Biol.*, **13**, 448–462.

35 Pacis, E., Yu, M., Autsen, J., Bayer, R., and Li, F. (2011) Effects of cell culture conditions on antibody N-linked glycosylation-what affects high mannose 5 glycoform. *Biotechnol. Bioeng.*, **108** (10), 2348–2358.

36 Stöckmann, H., Adamczyk, B., Hayes, J., and Rudd, P.M. (2013) Automated, high-throughput IgG-antibody glycoprofiling platform. *Anal. Chem.*, **85** (18), 8841–8849.

37 Aich, U., Lakbub, J., and Liu, A. (2016) State of the art technologies for rapid and high-throughput sample preparation and analysis of *n*-glycans from antibodies. *Electrophoresis*. doi: 10.1002/elps.201500551.

38 Dashivets, T., Thomann, M., Rueger, P., Knaupp, A., Buchner, J., and Schlothauer, T. (2015) Multi-angle effector function analysis of human monoclonal IgG glycovariants. *PLoS One*, **10** (12), e0143520.

39 Liu, J., Chen, X., Fan, L., Deng, X., Fai Poon, H., Tan, W.S., and Liu, X. (2015) Monitoring sialylation levels of Fc-fusion protein using size-exclusion

chromatography as a process analytical technology tool. *Biotechnol. Lett.*, **37** (7), 1371–1377.

40 Shubhakar, A., Reiding, K.R., Gardner, R.A., Spencer, D.I., Fernandes, D.L., and Wuhrer, M. (2015) High-throughput analysis and automation for glycomics studies. *Chromatographia*, **78**, 321–333.

41 Godawat, R. Konstantinov, K. Rohani, M. Warikoo, V. (2015) End-to-end integrated fully continuous production of recombinant monoclonal antibodies. *J Biotechnol.* **213**, 13–19.

42 ICH (2000) ICH Harmonized Tripartite Guideline: Good Manufacturing Practice Guide For Active Pharmaceutical Ingredients Q7. International Conference on Harmonization of Technical Requirements for Registration of Pharmaceuticals for Human Use (ICH), Geneva.

43 ICH (2009) ICH Harmonized Tripartite Guideline: Pharmaceutical Development Q8, Rev. 2. International Conference on Harmonization of Technical Requirements for Registration of Pharmaceuticals for Human Use (ICH), Geneva.

44 ICH (2005) ICH Harmonized Tripartite Guideline: Quality Risk Management Q9. International Conference on Harmonization of Technical Requirements for Registration of Pharmaceuticals for Human Use (ICH), Geneva.

45 ICH (2008) ICH Harmonized Tripartite Guideline: Pharmaceutical Quality System Q10. International Conference on Harmonization of Technical Requirements for Registration of Pharmaceuticals for Human Use, Geneva.

46 ICH (2012) ICH Harmonized Tripartite Guideline: Development and Manufacture of Drug Substances (Chemical Entities and Biotechnological/Biological Entities) Q11. International Conference on Harmonization of Technical Requirements for Registration of Pharmaceuticals for Human Use, Geneva.

47 FDA (US Food and Drugs Administration) (2011) Guidance for Industry: Process Validation – General Principles And Practices. US Department of Health and Human Services, Rockville, MA.

48 EMA (2016) Guideline on process validation for finished products: information and data to be provided in regulatory submissions. Rev1 Corr.1.

49 ASTM Standard 2537 (2008) Validation: Standard Guide for the Application of Continuous Quality Verification to Pharmaceutical and Biopharmaceutical, American Society for Testing and Material.

50 Klutz, S., Magnus, J., Lobedann, M., Schwan, P., Maiser, B., Niklas, J., Temming, M., and Schembecher, G. (2015) Developing the biofacility of the future based on continuous processing and single-use technology. *J. Biotechnol.*, **213**, 120–130.

51 Desai, S.G. (2015) Continuous and semi-continuous cell culture for production of blood clotting factors. *J. Biotechnol.*, **213**, 20–27.

52 Aranha, H. and Forbes, S. (2001) Viral clearance strategies for biopharmaceutical safety. *Pharm. Technol.*, **June**, 26–41.

53 Mazzer, A.R., Perraud, X., Halley, J., ÓHara, J., and Bracewell, D.G. (2015) Protein A chromatography increases monoclonal antibody aggregation rate during subsequent low pH virus inactivation hold. *J. Chromatogr. A*, **1415**, 83–90.

54 Bae, J.E., Eun, K.J., Jae, I.L., Jeong, I.L., In, S.K., and Jong-su, K. (2009) Evaluation of viral inactivation efficacy of continuous flow ultraviolet-C reactor (UVivatec). *Korean J. Microbiol. Biotechnol.*, **37** (4), 377–382.

55 Schmidberger, T., Posch, C., Sasse, A., Gülch, C., and Huber, R. (2015) Progress toward forecasting product quality and quantity of mammalian cell culture processes by performance-based modeling. *Biotechnol. Prog.*, **31** (4), 1119–1127.

56 ICH (1999) ICH Harmonized Tripartite Guideline: Viral Safety Evaluation Of Biotechnology Products Derived From Cell Lines Of Human Or Animal Origin Q5A, Rev. 1. International Conference on Harmonization of Technical Requirements for Registration of Pharmaceuticals for Human Use (ICH), Geneva.

57 Jayapal, K.P., Goudar, C.T. (2014) Transcriptomics as a tool for assessing the scalability of mammalian cell perfusion systems. *Adv. Biochem. Eng. Biotechnol.*, **139**, 227–243.

58 Shimoni, Y., Srinivasan, V., Jenne, M., and Goudar, C. (2014) Qualification of scale-down bioreactors: validation of process changes in commercial production of animal-cell-derived products, part 1 – concept. *Bioprocess Int.*, **12** (5), 38–45.

59 Shimoni, Y., Goudar, C., Jenne, M., and Srinivasan, V. (2014) Qualification of scale-down bioreactors validation of process changes in commercial production of animal-cell-derived products, part 2 — application. *Bioprocess Int.*, **12** (6), 54–61.

60 DiCesare, C., Yu, M., Yin, J., Zhou, W., Hwang, C., Tengtrakool, J., and Konstantinov, K. (2016) Development, qualification, and application of a bioreactor scale-down process: modeling large-scale microcarrier perfusion cell culture. *Bioprocess Int.*, **14** (1), 18–29.

61 Rader, R.A. and Langer, E.S. (2012) Upstream single-use bioprocessing systems: future market trends and growth assessment. *Bioprocess Int.*, **10**, 12–16.

62 Allison, G., Cain, Y.T., Cooney, C., Garcia, T., Bizjak, T.G., Holte, O., Jagota, N., Komas, B., Korakianiti, E., Kourti, D., Madurawe, R., Morefield, E., Montgomery, F., Nasr, M., Randolph, W., Robert, J.L., Rudd, D., and Zezza, D. (2015) Regulatory and quality considerations for continuous manufacturing. May 20–21, 2014. Continuous Manufacturing Symposium. *J. Pharm. Sci.*, **104** (3), 803–812.

19

Continuous Validation For Continuous Processing

Steven S. Kuwahara

GXP BioTechnology, LLC, Quality, 6336 Oracle Road, #326-313, Tucson, AZ 85704, USA

Continuous processing offers many advantages to the manufacturer. These come through improvements in efficiency and speed of processing. The automotive, petroleum, chemical, and food processing industries, among others, have been using it for some time and have developed many useful methods and concepts for dealing with the continuous flow of materials and intermediates through their processes. The pharmaceutical industry has been slow to adopt these methods primarily due to a fear of regulatory problems. This fear is no longer justified as regulatory agencies are strongly encouraging the adoption of more modern and efficient methods by the pharmaceutical industries.

One of the regulatory concerns has arisen from the need to closely monitor the properties of pharmaceutical products and their intermediates. A related problem is the need to maintain the monitoring methods in a validated state. This problem is being rapidly resolved through the adoption of equipment and techniques that have been developed for other industries, mainly in the chemical processing, petroleum, and food processing areas.

19.1 Quality Management

A basic principle of quality management as applied to manufacturing processes holds that "the earlier a problem is found and fixed, the lesser the cost to the company." The basic idea here is that this minimizes the amount of labor and materials that will be lost when the defective material becomes scrap. This is often a difficult principle to maintain when using a continuous process, as the defective material may progress through several additional processing steps before a problem is noticed and the process can be stopped. This, in turn, calls for the development of rapid or continuous measurement and detection systems to continuously verify the quality of the product as it progresses through the manufacturing system.

19.2 Regulatory Considerations

A transition from batch processing to continuous processing may require the submission of new documents to the FDA, and should be done only after a review of several relevant documents such as the regulations found in Title 21, US Code of Federal Regulations, Section 312, paragraph 30 (21 CFR 312.30) and 21 CFR 314.60 and related regulations concerning changes to approved applications (licenses). In addition, there are many guidance documents such as those for Scale-up and Post- Approval Changes (SUPAC) [1–6] and comparability protocols [7,8].[1] The continuous validation of continuous processing is a variant of Phase 3 (Continued Process Verification) of Process Validation [9,10] that is aimed at providing the assurance that a process remains in a validated state. To attain this state, a process validation needs to be performed and the critical specifications such as the Critical Quality Attributes (CQA) and the Critical Process Parameters (CPP) must be established and shown to be appropriate for maintaining the quality of the product. Normally, this is done during the development of the process [11] and the Annex of ICH Q8 (R2) on pharmaceutical development should be consulted, especially if a continuous process is planned from the beginning. In most instances, the development and design will initially be for a batch process with the expectation that the specifications can be maintained during the transition to a continuous process. This will not necessarily be true, and the worker must be careful to consider the changes that must be made to accommodate the needs of a continuous process. In some instances, the nature of the monitoring tests will change because of the needs of the process monitoring system and the specifications must change to match the output of the modified or changed tests.

Access to a statistician or statistically trained quality control worker who understands processes will help validation workers, as progress is made toward a continuous process. A full time professional statistician may not be required, but advice from a person who understands both processes and the statistical procedures that may be used for their development and control will provide vital support. The support should begin during the development stage as there are many lines of investigation or process design that are best introduced during the development phase. Introducing them at a later stage will disrupt progress toward the development of a final validated process and tempt workers to employ retrospective studies that are often unsatisfactory.

19.3 Setting Specifications

If the critical specifications for the process are established during the design process and maintained through the development work and Stages 1 and 2 of the

---

1 Note that the author is located in the United States and works primarily with FDA documents. Thus the references cited in this chapter are primarily to US documents; however, it should be noted that similar documents may, and usually do, exist for other jurisdictions, especially in the European Union and Japan, thus the reader should seek comparable documents that are applicable to the specific region of the world where the products will be used.

process validation, they may also be employed as the initial specifications for the continuous validation of the process. In reality, one should expect specifications to change during the early stages of establishing the process, as the understanding of the process should grow along with the maturity of the process. It is not useful to expect that early process parameters and their specifications will remain unchanged especially as an understanding of the process grows. Setting parameters and specifications early in the development of a process and insisting that they remain unchanged, especially during a transition from batch processing to continuous processing, is not reasonable. Thus it may not be possible to complete a full and definitive validation study until the development of a product is complete and commercial production begins.

While the continuous verification of the validated state of a process is a part of methods such as process analytical technology (PAT) [12] and real time release testing (RTRT), it does not constitute the whole of these methods. However, the terminology used for these methods will be employed to discuss continuous validation, and all of this validation work must be conducted within the context of risk management [13–16] requirements. If it is possible to work closely with the regulators, the work that will be described here can serve as a part of the data required to support the introduction of these methods.

19.4 Sequence of Events

In the normal sequence of events, processes are initially designed to be batch processes. This allows the establishment of the initial validation specifications that will evolve into the specifications for the final process validation. However, before a process validation is performed there are several other validations and qualification studies that must be performed. The cleaning of the processing area and its tools and equipment must be validated, the ability to maintain critical environmental properties, and, if an aseptic process is employed, the sterilization processes must also be validated. The procedures for bioburden monitoring should also be validated. It is especially important to validate the test methods that will be employed for the process and its supporting validations [17–20]. The worker must remember that before a process can be validated, the instruments and key equipment used in the process must be qualified. Thus the test instruments must be qualified before the test method can be validated, and the test method must be validated before the process can be validated. Similarly, the test methods and procedures used to maintain the processing environment must be validated before the process is validated. The process validation is then conducted using qualified instruments and equipment with validated test methods in a validated environment.

It is important to remember that when continuous verification of a process validation is conducted, there must also be a continual verification of the validity for validation of the supporting validations. (Note that continual is used to mean an action that is repeated intermittently as opposed to one that is constantly performed.) Thus, a test method such as reflectance near-infrared spectrometry that is used to constantly monitor product as it flows by, must be periodically checked either by testing it offline or by introducing a control into the product flow. Also, the state of cleanliness of the process environment must be verified by periodic checks.

19.5 Verification of Validated States

The need for verifications of the various validated states will be dependent on the length of the continuous process. A process that takes a day or less could have its validation status verified by checks before and after the process is run. This is analogous to the before and after bioburden checks that are done for aseptic processing. Also, in the situation where a continuous processing step is performed as a part of a batch process, it may not be necessary to continuously verify the status of the process, provided that the step does not require too much time. The decision of how often a process validation will need to be verified will be dependent upon knowledge of how variable the processing and the testing methods are. Production and analytical processes cannot be assumed to be stable forever, so the workers must understand the process and its potential sources of variation.

19.6 Choice of Test Methods

When designing a continuous process, it is important to consider the test methods that will be employed to monitor the process. Many test methods that are acceptable for batch processing will not be acceptable for continuous processing because of the time lag between the sampling of the process stream and the time when the test results are available for interpretation. The speed of the process stream can be such that a considerable amount of defective product or intermediate could be produced while testing is performed. To prevent this, it will be necessary to introduce rapid testing methods that can produce interpretable test results before an excessive amount of waste is produced. In addition to the rapid testing methods, it will be necessary to employ software and computers that will be able to take the signal from the instrument generating the signal from the rapid testing method, convert it into an appropriate result, interpret the result, and stop or adjust the process as appropriate. This system will need to be compliant with 21 CFR 11 and its guidance document [21].

19.7 Types of Monitoring

The FDA PAT Guidance [12] describes the following three types of monitoring for continuous processing:

- *At line:* Measurement where the sample is removed, isolated from, and analyzed in close proximity to the process stream.

- *Online:* Measurement where the sample is diverted from the manufacturing process, and may be returned to the process stream.
- *In-line:* Measurement where the sample is not removed from the process stream and can be invasive or noninvasive.

At line measurements are usually those that are destructive to the sample or damage it to the point where it cannot be returned to the process stream. It may not be possible to avoid this type of testing in situations where test method specificity or sensitivity is critical. The advantage here is that test method verification can be performed through the use of control preparations and standards. There are also some tests such as disintegration and dissolution tests that will need to be conducted offline. In these instances, a rapid method such as disintegration testing may be preferred over a slower method such as the dissolution test. This substitution should be acceptable as long as a clear correlation can be shown between the two measurements.

Online measurements offer the advantage of returning the sample to the process stream, and thereby avoiding the cost of losing expensive intermediates or final products. Its disadvantage comes from the need to closely monitor the timing between the location where the sample is removed and when the test result can be acted upon. The time delay must be known in order to stop unsuitable material from progressing too far down the process stream.

In-line measurements are to be preferred as the delay associated with the diversion and testing of material can be avoided. There will still be a delay between the time a measurement is made and the time when a reading showing an unacceptable product can stop the process. This may be minimized through the use of electronic data processing and electronic process controls.

In general, the speed for acting on unacceptable material increases on the order of at line, online, and in-line measurements. Consequently, in-line measurements are preferred over the others, especially since an in-line measurement usually involves less sample preparation or modifications to the process line. This is bolstered by the availability of spectroscopic methods such as reflectance ultraviolet, reflectance near-infrared, fluorescence, and Raman spectroscopy. In many cases, these techniques provide a large amount of information allowing for the simultaneous measurement of more than one parameter. This, often overwhelming, amount of data may be handled through the use of multivariate analysis techniques such as multivariate regression and partial least squares regression; chemometric methods and the multivariate design of experiments. In some instances, a finger printing method similar to the use of the finger print region of an infrared spectrum or a voice print may be employed. Individual substances may not be identified, but the maintenance of the finger print will tell the operator that the intermediate or product is in an acceptable state.

The use of computerized data analysis programs will be critical. In many situations, it will be necessary to allow the software to make decisions and give directions to the process equipment. The large increase in data processing speed that has been achieved with modern computers allows for rapid decision making and controls. In the food industry, it has been possible to select individual fruits, nuts, and vegetables on a conveyor by programming process analyzers to detect unsuitable raw material and reject them during high-speed processing. With cell sorters it is possible to measure more than one characteristic and divert cells into different flow paths. Similar actions may be taken with pharmaceutical raw material and intermediates.

A regulatory concern with some monitoring systems has been their proximity to the process stream. In batch processing, samples are normally taken to a separate laboratory for analysis; greatly reducing the possibility of contaminating the product or intermediates with substances such as laboratory reagents and solvents. With continuous process monitoring methods, it is very important to protect the process stream from contamination with any reagents or solvents that are used for testing. It is also important to ensure that the sampling process itself does not expose the process stream to contamination.

19.8 Process Stream Analyzers

In some situations, a conversion from batch processing to the use of process stream analyzers and continuous processing offers such great advantages in efficiency and cost that it is advantageous to consider the conversion from a batch to continuous process. For instance, in situations where processing steps need to be modified to accommodate variances in the properties of raw material, the traditional batch testing and qualification of raw material could be replaced with a process stream analyzer that measures the identity and properties of incoming raw material and directs the modification of the downstream processing steps or rejects the material completely. Although the process stream analyzer and its design will result in significant costs, the reduced expenses related to its use in a continuous process will compensate the company over time.

Unless a process is extremely simple, continuous processing will require the use of process stream analyzers. If this is done, the testing methods used by the analyzer will need to be shown to provide results equivalent to a validated control method for measuring the particular analyte [22,23]. These studies should provide the data needed to generate specifications for the verification of the process. However, the process stream analyzer will need to be qualified and validated before it can be employed. This will mean that its software and feedback loops will need to be validated (21 CFR 11 and [21]) along with its test methods. There are standards and guidelines that will aid in selecting and designing the sampling process [24] and correlating an analyzer's results with those of a primary test method [25].

19.9 Validation/Qualification of Process Stream Analyzers

There are four sequential activities required for the validation of a process stream analyzer system by the ASTM standard practice [26]. Some of these steps are similar to the instrument qualification steps described in General Chapter <1058> of the United States Pharmacopoeia (USP) [27]. Note also that the

USP states that processes are validated, while equipment and instruments are qualified. In the following discussion, the pharmaceutical quality control analyst will note that the ASTM method may be seen as a part of the USP instrument qualification study. The following are the steps for the ASTM validation of the analyzer:

1) *Analyzer calibration:* Performed when the analyzer is initially installed or after major maintenance, the calibration testing is done to show that the analyzer meets the manufacturer's specifications and historical performance standards. If the testing shows that it is needed, the analyzer response may be adjusted to meet predetermined output levels for reference material. Note that the design qualification (DQ) step should have provided the data on the manufacturer's specifications and historical performance standards. This work can be included as a part of an installation qualification (IQ) step.
2) *Correlation for the same material:* Process stream samples are analyzed using the process stream analyzer. For applications where the process stream analyzer results must agree with the results from a primary test method (PTM), an equation should be developed to relate the process stream analyzer results to the primary test method results (PTMR) [25]. The application of the mathematical function to the analyzer result should produce results that may be considered to be the predicted primary test method result (PPTMR) for the same material. The PTM used here may be the validated test method that was originally developed for use in the batch production method.
3) *Correlation for material including effect from additional treatment to the material:* The PPTMR from 2 is used as an input to a mathematical model to predict the effect of an additive and/or a blend stock added to a base stock material that was tested by the PTM.
4) *Probationary validation:* After the correlation between the process stream analyzer results and the PTMR are established, the probationary validation is performed using an independent but limited set of materials that were not a part of the correlation activity. The object is to show that the PPTMRs and the PTMRs agree within the user-defined limits for the process stream analyzer system.

If the set of materials used here covers the range of expected test results and includes variables to demonstrate variability, accuracy, specificity, and robustness of the analyzer system results, the data gathered in steps 2, 3, and 4 could be included in a operational qualification (OQ) study.

The PTM that is used to generate the PTMRs that validate the PPTMRs produced by the process stream analyzer must, itself, be validated. In addition, the PTM should be a fairly rapid method even though it does not need to be as rapid as the test method used by the process analyzer. The PTM should become a part of the standard repertory of the quality control laboratory as it may need to be employed with very little advanced warning. The reason for this is that if the result from the process stream analyzer, which will be seen as the PPTMR, falls outside of the specifications for that measurement, the first question would be if the result is due to a problem with the product or due to the test method. To answer this question, it will be necessary to take a sample from the process stream that matches the time point of the material that produced the problematic result from the process analyzer, and test it by the PTM to see if the PPTMR can be confirmed. This check will need to be performed before any subsequent decisions can be made. This will, in turn, mean that there will be a need to quickly execute this test.

5) *General and continual validation:* After an adequate amount (statistically significant) of pairs of PPTMRs and PTMRs have been acquired from material that was not a part of the correlation study or the probationary validation, a comprehensive statistical assessment is conducted to show that the PPTMRs agree with the PTMRs within the tolerances that were established using data collected in the correlation and probationary validation studies. If this study is successful in demonstrating the correlation between the PPTMRs and the PTMRs, a quality assurance control chart (Shewhart chart, SPC chart) may be employed to monitor the differences between the PPTMRs and the PTMRs during the normal operation of the process stream analyzer to demonstrate that the agreement between the PPTMR and the PTMR that was established during the general validation study is maintained over time.

In some situations where good standards or control materials are available, PTMs may not be necessary as the control or standard could be introduced into the testing stream and the PPTMRs compared against the values that should have been obtained from the unmodified standard or control material. This scheme may not be possible for in-line testing systems where the control or standard may contaminate the process stream, but with at line or online systems where a sampling shunt is used and the tested material can be diverted from the process stream, it could be considered.

19.10 Control Charting

When choosing the particular control chart to use, it should be noted that a sample that is isolated from the process stream and the test measurements for that sample represent only a single point in time and a single sample. Therefore, any replicate testing on the sample by the PTM will produce PTMRs whose variance will only represent the variance of the particular PTM not the product stream. Consequently, the control chart chosen should be a control chart designed for use with single-point data. One such chart is the X and mR or moving range (mR) chart [28].

Step 5 as already described and the subsequent control chart can be used as a part of the performance qualification (PQ) phase of the analytical instrument qualification procedure given in the USP [27]. The control chart can then be used to show that the instrument remains in a validated/qualified state. If the process stream analyzer monitors more than one parameter and/or has several output/feedback functions, the testing for each parameter will need to be monitored separately as the PTM and the resulting PPTMR will be different because the equations for converting the result of the raw measurements into the PPTMRs will be different. While the continuous validation of the process stream analyzer is important, it is only a part of the continuous validation of a continuous process. A continuous process may have several in-line process stream analyzers monitoring a multitude of parameters. However, the PPTMRs from the process stream can be used in the continuous validation.

19.11 The Moving Range Chart

The moving range chart (X, mR chart) assumes that the data are normally distributed and requires 25 previous observations before it can be constructed. In very early stages of the work, these observations could be then taken from those used for the precontrol charts (discussed further) but they should be replaced by 25 observations from later, more established, production runs as the implementation of the continuous process proceeds. As with most statistical process control (SPC) charts, there are really two charts that are plotted against common points in time.

The X chart is constructed by using the individual observations (X_i) in the following manner:

$\bar{x} = \sum X_i/n$ (centerline), where $n=$ the number of x_i observations.

The upper control limit (UCL) and lower control limit (LCL) are UCL = $\bar{x} + d_2(\overline{mR})$, LCL = $\bar{x} - d_2(\overline{mR})$, where d_2 is the control chart constant that is used to convert ranges into standard deviations (s). The reader should consult tables of control chart constants that are found in many textbooks on statistics.

The mR chart is constructed from moving ranges using the differences between successive values of X as follows:

$$mR = |X\_2 - X\_1| \cdot |X\_3 - X\_2| \ldots$$

$\overline{mR} = \sum mR/n - 1$ (Centerline), $n-1$ is used as the number of range pairs is one less than the total number of X_i.

The upper control limit for the moving range is UCL = $D_4(\overline{mR})$. The LCL for ranges does not exist as its calculation requires a constant D_3, which is zero for less than seven replicates. Note that the constants are case sensitive such that D_2, D_3, and d_2, d_3 all exist as different numbers.

An estimate of the standard deviation (s) of the data can be calculated as follows:

$$s = \overline{mR}/d_2.$$

Note that the control limits (UCL and LCL) for control charts are actually calculated from the data. In contrast, the UCL and LCL used for process capability studies are not statistically based, but are assigned as specification limits by the company.

19.12 Continuous Validation

With continuous process validations as with most validation studies, most of the work must be done in preparation for the study. Continuous validation of a continuous process is done after the first two phases of process validation are completed [9]. All of the previously described work will need to be completed and this will provide the worker with the data that will be needed to set the specifications for the continuous verification of the validated state of the process.

During the early stages of introducing the continuous process, the process stream analyzer and any other monitoring equipment will not have completed the PQ phase of their qualification/validation studies. The process itself will be new and there will not be sufficient manufacturing runs to provide the replicates needed to initiate monitoring via control charts. In some instances, the control charting is started by using data from the preceding batch processing, but it is best to allow the process to speak for itself. In this situation, the use of a precontrol chart (stoplight chart) [29] will be useful. These charts are not dependent on the form of the distribution and work with single-point data. The centerline is based on the target specification and allows a determination of whether or not the process is ready for control charting. The target specification (centerline) is chosen by the company and is usually the targeted center point of the specification range that is desired. These values are normally developed in the design and development phase of the process.

19.13 Choosing Other Control Charts

The most commonly used control charts are based on the assumption that the data will be normally distributed. However, this is not true for all types of process data and a determination should be made during the process development stage of the nature of the data distribution. Control charts have been developed for data that follow the binomial and Poisson distributions. Also, multivariate control charts have been generated for use with software packages that may be employed with process stream analyzers that simultaneously measure more than one parameter.

The continuous validation of a continuous process may be performed through the use of appropriate control charts that monitor CQAs and CPPs. However, these charts by themselves are not sufficient as a process validation requires attention to more than the CQA and CPP. The status of the manufacturing environment and the supporting validations should be periodically checked, especially with continuous processes that are run constantly and may only shut down once or twice a year for maintenance.

19.14 Information Awareness

The continuous validation of the continuous process itself is actually a continual process. At various times (discussed above and below) the workers will need to obtain a sample from the process stream and test it using the PTM. This test result (a PTMR) can then be used in two ways. First, it will be used to confirm the corresponding PPTMR that will be monitored by the process stream analyzer. Second, the difference between the PTMR and the PPTMR will be used for the control chart that monitors the performance of the process analyzer. The PPTMRs from the process analyzer system, themselves, should be monitored using software that can detect trends and violations of control limits. It will be wise to program alert limits into the system, as it is not feasible to wait for outright out-of-specification results before taking action on a problem. In this sense, the software should be continuously generating a control chart showing the status of the process stream. This will produce a continuous verification of the validation state of the process as is required for phase 3 of a process validation [9].

The quality control worker must retain an awareness of the status of factors such as the environment of the manufacturing area where the process is operating. Matters such as the bioburden levels and presence of objectionable organisms in raw material and containers and closures, and cleanliness levels in the processing area must be known. If aseptic processing is required, monitoring of the aseptic status is needed along with knowledge about the sterilization procedures. In reality, there are usually so many factors that must be monitored that an electronic system programmed to monitor all of them and provide alerts to workers will be needed. The speed of electronic systems for making simple decisions is advantageous and should be used to provide a rapid response to problems. For instance, a computer may receive continuous input from a laser particle monitor that checks the particle count of the air in an aseptic processing area. This computer can be programmed to alert workers when a certain particle level is noted and to immediately stop the process if an extreme level is surpassed.

19.15 Cost Issues

The amount of product that could be lost is also a major factor in deciding the intervals for revalidating or verifying the validation status of the process. When a failure is found, it will be necessary to sequester all of the product made before the problem was found, going back to the last point when the product was known to be within specifications. It will also be necessary to hold all of the product that was produced after the problem was found and before the process could be stopped. The sequestered product will need to be examined to determine how much of it is really a failing product and how much of it can be accepted for further processing. For a single processing stream, the associated costs will include not only the cost of the material and labor that is related to the lost material, but also the associated overhead costs to the company. There will be a cost associated with the time that the process stream is stopped and no product can be made (but the company is open for business). The company must determine the cost that it is willing to tolerate in these situations. If the cost is intolerable, the intervals between reverification points should be reduced until the cost becomes tolerable.

19.16 Revalidations

The continuous validation of a continuous process may be a continuous activity while using process stream analyzers and other in-line testing systems, but there will be associated revalidations on a continual basis for other validations and monitoring tests. Even if the process stream analyzers show that the process is functioning normally, a failure of a validated environmental control, such as airborne particle levels or background bioburden levels, will mean that the process must be stopped until the problem can be corrected. When a continuous process is stopped, losses will mount by the minute and a restoration of the validated state must be done as soon as possible.

The need to restore the validated state as rapidly as possible will mean the lengthy testing methods must be avoided. Even with the use of a process stream analyzer, the PTM that is used to verify the PPTMRs from the analyzer must be rapid, as these PTMRs will be needed to validate the output from the analyzer. For instance the use of a 14-day sterility test will not be acceptable for revalidating a sterilization process. Rapid microbiological testing methods should be developed and validated in conformance with 21 CFR 610.12. Similar situations exist with other testing methods and companies should work at developing and validating rapid test methods that will meet their needs.

19.17 Management and Personnel

The employment of continuous validation will require the ability of employees to rapidly respond to failures or alarm/alert conditions, where the alarm is set to warn workers of an impending out-of-specification event. Quality control and manufacturing personnel will need to be available any time the process is being operated. The reason for this is that a continuous process can produce a large amount of product in a short period of time. If a failure is noted or an alarm is raised, personnel will need to quickly correct the problem to minimize the amount of product that could be lost.

From the foregoing discussion, it should be clear that the continuous validation of continuous processes would require the formation of a team of individuals with an in-depth knowledge of several areas of manufacturing. For a company that does only continuous processing, this team will be a part of its normal management structure, but if batch processing and continuous processing are both done, a separate team at an appropriate management level will be needed. This team will need the authority to request rapid help from various specialists as well as the ability to request help from consultants. As a result, at least the head of the team should be at the upper levels of company management.

The use of continuous processing coupled with the continuous validation of the process offers many advantages that will lead to reductions in unit production costs. These will include reduced final product testing that will allow reductions in inventory and a more rapid recovery of production expenses. The advantage is greatest for companies that intend to produce a product over a long term. Careful planning combined with knowledgeable management, willing to work closely with regulators, will be beneficial for stockholders and patients.

References

1. CDER, (1995) Guidance for Industry: Immediate Release Solid Oral Dosage Forms. Scale-Up and Postapproval Changes: Chemistry, Manufacturing, and Controls; *In Vitro* Dissolution Testing, and *In Vivo* Bioequivalence Documentation. CMC 5, Center for Drug Evaluation and Research, FDA, USDHHS, November.
2. Willians, R.L. (1997) SUPAC-IR questions and answers about SUPAC-IR guidance. Center for Drug Evaluation and Research (CDER), FDA, USDHHS, February 18.
3. FDA (1997) Guidance for Industry: Nonsterile Semisolid Dosage Forms. Scale-Up and Postapproval Changes: Chemistry, Manufacturing, and Controls; *In Vitro* Release Testing and *In Vivo* Bioequivalence Documentation. SUPAC-SS. CMC 7, Center for Drug Evaluation and Research (CDER), FDA, USDHHS, May.
4. FDA (1997) Guidance for Industry: SUPAC-MR: Modified Release Solid Oral Dosage Forms. Scale-Up and Postapproval Changes: Chemistry, Manufacturing, and Controls; *In Vitro* Dissolution Testing, and *In Vivo* Bioequivalence Documentation. CMC 8, Center for Drug Evaluation and Research (CDER), FDA, USDHHS, September.
5. FDA (1998) Guidance for Industry: PAC – ATLS: Postapproval Changes – Analytical Testing Laboratory Sites. Center for Drug Evaluation and Research. CMC 10, CDER, FDA, USDHHS, April.
6. CDER (2014) Guidance for Industry: SUPAC: Manufacturing Equipment Addendum. (Pharmaceutical Quality/CMC) CDER, FDA, USDHHS, December.
7. CDER (2003) Draft Guidance for Industry: Comparability Protocols – Chemistry, Manufacturing, and Controls Information (CMC). Center for Drug Evaluation and Research, Center for Biologics Evaluation and Research (CBER), and the Center for Veterinary Medicine (CVM), FDA, USDHHS, February.
8. CDER (2005) Guidance for Industry: Q5E Comparability of Biotechnological/Biological Products Subject to Changes in Their Manufacturing Process. (ICH Q5E) Center for Drug Evaluation and Research, Center for Biologics Evaluation and Research (CBER), FDA, USDHHS, June.
9. CDER (2011) Guidance for Industry: Process Validation: General Principles and Practices. (CGMP, Rev. 1) Center for Drug Evaluation and Research, Center for Biologics Evaluation and Research (CBER), and the Center for Veterinary Medicine (CVM), FDA, USDHHS, January.
10. WHO (2006) Supplementary Guidelines on Good Manufacturing Practices: Validation. World Health Organization (WHO) Technical report series, No. 937 (1).

11 CDER (2009 Guidance for Industry: ICH Q8 (R2) Pharmaceutical Development. (ICH Rev. 2) CDER, (CBER), FDA, USDHHS, November.
12 CDER (2004 Guidance for Industry: PAT – A Framework for Innovative Pharmaceutical Development, Manufacturing, and Quality Assurance. (Pharmaceutical CGMPs) CDER, CVM, Office for Regulatory Affairs (ORA), FDA, USDHHS, September.
13 ICH (2006) Guidance for Industry: Q9 Quality Risk Management. CDER, CBER, FDA, USDHHS, June.
14 European Commission (2008) Annex 20: Quality Risk Management. EudraLex: The Rules Governing Medicinal Products in the European Union. Vol. 4, EU Guidelines to Good Manufacturing Practice: Medicinal Products for Human and Veterinary Use. February 14.
15 FDA (2004) Pharmaceutical CGMPs for the 21st Century – A Risk-Based Approach. Final report. FDA, USDHHS, September.
16 FDA (2004) Draft Guidance for Industry: Premarketing Risk Assessment. (Clinical, Medical) CDER, CBER, FDA, USDHHS, May.
17 ICH (2005) Validation of Analytical Procedures: Text and Methodology. ICH Harmonised Tripartite Guideline Q2(R1), (Current step 4 version) International Conference on Harmonization of Technical Requirements for Registration of Pharmaceuticals for Human Use, November.
18 OMCL Network of the Council of Europe General Document (2014) Validation of Analytical Procedures. PA/PH/OMCL (13) 82 2R, February.
19 CDER (2015) Guidance for Industry: Analytical Procedures and Methods Validation for Drugs and Biologics. (Pharmaceutical Quality/CMC) CDER, CBER, FDA, USDHHS, July.
20 CDER (2001 Guidance for Industry: Bioanalytical Method Validation. (BP) CDER, CVM, FDA, USDHHS, May.
21 CDRH (2002) General Principles of Software Validation: Final Guidance for Industry and FDA Staff. Center for Devices and Radiological Health, CBER, FDA, USDHHS, January.
22 CDER (2003 Guidance for Industry: Comparability Protocols – Chemistry, Manufacturing, and Controls Information. CMC, CDER, CBER, CVM, FDA, USDHHS, February.
23 ASTM (2015) Standard Practice for Conducting Equivalence Testing in Laboratory Applications. American Society for Testing and Materials International (ASTM), Designation: E2935 - 15. (In recent years, the FDA and other regulatory bodies have begun to accept standards and procedures prepared by well recognized third party standard setting bodies. Examples are organizations such as the USP, ISO, and ASTM. With ASTM standards the designation is for the document itself and the numbers following the dash represent the year of approval.)
24 ASTM (2010) Standard Practice for Sampling a Stream of Product by Variables Indexed by AQL. Designation: E2762 – 10 (reapproved in 2014).
25 ASTM (2014) Standard Guide for Establishing a Linear Correlation Between Analyzer and Primary Test Method Results Using Relevant ASTM Standard Practices. ASTM Designation: D7235-14.

26 ASTM (2015) Standard Practice for Validation of the Performance of Process Stream Analyzer Systems. ASTM Designation: D3764-15.
27 USP (2008) Analytical Instrument Qualification. General Chapter <1058 United States Pharmacopeia, vol. 31 (suppl. 1). (The chapter was first published in *Pharmacopeial Forum*, 32, (6) 1784–1794, 2006.)
28 Durivage, M. A. (2015) *Practical Engineering, Process, and Reliability Statistics*. American Society for Quality, QualityPress, Milwaukee, WI. pp. 131.
29 Durivage, M. A. (2015) *Practical Engineering, Process, and Reliability Statistics*. American Society for Quality, QualityPress, Milwaukee, WI pp. 134.

20

Validation, Quality, and Regulatory Considerations in Continuous Biomanufacturing

Laura Okhio-Seaman

Sartorius Stedim North America, Validation Services, 5 Orville Drive, Bohemia, NY 11716, USA

20.1 Introduction

20.1.1 What is Continuous Biomanufacturing?

The definition is a manufacturing process that is not composed of single batches[*], which may take 10 days or so, but instead is composed of a steady stream of product manufactured all in one sequence that can potentially last as long as 90 days. Deeper reflection revels that in continuous biomanufacturing the process design is such that the distinction and separation between upstream and downstream no longer exists. Consider a facility that can operate processes 24 hours a day, 7 days a week with only a week or two of downtime for system maintenance, and the resulting impact on the manufacture of critical drugs. The facility would be modern, streamlined, and efficient with little need of personnel except for system monitoring. Simply put, raw materials are delivered into a process continuously, while the processed product materials are removed continuously.

The conversion from batch to continuous manufacturing has already been effectively applied in industries such as steel casting, petrochemical, chemical, and food. Despite vast differences between the product types, the advantages of continuous over batch manufacturing are consistent and include a steady-state operation.

There is now a growing interest in realizing the benefits of continuous processing in biologics manufacturing. Many companies are actively investing in the development of continuous systems. This new technology, does however, poses it's own challenges, which need to be addressed before large scale implementation can be considered. Thus, in some cases a continous process has not yet been fully achieved. Instead, the manufacturing process consists of a combination or hybrid of batch and continuous process steps.

Within 10 to 15 years, by some estimates, continuous processing will become the most prevalent platform in biomanufacturing. Indeed the drive to reduce

---

[*]A batch is defined by 21CFR 210.3 as "a specific quantity of a drug or other material that is intended to have uniform character and quality, within specified limits and is produced according to a single manufacturing order during the same cycle of manufacture."

Continuous Biomanufacturing: Innovative Technologies and Methods, First Edition.
Edited by Ganapathy Subramanian.
© 2018 Wiley-VCH Verlag GmbH & Co. KGaA. Published 2018 by Wiley-VCH Verlag GmbH & Co. KGaA.

overall costs has leading companies pushing to implement new solutions and technologies. The following advantages are included:

- Improvement in product quality
- Manufacturing consistency
- Efficient facility and personnel utilization
- Reduction in capital expenditure and cost

20.1.2 Improvement in Product Quality

Continuous biomanufacturing brings into play an increased dependence on quality and consistency of raw materials in the supply chain. This is essential in order to minimize variance going into the process. Risk management of raw materials is an area that is often overlooked as a Company develops new products but depending on the product there can be anything from 15 to 60 raw materials that need to be sourced and qualified before the process can be moved from initiation to completion and contamination is a special cause of concern. If, in addition, single-use technologies are incorporated into the process, there is greater dependency on technology vendors. In these cases, extra rigor in quality assurance will be required because it takes on additional importance.

20.1.3 Manufacturing Consistency

Continuous operation also has the potential to positively influence the downstream side of the supply chain, as there is greater flexibility in meeting variable demands from the market since scaling is by time extension and not by the number of batches. This is further enhanced by the short process cycle time and the opportunity to reduce inventories at hands. Demand may also be met by the mobility of smaller production systems and the opportunity to decentralize production and distribution.

20.1.4 Efficient Facility and Personnel Utilization

One of the advantages of continuous processing equipment is that the physical size of the equipment does not change anywhere near the magnitude that batch equipment changes when the process is scaled up to larger quantities.

Continuous bioprocessing provides a system that implements standardization, that is, all biological drugs are manufactured in the same system. Companies that manufacture multiple products can be designed using a standard continuous process and because the amount of equipment can now be significantly reduced full standardization can also be implemented across the whole system from the beginning of drug development right through to commercial manufacturing.

20.1.5 Reduction in Capital Expenditure and Cost

Facilities that are primarily established with stainless steel tend to have a higher level of investment costs due to larger equipment, which means more space is utilized and also there downtime for this stainless steel equipment to be cleaned and sanitized before reuse. On the other hand, continuous biomanufacturing significantly reduces capital and operating costs. Similarly, the relatively high degree of automation in continuous biomanufacturing that takes place in a "closed" system ensures that the risk of bioburden contamination is reduced if not eliminated entirely in the downstream process.

Given all of these benefits, the likely first step of most pharma companies adapting the continuous manufacturing approach will be to convert as much of the existing batch operations into the continuous manufacturing system, but an important fact to remember is that a batch process is not only converted to continuous by simply connecting together the existing batch equipment, followed by a focused investment of equipment into the continuous manufacturing platform but also by designing new fit-for-purpose unit operations and processes. Moving away from the accepted batch process is a challenge and needs to be backed by a solid strategy

This chapter will go on to explore the quality, validation, and regulatory requirements necessary to constitute an effective and stable continuous manufacturing program.

20.2 Quality

For continuous manufacturing to be successful, there must essentially be a state of control. Control is established and maintained by a robust quality and validation program, which includes strict monitoring and full traceability of all materials used in the process. Exactly the same regulatory requirements apply for continuous manufacturing as for batch manufacturing, in which it is expected that a control strategy is put into place that ensures that the manufacturing process is reproducible to the same high standard.

ICH Q7 – Good Manufacturing Practice states:

> "Each manufacturer should establish, document, and implement an effective system for managing quality that involves the active participation of management and appropriate manufacturing personnel."

ICH Q9 – Quality Risk Management states:

> "Risk is defined as the combination of the probability of occurrence of harm and the severity of that harm . . . , the protection of the patient by managing the risk to quality should be considered of prime importance."

ICH Q10 states:

> "Management must provide a state of control whereby process conditions are established so that there is a confidence and assurance that quality is maintained."

Similarly CFR 211.110(b) states:

> "In-process controls shall monitor and validate the performance of the manufacturing processes that may cause variability in the drug product."

A change from batch to continuous manufacturing is likely to result in changes to equipment, process parameters, and facilities. Maintaining a state of control provides assurance that the desired product quality is consistently met; thus, certain criteria must be established for determining when the process is in or out of control. In particuar, stringent procedures should be implemented during the start-up and shutdown of the process and for taking care of an unexpected "glitch." This ability to detect and correct process deviations helps to ensure the consistency and quality of the manufacturing process over time. Documentation should be available that addresses critical points of prevention or reduction. Such documentation should include a review of the process including personnel, facilities, equipment, and raw materials. In particular, if multiple lots of a raw material are used in continuous manufacturing, the quality and traceability must be documented and recorded.

Risk management is at the core of continuous manufacturing. The objective is to minimize or eliminate issues before they happen. For example, there may be a risk of contamination of the product if it is susceptible to bioburden contamination. Since this can be detrimental, it should be identified and mitigated. The high degree of automation in continuous processing minimizes manual operations and the closed system for both upstream and downstream also reduces bioburden risk.

This then moves into the realm of Quality by Design (QbD). QbD presents an opportunity for companies to manage the manufacturing costs without compromising the quality of drug products. In order to do this, a thorough understanding of risk and how to manage risk is essential. Thus, QbD starts early in the drug development phase and continues throughout the product life cycle. By maintaining very high-quality standards (such as getting it right the first time) a company can take steps to control a process so that non-conformances are rare. Full implementation of QbD in the pharmaceutical industry provides long lasting benefits to industry, regulators, and patients. Tools such as risk assessment and Design of Experiments (DOE) provide scientific rationale for the choices made during the process. Using the QbD approach during analytical method development and qualification produces consistent and reliable data.

20.2.1 Other Considerations in Quality

20.2.1.1 Contract Manufacturing Organizations (CMO's)

In many indutrsies such as automative, food and beverage and personal care, there are companies that do not have the finance, facilities, or personnel to provide their own production efforts will turn to a contract manufacturing organization or CMO. This often provides a positive impact on the business that is enabled by the increase in flexibility, shortened scale-up and process development time, standardization, and reduced capital requirements for an operating plant. The pharmaceutical industry is no different and like the other industries the typical scenario involves the hiring organization approaching the CMO, stating necessary requirements. The CMO will then quote for labor, services, materials, and so on. Once a satisfactory arrangement has been made, the CMO will now act as the hiring organization's factory, producing and shipping units of the product on behalf of the hiring firm.

In a nutshell, the hiring organization has outsourced its entire manufacturing process, thus the position of trust that has to be established and maintained between the hiring organization and the CMO is crucial for success. Of course the benefits usually far out way the risks but still, hiring organizations need to tread warily. The hiring organization should be aware of its own competency. This core competency results from the production of a specific drug that delivers additional value to the customer, that is difficult to imitate by competitors. Assessing the capabilities of the selected CMO can be a long drawn out process involving numerous communications as the need for capability and quality are critical.

Following are the benefits of a CMO that the hiring organization receives:

- *Cost savings* – The hiring organization does not have to pay for a facility and the equipment needed for production. They also save on labor costs such as wages, training, and benefits and thus time.
- *Advanced skills* – The contract manufacturer has certain skills and expertise that the hiring organization does not.
- *Quality* – Contract manufacturers have their own methods of quality control in place.

The risks of working with a CMO can be alarming, some of the major concerns for the organizing company are as follows:

- *Lack of control* – When a company signs the contract allowing another company to produce their product, they, for all intents and purposes are giving up a significant amount of control over that product.
- *Changes in strategy* can only be suggested or recommended. Once the hiring company signs on the dotted line, they have accepted all of the terms and conditions so they cannot force the CMO to implement a new strategy after the fact.
- *Relationships* – As with any other customer, the hiring company has to bear in mind that they may be one of many customers working with the CMO so it is imperative that the company forms a good relationship with its contract manufacturer. Most companies mitigate this risk by working cohesively with the manufacturer and awarding good performance with additional business.
- *Quality concerns* – When entering into a contract, companies must make sure that the manufacturer's standards are congruent with their own. They should evaluate the methods in which they test products to make sure they are of good quality. The company has to rely on the contract manufacturer for having good suppliers that also meet these standards.
- *Capacity constraints* – This ties in with having a good relationship and always looking to future endeavors with the CMO. If a company does not make up a

large portion of the contract manufacturer's business, they may find that they are deprioritized over other companies during high production periods. Thus, they may not obtain the product they need when they need it.
- *Loss of flexibility* – Without direct control over the manufacturing facility, the company will lose some of its ability to respond to disruptions in the supply chain or a requirement for additional product in a certain time frame.

CMOs have been perhaps the most willing to adopt single-use systems in recent years because single use is efficient (up to 75% more so than standard stainless steel), as there is virtually no down time (down by 58%) and cleaning validation for most parts is eliminated. However, on the flipside and perhaps ironically single use can be seen by CMOs to be expensive and there is an additional concern about breakage (bags) and extractables and leachables (from filters, bags, tubing, etc.). Many CMO companies balk at the lack of clear regulatory guidelines for extractables and leachables and then further cringe as they are often responsible for conducting the appropriate extractables and leachables studies, as required by regulatory, to help support the integration of the single-use component. It is of course understandable that these concerns arise as scientific studies can get very expensive and potentially can eat away at the CMO gross profit margin.

A recent (2015) survey by Aspen Brook, of the single-use market in which 100 s of end users were interviewed in various processes (with both upstream and downstream decision making responsibilities for Process Development, Pilot and Manufacturing scale applications of single-use solutions) concluded the following:

How do you view the industry adoption of single-use solutions today?

Increasing 83.4%
Stabilizing 14.5%
Decreasing 1.9%

What do you consider to be the most important benefit of single-use solutions?

Eliminate cleaning requirements	45.6%
Increase capacity without capital improvements	30.0%
Mitigate risks	10.6%
Reduce operating costs	9.7%
Improve process performance	3.8%

Which of these has been your greatest source of frustration with single-use solutions?

Price	31.0%
Reliability	29.1%
Interoperability	11.6%
Scalability	9.7%
Performance	8.7%
Documentation	7.7%

Thus, the implementation of single-use technology by CMOs continues to grow by leaps and bounds as the overall benefits far outweigh the drawbacks.

20.2.1.2 Good Manufacturing Practices (GMP)

It is the manufacturers responsibility to demonstrate that its manufacturing process is consistent and meets applicable quality specifications and guidelines.

Good manufacturing practices (GMP) are the practices required in order to conform to these guidelines to ensure that the products do not pose any risk to the public. The four main parameters are, communication, accountability, traceability, and control.

The guidelines apply to the manufacture, testing, and quality assurance of the drug product to ensure that it is safe for human use and they are constructed as general principles that a company can then adapt to produce their own detailed instructions.

- Code of Federal (CFR) Regulations requires that
- Written records shall be maintained so that the data can be used for evaluating the quality standards of each product produced.

All guidelines follow a few basic principles:

Buildings and Facilities

- Manufacturing facilities must maintain a clean and hygienic manufacturing area.
- Cross contamination of drug product must be avoided.
- Manufacturing processes are clearly defined and controlled.
- Manufacturing processes are controlled, and any changes to the process are evaluated. Changes that have an impact on the quality of the drug are validated as necessary.
- Good documentation practices – clear concise language *production record review.*
- Laboratory records, distribution records, complaint files.
- Trained operators using established and verified standard operating procedures (SOPs).
- Records of manufacture (including distribution) that enable the complete history of a batch to be traced are retained in a comprehensible and accessible form.
- The distribution of the food or drugs minimizes any risk to their quality.
- A system in place that is available for recalling any batch from sale or supply.
- Design and construction features.
- Adequate lighting, ventilation, air filtration, air heating, and cooling.
- Plumbing, sewage and refuse, washing and sanitation facilities.
- General maintenance.
- Warehousing procedures and distribution procedures.

20.2.1.3 Supply Chains

In industries such as medicinals, food, and chemicals, which are highly regulated, there is an increased dependency on quality and consistency in the supply chain.

In particular quality of excipients, active pharmaceutical ingredients, and applicable sterilization methods have to meet regulatory requirements. With raw materials there is also the additional importance of limiting lot-to-lot variability. This variability must be monitored and controlled during development of a continuous biomanufacturing system. Thus, sourcing and management of raw material vendors is quite essential and it is important that a thorough risk assessment is conducted by the company in advance in order to manage production satisfactorily. Organized risk assessment can be used as a tool with which critical vendors can be identified. These vendors can, potentially, cause major disruption in the supply chain, resulting in detrimental effects on the process. The risk assessment should be used to develop safe guards and a plan to address or mitigate potential issues, especially if these issues are directly related to quality or regulatory compliance.

The risk assessment can be broken down into several parameters, which include but are not necessarily limited to the following:

1) How many materials are required for the process, the higher the number, the increasing risk and possibility of impact should a material become limited in supply.
2) Criticality of the materials to the process.
3) Availability of the material. Are the materials supplied by one supplier or by one sole supplier and does the sole supplier have alternate production facilities.
4) The financial impact to the company, should a supplier not be able to provide the material(s).

Considering the above listed factors, a critical vendor can be defined as

A vendor or supplier that provides critical materials, without which, there is significant impact on the buyer company.

Suppliers and vendors have to be monitored and re-evaluated on a regular basis and audited accordingly. This will ensure that consistency, quality, and change management is maintained. Documented evidence of the supplier compliance should be kept to demonstrate supplier capabilities. A list of compliance requirements can be compiled and should include the following:

1) Quality and quality systems
2) Capacity
3) Audit schedules on the facility
4) Policy and procedures

20.2.1.4 Change Management and Control

When a change occurs with any material of construction (MOC) or manufacturing process, the supplier or CMO must have an established qualification procedure and timely notification procedure that can be provided to the end user. An agreement for this should be established at the beginning of negotiations to do business.

If the result of a change notification determines that there is a potential for a regulatory cause for concern or issue, it is important to address the concern as speedily as possible. The key is to have a process in place to act on any pertinent information obtained so the company can take the appropriate action with its critical vendors if necessary.

Identifying and evaluating change is a crucial part of critical vendor management. Criteria should be agreed between the critical vendor and the hiring company regarding actions to be taken when a critical, major, or minor incident occurs. Timelines for the handling of the incident should be in writing and adhered to. Immediate evaluation of regulatory impact, if applicable, should be conducted. Traceability on actions and activities must be transparent and all items must be closed out once the investigative and disposition process is completed inclusive of corrective and preventative actions (CAPA).

20.3 Validation

20.3.1 Validate to Eliminate!

The FDA defines validation as "establishing documented evidence which provides a high degree of assurance that a specific process which will consistently produce a product meeting its pre-determined specifications and quality attributes."

Process validation is critical for pharmaceutical and biotech organizations and production over varying periods of time has to be validated to provide additional operational flexibility.

- *Installation Qualification (IQ)* – establishes that process equipment and ancillary systems are capable of consistently operating within specified limits.
- *Operation Qualification (OQ)* – provides the verification documents for the intended operation of the equipment (or system).
- *Performance Qualification (PQ)* – provides the verification documents for the conformity of the new process and related systems to the intended performance throughout all anticipated product ranges.
- *Process Validation* (PV) – complete batches manufactured to provide supporting evidence of the reproducibility and consistency of the process.

Companies that are operating without an effective process validation system in place, face significant regulatory consequences including "For Cause" audits by regulators, recalls, or even plant closure. It should be noted that the application of process validation is not only limited to the drug manufacturer but also extends to include suppliers of raw materials and equipment. Regulations have also been established that direct organizations to assess the adequacy of process validation activities at their suppliers. For example, regulators expect organizations to begin process validation activities before new equipment is installed.

The intergration of hybrid technology, in which disposable and stainless steel technologies are employed together is developing rapidly and is extremely beneficial because the end user can reduce cleaning costs, down time, and the number of personnel, ultimately making more money. While many companies have stainless steel tanks, they have also progressively moved to flexible facilities with disposable systems including tanks and bioreactors. This move to disposables has reduced capital expenditures and produced a facility that is more flexible. This kind of flexibility is obviously very important when producing a variety of products at varying scales. There are some major benefits of single-use, such as low capital cost and minimal downtime that fits in and complements very well with the concept of continuous processing. Systems may be smaller or more compact. Thus, further single-use technology is often seen as the forerunner for continuous closed processing, including both up and downstream operations. Single-use technology has been steadily growing for the past 20 years and the resulting flexibility and shortened manufacturing time is key. However, there can be issues and there is a need for collaboration and alignment between the manufacturers and suppliers of single-use systems and the end user. Regulatory requirements must be considered and standardization of the qualifying procedures is becoming more necessary, especially within the realms of extractables and leachables testing.

Currently, end users often put the onus on the single use manufacturer to provide validation guides based on studies conducted with the disposable component.

PDA Technical Report No. 66, "Application of Single-Use Systems in Pharmaceuticals Manufacturing" has been developed to provide comprehensive, high-level guidance on how to qualify single-use systems.

Unlike stainless steel, single-use components are made from polymers (plastics and elastomers) and they are therefore susceptible to physical attack by acids and bases that can if powerful enough literally degrade the polymers. The degradation is dependent on the concentration of the solvent, the exposure time, and the temperature. In addition, pretreatment methods such as steam sterilization, gamma irradiation, and ethylene oxide can also adversely affect the polymers so it is essential that pretreatment steps do not exceed the manufacturers recommended levels.

It is important that each company conducts a risk-based analysis of single-use components used in the manufacturing process in order to eliminate any potential safety concerns. The analysis should take into account some of the following parameters: chemical compatibility, leak test, particulates, endotoxin, and of course extractables.

Any contact between a drug product and a single-use component such as a filter or a bag, provides the opportunity for the two to interact. When this happens it is possible that the composition of the drug product or the single-use component may be altered. The alteration could produce adverse effects in the drug product or the contact material. Adverse effects can harm a patient. This is the basis for extractables and leachables testing.

So, what exactly are extractables and leachables?

Extractables are chemical compounds that migrate from any product-contact material (including elastomeric, plastic, glass, stainless steel, or coating components) when exposed to an appropriate solvent under exaggerated conditions of time and temperature.

Leachables are chemical compounds, typically a subset of extractables that migrate into a drug formulation from any product-contact material (including elastomeric, plastic, glass, stainless steel, or coating components) as a result of direct contact under normal process conditions or accelerated storage conditions. These are likely to be found in the final drug product.

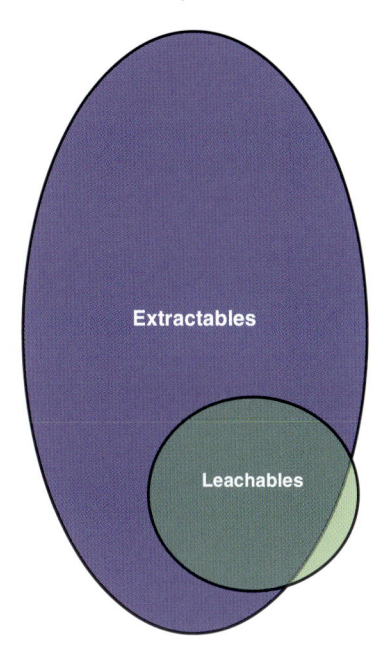

Any material that has the potential to affect the final product should be evaluated and a list should be compiled. The list should begin upstream with the buffers and should finish with materials used directly before the final fill of containers. The list should include tubing, bags, filters, O-rings, connectors, and so on.

Companies often put the onus on the disposable manufacturer to provide documents or guides based on studies conducted with the disposable component. Of course the guide cannot represent every possible scenario but they do help the end user to select a disposable component with some idea of how that component may perform in their process.

However, not every disposable manufacturer provides this basic information, so the end user does not always know how or where to start in the disposable validation process.

Many companies do not want to approach their local regulatory agency for guidance, for fear that they will turn the spotlight on themselves and be subject to audits and inspections. Few guidelines are available that tackle the validation of disposables in any great detail. So what should a company do?

Each company should conduct a risk-based analysis of disposable products used in the manufacturing process. If a risk assessment is done properly, it can enable a biopharm or biotech company to evaluate extractables data and assess worst-case toxicity. This is essential to eliminate any safety concerns.

Parenteral Technical Report 26 (PDA TR 26, revised 2008) states:

> "Given the various sources of extractables and the number of factors that affect them, it is recommended that users perform studies (when possible) with actual process fluids and the same filter type as will be used in production."

> "Once the extraction solution(s) (product, surrogate, or combination of solutions) are determined, extraction studies should be designed to mimic the worst-case actual process conditions with regard to key variables such as temperature, time, pH, and pretreatment (e.g., flushing, sterilization) steps.

> The Food and Drug Administration (FDA) has published *Guides to Inspections* that reference the requirement of manufacturers to establish an appropriate impurities profile for each Bulk Pharmaceutical Chemical."

These guidelines specifically state the need for validation procedures, which approximate routine production as closely as possible.

Similarly, the Code of Federal Regulations (CFR) mandates that materials used in the production of pharmaceutical products be compatible with the drug products. CFR Title 21, Part 211.65 states:

> "Equipment shall be constructed so that surfaces that contact components shall not be reactive, additive, or absorptive so as to alter the safety, identity strength, quality, or purity of the drug product beyond the official or other established requirements."

> Relatively new in the mix, is the BPSA, Bioprocess Systems Alliance, an organization established by disposable systems, components, and services suppliers to the Biopharm industry. The BPSA has provided two recommendation documents. Part 1 "Introduction, Regulatory Issues and Risk Assessment, and Part 2 "Executing a Program." Both documents are well worth the read as they set about clarifying the important steps in the validation of disposable products.

20.3.2 Test Conditions for Extractable and Leachable Analysis

When choosing the conditions and methods for extractables and leachables analysis, the regulatory guidelines recommend those conditions that are as close to the actual processing conditions as possible. Since it is difficult to perform an extraction test during an actual process, simulated conditions that are at least as severe as the actual processing conditions can be substituted. For instance, extracting a filter in only 1 Liter of formulation will lead to a higher level of leachables then if the extraction were performed in an entire process. This maximizes the ratio of component surface area to formulation volume and allows for detection of worst-case leachables concentrations.

Other test conditions that should simulate or exceed the actual processing conditions include pH, temperature, contact time, concentration, autoclave cycles, and flush procedures. The choice of test conditions should be the actual process conditions or close to the process conditions to accurately depict results expected with the formulation. If the conditions are too harsh, then unrealistically high levels of leachables may be present. If the conditions are too moderate, the reduced tendency to leach will give a false sense of security. The extraction conditions are unique to each application. For this reason, the limited testing performed by filter manufacturers may not be used in place of actual testing and rather be used as a guideline. The vital interaction of the formulation with the filter polymers is not necessarily mimicked with a less complex formulation. The choice of extraction medium is a fundamental decision with regulatory impact.

20.3.3 Test Solutions for Extractable and Leachable Analysis

Based on the regulatory references, it is apparent that the first choice of extraction medium and conditions should be the formulation and the respective process condition itself. Under circumstances where interfering substances preclude available analytical methods, then the use of a substitute solution is appropriate. Complex schemes for choosing these substitute solutions have been developed and published. As a result of these schemes, extractables analysis may have to be performed with multiple substitute solutions, with each one approximating the effect of one or more of the formulation's components. The "equivalency" of the substitute solutions is justified based on their likelihood to extract to a similar degree as the actual solution component.

The use of substitute solutions has been popularized by the perception that interfering substances in the actual formulation almost always preclude analytical methods. This theory is mainly due to the low levels of leachables expected when compared to the concentrations of components in the formulation.

20.3.4 Analytical Techniques for Leachables Analysis

When analyzing for leachables, it is important to choose the most appropriate analysis technique for a specific application. The common analytical techniques for the measurement of extractables and leachables fall into two categories, nonspecific and specific. Non-specific analytical techniques such as nonvolatile residue (NVR) and total organic carbon (TOC) measure the total nonvolatile weight (in the case of NVR) and total organic carbon. However, they do not differentiate between the components of a drug formulation and the leachables. A buffer or salt component in a formulation will cause an NVR weight that is much higher than the weight of the potential leachables. Likewise, the presence of an organic drug product in a formulation can lead to a TOC value that is much higher than the potential leachables. In these cases, the level of leachables is likely to fall within the error range of the test method.

Specific analytical techniques include standard analytical techniques such as high performance liquid chromatography with a UV detector (HPLC-UV), high performance liquid chromatography with a mass spectroscopy detector (LC-MS) and gas chromatography with a mass spectroscopy detector (GC-MS). Unlike nonspecific analytical methods, these methods do detect separate individual components. Thus, it is possible to detect a small level of a leachable in the presence of a much higher level of the drug product. In general, no single analytical technique should be relied upon. For instance, GC-MS is very valuable technique but will only detect volatile and semivolatile compounds. HPLC-UV can detect most compounds but may not detect certain compounds without a significant UV chromophore such as certain saturated aliphatic oligomers. The experienced analytical chemist will choose the appropriate methods depending on their knowledge of potential extractables from polymer and rubber materials.

Fourier transform infrared spectroscopy (FTIR) is another powerful analytical technique that can be used to identify individual leachables after fractionation with a specific analytical technique such as HPLC-UV. It has been suggested that FTIR can be used in combination with a nonspecific analysis technique such as NVR to identify leachables. However, FTIR does not adequately differentiate between identical chemical moieties from different compounds.

20.3.5 Description of the Model Approach

The model stream approach allows the use of a suitable model solution, or solutions, to approximate the leachables that would be present when filtering the actual pharmaceutical formulation. Model solutions are identified as those that have a similar effect on extractables and/or fall into the same main solvent group as the formulation components. For example, a formulation containing monosodium phosphate would be modeled using water, and a formulation containing polyethylene glycol would be modeled using 100% ethanol. Model solvents are used under the actual process conditions to simulate the actual effect on the polymeric equipment used. The rationale for these model solvents is that they have similar interactions with the filter device.

Once the pharmaceutical formulation has been classified and appropriate model solution(s) identified, separate filters are extracted in each model solution at worst-case conditions. A blank analysis is performed by subjecting the model solution(s) to identical test conditions without a filter. NVR levels in the blank and the extracts may then be compared to calculate weight of extractables. TOC may be performed to determine the organic content of the blank and extracts. Specific analytical methods such as FTIR, HPLC, and GC-MS may also be used to identify extractables.

The model approach may fall short of regulatory requirements in some cases since the guidances suggest that the actual formulation should be used, unless analytical interference precludes the use of standard techniques. Without first proving that the actual formulation causes significant interference, the model approach may be viewed as a rush to judgment. The use of a model solution should not be an automatic reaction to a complex pharmaceutical formulation without first demonstrating the inability to use the actual formulation.

Although the model approach gives only an approximation of what would be observed in actual product, there are circumstances when the model approach proves valuable. When the formulation contains biologically active or otherwise hazardous materials, the use of a model solution could be used to reduce safety concerns. When there is unavoidable interference with standard analytical techniques, a model solution is justified according to PDA Technical Report 26. Also, as mentioned previously, filter manufacturers regularly use model solvents to gather preliminary compatibility information for their products. Pharmaceutical manufacturers may use this information to assist in choosing filters for a particular application; however, they should avoid using this information as a replacement to performing their own analysis.

20.3.6 Actual Formulation Approach

When leachables analysis is performed with the actual pharmaceutical formulation, only one solution must be investigated. The individual components of the formulation do not need to be classified and evaluated for their effect on leachables. A justification is not required to defend the choices of model solution(s), and regulatory compliance is achieved.

During extraction, the actual formulation is placed in a container with a filter at worst-case processing conditions. From that point, the methodology mirrors the model approach. Analytical techniques are employed to detect, identify, and sometimes quantify extractables. As an initial level of detection, RP-HPLC with UV detection and GC-MS can be used. The chromatograms and spectra are compared to find peaks that are in the extract but not the blank. The following chromatograms indicate the expected result when no extractables are present.

In Figures 20.1 and 20.2, the expected elution time of leachables is from 5 to 55 min. The peaks prior to 5 min are generally highly hydrophilic and attributable to the formulation itself. Thus, there is no interference by the formulation in the region where leachable peaks are expected. Additionally, there are no peaks visible in the extract that are not visible in the blank, indicating a lack of leachables.

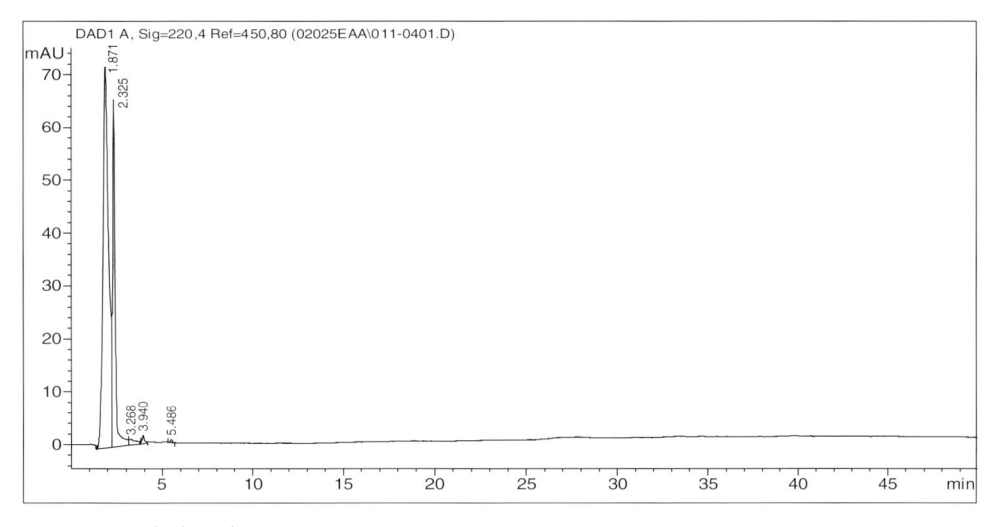

Figure 20.1 Blank analysis, HPLC-UV.

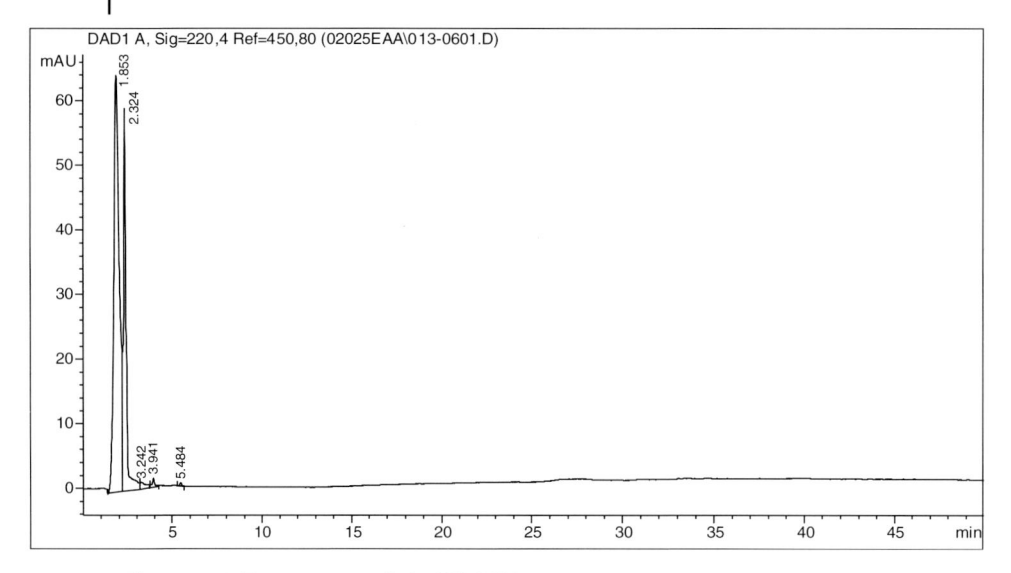

Figure 20.2 Test extract analysis, HPLC-UV.

If a leachable is detected, HPLC peaks can be collected and subsequently identified using FTIR or LC-MS. Once identification is complete, and the HPLC method can be validated to determine component concentration. If HPLC interference is a problem, sample preparation techniques may be employed to achieve the required analytical sensitivity. A variety of sample preparation techniques are available in this instance.

20.4 Regulatory

Even though the United States is considered a world leader in drug discovery and development, it is no longer in the forefront of drug manufacturing. In the past it was taken for granted that the production of medicines for the US population took place in the United States. However, now it is currently estimated that 40% of the drugs prescribed by the US doctors to the US patients and 80% of the active ingredients in these drugs, come from sources overseas, from countries such as India and China. The reasons for this shift toward an outsourced supply vary but perhaps predominantly it is because of the availability of a labor force that comes at a relatively low cost. The reliance on foreign-sourced materials makes the United States vulnerable, simply because there is a huge risk to the supply of Active Pharmaceutical ingredients (API) and thus drug shortages due to, for example, severe weather conditions in a supply location, which then disrupts on time transport and shipping delivery.

Much of the pharma industry has over the years viewed the FDA as generally against the risks of implementing new technology, if the new technology caused a decline in quality. However, this is far from the truth and regulatory expectations for reliance on quality assurance, is the same for both batch and continuous processing.

Unfortunately, despite the best efforts by the FDA to educate and support the industry in the bid to move to the continuous biomanufacturing methodology, there are still some misconceptions.

Two typical misconceptions are as follows:

1) The FDA is very stringent in their definition and interpretation when defining a batch versus a continuous operation.

   In actual fact the FDA allows companies to choose their own definition, whether by time, volume processed, and so on
2) The cost of batch manufacturing is so low compared to continuous manufacturing that it is not worth the extensive work needed to make the change and therefore the benefits are moot.

   While this may be somewhat true for certain companies that are deeply entrenched in batch manufacturing, it is not necessarily true in the case for new entrants to the market.

This is good news and an indication of the change to come as regulatory agencies are opening up the industry to the concept of continuous manufacturing and are working to break down perceived regulatory obstacles.

The success of continuous flow processing in the pharmaceutical industry has impressed regulatory agencies who are becoming increasingly supportive of this new manufacturing concept. FDA has recently provided guidance about the applicability of continuous manufacture of synthetic drugs with highly encouraging conclusions.

The FDA has issued a draft guidance entitled "Guidance for Industry "PAT A Framework for Innovative Pharmaceutical Manufacturing and Quality Assurance."" This draft guidance recommends Process Analytical Technology.

"Process Analytical Technology, or PAT, should help manufacturers develop and implement new efficient tools for use during pharmaceutical development, manufacturing, and quality assurance while maintaining or improving the current level of product quality assurance."

The main focus is on control of automated systems and procedures, for example, the automated regulation and monitoring of flow rate and pressure in a closed system. This is no different to the need to monitor these parameters in a batch system; however, it is more complex so an approach of sterile sampling can be implemented at specific points in the process. The FDA "Guidance for Industry PAT-A Framework for Innovative Pharmaceutical Development, Manufacturing, and Quality Assurance" specifically introduces continuouos biomanufacturing.

20.4.1 Current Regulatory References

Regulatory authorities are encouraging the industry to adapt to new technology.

ICH Q8(R2), Q9, Q10, and Q11 emphasize a risk-based approach.

ASTM E2537 Validation: Standard Guide for the Application of Continuous Quality Verification to Pharmaceutical and Biopharmaceutical provides a science-based approach for quality.

EU Guidelines that might be particularly relevant to continuous manufacturing include the *Guidelines for Process Validation* where the concept of continuous process verification is introduced

20.5 Conclusion

The design of new continuous processing equipment can maintain product quality while increasing the efficiency of the manufacturing process. These benefits have already been realized in other industries that have moved from batch to continuous processing.

As it is with any new technology, continuous bioprocessing is not yet mature and requires sustained innovation and guided evolution in terms of methodology and technology.

Continuous processing offers exciting new opportunities for improving the manufacture of biological products. For this technology to be successfully implemented, the joint efforts of industry, academia, regulatory authorities, and equipment vendors are needed. The future of drug manufacturing lies in high-technology, computer-controlled production facilities that can rapidly respond to changes in demand and are capable of seamlessly producing a variety of dosages and even dosage forms. This future should unfold within the United States, or it will take place elsewhere, forcing US patients to continue to rely on drugs produced on other continents.

> "It is not the strongest or the most intelligent who will survive but those who can best manage change."
>
> Charles Darwin

Further Reading

1 Thomas, H. (2008) Batch-to-continuous – coming out of age. *Chem. Eng.*, **805**, 38–40.
2 Fletcher, N. (2010) Turn batch to continuous processing. *Manufact. Chem.*, **2010**, 24–26.
3 Laird, T. (2007) Continuous processes in small-scale manufacture. *Org. Process Res. Dev.*, **11** (6), 927.
4 Utterback, J.M. (1994) *Mastering the Dynamics of Innovation*, Harvard Business School Press..
5 Bisson, W. (2008) Continuous manufacturing – the ultra lean way of manufacturing. ISPE Innovations in Process Technology for Manufacture of APIs and BPCs, Copenhagen, April 7–11.
6 Whitford, W. (2013) Single-use technology supporting continuous biomanufacturing. Integrated Continuous Biomanufacturing Conference, Barcelona, October 20–24,.
7 Konstantinov, K.B. and Cooney, C.L. (2014) Continuous bioprocessing. International Symposium on Continuous Manufacturing of Pharmaceuticals, MIT-Cambridge, USA, May 20–21.

8 Moore, C. (2011) Continuous manufacturing – FDA perspective on submissions and implementation. 3rd Symposium on Continuous Flow Reactor Technology for Industrial Applications, Lake Como, October 2–4.
9 Baker, J. (2013) Matching flows: the development of continuous bioprocessing, new initiatives in the approval of bioproducts, and assurance of product quality throughout the product lifecycle. Integrated Continuous Biomanufacturing Conference, Barcelona, October 20–24.
10 Badman, C. and Trout, B. (2014) Introductory white paper. Achieving Continuous Manufacturing. International Symposium on Continuous Manufacturing of Pharmaceuticals, MIT-Cambridge, USA, May 20–21.
11 Godwin, F. (2011) Continuous manufacturing, a regulatory perspective. INTERPHEX, New York, March, FDA.
12 Bioplan Associates (2015) 12th annual report and survey of Biopharmaceutical Manufacturing Capacity and Production, Rockville, MD.
13 PDA (2014) PDA Technical Report No. 66, Application of single-use systems in pharmaceuticals manufacturing.
14 ICH (2016), Q7: Good Manufacturing Practice Guide for Active Pharmaceutical Ingredients.
15 Weitzmann, C.J. (1997) The use of model solvents for evaluating extractables from filters used to process pharmaceutical products, part 1 – practical considerations. *Pharm. Technol.*, **10**, 44–60.
16 Stone, T.E. (1994) Methodology for analysis if filter extractables: a model stream approach. *Pharm. Technol.*, **18**, 116–130.
17 Reif, O.W., Soelkner, P., and Rupp, J. (1996) Analysis and evaluation of filter cartridge extractables for validation in pharmaceutical downstream processing. *PDA J. Pharm. Sci. Technol.*, **50**, 399–410.
18 Aspen Brook (2014), 6th Annual Survey of the Single Use Bioprocessing Market.

Part Nine

Industry Perspectives

21

Evaluation of Continuous Downstream Processing: Industrial Perspective

Venkatesh Natarajan,[1] John Pieracci,[1] and Sanchayita Ghose[2]

[1]Biogen, Engineering & Technology, 225 Binney Street, Cambridge, MA 02142, USA
[2]Bristol-Myers Squibb, Downstream Process Development, 38 Jackson Road, Danvers, MA, USA

21.1 Biogen mAb Downstream Platform Process

The downstream process utilized by Biogen for the purification of monoclonal antibodies (mAb) was developed to deliver purified drug substance and provide acceptable viral safety. The process consists of harvest steps, chromatographic columns, two virus inactivation steps, a viral filtration step, and one or two ultrafiltration steps (Figure 21.1). The platform process minimizes and controls intermediate pool volumes to allow processing very high titer cell culture without risk of exceeding tank volume constraints. The column steps were designed to operate at high adsorption capacities with minimal pH adjustments in between steps.

The downstream process begins with the clarification of the cell culture to remove cells and debris. The bioreactor is chilled to 2–12 °C and then pretreated with the addition of an acid to adjust the cell culture to a target pH of 4.8. This acidification improves clarification by inducing precipitation/flocculation of cells and debris [1]. The bioreactor is then processed through a harvest train consisting of a centrifuge followed by depth filtration and sterile filtration. The harvested cell culture pool is then adjusted to neutral pH and held for further processing. A viral inactivation step is performed on the neutralized harvest pool using the detergent Triton X-100 or lauryldimethylamine N-oxide (LDAO) [2]. The detergent is added into the hold tank and the contents are mixed and held for the validated hold time.

The inactivated pool is then processed through a Protein A chromatography step to capture and concentrate the mAb product and remove impurities. The Protein A ligand has specific affinity for mAbs, the primary binding site for Protein A being on the FC region at the junction of the $C\gamma 2$ and $C\gamma 3$ domains [3,4]. During the loading step, the mAb adsorbs to the adsorbent, while most impurities flow through the column. A post load wash step containing a high concentration of sodium chloride is used to further reduce impurities such as host cell proteins

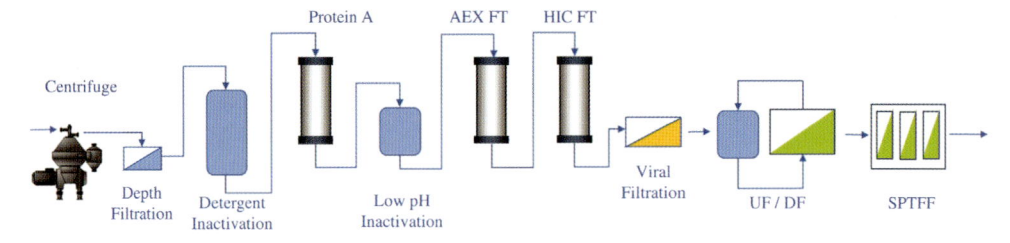

Figure 21.1 Platform downstream process for monoclonal antibodies.

(HCPs) and DNA that are bound to the adsorbent or mAb itself by electrostatic interactions. Product elution is achieved by using a low conductivity acidic solution. An elution strategy was designed to achieve several objectives: (1) minimize elution volume, (2) minimize pool conductivity, and (3) a target elution pool pH of <3.6 in order to eliminate the need for pH adjustment prior to the subsequent low pH viral inactivation step. The consequence of this design parameter is to minimize the intermediate pool volume post the Protein A column that was deemed critical as a large intermediate pool volume would then cascade through the subsequent flowthrough chromatography steps: anion exchange (AEX) and hydrophobic interaction chromatography (HIC).

The Protein A elution pool is mixed, the pH confirmed, and the pool held for the validated hold time to achieve the low pH viral inactivation. The inactivated pool is processed through the AEX and HIC steps in order to reduce the levels of HCPs, DNA, and high molecular weight product-related species (HMW). A pH adjustment is performed prior to the AEX chromatography step. The HIC step was designed without the need for a lyotropic salt to drive separation; selectivity is achieved based on a pH adjustment alone [5]. The HIC flowthrough pool is processed through a viral filter without any pH adjustment.

Two ultrafiltration (UF) process steps are used to concentrate the purified mAb pool and exchange it into the final formulation solution. The first UF process concentrates and diafilters the retentate pool into the formulation solution. A traditional UF system with a retentate tank is employed and the HIC flowthrough pool is concentrated to a high enough concentration to allow the entire pool to fit into the retentate tank prior to the diafiltration step [6]. The retentate stream is recirculated back to the retentate tank with each pass by the filter; multiple passes are required to achieve the desired concentration. The diafiltration (DF) is then performed with 10 diavolumes. The second UF process step is operated in single-pass tangential flow filtration (SPTFF) mode to further concentrate the formulated pool to its final concentration of >150 mg/ml. SPTFF utilizes a filter configuration based on multiple filters connected in series that allows the final concentration to be achieved in a single pass [6,7]. The final concentration is controlled by the number of filters placed in series and the inlet flowrate. SPTFF operation allows higher concentrations than conventional UF/DF systems because the system hold up is significantly smaller. Additional excipients could also be added to this concentrated pool to generate the drug substance (DS) prior to the final sterile filtration during the bottling process.

21.2 Potential Platform Process Bottlenecks Pertaining to Large Scale Manufacturing

The individual unit operations in antibody manufacture are executed in a batch mode. The mass of antibody drug substance produced per batch is dependent on the size of the bioreactor, the titer in the bioreactor and the overall yield of the downstream unit operations. In a well-designed antibody manufacturing process, the production bioreactor should be the rate-limiting step as all the potential DS is produced in it. Thus, the process imperatives for downstream operations are to increase yield and ensure that none of the unit operations are rate-limiting. The size of the downstream equipment would be affected by the operational philosophy – if the requirement is that only material from a single batch can be on the manufacturing floor at any given time, the entire downstream process (harvest to UF/DF) would need to be completed within the time between two successive harvests. In this case, the equipment of all the unit operations would need to increase to accommodate faster run rates. On other hand, if the operational philosophy allows multiple batches of the same campaign to coexist in the manufacturing space, then the requirement becomes that the slowest downstream operation should be completed between two successive harvests.

In spreadsheet calculations, the rate limiting step is usually identified in terms of the actual processing time. For instance, in the case of chromatography operations, the process time would simply be the time to complete one cycle multiplied by the number of cycles with the former calculated purely based on the residence time for a given step and the number of column volumes to be flushed in that step. However, in a manufacturing facility, the actual process time may not be indicative of the time needed to complete the unit operation. The number of auxiliary equipment, such as CIP skids, and the number and size of buffer hold vessels and product hold vessels can also have a significant impact on the duration of the unit operation. The sharing of buffer hold vessels between unit operations could have a seriously impairing effect on the run-rate. If multiple preps of a given buffer need to be prepared due to the loading levels and/or size of buffer hold vessels, the buffer prep turnaround time could significantly slow down the unit operation. If the product hold vessels are undersized, then considerable yield loss may be incurred or an additional operation such as SPTFF may need to be inserted to reduce elution/flow through volumes. Finally, the downstream run rate could be controlled not by a unit operation but by a piece of equipment, such as a product hold vessel or a CIP skid.

21.3 Continuous Downstream Process

Figure 21.2 illustrates a schematic of continuous downstream whose feed is harvested cell culture fluid. The schematic is similar to that reported in the literature by other investigators [8]. The flow rate into the system could be dictated either by the column sizes one wants to use in the multicolumn

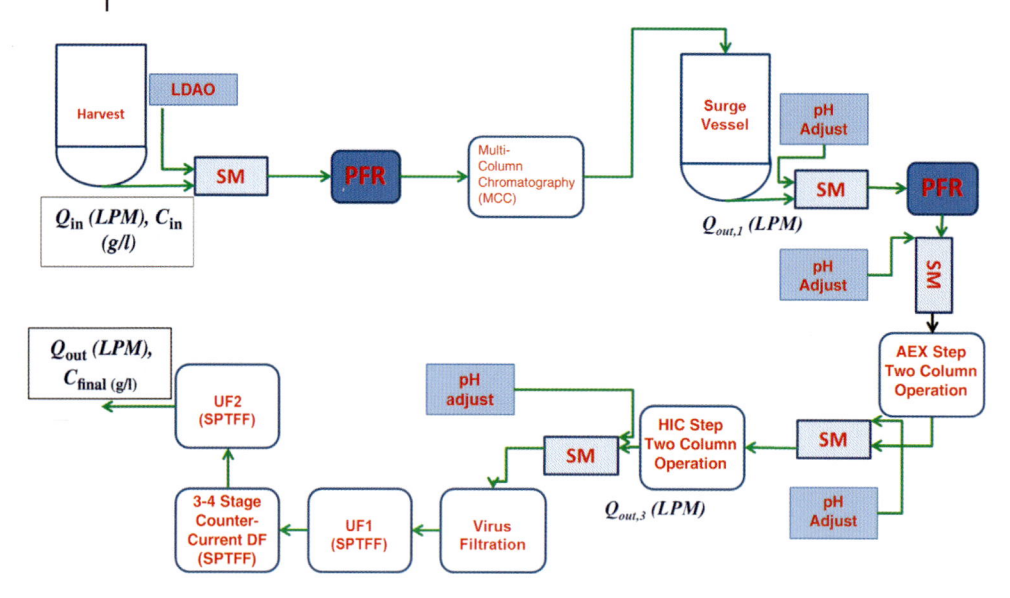

Figure 21.2 Schematic of a potential continuous downstream process.

system or by the harvest turnaround time. The advantage of the scheme outlined above is that it diminishes the size and number of product hold vessels as well as the purification equipment. The use of multicolumn chromatography (MCC) in protein A chromatography enables higher loading than can be achieved in standard batch chromatography. However, given the type of resin used, the increase may be modest (∼20–25%). Thus, the buffer volumes may not be reduced significantly (defined as at least a 2× improvement). However, recasting the buffer infrastructure to deliver buffers continuously could reduce the sizes of the prep and hold vessels. The sizing of the remaining unit operations is not expected to be significantly different from that of batch operations. Thus, the cost of consumables (resin, filters, and buffer components) to operate the continuous scheme may not be significantly different from a batch operation.

The process robustness of the continuous scheme outlined in Figure 21.2 is currently unknown. There is scant information in the reported literature on how continuous downstream would react to and recover from process upsets or how viral clearance can be validated. In conference meetings, these issues are admitted. In all probability, these issues are not show-stoppers and can be overcome with appropriate process design, and engineering. However, the efforts needed to achieve and demonstrate robustness should not be underestimated. For continuous downstream to supplant standard batch operations whose robustness is well established, there needs to be significant benefits to justify the time process development groups would need to spend to achieve levels of robustness comparable to standard batch operations. What constitutes a "significant benefit" would be dependent on the business drivers for a given company and a universal prescription may not exist.

21.3.1 Multicolumn Chromatography (MCC) for Continuous Capture

The technology of multicolumn chromatography is based on the concept of simulated moving bed (SMB) but is simpler in design and implementation complexity. The use of this technology for the protein A capture step has gained increasing popularity in the recent years [9]. There are several vendors in this space with slight variations in their product offering such as sequential multi-column chromatography (SMCC) from Novasep (Pompey, France), BioSMB from Pall Corporation (Worcester, MA), Semba Biosystem (Madison, WI), Conti-chrom (Knauer, Germany), and periodic counter current (PCC) system from GE Healthcare (Uppsala, Sweden). While each of these systems vary slightly from each other with respect to the number of column used and thereby their throughput improvement capabilities, they are all broadly based on the same concept described in the subsequent section.

21.3.1.1 Background

In any traditional batch chromatography process, due to mass transfer resistance, protein breakthrough from a column occurs well before all the protein-accessible binding sites are utilized. This results in underutilization of the available resin capacity and contributes to significant resin costs. Figure 21.3 shows a typical breakthrough curve and the amount of capacity that is usually unbound [9].

If we consider the state of a batch column before product leakage happens, the preceding layers of the stationary phase are loaded more than the layers close to the outlet of the column. Even though some sites are still available for binding, loading is usually stopped to prevent loss of valuable product. The technology of MCC tries to overcome some of this limitation by "slicing" one traditional batch column into four smaller columns. This is schematically shown in Figure 21.4. The effluent after the first column is fully saturated and can be easily captured by the subsequent column.

In a MCC process, multiple columns are used to perform the same steps in parallel such that continuous feed is achieved and all other process steps are discrete in time. In other words, the highly loaded layers are contained in the first

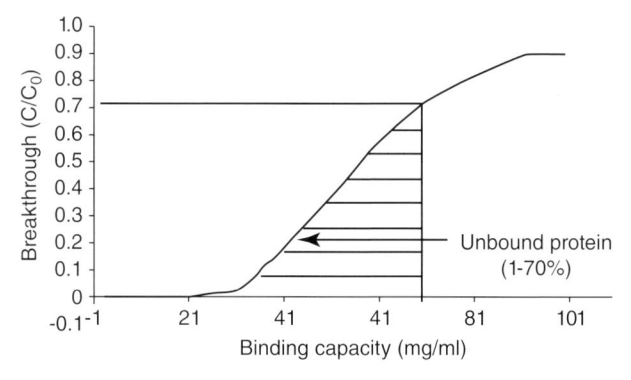

Figure 21.3 Breakthrough curve in a typical batch process. Unbound protein can be calculated by integrating the area under the curve.

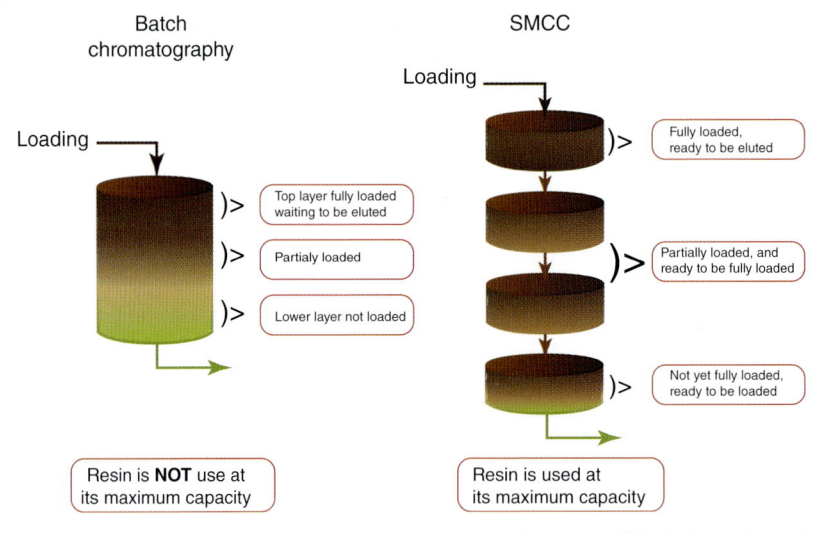

Figure 21.4 Comparison of batch chromatography and MCC (Modified after Ref. [10].)

column, which is isolated from the process and the latter steps of that chromatographic process (for e.g., wash, elution, regeneration, etc.) are applied independently. A wash step on the first column sweeps unbound target protein off the first column onto the second, thus, minimizing product loss and maximizing buffer usage. The first column is then eluted offline while loading restarts in linear sequence through the rest of the partially loaded column. The second column soon reaches a fully loaded stage as the first column, and the exact same procedure (as described for the first column) is repeated for each column completing a full MCC cycle.

21.3.1.2 Process Optimization

To optimize the parameters of a MCC operation, batch experiments are initially performed to determine breakthrough curves at different flow rates. This is usually performed for a given resin – antibody pair. The data is then fitted to a lumped rate transport model and an equilibrium model (typically employing the Langmuir isotherm) to calculate a transport parameter and the equilibrium static capacity, respectively. This information is then used to calculate key operating parameters for the MCC operation such as the number and bed height of each column, timing of feed switching, and flow-rates for each step and so forth. In silico simulations (using continuous chromatography simulation software developed by the respective vendor) are then performed to select the optimum MCC process conditions. The theoretical predictions are then experimentally verified using MCC equipment. The experimental results described in this chapter were obtained using the BioSMB system from Pall Corporation. Furthermore, the primary application of this technology was focused on Protein A capture even though it can be used for all modes of chromatography operated in the bind and elute mode.

Table 21.1 Comparison of Protein A adsorbent capacity and transport parameter.

	Q_{max} (mg/ml)	k_{oL} (h^{-1})
MAbSelect SuRe	60	648
MAbSelect SuRe LX	80	330
MAbSelect SuRe PCC	90	1200

21.3.1.3 Experimental Results

The main way to maximize throughput for the Protein A step using this technology would be to load the column to maximum capacity at the maximum possible flow rate. Hence, apart from the static binding capacity of the resin, the transport property of the resin becomes equally important. Three different versions of Protein A resin which have the same Protein A ligand were used for this evaluation and compared for optimal performance. Table 21.1 lists the equilibrium saturation capacity (Q_{max}) and lumped transport parameter (k_{oL}) calculated from breakthrough curve using the methodology described above.

As can be seen from Table 21.1, the three resins varied significantly in their equilibrium binding capacity and transport properties. MAbSelect SuRe (the most commonly used protein A resin in the biotech industry) had moderate Q_{max} and k_{oL} values. On the other hand, MAbSelect SuRe LX had a significant improvement in capacity but almost half of the k_{oL} value. This is consistent with recent findings in the literature [11]. The current Biogen batch platform has recently migrated from MAbSelect SuRe to MAbSelect SuRe LX to take advantage of the higher capacity; however, this might not be the most optimum resin for MCC application. To overcome some of the transport limitations of the LX resin, GE Healthcare recently has introduced the newest family of Protein A resin (MAbSelect SuRe PCC) that had both improved capacity as well as transport making it the most suitable for MCC application.

Optimum MCC conditions were developed on all three resins for a comprehensive and objective evaluation. It is to be mentioned that the optimization was heavily based on capacity maximization (and less on throughput) depending on facility bottlenecks mentioned in Section 21.2. Furthermore, the Protein A conditions used for the batch process were kept unchanged for the MCC process. Table 21.2 list compares the performance of the batch and MCC process across the three different resins mentioned above.

Table 21.2 clearly shows that operating the Protein A step in the MCC mode can significantly help improve step throughput without affecting step performance or product quality. A double or triple increase in productivity can be obtained based on the resin selected. It is to be further noted that the current batch Protein A step had several constraints (specific hold times for the wash and the cleaning step) that were not favorable for MCC application. Reoptimization of the Protein A step itself can further improve the throughput of the MCC process several fold, but were not included in the scope of this study.

Table 21.2 Comparison of batch and MCC performance for Protein A adsorbents.

			Process performance			Product quality			
			Loading (g/l)	Yield (%)	Throughput (g/l/d)	HMW (%)	HCP (ppm)	DNA (ppb)	Leached PrA (ppm)
MAbSelect SuRe	Batch	35	97	265	2.8	530	<2	39
	MCC	50	95	509	2.4	550	<2	4
MAbSelect SuRe LX	Batch	55	85	260	4.0	1934	2	20
	MCC	70	90	694	4.0	1330	2	4
MAbSelect SuRe PCC	Batch	60	96	369	2.8	5812	4	19
	MCC	83	92	900	2.3	4737	4	5

Differences in product quality seen between resins was primarily due to different lots of cell culture harvest used.

21.3.2 Continuous Viral Inactivation

Viral inactivation (VI), either by lowering the solution pH or by addition of detergent, is an integral part of the downstream processing of recombinant proteins produced in mammalian cell cultures. In the case of antibody processes, low pH viral inactivation is executed on protein A elution pools to leverage the elution conditions in that unit operation. The critical parameters in this unit operation are the solution pH, temperature, and the hold time. The kinetics of viral inactivation are strongly dependent on the solution pH and temperature – at pH<3.6 under ambient conditions, the kinetics are extremely rapid, resulting in effective inactivation in 5–10 min [12]. On the other hand, as the pH increases and/or temperature decreases, the kinetics become appreciably slower and hold times as high as 2 h may be needed at pH ~ 4. The stability of the molecule usually dictates the pH and temperature at which VI is executed. Thus, a hold time of 1–2 h is typically employed to account for worst-case conditions.

In current antibody manufacture, VI is carried out in the batch mode in a holding vessel. All operations involved in the inactivation process (pH adjustment, mixing, and inactivation hold) can be performed in a single tank. Two vessels are also sometimes used in which the pH adjustment and mixing are performed in one tank and the pool is then transferred to a second tank where it is held for the validated inactivation time. In addition, pH adjustment is currently a manual process. Recent improvements in the antibody platform have adjusted elution buffer conditions such that the final collected pool is at the appropriate pH for inactivation. However, the buffering capacity of the high concentration of antibody in the elution pool necessitates the collection of a considerably higher volume than that over which the product elutes. As titers continue to increase, the tankage requirements for batch VI could start presenting facility fit constraints, and technologies such as single pass UF may need to be employed to fit the process to available tankage.

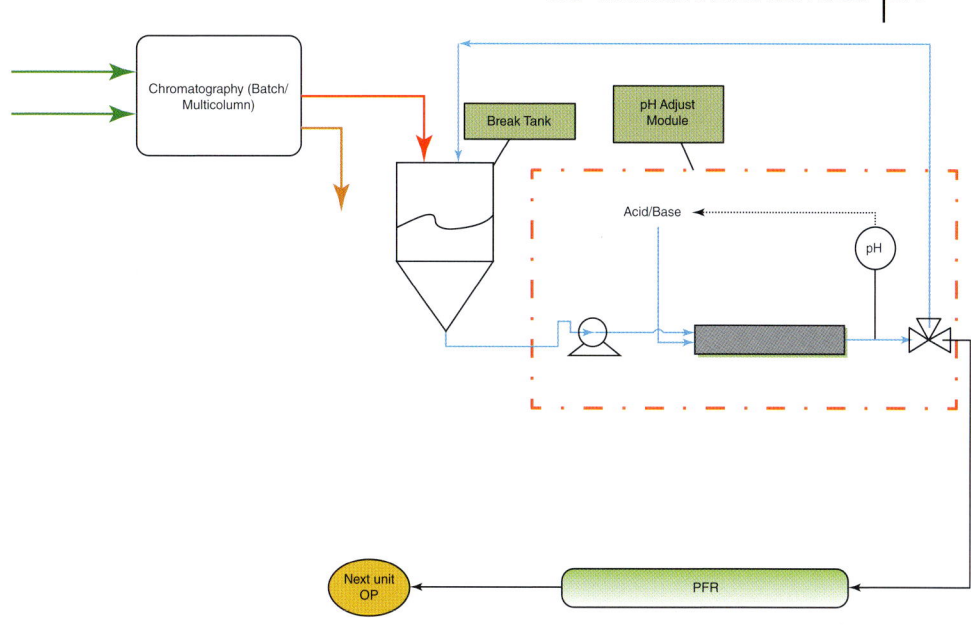

Figure 21.5 Set-up for continuous viral inactivation. PFR = plug flow reactor.

The facility fit constraints could potentially be overcome by performing VI in a continuous manner. Figure 21.5 illustrates one potential schematic for continuous VI. The system combines a pH adjustment module, a break tank, and a plug flow reactor (PFR) to provide the requisite hold time for inactivation.

The break tank would collect a finite volume of Protein A elution pool and the pH adjust module would ensure that the stream entering the PFR is at the validated pH for VI. The PFR could be a straight length of tube with mixing elements to promote radial mixing or a coiled length of tube that would leverage induced secondary flows to achieve plug flow [13]. In this setup, the critical temporal parameter would be the residence time in the PFR. The size of the PFR would be dependent on the flow rate and the validated residence time.

The LRV that can be claimed in this setup is related to the breakthrough curve of virus particles as follows:

$$\text{LRV} = -\log \frac{C}{C_o}.$$

In an ideal plug flow, all concentrations have the residence time. However, in real pipe flows, the presence of dispersion creates a distribution of residence times that needs to be accurately captured – especially at the very low concentrations – to ensure robustness of viral clearance claims. Dispersion in pipe flows is a function of Reynolds number and the LRV may be expected to depend on the Reynolds number. One way to overcome this uncertainty and eliminate extensive testing would be to oversize the PFR by increasing the residence time by a factor of 1.5 or 2. However, if the operating pH is >3.6, this may lead to long lengths of tubing.

Continuous VI could be integrated with existing batch processes or could be an element of a continuous downstream scheme. It has the potential to reduce the size and number of product hold vessels that may be needed in a facility.

21.3.3 Connected Chromatography Steps

As mentioned in Section 21.1, the typical Biogen mAb platform consists of two polishing steps usually operated in the flow-through mode. In the current manufacturing setup, the flow-through pool from the first polishing step is collected in a product pool tank, adjusted for pH and/or conductivity needed for the subsequent chromatography step and then processed through the next step. As cell culture titers (thereby product mass) increase, the volume of these intermediate pools can be a significant bottleneck in an existing manufacturing facility. One way of overcoming this limitation is through integration of multiple unit operations into a single fluid system thereby eliminating the need for expensive in-process product hold tanks. Such instances of semicontinuous connected operations do exist in the literature such as the use of tandem depth filtration followed by AEX membrane chromatography for harvesting of perfusion cell culture [14] and tandem use of Protein A and AEX chromatography [15]. At Biogen, we have found that it is most prudent to operate the two polishing steps in a connected fashion by using an appropriate in-line pH/conductivity adjustment technology (Figure 21.6) [16].

Several engineering solutions can be used to achieve the process schematic shown in Figure 21.6b. GE Healthcare currently has a skid called straight-through

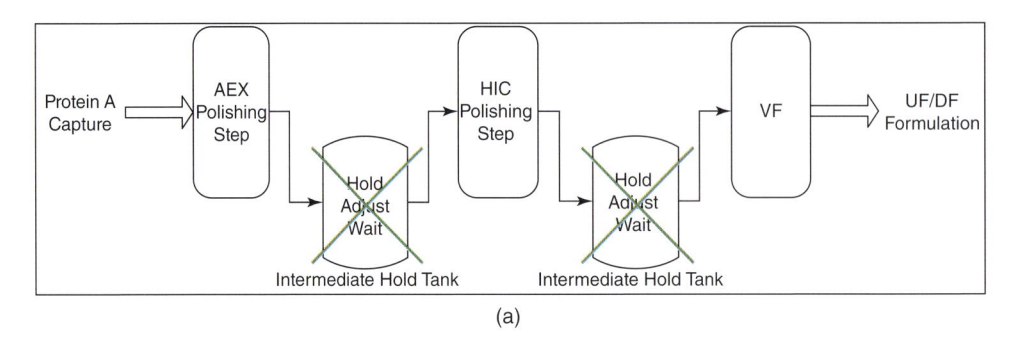

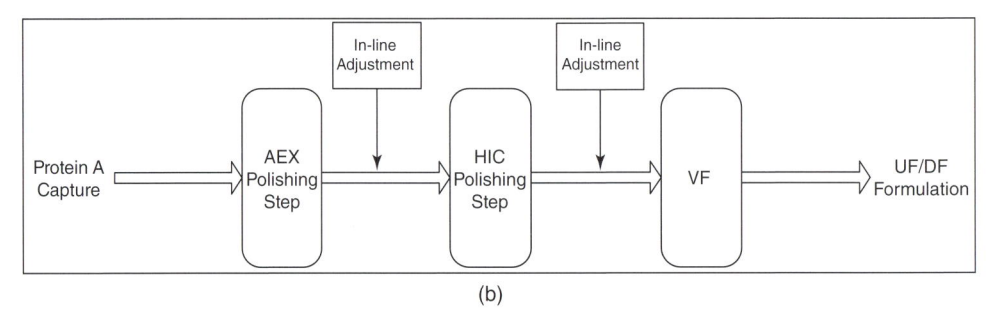

Figure 21.6 (a) Current manufacturing setup. (b) Proposed connected polishing steps.

processing (STP) capable of achieving the above mentioned process flow. To test the capability of this technology at lab scale and perform proof-of-concept (POC) studies, an existing AKTA PuRe purification system was retrofitted to achieve similar functionality as the STP skid.

21.3.3.1 Comparison of Current and Pool-Less Process

The current traditional and pool-less processes were run side-by-side for all three mAbs keeping other process parameters constant. Three of Biogen's high titer mAbs (A, B, and C) were used for this study. MAbs A and B used the typical Biogen platform where the second polishing step was HIC flow-through operated under no-salt conditions [5]. This new no-salt HIC strategy eliminated the need for addition of larger quantity of kosmotropic salts typically used to achieve selectivity on HIC and thus was an ideal candidate for this application. The AEX flow-through pool only needed to be pH adjusted by using an appropriate titration solution. On the other hand, mAb C was an atypical and challenging mAb that needed the use of a mixed mode (MM) resin (Capto Adhere) as the second polishing step to achieve the desired product quality. In this case, the AEX flow-through pool needed to be adjusted both in pH and conductivity for the subsequent Capto Adhere step.

Table 21.3 lists the step yields and key product quality data obtained from these runs. As shown in Table 21.3, comparable process performance and product quality was achieved between the traditional process and the connected process for all cases. Furthermore, mAb C demonstrates that this technology was able to support a more complicated process where a simultaneous pH and conductivity adjustment was needed for the mixed-mode step. Figure 21.7 shows a typical chromatogram from the pool-less process with the two flow-through columns running in tandem. The product flow-through peaks from the two individual columns are overlapping, which translates to a significant reduction in overall processing time at manufacturing. This is an additional benefit of this technology apart from process simplification and elimination of product hold tanks.

Table 21.3 Performance of tradition and pool-less process.

In-line adjustment	Chromatography mode	Molecule	Process	Cumulative yield (%)	HMW (%)	HCP (ppm)
pH only	AEX to HIC	A	Control	91	0.66	2.8
			Pool-less	90	0.60	2.2
		B	Control	82	0.54	0.8
			Pool-less	82	0.52	0.8
pH and conductivity	AEX to MM	C	Control	78	0.74	<LLOQ[a]
			Pool-less	78	0.71	<LLOQ

a) LLOQ: Lower level of quantitation.

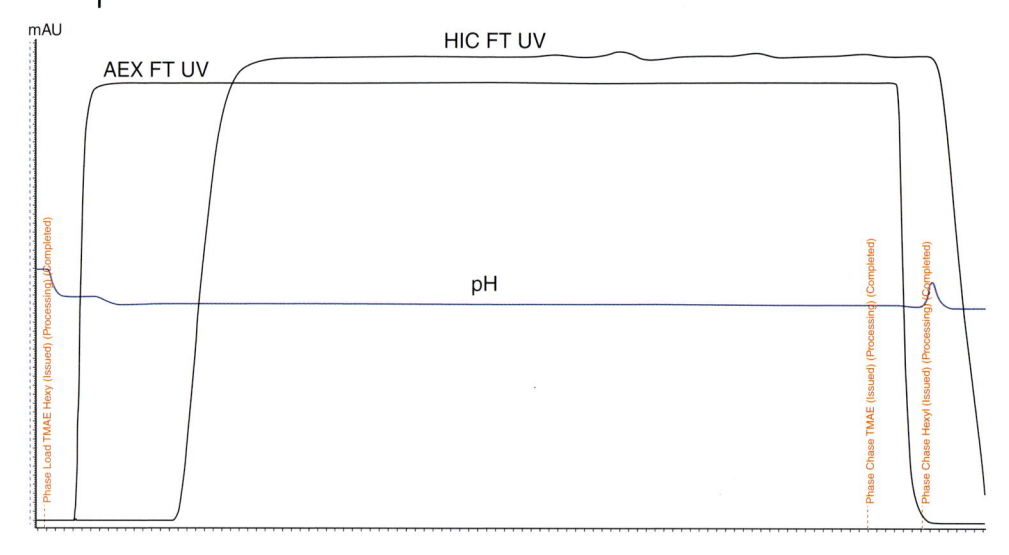

Figure 21.7 Chromatogram from the connected polishing steps.

21.3.4 Continuous UF/DF Processes

Batch UF/DF relies on a system configuration that achieves concentration and diafiltration by passing the feed pool across the filter bank multiples times and recirculating the retentate stream back to a holding tank (Figure 21.8) [6]. A typical process consists of an initial concentration step, followed by a diafiltration step and then a final concentration step to achieve the final concentration. The need for multiple passes and feed recirculation makes this configuration inherently non-continuous. The concentration steps could be made continuous through the use of SPTFF, in which multiple filters are connected in series such that concentration is achieved in a single pass through the filter (Figure 21.9). But this still leaves the diafiltration step as the obstacle to realizing a continuous UF/DF process.

Continuous diafiltration could be achieved repeating a sequence of operations that consist of the addition of diafiltration solution followed by a concentration

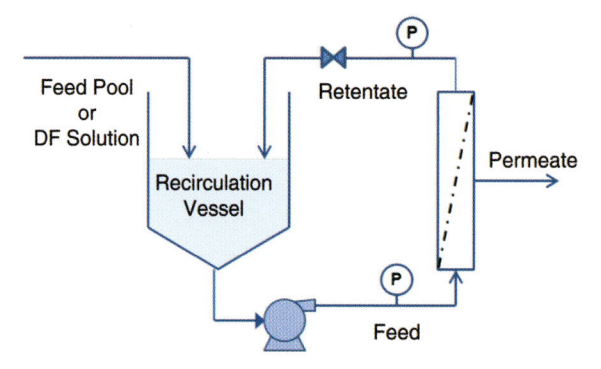

Figure 21.8 Batch UF/DF system.

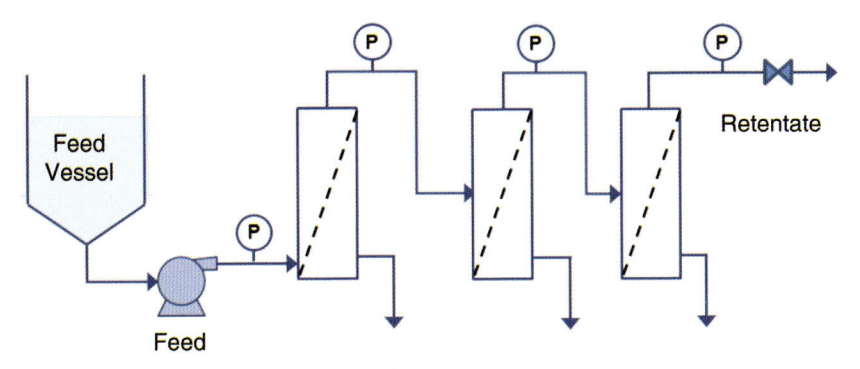

Figure 21.9 Single pass tangential flow filtration system.

step using SPTFF. The rate of addition of the diafiltration solution and the permeate rate at each stage would be matched. Two configurations are most common: cocurrent and countercurrent (Figure 21.10) [7]. In the cocurrent configuration, diafiltration solution is added into the feed stream of each filter stage and the permeate from each stage is discarded. In the countercurrent configuration, diafiltration solution is added to the final filter stage and the permeate from each stage acts as the diafiltration solution and is added back into the feed stream of the previous stage.

In order to compare the performance of the fed-batch, cocurrent, and countercurrent DF configurations, it is necessary to define the rate of exchange of solute that occurs during diafiltration. This rate of exchange can be derived from a simple mass balance on the solute being removed:

$$\text{Exchange Rate} = \frac{C_0}{C_R} = e^{(DV \cdot S)},$$

where C_o is the initial solute concentration of solute, C_R is the solute remaining in the retentate, DV is the number of diavolumes of solution (where 1 DV = the

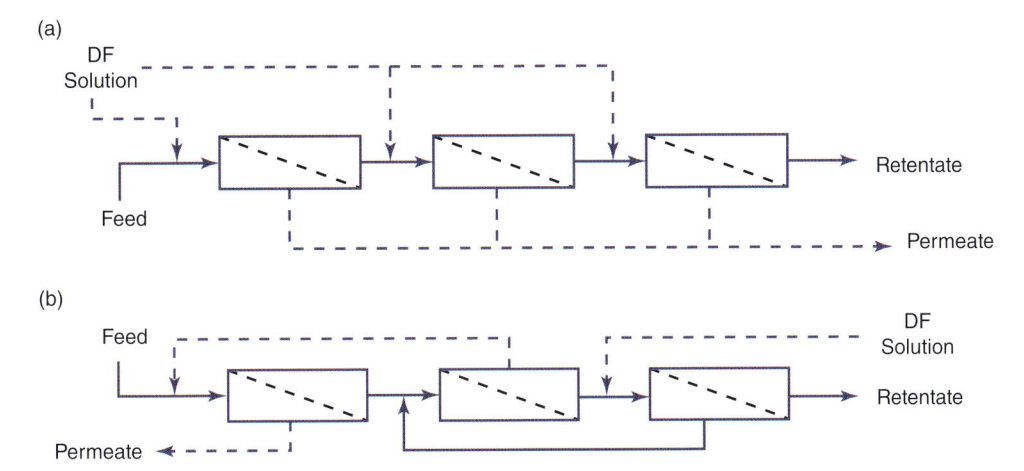

Figure 21.10 Continuous diafiltration configurations: (a) cocurrent and (b) countercurrent.

retentate volume), and S is the sieving coefficient (in this case equals to 1 as the salts in the feed are expected to fully permeate through the membrane). In the Biogen platform downstream process, the fed-batch diafiltration is performed with a 30 kDa membrane with a total of eight diavolumes, which translates into an expected exchange rate of 2981 or 99.967% solute removal [7].

Cocurrent and countercurrent configurations would have to be designed with at least four stages to provide the same exchange rate of the fed-batch system operated with eight DVs [17]. In a four-stage cocurrent system, ~3× higher DF solution consumption would be expected compared to a fed-batch system as a consequence of adding diafiltration at each of the multiple stage [17]. While DF solution consumption can be reduced by adding more stages, the overall DF solution consumption is still expected to be higher than fed-batch operation. A four-stage countercurrent configuration would have a DF solution consumption that is approximately equal to the fed-batch by virtue of the DF solution being fed to one stage and fed back through the preceding steps [17]. The DF solution consumption could be further significantly reduced by adding more stages, but at a cost of greater operational complexity.

Another option to consider for a continuous UF/DF process is the use of multiple fed-batch systems in tandem (Figure 21.11). While each fed-batch system itself would not be a continuous process, the tandem UF/DF systems could support an overall continuous process by accepting a continuous input from the preceding process step. The filter areas and retentate tank would be sized appropriately to match the input mass flow rate. Although the output of each UF system would not be continuous, the overall timing of operation would mimic a continuous operation. For a tandem two-system fed-batch operation, one system would perform the concentration step while the other is performing the diafiltration in an alternating fashion (Figure 21.11).

The choice of system (tandem versus countercurrent) would be guided by comparing system requirements (number of filter stages, filter area, tank sizes, etc.) to achieve target process productivity and DF solution consumption.

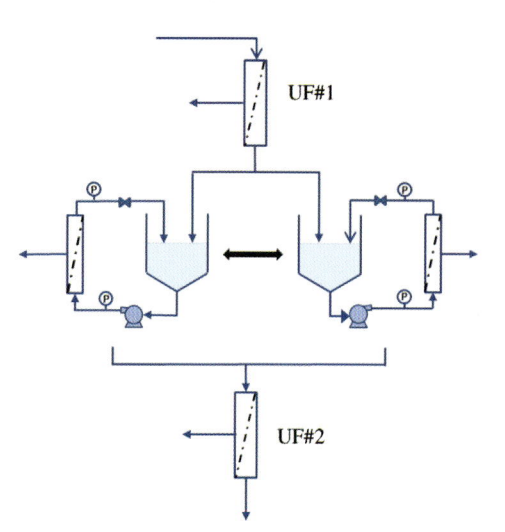

Figure 21.11 Tandem two fed-batch system.

21.4 Productivity Comparison of Batch and Continuous Downstream Process

The annual mass throughput that can be obtained from a given manufacturing facility is dependent on the rate-limiting step and the overall yield of the downstream operations. In an ideal scenario, the rate-limiting step would be the production bioreactor in which all the potential DS is actually generated. Thus, downstream operations would impact the throughput only through the yield term. If batch downstream operations are not rate-limiting, then, from a throughput perspective, continuous downstream operation would be beneficial only if it resulted in an increase in the overall yield when compared with standard discrete batch operation.

If downstream operations do become rate-limiting, continuous processing may provide an operational advantage. If DSP is rate-limiting because all the unit operations need to be completed before the subsequent harvest can be processed, the continuous scheme should allow linking of all unit operations. This situation would arise in facilities whose manufacturing philosophy forbids the coexistence of multiple batches of the same product. On the other hand, if multiple batches can coexist, then continuous processing needs only to be applied to the unit operation(s) that threaten to become rate-limiting.

If one defines the productive steps in an unit operation as those in which the product is processed, then in both chromatography and filtration, the productive time is a fraction of the actual processing time. For instance, in bind-and-elute chromatography, the typical steps are equilibration, load, chase, wash(es), elution, regeneration that are executed serially. The only productive steps are the load and elution steps wherein the product is processed. Depending on titer and loading, these could be a small fraction of the overall processing time. Now, both chromatography and filtration operations are inherently batch operations insofar as they can only process a finite quantity of material at a time. To render them continuous, requires the use of multiple devices – multiple columns in the case of chromatography and multiple filtration units in the case of filtration. The use of multiple devices enables the parallelization of productive and nonproductive steps resulting in reduced process times and, consequently, increased productivity. However, if the loading on the columns or the filtration units remains the same, transforming a batch chromatography/filtration step to a continuous mode of operation by using multiple devices will not result in any decrease in consumables costs.

References

1 Westoby, M., Chrostowski, J., de Vilmorin, P., Smelko, J.P., and Romero, J.K. (2011) Effects of solution environment on mammalian cell fermentation broth properties: enhanced impurity removal and clarification performance. *Biotechnol. Bioeng.*, **108** (1), 50–58.

2 Conley, L. and Tao, Y. (2014) Methods and compositions for inactivating enveloped viruses. WO 2014025771 A2, Feb. 13, 2014.

3 Diesenhofer, J. (1981) Crystallographic refinement and atomic models of a human Fc fragment and its complex with fragment B of protein A from Staphylococcus aureus at 2.9- and 2.8-A resolution. *Biochemistry*, **20** (9), 2361–2370.

4 Torigoe, H., Shimada, I., Saito, A., Sato, M., and Arata, Y. (1990) Sequential IH NMR assignments and secondary structure of the B domain of staphylococcal protein A: structural changes between the free B domain in solution and the Fc-bound B domain in crystal. *Biochemistry*, **29**, 8787–8793.

5 Ghose, S., Tao, Y., Conley, L., and Cecchini, D. (2013) Purification of monoclonal antibodies by hydrophobic interaction chromatography under no-salt conditions. *MAbs*, **5** (5), 795–800.

6 Zeman, L. and Zydney., A. (1996) *Microfiltration and Ultrafiltration: Principles and Applications*, Marcel Dekker, New York, NY.

7 Jungbauer, A. (2013) Continuous processing of biopharmaceuticals. *Trends Biotechnol.*, **31** (8), 479–492.

8 Walther, J., Godawat, R., Hwang, C., Abe, Y., Sinclair, A., and Konstantinov, K. (2015) The business impact of an integrated continuous biomanufacturing platform for recombinant protein production. *J. Biotechnol.*, **213**, 3–12.

9 Mahajan, E., George, A., and Wolk, B. (2012) Improving affinity chromatography resin efficiency using semi-continuous chromatography. *J. Chromatogr. A*, **1227**, 154–162.

10 Holzer, M., Osuna-Sanchez, H., and David, L. (2008) Multicolumn chromatography: a new approach to relieving capacity bottlenecks for downstream processing efficiency. *BioProcess Int.*, **6**, 74–80.

11 Ghose, S., Zhang, J., Conley, L., Caple, R., Williams, K.P., and Cecchini, D. (2014) Maximizing binding capacity for Protein A chromatography. *Biotech. Progress*, **30** (6), 1335–1340.

12 Mattila, J., Clark, M., Liu, S., Pieracci, J., Gervais, T., Wilson, E., Galperina, O., Li, X., Roush, D., Zoeller, K., and Brough, H. (2016) Retrospective evaluation of low pH viral inactivation and viral filtration data from multiple company collaboration. *PDA J. Pharm. Sci. Technol.*, **70** (3), 293–299.

13 Klutz, S., Lobedann, M., Bramsiepe, C., and Schembecker, G. (2016) Continuous viral inactivation at low pH value in antibody manufacturing. *Chem. Eng. Process.*, **102**, 88–101.

14 Vogel, J.H., Nguyen, H., Giovannini, R., Ignowski, J., Garger, S., Salgotra, A., and Tom, J. (2012) A new large-scale manufacturing platform for complex biopharmaceuticals. *Biotechnol. Bioeng.*, **109**, 3049–3058.

15 Shamashkin, M., Godavarti, R., Iskra, T., and Coffman, J. (2013) A tandem laboratory scale protein purification process using Protein A affinity and anion exchange chromatography operated in a weak partitioning mode. *Biotechnol. Bioeng.*, **110**, 2655–2663.

16 Zhang, J., Conley, L., Ghose, S., and Pieracci, J. (2017) Pool-less processing to streamline downstream purification of monoclonal antibodies. *Engineering in Life Sciences*, **17**, 117–124.

17 Westoby, M. and Brinkmann, A. (2015) Exploring options for continuous diafiltration. The 249th American Chemical Society Meeting, Denver, CO.

Index

a

ab initio cell design, 234
accelerostat, 234, 237, 240
acetate recycling, 242
acetyl-CoA synthetase (Acs), 242
acoustic wave separators (AWS), 220
active pharmaceutical ingredient (API), 564
adaptastat, 246
adaptive control techniques, 14
adsorption train, 467
advanced skills, 553
AEX. *see* anion exchange (AEX)
affinity chromatography, 329, 344, 417
– process design objective, 357
agile approach, 109
Air Wheel bioreactor, 37
ÄKTA Explorer 10 system, 54
ÄKTA FPLC system, 54
ÄKTApcc, 49, 467
AKTA PuRe purification system, 581
Akta purifier system, 352
ÄKTA system, for 3C-PCC, 50
alternating tangential filtration (ATF)
– application of, 211
– ATF™-4 system, 39
– details for different bioreactor sizes and scale-up, 211
– filtration module, 38
– physical bioreactor characterization, 465
– systems, 40, 208, 210, 211, 216, 273, 463, 470, 483, 491
amino acids, 329
– concentration, 518
analytical cation exchange chromatogram, 414
analyzer calibration, 539
ancillary systems, 525
animal cell culture, bioprocesses, 301
animal cell separation, hydrocyclone designed, 39
anion exchange (AEX), 572
– chromatography, 572

– flow-through pool, 581
– flow-through validation, 315
– membrane chromatography, 580
antibody, 474
– drug substance, 573
– isoform separation, 412
– manufacture, 578
– production phase, 178
– purification, 482
– volumetric productivity, 36
antibody drug conjugates (ADC), 414, 425
Applikon BioSep system, 212
aqueous two-phase extraction (ATPE), 271, 482
– purification effect, 498
aqueous two phase system (ATPS), 42, 51
– devices, 272
– DoE-based experiments, 497
– separation
– – countercurrent mixer-settler battery, schematic representation of, 52
ascorbic acid, 495
A-stat, 242, 246
– applications, 242, 243
– cultivation, 243
– integration with Omics methods, 243
– technology, 16
– yeast physiology studies, 243
– yields, 243
ATF. *see* alternating tangential filtration (ATF)
ATF2 device, 464
automated microbioreactors, 183
auxiliary electrodes, 139
auxoaccelerostats, 245, 246
– culture, 245
auxostats, 245
axial pumping stirrer, 141

b

Bachellier-impeller, 145
– mixing profile, 146
– prototype design, 146

bag film materials
– testing, 137
basal medium screening, 183
batch and continuous protein refolding processes, 277
batch bioprocesses, 445
batch chromatography, 325–327, 330, 393, 394, 399, 415
– with linear elution gradient, 331
– MCC, comparison of, 576
– processes, 55, 330
– system, 330
– trade-offs, 399–400
– transformation of, 335
– true moving bed/simulated moving bed/comparison, 370
batch downstream process, 274
batch/fed-batch processes, 207
– cell cultivation, 31
batch fermentation, 405
batch/lot, definition, 527
batch manufacturing, 565
batch operations, 574
– advantages and disadvantages, 460
batch paradigm, in synthetic drug industry, 31
batch polishing, 418
batch processing, 359, 538, 575
batch purification
– for monoclonal antibodies, 111
– resin chromatography, 31
batch-topped off filtration, 273
batch UF/DF system, 582
batch virus reduction filtration step, 316
batch-wise processing, 468
Bayer Technology Services, 161
Beer–Lambert law, 385
BEND Mast aseptic sampling system, 118
BeneFIX, 522, 527
binary mixture, 342
binary separations, 329, 349
– principle of operation, 44
bind/elute mechanism, 331
bioburden contamination, 551
bioburden filtration, 112
bioburden reduction filtration, 117
bioburden testing, 448
biochemical processes, steady-state flux patterns of, 233
– environmental conditions, relationship, 233
biochromatography, 369
biocomparability, 355
Bioflo 110, 37
Biogen mAb
– downstream platform process, 571
– platform, 580
biological contaminants, 289

biologic drugs contaminations, 290
biomanufacturing, 142, 301
biomass
– concentration, 245
– constant feeding profiles, effect of, 19
– exponential feeding profiles, effect of, 20
biomolecules, hydrophobic characteristics of, 329
biopharmaceutical companies, 149
biopharmaceutical downstream process, 380
biopharmaceutical landscape, 149
biopharmaceutical manufacturing, 149, 397
biopharmaceutical purification processes
– viral clearance robustness, 55
biopharmaceutical sector, 31, 178, 343
bioprocess complexity, 8
bioprocesses, 298
– intensification, 178
bioreactors, 172, 176, 206
– confirmation, 191
– – data, 184
– operating conditions, 476
– process, 171
– technologies for clinical/commercial scale manufacturing, 155
– volume, 175
BioSC®, 467
Biosep acoustic perfusion system, 212
bioseparation problem, schematic illustration, 407
BioSep device, 212
Biosep system applications, 212
BioSMB system, 50, 58, 575
– BioSMB™, 467
– CMCC system, 124
– protein A chromatography, 113
biotech
– continuous process models, 12
– – complexity, 11
– – control, 14
– – process monitoring, 13
– iterative process development, 14
– – fed-batch technology, 16
– – methods for development, 14
biotechnological process, 369
– characterization, 8
– upstream process, 33
black box models, 12
blank analysis, 562
bovine serum, 291
bovine serum albumin (BSA), 50, 344
bovine spongiform encephalopathy (BSE), 291
breakthrough (BT), 401
breakthrough curve, schematic illustration, 402
BSE. *see* bovine spongiform encephalopathy (BSE)

BT. *see* breakthrough (BT)
buffer compositions, 333, 357
buffer consumption, 337, 359, 401
buffer gradient, 338
buffer strength, 332
butyric acid, biphasic perfusion, 36

c

$CaCl_2$/ethanol continuous precipitation process, 53
calcitonin, 412
capacity utilization (CU), 401
capillary gel electrophoresis, 474
caprylic acid (CA), 53
capture applications, 400–406
– examples, 405
– process principle, 403
capture operation, cycle length of, 469
CaptureSMB
– configuration, 359, 360
– process, 337
– units, 353
carbon dioxide mass transfer coefficient, 142
Carman–Kozeny equation, 358
β-carotene, in microalgae *D. tertiolecta*, 243
carotenoids lutein, 243
cation exchange
– chromatogram, 415
– chromatography, 425
– MCSGP process, 46
caustic stability, 402
cell banks, 291, 293
cell cultivation processes, 5
cell cultures, 237, 295
– based products, 4
– broth, 177
– Cell Culture Engineering, 31
– contamination in, 291
– growth, 526
– media, 180, 293, 299, 300
– – NaCl addition, 518
– medium development, limitation, 180
– steady-state of, 233
cell cycle mechanisms, 234, 236
cell design, 235
cell geometry, 234, 236
cell-less biomass, 248
cell lines, 293
– genetic stability, 522
– selection and testing, 290
– specific requirements, 181
– stability, 474
cell lysis, 262
cell metabolism, 235
cell microenvironment
– various process parameters interaction, 144

cell physiologies theories, 233, 234
cell productivity, 462
cell retention, 11
– ATF /TFF system for, 464
cell retention device (CRD), 39, 171, 173, 176, 210, 213, 483
– classification, 176
– selection of, 176
– used for continuous upstream process, 208
cell separation devices, 35, 38, 134
cell settlers, 38
cell-specific perfusion rate (CSPR), 35, 173
CELL-tainer® bioreactor, 142
– application of, 142
– single-use, 137, 145
cellular behavior study, omics methods, 180
cellular metabolism, 178
Cellvento CHO-100 medium, 188
Center for Drug Evaluation and Research (CDER), 514
centrifuges, 214
– cell retention devices, used in perfusion process, 210
– continuous disk stack, 40
– rotor, 263
– SEC diagram of, 497
– semicontinuous perfusion (SCP), 187
CEX elution peak, 315
CFI. *see* coiled flow inverter (CFI)
changestat methods
– advanced continuous cultivation methods, 237–242
– categories, 240
– – chemostat-based methods, 240
– – turbidostat-based methods, 240
– cultivation methods, 237
– deceleration-stat, 244
– experiments, 238
– for GSA, benefits, 240
– review of results, 242–247
– tools for scanning steady-state growth space, 240
charge isoform profile, product quality, 475
CHEF1 cell line, 215
chemical engineering, 327
chemoenzymatic processes, 4
chemostats, 233, 236, 237, 240
– cultivation method, 236
– – disadvantages, 233
– cultures, 184
– – drawback, 233
– – tool for SSGSA, 236
– steady-states, 237
– – metabolism, 236
Chinese hamster ovary (CHO), 34, 181, 267
– based monoclonal antibody, 212

- cell lines, 185, 470
  - – DHFR system, 58, 519
  - – K1 cell line, 181, 189
  - – lineage and development history, 182
- cells, 40
  - – mAb produced, integrated perfusion/capture of, 57
  - – recombinant technology, 459
  - – shear stress impact, 145
- culture, gymetrics up-stream pH sensor, 140

chiral separations, 329
- batch separations, 330
- PR&D phase, 333

CHO. *see* Chinese hamster ovary (CHO)
chromatogram, 331, 396, 582
chromatographic capture process
- harming proteins, separation of, 466
- high product affinity, 466

chromatographic columns, 461
chromatographic interactions, 328
- types of, 326, 329

chromatographic polishing applications, 406–414
chromatographic technology, 268, 311, 324, 326, 369, 393, 516
- flow-through, 311
- interaction mechanisms used in, 328
- manufacturing aspects, 396–399
- multicolumn, 311
- separations, 54
  - – purification processes, 326
- simulated moving bed, 311
- steps, 393
  - – capture step, 393
  - – flow sheet of a possible continuous process, 501
  - – manufacturing constraints, 396
  - – polishing steps, 393

CIP. *see* clean-in-place (CIP)
classical detection methods, 385
classical single-column chromatography, 370
classical SMB (4-SMB) and 1-SMB with 7 stirred tanks, comparison, 377
cleanability, 355
clean-in-place (CIP), 42, 218
- cycles, 298
- lines, 355
- regimes, 450
- skids, 573
- steps, 355
- strategy, 357

CMOs. *see* contract manufacturing organizations (CMO's)
Code of Federal Regulation (CFR), 555, 560
coextruding, 137

coiled flow inverter (CFI), 265
- based reactor, 276
- continuous precipitation, 267
- reactor (*see also* coiled flow inverter reactor (CFIR))
- use of, 55

coiled flow inverter reactor (CFIR)
- case study, 276

CO_2 incubator, 34
column capacity, 357
column characteristics, influence separation performance, 342
column configurations, 348
- theoretical comparison, 359
column length, 347
6-column MCSGP, 379
column packing, 343, 355
3-column PCC (3C-PCC) system, 47
columns cycle, 334
column variability, 355
column volume (CV), 331
complex mixtures, 343
computational fluid dynamic (CFD), 13, 355
computer aided biology (CAB), 12
computer aided process engineering (CAPE), 12
concentrated fed-batch (CFB), 179
- processes, 175
concentration factor (CF), 433
connected chromatography steps, 580
- current and pool-less process, 581
conservatism, in biopharmaceutical sector, 31
constraint-based metabolic network models, 248
contaminants, 488
Contichrom® Lab-10, 467, 472
continuous affinity chromatography, 337
continuous annular chromatography, 394
continuous batch chromatography, 327, 331, 358
- using multiple columns, 336
continuous batch operation, 338
continuous biomanufacturing process, 289, 294, 513, 550, 551
- advantages, 550
- ASTM, 521
- capital expenditure/cost reduction, 550
- defined, 549
- efficient facility/personnel utilization, 550
- EMA Guideline, 521
- FDA support, 513, 520
- ICH, 520
- implementation of, 513
- manufacturing consistency, 550
- PAT applications
  - – as facilitator for implementation, 514
  - – in pharmaceutical industry, 516

- process validation, 519
- quality, 551
- regulatory documents, 520
- scale-down for continuous bioprocessing, 524
- single-use equipment/components, 217
- single-use systems impact in process validation, 526
- validation, 521
continuous bioprocessing, 150, 154, 303, 550
- advantages of, 3
- challenges, 5
- – in continuous biomanufacturing, 5
- – engineering approach to complex systems, 8
- – inherent changes in microbial system, 6
- – lack of process information, 7
- – models-based process development, 8
- computer-aided operation of robotic facilities, 20
- concerns for, 158
- continuous downstream processing, 155
- continuous-like fed-batch cultivations, 18
- continuous upstream processing, 154
- development, 3, 302
- limited control strategies, 9
- mimicking industrial scale conditions in lab, 17
- model building/experimental validation, 21
- operational issues, 158
- platform, end-to-end, 278
- technologies, regulatory aspects, 164
- traditional control strategies, 9
continuous bioproduction (CB), 218
continuous capture development, 466
- continuous two-column capture process, 467
- objectives/requirements, 466
- process control, 469
- process performance, 468
continuous capture step, process design of, 357
continuous cell cultivation processes, 32
- cell retention devices, 36
- continuous bioprocesses, 33
- early/scale-down perfusion development, 34
- feeding/operational strategies in perfusion processes, 35
continuous chromatography processes, 340, 370, 393–396, 485
- comparison, 387
- countercurrent multicolumn gradient chromatography, 378
- integrated countercurrent chromatography, 379
- risk assessment of, 353
- serial multicolumn continuous chromatography, 377
- SMB, 370–377
- use in biopharmaceutical industry, 394

Continuous cocurrent diafiltration, 273
continuous countercurrent tangential chromatography (CCTC), 50
- multicolumn gradient, 378
- for protein A purification of mAbs, 51
continuous cultivation methods, 236
- accelerostat, 237
- advanced, 237–242
continuous culture, 237
continuous diafiltration, 274, 582
- configurations, 583
continuous downstream process, 41, 155, 275, 573
- batch, productivity comparison of, 585
- continuous annular chromatography (CAC), 42
- continuous chromatography, 157
- continuous countercurrent tangential chromatography (CCTC), 50
- continuous liquid chromatography (CLC), 42
- multicolumn countercurrent solvent gradient purification (MCSGP), 45
- periodic countercurrent chromatography (PCC), 47
- tangential flow filtration, 156
- true and simulated moving bed chromatography (TMB/SMB, 43
continuous feed stream, 342
- flow rate, 398
continuous fermentation, 397, 419
continuous flow UVC reactor (UVivatec), 523
continuous integrated process
- bioreactor operation, 470
- cell growth, 470
- monoclonal antibody
- – capture, 472
- – production, 471
- perfusion bioreactor, 460
- process performance, 473
- product quality, 474
continuous liquid chromatography (CLC), 42
continuous manufacturing, 419
- approach, pharma companies, 551
- framework for, 323
- platform, 262
continuous monoclonal antibody laboratory
- horseshoe lay-out of, 160
continuous multicolumn chromatography (CMCC), 124
- conceptual diagram of, 157
continuous operation, 328, 356
continuous over batch manufacturing, 549
continuous perfusion bioreactors, 191
continuous processing (CP), 218
- perfusion, 141, 145
- requirements, mass transfer and mixing, 141

continuous process verification (CPV), 447, 448
continuous protein refolding, 264
continuous quality verification (CQV), 521
continuous seed train, 145
continuous separation, of BSA and myoglobin, 352
continuous stirred tank reactor (CSTR), 5, 53, 265
continuous ternary separations, 351
continuous UF/DF processes, 582
continuous upstream processing, 154, 207
continuous validation, 541, 542
– management and personnel, 544
– of process, 533, 544
continuous viral inactivation, 578
– set-up, 579
– virus clearance processes, 54
contract manufacturing organizations (CMO's), 552
– actual formulation approach, 563
– benefits, 553
– buildings/facilities, 555
– capacity constraints, 553
– change management/control, 556
– compliance, 556
– extractable and leachable analysis, 560, 561
– leachables analysis, 561
– loss of flexibility, 554
– model approach, 562
– quality concerns, 553
– relationships, 553
– risk assessment, 556
– risks of working, 553
– single-use technology, 555
– supply chains, 555
– validate to eliminate, 557
control charting, 540
– choosing, 542
conventional batch bioprocesses, 294
conventional SMB process, 375
conveyor belt, 394
copper sulfate, crystallization, 267
corrective and preventative actions (CAPA), 557
Corynebacterium glutamicum, 244
cost distribution, in USP/DSP, 486
cost issues, process validation, 543
cost of good (COG), 203, 209, 487
cost–risk assessment, 356
cost savings, 553
costs of goods (COGs), 203
countercurrent, 340
– chromatographic processes, 395, 396, 399, 409
– diafiltration, 438
– flow, 351
– process analogy, 395
– single-pass diafiltration, 438

Crabtree effect, in yeast, 243
CRD. *see* cell retention device (CRD)
Cricetulus griseus, 181
critical material attribute (CMA), 448
critical process parameter (CPP), 221, 327, 522, 534, 542
critical quality attribute (CQA), 3, 327, 490, 522, 534, 542
cryopreserved cells, 186
CSEP process, 349
– from Carbon Calgon, 350
CSPR. *see* cell-specific perfusion rate (CSPR)
CU. *see* capacity utilization (CU)
cultivation vessels, 186
current manufacturing setup, 580
cycle durations, yield of capture process, 473

d

DASGIP 2.5L bioreactor system, 463
DBC. *see* dynamic binding capacity (DBC)
dead-end filtration, 304
DEAE membrane radial column, 271
deceleration-stat (De-stat), 244
DeltaV controller, 114
denatured and reduced (D&R), 266
design batch chromatogram, 411
design of experiment (DOE), 17, 276, 426, 492, 521, 552
design qualification (DQ), 539
De-stat, 244
diafiltration, 583
diafiltration (DF), 572
dihydrofolate reductase (DHFR), 181
– transfection system, 186
dilution rate stat (D-Stat), 244
diode array detector (DAD), 385
disposable, 557
DNA microarray-based transcriptomics, 525
DNA replication, 249
downstream (DSP), 482
– bioprocessing production, 261
– case study, 494
– components
– – stainless steel and single-use, 209
– development, 489, 491
– – in continuous mAb manufacturing, 484
– methodical approach regarding impurity processing, 492
– processing, 486, 551, 571, 573
– – case studies, related to continuous manufacturing, 276
– – cell lysis, 262
– – centrifugation, 263
– – chromatography, 267
– – continuous annular chromatography (CAC), 267

– – continuous extraction, 271
– – continuous filtration, 272
– – continuous manufacturing technologies, 262
– – continuous process development, 274
– – countercurrent tangential (CCT), 269
– – design, 41
– – expanded bed chromatography (EBC), 269
– – experimental presentation of the methodical approach, 493
– – periodic countercurrent chromatography (PCC), 269
– – precipitation, 267
– – purification, 393
– – refolding, 264
– – schematic presentation of the methodical approach, 492
– – step, 396
– scheme, 324, 325
– trends/integration approaches, 490
– unit operations, 486
drug–antibody ratio (DAR), 414
drug development, 311
drug product (DP), 207, 298
drug proteins production, 181
drug substance (DS), 207, 572
dry sensors, 140
D-stat, 246
Dunaliella tertiolecta, 243
dynamic binding capacity (DBC), 401
dynamic breakthrough (DBC), 336
– curves, 358

e
Ebola virus, 33
EcoPrime Twin, 337
– LPLC system, 338
– process, 354
– 100 prototype, 325
electrode reuse, disadvantages, 139
electron beam irradiation (β-irradiation), 138
eluent consumption, 44
elution, 407
EMD Millipore Prostak™-microfiltration series, 213
EMD Millipore's Pellicon® 3, 439
EnBase® technology, 16
endogenous retroviruses, 303, 306
endogenous virus, 290
Engage®, 138
enzyme capture, using a HIC resin, 57
equilibrium processes, 234
equilibrium thermodynamics, 234
equipment components, risk assessments of, 356

Escherichia Coli, 7, 33, 134, 243, 264
– acetate overflow metabolism in, 242
ethylene vinyl acetate, 355
European Medicines Agency (EMA), 261
Evatane®, 138
exogenous/adventitious virus, 290
expanded bed chromatography (EBA), 43
experimental validation, 385
extracapillary space (ECS), 219
extractables, 153, 558
– risk assessments, 452
extraneous contamination, 290

f
fast moving consumer good (FMCG), 261
FBA. *see* flux balance analysis (FBA)
FBS. *see* fetal bovine serum (FBS)
fed-batch
– based processes, 420
– batch upstream production process, 206
– bioreactors, 297
– and continuous/perfusion upstream process, 204
– cultivation *vs.* perfusion, 172
– culture processes, 32
– fermentation, 419
– manufacturing processes, 216
– media, 180
– moving to integrated continuous operation, 463
– processes, 3, 20, 172, 483
– reactors, 459, 461
– system, 584
– volumetric productivity
– – *vs.* perfusion process, 143
feedback controller, 514
feed concentrations, 397
– yield of capture process, 473
feed continuity constraint, 49
feed flow excursion experiments, 440
fermentation
– broths, 493
– industrial relevant analytical methods, 488
fetal bovine serum (FBS), 290
fibrinopeptide, 415
film polyethylene type materials, 137
filtration, 38, 324
– adsorption system, 56
– relative yield of antibodies, 496
five-zone continuous process for ternary separations, 351
flat sheet cassettes, 428
flow cytometry, 144
flow design, 355
flow meter, 328

flow rates, 335, 336, 342, 348, 358
– determination for different SMB zones, 347
fluorescence spectroscopy, 276
fluorophoric sensors, 140
flux balance analysis (FBA), 234
FMEA analysis, 345
FMEA, risk category, 452
folic acid, 495
Food and Drug Administration (FDA), 3, 181, 290, 534, 565
– FDA PAT Guidance, 536
Fourier transform infrared spectroscopy (FTIR), 516, 562
fragile proteins, 171
fuzzy logic controls, 14

g

galactosylation, 475
gamma irradiation, 313
gas chromatography with mass spectroscopy detector (GC-MS), 561, 562
gas dispersing impellers, 141
gas–liquid mass transfer, 135, 463
– without harming cellular integrity, 470
GE Healthcare, 575
gene regulatory networks, 12
generic antibodies, 481
generic assessment, 356
generic integrated continuous downstream process, 324
generic manufacturing process, for monoclonal antibodies, 485
genome-scale metabolic models, 249
Genzyme, 159, 297
glutamine synthetase (GS), 181
glycine, 495
glycoforms, 461
N-glycosylation, 517
glycosylation pattern, 517
GMP. see good manufacturing practice (GMP)
good manufacturing practice (GMP), 555
– environment, 117, 461
– GAMP5 directions, 356
– manufacturing, 134
– requirements, 444
gradient chromatography, 331, 333
gradient processes, 332
gradient SMB processes, isolation of proteins and antibodies using, 352
granulocyte colony stimulating factor (GCSF), 271
gravimetric control, of bioreactor volume, 472
gravimetric settlers, 172
gravity-based cell settlers, 483
growth-limiting substrate, 236

growth spaces, 234
– scanning of, environmental parameter, 241
– three-dimensional, 235
Gymetrics pH prototype material, 140

h

Hanseniaspora guilliermondii, 243
HCIC systems, 271
HCP. see host cell protein (HCP)
heat sealers, 138
HIC. see hydrophobic interaction chromatography (HIC)
high-density seed bioreactors, 483
high molecular weight product-related species (HMW), 572
high performance liquid chromatography with a UV detector (HPLC-UV), 561
high-pressure limitations, 359
high-pressure liquid chromatography (HPLC), 328, 562
– HPLC-UV, 562
– – blank analysis, 563
– – test extract analysis, 564
– interference, 564
high-temperature short-time (HTST)
– media exposure, 303
– techniques, 313
– treatment, 298
high-throughput (HT), 21
Hollow Fiber Perfusion Bioreactor (HFPB), 219
hollow fibers, 426
– media exchange, 220
– microporous hollow fiber membranes retain, 50
– modules, cross flow filtration, 463
– perfusion culture, 220
homogenizer, high-pressure, 262
host cell protein (HCP), 41, 309, 393, 482, 488, 571
– composition, 489, 493
– concentration, 493
– profiles, 494
– spectrum, 482
HPLC. see high-pressure liquid chromatography (HPLC)
HTP methods, 498
HTST. see high-temperature short-time (HTST)
human-like glycosylation, 181
human serum proteins, 267
hybridoma cells, 178, 303
hydrocarbons, 324
hydrodynamic resistance
– for the nonvolumetric centrifugal pump, 465
hydrodynamic stress, 180
hydrophilic amino acid residues, 52

hydrophobic interaction chromatography (HIC), 227, 266, 311, 329, 344, 385, 403, 405, 412, 572
– chromatogram, 382
– resins, 487

i

15 iCCC cycles, simulation results, 384
iCCC process scheme, 382, 500
ICH Q7, 527
ICH-Q5A, 524
ICH Q7 - Good Manufacturing Practice states, 551
ICH Q9 - Quality Risk Management states, 551
ICH Q10 states, 551
IgG4 antibody, 523
IgG capture step, 352
IgG concentrations, 495, 497
IgG production, 495
Immobilized metal ion affinity chromatography (IMAC), 266
immunological diseases, 481
impurities, from host cells, 488
independent qualified (IQ), 450
in flow-through mode, 393
information awareness, 542
injection, 333
innovative downstream approaches, 323
innovative negative elution gradient, 332
in silico ab initio "first time right" cell design, 250
installation and operational qualification (IQ/OQ) procedures, 134
installation qualification (IQ), 539, 557
insulin, 329
integrated continuous processes, 55
– comparison of, 163
integrated countercurrent chromatography (MCSGP), 370, 379
– basic idea of, 380
– countercurrent behavior, 380
– fermentation broth, 383
– ion exchange binding capacity, 385
– mathematical peak identification methods, 386
– parameters for, 379
– steps of, 383
integrated single-use continuous bioprocessing, 159
interaction mechanisms, 328
interconnected operation, 337
intermittent SMB, 374
internal recycling, 398
International Conference on Harmonisation (ICH), 490
International Symposium on Continuous Manufacturing of Pharmaceuticals, 324

in'vitro detection techniques, 291
ion exchange, 344
ion exchange chromatography (IEC), 266
iSMB process scheme, 375
isocratic batch chromatography, 372
isocratic chromatography, 330, 332, 333

k

Kalman filters, 13
keep it simple and stupid (KISS) rule, 32
kinetic principles, 342
knowledge discovery of data (KDD), 12

l

lab-scale systems, 353
lack of control, 553
lactate metabolism, 191
Lactobacillus paracasei, 246
Lactococcus lactis, 241, 243, 246
Langmuirian-type isotherm, 342
lauryldimethylamine N-oxide (LDAO), 571
leachables, 559
linear adsorption models, 348
linear feed flow rate, 397
linear range of Langmuir, 372
linear retention factors, 348
linear velocities, 336, 359, 360
liquid chromatography (LC), 42
liquid chromatography with a mass spectroscopy detector (LC-MS), 561
liquid flow distribution, 397
liquid–liquid extraction (LLE), 51, 271, 394, 482
loading, to column capacity, 335
load volumes, 331
log reduction value (LRV), 293, 303, 304
lower control limit (LCL), 541
lower fermentation temperatures, advantage, 178
low pH inactivation, 55
– low-pH virus, 425
low-pressure liquid chromatography (LPLC), 328
lysine, fermentation, 135
lysozyme purification, 409, 410

m

mAB. *see* monoclonal antibodies (mAB)
MabSelect SuRe, 227
– protein A columns, 472
mammalian cell culture processes, 141, 182
mammalian cell perfusion cultures, 176, 463
manufacturing process, 556
marine-type impellers, 141
mass balancing, 174
– equations, 342
mass flow ratio, 373
massive parallel sequencing (MSP), 291

mass transfer, 401
– correlations, 348
– effects, 342
master cell bank (MCB), 291
– process, 356, 575, 577
material of construction (MOC), 355, 556
MCB. see master cell bank (MCB)
MCC. see multicolumn chromatography (MCC)
MCSGP. see multicolumn countercurrent solvent gradient purification (MCSGP)
mechanical stress, 395
media screening, 189
– Cellvento CHO-100 Medium, 189
– cultivation process illustration, 186
membranes, 324
– based adsorption, 482
– fouling, 432
Merck & Co., 160, 425
metabolic flux analysis (MFA), 15, 234, 243
metabolic modeling, 243
metabolic network models, 249
metabolic rates, determination, 174
metabolomics, 236
MFA. see metabolic flux analysis (MFA)
M. genitalium, 249
microbioreactors, 15
microchip-based circulating loop bioreactor, 15
microfiltration processes, 179, 423
microfluidic channel continuously, 263
microfluidic devices, 263
micro-Matrix (Applikon), 35
Micro-24 MicroReactor system, 35
microsparger, 141
Milli-Q® water, 186
mixed suspension and mixed product removal (MSMPR) reactors, 53, 267
mixer-settler devices, 272
MLV. see murine leukemia virus (MLV)
mobile-phase, 342
modeling in steady-state growth analysis, 248–250
model solvents, 562
model viruses, choice of, 293
modified batch chromatography system, 351
modular design, 353
modular facilities, 491
molecular mass (MM) contaminants, 51
molecules measurement, analytical methods, 236
monitoring, for continuous processing, 536
monoclonal antibodies (mAB), 4, 32, 34, 46, 178, 221, 301, 306, 323, 369, 401, 571
– capture system, 56
– DSP scheme of, 337
– isoform separation, 413
– molecules, 329
– platform downstream process, 572
– precipitation techniques, 53
– production, 181, 289
– – for human use, 289
– purification, 329
– therapeutic, 304
Monod equation, 13, 235, 236
moving range (mR), 540
– chart, 541
mPES membrane, 123
MSP. see massive parallel sequencing (MSP)
multicolumn, 340
multicolumn chromatography (MCC), 114, 405, 574, 575
– background, 575
– continuous capture, 575
– continuous process, 323, 325, 398
– – technical challenges, 353
– countercurrent chromatography, 396, 401, 414
– – continuous, 327
– – continuous upstream, 418–419
– – discovery and development applications, 414–415
– – as replacement for batch chromatography unit operations, 417–418
– – scale-up, 415–417
– experimental results, 577
– process optimization, 576
– systems, 222, 224
multicolumn countercurrent processes, 417, 418
– regulatory aspects and control, 419–420
multicolumn countercurrent solvent gradient purification (MCSGP), 45, 224, 270, 370, 399
– case study, 412
– multicolumn, 399
– principle, 407
– process design, 46, 340, 409
– – schematic, 411
– – with three consecutive cycles, 340
– standard process cycle of, 339
multicolumn parallel operation, 333, 337
multicolumn solvent gradient purification (MCSGP), 338
multicolumn systems, 327
multicomponent separations, 344
multicomponent SMB system, schematic design of, 349
Multifors 2 (Infors), 35
multimodal chromatography, 271
– cation exchange, 271
multiple protein A elutions, 315
multiple sclerosis, 481
multivariate data analysis (MVDA), 113, 516
– software, 127

murine leukemia virus (MLV), 303
murine retroviruses, 303
MuScan, 121

n
N-acetyl-DL-tryptophan, 313
nanofiltration, for virus removal, 112
new mixer designs, 145
next-generation sequencing (NGS), 180
NFF. see normal flow filtration (NFF)
NIR spectroscopy, 118
N-linked glycosylation pattern, 475
non-chromatographic continuous processes, 51
– aqueous two-phase systems(ATPS), 51, 498
– DSP operation, 497
– – one step toward a chromatography free purification process, 500
– – precipitation, 499
– protein precipitation, 52
nonchromatographic steps, 418
nongrowth spaces, 234
nonideal separation conditions, 343
nonsteady physiological states, 234
nonvolatile residue (NVR), 561
normal flow filtration (NFF), 304
normalized water permeability measurements (NWP), 450
– measurement, 451
normal phase (NP), 329
Novartis-MIT Center for Continuous Manufacturing, 204, 261
N-Rich process principle, 416

o
OCTAVE™ SMB (Tarpon) systems, 467
offline CEDEX cell counter measurements, 124
omic methods, 234, 236, 237
– and modeling, 243
on-column refolding, 265
– protein refolding technique, 266
one-cell clone, 7
open retention devices, 176
operational qualification (OQ), 539
operation diagram, for feed concentration, 373
operation qualification (OQ), 557
optimal chromatography process, 399
optimal dissolved oxygen, 141
optimal downstream process, decision tree, 400
optimal experimental design (OED), 17, 21
optimal operation performance, 469
ordinary differential equation (ODE), 13
organic solvents, 326
organized risk assessment, 556
OTR-limited reactors, 178
oxygen mass transfer coefficient, 141, 465

p
Packed bed (PB), 219
packing materials, 326, 345
packing utilization, 325
parallel columns, 334
parallel operation, 334
parallel-sequential experiments, generic process, 248
parallel-sequential fermentation (PSF), 247
– scheme, 248
Parenteral Drug Association (PDA), 490
– Technical Report 26, 563
– Technical Report No. 66, 558
Parenteral Technical Report 26 (PDA TR 26, revised 2008) states, 560
partial least squares (PLS), 12
parvoviruses, 290, 306
PAT. see process analytical technology (PAT)
pathogen safety, 301
– risk mitigation, 293, 300
PCC. see periodic countercurrent chromatography (PCC)
PDA. see Parenteral Drug Association (PDA)
PEG400-phosphate-based ATPS, 495
perforated rotating disk contractor (PRDC), 272
performance qualification (PQ) phase, 540, 542, 557
perfused expansion bioreactor, 179
perfusion, 171
– advantages of, 171
– batch, 215
– bioreactors, 36, 175, 467, 470
– – downstream of, 58
– – process, 188
– cell culture, 11
– – bioreactor setup, 463
– – cell free harvest, 466
– – development, 463
– – objectives/requirements, 463
– – physical bioreactor characterization, 464
– comeback of, 172
– devices, failure, 214
– history, 172
– media, 173
– – development strategies, 179–184
– medium formulation, 179
– microbioreactor platform, 184
– processes, 34, 210
– – biochemical modification, 177
– – bioreactor temperature, 178
– – characterization of, 172–177
– – designing, 177
– – formats, 177–179
– – innovative formats, 178
– – limiting factors, 185
– – process development for, 185

– – productivity, 175
– – required bioreactor sizes, 178
– – scale-down models, 181
– – segmentation, 177
– – sequential phases, 177
– – steady-state definition, 176
– rate, definition, 173
– reactors, 459, 461
– scale-down applications, examples, 184
– scale-down model development, case study, 185–192
– – material & methods, 186
– – results, 187
– systems
– – external cell retention devices, 37
– – internal cell retention devices, 37
periodic countercurrent chromatography (PCC), 47, 227, 274, 378, 575
– CaptureSMB system, 49
– cyclic operation, 270
– schematic comparison, 47
– system, 225
pharma area, continuous cell culture processes, 5
Pharmaceutical cGMP for the 21st Century – A Risk-Based Approach, 513
pharmaceutical industry, 327
– single-use technologies, 150
pharmaceutical manufacturers, 563
pH auxostat, 10, 245
pH sensors, 139
physiological state of cells, 234, 237
– environmental parameters, 234
pillow bags, 150
plasmid DNA, 267
plug flow model, 377
plug flow reactor (PFR), 5, 579
PMMA particulate system, 466
polishing chromatography, 399
polyamide, 355
polyethylene (PE), 138, 355
polyethylene glycol (PEG), 51, 53, 271, 498
polymeric resins, 328
polystyrene, 355
polysulfone, 355
pool-less process, performance, 581
porcine circovirus (PCV-1), 291
porcine trypsin, virus-contaminated, 291
posttranslational modifications, 181
potential continuous downstream process, schematic of, 574
PRD/GMP manufacturing environments, 355
precipitate recovery, 53
precipitation processes, 499
predicted primary test method result (PPTMR), 539, 540, 543

preparative batch chromatography, 406
preparative continuous annular chromatography (P-CAC), 265
pressure ratings, of chromatographic systems, 328
pressure regulators, 355
pressure sensors monitor, single-use, 115
presterilized single-use wave-action bioreactors, 219
preuse integrity test, 299
primary test method (PTM), 539
primary test method result (PTMR), 539
principal component analysis (PCA), 12, 126
probationary validation, 539
process analytical technology (PAT), 113, 171, 222, 323, 447, 448, 515, 535, 565
– to cell culture operations, 517
– Framework for Innovative Pharmaceutical Manufacturing and Quality Assurance, 565
– implementation, 516
– to purification operations, 518
process development, challenges, 485
– changing titers/impurities impact on cost structures, 485
– host cell proteins, 488
– impurities, as critical parameters, 487
– regulatory aspects, 490
process monitoring system, 534
process performance parameters yield, simulated, 469
Process qualification, 521
process stream analyzers, validation/qualification of, 538
process validation (PV), 557
– sequence of events, 535
– setting specifications, 535
production rate, 358
production record review, 555
productivity, 336, 352, 357, 358, 360
product life cycle management, 356
product quality, improvement, 550
product recalls, 290
product-related impurities, 262
Profibus, 114
proof-of-concept (POC), 581
ProSep rA (Millipore), 352
Prostak TFF cell retention device, 213
proteases, 482
protein A, 358
– adsorbent capacity
– – batch vs. MCC performance, 578
– – transport parameter, comparison, 577
– affinity chromatography, 51, 127, 308, 425
– based conventional technology, 53
– CAC processes, 43
– chromatography, 574

– ligand, 56
– reoptimization, 577
protein refinery operations lab, 111
– automation user control layouts, 116
– biologics PAT toolkit, 119
– case studies, 122
– – concept automated handling of deliberate process deviations, 127
– – continuous purification, 124
– – perfusion, 122
– challenges impacting biotech and biopharma industry, 110
– CHO mAb continuous processing, 128
– CHO mAb manufacturing improvement, 110
– continuous mAb capture using continuous multicolumn chromatography, 125
– design and implementation, 112
– design and operating principles, 115
– downstream process deviation analysis, 128
– horseshoe layout, 114
– mAb production, 112
– modular facility for bioprocessing manufacturing, 111
– multiattribute method via peptide mapping/LC-MS, 122
– perfusion processing, 123
– – deviation analysis, 127
– process analytical technology (PAT), 117
– – technologies, 118
– – tools to support key bioprocess parameters, 120
– real-time release (RTR), 117
– – via PAT and parametric models, 117
– UPSEC multiattribute process monitoring, 120
– UV data from automated perfusion, 126
– UV elution profiles, 125
– viable microorganism analysis using MuScan, 121
Protein Refinery Operations Lab (PRO lab), 112, 113
proteins, 329
– extraction of, 498
– online peak deconvolution of, 386
– polymerization, 249
– refinery (see also protein refinery operations lab)
– separation, 438
– solution with refolding buffer, 265
proteomics, 236
pseudoaffinity, 405
pump design, 355
purified bovine IgG, by three-zone SMB process, 45
purity requirement, 342
purity trade-off, 407

push-to-low strategy, 35
PV. see process validation (PV)

q
quality, 553
quality by design (QbD), 3, 171, 221, 426, 448, 481, 552
– starts, 552
quality control, 543
quality management, 533
quantitative infectivity assays, 294
quantitative polymerase chain reaction (qPCR), 291
quasi steady-state (QSS), 248
– parallel-sequential scheme, bioreactor system, 247

r
racemic mixtures of small molecules, 329
racemic pharmaceutical enantiomer, 44
radial flow chromatography, 270
Raman spectroscopy, 469, 537
reactor design, 141
– material aspects, 136
readsorption, 404
real time release testing (RTRT), 535
recombinant biologics, 301
recombineering methods, 234
– CRISPR Cas9, 234
– MAGE, 234
recovery/regeneration operation (REC/REG), 48
recycling process, 333
reduced cycle time, 261
reference electrodes, 139
Refine technology, 223
reflectance near-infrared spectrometry, 535
regular network models, 249
regulatory, 564
– authorities, 565
– considerations, 534
– current regulatory references, 565
– misconceptions, 565
– requirements, 558
repeated batch (RB), 184, 187
residence time (RT), 185, 409
residual Protein A, 57
residual substrate concentration, 235
resin, 381
– static binding capacity, 577
retentate pressure impact, 440
retrovirus-like particles (RVLP), 290, 293, 298
revalidations, 544
reverse osmosis (RO), 423
reverse phase (RP), 329, 344
reverse phase chromatography, 412
Reynolds number, 144, 579

rhEnzyme proteins, 226
Rhodobacter capsulatus, 244
risk assessment, 559
risk management, 552
– of raw materials, 550
robust optimization program, 22
rocking platform-based single-use bioreactor, types of, 135
Rogers' bell curve, 166
rotating circular system, prototype of CAC made, 43
rotavirus vaccine
– contamination, 291
– live attenuated, 291
RP-HPLC, with UV detection, 563
RS-232 interface, 114
Rushton turbine impeller, 464
RVLP. *see* retrovirus-like particles (RVLP)

s

Saccharomyces cerevisiae, 7, 243, 244, 246
Saccharomyces uvarum, 245
sanitization agent, validating methods, 450
sanitization step, 449
SBC. *see* static binding capacity (SBC)
scale-down model, 524
– comparison, 188
– cultivation methods, 182
– metabolic behavior, 191
Scale-up and Post- Approval Change (SUPAC), 534
Scale-Up Aspects, 142
scaling-up parameters, 143
– critical process parameters variation, 143
SEC. *see* size-exclusion chromatography (SEC)
segment-wise sterilization, 15
Semba Octave Chromatography System, 269
semicontinuous batch chromatography, 331
semicontinuous chemostat (SCC), 184, 187
– mode, 187
semicontinuous cultivation, 183
semicontinuous perfusion
– VCD, CSPR, and q_{IgG}, 189
– VCD, viability, and q_{IgG}, 190
semicontinuous perfusion (SCP), 184, 187
– bioreactor confirmation, 192
sensors, 139, 444
– storage, condition, 140
sequence-based polymer composition data, 249
sequential multicolumn chromatography (SMCC), 378, 575
serial column setup, example for, 378
serial multicolumn continuous chromatography, 377
settling, 38
shake flasks, 184

shear sensitive cells, 146
silver-stained SDS-PAGE analysis, 57
SIMCA batch, 118
simulated moving bed (SMB), 44, 265, 295, 327, 369, 370, 395, 575
– advantages, 372
– approach, 296
– binary SMB system
– – with buffer gradient, 346
– – chromatography using step gradients, 347
– bind and elute cation exchange chromatography, 297
– chromatography, 268, 297, 324, 329, 341, 352
– in column concentration profiles, 371
– column dimensions, 372
– column height and switch time, 371
– eight-column four-zone, 395
– examples of, 374
– Henry coefficient, 372
– linear velocity of the solid phase, 371
– modification, 352
– parameters for, 372
– purifications, 349
– raffinate stream, 374
– semicontinuous SMB-like unit operation, 375
– separation factors, 372
– separation zones, 371
– 1-SMB process, zones of, 376
– systems, 345
– tandem SMB configurations, 345
– tandem SMB for SEC purification of insulin, 346
– technology, 324, 341, 345, 352
– – for biomolecules, 343
– Varicol-SMB switching times, schematic visualization of, 374
– velocities of zones, 372
– zones for ternary separations, 344
single cell model (SCM), 234, 249
single column, 331
– batch chromatogram, 409
– batch process, 327, 359, 398, 399
– chromatography, 393, 402
– recycling schemes, 333
single-pass tangential flow filtration (SPTFF), 156, 273, 297, 423, 424, 428, 430, 572
– application, 442
– biopharmaceutical industry, 424
– concentration using batch, 431
– continuous mAb template, 429
– continuous processing, 425
– conversion dependency on retentate pressure, 435
– conversion *vs.* concentration, 434, 435
– conversion *vs.* feed flow, 440
– – graph, 436, 441

- experimental setup, 439
- feed channel, 432
- multilevel sections configuration, 443
- multiple sections, 443
- process, 431
- single holder configuration, 443
- single section configuration, 443
- single-section installations, 442
- sizing example, 442
- system, 298
- system setup, 430, 583
- ultrafiltration, 437
- unit installation, 444
single-use (SU), 216
- bags, 136
- - manufacturing materials, 137
- - production and assembly of, 138
- bioreactor (see single use bioreactor)
- continuous bioprocessing, adoption rate of, 165
- disposable technology, 297
- downstream processing, 151
- - chromatography steps, 152
- - tangential flow filtration, 151
- equipment, 58
- - vs. stainless steel equipment, 134
- market, 554
- reactor types, 135
- solutions, 554
- systems, 149
- technologies, 150, 172, 484, 558
- - current trends and future predictions, 153
- - early skepticism, 152
- - history of, 150
- - single-use downstream processing, 151
- - single-use upstream processing, 151
- upstream processing, 151
single use bioreactor (SUB), 133, 136, 216
- classification, 135
- overview of, 137
- parts, 133
- scale-up aspects, 142
- shear effects, 144
size-exclusion chromatography (SEC), 44, 266, 329
- diagram, top and bottom phase, 497
- SMB system, for insulin purification, 345
SMB. see simulated moving bed (SMB)
solid-state fermentation processes, 6
solvent compositions, 345
solvent consumption, 332, 343
space time yield (STY), 175
spectroscopy techniques, 516
spiking techniques, 295
- inline spiking, 295
spin-filter technology, 40, 172, 210

SpinTubes, 184
- bioreactors, 184, 186
- semicontinuous perfusion, 188
- using, advantage, 34
spiral wounds, 426
SPTFF. see single-pass tangential flow filtration (SPTFF)
SSGSA. see steady-state growth space analysis (SSGSA)
stack injection, 332
stainless steel technologies, 557
standard feedback control methods, 11
standard operating procedure (SOP), 555
starting cell density, 188
- determination, 187
state-of-the-art foci, 493
static binding capacity (SBC), 401
statistical process control (SPC), 541
steady-state cell physiology, 240
steady-state cultivation methods, 234
steady-state growth space analysis (SSGSA), 233
- chemostat culture as tool for, 236
- comprehensive quantitative, challenges, 236
- high-resolution, 241
- physiological state of cells, 234
steady-state GSA, using parallel-sequential cultivations, 247
steady-state metabolism, 233, 240
- of cells, 236
- chemostat cultivation, disadvantages, 233
- intracellular, modeling methods, 234
- - flux balance analysis (FBA), 234
- - metabolic flux analysis (MFA), 234
steady-state recycling (SSR), 333
- process, flow diagram of, 334
steady-states of metabolism, 236
steel casting, 3
sterile filtration, 425
sterile tube-welder, 138
sterilization processes, 535
stirred bioreactors, 34
stirred tank bioreactor (STR), 34, 219
stirred tank perfusion bioreactor setup, 464
straight-through processes (STP), 53, 54, 581
Streptococcus thermophilus, 246
SUB. see single use bioreactor (SUB)
substitute solutions, use of, 561
surface-enhanced Raman scattering (SERS), 516
switches
- pairs of tasks, 408
- time, 335
synthetic peptides, 329
system behavior analysis, 9
system control method, 9
system design method, 9

t

tandem configurations, 344
tangential flow filter (TFF), 40, 151, 156, 424, 426, 463, 464
– advantages, 429
– applications, 424
– batch
– – fed-batch, 433
– – *vs.* single-pass, 426
– cell retention device, 213, 470
– disadvantages, 429
– equipment configuration/requirements, 442
– flux excursion batch, 433
– laboratory-scale process development example, 438
– membrane type and format for applications, 426
– modes of operation, 427
– physical bioreactor characterization, 465
– pressure-independent region, 436
– process design, 430
– Prostak system, 213
– schemial drawing, 37
– serial and parallel configuration, 431
– single-pass tangential flow filtration (SPTFF), 428
– systems, 470
– UF holder, 442
Tarpon Biosystems BioSMB, 269
ternary separations, 344
TFF. *see* tangential flow filter (TFF)
therapeutic oligonucleotides, purification, 352
therapeutic proteins, 109
thermodynamics, 234, 342
thermoplastic elastomers, 138
three-column capture process (3C-PCC), 47
– breakthrough curve, 48
– protein A columns of, 55
TMP. *see* transmembrane pressure (TMP)
TOC flush residuals, 450
total organic carbon (TOC), 561
traditional batch processes, 294, 296
traditional batch processing, 295
traditional steel or glass bioreactors, 134
transcription-mediated amplification (TMA), 291
transcriptome analysis, 180
transcriptomics, 180, 236
transit time, 397
transmembrane pressure (TMP)
– feed channel, 432
– filtrate flux dependency, 432
transmissible spongiform encephalopathies (TSE), 289, 300
triangle theory, 342, 347, 352
tri-*n*-Butyl phosphate (TnBP), 310

Triton X-100, 310
true moving bed (TMB), 44, 341, 371, 395
– chromatography, 43
TSE. *see* transmissible spongiform encephalopathies (TSE)
tubing, proper guiding, 136
Tunable Aqueous Polymer Phase Impregnated Resins (TAPPIR®), 52
turbidostat, 9, 245
twin-column CaptureSMB process, schematic illustration, 403
twin-column countercurrent processes, 400
twin-column MCSGP process principle, schematic illustration, 408
twin-column process, 337, 355, 398
– capture process, flow chart of, 468
two one-sided test (TOST), 525
two-phase liquid extraction, 369
Tygon tubing, 277

u

ultrafiltration (UF), 423, 572
– cassette modules, 439
DF systems, 572
– membrane, 438
– – configuration, 428
– processes, 423
ultrasonic separators, 220
ultraviolet
– detectors, 330, 444
– elution profiles, 125
– irradiation, 55
– signal, 48
– vis spectroscopy, 276
ultraviolet-C (UV-C)
– gamma irradiation exposure, 303
– radiation, 312
– viral inactivation devices, 312
United States Pharmacopoeia (USP), 538
UPLC tools, 118
upper control limit (UCL), 541
upstream continuous/perfusion process, 207
– acoustic wave separation, 220
– alternating tangential filtration
– – application of, 211
– – details for different bioreactor sizes and scale-up, 211
– – system, 210, 211
– batch and fed-batch processes, 222
– case study-perfusion using ATF cell retention device, 212
– cell line stability, 215
– cell retention devices, 210
– continuous downstream processing, 223
– continuous flow centrifugation, 220

- costs and benefits of continuous manufacturing, 222
- costs of adoption, 223
- current biotherapeutics, 217
- design objective for integrated continuous processing, 226
- FDA supports continuous processing, 221
- fixed/floating filter bioreactors, 219
- hollow fiber media exchange, 220
- hollow fiber perfusion bioreactors, 219
- integrated continuous manufacturing, 224
- manufacturing scale-up challenges, 214
- packed bed bioreactors, 219
- process complexity and control, 214
- single-use accessories supporting perfusion culture, 220
- single-use bioreactors for perfusion culture, 219
- single-use continuous bioproduction, 218
- single-use perfusion bioreactors, 219
- single-use technologies, 215
- spin filters, 210, 220
- Stainless Steel and Single-Use, 209
- stirred tank suspension reactors, 219
- SUBs application in continuous processing, 218
- traditional fed-batch, traditional perfusion culture, 225
- upstream process-type selection, 209
- validation, 215

upstream process (USP), 203, 481, 489
- application, sensor for, 139
- biomanufacturing, 205
- bioprocessing, 302
- case study, 494
- continuous or perfusion, 208
- development, 492
- – in continuous manufacturing, 483
- experimental presentation of methodical approach, 493
- methodical approach regarding impurity processing, 492
- operating modes, 206
- – continuous/perfusion process, 207
- – continuous upstream/perfusion, advantage of, 207
- – fed-batch process, 206
- optimization, 482
- trends/integration approaches, 490

v

vaccines, 109
validated states, verification of, 536
validation, of bioprocess, 355
- between batch and continuous processing, 445
- single-pass TFF, 449
- virus injection, 318

valve configurations, 355
Varicol process, 375
Varicol-SMB switching times, schematic visualization of, 374
VCD time course, illustration of, 183
velocity, 342
viable cell density
- productivity, media consumption, buffer consumption, and overall yield, 473
- steady-state operation, 471
viral clearance, 54, 293, 294, 304, 311
- validation, 299
viral contamination, 289
- of biotechnology products, 290
- sources, 293
viral filters, 307
viral genome, 290, 292
viral inactivation (VI), 310, 522, 578
- chemical, 310
- low pH, 523
- methods, 292
- reduction, 293
- – mechanism of action, 293
viral reduction
- filters, 306, 307
- filtration, 305, 306, 308
- inactivation, orthogonal, 304
viral safety, 290, 292–294, 296
- batch vs. continuous, 297
- of biologics, 289
- chemical inactivation, 308, 315
- – in continuous processing, 310
- chromatography, 311
- continuous processing, 295
- downstream virus removal strategies, 304
- gamma irradiation, 313
- heat treatment/pasteurization, 313
- inactivation by ultraviolet radiation (UV-C), 312
- in-line virus injection validation method, 317
- intermediate and polishing chromatography, 315
- limitations, 292
- low-pH inactivation, 308
- of medicinal product, 292
- normal flow filtration (NFF), 304
- pH/S/D viral inactivation manufacturing set-up for batch manufacturing, 309
- protein a capture chromatography, 314
- raw material safety/testing, 299
- risk mitigation strategies, 300
- solvent detergent (S/D) inactivation, 310
- typical batch vs. fed-batch vs. continuous processing modes of operation, 302

- upstream and bioreactor safety, 301
- validation in continuous manufacturing processes, 313
- viral reduction and inactivation unit operations, 304
- virus filtration
  - – flow decay *vs.* mass throughput, 318
  - – manifold for continuous processing, 308
  - – reduction, 316
viral transmission, 290
viral validation, 293
viruses
- barrier filtration, 298, 303
- barrier technologies, 291, 301
- classification, 290
  - – endogenous virus, 290
  - – exogenous/adventitious virus, 290
- clearance, 301
- contamination, 290, 293, 298
- detection methods, 291
  - – analytical methods, 292
  - – *in vitro* detection techniques, 291
  - – quantitative polymerase chain reaction (qPCR), 291
  - – transcription-mediated amplification (TMA), 291
- filtration, 299, 306, 307, 425
- log reduction value, 316
virus filters, 307, 317
- integrity test, 299
- log reduction value, 307
- maximum operating pressure, 442
- postuse integrity status, 307
- validation, 307
virus inactivation, 309, 459
- solvent detergent (S/D) treatment for, 310
virus reduction methods, 298, 301
- filters, 299
- filtration, 305
  - – buffer compatibility, 305

- – facility fit, 305
- – incoming raw materials, 305
- – protein concentration, 305
virus removal, 301
- efficacy, 301
- mechanism of action
  - – adsorption, 293
  - – chemical inactivation, 293
  - – size exclusion, 293
- steps, 55
- validation studies, 293, 294
  - – model virus, 293
  - – using scale-down models, 293
virus retention, 305
virus safety, 295, 296
- risk mitigation, 302
- risk reduction, 303
  - – in continuous upstream bioprocessing, 303
- risk strategies, 303
- strategy in continuous processing, 313
virus spiking, 295
virus validation, 314
vitamins C and E, 243
volumetric concentration factor (VCF), 430
volumetric flow rates, 463
volumetric productivity (VP), 175
vortex flow filters, 208, 483

w

waste treatment, 486
wave bioreactor, 35, 135, 151
weak cation exchange chromatography, 474
whole-cell models, 249

y

Yarrowia lipolytica, 245
yield/purity chart, 413

z

Zika, 33